Mastering Algebra
An Introduction

By

SAID HAMILTON

Book Title: Mastering Algebra - An Introduction

Author: Said Hamilton

Editor: Pat Eblen

Cover design by: Kathleen Myers

First published in 1997

Hamilton Education Guides
P.O. Box 681
Vienna, Va. 22183

Library of Congress Catalog Card Number 97-93529
Library of Congress Cataloging-in-Publication Data

ISBN 0-9649954-1-7

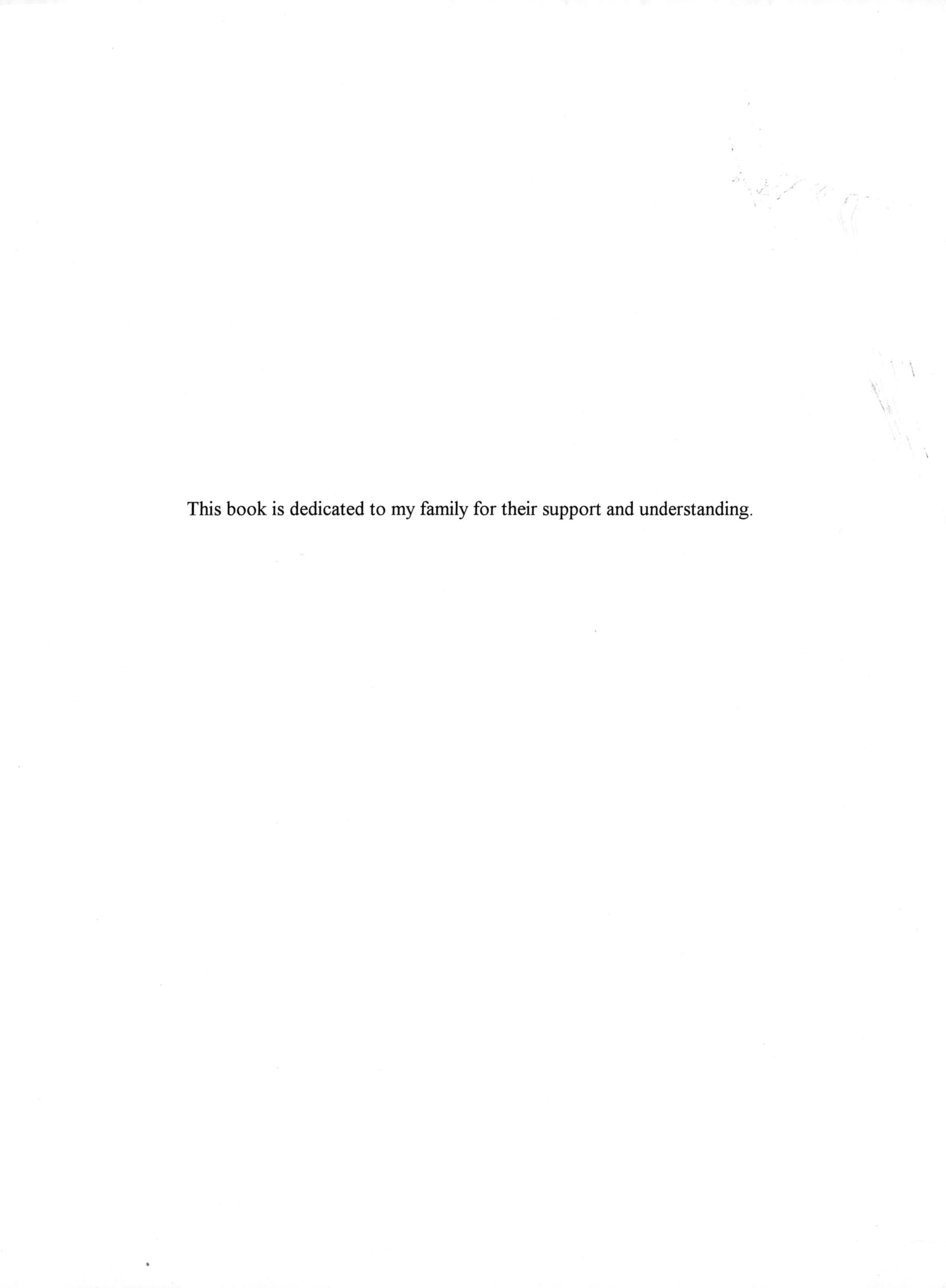

This book is dedicated to my family for their support and understanding.

General Contents

Detailed Contents

Chapter 1 - Parentheses and Brackets

Chapter 2 - Integer Fractions

Chapter 3 - Exponents

Chapter 4 - Radicals

Chapter 5 - Fractional Exponents

Chapter 6 - Polynomials

Appendix - Exercise Solutions

Acknowledgments

The primary motivating factor in writing the Hamilton Education Guides is observing the difficulty my children have in following the math concepts presented in the books they use in their school programs. I therefore, would like to acknowledge my children for giving me the inspiration to proceed with the writing of these books. I am grateful to Pat Eblen for his editorial comments. His constructive comments and suggestions on more precise and easier presentation of the topics truly elevated the usefulness of this book. I would also like to acknowledge and give my thanks to the following education professionals who reviewed and provided comments to further enhance this book: Mrs. Linda Clark, Mrs. Bodil Nadler, Mr. Omar Spaulding, and various other contributors. My special thanks to Kathleen Myers for her outstanding cover design. Finally, I would like to thank my family for their understanding and patience in allowing me to take on the task of writing this book. I hope users of this book will find it valuable.

Introduction and Overview

In reminiscing of my teaching career, I am frequently reminded of how unwillingly many of my students were to listen to me during the first few class lectures. The lack of interest and frankly not knowing how to study was very evident. I could only hope to keep them interested enough to stay in the course. My approach in teaching mathematics and other technical subjects has always been in the form of trying to present any topic in the simplest way possible. I have always believed that many of our academic texts, particularly those that lay the foundation for learning math and science, are either written in an abstract and difficult to follow language, lack sufficient number of detailed sample problems, or are not explained adequately for a student to become interested in the subject.

It is my belief that the key to learning mathematics is through positive motivation. Students can be greatly motivated if subjects are presented concisely and the problems are solved in a detailed step by step approach. This keeps the student motivated and provides a great deal of encouragement in wanting to learn the next subject or to solve the next problem. During my teaching career, I found this method to be an effective way of teaching. I never forget the expressions of gratitude I have received from students for helping them realize mathematics is indeed an interesting subject to pursue. The fact that they could truly learn math and develop a positive interest in taking the next math course was gratifying for me. I hope by presenting subjects with the methods used in this book, more students will become interested in the subject of mathematics and can carry this approach and philosophy to future generations.

The scope of this book is intended for educational levels ranging from the 8th grade to adult. The book can also be used by students in home study programs, parents, teachers, special education programs, preparatory schools, and adult educational programs including colleges and universities as a main text, a thorough reference, or a supplementary book. A fundamental understanding of basic mathematical operations such as addition, subtraction, multiplication, and division is required.

"Mastering Algebra: An Introduction" is the first in a series of three books on algebra. It addresses the basics of algebra by introducing the student to topics such as integer fractions, exponents, radicals, fractional exponents, and polynomials. The second book, *"Mastering Algebra: Intermediate Level"* addresses topics such as factoring, algebraic fractions, functions of variables, and graphing. The third book, *"Mastering Algebra: Advanced Level"* addresses areas such as solving and graphing quadratic equations, parabolas, circles, ellipses, hyperbolas, and vectors.

This book is divided into six chapters. Chapter 1 introduces the student to the concept and use of signed numbers, parentheses, and brackets as math expressions in solving mathematical operations. How integer fractions are simplified, added, subtracted, multiplied, and divided are described in Chapter 2. (It is essential that students be thoroughly familiar with sign numbers, the use of parentheses, brackets, and fractional operations before proceeding with other chapters.) The subject of exponents is addressed in Chapter 3. Students learn how to solve and simplify numbers and variables that are raised to integer exponents. Radical expressions and how they are simplified and mathematically operated on are addressed in Chapter 4. In this section students learn how to add, subtract, multiply, divide, and rationalize radical expressions. Chapter 5 addresses a more difficult class of exponents by showing how to simplify and solve fractional exponents. The subject of polynomials and how different types of polynomials are added, subtracted, multiplied, and divided are addressed in Chapter 6. Finally, detailed solutions to the exercises are provided in the Appendix. Students are encouraged to solve each problem in the same detail and step by step format as shown in the text.

It is my hope that all the Hamilton Education Guides stand apart in their understandable treatment of the presented subjects and for their clarity and special attention to detail. I hope readers of these books will find them useful. Any comments and suggestions for improvement of this book will be appreciated.

With best wishes,

Said Hamilton

Chapter 1
Parentheses and Brackets

Quick Reference to Chapter 1 Case Problems

Case II c - Using Brackets to Add Two and Three Integer Numbers Sub-grouped by Parentheses, p. 16

$\boxed{4+[(3+12)+(9+15+23)]}=$; $\boxed{[(3+5)+(4+9+11)+6]+3}=$; $\boxed{(4+3)+[(6+9+12)+(30+5)+1]}=$

1.3 Using Parentheses and Brackets in Subtraction 19

Case I - Use of Parentheses in Subtraction, *p. 19*

Case I a - Subtracting Integer Numbers Without Using Parentheses, p. 19

$\boxed{24-5-13-7-8}=$; $\boxed{35-4-6-3}=$; $\boxed{-12-20-5}=$

Case I b - Subtracting Two Integer Numbers Grouped by Parentheses, p. 20

$\boxed{20-(15-45)}=$; $\boxed{(20-25)-(7-5)}=$; $\boxed{25-(35-12)-(8-3)}=$

Case I c - Subtracting Three Integer Numbers Grouped by Parentheses, p. 21

$\boxed{6-(22-16-8)}=$; $\boxed{(15-3-8)-(40-9-34)}=$; $\boxed{(-12-3-6)-5}=$

Case I d - Subtracting Two and Three Integer Numbers Grouped by Parentheses, p. 21

$\boxed{(43-6)-(54-13-7)-19}=$; $\boxed{(8-13-10)-(6-36)}=$; $\boxed{(-12-3-4)-(-15-5)}=$

Case II - Use of Brackets in Subtraction, *p. 22*

Case II a - Using Brackets to Subtract Two Integer Numbers Sub-grouped by Parentheses, p. 22

$\boxed{[(9-23)-12]-40}=$; $\boxed{5-[(18-7)-27]}=$; $\boxed{26-[(10-6)-(4-9)]}=$

Case II b - Using Brackets to Subtract Three Integer Numbers Sub-grouped by Parentheses, p. 24

$\boxed{[(45-13-7)-15]-20}=$; $\boxed{50-[(5-25-7)-(36-12-5)]}=$; $\boxed{[12-(-12-3-5)]-3}=$

Case II c - Using Brackets to Subtract Two and Three Integer Numbers Sub-grouped by Parentheses, p. 25

$\boxed{[(300-450)-(100-35-55)]-12}=$; $\boxed{34-[(324-130)-(250-39-85)]}=$; $\boxed{[(13-8)-(24-9-15)-6]-30}=$

1.4 Using Parentheses and Brackets in Multiplication 28

Case I - Use of Parentheses in Multiplication, *p. 28*

Case I a - Multiplying Integer Numbers Without Using Parentheses, p. 28

$\boxed{3\times 5\times 7\times 2\times 4}=$; $\boxed{3\times 8\times 2}=$; $\boxed{12\times 5\times 4\times 3}=$

Case I b - Multiplying Two Integer Numbers Grouped by Parentheses, p. 29

$\boxed{(2\times 5)\times(7\times 4)\times(1\times 3)}=$; $\boxed{2\times(5\times 3)\times(6\times 4)\times 7}=$; $\boxed{(3\times 5)\times(2\times 3)}=$

Case I c - Multiplying Three Integer Numbers Grouped by Parentheses, p. 29

$\boxed{2\times(3\times 8\times 10)}=$; $\boxed{(5\times 3\times 2)\times(10\times 4\times 7)}=$; $\boxed{(3\times 2\times 6)\times 4}=$

Case I d - Multiplying Two and Three Integer Numbers Grouped by Parentheses, p. 30

$\boxed{(3\times 1)\times(4\times 5\times 11)\times 2}=$; $\boxed{(2\times 9\times 8)\times(6\times 4)}=$; $\boxed{3\times(3\times 2\times 5)\times(2\times 8)}=$

Case II - Use of Brackets in Multiplication, *p. 30*

Case II a - Using Brackets to Multiply Two Integer Numbers Sub-grouped by Parentheses, p. 30

$\boxed{6\times[(12\times 3)\times(4\times 1)]}=$; $\boxed{[(4\times 1)\times(5\times 9)]\times(2\times 3)}=$; $\boxed{(7\times 4)\times[(13\times 2)\times(6\times 1)]}=$

Case II b - Using Brackets to Multiply Three Integer Numbers Sub-grouped by Parentheses, p. 31

$\boxed{[(7\times 3\times 10)\times 4]\times 2}=$; $\boxed{2\times[(5\times 1\times 6)\times(3\times 8\times 4)]}=$; $\boxed{3\times[10\times(2\times 5\times 4)]}=$

Case II c - Using Brackets to Multiply Two and Three Integer Numbers Sub-grouped by Parentheses, p. 32

$\boxed{2\times[(3\times 7)\times(1\times 10\times 5)]}=$; $\boxed{[(3\times 5)\times(4\times 1\times 7)\times 6]\times 2}=$; $\boxed{(5\times 3)\times[(6\times 2\times 8)\times(7\times 4)\times 1]}=$

1.5 Using Parentheses and Brackets in Division 35

Case I - Use of Parentheses in Division, *p. 35*

Case I a - Dividing Two Integer Numbers, p. 35

$\boxed{135\div 20}=$; $\boxed{25\div 5}=$; $\boxed{126\div 6}=$

Case I b - Dividing Two Integer Numbers Grouped by Parentheses, p. 35

$\boxed{38\div(12\div 3)}=$; $\boxed{(125\div 5)\div 4}=$; $\boxed{(15\div 4)\div(8\div 3)}=$

Case II - Use of Brackets in Division, *p. 36*

$\boxed{238\div[24\div(15\div 5)]}=$; $\boxed{[(28\div 13)\div(15\div 4)]\div 2}=$; $\boxed{238\div[(35\div 5)\div(14\div 7)]}=$

1.6 Using Parentheses and Brackets in Mixed Operations 41

Case I - Use of Parentheses in Addition, Subtraction, Multiplication, and Division, *p. 41*

$\boxed{30+(50\div 4)}=$; $\boxed{(23+5)\div(20-8)}=$; $\boxed{(49\div 5)-(12\times 4)}=$

Case II - Use of Brackets in Addition, Subtraction, Multiplication, and Division, *p. 43*

$\boxed{[(23-6)-8]+(12+7)}=$; $\boxed{35+[(12+5)-(4\times 2)]}=$; $\boxed{[(45\div 9)+(12\div 4)]\times(10+5)}=$

Chapter 1 - Parentheses and Brackets

The objective of this chapter is to teach the student the concept of grouping numbers. This is achieved by introduction of parentheses and brackets as tools for solving mathematical problems. In section 1.1 signed numbers are introduced and their use in addition, subtraction, multiplication, and division are discussed. Section 1.2 shows how numbers are grouped and solved in addition. Sections 1.3, 1.4, and 1.5 show how parentheses and brackets are used in subtraction, multiplication, and division, respectively. Section 1.6 shows the use of parentheses and brackets in solving mixed operations. The general algebraic approach for grouping numbers using parentheses and brackets is provided in each section. Additional examples are provided at the end of each section to help meet the objective of this chapter.

1.1 Signed Numbers

In mathematics, "$+$" and "$-$" symbols are used to indicate the use of positive and negative numbers, respectively. If a signed number has no symbol it is understood to be a positive number. Signed numbers are added, subtracted, multiplied, and divided as exemplified in the following cases:

Case I - Addition of Signed Numbers

When two numbers are added, the numbers are called **addends** and the result is called a **sum.** The sign of the sum dependents on the sign of the numbers. This is shown in the following cases with the sign change of two real numbers a and b:

Case I a.

$\boxed{a+b} = \boxed{A}$

For example,

1. $\boxed{5+6} = \boxed{\mathbf{11}}$
2. $\boxed{7+8} = \boxed{\mathbf{15}}$
3. $\boxed{1+0} = \boxed{\mathbf{1}}$
4. $\boxed{3+15} = \boxed{\mathbf{18}}$
5. $\boxed{15+9} = \boxed{\mathbf{24}}$

Case I b.

$\boxed{-a+b} = \boxed{B}$

For example,

1. $\boxed{-7+3} = \boxed{\mathbf{-4}}$
2. $\boxed{-9+0} = \boxed{\mathbf{-9}}$
3. $\boxed{-15+40} = \boxed{\mathbf{25}}$
4. $\boxed{-35+18} = \boxed{\mathbf{-17}}$
5. $\boxed{-8+30} = \boxed{\mathbf{22}}$

Case I c.

$\boxed{a+(-b)} = \boxed{a-b} = \boxed{C}$

For example,

1. $\boxed{2+(-5)} = \boxed{2-5} = \boxed{\mathbf{-3}}$
2. $\boxed{7+(-9)} = \boxed{7-9} = \boxed{\mathbf{-2}}$
3. $\boxed{0+(-1)} = \boxed{0-1} = \boxed{\mathbf{-1}}$
4. $\boxed{8+(-45)} = \boxed{8-45} = \boxed{\mathbf{-37}}$
5. $\boxed{40+(-9)} = \boxed{40-9} = \boxed{\mathbf{31}}$

Case I d.

$\boxed{(-a)+b} = \boxed{-a+b} = \boxed{D}$ *Note:* $(-a) = -a$

For example,

1. $\boxed{(-3)+9} = \boxed{-3+9} = \boxed{\mathbf{6}}$
2. $\boxed{(-12)+8} = \boxed{-12+8} = \boxed{\mathbf{-4}}$
3. $\boxed{(-7)+25} = \boxed{-7+25} = \boxed{\mathbf{18}}$
4. $\boxed{(-34)+10} = \boxed{-34+10} = \boxed{\mathbf{-24}}$
5. $\boxed{(-1)+0} = \boxed{-1+0} = \boxed{\mathbf{-1}}$

Case I e.

$\boxed{(-a)+(-b)} = \boxed{-a-b} = \boxed{E}$

For example,

1. $\boxed{(-6)+(-9)} = \boxed{-6-9} = \boxed{\mathbf{-15}}$
2. $\boxed{(-45)+(-6)} = \boxed{-45-6} = \boxed{\mathbf{-51}}$
3. $\boxed{(-10)+(-55)} = \boxed{-10-55} = \boxed{\mathbf{-65}}$
4. $\boxed{(-35)+(-20)} = \boxed{-35-20} = \boxed{\mathbf{-55}}$
5. $\boxed{(-5)+(-5)} = \boxed{-5-5} = \boxed{\mathbf{-10}}$

Case II - Subtraction of Signed Numbers

When two numbers are subtracted the result is called the **difference**. The sign of the difference depends on the sign of the numbers. This is shown in the following cases with the sign change of two real numbers a and b:

Case II a.

$\boxed{a-b} = \boxed{A}$

For example,

1. $\boxed{15-6} = \boxed{\mathbf{9}}$
2. $\boxed{17-47} = \boxed{\mathbf{-30}}$
3. $\boxed{1-0} = \boxed{\mathbf{1}}$
4. $\boxed{3-15} = \boxed{\mathbf{-12}}$
5. $\boxed{45-9} = \boxed{\mathbf{36}}$

Case II b.

$\boxed{-a-b}=\boxed{B}$

For example,

1. $\boxed{-7-3}=\boxed{\mathbf{-10}}$
2. $\boxed{-1+0}=\boxed{\mathbf{-1}}$
3. $\boxed{-15-45}=\boxed{\mathbf{-60}}$
4. $\boxed{-35-8}=\boxed{\mathbf{-43}}$
5. $\boxed{-8-30}=\boxed{\mathbf{-38}}$

Case II c.

$\boxed{a-(-b)}=\boxed{a+(b)}=\boxed{a+b}=\boxed{C}$

For example,

1. $\boxed{12-(-5)}=\boxed{12+(5)}=\boxed{12+5}=\boxed{\mathbf{17}}$
2. $\boxed{7-(-9)}=\boxed{7+(9)}=\boxed{7+9}=\boxed{\mathbf{16}}$
3. $\boxed{0-(-1)}=\boxed{0+(1)}=\boxed{0+1}=\boxed{\mathbf{1}}$
4. $\boxed{30-(-45)}=\boxed{30+(45)}=\boxed{30+45}=\boxed{\mathbf{75}}$
5. $\boxed{10-(-39)}=\boxed{10+(39)}=\boxed{10+39}=\boxed{\mathbf{49}}$

Case II d.

$\boxed{(-a)-(-b)}=\boxed{(-a)+(b)}=\boxed{-a+b}=\boxed{D}$

For example,

1. $\boxed{(-3)-(-9)}=\boxed{(-3)+(9)}=\boxed{-3+9}=\boxed{\mathbf{6}}$
2. $\boxed{(-32)-(-8)}=\boxed{(-32)+(8)}=\boxed{-32+8}=\boxed{\mathbf{-24}}$
3. $\boxed{(-17)-(-25)}=\boxed{(-17)+(25)}=\boxed{-17+25}=\boxed{\mathbf{8}}$
4. $\boxed{(-35)-(-10)}=\boxed{(-35)+(10)}=\boxed{-35+10}=\boxed{\mathbf{-25}}$
5. $\boxed{(-1)-(-6)}=\boxed{(-1)+(6)}=\boxed{-1+6}=\boxed{\mathbf{5}}$

Case III - Multiplication of Signed Numbers

When two numbers are multiplied, the numbers are called **factors** and the result is called a **product**. For example, when 12 is multiplied by 2 the result is 24.

$$\boxed{12\,(factor)\times 2\,(factor)} = \boxed{\mathbf{24}\,(product)}$$

Thus, 12 and 2 are the factors, and 24 is the product.

The sign of the product is positive if the factors have the same sign and is negative if the factors have different signs. This is shown in the following cases with the sign change of two real numbers a and b:

Case III a.

$\boxed{a\times b} = \boxed{ab}$

For example,

1. $\boxed{5\times 6} = \boxed{\mathbf{30}}$
2. $\boxed{7\times 8} = \boxed{\mathbf{56}}$
3. $\boxed{1\times 0} = \boxed{\mathbf{0}}$
4. $\boxed{10\times 7} = \boxed{\mathbf{70}}$
5. $\boxed{15\times 7} = \boxed{\mathbf{105}}$

Case III b.

$\boxed{(-a)\times b} = \boxed{-a\times b} = \boxed{-ab}$

For example,

1. $\boxed{(-7)\times 3} = \boxed{-7\times 3} = \boxed{\mathbf{-21}}$
2. $\boxed{(-1)\times 0} = \boxed{-1\times 0} = \boxed{\mathbf{0}}$
3. $\boxed{(-15)\times 40} = \boxed{-15\times 40} = \boxed{\mathbf{-600}}$
4. $\boxed{(-25)\times 16} = \boxed{-25\times 16} = \boxed{\mathbf{-400}}$
5. $\boxed{(-8)\times 20} = \boxed{-8\times 20} = \boxed{\mathbf{-160}}$

Case III c.

$\boxed{a\times(-b)} = \boxed{-a\times b} = \boxed{-ab}$

For example,

1. $\boxed{2\times(-5)} = \boxed{-2\times 5} = \boxed{\mathbf{-10}}$
2. $\boxed{7\times(-9)} = \boxed{-7\times 9} = \boxed{\mathbf{-63}}$
3. $\boxed{0\times(-1)} = \boxed{\mathbf{0}}$
4. $\boxed{30\times(-25)} = \boxed{-30\times 25} = \boxed{\mathbf{-750}}$
5. $\boxed{40\times(-9)} = \boxed{-40\times 9} = \boxed{\mathbf{-360}}$

Case III d.

$\boxed{(-a)\times(-b)} = \boxed{+ab} = \boxed{ab}$

For example,

1. $\boxed{(-3)\times(-9)}=\boxed{+27}=\boxed{\mathbf{27}}$

2. $\boxed{(-12)\times(-4)}=\boxed{+48}=\boxed{\mathbf{48}}$

3. $\boxed{(-8)\times(-150)}=\boxed{+1200}=\boxed{\mathbf{1200}}$

4. $\boxed{(-30)\times(-10)}=\boxed{+300}=\boxed{\mathbf{300}}$

5. $\boxed{(-5)\times(-25)}=\boxed{+125}=\boxed{\mathbf{125}}$

Case IV - Division of Signed Numbers

When one number is divided by another, the first number is called the **dividend**, the second number the **divisor**, and the result a **quotient**. For example, when 12 is divided by 2 the result is 6.

$$\boxed{\frac{12\,(dividend)}{2\,(divisor)}}=\boxed{6\,(quotient)}$$

Thus, 12 is the dividend, 2 is the divisor, and 6 is the quotient.

The sign of the quotient is positive if the divisor and the dividend have the same sign and is negative if the divisor and the dividend have different signs. This is shown in the following cases with the sign change of two real numbers a and b:

Case IV a.

$$\boxed{\frac{a}{b}}=\boxed{A}$$

For example,

1. $\boxed{\frac{9}{3}}=\boxed{\mathbf{3}}$

2. $\boxed{\frac{27}{3}}=\boxed{\mathbf{9}}$

3. $\boxed{\frac{75}{5}}=\boxed{\mathbf{15}}$

4. $\boxed{\frac{18}{4}}=\boxed{\mathbf{4.5}}$

5. $\boxed{\frac{36}{6}}=\boxed{\mathbf{6}}$

Case IV b.

$$\boxed{\frac{-a}{b}}=\boxed{-\frac{a}{b}}=\boxed{B}$$

For example,

1. $\boxed{\frac{-10}{2}}=\boxed{-\frac{10}{2}}=\boxed{\mathbf{-5}}$

2. $\boxed{\frac{-66}{3}}=\boxed{-\frac{66}{3}}=\boxed{\mathbf{-22}}$

3. $\boxed{\frac{-75}{5}}=\boxed{-\frac{75}{5}}=\boxed{\mathbf{-15}}$

4. $\boxed{\frac{-8}{2}}=\boxed{-\frac{8}{2}}=\boxed{\mathbf{-4}}$

5. $\boxed{\frac{-5}{3}}=\boxed{-\frac{5}{3}}=\boxed{\mathbf{-1.67}}$

Case IV c.

$$\boxed{\frac{a}{-b}} = \boxed{-\frac{a}{b}} = \boxed{C}$$

For example,

1. $\boxed{\frac{30}{-2}} = \boxed{-\frac{30}{2}} = \boxed{\mathbf{-15}}$ 2. $\boxed{\frac{88}{-8}} = \boxed{-\frac{88}{8}} = \boxed{\mathbf{-11}}$ 3. $\boxed{\frac{45}{-9}} = \boxed{-\frac{45}{9}} = \boxed{\mathbf{-5}}$

4. $\boxed{\frac{18}{-5}} = \boxed{-\frac{18}{5}} = \boxed{\mathbf{-3.6}}$ 5. $\boxed{\frac{35}{-7}} = \boxed{-\frac{35}{7}} = \boxed{\mathbf{-5}}$

Case IV d.

$$\boxed{\frac{-a}{-b}} = \boxed{+\frac{a}{b}} = \boxed{\frac{a}{b}} = \boxed{D}$$

For example,

1. $\boxed{\frac{-40}{-2}} = \boxed{\frac{40}{2}} = \boxed{\mathbf{20}}$ 2. $\boxed{\frac{-66}{-3}} = \boxed{\frac{66}{3}} = \boxed{\mathbf{22}}$ 3. $\boxed{\frac{-7}{-7}} = \boxed{\frac{7}{7}} = \boxed{\mathbf{1}}$

4. $\boxed{\frac{-28}{-4}} = \boxed{\frac{28}{4}} = \boxed{\mathbf{7}}$ 5. $\boxed{\frac{-8}{-3}} = \boxed{+\frac{8}{3}} = \boxed{\frac{8}{3}} = \boxed{\mathbf{2.67}}$

Signed Numbers - General Rules

Addition: $\boxed{(-)+(-)=(-)}$; $\boxed{(-)+(+)=(-) \text{ if negative No. is } \rangle \text{ positive No.}}$; $\boxed{(+)+(+)=(+)}$;

$\boxed{(+)+(-)=(+) \text{ if positive No. is } \rangle \text{ negative No.}}$ *Note: The symbol " $\rangle$ " means greater than.*

Subtraction: $\boxed{(-)-(-)=(-)+(+)=(-) \text{ if the 1st. negative No. is } \rangle \text{ the 2nd. negative No.}}$;

$\boxed{(+)-(+)=(+)+(-)=(+) \text{ if the 1st. positive No. is } \rangle \text{ the 2nd. positive No.}}$;

$\boxed{(+)-(-)=(+)+(+)=(+)}$; $\boxed{(-)-(+)=(-)+(-)=(-)}$

Multiplication: $\boxed{(-)\times(-)=(+)}$; $\boxed{(-)\times(+)=(-)}$; $\boxed{(+)\times(+)=(+)}$, $\boxed{(+)\times(-)=(-)}$

Division: $\boxed{\frac{(-)}{(-)}=(+)}$; $\boxed{\frac{(-)}{(+)}=(-)}$; $\boxed{\frac{(+)}{(+)}=(+)}$; $\boxed{\frac{(+)}{(-)}=(-)}$

Signed Numbers - Summary of Cases

1. Addition and Subtraction:

I a. $\boxed{a+b}=\boxed{A}$

I b. $\boxed{-a+b}=\boxed{B}$

I c. $\boxed{a+(-b)}=\boxed{a-b}=\boxed{C}$

I d. $\boxed{(-a)+b}=\boxed{-a+b}=\boxed{D}$

I e. $\boxed{(-a)+(-b)}=\boxed{-a-b}=\boxed{E}$

II a. $\boxed{a-b}=\boxed{A}$

II b. $\boxed{-a-b}=\boxed{B}$

II c. $\boxed{a-(-b)}=\boxed{a+(b)}$
$=\boxed{a+b}=\boxed{C}$

II d. $\boxed{(-a)-(-b)}=\boxed{(-a)+(b)}$
$=\boxed{-a+b}=\boxed{D}$

2. Multiplication and Division:

III a. $\boxed{a\times b}=\boxed{ab}$

III b. $\boxed{(-a)\times b}=\boxed{-a\times b}=\boxed{-ab}$

III c. $\boxed{a\times(-b)}=\boxed{-a\times b}=\boxed{-ab}$

III d. $\boxed{(-a)\times(-b)}=\boxed{+ab}=\boxed{ab}$

IV a. $\boxed{\frac{a}{b}}=\boxed{A}$

IV b. $\boxed{\frac{-a}{b}}=\boxed{-\frac{a}{b}}=\boxed{B}$

IV c. $\boxed{\frac{a}{-b}}=\boxed{-\frac{a}{b}}=\boxed{C}$

IV d. $\boxed{\frac{-a}{-b}}=\boxed{+\frac{a}{b}}=\boxed{\frac{a}{b}}=\boxed{D}$

Practice Problems - Signed Numbers

Section 1.1 Practice Problems - Show the correct sign by performing the following operations:

1. $\frac{-95}{-5}=$
2. $(-20)\times(-8)=$
3. $(-33)+(-14)=$
4. $(-18)-(-5)=$
5. $(-20)+8=$
6. $\frac{48}{-4}=$
7. $-15-32=$
8. $30+(-9)=$
9. $55-(-6)=$
10. $8\times(-35)=$

1.2 Using Parentheses and Brackets in Addition

Parentheses and brackets are used to group numbers as a means to minimize mistakes in solving mathematical operations. In this section the use of parentheses and brackets is discussed in detail. Two properties associated with addition are discussed first as follows:

Commutative and Associative Property of Addition

1. Changing the order in which two numbers are added does not change the final answer. This property of real numbers is called the **Commutative Property of Addition**, e.g., for any two real numbers a and b

   $\boxed{a+b}=\boxed{b+a}$

   For example, $\boxed{9+7}=\boxed{\mathbf{16}}$ and $\boxed{7+9}=\boxed{\mathbf{16}}$

2. Re-grouping numbers does not change the final answer. This property of real numbers is called the **Associative Property of Addition**, e.g., for any real numbers a, b, and c

   $\boxed{(a+b)+c}=\boxed{a+(b+c)}$

   For example,

   $\boxed{(5+4)+7}=\boxed{(9)+7}=\boxed{9+7}=\boxed{\mathbf{16}}$

   $\boxed{5+(4+7)}=\boxed{5+(11)}=\boxed{5+11}=\boxed{\mathbf{16}}$

Note that, although changing the order in which numbers are added or grouped does not affect the final answer, it is important to learn how to solve math operations in the exact order in which parentheses or brackets are used. Learning how to use parentheses and brackets properly will minimize mistakes in solving mathematical problems. Parentheses and brackets are used in different ways to group numbers. The use of parentheses and brackets in addition, using integer numbers, is discussed in the following cases:

Case I - Use of Parentheses in Addition

In addition, parentheses can be grouped in different ways as shown in the following example cases:

Case I a - *Adding Integer Numbers Without Using Parentheses*

Integer numbers are added without the use of parentheses, as shown in the following general and specific examples:

$\boxed{a+b+c+d+e}=$

Let $\boxed{a+b+c+d+e=A}$, then

$\boxed{a+b+c+d+e}=\boxed{A}$

Example 1.2-1

$\boxed{2+3+5+6+10}=$

Solution:

$\boxed{2+3+5+6+10} = \boxed{\mathbf{26}}$

Case I b - *Adding Two Integer Numbers Grouped by Parentheses*
Two integer numbers that are grouped by parentheses are added in the following ways, as shown by general and specific example cases:

Case I b-1.

$\boxed{(a+b)+(c+d)} =$

Let $\boxed{k_1 = a+b}$, $\boxed{k_2 = c+d}$, and $\boxed{k_1 + k_2 = A}$, then

$\boxed{(a+b)+(c+d)} = \boxed{(k_1)+(k_2)} = \boxed{k_1 + k_2} = \boxed{A}$

Example 1.2-2

$\boxed{(12+35)+(8+10)} =$

Solution:

$\boxed{(12+35)+(8+10)} = \boxed{(47)+(18)} = \boxed{47+18} = \boxed{\mathbf{65}}$

Case I b-2.

$\boxed{a+(b+c)+(d+e)+f} =$

Let $\boxed{k_1 = b+c}$, $\boxed{k_2 = d+e}$, and $\boxed{a + k_1 + k_2 + f = B}$, then

$\boxed{a+(b+c)+(d+e)+f} = \boxed{a+(k_1)+(k_2)+f} = \boxed{a+k_1+k_2+f} = \boxed{B}$

Example 1.2-3

$\boxed{6+(5+12)+(8+7)+23} =$

Solution:

$\boxed{6+(5+12)+(8+7)+23} = \boxed{6+(17)+(15)+23} = \boxed{6+17+15+23} = \boxed{\mathbf{61}}$

Case I b-3.

$\boxed{(a+b)+(c+d)+(e+f)} =$

Let $\boxed{k_1 = a+b}$, $\boxed{k_2 = c+d}$, $\boxed{k_3 = e+f}$, and $\boxed{k_1 + k_2 + k_3 = C}$, then

$\boxed{(a+b)+(c+d)+(e+f)} = \boxed{(k_1)+(k_2)+(k_3)} = \boxed{k_1+k_2+k_3} = \boxed{C}$

Example 1.2-4

$\boxed{(2+5)+(7+10)+(9+12)} =$

Solution:

$\boxed{(2+5)+(7+10)+(9+12)} = \boxed{(7)+(17)+(21)} = \boxed{7+17+21} = \boxed{\mathbf{45}}$

Case I c - *Adding Three Integer Numbers Grouped by Parentheses*

Three integer numbers that are grouped by parentheses are added in the following ways, as shown by general and specific example cases:

Case I c-1.

$\boxed{a+(b+c+d)} =$

Let $\boxed{k_1 = b+c+d}$, and $\boxed{a+k_1 = A}$, then

$\boxed{a+(b+c+d)} = \boxed{a+(k_1)} = \boxed{a+k_1} = \boxed{A}$

Example 1.2-5

$\boxed{6+(22+16+5)} =$

Solution:

$\boxed{6+(22+16+5)} = \boxed{6+(43)} = \boxed{6+43} = \boxed{\mathbf{49}}$

Case I c-2.

$\boxed{(a+b+c)+(d+e+f)} =$

Let $\boxed{k_1 = a+b+c}$, $\boxed{k_2 = d+e+f}$, and $\boxed{k_1+k_2 = B}$, then

$\boxed{(a+b+c)+(d+e+f)} = \boxed{(k_1)+(k_2)} = \boxed{k_1+k_2} = \boxed{B}$

Example 1.2-6

$\boxed{(10+3+7)+(20+6+13)} =$

Solution:

$\boxed{(10+3+7)+(20+6+13)} = \boxed{(20)+(39)} = \boxed{20+39} = \boxed{\mathbf{59}}$

Case I d - *Adding Two and Three Integer Numbers Grouped by Parentheses*

Two and three integer numbers that are grouped by parentheses are added in the following ways, as shown by general and specific example cases:

Case I d-1.

$\boxed{(a+b+c)+(d+e)} =$

Let $\boxed{k_1 = a+b+c}$, $\boxed{k_2 = d+e}$, and $\boxed{k_1+k_2 = A}$, then

$\boxed{(a+b+c)+(d+e)} = \boxed{(k_1)+(k_2)} = \boxed{k_1+k_2} = \boxed{A}$

Example 1.2-7

$\boxed{(22+13+8)+(6+24)} =$

Solution:

$\boxed{(22+13+8)+(6+24)} = \boxed{(43)+(30)} = \boxed{43+30} = \boxed{\mathbf{73}}$

Case I d-2.

$\boxed{(a+b)+(c+d+e)+f} =$

Let $\boxed{k_1 = a+b}$, $\boxed{k_2 = c+d+e}$, and $\boxed{k_1 + k_2 + f = B}$, then

$\boxed{(a+b)+(c+d+e)+f} = \boxed{(k_1)+(k_2)+f} = \boxed{k_1 + k_2 + f} = \boxed{B}$

Example 1.2-8

$\boxed{(43+6)+(4+13+7)+9} =$

Solution:

$\boxed{(43+6)+(4+13+7)+9} = \boxed{(49)+(24)+9} = \boxed{49+24+9} = \boxed{\mathbf{82}}$

Case II - Use of Brackets in Addition

In addition, brackets are used in a similar way as parentheses. However, brackets are used to separate mathematical operations that contain integer numbers already grouped by parentheses. Brackets are also used to group numbers in different ways, as shown in the following example cases:

Case II a - *Using Brackets to Add Two Integer Numbers Sub-grouped by Parentheses*
Two integer numbers, already grouped by parentheses, are regrouped by brackets and are added as in the following general and specific example cases:

Case II a-1.

$\boxed{a+[(b+c)+(d+e)]} =$

Let $\boxed{k_1 = b+c}$, $\boxed{k_2 = d+e}$, $\boxed{k_1 + k_2 = k_3}$, and $\boxed{a + k_3 = A}$, then

$\boxed{a+[(b+c)+(d+e)]} = \boxed{a+[(k_1)+(k_2)]} = \boxed{a+[k_1 + k_2]} = \boxed{a+[k_3]} = \boxed{a+k_3} = \boxed{A}$

Example 1.2-9

$\boxed{6+[(10+3)+(4+5)]} =$

Solution:

$\boxed{6+[(10+3)+(4+5)]} = \boxed{6+[(13)+(9)]} = \boxed{6+[13+9]} = \boxed{6+[22]} = \boxed{6+22} = \boxed{\mathbf{28}}$

Case II a-2.

$\boxed{[(a+b)+(c+d)]+(e+f)}=$

Let $\boxed{k_1=a+b}$, $\boxed{k_2=c+d}$, $\boxed{k_3=e+f}$, $\boxed{k_1+k_2=k_4}$, and $\boxed{k_4+k_3=B}$, then

$\boxed{[(a+b)+(c+d)]+(e+f)}=\boxed{[(k_1)+(k_2)]+(k_3)}=\boxed{[k_1+k_2]+k_3}=\boxed{[k_4]+k_3}=\boxed{k_4+k_3}=\boxed{B}$

Example 1.2-10

$\boxed{[(4+7)+(5+9)]+(20+3)}=$

Solution:

$\boxed{[(4+7)+(5+9)]+(20+3)}=\boxed{[(11)+(14)]+(23)}=\boxed{[11+14]+23}=\boxed{[25]+23}=\boxed{25+23}=\boxed{\mathbf{48}}$

Case II a-3.

$\boxed{(a+b)+[(c+d)+(e+f)]}=$

Let $\boxed{k_1=a+b}$, $\boxed{k_2=c+d}$, $\boxed{k_3=e+f}$, $\boxed{k_2+k_3=k_4}$, and $\boxed{k_1+k_4=C}$, then

$\boxed{(a+b)+[(c+d)+(e+f)]}=\boxed{(k_1)+[(k_2)+(k_3)]}=\boxed{k_1+[k_2+k_3]}=\boxed{k_1+[k_4]}=\boxed{k_1+k_4}=\boxed{C}$

Example 1.2-11

$\boxed{(7+12)+[(13+5)+(6+34)]}=$

Solution:

$\boxed{(7+12)+[(13+5)+(6+34)]}=\boxed{(19)+[(18)+(40)]}=\boxed{19+[18+40]}=\boxed{19+[58]}=\boxed{19+58}=\boxed{\mathbf{77}}$

Case II b - *Using Brackets to Add Three Integer Numbers Sub-grouped by Parentheses*
Three integer numbers, already grouped by parentheses, are regrouped by brackets and are added as in the following general and specific example cases:

Case II b-1.

$\boxed{[(a+b+c)+d]+e}=$

Let $\boxed{k_1=a+b+c}$, $\boxed{k_2=k_1+d}$, and $\boxed{k_2+e=A}$, then

$\boxed{[(a+b+c)+d]+e}=\boxed{[(k_1)+d]+e}=\boxed{[k_1+d]+e}=\boxed{[k_2]+e}=\boxed{k_2+e}=\boxed{A}$

Example 1.2-12

$\boxed{[(7+3+25)+4]+6}=$

Solution:

$\boxed{[(7+3+25)+4]+6}=\boxed{[(35)+4]+6}=\boxed{[35+4]+6}=\boxed{[39]+6}=\boxed{39+6}=\boxed{\mathbf{45}}$

Case II b-2.

$\boxed{a+[(b+c+d)+(e+f+g)]}=$

Let $\boxed{k_1=b+c+d}$, $\boxed{k_2=e+f+g}$, $\boxed{k_1+k_2=k_3}$, and $\boxed{a+k_3=B}$, then

$\boxed{a+[(b+c+d)+(e+f+g)]}=\boxed{a+[(k_1)+(k_2)]}=\boxed{a+[k_1+k_2]}=\boxed{a+[k_3]}=\boxed{a+k_3}=\boxed{B}$

Example 1.2-13

$\boxed{20+[(5+12+6)+(3+8+4)]}=$

Solution:

$\boxed{20+[(5+12+6)+(3+8+4)]}=\boxed{20+[(23)+(15)]}=\boxed{20+[23+15]}=\boxed{20+[38]}=\boxed{20+38}=\boxed{\mathbf{58}}$

Case II c - *Using Brackets to Add Two and Three Integer Numbers Sub-grouped by Parentheses*
Two and three integer numbers, already grouped by parentheses, are regrouped by brackets and are added as in the following general and specific example cases:

Case II c-1.

$\boxed{a+[(b+c)+(d+e+f)]}=$

Let $\boxed{k_1=b+c}$, $\boxed{k_2=d+e+f}$, $\boxed{k_1+k_2=k_3}$, and $\boxed{a+k_3=A}$, then

$\boxed{a+[(b+c)+(d+e+f)]}=\boxed{a+[(k_1)+(k_2)]}=\boxed{a+[k_1+k_2]}=\boxed{a+[k_3]}=\boxed{a+k_3}=\boxed{A}$

Example 1.2-14

$\boxed{4+[(3+12)+(9+15+23)]}=$

Solution:

$\boxed{4+[(3+12)+(9+15+23)]}=\boxed{4+[(15)+(47)]}=\boxed{4+[15+47]}=\boxed{4+[62]}=\boxed{4+62}=\boxed{\mathbf{66}}$

Case II c-2.

$\boxed{[(a+b)+(c+d+e)+f]+g}=$

Let $\boxed{k_1=a+b}$, $\boxed{k_2=c+d+e}$, $\boxed{k_1+k_2+f=k_3}$, and $\boxed{k_3+g=B}$, then

$\boxed{[(a+b)+(c+d+e)+f]+g}=\boxed{[(k_1)+(k_2)+f]+g}=\boxed{[k_1+k_2+f]+g}=\boxed{[k_3]+g}=\boxed{k_3+g}=\boxed{B}$

Example 1.2-15

$\boxed{[(3+5)+(4+9+11)+6]+3}=$

Solution:

$\boxed{[(3+5)+(4+9+11)+6]+3}=\boxed{[(8)+(24)+6]+3}=\boxed{[8+24+6]+3}=\boxed{[38]+3}=\boxed{38+3}=\boxed{\mathbf{41}}$

Case II c-3.

$\boxed{(a+b)+[(c+d+e)+(f+g)+h]} =$

Let $\boxed{k_1 = a+b}$, $\boxed{k_2 = c+d+e}$, $\boxed{k_3 = f+g}$, $\boxed{k_2+k_3+h = k_4}$, and $\boxed{k_1+k_4 = C}$, then

$\boxed{(a+b)+[(c+d+e)+(f+g)+h]} = \boxed{(k_1)+[(k_2)+(k_3)+h]} = \boxed{k_1+[k_2+k_3+h]} = \boxed{k_1+[k_4]} = \boxed{k_1+k_4} = \boxed{C}$

Example 1.2-16

$\boxed{(4+3)+[(6+9+12)+(30+5)+1]} =$

Solution:

$\boxed{(4+3)+[(6+9+12)+(30+5)+1]} = \boxed{7+[(27)+(35)+1]} = \boxed{7+[27+35+1]} = \boxed{7+[63]} = \boxed{7+63} = \boxed{\mathbf{70}}$

Additional Examples - Use of Parentheses and Brackets in Addition

The following examples further illustrate how to use parentheses and brackets in addition:

Example 1.2-17

$\boxed{5+(2+13+8)+(8+20)} = \boxed{5+(23)+(28)} = \boxed{5+23+28} = \boxed{\mathbf{56}}$

Example 1.2-18

$\boxed{(25+33)+(8+13)+7} = \boxed{(58)+(21)+7} = \boxed{58+21+7} = \boxed{\mathbf{86}}$

Example 1.2-19

$\boxed{18+(52+10+7)+15+(6+24)} = \boxed{18+(69)+15+(30)} = \boxed{18+69+15+30} = \boxed{\mathbf{132}}$

Example 1.2-20

$\boxed{12+[3+(16+4)+(2+13+5)]} = \boxed{12+[3+(20)+(20)]} = \boxed{12+[3+20+20]} = \boxed{12+[43]} = \boxed{12+43} = \boxed{\mathbf{55}}$

Example 1.2-21

$\boxed{(26+11+7)+[(2+13)+(23+8)+20]} = \boxed{(44)+[(15)+(31)+20]} = \boxed{44+[15+31+20]} = \boxed{44+[66]}$

$= \boxed{44+66} = \boxed{\mathbf{110}}$

Example 1.2-22

$\boxed{[12+(12+6+10)+(18+4+9)]+(16+5)} = \boxed{[12+(28)+(31)]+(21)} = \boxed{[12+28+31]+21} = \boxed{[71]+21}$

$= \boxed{71+21} = \boxed{\mathbf{92}}$

Example 1.2-23

$\boxed{23+(12+5)+[7+(12+9)]} = \boxed{23+(17)+[7+(21)]} = \boxed{23+17+[7+21]} = \boxed{23+17+[28]} = \boxed{23+17+28}$

$= \boxed{\mathbf{68}}$

Example 1.2-24

$\boxed{[(12+3+8)+(32+4)+3]+(5+20)} = \boxed{[(23)+(36)+3]+(25)} = \boxed{[23+36+3]+25} = \boxed{[62]+25} = \boxed{62+25}$

$= \boxed{\mathbf{87}}$

Example 1.2-25

$\boxed{(23+13)+7+[23+(12+9)]} = \boxed{(36)+7+[23+(21)]} = \boxed{36+7+[23+21]} = \boxed{43+[44]} = \boxed{43+44} = \boxed{\mathbf{87}}$

Example 1.2-26

$\boxed{[(25+13+2)+(16+84)]+(10+3)+5} = \boxed{[(40)+(100)]+(13)+5} = \boxed{[40+100]+13+5} = \boxed{[140]+18}$

$= \boxed{140+18} = \boxed{\mathbf{158}}$

Practice Problems - Use of Parentheses and Brackets in Addition

Section 1.2 Practice Problems - Add the following numbers in the order grouped:

1. $2+3+5+6 =$
2. $(2+5)+(6+3)+9 =$
3. $(6+3+8)+(2+3)+4 =$
4. $8+[(1+3+4)+(1+2)] =$
5. $[(18+4)+9]+[1+(2+3)] =$
6. $8+[(2+3)+(6+3)+15] =$
7. $(7+3+8)+[(7+2+3)+5] =$
8. $[(3+9+4)+1+(1+8)]+(8+2) =$
9. $[(2+3+6)+(1+8)]+[(1+3)+4] =$
10. $[[(3+5)+(4+3)+5]+(2+3+5)]+6 =$

1.3 Using Parentheses and Brackets in Subtraction

In this section the use of parentheses and brackets as applied to subtraction is discussed. Changing the order in which numbers are subtracted or grouped does alter the final answer. The following two properties associated with subtraction are discussed first and are as follows:

Properties Associated with Subtraction

1. Changing the order in which two numbers are subtracted does change the final answer. For example, for any two real numbers a and b

   $\boxed{a-b \neq b-a}$ *Note:* The symbol "$\neq$" means not equal.

   For example, $\boxed{20-8} = \boxed{12}$, but $\boxed{8-20} = \boxed{-12}$

2. Re-grouping numbers does change the final answer. For example, for any real numbers a, b, and c

   $\boxed{(a-b)-c \neq a-(b-c)}$

   For example,

   $\boxed{(25-6)-8} = \boxed{(19)-8} = \boxed{19-8} = \boxed{11}$, however

   $\boxed{25-(6-8)} = \boxed{25-(-2)} = \boxed{25+(2)} = \boxed{25+2} = \boxed{27}$

In the following cases the use of parentheses and brackets in subtraction, using integer numbers, is discussed:

Case I - Use of Parentheses in Subtraction

In subtraction, parentheses can be grouped in different ways as shown in the following example cases:

Case I a - *Subtracting Integer Numbers Without Using Parentheses*
Integer numbers are subtracted without the use of parentheses, as shown in the following general and specific examples:

$\boxed{a-b-c-d-e} =$

Let $\boxed{a-b-c-d-e=A}$, then

$\boxed{a-b-c-d-e} = \boxed{A}$

Example 1.3-1

$\boxed{24-5-13-7-8} =$

Solution:

$\boxed{24-5-13-7-8} = \boxed{-9}$

Case I b - *Subtracting Two Integer Numbers Grouped by Parentheses*
Two integer numbers that are grouped by parentheses are subtracted in the following ways as shown by general and specific example cases:

Case I b-1.

$\boxed{a-(b-c)} =$

Let $\boxed{k_1 = b-c}$, and $\boxed{a-k_1 = A}$, then

$\boxed{a-(b-c)} = \boxed{a-(k_1)} = \boxed{a-k_1} = \boxed{A}$

Example 1.3-2

$\boxed{20-(15-45)} =$

Solution:

$\boxed{20-(15-45)} = \boxed{20-(-30)} = \boxed{20+(30)} = \boxed{20+30} = \boxed{\mathbf{50}}$

Case I b-2.

$\boxed{(a-b)-(c-d)} =$

Let $\boxed{k_1 = a-b}$, $\boxed{k_2 = c-d}$, and $\boxed{k_1 - k_2 = B}$, then

$\boxed{(a-b)-(c-d)} = \boxed{(k_1)-(k_2)} = \boxed{k_1 - k_2} = \boxed{B}$

Example 1.3-3

$\boxed{(20-25)-(7-5)} =$

Solution:

$\boxed{(20-25)-(7-5)} = \boxed{(-5)-(2)} = \boxed{-5-2} = \boxed{-7}$

Case I b-3.

$\boxed{a-(b-c)-(d-e)} =$

Let $\boxed{k_1 = b-c}$, $\boxed{k_2 = d-e}$, and $\boxed{a-k_1-k_2 = C}$, then

$\boxed{a-(b-c)-(d-e)} = \boxed{a-(k_1)-(k_2)} = \boxed{a-k_1-k_2} = \boxed{C}$

Example 1.3-4

$\boxed{25-(35-12)-(8-3)} =$

Solution:

$\boxed{25-(35-12)-(8-3)} = \boxed{25-(23)-(5)} = \boxed{25-23-5} = \boxed{\mathbf{-3}}$

Case I c - *Subtracting Three Integer Numbers Grouped by Parentheses*
Three integer numbers that are grouped by parentheses are subtracted in the following ways, as shown by general and specific example cases:

Case I c-1.

$\boxed{a-(b-c-d)}=$

Let $\boxed{k_1=b-c-d}$, and $\boxed{a-k_1=A}$, then

$\boxed{a-(b-c-d)}=\boxed{a-(k_1)}=\boxed{a-k_1}=\boxed{A}$

Example 1.3-5

$\boxed{6-(22-16-8)}=$

Solution:

$\boxed{6-(22-16-8)}=\boxed{6-(-2)}=\boxed{6+(2)}=\boxed{6+2}=\boxed{\mathbf{8}}$

Case I c-2.

$\boxed{(a-b-c)-(d-e-f)}=$

Let $\boxed{k_1=a-b-c}$, $\boxed{k_2=d-e-f}$, and $\boxed{k_1-k_2=B}$, then

$\boxed{(a-b-c)-(d-e-f)}=\boxed{(k_1)-(k_2)}=\boxed{k_1-k_2}=\boxed{B}$

Example 1.3-6

$\boxed{(15-3-8)-(40-9-34)}=$

Solution:

$\boxed{(15-3-8)-(40-9-34)}=\boxed{(4)-(-3)}=\boxed{4+(3)}=\boxed{4+3}=\boxed{7}$

Case I d - *Subtracting Two and Three Integer Numbers Grouped by Parentheses*
Two and three integer numbers that are grouped by parentheses are subtracted in the following ways, as shown by general and specific example cases:

Case I d-1.

$\boxed{(a-b)-(c-d-e)-f}=$

Let $\boxed{k_1=a-b}$, $\boxed{k_2=c-d-e}$, and $\boxed{k_1-k_2-f=A}$, then

$\boxed{(a-b)-(c-d-e)-f}=\boxed{(k_1)-(k_2)-f}=\boxed{k_1-k_2-f}=\boxed{A}$

Example 1.3-7

$\boxed{(43-6)-(54-13-7)-19}=$

Solution:

$\boxed{(43-6)-(54-13-7)-19} = \boxed{(37)-(34)-19} = \boxed{37-34-19} = \boxed{3-19} = \boxed{\mathbf{-16}}$

Case I d-2.

$\boxed{(a-b-c)-(d-e)} =$

Let $\boxed{k_1 = a-b-c}$, $\boxed{k_2 = d-e}$, and $\boxed{k_1 - k_2 = B}$, then

$\boxed{(a-b-c)-(d-e)} = \boxed{(k_1)-(k_2)} = \boxed{k_1 - k_2} = \boxed{B}$

Example 1.3-8

$\boxed{(8-13-10)-(6-36)} =$

Solution:

$\boxed{(8-13-10)-(6-36)} = \boxed{(-15)-(-30)} = \boxed{-15+(30)} = \boxed{-15+30} = \boxed{\mathbf{15}}$

Case II - Use of Brackets in Subtraction

In subtraction, brackets are used in a similar way as parentheses. However, brackets are used to separate mathematical operations that contain integer numbers already grouped by parentheses. Brackets are also used to group numbers in different ways, as shown in the following example cases:

Case II a - *Using Brackets to Subtract Two Integer Numbers Sub-grouped by Parentheses*
Two integer numbers, already grouped by parentheses, are regrouped by brackets and subtracted as in the following general and specific example cases:

Case II a-1.

$\boxed{[a-(b-c)]-d} =$

Let $\boxed{k_1 = b-c}$, $\boxed{a-k_1 = k_2}$, and $\boxed{k_2 - d = A}$, then

$\boxed{[a-(b-c)]-d} = \boxed{[a-k_1]-d} = \boxed{[k_2]-d} = \boxed{k_2 - d} = \boxed{A}$

Example 1.3-9

$\boxed{[38-(12-9)]-30} =$

Solution:

$\boxed{[38-(12-9)]-30} = \boxed{[38-(3)]-30} = \boxed{[38-3]-30} = \boxed{[35]-30} = \boxed{35-30} = \boxed{\mathbf{5}}$

Case II a-2.

$\boxed{[(a-b)-c]-d} =$

Let $\boxed{k_1 = a - b}$, $\boxed{k_2 = k_1 - c}$, and $\boxed{k_2 - d = B}$, then

$\boxed{[(a-b)-c]-d} = \boxed{[(k_1)-c]-d} = \boxed{[k_1 - c]-d} = \boxed{[k_2]-d} = \boxed{k_2 - d} = \boxed{B}$

Example 1.3-10

$\boxed{[(9-23)-12]-40} =$

Solution:

$\boxed{[(9-23)-12]-40} = \boxed{[(-14)-12]-40} = \boxed{[-14-12]-40} = \boxed{[-26]-40} = \boxed{-26-40} = \boxed{\mathbf{-66}}$

Case II a-3.

$\boxed{a-[(b-c)-d]} =$

Let $\boxed{k_1 = b - c}$, $\boxed{k_2 = k_1 - d}$, and $\boxed{a - k_2 = C}$, then

$\boxed{a-[(b-c)-d]} = \boxed{a-[(k_1)-d]} = \boxed{a-[k_1 - d]} = \boxed{a-[k_2]} = \boxed{a-k_2} = \boxed{C}$

Example 1.3-11

$\boxed{5-[(18-7)-27]} =$

Solution:

$\boxed{5-[(18-7)-27]} = \boxed{5-[(11)-27]} = \boxed{5-[11-27]} = \boxed{5-[-16]} = \boxed{5+[16]} = \boxed{5+16} = \boxed{\mathbf{21}}$

Case II a-4.

$\boxed{a-[(b-c)-(d-e)]} =$

Let $\boxed{k_1 = b - c}$, $\boxed{k_2 = d - e}$, $\boxed{k_1 - k_2 = k_3}$, and $\boxed{a - k_3 = D}$, then

$\boxed{a-[(b-c)-(d-e)]} = \boxed{a-[(k_1)-(k_2)]} = \boxed{a-[k_1 - k_2]} = \boxed{a-[k_3]} = \boxed{a-k_3} = \boxed{D}$

Example 1.3-12

$\boxed{26-[(10-6)-(4-9)]} =$

Solution:

$\boxed{26-[(10-6)-(4-9)]} = \boxed{26-[(4)-(-5)]} = \boxed{26-[4+(5)]} = \boxed{26-[4+5]} = \boxed{26-[9]} = \boxed{26-9} = \boxed{\mathbf{17}}$

Case II a-5.

$\boxed{(a-b)-[(c-d)-(e-f)]} =$

Let; $\boxed{k_1 = a - b}$, $\boxed{k_2 = c - d}$, $\boxed{k_3 = e - f}$, $\boxed{k_2 - k_3 = k_4}$, and $\boxed{k_1 - k_4 = E}$, then

$\boxed{(a-b)-[(c-d)-(e-f)]} = \boxed{(k_1)-[(k_2)-(k_3)]} = \boxed{k_1-[k_2-k_3]} = \boxed{k_1-[k_4]} = \boxed{k_1-k_4} = \boxed{E}$

Example 1.3-13

$\boxed{(27-14)-[(13-9)-(26-8)]} =$

Solution:

$\boxed{(27-14)-[(13-9)-(26-8)]} = \boxed{(13)-[(4)-(18)]} = \boxed{13-[4-18]} = \boxed{13-[-14]} = \boxed{13+[14]} = \boxed{13+14} = \boxed{\mathbf{27}}$

Case II b - *Using Brackets to Subtract Three Integer Numbers Sub-grouped by Parentheses*
Three integer numbers, already grouped by parentheses, are regrouped by brackets and subtracted as in the following general and specific example cases:

Case II b-1.

$\boxed{[(a-b-c)-d]-e} =$

Let $\boxed{k_1=a-b-c}$, $\boxed{k_2=k_1-d}$, and $\boxed{k_2-e=A}$, then

$\boxed{[(a-b-c)-d]-e} = \boxed{[(k_1)-d]-e} = \boxed{[k_1-d]-e} = \boxed{[k_2]-e} = \boxed{k_2-e} = \boxed{A}$

Example 1.3-14

$\boxed{[(45-13-7)-15]-20} =$

Solution:

$\boxed{[(45-13-7)-15]-20} = \boxed{[(25)-15]-20} = \boxed{[25-15]-20} = \boxed{[10]-20} = \boxed{10-20} = \boxed{\mathbf{-10}}$

Case II b-2.

$\boxed{a-[(b-c-d)-(e-f-g)]} =$

Let $\boxed{k_1=b-c-d}$, $\boxed{k_2=e-f-g}$, $\boxed{k_1-k_2=k_3}$, and $\boxed{a-k_3=B}$, then

$\boxed{a-[(b-c-d)-(e-f-g)]} = \boxed{a-[(k_1)-(k_2)]} = \boxed{a-[k_1-k_2]} = \boxed{a-[k_3]} = \boxed{a-k_3} = \boxed{B}$

Example 1.3-15

$\boxed{50-[(5-25-7)-(36-12-5)]} =$

Solution:

$\boxed{50-[(5-25-7)-(36-12-5)]} = \boxed{50-[(-27)-(19)]} = \boxed{50-[-27-19]} = \boxed{50-[-46]} = \boxed{50+[46]}$

$= \boxed{50+46} = \boxed{\mathbf{96}}$

Case II c - *Using Brackets to Subtract Two and Three Integer Numbers Sub-grouped by Parentheses*

Two and three integer numbers, already grouped by parentheses, are regrouped by brackets and subtracted as in the following general and specific example cases:

Case II c-1.

$\boxed{[(a-b)-(c-d-e)]-f} =$

Let $\boxed{k_1 = a-b}$, $\boxed{k_2 = c-d-e}$, $\boxed{k_1 - k_2 = k_3}$, and $\boxed{k_3 - f = A}$, then

$\boxed{[(a-b)-(c-d-e)]-f} = \boxed{[(k_1)-(k_2)]-f} = \boxed{[k_1 - k_2]-f} = \boxed{[k_3]-f} = \boxed{k_3 - f} = \boxed{A}$

Example 1.3-16

$\boxed{[(300-450)-(100-35-55)]-12} =$

Solution:

$\boxed{[(300-450)-(100-35-55)]-12} = \boxed{[(-150)-(10)]-12} = \boxed{[-150-10]-12} = \boxed{[-160]-12} = \boxed{-160-12}$

$= \boxed{\mathbf{-172}}$

Case II c-2.

$\boxed{a-[(b-c)-(d-e-f)]} =$

Let $\boxed{k_1 = b-c}$, $\boxed{k_2 = d-e-f}$, $\boxed{k_1 - k_2 = k_3}$, and $\boxed{a - k_3 = B}$, then

$\boxed{a-[(b-c)-(d-e-f)]} = \boxed{a-[(k_1)-(k_2)]} = \boxed{a-[k_1-k_2]} = \boxed{a-[k_3]} = \boxed{a-k_3} = \boxed{B}$

Example 1.3-17

$\boxed{34-[(324-130)-(250-39-85)]} =$

Solution:

$\boxed{34-[(324-130)-(250-39-85)]} = \boxed{34-[(194)-(126)]} = \boxed{34-[194-126]} = \boxed{34-[68]} = \boxed{34-68} = \boxed{\mathbf{-34}}$

Case II c-3.

$\boxed{[(a-b)-(c-d-e)-f]-g} =$

Let $\boxed{k_1 = a-b}$, $\boxed{k_2 = c-d-e}$, $\boxed{k_1 - k_2 - f = k_3}$, and $\boxed{k_3 - g = C}$, then

$\boxed{[(a-b)-(c-d-e)-f]-g} = \boxed{[(k_1)-(k_2)-f]-g} = \boxed{[k_1-k_2-f]-g} = \boxed{[k_3]-g} = \boxed{k_3-g} = \boxed{C}$

Example 1.3-18

$\boxed{[(13-8)-(24-9-15)-6]-30} =$

Solution:

$[(13-8)-(24-9-15)-6]-30 = [(5)-(0)-6]-30 = [5-0-6]-30 = [-1]-30 = -1-30 = \mathbf{-31}$

Case II c-4.

$(a-b)-[(c-d-e)-(f-g)] =$

Let $k_1 = a-b$, $k_2 = c-d-e$, $k_3 = f-g$, $k_2 - k_3 = k_4$, and $k_1 - k_4 = D$, then

$(a-b)-[(c-d-e)-(f-g)] = (k_1)-[(k_2)-(k_3)] = k_1 - [k_2 - k_3] = k_1 - [k_4] = k_1 - k_4 = D$

Example 1.3-19

$(135-12)-[(8-6-4)-(20-18)] =$

Solution:

$(135-12)-[(8-6-4)-(20-18)] = (123)-[(-2)-(2)] = (123)-[-2-2] = (123)-[-4] = 123+[4]$

$= 123+4 = \mathbf{127}$

Additional Examples - Use of Parentheses and Brackets in Subtraction

Example 1.3-20

$(-35-3)-(10-3) = (-38)-(7) = -38-7 = \mathbf{-45}$

Example 1.3-21

$35-(12-3)-8 = 35-(9)-8 = 35-9-8 = 35-17 = \mathbf{18}$

Example 1.3-22

$-(38-6)-(-12+4) = -(32)-(-8) = -32+8 = \mathbf{-24}$

Example 1.3-23

$[(-8-12)-3]-(-3+2-4) = [(-20)-3]-(-5) = [-20-3]+5 = [-23]+5 = -23+5 = \mathbf{-18}$

Example 1.3-24

$-[(18-6)-(20-2)]-5 = -[(12)-(18)]-5 = -[12-18]-5 = -[-6]-5 = +6-5 = +1 = \mathbf{1}$

Example 1.3-25

$[-18-(22-4)]-(20-5-4) = [-18-(18)]-(20-9) = [-18-18]-(11) = [-36]-11 = -36-11 = \mathbf{-47}$

Example 1.3-26

$-(25-4)-[(8-3)-6] = -(21)-[(5)-6] = -21-[5-6] = -21-[-1] = -21+[1] = -21+1 = \mathbf{-20}$

Example 1.3-27

$$\boxed{-16-(-23+5-2)+6}=\boxed{-16-(-23+3)+6}=\boxed{-16-(-20)+6}=\boxed{-16+20+6}=\boxed{-16+26}=\boxed{+10}=\boxed{\mathbf{10}}$$

Example 1.3-28

$$\boxed{-[(5-8-2)-(2-3)]-(8-5)}=\boxed{-[(5-10)-(-1)]-(3)}=\boxed{-[(-5)+(1)]-3}=\boxed{-[-5+1]-3}=\boxed{-[-4]-3}$$

$$=\boxed{+[4]-3}=\boxed{4-3}=\boxed{\mathbf{1}}$$

Example 1.3-29

$$\boxed{(58-20)-[(18-6)-(8+2)]}=\boxed{(38)-[(12)-(10)]}=\boxed{38-[12-10]}=\boxed{38-[2]}=\boxed{38-2}=\boxed{\mathbf{36}}$$

Practice Problems - Use of Parentheses and Brackets in Subtraction

Section 1.3 Practice Problems - Subtract the following numbers in the order grouped:

1. $(55-5)-3-8=$
2. $59-38-12-(20-5)=$
3. $(20-5)-(11-2)=$
4. $[-25-(4-13)]-5=$
5. $350-(25-38)-30=$
6. $[(-30-3)-8]-(16-9)=$
7. $[(40-4)-(8-10)]-9=$
8. $(35-56)-[(20-15)-8]=$
9. $[(-175-55)-245]-(5-6)=$
10. $(48-80)-[(12-2)-(15-37)]=$

1.4 Using Parentheses and Brackets in Multiplication

Parentheses and brackets are the tools used for grouping numbers. In this section the use of parentheses and brackets as applied to multiplication is discussed. The following properties associated with multiplication are discussed first and are as follows:

Commutative, Associative, and Distributive Property of Multiplication

1. Changing the order in which two numbers are multiplied does not change the final answer. This property of real numbers is called the **Commutative Property of Multiplication**, e.g., for any two real numbers a and b

   $\boxed{a \times b} = \boxed{b \times a}$

   For example, $\boxed{3 \times 15} = \boxed{\mathbf{45}}$ and $\boxed{15 \times 3} = \boxed{\mathbf{45}}$

2. Re-grouping numbers does not change the final answer. This property of real numbers is called the **Associative Property of Multiplication**, e.g., for any real numbers a, b, and c

   $\boxed{(a \times b) \times c} = \boxed{a \times (b \times c)}$

   For example,

   $\boxed{(4 \times 8) \times 5} = \boxed{(32) \times 5} = \boxed{32 \times 5} = \boxed{\mathbf{160}}$

   $\boxed{4 \times (8 \times 5)} = \boxed{4 \times (40)} = \boxed{4 \times 40} = \boxed{\mathbf{160}}$

3. Multiplication can be distributed over addition. This property is called the **Distributive Property of Multiplication**, e.g., for any real numbers a, b, and c

   $\boxed{a \times (b + c)} = \boxed{ab + ac}$

   For example,

   $\boxed{9 \times (4 + 5)} = \boxed{(9 \times 4) + (9 \times 5)} = \boxed{36 + 45} = \boxed{\mathbf{81}}$

Similar to addition (see Section 1.2), changing the order in which numbers are multiplied or grouped does not affect the final answer. However, again, it is important to learn how to solve math operations in the exact order in which parentheses or brackets are used in grouping numbers. The use of parentheses and brackets in multiplication, using integer numbers, is discussed in the following cases:

Case I - Use of Parentheses in Multiplication

In multiplication, parentheses can be grouped in different ways, as shown in the following example cases:

Case I a - *Multiplying Integer Numbers Without Using Parentheses*

Integer numbers are multiplied without the use of parentheses, as shown in the following general and specific examples:

$\boxed{a \times b \times c \times d \times e} = \boxed{abcde}$

Example 1.4-1

$\boxed{3 \times 5 \times 7 \times 2 \times 4} =$

Solution:

$\boxed{3 \times 5 \times 7 \times 2 \times 4} = \boxed{\mathbf{840}}$

Case I b - *Multiplying Two Integer Numbers Grouped by Parentheses*

Two integer numbers that are grouped by parentheses are multiplied in the following ways, as shown by general and specific example cases:

Case I b-1.

$\boxed{(a \times b) \times (c \times d) \times (e \times f)} = \boxed{(ab) \times (cd) \times (ef)} = \boxed{ab \times cd \times ef} = \boxed{abcdef}$

Example 1.4-2

$\boxed{(2 \times 5) \times (7 \times 4) \times (1 \times 3)} =$

Solution:

$\boxed{(2 \times 5) \times (7 \times 4) \times (1 \times 3)} = \boxed{(10) \times (28) \times (3)} = \boxed{10 \times 28 \times 3} = \boxed{\mathbf{840}}$

Case I b-2.

$\boxed{a \times (b \times c) \times (d \times e) \times f} = \boxed{a \times (bc) \times (de) \times f} = \boxed{a \times bc \times de \times f} = \boxed{abcdef}$

Example 1.4-3

$\boxed{2 \times (5 \times 3) \times (6 \times 4) \times 7} =$

Solution:

$\boxed{2 \times (5 \times 3) \times (6 \times 4) \times 7} = \boxed{2 \times (15) \times (24) \times 7} = \boxed{2 \times 15 \times 24 \times 7} = \boxed{\mathbf{5040}}$

Case I c - *Multiplying Three Integer Numbers Grouped by Parentheses*

Three integer numbers that are grouped by parentheses are multiplied in the following ways, as shown by general and specific example cases:

Case I c-1.

$\boxed{a \times (b \times c \times d)} = \boxed{a \times (bcd)} = \boxed{a \times bcd} = \boxed{abcd}$

Example 1.4-4

$\boxed{2 \times (3 \times 8 \times 10)} =$

Solution:

$\boxed{2 \times (3 \times 8 \times 10)} = \boxed{2 \times (240)} = \boxed{2 \times 240} = \boxed{\mathbf{480}}$

Case I c-2.

$\boxed{(a \times b \times c) \times (d \times e \times f)} = \boxed{(abc) \times (def)} = \boxed{abc \times def} = \boxed{abcdef}$

Example 1.4-5

$\boxed{(5\times3\times2)\times(10\times4\times7)}=$

Solution:

$\boxed{(5\times3\times2)\times(10\times4\times7)}=\boxed{(30)\times(280)}=\boxed{30\times280}=\boxed{\mathbf{8400}}$

Case I d - *Multiplying Two and Three Integer Numbers Grouped by Parentheses*

Two and three integer numbers that are grouped by parentheses are multiplied in the following ways, as shown by general and specific example cases:

Case I d-1.

$\boxed{(a\times b)\times(c\times d\times e)\times f}=\boxed{(ab)\times(cde)\times f}=\boxed{abcdef}$

Example 1.4-6

$\boxed{(3\times1)\times(4\times5\times11)\times2}=$

Solution:

$\boxed{(3\times1)\times(4\times5\times11)\times2}=\boxed{(3)\times(220)\times2}=\boxed{3\times220\times2}=\boxed{\mathbf{1320}}$

Case I d-2.

$\boxed{(a\times b\times c)\times(d\times e)}=\boxed{(abc)\times(de)}=\boxed{abc\times de}=\boxed{abcde}$

Example 1.4-7

$\boxed{(2\times9\times8)\times(6\times4)}=$

Solution:

$\boxed{(2\times9\times8)\times(6\times4)}=\boxed{(144)\times(24)}=\boxed{144\times24}=\boxed{\mathbf{3456}}$

Case II - Use of Brackets in Multiplication

In multiplication, brackets are used in a similar way as parentheses. However, brackets are used to separate mathematical operations that contain integer numbers already grouped by parentheses. Brackets are also used to group numbers in different ways, as shown in the following example cases:

Case II a - *Using Brackets to Multiply Two Integer Numbers Sub-grouped by Parentheses*
Two integer numbers already grouped by parentheses are regrouped by brackets and are multiplied as in the following general and specific example cases:

Case II a-1.

$\boxed{a\times[(b\times c)\times(d\times e)]}=\boxed{a\times[(bc)\times(de)]}=\boxed{a\times[bc\times de]}=\boxed{a\times[bcde]}=\boxed{a\times bcde}=\boxed{abcde}$

Example 1.4-8

$\boxed{6\times[(12\times3)\times(4\times1)]}=$

Solution:

$\boxed{6\times\left[(12\times 3)\times(4\times 1)\right]} = \boxed{6\times\left[(36)\times(4)\right]} = \boxed{6\times\left[36\times 4\right]} = \boxed{6\times\left[144\right]} = \boxed{6\times 144} = \boxed{\mathbf{864}}$

Case II a-2.

$\boxed{\left[(a\times b)\times(c\times d)\right]\times(e\times f)} = \boxed{\left[(ab)\times(cd)\right]\times(ef)} = \boxed{\left[ab\times cd\right]ef} = \boxed{\left[abcd\right]ef} = \boxed{abcdef}$

Example 1.4-9

$\boxed{\left[(4\times 1)\times(5\times 9)\right]\times(2\times 3)} =$

Solution:

$\boxed{\left[(4\times 1)\times(5\times 9)\right]\times(2\times 3)} = \boxed{\left[(4)\times(45)\right]\times(6)} = \boxed{\left[4\times 45\right]\times 6} = \boxed{\left[180\right]\times 6} = \boxed{180\times 6} = \boxed{\mathbf{1080}}$

Case II a-3.

$\boxed{(a\times b)\times\left[(c\times d)\times(e\times f)\right]} = \boxed{(ab)\times\left[(cd)\times(ef)\right]} = \boxed{ab\times\left[cd\times ef\right]} = \boxed{ab\times\left[cdef\right]} = \boxed{ab\times cdef} = \boxed{abcdef}$

Example 1.4-10

$\boxed{(7\times 4)\times\left[(13\times 2)\times(6\times 1)\right]} =$

Solution:

$\boxed{(7\times 4)\times\left[(13\times 2)\times(6\times 1)\right]} = \boxed{(28)\times\left[(26)\times(6)\right]} = \boxed{28\times\left[156\right]} = \boxed{28\times 156} = \boxed{\mathbf{4368}}$

Case II b - *Using Brackets to Multiply Three Integer Numbers Sub-grouped by Parentheses*
Three integer numbers, already grouped by parentheses, are regrouped by brackets and are multiplied as in the following general and specific example cases:

Case II b-1.

$\boxed{\left[(a\times b\times c)\times d\right]\times e} = \boxed{\left[(abc)\times d\right]\times e} = \boxed{\left[abc\times d\right]\times e} = \boxed{\left[abcd\right]\times e} = \boxed{abcd\times e} = \boxed{abcde}$

Example 1.4-11

$\boxed{\left[(7\times 3\times 10)\times 4\right]\times 2} =$

Solution:

$\boxed{\left[(7\times 3\times 10)\times 4\right]\times 2} = \boxed{\left[(210)\times 4\right]\times 2} = \boxed{\left[210\times 4\right]\times 2} = \boxed{\left[840\right]\times 2} = \boxed{840\times 2} = \boxed{\mathbf{1680}}$

Case II b-2.

$\boxed{a\times\left[(b\times c\times d)\times(e\times f\times g)\right]} = \boxed{a\times\left[(bcd)\times(efg)\right]} = \boxed{a\times\left[bcd\times efg\right]} = \boxed{a\times\left[bcdefg\right]} = \boxed{a\times bcdefg} = \boxed{abcdefg}$

Example 1.4-12

$\boxed{2\times\left[(5\times 1\times 6)\times(3\times 8\times 4)\right]} =$

Solution:

$\boxed{2\times\left[(5\times 1\times 6)\times(3\times 8\times 4)\right]} = \boxed{2\times\left[(30)\times(96)\right]} = \boxed{2\times\left[30\times 96\right]} = \boxed{2\times\left[2880\right]} = \boxed{2\times 2880} = \boxed{\mathbf{5760}}$

Case II c - *Using Brackets to Multiply Two and Three Integer Numbers Sub-grouped by Parentheses*

Two and three integer numbers, already grouped by parentheses, are regrouped by brackets and are multiplied as in the following general and specific example cases:

Case II c-1.

$$\boxed{a\times[(b\times c)\times(d\times e\times f)]}=\boxed{a\times[(bc)\times(def)]}=\boxed{a\times[bc\times def]}=\boxed{a\times[bcdef]}=\boxed{a\times bcdef}=\boxed{abcdef}$$

Example 1.4-13

$$\boxed{2\times[(3\times 7)\times(1\times 10\times 5)]}=$$

Solution:

$$\boxed{2\times[(3\times 7)\times(1\times 10\times 5)]}=\boxed{2\times[(21)\times(50)]}=\boxed{2\times[21\times 50]}=\boxed{2\times[1050]}=\boxed{2\times 1050}=\boxed{\mathbf{2100}}$$

Case II c-2.

$$\boxed{[(a\times b)\times(c\times d\times e)\times f]\times g}=\boxed{[(ab)\times(cde)\times f]\times g}=\boxed{[ab\times cde\times f]\times g}=\boxed{[abcdef]\times g}=\boxed{abcdef\times g}$$

$$=\boxed{abcdefg}$$

Example 1.4-14

$$\boxed{[(3\times 5)\times(4\times 1\times 7)\times 6]\times 2}=$$

Solution:

$$\boxed{[(3\times 5)\times(4\times 1\times 7)\times 6]\times 2}=\boxed{[(15)\times(28)\times 6]\times 2}=\boxed{[15\times 28\times 6]\times 2}=\boxed{[2520]\times 2}=\boxed{2520\times 2}=\boxed{\mathbf{5040}}$$

Case II c-3.

$$\boxed{(a\times b)\times[(c\times d\times e)\times(f\times g)\times h]}=\boxed{(ab)\times[(cde)\times(fg)\times h]}=\boxed{ab\times[cde\times fg\times h]}=\boxed{ab\times[cdefgh]}$$

$$=\boxed{ab\times cdefgh}=\boxed{abcdefgh}$$

Example 1.4-15

$$\boxed{(5\times 3)\times[(6\times 2\times 8)\times(7\times 4)\times 1]}=$$

Solution:

$$\boxed{(5\times 3)\times[(6\times 2\times 8)\times(7\times 4)\times 1]}=\boxed{(15)\times[(96)\times(28)\times 1]}=\boxed{15\times[96\times 28\times 1]}=\boxed{15\times[2688]}=\boxed{15\times 2688}$$

$$=\boxed{\mathbf{40320}}$$

Additional Examples - Use of Parentheses and Brackets in Multiplication

The following examples further illustrate how to use parentheses and brackets in multiplication:

Example 1.4-16

$\boxed{5\times(2\times11\times8)\times(4\times6)}=\boxed{5\times(176)\times(24)}=\boxed{5\times176\times24}=\boxed{\mathbf{21120}}$

Example 1.4-17

$\boxed{(6\times5)\times(8\times10)\times3}=\boxed{(30)\times(80)\times3}=\boxed{30\times80\times3}=\boxed{\mathbf{7200}}$

Example 1.4-18

$\boxed{(2\times10\times7)\times(6\times2)\times4}=\boxed{(140)\times(12)\times4}=\boxed{140\times12\times4}=\boxed{\mathbf{6720}}$

Example 1.4-19

$\boxed{9\times[3\times(10\times4)\times(2\times7\times5)]}=\boxed{9\times[3\times(40)\times(70)]}=\boxed{9\times[3\times40\times70]}=\boxed{9\times[8400]}=\boxed{9\times8400}=\boxed{\mathbf{75600}}$

Example 1.4-20

$\boxed{(20\times1\times5)\times[(2\times6)\times(4\times8)\times3]}=\boxed{(100)\times[(12)\times(32)\times3]}=\boxed{100\times[12\times32\times3]}=\boxed{100\times[1152]}$

$=\boxed{100\times1152}=\boxed{\mathbf{115200}}$

Example 1.4-21

$\boxed{[(5\times3)\times(11\times4\times2)]\times(6\times5)}=\boxed{[(15)\times(88)]\times(30)}=\boxed{[15\times88]\times30}=\boxed{[1320]\times30}=\boxed{1320\times30}=\boxed{\mathbf{39600}}$

Example 1.4-22

$\boxed{5\times(8\times5)\times[7\times(4\times9)]}=\boxed{5\times(40)\times[7\times(36)]}=\boxed{5\times40\times[7\times36]}=\boxed{5\times40\times[252]}=\boxed{5\times40\times252}=\boxed{\mathbf{50400}}$

Example 1.4-23

$\boxed{[(12\times3\times1)\times(2\times4)\times3]\times(5\times8)}=\boxed{[(36)\times(8)\times3]\times(40)}=\boxed{[36\times8\times3]\times40}=\boxed{[864]\times40}=\boxed{864\times40}$

$=\boxed{\mathbf{34560}}$

Example 1.4-24

$\boxed{(5\times13\times3)\times[8\times(10\times2)]\times3}=\boxed{(195)\times[8\times(20)]\times3}=\boxed{195\times[8\times20]\times3}=\boxed{195\times[160]\times3}=\boxed{195\times160\times3}$

$=\boxed{\mathbf{93600}}$

Example 1.4-25

$\boxed{[(2\times7\times4)\times(6\times8)]\times(2\times3)\times4}=\boxed{[(56)\times(48)]\times(6)\times4}=\boxed{[56\times48]\times6\times4}=\boxed{[2688]\times6\times4}$

$=\boxed{2688\times6\times4}=\boxed{\mathbf{64512}}$

Practice Problems - Use of Parentheses and Brackets in Multiplication

Section 1.4 Practice Problems - Multiply the following numbers in the order grouped:

1. $5 \times 2 \times 7 \times 4 =$
2. $(3 \times 5) \times (4 \times 2) \times 7 =$
3. $(20 \times 3 \times 4) \times (1 \times 2 \times 6) =$
4. $8 \times [(1 \times 5 \times 6) \times (7 \times 2)] =$
5. $[(2 \times 7) \times 4] \times [6 \times (5 \times 3)] =$
6. $(6 \times 8) \times [(2 \times 3) \times 5] \times 10 =$
7. $(2 \times 3 \times 9) \times [(4 \times 5) \times 0] \times 7 =$
8. $[(1 \times 6 \times 3) \times [(7 \times 3) \times 5]] \times 3 =$
9. $[(2 \times 3) \times (6 \times 5 \times 2)] \times [4 \times (2 \times 4)] =$
10. $[(2 \times 3) \times (6 \times 7) \times 2] \times [(4 \times 2) \times 5] =$

1.5 Using Parentheses and Brackets in Division

In this section the use of parentheses and brackets as applied to division is discussed. Similar to subtraction, discussed in Section 1.3, changing the order in which numbers are divided or grouped does alter the final answer. These two properties associated with division are discussed below:

Properties Associated with Division

1. Changing the order in which two numbers are divided does change the final answer. For example, for any two real numbers a and b

   $\boxed{a \div b \neq b \div a}$ *Note 1:* $\frac{a}{b}$, $b \neq 0$ *and* $\frac{b}{a}$, $a \neq 0$ *Note 2:* $\frac{a}{0}$ *is not defined.*

   For example, $\boxed{15 \div 5} = \boxed{3}$, but $\boxed{5 \div 15} = \boxed{0.33}$

2. Re-grouping numbers does change the final answer. For example, for any real numbers a, b, and c

   $\boxed{(a \div b) \div c \neq a \div (b \div c)}$

   For example,

   $\boxed{(28 \div 4) \div 2} = \boxed{(7) \div 2} = \boxed{7 \div 2} = \boxed{3.5}$, however

   $\boxed{28 \div (4 \div 2)} = \boxed{28 \div (2)} = \boxed{28 \div 2} = \boxed{14}$

In the following cases the use of parentheses and brackets in division is discussed:

Case I - Use of Parentheses in Division

In division, parentheses can be grouped in different ways, as shown in the following example cases:

Case I a - *Dividing Two Integer Numbers*

Two integer numbers are divided using the general division process. Following is a general and a specific example as to how two integer numbers are divided:

$\boxed{a \div b} = \boxed{A}$

Example 1.5-1

$\boxed{135 \div 15} =$

Solution:

$\boxed{135 \div 15} = \boxed{9}$

Case I b - *Dividing Two Integer Numbers Grouped by Parentheses*

Two integer numbers that are grouped by parentheses are divided in the following ways, as shown by general and specific example cases:

Case I b-1.

$\boxed{a \div (b \div c)} =$

Let $\boxed{b \div c = k_1}$ and $\boxed{a \div k_1 = B}$, then

$\boxed{a \div (b \div c)} = \boxed{a \div (k_1)} = \boxed{a \div k_1} = \boxed{B}$

Example 1.5-2

$\boxed{38 \div (12 \div 3)} =$

Solution:

$\boxed{38 \div (12 \div 3)} = \boxed{38 \div (4)} = \boxed{38 \div 4} = \boxed{\mathbf{9.5}}$

Case I b-2.

$\boxed{(a \div b) \div c} =$

Let $\boxed{a \div b = k_1}$ and $\boxed{k_1 \div c = C}$, then

$\boxed{(a \div b) \div c} = \boxed{(k_1) \div c} = \boxed{k_1 \div c} = \boxed{C}$

Example 1.5-3

$\boxed{(125 \div 5) \div 4} =$

Solution:

$\boxed{(125 \div 5) \div 4} = \boxed{(25) \div 4} = \boxed{25 \div 4} = \boxed{\mathbf{6.25}}$

Case I b-3.

$\boxed{(a \div b) \div (c \div d)} =$

Let $\boxed{a \div b = k_1}$, $\boxed{c \div d = k_2}$, and $\boxed{k_1 \div k_2 = D}$, then

$\boxed{(a \div b) \div (c \div d)} = \boxed{(k_1) \div (k_2)} = \boxed{k_1 \div k_2} = \boxed{D}$

Example 1.5-4

$\boxed{(15 \div 4) \div (8 \div 3)} =$

Solution:

$\boxed{(15 \div 4) \div (8 \div 3)} = \boxed{(3.75) \div (2.67)} = \boxed{3.75 \div 2.67} = \boxed{\mathbf{1.41}}$

Case II - Use of Brackets in Division

In division, brackets are used in a similar way as parentheses. However, brackets are used to separate mathematical operations that contain integer numbers already grouped by parentheses. Brackets are used to group numbers in different ways, as shown in the following general and specific example cases:

Case II-1.

$\boxed{[a \div (b \div c)] \div d} =$

Let $\boxed{b \div c = k_1}$, $\boxed{a \div k_1 = k_2}$, and $\boxed{k_2 \div d = A}$, then

$\boxed{[a \div (b \div c)] \div d} = \boxed{[a \div (k_1)] \div d} = \boxed{[a \div k_1] \div d} = \boxed{[k_2] \div d} = \boxed{k_2 \div d} = \boxed{A}$

Example 1.5-5

$\boxed{[15 \div (6 \div 4)] \div 2} =$

Solution:

$\boxed{[15 \div (6 \div 4)] \div 2} = \boxed{[15 \div (1.5)] \div 2} = \boxed{[15 \div 1.5] \div 2} = \boxed{[10] \div 2} = \boxed{10 \div 2} = \boxed{\mathbf{5}}$

Case II-2.

$\boxed{[(a \div b) \div c] \div d} =$

Let $\boxed{a \div b = k_1}$, $\boxed{k_1 \div c = k_2}$, and $\boxed{k_2 \div d = B}$, then

$\boxed{[(a \div b) \div c] \div d} = \boxed{[(k_1) \div c] \div d} = \boxed{[k_1 \div c] \div d} = \boxed{[k_2] \div d} = \boxed{k_2 \div d} = \boxed{B}$

Example 1.5-6

$\boxed{[(236 \div 12) \div 5] \div 3} =$

Solution:

$\boxed{[(236 \div 12) \div 5] \div 3} = \boxed{[(19.67) \div 5] \div 3} = \boxed{[19.67 \div 5] \div 3} = \boxed{[3.93] \div 3} = \boxed{3.93 \div 3} = \boxed{\mathbf{1.31}}$

Case II-3.

$\boxed{a \div [b \div (c \div d)]} =$

Let $\boxed{c \div d = k_1}$, $\boxed{b \div k_1 = k_2}$, and $\boxed{a \div k_2 = C}$, then

$\boxed{a \div [b \div (c \div d)]} = \boxed{a \div [b \div (k_1)]} = \boxed{a \div [b \div k_1]} = \boxed{a \div [k_2]} = \boxed{a \div k_2} = \boxed{C}$

Example 1.5-7

$\boxed{238 \div [24 \div (15 \div 5)]} =$

Solution:

$\boxed{238 \div [24 \div (15 \div 5)]} = \boxed{238 \div [24 \div (3)]} = \boxed{238 \div [24 \div 3]} = \boxed{238 \div [8]} = \boxed{238 \div 8} = \boxed{\mathbf{29.75}}$

Case II-4.

$\boxed{[(a \div b) \div (c \div d)] \div e} =$

Let $a \div b = k_1$, $c \div d = k_1$, $k_1 \div k_2 = k_3$, and $k_3 \div e = D$, then

$$[(a \div b) \div (c \div d)] \div e = [(k_1) \div (k_2)] \div e = [k_1 \div k_2] \div e = [k_3] \div e = k_3 \div e = D$$

Example 1.5-8

$[(28 \div 13) \div (15 \div 4)] \div 2 =$

Solution:

$$[(28 \div 13) \div (15 \div 4)] \div 2 = [(2.15) \div (3.75)] \div 2 = [2.15 \div 3.75] \div 2 = [0.57] \div 2 = 0.57 \div 2 = \mathbf{0.285}$$

Case II-5.

$[(a \div b) \div c] \div (d \div e) =$

Let $a \div b = k_1$, $k_1 \div c = k_2$, $d \div e = k_3$, and $k_2 \div k_3 = E$, then

$$[(a \div b) \div c] \div (d \div e) = [(k_1) \div c] \div (k_3) = [k_1 \div c] \div k_3 = [k_2] \div k_3 = k_2 \div k_3 = E$$

Example 1.5-9

$[(29 \div 5) \div 2] \div (15 \div 6) =$

Solution:

$$[(29 \div 5) \div 2] \div (15 \div 6) = [(5.8) \div 2] \div (2.5) = [5.8 \div 2] \div 2.5 = [2.9] \div 2.5 = 2.9 \div 2.5 = \mathbf{1.16}$$

Case II-6.

$a \div [(b \div c) \div (d \div e)] =$

Let $b \div c = k_1$, $d \div e = k_2$, $k_1 \div k_2 = k_3$, and $a \div k_3 = F$, then

$$a \div [(b \div c) \div (d \div e)] = a \div [(k_1) \div (k_2)] = a \div [k_1 \div k_2] = a \div [k_3] = a \div k_3 = F$$

Example 1.5-10

$238 \div [(35 \div 5) \div (14 \div 7)] =$

Solution:

$$238 \div [(35 \div 5) \div (14 \div 7)] = 238 \div [(7) \div (2)] = 238 \div [7 \div 2] = 238 \div [3.5] = 238 \div 3.5 = \mathbf{68}$$

Case II-7.

$[(a \div b) \div (c \div d)] \div (e \div f) =$

Let $a \div b = k_1$, $c \div d = k_2$, $e \div f = k_3$, $k_1 \div k_2 = k_4$, and $k_4 \div k_3 = G$, then

$\boxed{[(a \div b) \div (c \div d)] \div (e \div f)} = \boxed{[(k_1) \div (k_2)] \div (k_3)} = \boxed{[k_1 \div k_2] \div k_3} = \boxed{[k_4] \div k_3} = \boxed{k_4 \div k_3} = \boxed{G}$

Example 1.5-11

$\boxed{[(230 \div 5) \div (36 \div 4)] \div (25 \div 6)} =$

Solution:

$\boxed{[(230 \div 5) \div (36 \div 4)] \div (25 \div 6)} = \boxed{[(46) \div (9)] \div (4.17)} = \boxed{[46 \div 9] \div 4.17} = \boxed{[5.11] \div 4.17} = \boxed{5.11 \div 4.17} = \boxed{\mathbf{1.23}}$

Case II-8.

$\boxed{(a \div b) \div [(c \div d) \div (e \div f)]} =$

Let $\boxed{a \div b = k_1}$, $\boxed{c \div d = k_2}$, $\boxed{e \div f = k_3}$, $\boxed{k_2 \div k_3 = k_4}$, and $\boxed{k_1 \div k_4 = H}$, then

$\boxed{(a \div b) \div [(c \div d) \div (e \div f)]} = \boxed{(k_1) \div [(k_2) \div (k_3)]} = \boxed{k_1 \div [k_2 \div k_3]} = \boxed{k_1 \div [k_4]} = \boxed{k_1 \div k_4} = \boxed{H}$

Example 1.5-12

$\boxed{(358 \div 12) \div [(35 \div 7) \div (25 \div 2)]} =$

Solution:

$\boxed{(358 \div 12) \div [(35 \div 7) \div (25 \div 2)]} = \boxed{(29.83) \div [(5) \div (12.5)]} = \boxed{29.83 \div [5 \div 12.5]} = \boxed{29.83 \div [0.4]} = \boxed{29.83 \div 0.4}$

$= \boxed{\mathbf{74.58}}$

Additional Examples - Use of Parentheses and Brackets in Division

The following examples further illustrate how to use parentheses and brackets in division:

Example 1.5-13

$\boxed{(35 \div 5) \div 3} = \boxed{(7) \div 3} = \boxed{7 \div 3} = \boxed{\mathbf{2.33}}$

Example 1.5-14

$\boxed{240 \div (16 \div 2)} = \boxed{240 \div (8)} = \boxed{240 \div 8} = \boxed{\mathbf{30}}$

Example 1.5-15

$\boxed{(40 \div 2) \div (165 \div 15)} = \boxed{(20) \div (11)} = \boxed{20 \div 11} = \boxed{\mathbf{1.82}}$

Example 1.5-16

$\boxed{28 \div [15 \div (36 \div 3)]} = \boxed{28 \div [15 \div (12)]} = \boxed{28 \div [15 \div 12]} = \boxed{28 \div [1.25]} = \boxed{28 \div 1.25} = \boxed{\mathbf{22.4}}$

Example 1.5-17

$\boxed{[(80 \div 2) \div 5] \div 4} = \boxed{[(40) \div 5] \div 4} = \boxed{[40 \div 5] \div 4} = \boxed{[8] \div 4} = \boxed{8 \div 4} = \boxed{\mathbf{2}}$

Example 1.5-18

$\boxed{(238 \div 4) \div [16 \div (8 \div 2)]} = \boxed{(59.5) \div [16 \div (4)]} = \boxed{59.5 \div [16 \div 4]} = \boxed{59.5 \div [4]} = \boxed{59.5 \div 4} = \boxed{\mathbf{14.88}}$

Example 1.5-19

$\boxed{[(30 \div 3) \div (28 \div 2)] \div 5} = \boxed{[(10) \div (14)] \div 5} = \boxed{[10 \div 14] \div 5} = \boxed{[0.71] \div 5} = \boxed{0.71 \div 5} = \boxed{\mathbf{0.14}}$

Example 1.5-20

$\boxed{[(81 \div 3) \div 3] \div (18 \div 2)} = \boxed{[(27) \div 3] \div (9)} = \boxed{[27 \div 3] \div 9} = \boxed{[9] \div 9} = \boxed{9 \div 9} = \boxed{\mathbf{1}}$

Example 1.5-21

$\boxed{45 \div [25 \div (15 \div 5)]} = \boxed{45 \div [25 \div (3)]} = \boxed{45 \div [25 \div 3]} = \boxed{45 \div [8.33]} = \boxed{45 \div 8.33} = \boxed{\mathbf{5.4}}$

Example 1.5-22

$\boxed{(230 \div 10) \div [48 \div (24 \div 2)]} = \boxed{(23) \div [48 \div (12)]} = \boxed{23 \div [48 \div 12]} = \boxed{23 \div [4]} = \boxed{23 \div 4} = \boxed{\mathbf{5.75}}$

Practice Problems - Use of Parentheses and Brackets in Division

Section 1.5 Practice Problems - Divide the following numbers in the order grouped:

1. $(16 \div 2) \div 4 =$
2. $(125 \div 5) \div (15 \div 5) =$
3. $[25 \div (8 \div 2)] \div 3 =$
4. $[(140 \div 10) \div 2] \div 6 =$
5. $[155 \div (15 \div 3)] \div 9 =$
6. $250 \div [(48 \div 2) \div 4] =$
7. $[(28 \div 4) \div (16 \div 3)] \div 8 =$
8. $66 \div [48 \div (14 \div 2)] =$
9. $(180 \div 2) \div [(88 \div 2) \div 4] =$
10. $[(48 \div 4) \div 2] \div (18 \div 3) =$

1.6 Using Parentheses and Brackets in Mixed Operations

In this section the use of parentheses and brackets as applied to addition, subtraction, multiplication, and division, using integer numbers, is discussed. Similar to subtraction and division, the order in which mixed operations are grouped does change the final answer. This is discussed in the following cases:

Case I - Use of Parentheses in Addition, Subtraction, Multiplication, and Division

In mixed mathematical operations, parentheses can be grouped in different ways, as shown in the following example cases:

Case I-1.

$\boxed{a+(b\div c)}=$

Let $\boxed{b\div c=k_1}$ and $\boxed{a+k_1=A}$, then

$\boxed{a+(b\div c)}=\boxed{a+(k_1)}=\boxed{a+k_1}=\boxed{A}$

Example 1.6-1

$\boxed{30+(50\div 5)}=$

Solution:

$\boxed{30+(50\div 5)}=\boxed{30+(10)}=\boxed{30+10}=\boxed{\mathbf{40}}$

Case I-2.

$\boxed{a\div(b\times c)}=$

Let $\boxed{b\times c=k_1}$ and $\boxed{a\div k_1=B}$, then

$\boxed{a\div(b\times c)}=\boxed{a\div(k_1)}=\boxed{a\div k_1}=\boxed{B}$

Example 1.6-2

$\boxed{18\div(4\times 2)}=$

Solution:

$\boxed{18\div(4\times 2)}=\boxed{18\div(8)}=\boxed{18\div 8}=\boxed{\mathbf{2.25}}$

Case I-3.

$\boxed{(a\times b)\div c}=$

Let $\boxed{a\times b=k_1}$ and $\boxed{k_1\div c=C}$, then

$\boxed{(a\times b)\div c}=\boxed{(k_1)\div c}=\boxed{k_1\div c}=\boxed{C}$

Example 1.6-3

$\boxed{(20 \times 5) \div 8} =$

Solution:

$\boxed{(20 \times 5) \div 8} = \boxed{(100) \div 8} = \boxed{100 \div 8} = \boxed{\mathbf{12.5}}$

Case I-4.

$\boxed{(a \div b) + c} =$

Let $\boxed{a \div b = k_1}$ and $\boxed{k_1 + c = D}$, then

$\boxed{(a \div b) + c} = \boxed{(k_1) + c} = \boxed{k_1 + c} = \boxed{D}$

Example 1.6-4

$\boxed{(45 \div 5) + 25} =$

Solution:

$\boxed{(45 \div 5) + 25} = \boxed{(9) + 25} = \boxed{9 + 25} = \boxed{\mathbf{34}}$

Case I-5.

$\boxed{(a + b) \div (c - d)} =$

Let $\boxed{a + b = k_1}$, $\boxed{c - d = k_2}$, and $\boxed{k_1 \div k_2 = E}$, then

$\boxed{(a + b) \div (c - d)} = \boxed{(k_1) \div (k_2)} = \boxed{k_1 \div k_2} = \boxed{E}$

Example 1.6-5

$\boxed{(23 + 5) \div (20 - 8)} =$

Solution:

$\boxed{(23 + 5) \div (20 - 8)} = \boxed{(28) \div (12)} = \boxed{28 \div 12} = \boxed{\mathbf{2.33}}$

Case I-6.

$\boxed{(a \div b) - (c \times d)} =$

Let $\boxed{a \div b = k_1}$, $\boxed{c \times d = k_2}$, and $\boxed{k_1 - k_2 = F}$, then

$\boxed{(a \div b) - (c \times d)} = \boxed{(k_1) - (k_2)} = \boxed{k_1 - k_2} = \boxed{F}$

Example 1.6-6

$\boxed{(49 \div 5) - (12 \times 4)} =$

Solution:

$\boxed{(49 \div 5)-(12\times 4)} = \boxed{(9.8)-(48)} = \boxed{9.8-48} = \boxed{\mathbf{-38.2}}$

Case II - Use of Brackets in Addition, Subtraction, Multiplication, and Division

In mixed operations, brackets are used in a similar way as parentheses. However, brackets are used to separate mathematical operations that contain integer numbers already grouped by parentheses. Brackets are used to group numbers in different ways, as shown in the following general and specific example cases:

Case II-1.

$\boxed{[a \div (b+c)] \div d} =$

Let $\boxed{b+c=k_1}$, $\boxed{a \div k_1 = k_2}$, and $\boxed{k_2 \div d = A}$, then

$\boxed{[a \div (b+c)] \div d} = \boxed{[a \div (k_1)] \div d} = \boxed{[a \div k_1] \div d} = \boxed{[k_2] \div d} = \boxed{k_2 \div d} = \boxed{A}$

Example 1.6-7

$\boxed{[350 \div (12+8)] \div 4} =$

Solution:

$\boxed{[350 \div (12+8)] \div 4} = \boxed{[350 \div (20)] \div 4} = \boxed{[350 \div 20] \div 4} = \boxed{[17.5] \div 4} = \boxed{17.5 \div 4} = \boxed{\mathbf{4.38}}$

Case II-2.

$\boxed{[(a\times b) \div c]+d} =$

Let $\boxed{a\times b = k_1}$, $\boxed{k_1 \div c = k_2}$, and $\boxed{k_2 + d = B}$, then

$\boxed{[(a\times b) \div c]+d} = \boxed{[(k_1) \div c]+d} = \boxed{[k_1 \div c]+d} = \boxed{[k_2]+d} = \boxed{k_2+d} = \boxed{B}$

Example 1.6-8

$\boxed{[(12\times 4) \div 2]+46} =$

Solution:

$\boxed{[(12\times 4) \div 2]+46} = \boxed{[(48) \div 2]+46} = \boxed{[48 \div 2]+46} = \boxed{[24]+46} = \boxed{24+46} = \boxed{\mathbf{70}}$

Case II-3.

$\boxed{a\times[b-(c+d)]} =$

Let $\boxed{c+d=k_1}$, $\boxed{b-k_1=k_2}$, and $\boxed{ak_2 = C}$, then

$\boxed{a\times[b-(c+d)]} = \boxed{a\times[b-(k_1)]} = \boxed{a\times[b-k_1]} = \boxed{a\times[k_2]} = \boxed{a\times k_2} = \boxed{ak_2} = \boxed{C}$

Example 1.6-9

$\boxed{8\times[10-(5+9)]}=$

Solution:

$\boxed{8\times[10-(5+9)]}=\boxed{8\times[10-(14)]}=\boxed{8\times[10-14]}=\boxed{8\times[-4]}=\boxed{8\times-4}=\boxed{\mathbf{-32}}$

Case II-4.

$\boxed{[(a\times b)\div(c+d)]\div e}=$

Let $\boxed{a\times b=k_1}$, $\boxed{c+d=k_2}$, $\boxed{k_1\div k_2=k_3}$, and $\boxed{k_3\div e=D}$, then

$\boxed{[(a\times b)\div(c+d)]\div e}=\boxed{[(k_1)\div(k_2)]\div e}=\boxed{[k_1\div k_2]\div e}=\boxed{[k_3]\div e}=\boxed{k_3\div e}=\boxed{D}$

Example 1.6-10

$\boxed{[(4\times 5)\div(28+9)]\div 5}=$

Solution:

$\boxed{[(4\times 5)\div(28+9)]\div 5}=\boxed{[(20)\div(37)]\div 5}=\boxed{[20\div 37]\div 5}=\boxed{[0.54]\div 5}=\boxed{0.54\div 5}=\boxed{\mathbf{0.108}}$

Case II-5.

$\boxed{[(a-b)-c]+(d+e)}=$

Let $\boxed{a-b=k_1}$, $\boxed{k_1-c=k_2}$, $\boxed{d+e=k_3}$, and $\boxed{k_2+k_3=E}$, then

$\boxed{[(a-b)-c]+(d+e)}=\boxed{[(k_1)-c]+(k_3)}=\boxed{[k_1-c]+k_3}=\boxed{[k_2]+k_3}=\boxed{k_2+k_3}=\boxed{E}$

Example 1.6-11

$\boxed{[(23-6)-8]+(12+7)}=$

Solution:

$\boxed{[(23-6)-8]+(12+7)}=\boxed{[(17)-8]+(19)}=\boxed{[17-8]+19}=\boxed{[9]+19}=\boxed{9+19}=\boxed{\mathbf{28}}$

Case II-6.

$\boxed{a+[(b+c)-(d\times e)]}=$

Let $\boxed{b+c=k_1}$, $\boxed{d\times e=k_2}$, $\boxed{k_1-k_2=k_3}$, and $\boxed{a+k_3=F}$, then

$\boxed{a+[(b+c)-(d\times e)]}=\boxed{a+[(k_1)-(k_2)]}=\boxed{a+[k_1-k_2]}=\boxed{a+[k_3]}=\boxed{a+k_3}=\boxed{F}$

Example 1.6-12

$\boxed{35+[(12+5)-(4\times 2)]}=$

Solution:

$\boxed{35+[(12+5)-(4\times 2)]} = \boxed{35+[(17)-(8)]} = \boxed{35+[17-8]} = \boxed{35+[9]} = \boxed{35+9} = \boxed{\mathbf{44}}$

Case II-7.

$\boxed{[(a \div b)+(c \div d)] \times (e+f)} =$

Let $\boxed{a \div b = k_1}$, $\boxed{c \div d = k_2}$, $\boxed{e+f = k_3}$, $\boxed{k_1 + k_2 = k_4}$, and $\boxed{k_4 k_3 = G}$, then

$\boxed{[(a \div b)+(c \div d)] \times (e+f)} = \boxed{[(k_1)+(k_2)] \times (k_3)} = \boxed{[k_1 + k_2] \times k_3} = \boxed{[k_4] \times k_3} = \boxed{k_4 \times k_3} = \boxed{k_4 k_3} = \boxed{G}$

Example 1.6-13

$\boxed{[(45 \div 9)+(12 \div 4)] \times (10+5)} =$

Solution:

$\boxed{[(45 \div 9)+(12 \div 4)] \times (10+5)} = \boxed{[(5)+(3)] \times (15)} = \boxed{[5+3] \times 15} = \boxed{[8] \times 15} = \boxed{8 \times 15} = \boxed{\mathbf{120}}$

Case II-8.

$\boxed{(a-b)+[(c \div d) \times (e \div f)]} =$

Let $\boxed{a-b = k_1}$, $\boxed{c \div d = k_2}$, $\boxed{e \div f = k_3}$, $\boxed{k_2 k_3 = k_4}$, and $\boxed{k_1 + k_4 = H}$, then

$\boxed{(a-b)+[(c \div d) \times (e \div f)]} = \boxed{(k_1)+[(k_2) \times (k_3)]} = \boxed{k_1 + [k_2 \times k_3]} = \boxed{k_1 + [k_2 k_3]} = \boxed{k_1 + [k_4]} = \boxed{k_1 + k_4}$

$= \boxed{H}$

Example 1.6-14

$\boxed{(45-6)+[(12 \div 4) \times (34 \div 4)]} =$

Solution:

$\boxed{(45-6)+[(12 \div 4) \times (34 \div 4)]} = \boxed{(39)+[(3) \times (8.5)]} = \boxed{39+[3 \times 8.5]} = \boxed{39+[25.5]} = \boxed{39+25.5} = \boxed{\mathbf{64.5}}$

Case II-9.

$\boxed{(a+b+c) \div [d \times (e-f)]} =$

Let $\boxed{a+b+c = k_1}$, $\boxed{e-f = k_2}$, $\boxed{dk_2 = k_3}$, and $\boxed{k_1 \div k_3 = I}$, then

$\boxed{(a+b+c) \div [d \times (e-f)]} = \boxed{(k_1) \div [d \times (k_2)]} = \boxed{k_1 \div [d \times k_2]} = \boxed{k_1 \div [dk_2]} = \boxed{k_1 \div dk_2} = \boxed{k_1 \div k_3} = \boxed{I}$

Example 1.6-15

$\boxed{(8+50+5) \div [3 \times (25-12)]} =$

Solution:

$\boxed{(8+50+5)\div[3\times(25-12)]} = \boxed{(63)\div[3\times(13)]} = \boxed{63\div[3\times13]} = \boxed{63\div[39]} = \boxed{63\div39} = \boxed{\mathbf{1.62}}$

Additional Examples - Use of Parentheses and Brackets in Addition, Subtraction, Multiplication, and Division

The following examples further illustrate how to use parentheses and brackets in mixed operations:

Example 1.6-16

$\boxed{(39+5)\div4} = \boxed{(44)\div4} = \boxed{44\div4} = \boxed{\mathbf{11}}$

Example 1.6-17

$\boxed{36\times(12+3)} = \boxed{36\times(15)} = \boxed{36\times15} = \boxed{\mathbf{540}}$

Example 1.6-18

$\boxed{(23+5)\div(8\times2)} = \boxed{(28)\div(16)} = \boxed{28\div16} = \boxed{\mathbf{1.75}}$

Example 1.6-19

$\boxed{38+[15\times(20\div2)]} = \boxed{38+[15\times(10)]} = \boxed{38+[15\times10]} = \boxed{38+[150]} = \boxed{38+150} = \boxed{\mathbf{188}}$

Example 1.6-20

$\boxed{[(35\times2)+5]\div3} = \boxed{[(70)+5]\div3} = \boxed{[70+5]\div3} = \boxed{[75]\div3} = \boxed{75\div3} = \boxed{\mathbf{25}}$

Example 1.6-21

$\boxed{(28-18)\times[16-(8-3)]} = \boxed{(10)\times[16-(5)]} = \boxed{10\times[16-5]} = \boxed{10\times[11]} = \boxed{10\times11} = \boxed{\mathbf{110}}$

Example 1.6-22

$\boxed{[(20-4)+(15-5)]\div2} = \boxed{[(16)+(10)]\div2} = \boxed{[16+10]\div2} = \boxed{[26]\div2} = \boxed{26\div2} = \boxed{\mathbf{13}}$

Example 1.6-23

$\boxed{[(15+6)\div3]\times(8\div2)} = \boxed{[(21)\div3]\times(4)} = \boxed{[21\div3]\times4} = \boxed{[7]\times4} = \boxed{7\times4} = \boxed{\mathbf{28}}$

Example 1.6-24

$\boxed{30-[15\times(30+2)]} = \boxed{30-[15\times(32)]} = \boxed{30-[15\times32]} = \boxed{30-[480]} = \boxed{30-480} = \boxed{\mathbf{-450}}$

Example 1.6-25

$\boxed{(85\div5)\times[20+(13-8)]} = \boxed{(17)\times[20+(5)]} = \boxed{17\times[20+5]} = \boxed{17\times[25]} = \boxed{17\times25} = \boxed{\mathbf{425}}$

Example 1.6-26

$\boxed{\{(236\div4)\times[35-(24-5)+18]\}+8} = \boxed{\{(59)\times[35-(19)+18]\}+8} = \boxed{\{59\times[35-19+18]\}+8}$

$= \boxed{\{59\times[34]\}+8} = \boxed{\{59\times34\}+8} = \boxed{\{2006\}+8} = \boxed{2006+8} = \boxed{\mathbf{2014}}$

Practice Problems - Use of Parentheses and Brackets in Addition, Subtraction, Multiplication, and Division

Section 1.6 Practice Problems - Perform the indicated operations in the order grouped:

1. $(28 \div 4) \times 3 =$
2. $250 + (15 \div 3) =$
3. $28 \div [(23 + 5) \times 8] =$
4. $[(255 - 15) \div 20] + 8 =$
5. $[230 \div (15 \times 2)] + 12 =$
6. $55 \times [(28 + 2) \div 3] =$
7. $[(55 \div 5) + (18 - 4)] \times 4 =$
8. $35 - [400 \div (16 + 4)] =$
9. $(230 + 5) \div [2 \times (18 + 2)] =$
10. $[(38 \div 4) + 2] \times (15 - 3) =$

Chapter 2
Integer Fractions

Quick Reference to Chapter 2 Case Problems

$\boxed{\frac{25}{3}-\frac{2}{3}}=$; $\boxed{\frac{5}{10}-\frac{14}{10}}=$; $\boxed{\frac{15}{6}-\frac{53}{6}}=$

$\boxed{\frac{7}{4}-\frac{3}{4}-\frac{1}{4}}=$; $\boxed{\frac{25}{6}-\frac{4}{6}-\frac{1}{6}}=$; $\boxed{\frac{12}{7}-\frac{28}{7}-\frac{13}{7}}=$

$\boxed{\frac{4}{5}-\frac{3}{8}}=$; $\boxed{\frac{9}{8}-\frac{3}{4}}=$; $\boxed{\frac{10}{6}-35}=$

$\boxed{\frac{4}{5}-\frac{1}{3}-\frac{2}{6}}=$; $\boxed{\frac{4}{7}-\frac{2}{5}-\frac{3}{4}}=$; $\boxed{25-\frac{3}{4}-\frac{32}{5}}=$

$\boxed{\frac{4}{5}\times\frac{3}{8}}=$; $\boxed{25\times\frac{5}{8}}=$; $\boxed{\frac{140}{3}\times\frac{1}{5}}=$

$\boxed{\frac{25}{3}\times\frac{4}{7}\times\frac{6}{5}}=$; $\boxed{\frac{9}{8}\times\frac{33}{5}\times\frac{5}{48}}=$; $\boxed{\frac{125}{4}\times\frac{28}{13}\times 39}=$

$\boxed{\frac{3}{5}\div\frac{8}{15}}=$; $\boxed{\frac{125}{65}\div 230}=$; $\boxed{\frac{32}{18}\div\frac{50}{12}}=$

$\boxed{\left(\frac{3}{5}\div 4\right)\div\frac{9}{25}}=$; $\boxed{235\div\left(\frac{68}{15}\div\frac{33}{12}\right)}=$; $\boxed{\frac{12}{30}\div\left(\frac{15}{6}\div\frac{12}{5}\right)}=$

Chapter 2 - Integer Fractions

The objective of this chapter is to review fractions. It is essential that students be thoroughly familiar with the subject of fractions in order to understand the topics and be able to work the problems presented in the following chapters. Simplifying integer fractions is addressed in Section 2.1. Section 2.1 Appendix shows the steps for changing improper fractions to mixed fractions. How to add, subtract, multiply, and divide two or more integer fractions is addressed in Sections 2.2, 2.3, 2.4, and 2.5, respectively. The general algebraic approach in solving integer fractional operations is provided in each indicated section. The student, depending on his or her grade level and ability, can skip the algebraic approach to integer fractions and learn only the techniques that are followed by examples. Focusing on the examples, and the steps shown to solve each problem, should be adequate to teach the student the mechanics of how integer fractions are mathematically operated upon. (Students are encouraged to review the companion Hamilton Education Guide, *Mastering Fractions*, book for a more in-depth treatment of this subject.)

2.1 Simplifying Integer Fractions

Integer fractions of the form $\frac{a}{b}$, where both the numerator a and the denominator b are integer numbers, are simplified as in the following cases:

Case I - The Numerator and the Denominator are Even Numbers

Use the following steps to simplify the integer fractions if the numerator and the denominator are even numbers:

Step 1 Check the numerator and the denominator of the integer fraction to see if it is an $\frac{even}{even}$ type of fraction.

Step 2 Simplify the fraction to its lowest term by dividing the numerator and the denominator by their Greatest Common Factor (G.C.F.) which is an even number, i.e., $(2, 4, 6, 8, 10, 12, 14, ...)$. See the methods introduced in finding G.C.F. at the end of this section.

Step 3 Change the improper fraction to a mixed fraction if the fraction obtained from Step 2 is an improper fraction (see Section 2.1 Appendix).

Examples with Steps

The following examples show the steps as to how integer fractions with even numerator and denominator are simplified:

Example 2.1-1

$$\boxed{-\frac{366}{64}} =$$

Solution:

Step 1 $\boxed{-\frac{366}{64}} = \boxed{-\frac{366\,(\textit{is an even No.})}{64\,(\textit{is an even No.})}}$

Step 2 $\boxed{-\frac{366\,(\textit{is an even No.})}{64\,(\textit{is an even No.})}} = \boxed{-\frac{366 \div 2}{64 \div 2}} = \boxed{-\frac{183}{32}}$

Step 3 $\boxed{-\frac{183}{32}} = \boxed{-\left(5\frac{23}{32}\right)}$

Example 2.1-2

$\boxed{\frac{400}{350}} =$

Solution:

Step 1 $\boxed{\frac{400}{350}} = \boxed{\frac{400\,(\textit{is an even No.})}{350\,(\textit{is an even No.})}}$

Step 2 $\boxed{\frac{400\,(\textit{is an even No.})}{350\,(\textit{is an even No.})}} = \boxed{\frac{400 \div 50}{350 \div 50}} = \boxed{\frac{8}{7}}$

Step 3 $\boxed{\frac{8}{7}} = \boxed{1\frac{1}{7}}$

Example 2.1-3

$\boxed{\frac{2}{8}} =$

Solution:

Step 1 $\boxed{\frac{2}{8}} = \boxed{\frac{2\,(\textit{is an even No.})}{8\,(\textit{is an even No.})}}$

Step 2 $\boxed{\frac{2\,(\textit{is an even No.})}{8\,(\textit{is an even No.})}} = \boxed{\frac{2 \div 2}{8 \div 2}} = \boxed{\frac{1}{4}}$

Step 3 $\boxed{\textit{Not Applicable}}$

Note: See definition of *Not Applicable* in the glossary section.

Example 2.1-4

$\boxed{-\frac{18}{12}} =$

Solution:

Step 1 $\boxed{-\dfrac{18}{12}} = \boxed{-\dfrac{18\,(is\ an\ even\ No.)}{12\,(is\ an\ even\ No.)}}$

Step 2 $\boxed{-\dfrac{18\,(is\ an\ even\ No.)}{12\,(is\ an\ even\ No.)}} = \boxed{-\dfrac{18 \div 6}{12 \div 6}} = \boxed{-\dfrac{3}{2}}$

Step 3 $\boxed{-\dfrac{3}{2}} = \boxed{-\left(1\dfrac{1}{2}\right)}$

Example 2.1-5A

$\boxed{\dfrac{16}{32}} =$

Solution:

Step 1 $\boxed{\dfrac{16}{32}} = \boxed{\dfrac{16\,(is\ an\ even\ No.)}{32\,(is\ an\ even\ No.)}}$

Step 2 $\boxed{\dfrac{16\,(is\ an\ even\ No.)}{32\,(is\ an\ even\ No.)}} = \boxed{\dfrac{16 \div 16}{32 \div 16}} = \boxed{\dfrac{1}{2}}$

Step 3 *Not Applicable*

Example 2.1-5B

$\boxed{\dfrac{48}{-36}} =$

Solution:

Step 1 $\boxed{\dfrac{48}{-36}} = \boxed{-\dfrac{48\,(is\ an\ even\ No.)}{36\,(is\ an\ even\ No.)}}$

Step 2 $\boxed{-\dfrac{48\,(is\ an\ even\ No.)}{36\,(is\ an\ even\ No.)}} = \boxed{-\dfrac{48 \div 12}{36 \div 12}} = \boxed{-\dfrac{4}{3}}$

Step 3 $\boxed{-\dfrac{4}{3}} = \boxed{-\left(1\dfrac{1}{3}\right)}$

Case II - The Numerator and the Denominator are Odd Numbers

Use the following steps to simplify the integer fractions if the numerator and the denominator are odd numbers:

Step 1 Check the numerator and the denominator of the integer fraction to see if it is an $\frac{odd}{odd}$ type of fraction.

Step 2 Simplify the fraction to its lowest term by dividing the numerator and the denominator by their Greatest Common Factor (G.C.F.) which is an odd number, i.e., $(3, 5, 7, 9, 11, 13, 15, \ldots)$. See the methods introduced in finding G.C.F. at the end of this section.

Step 3 Change the improper fraction to a mixed fraction if the fraction obtained from Step 2 is an improper fraction (see Section 2.1 Appendix).

Examples with Steps

The following examples show the steps as to how integer fractions with odd numerator and denominator are simplified:

Example 2.1-6

$$\boxed{-\frac{3}{15}} =$$

Solution:

Step 1 $\boxed{-\frac{3}{15}} = \boxed{-\frac{3\,(is\ an\ odd\ No.)}{15\,(is\ an\ odd\ No.)}}$

Step 2 $\boxed{-\frac{3\,(is\ an\ odd\ No.)}{15\,(is\ an\ odd\ No.)}} = \boxed{-\frac{3 \div 3}{15 \div 3}} = \boxed{-\frac{1}{5}}$

Step 3 *Not Applicable*

Example 2.1-7

$$\boxed{\frac{7}{21}} =$$

Solution:

Step 1 $\boxed{\frac{7}{21}} = \boxed{\frac{7\,(is\ an\ odd\ No.)}{21\,(is\ an\ odd\ No.)}}$

Step 2 $\boxed{\frac{7\,(is\ an\ odd\ No.)}{21\,(is\ an\ odd\ No.)}} = \boxed{\frac{7 \div 7}{21 \div 7}} = \boxed{\frac{1}{3}}$

Step 3 $\boxed{\textit{Not Applicable}}$

Example 2.1-8

$$\boxed{\frac{17}{21}} =$$

Solution:

Step 1 $\boxed{\frac{17}{21}} = \boxed{\frac{\mathbf{17}\,(\textit{is an odd No.})}{\mathbf{21}\,(\textit{is an odd No.})}}$

Step 2 $\boxed{\textit{Not Applicable}}$

Step 3 $\boxed{\textit{Not Applicable}}$

Note - In cases where the answer to Steps 2 and 3 are stated as "Not Applicable" this indicates that the fraction is in its lowest term and can not be simplified any further.

Example 2.1-9

$$\boxed{-\frac{305}{35}} =$$

Solution:

Step 1 $\boxed{-\frac{305}{35}} = \boxed{-\frac{305\,(\textit{is an odd No.})}{35\,(\textit{is an odd No.})}}$

Step 2 $\boxed{-\frac{305\,(\textit{is an odd No.})}{35\,(\textit{is an odd No.})}} = \boxed{-\frac{305 \div 5}{35 \div 5}} = \boxed{-\frac{61}{7}}$

Step 3 $\boxed{-\frac{61}{7}} = \boxed{-\left(\mathbf{8}\frac{\mathbf{5}}{\mathbf{7}}\right)}$

Example 2.1-10

$$\boxed{\frac{105}{33}} =$$

Solution:

Step 1 $\boxed{\frac{105}{33}} = \boxed{\frac{105\,(\textit{is an odd No.})}{33\,(\textit{is an odd No.})}}$

Step 2 $\boxed{\frac{105\,(\textit{is an odd No.})}{33\,(\textit{is an odd No.})}} = \boxed{\frac{105 \div 3}{33 \div 3}} = \boxed{\frac{35}{11}}$

Step 3 $\boxed{\frac{35}{11}} = \boxed{\mathbf{3}\frac{\mathbf{2}}{\mathbf{11}}}$

Case III - The Numerator is an Even Number and the Denominator is an Odd Number

Use the following steps to simplify the integer fractions if the numerator is an even number and the denominator is an odd number:

Step 1 Check the numerator and the denominator of the integer fraction to see if it is an $\frac{even}{odd}$ type of fraction.

Step 2 Simplify the fraction to its lowest term by dividing the numerator and the denominator by their Greatest Common Factor (G.C.F.) which is an odd number, i.e., $(3, 5, 7, 9, 11, 13, 15, \ldots)$. See the methods introduced in finding G.C.F. at the end of this section.

Step 3 Change the improper fraction to a mixed fraction if the fraction obtained from Step 2 is an improper fraction (see Section 2.1 Appendix).

Examples with Steps

The following examples show the steps as to how integer fractions with an even numerator and an odd denominator are simplified:

Example 2.1-11

$$\boxed{\frac{18}{27}} =$$

Solution:

Step 1 $\boxed{\frac{18}{27}} = \boxed{\frac{18\,(is\ an\ even\ No.)}{27\,(is\ an\ odd\ No.)}}$

Step 2 $\boxed{\frac{18\,(is\ an\ even\ No.)}{27\,(is\ an\ odd\ No.)}} = \boxed{\frac{18 \div 9}{27 \div 9}} = \boxed{\frac{2}{3}}$

Step 3 *Not Applicable*

Example 2.1-12

$$\boxed{\frac{14}{25}} =$$

Solution:

Step 1 $\boxed{\frac{14}{25}} = \boxed{\frac{14\,(is\ an\ even\ No.)}{25\,(is\ an\ odd\ No.)}}$

Step 2 *Not Applicable*

Step 3 *Not Applicable*

Example 2.1-13

$$\boxed{\frac{334}{15}} =$$

Solution:

Step 1 $\boxed{\frac{334}{15}} = \boxed{\frac{334\,(is\ an\ even\ No.)}{15\,(is\ an\ odd\ No.)}}$

Step 2 $\boxed{Not\ Applicable}$

Step 3 $\boxed{\frac{334\,(is\ an\ even\ No.)}{15\,(is\ an\ odd\ No.)}} = \boxed{\mathbf{22\frac{4}{15}}}$

Example 2.1-14

$$\boxed{-\frac{108}{27}} =$$

Solution:

Step 1 $\boxed{-\frac{108}{27}} = \boxed{-\frac{108\,(is\ an\ even\ No.)}{27\,(is\ an\ odd\ No.)}}$

Step 2 $\boxed{-\frac{108\,(is\ an\ even\ No.)}{27\,(is\ an\ odd\ No.)}} = \boxed{-\frac{108 \div 27}{27 \div 27}} = \boxed{-\frac{4}{1}} = \boxed{\mathbf{-4}}$

Step 3 $\boxed{Not\ Applicable}$

Example 2.1-15

$$\boxed{\frac{386}{13}} =$$

Solution:

Step 1 $\boxed{\frac{386}{13}} = \boxed{\frac{386\,(is\ an\ even\ No.)}{13\,(is\ an\ odd\ No.)}}$

Step 2 $\boxed{Not\ Applicable}$

Step 3 $\boxed{\frac{386\,(is\ an\ even\ No.)}{13\,(is\ an\ odd\ No.)}} = \boxed{\mathbf{29\frac{9}{13}}}$

Case IV - The Numerator is an Odd Number and the Denominator is an Even Number

Use the following steps to simplify the integer fractions if the numerator is an odd number and the denominator is an even number:

Step 1 Check the numerator and the denominator of the integer fraction to see if it is an $\frac{odd}{even}$ type of fraction.

Step 2 Simplify the fraction to its lowest term by dividing the numerator and the denominator by their Greatest Common Factor (G.C.F.) which is an odd number, i.e., $(3, 5, 7, 9, 11, 13, 15, \ldots)$. See the methods introduced in finding G.C.F. at the end of this section.

Step 3 Change the improper fraction to a mixed fraction if the fraction obtained from Step 2 is an improper fraction (see Section 2.1 Appendix).

Examples with Steps

The following examples show the steps as to how integer fractions with an odd numerator and an even denominator are simplified:

Example 2.1-16

$$\boxed{\frac{15}{60}} =$$

Solution:

Step 1 $$\boxed{\frac{15}{60}} = \boxed{\frac{15\,(\text{is an odd No.})}{60\,(\text{is an even No.})}}$$

Step 2 $$\boxed{\frac{15\,(\text{is an odd No.})}{60\,(\text{is an even No.})}} = \boxed{\frac{15 \div 15}{60 \div 15}} = \boxed{\frac{1}{4}}$$

Step 3 *Not Applicable*

Example 2.1-17

$$\boxed{\frac{333}{36}} =$$

Solution:

Step 1 $$\boxed{\frac{333}{36}} = \boxed{\frac{333\,(\text{is an odd No.})}{36\,(\text{is an even No.})}}$$

Step 2 $$\boxed{\frac{333\,(\text{is an odd No.})}{36\,(\text{is an even No.})}} = \boxed{\frac{333 \div 9}{36 \div 9}} = \boxed{\frac{37}{4}}$$

Step 3 $\boxed{\frac{37}{4}} = \boxed{\mathbf{9\frac{1}{4}}}$

Example 2.1-18

$\boxed{\frac{305}{200}} =$

Solution:

Step 1 $\boxed{\frac{305}{200}} = \boxed{\frac{305\,(is\ an\ odd\ No.)}{200\,(is\ an\ even\ No.)}}$

Step 2 $\boxed{\frac{305\,(is\ an\ odd\ No.)}{200\,(is\ an\ even\ No.)}} = \boxed{\frac{305 \div 5}{200 \div 5}} = \boxed{\frac{61}{40}}$

Step 3 $\boxed{\frac{61}{40}} = \boxed{\mathbf{1\frac{21}{40}}}$

Example 2.1-19

$\boxed{\frac{25}{10}} =$

Solution:

Step 1 $\boxed{\frac{25}{10}} = \boxed{\frac{25\,(is\ an\ odd\ No.)}{10\,(is\ an\ even\ No.)}}$

Step 2 $\boxed{\frac{25\,(is\ an\ odd\ No.)}{10\,(is\ an\ even\ No.)}} = \boxed{\frac{25 \div 5}{10 \div 5}} = \boxed{\frac{5}{2}}$

Step 3 $\boxed{\frac{5}{2}} = \boxed{\mathbf{2\frac{1}{2}}}$

Example 2.1-20

$\boxed{-\frac{327}{24}} =$

Solution:

Step 1 $\boxed{-\frac{327}{24}} = \boxed{-\frac{327\,(is\ an\ odd\ No.)}{24\,(is\ an\ even\ No.)}}$

Step 2 $\boxed{-\frac{327\,(is\ an\ odd\ No.)}{24\,(is\ an\ even\ No.)}} = \boxed{-\frac{327 \div 3}{24 \div 3}} = \boxed{-\frac{109}{8}}$

Step 3 $\boxed{-\frac{109}{8}} = \boxed{\mathbf{-\left(13\frac{5}{8}\right)}}$

Note that in Cases II, III, and IV where the integer fractions are $\frac{odd}{odd}$, $\frac{even}{odd}$, and $\frac{odd}{even}$ respectively, odd numbers are always used to simplify the fractions.

Additional Examples - Simplifying Integer Fractions

The following examples further illustrate how to simplify integer fractions:

Example 2.1-21

$$\boxed{\frac{15}{3}} = \boxed{\frac{15\,(is\ an\ odd\ No.)}{3\,(is\ an\ odd\ No.)}} = \boxed{\frac{15 \div 3}{3 \div 3}} = \boxed{\frac{5}{1}} = \boxed{5}$$

Example 2.1-22

$$\boxed{-\frac{6}{8}} = \boxed{\frac{6\,(is\ an\ even\ No.)}{8\,(is\ an\ even\ No.)}} = \boxed{-\frac{6 \div 2}{8 \div 2}} = \boxed{-\frac{3}{4}}$$

Example 2.1-23

$$\boxed{\frac{12}{3}} = \boxed{\frac{12\,(is\ an\ even\ No.)}{3\,(is\ an\ odd\ No.)}} = \boxed{\frac{12 \div 3}{3 \div 3}} = \boxed{\frac{4}{1}} = \boxed{4}$$

Example 2.1-24

$$\boxed{\frac{35}{7}} = \boxed{\frac{35\,(is\ an\ odd\ No.)}{7\,(is\ an\ odd\ No.)}} = \boxed{\frac{35 \div 7}{7 \div 7}} = \boxed{\frac{5}{1}} = \boxed{5}$$

Example 2.1-25

$$\boxed{\frac{100}{3}} = \boxed{\frac{100\,(is\ an\ even\ No.)}{3\,(is\ an\ odd\ No.)}} = \boxed{33\frac{1}{3}}$$

Example 2.1-26

$$\boxed{\frac{112}{2}} = \boxed{\frac{112\,(is\ an\ even\ No.)}{2\,(is\ an\ even\ No.)}} = \boxed{\frac{112 \div 2}{2 \div 2}} = \boxed{\frac{56}{1}} = \boxed{56}$$

Example 2.1-27

$$\boxed{-\frac{325}{40}} = \boxed{-\frac{325\,(is\ an\ odd\ No.)}{40\,(is\ an\ even\ No.)}} = \boxed{-\frac{325 \div 5}{40 \div 5}} = \boxed{-\frac{65}{8}} = \boxed{-\left(8\frac{1}{8}\right)}$$

Example 2.1-28

$$\boxed{\frac{22}{6}} = \boxed{\frac{22\,(is\ an\ even\ No.)}{6\,(is\ an\ even\ No.)}} = \boxed{\frac{22 \div 2}{6 \div 2}} = \boxed{\frac{11}{3}} = \boxed{3\frac{2}{3}}$$

Example 2.1-29

$$\boxed{\frac{36}{3}} = \boxed{\frac{36\,(is\ an\ even\ No.)}{3\,(is\ an\ odd\ No.)}} = \boxed{\frac{36 \div 3}{3 \div 3}} = \boxed{\frac{12}{1}} = \boxed{12}$$

Example 2.1-30

$$\boxed{\frac{6}{39}} = \boxed{\frac{6\,(is\ an\ even\ No.)}{39\,(is\ an\ odd\ No.)}} = \boxed{\frac{6 \div 3}{39 \div 3}} = \boxed{\frac{\mathbf{2}}{\mathbf{13}}}$$

Greatest Common Factor

Greatest Common Factor (G.C.F.) can be found in two ways: 1. Trial and error method, and 2. Prime factoring method.

1. ***Trial and Error Method***: In the trial and error method the numerator and the denominator are divided by odd or even numbers until the largest divisor for both the numerator and the denominator is found.
2. ***Prime Factoring Method***: The steps in using the prime factoring method are:
   a. Rewrite both the numerator and the denominator by their equivalent prime number products.
   b. Identify the prime numbers that are common in both the numerator and the denominator.
   c. Multiply the common prime numbers to obtain the G.C.F.

   The following are examples of how G.C.F. can be found using the prime factoring method:

   1. $\frac{24}{45} = \frac{8 \times 3}{9 \times 5} = \frac{4 \times 2 \times 3}{3 \times 3 \times 5} = \frac{2 \times 2 \times 2 \times 3}{3 \times 3 \times 5}$. The common prime number in both the numerator and the denominator is 3. Therefore, $G.C.F. = 3$.
   2. $\frac{400}{350} = \frac{4 \times 100}{35 \times 10} = \frac{2 \times 2 \times 4 \times 25}{7 \times 5 \times 5 \times 2} = \frac{2 \times 2 \times 2 \times 2 \times 5 \times 5}{7 \times 5 \times 5 \times 2}$. The common prime numbers in both the numerator and the denominator are 2, 5, and 5. Therefore, $G.C.F. = 2 \times 5 \times 5 = 50$.
   3. $\frac{15}{60} = \frac{5 \times 3}{6 \times 10} = \frac{5 \times 3}{2 \times 3 \times 5 \times 2}$. The common prime numbers in both the numerator and the denominator are 3 and 5. Therefore, $G.C.F. = 3 \times 5 = 15$.
   4. $\frac{108}{27} = \frac{12 \times 9}{9 \times 3} = \frac{6 \times 2 \times 3 \times 3}{3 \times 3 \times 3} = \frac{2 \times 3 \times 2 \times 3 \times 3}{3 \times 3 \times 3}$. The common prime numbers in both the numerator and the denominator are 3, 3, and 3. Therefore, $G.C.F. = 3 \times 3 \times 3 = 27$.

Practice Problems - Simplifying Integer Fractions

Section 2.1 Practice Problems - Simplify the following integer fractions:

1. $\frac{60}{150} =$ 2. $\frac{8}{18} =$ 3. $\frac{355}{15} =$ 4. $\frac{3}{8} =$ 5. $\frac{27}{6} =$

6. $\frac{33}{6} =$ 7. $\frac{250}{1000} =$ 8. $\frac{4}{32} =$ 9. $\frac{284}{568} =$ 10. $\frac{45}{75} =$

2.1 Appendix: Changing Improper Fractions to Mixed Fractions

Improper fractions of the form $\frac{c}{b}$ with absolute values of greater than one are changed to mixed fractions of the form $k\frac{a}{b}$, where k is a positive or negative whole number and $\frac{a}{b}$ is an integer fraction with value of less than one, using the following steps:

Step 1 Divide the dividend, i.e., the numerator of the improper fraction by the divisor, i.e., the denominator of the improper fraction using the general division process.

Step 2 a. Use the whole number portion of the quotient as the whole number portion of the mixed fraction.

b. Use the dividend of the remainder as the dividend (numerator) in the remainder portion of the quotient.

c. Use the divisor of the improper fraction as the divisor (denominator) in the remainder portion of the quotient.

Examples with Steps

The following examples show the steps as to how improper fractions are changed to integer fractions:

Example 2.1A-1

$$\boxed{\frac{86}{5}} =$$

Solution:

Step 1

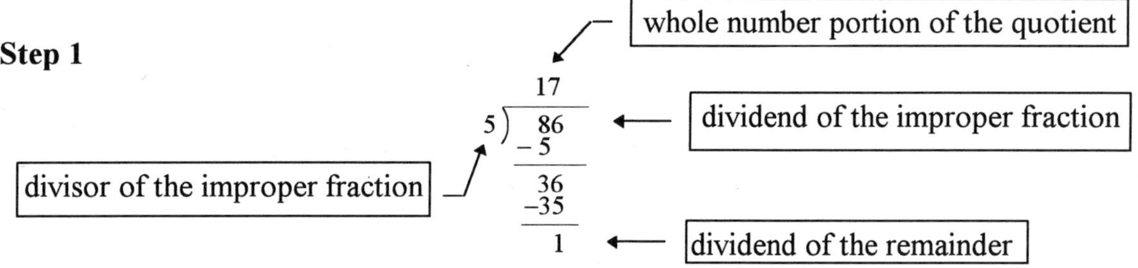

Step 2

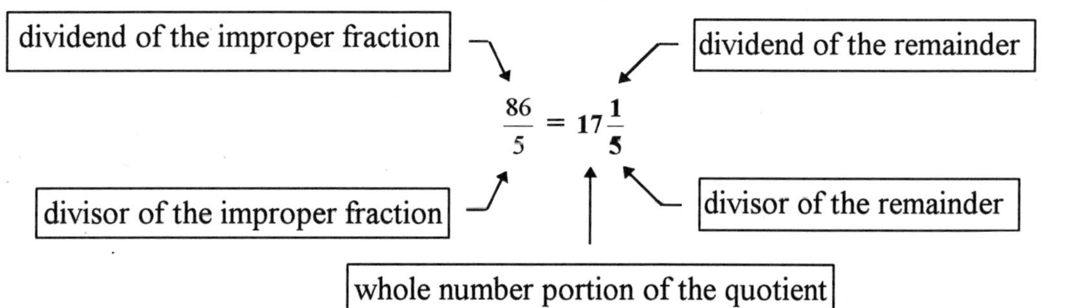

Example 2.1A-2

$$\boxed{\frac{506}{3}} =$$

Solution:

Step 1

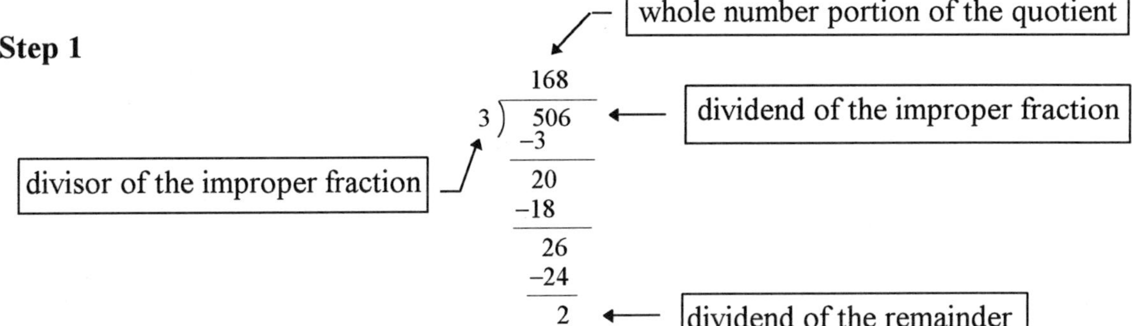

Step 2

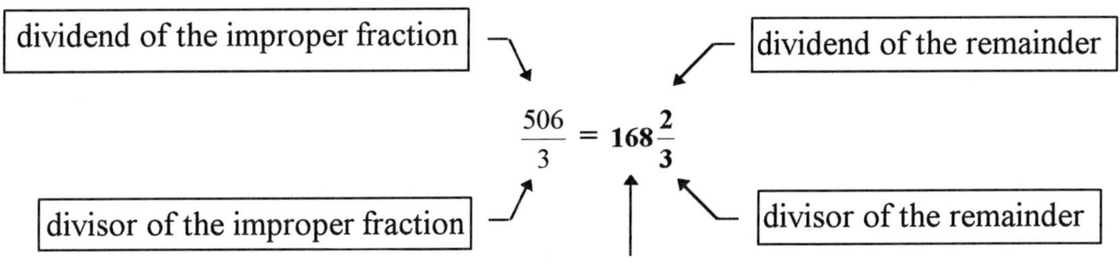

Example 2.1A-3

$$\boxed{\frac{296}{5}} =$$

Solution:

Step 1

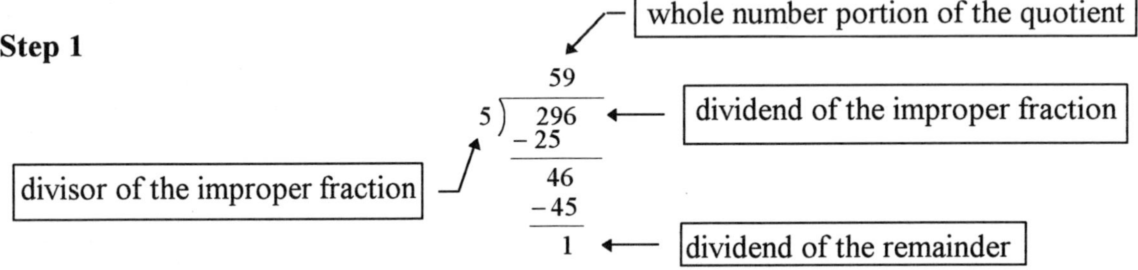

Step 2

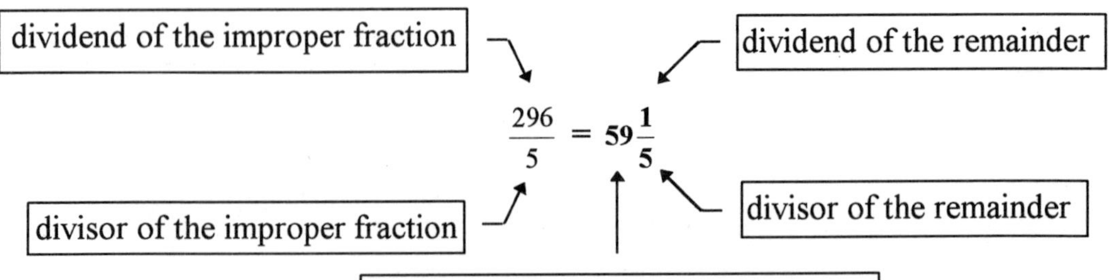

Example 2.1A-4

$$\boxed{-\frac{597}{10}} =$$

Solution:

Step 1

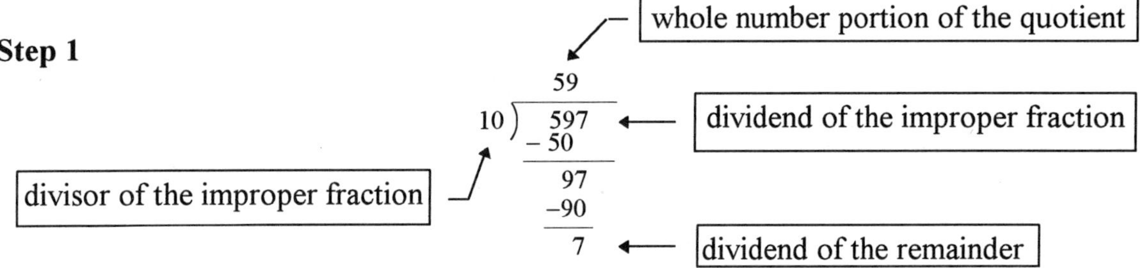

Step 2

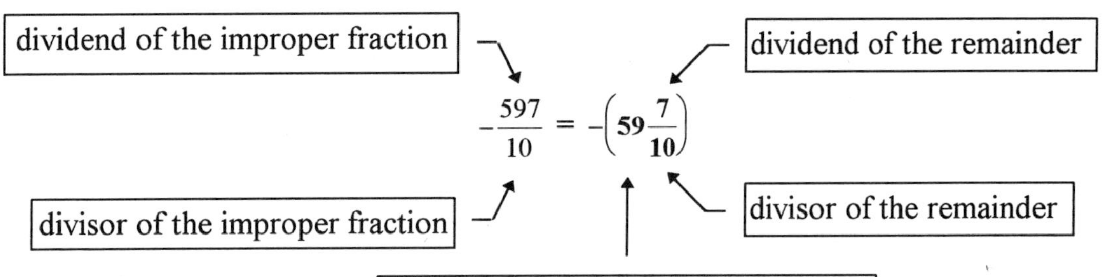

Example 2.1A-5

$$\boxed{\frac{1428}{45}} =$$

Solution:

Step 1

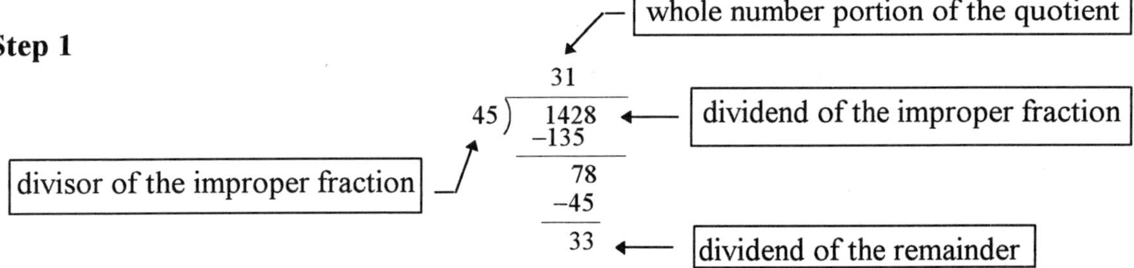

Step 2

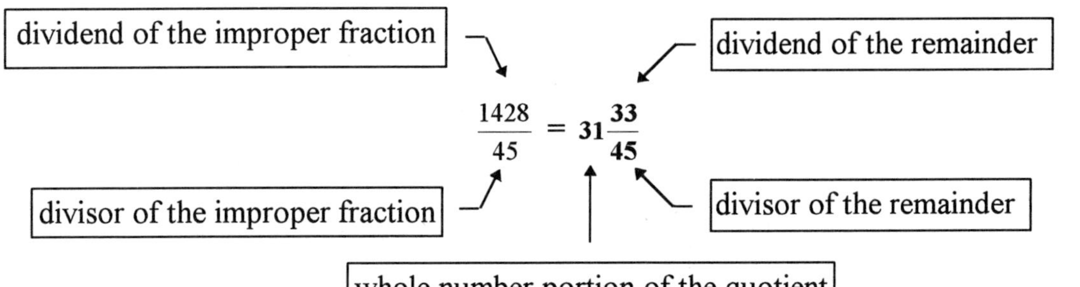

Example 2.1A-6

$$\boxed{-\frac{38}{3}} =$$

Solution:

Step 1

whole number portion of the quotient

$$3\overline{)38}$$ quotient 12

38 ← dividend of the improper fraction

divisor of the improper fraction → 3

$$\begin{array}{r} 12 \\ 3\overline{)\ 38} \\ -3 \\ \hline 08 \\ -6 \\ \hline 2 \end{array}$$

2 ← dividend of the remainder

Step 2

$$-\frac{38}{3} = -\left(12\frac{2}{3}\right)$$

dividend of the improper fraction (38); divisor of the improper fraction (3); whole number portion of the quotient (12); dividend of the remainder (2); divisor of the remainder (3)

Example 2.1A-7

$$\boxed{\frac{1967}{60}} =$$

Solution:

Step 1

$$\begin{array}{r} 32 \\ 60\overline{)\ 1967} \\ -180 \\ \hline 167 \\ -120 \\ \hline 47 \end{array}$$

whole number portion of the quotient (32); dividend of the improper fraction (1967); divisor of the improper fraction (60); dividend of the remainder (47)

Step 2

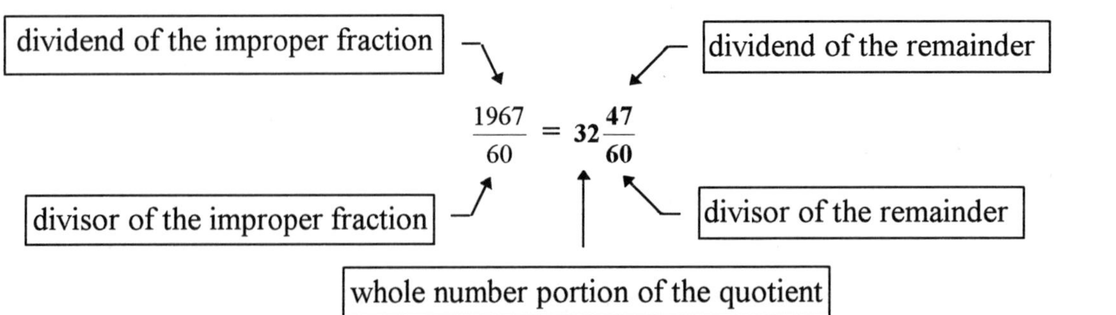

Example 2.1A-8

$$\boxed{-\frac{28}{13}} =$$

Solution:

Step 1

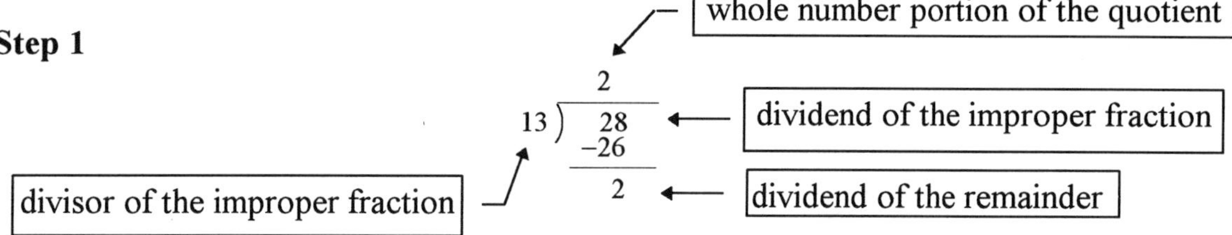

Step 2

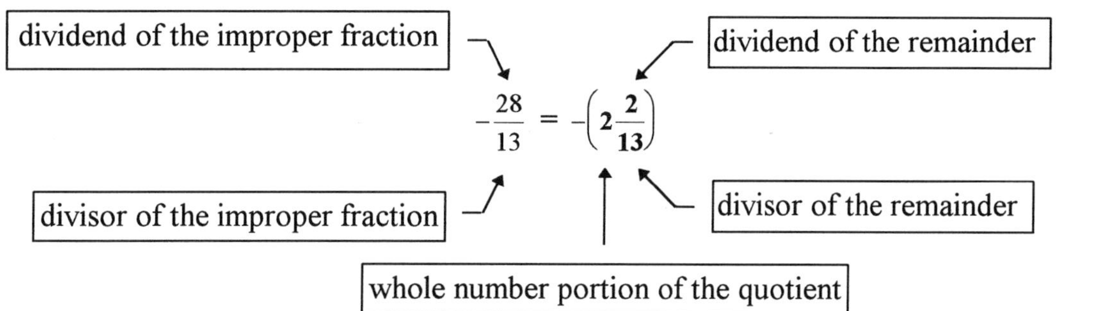

Example 2.1A-9

$$\boxed{\frac{273}{8}} =$$

Solution:

Step 1

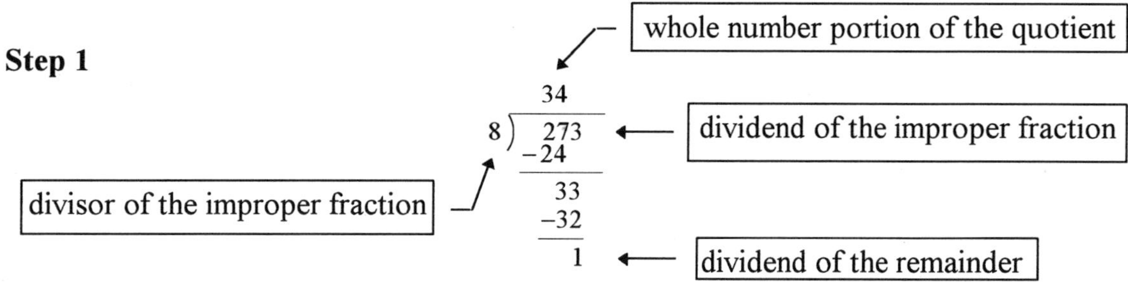

Step 2

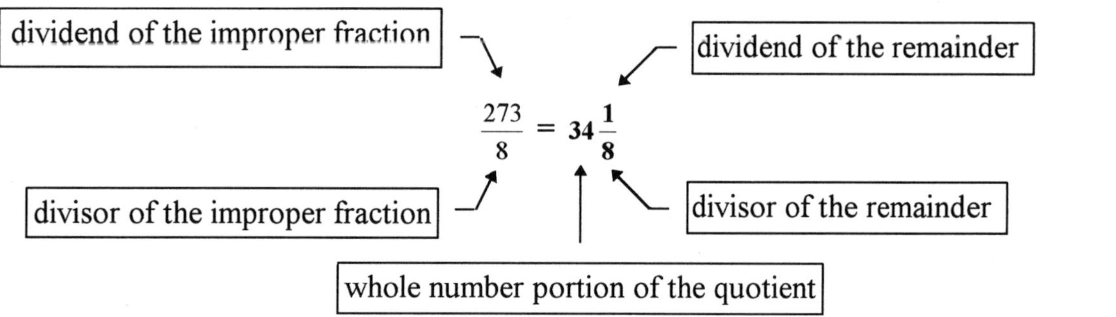

Example 2.1A-10

$$\boxed{-\frac{355}{102}} =$$

Solution:

Step 1

Step 2

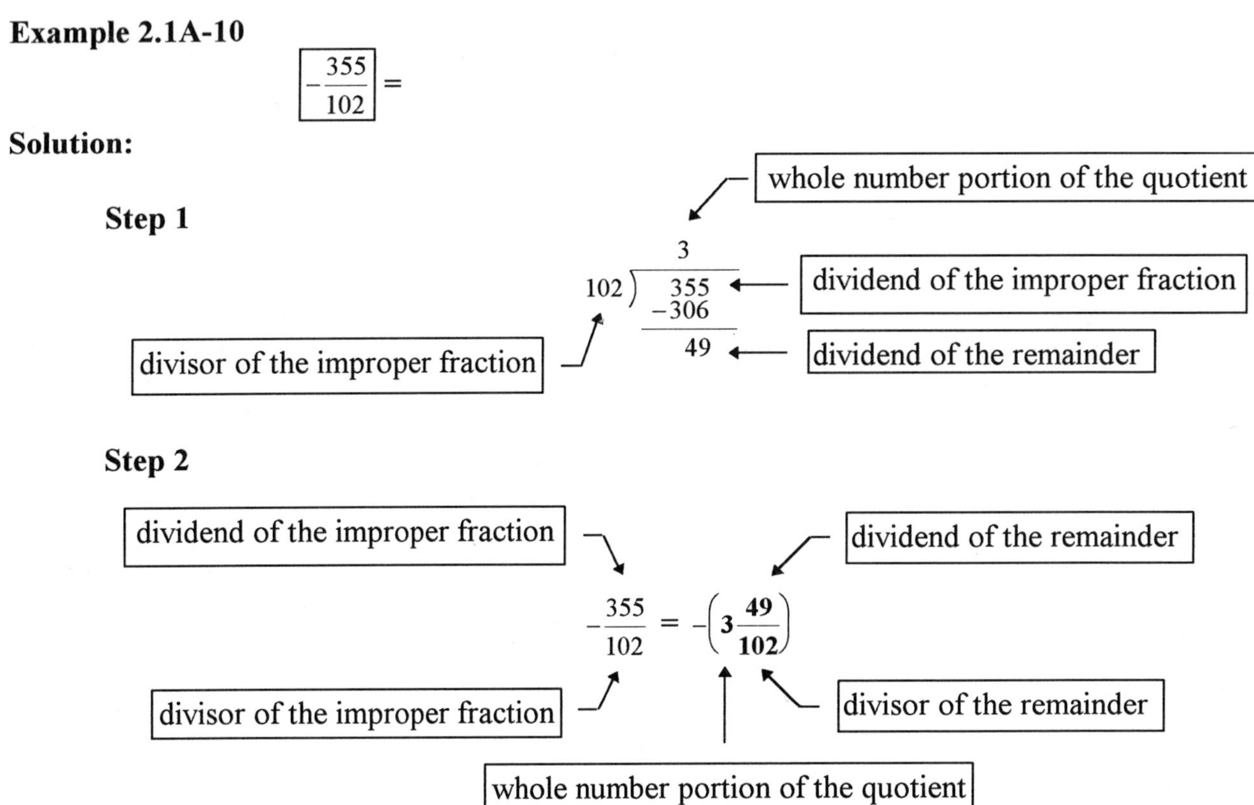

$$\begin{array}{r} 3 \\ 102 \overline{) \; 355} \\ -306 \\ \hline 49 \end{array}$$

$$-\frac{355}{102} = -\left(3\frac{49}{102}\right)$$

In general, an improper integer fraction $\frac{c}{b}$, where c is bigger than b, is changed to a mixed fraction in the following way:

1. divide the numerator c by its denominator b using the general division process.

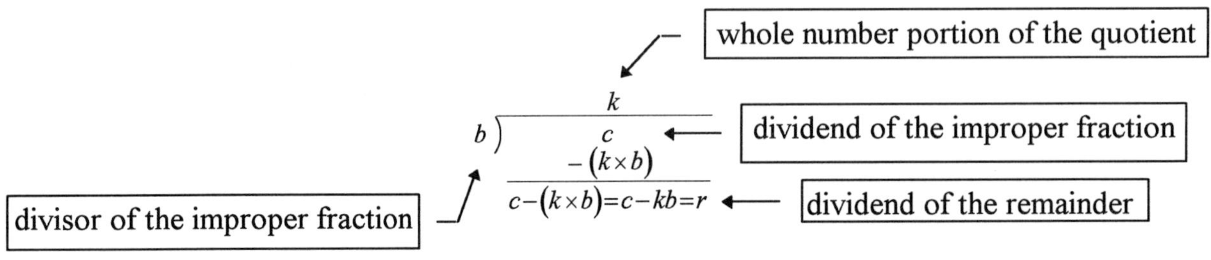

$$\begin{array}{r} k \\ b \overline{) \; c} \\ -(k \times b) \\ \hline c-(k \times b)=c-kb=r \end{array}$$

2. Use the whole number portion of the quotient k, the dividend of the remainder r, and the divisor of the improper fraction b to represent the mixed fraction as:

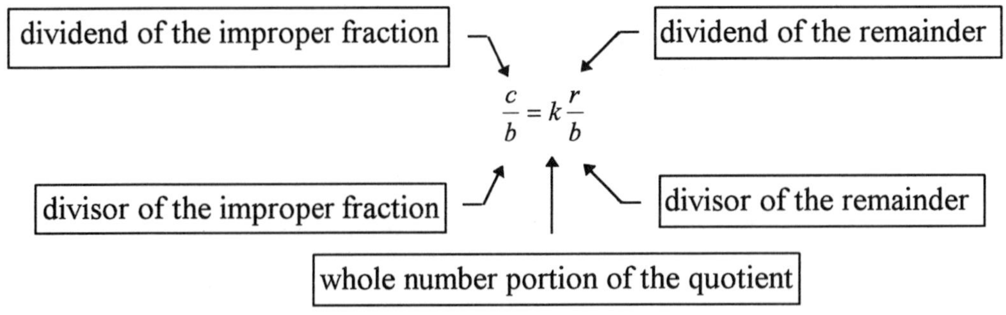

$$\frac{c}{b} = k\frac{r}{b}$$

Note 1 - In the general equation $\left(\frac{c}{b} = k\frac{r}{b}\right)$; $\frac{c}{b}$ is the improper fraction, $k\frac{r}{b}$ is the quotient, k is the whole number portion of the quotient, and $\frac{r}{b}$ is the remainder portion of the quotient.

Note 2 - The divisor of the improper fraction is always used as the divisor of the remainder. This is shown in Step 2 of examples above.

2.1 Appendix Practice Problems - Changing Improper Fractions to Mixed Fractions

2.1 Appendix Practice Problems - Change the following improper fractions to mixed fractions:

1. $\frac{83}{4} =$
2. $\frac{13}{3} =$
3. $-\frac{26}{5} =$
4. $\frac{67}{10} =$
5. $\frac{9}{2} =$
6. $-\frac{332}{113} =$
7. $\frac{205}{9} =$
8. $-\frac{235}{14}$
9. $\frac{207}{11} =$
10. $-\frac{523}{101} =$

2.2 Adding Integer Fractions

Integer fractions, i.e., fractions where both the numerator and the denominator are integers, are added as in the following cases:

Case I Adding Two or More Integer Fractions with Common Denominators

Integer fractions with two or more common denominators are added using the steps given as in each case below:

Case I-A Adding Two Integer Fractions with Common Denominators

Add two integer fractions with common denominators using the following steps:

Step 1 a. Use the common denominator between the first and second fractions as the new denominator.

b. Add the numerators of the first and second fractions to obtain the new numerator.

Step 2 Simplify the fraction to its lowest term (see Section 2.1).

Step 3 Change the improper fraction to a mixed fraction if the fraction obtained from Step 2 is an improper fraction (see Section 2.1 Appendix).

Examples with Steps

The following examples show the steps as to how two integer fractions with common denominators are added:

Example 2.2-1

$$\boxed{\frac{2}{3}+\frac{8}{3}} =$$

Solution:

Step 1 $\boxed{\frac{2}{3}+\frac{8}{3}} = \boxed{\frac{2+8}{3}} = \boxed{\frac{10}{3}}$

Step 2 *Not Applicable*

Step 3 $\boxed{\frac{10}{3}} = \boxed{3\frac{1}{3}}$

Example 2.2-2

$$\boxed{\frac{15}{4}+\frac{9}{4}} =$$

Solution:

Step 1 $\boxed{\frac{15}{4}+\frac{9}{4}} = \boxed{\frac{15+9}{4}} = \boxed{\frac{24}{4}}$

Step 2 $\boxed{\frac{24}{4}} = \boxed{\frac{24 \div 4}{4 \div 4}} = \frac{6}{1} = \boxed{6}$

Step 3 *Not Applicable*

Example 2.2-3

$$\boxed{\frac{5}{9}+\frac{2}{9}} =$$

Solution:

Step 1 $\boxed{\frac{5}{9}+\frac{2}{9}} = \boxed{\frac{5+2}{9}} = \boxed{\frac{7}{9}}$

Step 2 *Not Applicable*

Step 3 *Not Applicable*

Example 2.2-4

$$\boxed{\frac{4}{7}+\frac{15}{7}} =$$

Solution:

Step 1 $\boxed{\frac{4}{7}+\frac{15}{7}} = \boxed{\frac{4+15}{7}} = \boxed{\frac{19}{7}}$

Step 2 *Not Applicable*

Step 3 $\boxed{\frac{19}{7}} = \boxed{2\frac{5}{7}}$

Example 2.2-5

$$\boxed{\frac{12}{5}+\frac{33}{5}} =$$

Solution:

Step 1 $\boxed{\frac{12}{5}+\frac{33}{5}} = \boxed{\frac{12+33}{5}} = \boxed{\frac{45}{5}}$

Step 2 $\boxed{\frac{45}{5}} = \boxed{\frac{45 \div 5}{5 \div 5}} = \boxed{\frac{9}{1}} = \boxed{9}$

Step 3 *Not Applicable*

In general, two integer fractions with a common denominator are added in the following way:

$$\boxed{\frac{a}{d}+\frac{b}{d}} = \boxed{\frac{a+b}{d}}$$

Example 2.2-6

$$\boxed{\frac{5}{3}+\frac{13}{3}} = \boxed{\frac{5+13}{3}} = \boxed{\frac{\overset{6}{\cancel{18}}}{\underset{1}{\cancel{3}}}} = \boxed{\frac{6}{1}} = \boxed{6}$$

Case I-B Adding Three Integer Fractions with Common Denominators

Add three integer fractions with common denominators using the following steps:

Step 1 a. Use the common denominator between the first, second, and third fractions as the new denominator.

b. Add the numerators of the first, second, and third fractions to obtain the new denominator.

Step 2 Simplify the fraction to its lowest term (see Section 2.1).

Step 3 Change the improper fraction to a mixed fraction if the fraction obtained from Step 2 is an improper fraction (see Section 2.1 Appendix).

Examples with Steps

The following examples show the steps as to how three integer fractions with common denominators are added:

Example 2.2-7

$$\boxed{\frac{3}{5}+\frac{4}{5}+\frac{1}{5}} =$$

Solution:

Step 1 $\boxed{\frac{3}{5}+\frac{4}{5}+\frac{1}{5}} = \boxed{\frac{3+4+1}{5}} = \boxed{\frac{8}{5}}$

Step 2 *Not Applicable*

Step 3 $\boxed{\frac{8}{5}} = \boxed{\mathbf{1\frac{3}{5}}}$

Example 2.2-8

$$\boxed{\frac{5}{8}+\frac{2}{8}+\frac{14}{8}} =$$

Solution:

Step 1 $\boxed{\frac{5}{8}+\frac{2}{8}+\frac{14}{8}} = \boxed{\frac{5+2+14}{8}} = \boxed{\frac{21}{8}}$

Step 2 *Not Applicable*

Step 3 $\boxed{\frac{21}{8}} = \boxed{\mathbf{2\frac{5}{8}}}$

Example 2.2-9

$$\boxed{\frac{17}{3}+\frac{12}{3}+\frac{10}{3}} =$$

Solution:

Step 1 $\boxed{\frac{17}{3}+\frac{12}{3}+\frac{10}{3}} = \boxed{\frac{17+12+10}{3}} = \boxed{\frac{39}{3}}$

Step 2 $\boxed{\frac{39}{3}} = \boxed{\frac{39 \div 3}{3 \div 3}} = \boxed{\frac{13}{1}} = \boxed{\mathbf{13}}$

Step 3 $\boxed{\textit{Not Applicable}}$

Example 2.2-10

$\boxed{\frac{5}{4}+\frac{23}{4}+\frac{38}{4}} =$

Solution:

Step 1 $\boxed{\frac{5}{4}+\frac{23}{4}+\frac{38}{4}} = \boxed{\frac{5+23+38}{4}} = \boxed{\frac{66}{4}}$

Step 2 $\boxed{\frac{66}{4}} = \boxed{\frac{66 \div 2}{4 \div 2}} = \boxed{\frac{33}{2}}$

Step 3 $\boxed{\frac{33}{2}} = \boxed{\mathbf{16\frac{1}{2}}}$

Example 2.2-11

$\boxed{\frac{5}{12}+\frac{14}{12}+\frac{25}{12}} =$

Solution:

Step 1 $\boxed{\frac{5}{12}+\frac{14}{12}+\frac{25}{12}} = \boxed{\frac{5+14+25}{12}} = \boxed{\frac{44}{12}}$

Step 2 $\boxed{\frac{44}{12}} = \boxed{\frac{44 \div 4}{12 \div 4}} = \boxed{\frac{11}{3}}$

Step 3 $\boxed{\frac{11}{3}} = \boxed{\mathbf{3\frac{2}{3}}}$

In general, three integer fractions with a common denominator are added in the following way:

$\boxed{\frac{a}{d}+\frac{b}{d}+\frac{c}{d}} = \boxed{\frac{a+b+c}{d}}$

Example 2.2-12

$\boxed{\frac{3}{5}+\frac{2}{5}+\frac{5}{5}} = \boxed{\frac{3+2+5}{5}} = \boxed{\frac{\overset{2}{\cancel{10}}}{\underset{1}{\cancel{5}}}} = \boxed{\frac{2}{1}} = \boxed{\mathbf{2}}$

Case II Adding Two or More Integer Fractions Without a Common Denominator

Two or more integer fractions without a common denominator are added using the steps given as in each case below:

Case II-A Adding Two Integer Fractions Without a Common Denominator

Add two integer fractions without a common denominator using the following steps:

Step 1 Change the integer number a to an integer fraction of the form $\frac{a}{1}$, e.g., change 5 to $\frac{5}{1}$.

Step 2 a. Multiply the denominators of the first and second fractions to obtain the new denominator.

b. Cross multiply the numerator of the first fraction with the denominator of the second fraction.

c. Cross multiply the numerator of the second fraction with the denominator of the first fraction.

d. Add the results from the steps 2b and 2c above to obtain the new numerator.

Step 3 Simplify the fraction to its lowest term (see Section 2.1).

Step 4 Change the improper fraction to a mixed fraction if the fraction obtained from Step 3 is an improper fraction (see Section 2.1 Appendix).

Examples with Steps

The following examples show the steps as to how two integer fractions without a common denominator are added:

Example 2.2-13

$$\boxed{\frac{2}{5}+\frac{3}{4}} =$$

Solution:

Step 1 *Not Applicable*

Step 2 $\boxed{\frac{2}{5}+\frac{3}{4}} = \boxed{\frac{(2\times 4)+(3\times 5)}{5\times 4}} = \boxed{\frac{8+15}{20}} = \boxed{\frac{23}{20}}$

Step 3 *Not Applicable*

Step 4 $\boxed{\frac{23}{20}} = \boxed{\mathbf{1\frac{3}{20}}}$

Example 2.2-14

$$\boxed{40+\frac{4}{3}} =$$

Solution:

Step 1 $\boxed{40+\frac{4}{3}} = \boxed{\frac{40}{1}+\frac{4}{3}}$

Step 2 $\boxed{\frac{40}{1}+\frac{4}{3}} = \boxed{\frac{(40\times 3)+(4\times 1)}{1\times 3}} = \boxed{\frac{120+4}{3}} = \boxed{\frac{124}{3}}$

Step 3 Not Applicable

Step 4 $\boxed{\frac{124}{3}} = \boxed{\mathbf{41\frac{1}{3}}}$

Example 2.2-15

$\boxed{\frac{3}{5}+\frac{2}{7}} =$

Solution:

Step 1 Not Applicable

Step 2 $\boxed{\frac{3}{5}+\frac{2}{7}} = \boxed{\frac{(3\times 7)+(2\times 5)}{5\times 7}} = \boxed{\frac{21+10}{35}} = \boxed{\mathbf{\frac{31}{35}}}$

Step 3 Not Applicable

Step 4 Not Applicable

Example 2.2-16

$\boxed{\frac{8}{15}+\frac{3}{5}} =$

Solution:

Step 1 Not Applicable

Step 2 $\boxed{\frac{8}{15}+\frac{3}{5}} = \boxed{\frac{(8\times 5)+(3\times 15)}{15\times 5}} = \boxed{\frac{40+45}{75}} = \boxed{\frac{85}{75}}$

Step 3 $\boxed{\frac{85}{75}} = \boxed{\frac{85\div 5}{75\div 5}} = \boxed{\frac{17}{15}}$

Step 4 $\boxed{\frac{17}{15}} = \boxed{\mathbf{1\frac{2}{15}}}$

Example 2.2-17

$\boxed{\frac{5}{6}+3} =$

Solution:

Step 1 $\boxed{\frac{5}{6}+3} = \boxed{\frac{5}{6}+\frac{3}{1}}$

Step 2 $\boxed{\frac{5}{6}+\frac{3}{1}} = \boxed{\frac{(5\times 1)+(3\times 6)}{6\times 1}} = \boxed{\frac{5+18}{6}} = \boxed{\frac{23}{6}}$

Step 3 *Not Applicable*

Step 4 $\boxed{\frac{23}{6}} = \boxed{\mathbf{3\frac{5}{6}}}$

In general, two integer fractions without a common denominator are added in the following way:

$$\boxed{\frac{a}{b}+\frac{c}{d}} = \boxed{\frac{(a\times d)+(c\times b)}{(b\times d)}} = \boxed{\frac{ad+cb}{bd}}$$

Example 2.2-18

$$\boxed{\frac{6}{3}+\frac{9}{4}} = \boxed{\frac{(6\times 4)+(3\times 9)}{3\times 4}} = \boxed{\frac{24+27}{12}} = \boxed{\frac{\overset{17}{\cancel{51}}}{\underset{4}{\cancel{12}}}} = \boxed{\frac{17}{4}} = \boxed{\mathbf{4\frac{1}{4}}}$$

Case II-B Adding Three Integer Fractions Without a Common Denominator

Add three integer fractions without a common denominator using the following steps:

Step 1 Use parentheses to group the first and second fractions.

Step 2 Change the integer number a to an integer fraction of the form $\frac{a}{1}$, e.g., change 28 to $\frac{28}{1}$.

Step 3 a. Add the grouped fractions following Steps 2a through 2d, outlined in Section 2.2, Case II-A above, to obtain a new integer fraction.

b. Add the new integer fraction to the third fraction by repeating Steps 2a through 2d outlined in Section 2.2, Case II-A above.

Step 4 Simplify the fraction to its lowest term (see Section 2.1).

Step 5 Change the improper fraction to a mixed fraction if the fraction obtained from Step 4 is an improper fraction (see Section 2.1 Appendix).

Examples with Steps

The following examples show the steps as to how three integer fractions without a common denominator are added:

Example 2.2-19

$$\boxed{\frac{3}{5}+\frac{4}{3}+\frac{1}{6}} =$$

Solution:

Step 1 $\boxed{\frac{3}{5}+\frac{4}{3}+\frac{1}{6}} = \boxed{\left(\frac{3}{5}+\frac{4}{3}\right)+\frac{1}{6}}$

Step 2 Not Applicable

Step 3 $\left(\frac{3}{5}+\frac{4}{3}\right)+\frac{1}{6}=\left(\frac{(3\times3)+(4\times5)}{5\times3}\right)+\frac{1}{6}=\left(\frac{9+20}{15}\right)+\frac{1}{6}=\left(\frac{29}{15}\right)+\frac{1}{6}=\frac{29}{15}+\frac{1}{6}$

$=\frac{(29\times6)+(1\times15)}{15\times6}=\frac{174+15}{90}=\frac{189}{90}$

Step 4 $\frac{189}{90}=\frac{189\div9}{90\div9}=\frac{21}{10}$

Step 5 $\frac{21}{10}=\mathbf{2\frac{1}{10}}$

Example 2.2-20

$\frac{4}{6}+\frac{2}{5}+\frac{1}{8}=$

Solution:

Step 1 $\frac{4}{6}+\frac{2}{5}+\frac{1}{8}=\left(\frac{4}{6}+\frac{2}{5}\right)+\frac{1}{8}$

Step 2 Not Applicable

Step 3 $\left(\frac{4}{6}+\frac{2}{5}\right)+\frac{1}{8}=\left(\frac{(4\times5)+(2\times6)}{6\times5}\right)+\frac{1}{8}=\left(\frac{20+12}{30}\right)+\frac{1}{8}=\left(\frac{32}{30}\right)+\frac{1}{8}=\frac{32}{30}+\frac{1}{8}$

$=\frac{(32\times8)+(1\times30)}{30\times8}=\frac{256+30}{240}=\frac{286}{240}$

Step 4 $\frac{286}{240}=\frac{286\div2}{240\div2}=\frac{143}{120}$

Step 5 $\frac{143}{120}=\mathbf{1\frac{23}{120}}$

Example 2.2-21

$\frac{3}{5}+12+\frac{5}{8}=$

Solution:

Step 1 $\frac{3}{5}+12+\frac{5}{8}=\left(\frac{3}{5}+12\right)+\frac{5}{8}$

Step 2 $\left(\frac{3}{5}+12\right)+\frac{5}{8}=\left(\frac{3}{5}+\frac{12}{1}\right)+\frac{5}{8}$

Step 3 $\boxed{\left(\frac{3}{5}+\frac{12}{1}\right)+\frac{5}{8}} = \boxed{\left(\frac{(3\times 1)+(12\times 5)}{5\times 1}\right)+\frac{5}{8}} = \boxed{\left(\frac{3+60}{5}\right)+\frac{5}{8}} = \boxed{\left(\frac{63}{5}\right)+\frac{5}{8}} = \boxed{\frac{63}{5}+\frac{5}{8}}$

$= \boxed{\frac{(63\times 8)+(5\times 5)}{5\times 8}} = \boxed{\frac{504+25}{40}} = \boxed{\frac{529}{40}}$

Step 4 *Not Applicable*

Step 5 $\boxed{\frac{529}{40}} = \boxed{\mathbf{13\frac{9}{40}}}$

Example 2.2-22

$\boxed{15+\frac{3}{4}+\frac{5}{6}} =$

Solution:

Step 1 $\boxed{15+\frac{3}{4}+\frac{5}{6}} = \boxed{\left(15+\frac{3}{4}\right)+\frac{5}{6}}$

Step 2 $\boxed{\left(15+\frac{3}{4}\right)+\frac{5}{6}} = \boxed{\left(\frac{15}{1}+\frac{3}{4}\right)+\frac{5}{6}}$

Step 3 $\boxed{\left(\frac{15}{1}+\frac{3}{4}\right)+\frac{5}{6}} = \boxed{\left(\frac{(15\times 4)+(3\times 1)}{1\times 4}\right)+\frac{5}{6}} = \boxed{\left(\frac{60+3}{4}\right)+\frac{5}{6}} = \boxed{\left(\frac{63}{4}\right)+\frac{5}{6}} = \boxed{\frac{63}{4}+\frac{5}{6}}$

$= \boxed{\frac{(63\times 6)+(5\times 4)}{4\times 6}} = \boxed{\frac{378+20}{24}} = \boxed{\frac{398}{24}}$

Step 4 $\boxed{\frac{398}{24}} = \boxed{\frac{398\div 2}{24\div 2}} = \boxed{\frac{199}{12}}$

Step 5 $\boxed{\frac{199}{12}} = \boxed{\mathbf{16\frac{7}{12}}}$

Example 2.2-23

$\boxed{25+\frac{4}{5}+\frac{2}{3}} =$

Solution:

Step 1 $\boxed{25+\frac{4}{5}+\frac{2}{3}} = \boxed{\left(25+\frac{4}{5}\right)+\frac{2}{3}}$

Step 2 $\boxed{\left(25+\frac{4}{5}\right)+\frac{2}{3}} = \boxed{\left(\frac{25}{1}+\frac{4}{5}\right)+\frac{2}{3}}$

Step 3

$$\boxed{\left(\frac{25}{1}+\frac{4}{5}\right)+\frac{2}{3}} = \boxed{\left(\frac{(25\times 5)+(4\times 1)}{1\times 5}\right)+\frac{2}{3}} = \boxed{\left(\frac{125+4}{5}\right)+\frac{2}{3}} = \boxed{\frac{129}{5}+\frac{2}{3}}$$

$$= \boxed{\frac{(129\times 3)+(2\times 5)}{5\times 3}} = \boxed{\frac{387+10}{15}} = \boxed{\frac{397}{15}}$$

Step 4 *Not Applicable*

Step 5

$$\boxed{\frac{397}{15}} = \boxed{\mathbf{26\frac{7}{15}}}$$

In general, three integer fractions without a common denominator are added as in the following cases:

Case I.

$$\boxed{\frac{a}{b}+\frac{c}{d}+\frac{e}{f}} = \boxed{\left(\frac{a}{b}+\frac{c}{d}\right)+\frac{e}{f}} = \boxed{\left(\frac{(a\times d)+(c\times b)}{b\times d}\right)+\frac{e}{f}} = \boxed{\left(\frac{ad+cb}{bd}\right)+\frac{e}{f}} = \boxed{\frac{[(ad+cb)\times f]+(e\times bd)}{bd\times f}}$$

$$= \boxed{\frac{(ad+cb)f+ebd}{bdf}} = \boxed{\frac{adf+cbf+ebd}{bdf}}$$

Example 2.2-24

$$\boxed{\frac{1}{2}+\frac{3}{4}+\frac{2}{5}} = \boxed{\left(\frac{1}{2}+\frac{3}{4}\right)+\frac{2}{5}} = \boxed{\left(\frac{(1\times 4)+(3\times 2)}{2\times 4}\right)+\frac{2}{5}} = \boxed{\left(\frac{4+6}{8}\right)+\frac{2}{5}} = \boxed{\left(\frac{10}{8}\right)+\frac{2}{5}} = \boxed{\frac{10}{8}+\frac{2}{5}}$$

$$= \boxed{\frac{(10\times 5)+(2\times 8)}{8\times 5}} = \boxed{\frac{50+16}{40}} = \boxed{\frac{\overset{33}{\cancel{66}}}{\underset{20}{\cancel{40}}}} = \boxed{\frac{33}{20}} = \boxed{\mathbf{1\frac{13}{20}}}$$

Case II.

$$\boxed{\frac{a}{b}+\frac{c}{d}+\frac{e}{f}} = \boxed{\frac{a}{b}+\left(\frac{c}{d}+\frac{e}{f}\right)} = \boxed{\frac{a}{b}+\left(\frac{(c\times f)+(e\times d)}{d\times f}\right)} = \boxed{\frac{a}{b}+\left(\frac{cf+ed}{df}\right)} = \boxed{\frac{(a\times df)+[(cf+ed)\times b]}{b\times df}}$$

$$= \boxed{\frac{adf+(cf+ed)b}{bdf}} = \boxed{\frac{adf+cfb+edb}{bdf}}$$

Example 2.2-25

$$\boxed{\frac{1}{2}+\frac{3}{4}+\frac{2}{5}} = \boxed{\frac{1}{2}+\left(\frac{3}{4}+\frac{2}{5}\right)} = \boxed{\frac{1}{2}+\left(\frac{(3\times 5)+(2\times 4)}{4\times 5}\right)} = \boxed{\frac{1}{2}+\left(\frac{15+8}{20}\right)} = \boxed{\frac{1}{2}+\left(\frac{23}{20}\right)} = \boxed{\frac{1}{2}+\frac{23}{20}}$$

$$= \boxed{\frac{(1\times 20)+(23\times 2)}{2\times 20}} = \boxed{\frac{20+46}{40}} = \boxed{\frac{\overset{33}{\cancel{66}}}{\underset{20}{\cancel{40}}}} = \boxed{\frac{33}{20}} = \boxed{\mathbf{1\frac{13}{20}}}$$

Note - In addition the use of parentheses does not change the final answer; the two examples above have the same answer (see Section 1.2).

Additional Examples - Adding Integer Fractions

The following examples further illustrate how to add integer fractions: Note that fractional operations do not necessarily have to be solved in the exact "step" order as is given in this chapter. For example, in many instances, the process of adding, subtracting, multiplying, and dividing fractions is greatly simplified if fractions are reduced to their lowest terms first. In some instances, fractions are simplified several times at various steps of an operation.

Example 2.2-26

$$\boxed{\frac{3}{6}+\frac{4}{5}}=\boxed{\frac{(3\times5)+(4\times6)}{6\times5}}=\boxed{\frac{15+24}{30}}=\boxed{\frac{\overset{13}{\cancel{39}}}{\underset{10}{\cancel{30}}}}=\boxed{\frac{13}{10}}=\boxed{\mathbf{1\frac{3}{10}}}$$

Example 2.2-27

$$\boxed{\frac{3}{5}+\frac{8}{5}+\frac{4}{5}}=\boxed{\frac{3+8+4}{5}}=\boxed{\frac{\overset{3}{\cancel{15}}}{\underset{1}{\cancel{5}}}}=\boxed{\frac{3}{1}}=\boxed{\mathbf{3}}$$

Example 2.2-28

$$\boxed{\frac{2}{5}+\frac{1}{4}+\frac{4}{3}}=\boxed{\left(\frac{2}{5}+\frac{1}{4}\right)+\frac{4}{3}}=\boxed{\left(\frac{(2\times4)+(1\times5)}{5\times4}\right)+\frac{4}{3}}=\boxed{\left(\frac{8+5}{20}\right)+\frac{4}{3}}=\boxed{\left(\frac{13}{20}\right)+\frac{4}{3}}=\boxed{\frac{13}{20}+\frac{4}{3}}$$

$$=\boxed{\frac{(13\times3)+(4\times20)}{20\times3}}=\boxed{\frac{39+80}{60}}=\boxed{\frac{119}{60}}=\boxed{\mathbf{1\frac{59}{60}}}$$

Example 2.2-29

$$\boxed{\frac{1}{2}+\left(\frac{2}{3}+\frac{1}{5}\right)}=\boxed{\frac{1}{2}+\left(\frac{(2\times5)+(1\times3)}{3\times5}\right)}=\boxed{\frac{1}{2}+\left(\frac{10+3}{15}\right)}=\boxed{\frac{1}{2}+\left(\frac{13}{15}\right)}=\boxed{\frac{1}{2}+\frac{13}{15}}=\boxed{\frac{(1\times15)+(13\times2)}{2\times15}}$$

$$=\boxed{\frac{15+26}{30}}=\boxed{\frac{41}{30}}=\boxed{\mathbf{1\frac{11}{30}}}$$

Example 2.2-30

$$\boxed{6+\frac{4}{3}+\frac{8}{1}+\frac{9}{6}}=\boxed{\left(\frac{6}{1}+\frac{4}{3}\right)+\left(\frac{8}{1}+\frac{\overset{3}{\cancel{9}}}{\underset{2}{\cancel{6}}}\right)}=\boxed{\left(\frac{(6\times3)+(4\times1)}{1\times3}\right)+\left(\frac{8}{1}+\frac{3}{2}\right)}=\boxed{\left(\frac{18+4}{3}\right)+\left(\frac{(8\times2)+(3\times1)}{1\times2}\right)}$$

$$=\boxed{\left(\frac{22}{3}\right)+\left(\frac{16+3}{2}\right)}=\boxed{\frac{22}{3}+\left(\frac{19}{2}\right)}=\boxed{\frac{22}{3}+\frac{19}{2}}=\boxed{\frac{(22\times2)+(19\times3)}{3\times2}}=\boxed{\frac{44+57}{6}}=\boxed{\frac{101}{6}}=\boxed{\mathbf{16\frac{5}{6}}}$$

Example 2.2-31

$$\boxed{\left(\frac{2}{3}+\frac{6}{3}\right)+\left(\frac{8}{6}+\frac{2}{6}+\frac{1}{6}\right)}=\boxed{\left(\frac{2+6}{3}\right)+\left(\frac{8+2+1}{6}\right)}=\boxed{\left(\frac{8}{3}\right)+\left(\frac{11}{6}\right)}=\boxed{\frac{8}{3}+\frac{11}{6}}=\boxed{\frac{(8\times6)+(11\times3)}{3\times6}}=\boxed{\frac{48+33}{18}}$$

$$=\boxed{\frac{\overset{9}{\cancel{81}}}{\underset{2}{\cancel{18}}}}=\boxed{\frac{9}{2}}=\boxed{\mathbf{4\frac{1}{2}}}$$

Example 2.2-32

$$\boxed{\frac{2}{3}+\frac{1}{5}+\frac{7}{10}+\frac{4}{5}+\frac{3}{8}}=\boxed{\left(\frac{2}{3}+\frac{1}{5}\right)+\left(\frac{7}{10}+\frac{4}{5}\right)+\frac{3}{8}}=\boxed{\left(\frac{(2\times 5)+(1\times 3)}{3\times 5}\right)+\left(\frac{(7\times 5)+(4\times 10)}{10\times 5}\right)+\frac{3}{8}}$$

$$=\boxed{\left(\frac{10+3}{15}\right)+\left(\frac{35+40}{50}\right)+\frac{3}{8}}=\boxed{\left(\frac{13}{15}\right)+\left(\frac{75}{50}\right)+\frac{3}{8}}=\boxed{\frac{13}{15}+\frac{\overset{3}{\cancel{75}}}{\underset{2}{\cancel{50}}}+\frac{3}{8}}=\boxed{\frac{13}{15}+\frac{3}{2}+\frac{3}{8}}=\boxed{\left(\frac{13}{15}+\frac{3}{2}\right)+\frac{3}{8}}$$

$$=\boxed{\left(\frac{(13\times 2)+(3\times 15)}{15\times 2}\right)+\frac{3}{8}}=\boxed{\left(\frac{26+45}{30}\right)+\frac{3}{8}}=\boxed{\left(\frac{71}{30}\right)+\frac{3}{8}}=\boxed{\frac{71}{30}+\frac{3}{8}}=\boxed{\frac{(71\times 8)+(3\times 30)}{30\times 8}}=\boxed{\frac{568+90}{240}}$$

$$=\boxed{\frac{\overset{329}{\cancel{658}}}{\underset{120}{\cancel{240}}}}=\boxed{\frac{329}{120}}=\boxed{\mathbf{2\frac{89}{120}}}$$

Example 2.2-33

$$\boxed{2+\frac{0}{200}+\frac{5}{10}+\frac{4}{5}+6}=\boxed{\frac{2}{1}+0+\frac{5}{10}+\frac{4}{5}+\frac{6}{1}}=\boxed{\frac{2}{1}+\frac{5}{10}+\frac{4}{5}+\frac{6}{1}}=\boxed{\left(\frac{2}{1}+\frac{5}{10}\right)+\left(\frac{4}{5}+\frac{6}{1}\right)}$$

$$=\boxed{\left(\frac{(2\times 10)+(5\times 1)}{1\times 10}\right)+\left(\frac{(4\times 1)+(6\times 5)}{5\times 1}\right)}=\boxed{\left(\frac{20+5}{10}\right)+\left(\frac{4+30}{5}\right)}=\boxed{\left(\frac{25}{10}\right)+\left(\frac{34}{5}\right)}=\boxed{\frac{25}{10}+\frac{34}{5}}$$

$$=\boxed{\frac{(25\times 5)+(34\times 10)}{10\times 5}}=\boxed{\frac{125+340}{50}}=\boxed{\frac{\overset{93}{\cancel{465}}}{\underset{10}{\cancel{50}}}}=\boxed{\frac{93}{10}}=\boxed{\mathbf{9\frac{3}{10}}}$$

Example 2.2-34

$$\boxed{\frac{4}{5}+\left[\left(\frac{3}{4}+\frac{1}{5}\right)+\left(5+\frac{2}{3}\right)\right]}=\boxed{\frac{4}{5}+\left[\left(\frac{(3\times 5)+(1\times 4)}{4\times 5}\right)+\left(\frac{5}{1}+\frac{2}{3}\right)\right]}=\boxed{\frac{4}{5}+\left[\left(\frac{15+4}{20}\right)+\left(\frac{(5\times 3)+(2\times 1)}{1\times 3}\right)\right]}$$

$$=\boxed{\frac{4}{5}+\left[\left(\frac{19}{20}\right)+\left(\frac{15+2}{3}\right)\right]}=\boxed{\frac{4}{5}+\left[\frac{19}{20}+\left(\frac{17}{3}\right)\right]}=\boxed{\frac{4}{5}+\left[\frac{19}{20}+\frac{17}{3}\right]}=\boxed{\frac{4}{5}+\left[\frac{(19\times 3)+(17\times 20)}{20\times 3}\right]}$$

$$=\boxed{\frac{4}{5}+\left[\frac{57+340}{60}\right]}=\boxed{\frac{4}{5}+\left[\frac{397}{60}\right]}=\boxed{\frac{4}{5}+\frac{397}{60}}=\boxed{\frac{(4\times 60)+(397\times 5)}{5\times 60}}=\boxed{\frac{240+1985}{300}}=\boxed{\frac{\overset{89}{\cancel{2225}}}{\underset{12}{\cancel{300}}}}=\boxed{\frac{89}{12}}=\boxed{\mathbf{7\frac{5}{12}}}$$

Example 2.2-35

$$\left[\left(\frac{1}{4}+\frac{1}{2}\right)+\left(\frac{3}{4}+\frac{5}{4}\right)\right]+\left(\frac{1}{5}+\frac{3}{5}+\frac{4}{5}\right)=\left[\left(\frac{(1\times 2)+(1\times 4)}{4\times 2}\right)+\left(\frac{3+5}{4}\right)\right]+\left(\frac{1+3+4}{5}\right)$$

$$=\left[\left(\frac{2+4}{8}\right)+\left(\frac{\overset{2}{\cancel{8}}}{\underset{1}{\cancel{4}}}\right)\right]+\left(\frac{8}{5}\right)=\left[\left(\frac{6}{8}\right)+\left(\frac{2}{1}\right)\right]+\left(\frac{8}{5}\right)=\left[\frac{6}{8}+\frac{2}{1}\right]+\frac{8}{5}=\left[\frac{(6\times 1)+(2\times 8)}{8\times 1}\right]+\frac{8}{5}=\left[\frac{6+16}{8}\right]+\frac{8}{5}$$

$$=\frac{22}{8}+\frac{8}{5}=\frac{(22\times 5)+(8\times 8)}{8\times 5}=\frac{110+64}{40}=\frac{\overset{87}{\cancel{174}}}{\underset{20}{\cancel{40}}}=\frac{87}{20}=4\frac{7}{20}$$

Practice Problems - Adding Integer Fractions

Section 2.2 Practice Problems - Add the following integer fractions:

1. $\frac{4}{9}+\frac{2}{9}=$
2. $\frac{3}{8}+\frac{2}{5}=$
3. $\frac{3}{8}+\frac{2}{4}+\frac{5}{6}=$
4. $\frac{4}{5}+\frac{2}{5}+\frac{3}{5}=$
5. $5+\frac{0}{10}+\frac{6}{1}+\frac{4}{8}=$
6. $\left(\frac{3}{16}+\frac{1}{8}\right)+\frac{1}{6}=$
7. $\left(\frac{4}{5}+\frac{2}{8}\right)+\left(\frac{2}{4}+\frac{1}{4}+\frac{3}{4}\right)=$
8. $\frac{2}{5}+\left(\frac{4}{9}+\frac{2}{9}+\frac{1}{9}\right)=$
9. $\frac{2}{5}+\frac{1}{2}+\frac{4}{5}+\frac{2}{3}+12=$
10. $\left[\frac{5}{8}+\left(\frac{3}{5}+\frac{1}{8}\right)\right]+\left(\frac{1}{8}+\frac{3}{8}\right)=$

2.3 Subtracting Integer Fractions

Integer fractions, i.e., fractions where both the numerator and the denominator are integers, are subtracted as in the following cases:

Case I Subtracting Two or More Integer Fractions with Common Denominators

Integer fractions with two or more common denominators are subtracted using the steps given as in each case below:

Case I-A Subtracting Two Integer Fractions with Common Denominators

Subtract two integer fractions with common denominators using the following steps:

Step 1 a. Use the common denominator between the first and second fractions as the new denominator.

b. Subtract the numerators of the first and second fractions to obtain the new numerator.

Step 2 Simplify the fraction to its lowest term (see Section 2.1).

Step 3 Change the improper fraction to a mixed fraction if the fraction obtained from Step 2 is an improper fraction (see Section 2.1 Appendix).

Examples with Steps

The following examples show the steps as to how two integer fractions with common denominators are subtracted:

Example 2.3-1

$$\boxed{\frac{25}{3}-\frac{2}{3}}=$$

Solution:

Step 1 $\boxed{\frac{25}{3}-\frac{2}{3}}=\boxed{\frac{25-2}{3}}=\boxed{\frac{23}{3}}$

Step 2 $\boxed{\textit{Not Applicable}}$

Step 3 $\boxed{\frac{23}{3}}=\boxed{7\frac{2}{3}}$

Example 2.3-2

$$\boxed{\frac{40}{4}-\frac{10}{4}}=$$

Solution:

Step 1 $\boxed{\frac{40}{4}-\frac{10}{4}}=\boxed{\frac{40-10}{4}}=\boxed{\frac{30}{4}}$

Step 2 $\boxed{\frac{30}{4}} = \boxed{\frac{30 \div 2}{4 \div 2}} = \boxed{\frac{15}{2}}$

Step 3 $\boxed{\frac{15}{2}} = \boxed{7\frac{1}{2}}$

Example 2.3-3

$\boxed{\frac{9}{12} - \frac{22}{12}} =$

Solution:

Step 1 $\boxed{\frac{9}{12} - \frac{22}{12}} = \boxed{\frac{9-22}{12}} = \boxed{\frac{-13}{12}}$

Step 2 *Not Applicable*

Step 3 $\boxed{\frac{-13}{12}} = \boxed{-\left(1\frac{1}{12}\right)}$

Example 2.3-4

$\boxed{\frac{5}{10} - \frac{14}{10}} =$

Solution:

Step 1 $\boxed{\frac{5}{10} - \frac{14}{10}} = \boxed{\frac{5-14}{10}} = \boxed{-\frac{9}{10}}$

Step 2 *Not Applicable*

Step 3 *Not Applicable*

Example 2.3-5

$\boxed{\frac{15}{6} - \frac{53}{6}} =$

Solution:

Step 1 $\boxed{\frac{15}{6} - \frac{53}{6}} = \boxed{\frac{15-53}{6}} = \boxed{\frac{-38}{6}}$

Step 2 $\boxed{\frac{-38}{6}} = \boxed{\frac{-38 \div 2}{6 \div 2}} = \boxed{\frac{-19}{3}}$

Step 3 $\boxed{\frac{-19}{3}} = \boxed{-\left(6\frac{1}{3}\right)}$

In general, two integer fractions with a common denominator are subtracted in the following way:

$$\boxed{\frac{a}{d}-\frac{b}{d}}=\boxed{\frac{a-b}{d}}$$

Example 2.3-6

$$\boxed{\frac{6}{8}-\frac{4}{8}}=\boxed{\frac{6-4}{8}}=\boxed{\frac{\overset{1}{\cancel{2}}}{\underset{4}{\cancel{8}}}}=\boxed{\frac{1}{4}}$$

Case I-B Subtracting Three Integer Fractions with Common Denominators

Subtract three integer fractions with common denominators using the following steps:

Step 1 a. Use the common denominator between the first, second, and third fractions as the new denominator.

b. Subtract the numerators of the first, second, and third fractions to obtain the new numerator.

Step 2 Simplify the fraction to its lowest term (see Section 2.1).

Step 3 Change the improper fraction to a mixed fraction if the fraction obtained from Step 2 is an improper fraction (see Section 2.1 Appendix).

Examples with Steps

The following examples show the steps as to how three integer fractions with common denominators are subtracted:

Example 2.3-7

$$\boxed{\frac{7}{4}-\frac{3}{4}-\frac{1}{4}}=$$

Solution:

Step 1 $\boxed{\frac{7}{4}-\frac{3}{4}-\frac{1}{4}}=\boxed{\frac{7-3-1}{4}}=\boxed{\frac{7-4}{4}}=\boxed{\frac{3}{4}}$

Step 2 *Not Applicable*

Step 3 *Not Applicable*

Example 2.3-8

$$\boxed{\frac{25}{8}-\frac{3}{8}-\frac{4}{8}}=$$

Solution:

Step 1 $\boxed{\frac{25}{8}-\frac{3}{8}-\frac{4}{8}}=\boxed{\frac{25-3-4}{8}}=\boxed{\frac{25-7}{8}}=\boxed{\frac{18}{8}}$

Step 2 $\boxed{\dfrac{18}{8}} = \boxed{\dfrac{18 \div 2}{8 \div 2}} = \boxed{\dfrac{9}{4}}$

Step 3 $\boxed{\dfrac{9}{4}} = \boxed{\mathbf{2\dfrac{1}{4}}}$

Example 2.3-9

$\boxed{\dfrac{25}{6} - \dfrac{4}{6} - \dfrac{1}{6}} =$

Solution:

Step 1 $\boxed{\dfrac{25}{6} - \dfrac{4}{6} - \dfrac{1}{6}} = \boxed{\dfrac{25 - 4 - 1}{6}} = \boxed{\dfrac{25 - 5}{6}} = \boxed{\dfrac{20}{6}}$

Step 2 $\boxed{\dfrac{20}{6}} = \boxed{\dfrac{20 \div 2}{6 \div 2}} = \boxed{\dfrac{10}{3}}$

Step 3 $\boxed{\dfrac{10}{3}} = \boxed{\mathbf{3\dfrac{1}{3}}}$

Example 2.3-10

$\boxed{\dfrac{12}{7} - \dfrac{28}{7} - \dfrac{13}{7}} =$

Solution:

Step 1 $\boxed{\dfrac{12}{7} - \dfrac{28}{7} - \dfrac{13}{7}} = \boxed{\dfrac{12 - 28 - 13}{7}} = \boxed{\dfrac{12 - 41}{7}} = \boxed{\dfrac{-29}{7}}$

Step 2 *Not Applicable*

Step 3 $\boxed{\dfrac{-29}{7}} = \boxed{\mathbf{-\left(4\dfrac{1}{7}\right)}}$

Example 2.3-11

$\boxed{\dfrac{125}{12} - \dfrac{25}{12} - \dfrac{360}{12}} =$

Solution:

Step 1 $\boxed{\dfrac{125}{12} - \dfrac{25}{12} - \dfrac{360}{12}} = \boxed{\dfrac{125 - 25 - 360}{12}} = \boxed{\dfrac{125 - 385}{12}} = \boxed{\dfrac{-260}{12}}$

Step 2 $\boxed{\dfrac{-260}{12}} = \boxed{\dfrac{-260 \div 4}{12 \div 4}} = \boxed{\dfrac{-65}{3}}$

Step 3 $\boxed{\dfrac{-65}{3}} = \boxed{\mathbf{-\left(21\dfrac{2}{3}\right)}}$

In general, three integer fractions with a common denominator are subtracted in the following way:

$$\boxed{\frac{a}{d}-\frac{b}{d}-\frac{c}{d}} = \boxed{\frac{a-b-c}{d}}$$

Example 2.3-12

$$\boxed{\frac{5}{6}-\frac{2}{6}-\frac{1}{6}} = \boxed{\frac{5-2-1}{6}} = \boxed{\frac{5-3}{6}} = \boxed{\frac{\overset{1}{\cancel{2}}}{\underset{3}{\cancel{6}}}} = \boxed{\mathbf{\frac{1}{3}}}$$

Case II Subtracting Two or More Integer Fractions Without a Common Denominator

Two or more integer fractions without a common denominator are subtracted using the steps given as in each case below:

Case II-A Subtracting Two Integer Fractions Without a Common Denominator

Subtract two integer fractions without a common denominator using the following steps:

Step 1 Change the integer number a to an integer fraction of the form $\frac{a}{1}$, e.g., change 358 to $\frac{358}{1}$.

Step 2 a. Multiply the denominators of the first and second fractions to obtain the new denominator.

b. Cross multiply the numerator of the first fraction with the denominator of the second fraction.

c. Cross multiply the numerator of the second fraction with the denominator of the first fraction.

d. Subtract the results from steps 2b and 2c above to obtain the new numerator.

Step 3 Simplify the fraction to its lowest term (see Section 2.1).

Step 4 Change the improper fraction to a mixed fraction if the fraction obtained from Step 3 is an improper fraction (see Section 2.1 Appendix).

Examples with Steps

The following examples show the steps as to how two integer fractions without a common denominator are subtracted:

Example 2.3-13

$$\boxed{5-\frac{12}{8}} =$$

Solution:

Step 1 $\boxed{5-\frac{12}{8}} = \boxed{\frac{5}{1}-\frac{12}{8}}$

Step 2 $\boxed{\frac{5}{1}-\frac{12}{8}} = \boxed{\frac{(5\times 8)-(12\times 1)}{1\times 8}} = \boxed{\frac{40-12}{8}} = \boxed{\frac{28}{8}}$

Step 3 $\boxed{\frac{28}{8}} = \boxed{\frac{28\div 4}{8\div 4}} = \boxed{\frac{7}{2}}$

Step 4 $\boxed{\frac{7}{2}} = \boxed{\mathbf{3\frac{1}{2}}}$

Example 2.3-14

$\boxed{\frac{4}{5}-\frac{3}{8}} =$

Solution:

Step 1 *Not Applicable*

Step 2 $\boxed{\frac{4}{5}-\frac{3}{8}} = \boxed{\frac{(4\times 8)-(3\times 5)}{5\times 8}} = \boxed{\frac{32-15}{40}} = \boxed{\mathbf{\frac{17}{40}}}$

Step 3 *Not Applicable*

Step 4 *Not Applicable*

Example 2.3-15

$\boxed{\frac{9}{8}-\frac{3}{4}} =$

Solution:

Step 1 *Not Applicable*

Step 2 $\boxed{\frac{9}{8}-\frac{3}{4}} = \boxed{\frac{(9\times 4)-(3\times 8)}{8\times 4}} = \boxed{\frac{36-24}{32}} = \boxed{\frac{12}{32}}$

Step 3 $\boxed{\frac{12}{32}} = \boxed{\frac{12\div 4}{32\div 4}} = \boxed{\mathbf{\frac{3}{8}}}$

Step 4 *Not Applicable*

Example 2.3-16

$\boxed{\frac{10}{6}-35} =$

Solution:

Step 1 $\boxed{\frac{10}{6}-35} = \boxed{\frac{10}{6}-\frac{35}{1}}$

Step 2 $\frac{10}{6}-\frac{35}{1} = \frac{(10\times 1)-(35\times 6)}{6\times 1} = \frac{10-210}{6} = \frac{-200}{6}$

Step 3 $\frac{-200}{6} = \frac{-200\div 2}{6\div 2} = \frac{-100}{3}$

Step 4 $\frac{-100}{3} = -\left(\mathbf{33\frac{1}{3}}\right)$

Example 2.3-17

$\frac{3}{9}-\frac{4}{15} =$

Solution:

Step 1 *Not Applicable*

Step 2 $\frac{3}{9}-\frac{4}{15} = \frac{(3\times 15)-(4\times 9)}{9\times 15} = \frac{45-36}{135} = \frac{9}{135}$

Step 3 $\frac{9}{135} = \frac{9\div 9}{135\div 9} = \mathbf{\frac{1}{15}}$

Step 4 *Not Applicable*

In general, two integer fractions without a common denominator are subtracted in the following way:

$$\frac{a}{b}-\frac{c}{d} = \frac{(a\times d)-(c\times b)}{b\times d} = \frac{ad-cb}{bd}$$

Example 2.3-18

$$\frac{3}{4}-\frac{1}{8} = \frac{(3\times 8)-(1\times 4)}{4\times 8} = \frac{24-4}{32} = \frac{\overset{5}{\cancel{20}}}{\underset{8}{\cancel{32}}} = \mathbf{\frac{5}{8}}$$

Case II-B Subtracting Three Integer Fractions Without a Common Denominator

Subtract three integer fractions without a common denominator using the following steps:

Step 1 Use parentheses to group the first and second fractions.

Step 2 Change the integer number a to an integer fraction of the form $\frac{a}{1}$, e.g., change 12 to $\frac{12}{1}$

Step 3 a. Subtract the grouped fraction following Steps 2a through 2d, outlined in Section 2.3, Case II-A above, to obtain a new integer fraction.

b. Subtract the new integer fraction from the third fraction by repeating Steps 2a through 2d, outlined in Section 2.3, Case II-A above.

Step 4 Simplify the fraction to its lowest term (see Section 2.1).

Step 5 Change the improper fraction to a mixed fraction if the fraction obtained from Step 4 is an improper fraction (see Section 2.1 Appendix).

Examples with Steps

The following examples show the steps as to how three integer fractions without a common denominator are subtracted:

Example 2.3-19

$$\boxed{\frac{4}{5}-\frac{1}{3}-\frac{2}{6}} =$$

Solution:

Step 1 $\boxed{\frac{4}{5}-\frac{1}{3}-\frac{2}{6}} = \boxed{\left(\frac{4}{5}-\frac{1}{3}\right)-\frac{2}{6}}$

Step 2 *Not Applicable*

Step 3 $\boxed{\left(\frac{4}{5}-\frac{1}{3}\right)-\frac{2}{6}} = \boxed{\left(\frac{(4\times 3)-(1\times 5)}{5\times 3}\right)-\frac{2}{6}} = \boxed{\left(\frac{12-5}{15}\right)-\frac{2}{6}} = \boxed{\left(\frac{7}{15}\right)-\frac{2}{6}} = \boxed{\frac{7}{15}-\frac{2}{6}}$

$= \boxed{\frac{(7\times 6)-(2\times 15)}{15\times 6}} = \boxed{\frac{42-30}{90}} = \boxed{\frac{12}{90}}$

Step 4 $\boxed{\frac{12}{90}} = \boxed{\frac{12\div 6}{90\div 6}} = \boxed{\mathbf{\frac{2}{15}}}$

Step 5 *Not Applicable*

Example 2.3-20

$$\boxed{\frac{4}{7}-\frac{2}{5}-\frac{3}{4}} =$$

Solution:

Step 1 $\boxed{\frac{4}{7}-\frac{2}{5}-\frac{3}{4}} = \boxed{\left(\frac{4}{7}-\frac{2}{5}\right)-\frac{3}{4}}$

Step 2 *Not Applicable*

Step 3 $\boxed{\left(\frac{4}{7}-\frac{2}{5}\right)-\frac{3}{4}} = \boxed{\left(\frac{(4\times 5)-(2\times 7)}{7\times 5}\right)-\frac{3}{4}} = \boxed{\left(\frac{20-14}{35}\right)-\frac{3}{4}} = \boxed{\left(\frac{6}{35}\right)-\frac{3}{4}} = \boxed{\frac{6}{35}-\frac{3}{4}}$

$= \boxed{\dfrac{(6\times 4)-(3\times 35)}{35\times 4}} = \boxed{\dfrac{24-105}{140}} = \boxed{\mathbf{-\dfrac{81}{140}}}$

Step 4 $\boxed{\textit{Not Applicable}}$

Step 5 $\boxed{\textit{Not Applicable}}$

Example 2.3-21

$\boxed{15-\dfrac{5}{8}-\dfrac{2}{3}} =$

Solution:

Step 1 $\boxed{15-\dfrac{5}{8}-\dfrac{2}{3}} = \boxed{\left(15-\dfrac{5}{8}\right)-\dfrac{2}{3}}$

Step 2 $\boxed{\left(15-\dfrac{5}{8}\right)-\dfrac{2}{3}} = \boxed{\left(\dfrac{15}{1}-\dfrac{5}{8}\right)-\dfrac{2}{3}}$

Step 3 $\boxed{\left(\dfrac{15}{1}-\dfrac{5}{8}\right)-\dfrac{2}{3}} = \boxed{\left(\dfrac{(15\times 8)-(5\times 1)}{1\times 8}\right)-\dfrac{2}{3}} = \boxed{\left(\dfrac{120-5}{8}\right)-\dfrac{2}{3}} = \boxed{\left(\dfrac{115}{8}\right)-\dfrac{2}{3}} = \boxed{\dfrac{115}{8}-\dfrac{2}{3}}$

$= \boxed{\dfrac{(115\times 3)-(2\times 8)}{8\times 3}} = \boxed{\dfrac{345-16}{24}} = \boxed{\dfrac{329}{24}}$

Step 4 $\boxed{\textit{Not Applicable}}$

Step 5 $\boxed{\dfrac{329}{24}} = \boxed{\mathbf{13\dfrac{17}{24}}}$

Example 2.3-22

$\boxed{25-\dfrac{3}{4}-\dfrac{32}{5}} =$

Solution:

Step 1 $\boxed{25-\dfrac{3}{4}-\dfrac{32}{5}} = \boxed{\left(25-\dfrac{3}{4}\right)-\dfrac{32}{5}}$

Step 2 $\boxed{\left(25-\dfrac{3}{4}\right)-\dfrac{32}{5}} = \boxed{\left(\dfrac{25}{1}-\dfrac{3}{4}\right)-\dfrac{32}{5}}$

Step 3 $\boxed{\left(\dfrac{25}{1}-\dfrac{3}{4}\right)-\dfrac{32}{5}} = \boxed{\left(\dfrac{(25\times 4)-(3\times 1)}{1\times 4}\right)-\dfrac{32}{5}} = \boxed{\left(\dfrac{100-3}{4}\right)-\dfrac{32}{5}} = \boxed{\left(\dfrac{97}{4}\right)-\dfrac{32}{5}}$

$= \boxed{\dfrac{97}{4}-\dfrac{32}{5}} = \boxed{\dfrac{(97\times 5)-(32\times 4)}{4\times 5}} = \boxed{\dfrac{485-128}{20}} = \boxed{\dfrac{357}{20}}$

Step 4 *Not Applicable*

Step 5 $\dfrac{357}{20} = \mathbf{17\dfrac{17}{20}}$

Example 2.3-23

$$\frac{4}{5}-\frac{1}{4}-7=$$

Solution:

Step 1 $\dfrac{4}{5}-\dfrac{1}{4}-7 = \left(\dfrac{4}{5}-\dfrac{1}{4}\right)-7$

Step 2 $\left(\dfrac{4}{5}-\dfrac{1}{4}\right)-7 = \left(\dfrac{4}{5}-\dfrac{1}{4}\right)-\dfrac{7}{1}$

Step 3 $\left(\dfrac{4}{5}-\dfrac{1}{4}\right)-\dfrac{7}{1} = \left(\dfrac{(4\times4)-(1\times5)}{5\times4}\right)-\dfrac{7}{1} = \left(\dfrac{16-5}{20}\right)-\dfrac{7}{1} = \left(\dfrac{11}{20}\right)-\dfrac{7}{1} = \dfrac{11}{20}-\dfrac{7}{1}$

$$= \frac{(11\times1)-(7\times20)}{20\times1} = \frac{11-140}{20} = \frac{-129}{20}$$

Step 4 *Not Applicable*

Step 5 $\dfrac{-129}{20} = \mathbf{-\left(6\dfrac{9}{20}\right)}$

In general, three integer fractions without a common denominator are subtracted as in the following cases:

Case I.

$$\frac{a}{b}-\frac{c}{d}-\frac{e}{f} = \left(\frac{a}{b}-\frac{c}{d}\right)-\frac{e}{f} = \left(\frac{(a\times d)-(c\times b)}{b\times d}\right)-\frac{e}{f} = \left(\frac{ad-cb}{bd}\right)-\frac{e}{f} = \frac{[(ad-cb)\times f]-(e\times bd)}{bd\times f}$$

$$= \frac{[adf-cbf]-ebd}{bdf} = \frac{adf-cbf-ebd}{bdf}$$

Example 2.3-24

$$\frac{7}{4}-\frac{1}{2}-\frac{2}{3} = \left(\frac{7}{4}-\frac{1}{2}\right)-\frac{2}{3} = \left(\frac{(7\times2)-(1\times4)}{4\times2}\right)-\frac{2}{3} = \left(\frac{14-4}{8}\right)-\frac{2}{3} = \left(\frac{10}{8}\right)-\frac{2}{3} = \frac{10}{8}-\frac{2}{3}$$

$$= \frac{(10\times3)-(2\times8)}{8\times3} = \frac{30-16}{24} = \frac{\overset{7}{\cancel{14}}}{\underset{12}{\cancel{24}}} = \mathbf{\frac{7}{12}}$$

Case II.

$$\boxed{\frac{a}{b}-\frac{c}{d}-\frac{e}{f}} = \boxed{\frac{a}{b}+\left(-\frac{c}{d}-\frac{e}{f}\right)} = \boxed{\frac{a}{b}+\left(\frac{-(c\times f)-(e\times d)}{d\times f}\right)} = \boxed{\frac{a}{b}+\left(\frac{-cf-ed}{df}\right)}$$

$$= \boxed{\frac{(a\times df)+[b\times(-cf-ed)]}{b\times df}} = \boxed{\frac{adf+[-bcf-bed]}{bdf}} = \boxed{\frac{adf-bcf-bed}{bdf}}$$

Example 2.3-25

$$\boxed{\frac{7}{4}-\frac{1}{2}-\frac{2}{3}} = \boxed{\frac{7}{4}+\left(-\frac{1}{2}-\frac{2}{3}\right)} = \boxed{\frac{7}{4}+\left(\frac{-(1\times 3)-(2\times 2)}{2\times 3}\right)} = \boxed{\frac{7}{4}+\left(\frac{-3-4}{6}\right)} = \boxed{\frac{7}{4}+\left(\frac{-7}{6}\right)} = \boxed{\frac{7}{4}-\frac{7}{6}}$$

$$= \boxed{\frac{(7\times 6)-(7\times 4)}{4\times 6}} = \boxed{\frac{42-28}{24}} = \boxed{\frac{\overset{7}{\cancel{14}}}{\underset{12}{\cancel{24}}}} = \boxed{\frac{7}{12}}$$

Additional Examples - Subtracting Integer Fractions

The following examples further illustrate how to subtract integer fractions:

Example 2.3-26

$$\boxed{\frac{45}{8}-\frac{5}{8}} = \boxed{\frac{45-5}{8}} = \boxed{\frac{\overset{5}{\cancel{40}}}{\underset{1}{\cancel{8}}}} = \boxed{\frac{5}{1}} = \boxed{5}$$

Example 2.3-27

$$\boxed{\frac{3}{6}-\frac{5}{8}} = \boxed{\frac{\overset{1}{\cancel{3}}}{\underset{2}{\cancel{6}}}-\frac{5}{8}} = \boxed{\frac{1}{2}-\frac{5}{8}} = \boxed{\frac{(1\times 8)-(5\times 2)}{2\times 8}} = \boxed{\frac{8-10}{16}} = \boxed{-\frac{\overset{1}{\cancel{2}}}{\underset{8}{\cancel{16}}}} = \boxed{-\frac{1}{8}}$$

Example 2.3-28

$$\boxed{\frac{8}{3}-\frac{1}{6}-\frac{2}{5}} = \boxed{\left(\frac{8}{3}-\frac{1}{6}\right)-\frac{2}{5}} = \boxed{\left(\frac{(8\times 6)-(1\times 3)}{3\times 6}\right)-\frac{2}{5}} = \boxed{\left(\frac{48-3}{18}\right)-\frac{2}{5}} = \boxed{\left(\frac{45}{18}\right)-\frac{2}{5}} = \boxed{\frac{\overset{15}{\cancel{45}}}{\underset{6}{\cancel{18}}}-\frac{2}{5}} = \boxed{\frac{15}{6}-\frac{2}{5}}$$

$$= \boxed{\frac{(15\times 5)-(2\times 6)}{6\times 5}} = \boxed{\frac{75-12}{30}} = \boxed{\frac{\overset{21}{\cancel{63}}}{\underset{10}{\cancel{30}}}} = \boxed{\frac{21}{10}} = \boxed{2\frac{1}{10}}$$

Example 2.3-29

$$\boxed{\frac{16}{4}-\frac{2}{4}-\frac{4}{4}} = \boxed{\frac{16-2-4}{4}} = \boxed{\frac{16-6}{4}} = \boxed{\frac{\overset{5}{\cancel{10}}}{\underset{2}{\cancel{4}}}} = \boxed{\frac{5}{2}} = \boxed{2\frac{1}{2}}$$

Example 2.3-30

$$\boxed{\frac{3}{5}-\frac{2}{3}-9} = \boxed{\left(\frac{3}{5}-\frac{2}{3}\right)-\frac{9}{1}} = \boxed{\left(\frac{(3\times 3)-(2\times 5)}{5\times 3}\right)-\frac{9}{1}} = \boxed{\left(\frac{9-10}{15}\right)-\frac{9}{1}} = \boxed{\left(\frac{-1}{15}\right)-\frac{9}{1}} = \boxed{\frac{-1}{15}-\frac{9}{1}}$$

$$= \boxed{\frac{(-1\times 1)-(9\times 15)}{15\times 1}} = \boxed{\frac{-1-135}{15}} = \boxed{\frac{-136}{15}} = \boxed{-\left(9\frac{1}{15}\right)}$$

Example 2.3-31

$$\boxed{\left(\frac{13}{8}-\frac{4}{3}\right)-\frac{1}{5}} = \boxed{\left(\frac{(13\times 3)-(4\times 8)}{8\times 3}\right)-\frac{1}{5}} = \boxed{\left(\frac{39-32}{24}\right)-\frac{1}{5}} = \boxed{\left(\frac{7}{24}\right)-\frac{1}{5}} = \boxed{\frac{7}{24}-\frac{1}{5}} = \boxed{\frac{(7\times 5)-(1\times 24)}{24\times 5}}$$

$$= \boxed{\frac{35-24}{120}} = \boxed{\frac{11}{120}}$$

Example 2.3-32

$$\boxed{\frac{2}{4}-\left(\frac{1}{3}-\frac{1}{5}\right)} = \boxed{\frac{\overset{1}{\cancel{2}}}{\underset{2}{\cancel{4}}}-\left(\frac{(5\times 1)-(1\times 3)}{3\times 5}\right)} = \boxed{\frac{1}{2}-\left(\frac{5-3}{15}\right)} = \boxed{\frac{1}{2}-\left(\frac{2}{15}\right)} = \boxed{\frac{1}{2}-\frac{2}{15}} = \boxed{\frac{(1\times 15)-(2\times 2)}{2\times 15}} = \boxed{\frac{15-4}{30}}$$

$$= \boxed{\frac{11}{30}}$$

Example 2.3-33

$$\boxed{\left(\frac{20}{3}-\frac{1}{5}\right)-\left(\frac{4}{7}-\frac{6}{7}\right)} = \boxed{\left(\frac{(20\times 5)-(1\times 3)}{3\times 5}\right)-\left(\frac{4-6}{7}\right)} = \boxed{\left(\frac{100-3}{15}\right)-\left(\frac{-2}{7}\right)} = \boxed{\left(\frac{97}{15}\right)+\left(\frac{2}{7}\right)} = \boxed{\frac{97}{15}+\frac{2}{7}}$$

$$= \boxed{\frac{(97\times 7)+(2\times 15)}{15\times 7}} = \boxed{\frac{679+30}{105}} = \boxed{\frac{709}{105}} = \boxed{6\frac{79}{105}}$$

Example 2.3-34

$$\boxed{\frac{4}{5}-\left[\left(\frac{3}{2}-\frac{1}{4}\right)-\frac{1}{5}\right]} = \boxed{\frac{4}{5}-\left[\left(\frac{(3\times 4)-(1\times 2)}{2\times 4}\right)-\frac{1}{5}\right]} = \boxed{\frac{4}{5}-\left[\left(\frac{12-2}{8}\right)-\frac{1}{5}\right]} = \boxed{\frac{4}{5}-\left[\left(\frac{10}{8}\right)-\frac{1}{5}\right]} = \boxed{\frac{4}{5}-\left[\frac{\overset{5}{\cancel{10}}}{\underset{4}{\cancel{8}}}-\frac{1}{5}\right]}$$

$$= \boxed{\frac{4}{5}-\left[\frac{5}{4}-\frac{1}{5}\right]} = \boxed{\frac{4}{5}-\left[\frac{(5\times 5)-(1\times 4)}{4\times 5}\right]} = \boxed{\frac{4}{5}-\left[\frac{25-4}{20}\right]} = \boxed{\frac{4}{5}-\left[\frac{21}{20}\right]} = \boxed{\frac{4}{5}-\frac{21}{20}} = \boxed{\frac{(4\times 20)-(21\times 5)}{5\times 20}}$$

$$= \boxed{\frac{80-105}{100}} = \boxed{-\frac{\overset{1}{\cancel{25}}}{\underset{4}{\cancel{100}}}} = \boxed{-\frac{1}{4}}$$

Example 2.3-35

$$\boxed{\left[\left(20-\frac{1}{5}\right)-\left(\frac{4}{3}-\frac{3}{3}\right)\right]-4} = \boxed{\left[\left(\frac{20}{1}-\frac{1}{5}\right)-\left(\frac{4-3}{3}\right)\right]-\frac{4}{1}} = \boxed{\left[\left(\frac{(20\times 5)-(1\times 1)}{1\times 5}\right)-\left(\frac{1}{3}\right)\right]-\frac{4}{1}}$$

$$= \boxed{\left[\left(\frac{100-1}{5}\right)-\frac{1}{3}\right]-\frac{4}{1}} = \boxed{\left[\left(\frac{99}{5}\right)-\frac{1}{3}\right]-\frac{4}{1}} = \boxed{\left[\frac{99}{5}-\frac{1}{3}\right]-\frac{4}{1}} = \boxed{\left[\frac{(99\times 3)-(1\times 5)}{5\times 3}\right]-\frac{4}{1}} = \boxed{\left[\frac{297-5}{15}\right]-\frac{4}{1}}$$

$$= \boxed{\left[\frac{292}{15}\right] - \frac{4}{1}} = \boxed{\frac{292}{15} - \frac{4}{1}} = \boxed{\frac{(292 \times 1) - (4 \times 15)}{15 \times 1}} = \boxed{\frac{292 - 60}{15}} = \boxed{\frac{232}{15}} = \boxed{\mathbf{15\frac{7}{15}}}$$

Practice Problems - Subtracting Integer Fractions

Section 2.3 Practice Problems - Subtract the following integer fractions:

1. $\frac{3}{5} - \frac{2}{5} =$
2. $\frac{2}{5} - \frac{3}{4} =$
3. $\frac{12}{15} - \frac{3}{15} - \frac{6}{15} =$
4. $\frac{5}{8} - \frac{3}{4} - \frac{1}{3} =$
5. $\left(\frac{2}{8} - \frac{1}{6}\right) - \frac{2}{5} =$
6. $28 - \left(\frac{1}{8} - \frac{2}{3}\right) =$
7. $\left(\frac{4}{6} - \frac{1}{8}\right) - \left(\frac{4}{5} - \frac{1}{2}\right) =$
8. $\left(20 - \frac{1}{6}\right) - \left(\frac{3}{4} - \frac{1}{2}\right) =$
9. $\left[\frac{18}{5} - \left(\frac{4}{3} - \frac{2}{3}\right)\right] - 2 =$
10. $\left[\left(18 - \frac{1}{2}\right) - \left(\frac{16}{2} - 2\right)\right] - \frac{1}{5} =$

2.4 Multiplying Integer Fractions

Two or more integer fractions with or without a common denominator are multiplied using the steps given in each case below:

Case I - Multiplying Two Integer Fractions with or Without a Common Denominator

Multiply two integer fractions using the following steps:

Step 1 Change the integer number a to an integer fraction of the form $\frac{a}{1}$, e.g., change 300 to $\frac{300}{1}$.

Step 2 a. Multiply the numerator of the first fraction with the numerator of the second fraction to obtain the new numerator.

b. Multiply the denominator of the first fraction with the denominator of the second fraction to obtain the new denominator.

Step 3 Simplify the fraction to its lowest term (see Section 2.1).

Step 4 Change the improper fraction to a mixed fraction if the fraction obtained from Step 3 is an improper fraction (see Section 2.1 Appendix).

Examples with Steps

The following examples show the steps as to how two integer fractions with or without a common denominator are multiplied:

Example 2.4-1

$$\frac{4}{5} \times \frac{3}{8} =$$

Solution:

Step 1 *Not Applicable*

Step 2 $\frac{4}{5} \times \frac{3}{8} = \frac{4 \times 3}{5 \times 8} = \frac{12}{40}$

Step 3 $\frac{12}{40} = \frac{12 \div 4}{40 \div 4} = \mathbf{\frac{3}{10}}$

Step 4 *Not Applicable*

Example 2.4-2

$$25 \times \frac{5}{8} =$$

Solution:

Step 1 $25 \times \frac{5}{8} = \frac{25}{1} \times \frac{5}{8}$

Step 2 $\boxed{\frac{25}{1}\times\frac{5}{8}}=\boxed{\frac{25\times 5}{1\times 8}}=\boxed{\frac{125}{8}}$

Step 3 $\boxed{\textit{Not Applicable}}$

Step 4 $\boxed{\frac{125}{8}}=\boxed{\mathbf{15\frac{5}{8}}}$

Example 2.4-3

$\boxed{\frac{140}{3}\times\frac{1}{5}}=$

Solution:

Step 1 $\boxed{\textit{Not Applicable}}$

Step 2 $\boxed{\frac{140}{3}\times\frac{1}{5}}=\boxed{\frac{140\times 1}{3\times 5}}=\boxed{\frac{140}{15}}$

Step 3 $\boxed{\frac{140}{15}}=\boxed{\frac{140\div 5}{15\div 5}}=\boxed{\frac{28}{3}}$

Step 4 $\boxed{\frac{28}{3}}=\boxed{\mathbf{9\frac{1}{3}}}$

Example 2.4-4

$\boxed{36\times\frac{4}{28}}=$

Solution:

Step 1 $\boxed{36\times\frac{4}{28}}=\boxed{\frac{36}{1}\times\frac{4}{28}}$

Step 2 $\boxed{\frac{36}{1}\times\frac{4}{28}}=\boxed{\frac{36\times 4}{1\times 28}}=\boxed{\frac{144}{28}}$

Step 3 $\boxed{\frac{144}{28}}=\boxed{\frac{144\div 4}{28\div 4}}=\boxed{\frac{36}{7}}$

Step 4 $\boxed{\frac{36}{7}}=\boxed{\mathbf{5\frac{1}{7}}}$

Example 2.4-5

$\boxed{\frac{9}{38}\times 12}=$

Solution:

Step 1 $\boxed{\frac{9}{38}\times 12}=\boxed{\frac{9}{38}\times\frac{12}{1}}$

Step 2 $\boxed{\frac{9}{38} \times \frac{12}{1}} = \boxed{\frac{9 \times 12}{38 \times 1}} = \boxed{\frac{108}{38}}$

Step 3 $\boxed{\frac{108}{38}} = \boxed{\frac{108 \div 2}{38 \div 2}} = \boxed{\frac{54}{19}}$

Step 4 $\boxed{\frac{54}{19}} = \boxed{\mathbf{2\frac{16}{19}}}$

In general, two integer fractions with or without a common denominator are multiplied in the following way:

$\boxed{\frac{a}{b} \times \frac{c}{d}} = \boxed{\frac{a \times c}{b \times d}} = \boxed{\frac{ac}{bd}}$

Example 2.4-6

$\boxed{\frac{2}{5} \times \frac{3}{4}} = \boxed{\frac{2 \times 3}{5 \times 4}} = \boxed{\frac{\overset{3}{\cancel{6}}}{\underset{10}{\cancel{20}}}} = \boxed{\mathbf{\frac{3}{10}}}$

Case II - Multiplying Three Integer Fractions with or Without a Common Denominator

Multiply three integer fractions using the following steps:

Step 1 Change the integer number a to an integer fraction of the form $\frac{a}{1}$, e.g., change 25 to $\frac{25}{1}$.

Step 2 a. Multiply the numerators of the first, second, and third fractions to obtain the new numerator (see Section 1.4).

b. Multiply the denominator of the first, second, and third fractions to obtain the new denominator (see Section 1.4).

Step 3 Simplify the fraction to its lowest term (see Section 2.1).

Step 4 Change the improper fraction to a mixed fraction if the fraction obtained from Step 3 is an improper fraction(see Section 2.1 Appendix).

Examples with Steps

The following examples show the steps as to how three integer fractions with or without a common denominator are multiplied:

Example 2.4-7

$\boxed{12 \times \frac{3}{5} \times \frac{1}{8}} =$

Solution:

Step 1 $\boxed{12 \times \frac{3}{5} \times \frac{1}{8}} = \boxed{\frac{12}{1} \times \frac{3}{5} \times \frac{1}{8}}$

Step 2 $\boxed{\frac{12}{1}\times\frac{3}{5}\times\frac{1}{8}} = \boxed{\frac{12\times3\times1}{1\times5\times8}} = \boxed{\frac{36}{40}}$

Step 3 $\boxed{\frac{36}{40}} = \boxed{\frac{36\div4}{40\div4}} = \boxed{\frac{9}{10}}$

Step 4 *Not Applicable*

Example 2.4-8

$\boxed{\frac{25}{3}\times\frac{4}{7}\times\frac{6}{5}} =$

Solution:

Step 1 *Not Applicable*

Step 2 $\boxed{\frac{25}{3}\times\frac{4}{7}\times\frac{6}{5}} = \boxed{\frac{25\times4\times6}{3\times7\times5}} = \boxed{\frac{600}{105}}$

Step 3 $\boxed{\frac{600}{105}} = \boxed{\frac{600\div15}{105\div15}} = \boxed{\frac{40}{7}}$

Step 4 $\boxed{\frac{40}{7}} = \boxed{5\frac{5}{7}}$

Example 2.4-9

$\boxed{\frac{25}{3}\times14\times\frac{9}{50}} =$

Solution:

Step 1 $\boxed{\frac{25}{3}\times14\times\frac{9}{50}} = \boxed{\frac{25}{3}\times\frac{14}{1}\times\frac{9}{50}}$

Step 2 $\boxed{\frac{25}{3}\times\frac{14}{1}\times\frac{9}{50}} = \boxed{\frac{25\times14\times9}{3\times1\times50}} = \boxed{\frac{3150}{150}}$

Step 3 $\boxed{\frac{3150}{150}} = \boxed{\frac{3150\div150}{150\div150}} = \boxed{\frac{21}{1}} = \boxed{21}$

Step 4 *Not Applicable*

Example 2.4-10

$\boxed{\frac{9}{8}\times\frac{33}{5}\times\frac{5}{48}} =$

Solution:

Step 1 *Not Applicable*

Step 2 $\boxed{\frac{9}{8}\times\frac{33}{5}\times\frac{5}{48}} = \boxed{\frac{9\times 33\times 5}{8\times 5\times 48}} = \boxed{\frac{1485}{1920}}$

Step 3 $\boxed{\frac{1485}{1920}} = \boxed{\frac{1485\div 15}{1920\div 15}} = \boxed{\frac{\mathbf{99}}{\mathbf{128}}}$

Step 4 *Not Applicable*

Example 2.4-11

$\boxed{\frac{125}{4}\times\frac{28}{13}\times 39} =$

Solution:

Step 1 $\boxed{\frac{125}{4}\times\frac{28}{13}\times 39} = \boxed{\frac{125}{4}\times\frac{28}{13}\times\frac{39}{1}}$

Step 2 $\boxed{\frac{125}{4}\times\frac{28}{13}\times\frac{39}{1}} = \boxed{\frac{125\times 28\times 39}{4\times 13\times 1}} = \boxed{\frac{136500}{52}}$

Step 3 $\boxed{\frac{136500}{52}} = \boxed{\frac{136500\div 52}{52\div 52}} = \boxed{\frac{2625}{1}} = \boxed{\mathbf{2625}}$

Step 4 *Not Applicable*

In general, three integer fractions with or without a common denominator are multiplied as in the following cases:

Case I.

$\boxed{\frac{a}{b}\times\frac{c}{d}\times\frac{e}{f}} = \boxed{\frac{a\times c\times e}{b\times d\times f}} = \boxed{\frac{ace}{bdf}}$

Example 2.4-12

$\boxed{\frac{2}{3}\times\frac{3}{15}\times\frac{5}{2}} = \boxed{\frac{\overset{1}{\cancel{2}}\times\overset{1}{3}\times\overset{1}{\cancel{5}}}{\underset{1}{\cancel{3}}\times\underset{3}{\cancel{15}}\times\underset{1}{\cancel{2}}}} = \boxed{\frac{1\times 1\times 1}{1\times 3\times 1}} = \boxed{\frac{\mathbf{1}}{\mathbf{3}}}$

Case II.

$\boxed{\frac{a}{b}\times\frac{c}{d}\times\frac{e}{f}} = \boxed{\left(\frac{a}{b}\times\frac{c}{d}\right)\times\frac{e}{f}} = \boxed{\left(\frac{a\times c}{b\times d}\right)\times\frac{e}{f}} = \boxed{\left(\frac{ac}{bd}\right)\times\frac{e}{f}} = \boxed{\frac{ac}{bd}\times\frac{e}{f}} = \boxed{\frac{ac\times e}{bd\times f}} = \boxed{\frac{ace}{bdf}}$

Example 2.4-13

$\boxed{\frac{2}{3}\times\frac{3}{15}\times\frac{5}{2}} = \boxed{\left(\frac{2}{3}\times\frac{3}{15}\right)\times\frac{5}{2}} = \boxed{\left(\frac{2\times\overset{1}{\cancel{3}}}{\underset{1}{\cancel{3}}\times 15}\right)\times\frac{5}{2}} = \boxed{\left(\frac{2\times 1}{1\times 15}\right)\times\frac{5}{2}} = \boxed{\left(\frac{2}{15}\right)\times\frac{5}{2}} = \boxed{\frac{2}{15}\times\frac{5}{2}} = \boxed{\frac{2\times\overset{1}{\cancel{5}}}{\underset{3}{\cancel{15}}\times 2}} = \boxed{\frac{1\times 1}{3\times 1}} = \boxed{\frac{\mathbf{1}}{\mathbf{3}}}$

Case III.

$$\boxed{\frac{a}{b}\times\frac{c}{d}\times\frac{e}{f}} = \boxed{\frac{a}{b}\times\left(\frac{c}{d}\times\frac{e}{f}\right)} = \boxed{\frac{a}{b}\times\left(\frac{c\times e}{d\times f}\right)} = \boxed{\frac{a}{b}\times\left(\frac{ce}{df}\right)} = \boxed{\frac{a}{b}\times\frac{ce}{df}} = \boxed{\frac{a\times ce}{b\times df}} = \boxed{\frac{ace}{bdf}}$$

Example 2.4-14

$$\boxed{\frac{2}{3}\times\frac{3}{15}\times\frac{5}{2}} = \boxed{\frac{2}{3}\times\left(\frac{3}{15}\times\frac{5}{2}\right)} = \boxed{\frac{2}{3}\times\left(\frac{3\times\overset{1}{\cancel{5}}}{\underset{3}{\cancel{15}}\times 2}\right)} = \boxed{\frac{2}{3}\times\left(\frac{\overset{1}{\cancel{3}}\times 1}{\underset{1}{\cancel{3}}\times 2}\right)} = \boxed{\frac{2}{3}\times\left(\frac{1\times 1}{1\times 2}\right)} = \boxed{\frac{2}{3}\times\left(\frac{1}{2}\right)} = \boxed{\frac{2}{3}\times\frac{1}{2}}$$

$$= \boxed{\frac{\overset{1}{\cancel{2}}\times 1}{3\times\underset{1}{\cancel{2}}}} = \boxed{\frac{1\times 1}{3\times 1}} = \boxed{\frac{1}{3}}$$

Note - In multiplication the use of parentheses does not change the final answer; the three examples above have the same answer (see Section 1.4).

Additional Examples - Multiplying Integer Fractions

The following examples further illustrate how to multiply integer fractions:

Example 2.4-15

$$\boxed{\frac{3}{5}\times\frac{2}{6}} = \boxed{\frac{\overset{1}{\cancel{3}}\times 2}{5\times\underset{2}{\cancel{6}}}} = \boxed{\frac{1\times\overset{1}{\cancel{2}}}{5\times\underset{1}{\cancel{2}}}} = \boxed{\frac{1\times 1}{5\times 1}} = \boxed{\frac{1}{5}}$$

Example 2.4-16

$$\boxed{\frac{2}{3}\times 24} = \boxed{\frac{2}{3}\times\frac{24}{1}} = \boxed{\frac{2\times\overset{8}{\cancel{24}}}{\underset{1}{\cancel{3}}\times 1}} = \boxed{\frac{2\times 8}{1\times 1}} = \boxed{\frac{16}{1}} = \boxed{16}$$

Example 2.4-17

$$\boxed{\frac{2}{5}\times\frac{4}{5}\times\frac{25}{8}} = \boxed{\frac{2\times\overset{1}{\cancel{4}}\times\overset{5}{\cancel{25}}}{\underset{1}{\cancel{5}}\times 5\times\underset{2}{\cancel{8}}}} = \boxed{\frac{\overset{1}{\cancel{2}}\times 1\times\overset{1}{\cancel{5}}}{1\times\underset{1}{\cancel{5}}\times\underset{1}{\cancel{2}}}} = \boxed{\frac{1\times 1\times 1}{1\times 1\times 1}} = \boxed{\frac{1}{1}} = \boxed{1}$$

Example 2.4-18

$$\boxed{\frac{6}{3}\times\frac{1}{3}\times\frac{0}{1}} = \boxed{\frac{6\times 1\times 0}{3\times 3\times 1}} = \boxed{\frac{0}{1}} = \boxed{0}$$

Example 2.4-19

$$\boxed{1000\times\frac{2}{100}\times\frac{1}{10}\times\frac{1}{2}} = \boxed{\frac{1000}{1}\times\frac{2}{100}\times\frac{1}{10}\times\frac{1}{2}} = \boxed{\frac{\overset{10}{\cancel{1000}}\times\overset{1}{\cancel{2}}\times 1\times 1}{1\times\underset{1}{\cancel{100}}\times 10\times\underset{1}{\cancel{2}}}} = \boxed{\frac{\overset{1}{\cancel{10}}\times 1\times 1\times 1}{1\times 1\times\underset{1}{\cancel{10}}\times 1}} = \boxed{\frac{1\times 1\times 1\times 1}{1\times 1\times 1\times 1}} = \boxed{\frac{1}{1}} = \boxed{1}$$

Example 2.4-20

$$\boxed{\frac{3}{8}\times\frac{4}{5}\times\frac{6}{10}\times\frac{1}{3}} = \boxed{\frac{\overset{1}{\cancel{3}}\times\overset{1}{\cancel{4}}\times 6\times 1}{\underset{2}{\cancel{8}}\times 5\times 10\times\underset{1}{\cancel{3}}}} = \boxed{\frac{1\times 1\times 6\times 1}{2\times 5\times 10\times 1}} = \boxed{\frac{\overset{3}{\cancel{6}}}{\underset{50}{\cancel{100}}}} = \boxed{\mathbf{\frac{3}{50}}}$$

Example 2.4-21

$$\boxed{\frac{3}{8}\times\left(\frac{5}{12}\times\frac{6}{7}\times 36\right)} = \boxed{\frac{3}{8}\times\left(\frac{5}{12}\times\frac{6}{7}\times\frac{36}{1}\right)} = \boxed{\frac{3}{8}\times\left(\frac{5\times 6\times\overset{3}{\cancel{36}}}{\underset{1}{\cancel{12}}\times 7\times 1}\right)} = \boxed{\frac{3}{8}\times\left(\frac{5\times 6\times 3}{1\times 7\times 1}\right)} = \boxed{\frac{3}{8}\times\left(\frac{90}{7}\right)} = \boxed{\frac{3}{8}\times\frac{90}{7}}$$

$$= \boxed{\frac{3\times 90}{8\times 7}} = \boxed{\frac{\overset{135}{\cancel{270}}}{\underset{28}{\cancel{56}}}} = \boxed{\frac{135}{28}} = \boxed{\mathbf{4\frac{23}{28}}}$$

Example 2.4-22

$$\boxed{\left(\frac{3}{10}\times 24\right)\times\left(\frac{3}{8}\times\frac{25}{6}\right)} = \boxed{\left(\frac{3}{10}\times\frac{24}{1}\right)\times\left(\frac{3}{8}\times\frac{25}{6}\right)} = \boxed{\left(\frac{3\times\overset{12}{\cancel{24}}}{\underset{5}{\cancel{10}}\times 1}\right)\times\left(\frac{\overset{1}{\cancel{3}}\times 25}{8\times\underset{2}{\cancel{6}}}\right)} = \boxed{\left(\frac{3\times 12}{5\times 1}\right)\times\left(\frac{1\times 25}{8\times 2}\right)} = \boxed{\left(\frac{36}{5}\right)\times\left(\frac{25}{16}\right)}$$

$$= \boxed{\frac{36}{5}\times\frac{25}{16}} = \boxed{\frac{\overset{9}{\cancel{36}}\times\overset{5}{\cancel{25}}}{\underset{1}{\cancel{5}}\times\underset{4}{\cancel{16}}}} = \boxed{\frac{9\times 5}{1\times 4}} = \boxed{\frac{45}{4}} = \boxed{\mathbf{11\frac{1}{4}}}$$

Example 2.4-23

$$\boxed{\left(\frac{2}{5}\times\frac{3}{4}\right)\times\left(\frac{1}{3}\times\frac{20}{33}\times 121\right)} = \boxed{\left(\frac{\overset{1}{\cancel{2}}\times 3}{5\times\underset{2}{\cancel{4}}}\right)\times\left(\frac{1}{3}\times\frac{20}{33}\times\frac{121}{1}\right)} = \boxed{\left(\frac{1\times 3}{5\times 2}\right)\times\left(\frac{1\times 20\times 121}{3\times 33\times 1}\right)} = \boxed{\left(\frac{3}{10}\right)\times\left(\frac{2420}{99}\right)}$$

$$= \boxed{\frac{3}{10}\times\frac{2420}{99}} = \boxed{\frac{\overset{1}{\cancel{3}}\times\overset{242}{\cancel{2420}}}{\underset{1}{\cancel{10}}\times\underset{33}{\cancel{99}}}} = \boxed{\frac{1\times\overset{22}{\cancel{242}}}{1\times\underset{3}{\cancel{33}}}} = \boxed{\frac{1\times 22}{1\times 3}} = \boxed{\frac{22}{3}} = \boxed{\mathbf{7\frac{1}{3}}}$$

Example 2.4-24

$$\boxed{\left[\left(\frac{35}{5}\times\frac{1}{7}\right)\times 5\right]\times\left(\frac{1}{8}\times 3\right)} = \boxed{\left[\left(\frac{35\times 1}{5\times 7}\right)\times\frac{5}{1}\right]\times\left(\frac{1}{8}\times\frac{3}{1}\right)} = \boxed{\left[\left(\frac{\overset{1}{\cancel{35}}}{\underset{1}{\cancel{35}}}\right)\times\frac{5}{1}\right]\times\left(\frac{1\times 3}{8\times 1}\right)} = \boxed{\left[\left(\frac{1}{1}\right)\times\frac{5}{1}\right]\times\left(\frac{3}{8}\right)}$$

$$= \boxed{\left[\frac{1}{1}\times\frac{5}{1}\right]\times\frac{3}{8}} = \boxed{\left[\frac{1\times 5}{1\times 1}\right]\times\frac{3}{8}} = \boxed{\left[\frac{5}{1}\right]\times\frac{3}{8}} = \boxed{\frac{5}{1}\times\frac{3}{8}} = \boxed{\frac{5\times 3}{1\times 8}} = \boxed{\frac{15}{8}} = \boxed{\mathbf{1\frac{7}{8}}}$$

Example 2.4-25

$$\boxed{\left(\frac{9}{80}\times 80\right)\times\left(\frac{1}{50}\times 5\right)\times\left(\frac{100}{2}\times\frac{50}{1}\right)} = \boxed{\left(\frac{9}{80}\times\frac{80}{1}\right)\times\left(\frac{1}{50}\times\frac{5}{1}\right)\times\left(\frac{\overset{50}{\cancel{100}}}{\underset{1}{\cancel{2}}}\times\frac{50}{1}\right)} = \boxed{\left(\frac{9\times\overset{1}{\cancel{80}}}{\underset{1}{\cancel{80}}\times 1}\right)\times\left(\frac{1\times\overset{1}{\cancel{5}}}{\underset{10}{\cancel{50}}\times 1}\right)\times\left(\frac{50\times 50}{1\times 1}\right)}$$

$$= \boxed{\left(\frac{9\times1}{1\times1}\right)\times\left(\frac{1\times1}{10\times1}\right)\times\left(\frac{2500}{1}\right)} = \boxed{\left(\frac{9}{1}\right)\times\left(\frac{1}{10}\right)\times\frac{2500}{1}} = \boxed{\frac{9}{1}\times\frac{1}{10}\times\frac{2500}{1}} = \boxed{\frac{9\times1\times\overset{250}{\cancel{2500}}}{1\times\underset{1}{\cancel{10}}\times1}} = \boxed{\frac{9\times1\times250}{1\times1\times1}}$$

$$= \boxed{\frac{2250}{1}} = \boxed{2250}$$

Practice Problems - Multiplying Integer Fractions

Section 2.4 Practice Problems - Multiply the following integer fractions:

1. $\frac{4}{8}\times\frac{3}{5} =$
2. $\frac{4}{8}\times\frac{5}{6}\times100 =$
3. $\frac{7}{3}\times\frac{9}{4}\times\frac{6}{3} =$
4. $34\times\frac{1}{5}\times\frac{3}{17}\times\frac{1}{8}\times20 =$
5. $\left(\frac{2}{55}\times3\right)\times\left(\frac{4}{5}\times\frac{25}{8}\right) =$
6. $\left(1000\times\frac{1}{5}\right)\times\left(\frac{25}{5}\times\frac{1}{8}\right)\times\frac{0}{100} =$
7. $\frac{2}{6}\times\frac{36}{1}\times\frac{1}{100}\times10\times\frac{1}{6} =$
8. $\left(\frac{7}{8}\times\frac{9}{4}\right)\times\left(\frac{4}{18}\times\frac{1}{14}\times\frac{1}{9}\right) =$
9. $\left[\left(18\times\frac{2}{8}\right)\times\left(\frac{1}{5}\times\frac{25}{3}\right)\right]\times\frac{2}{9} =$
10. $\left(\frac{3}{8}\times\frac{4}{49}\times\frac{6}{5}\right)\times\left(\frac{7}{3}\times\frac{4}{8}\right)\times\frac{7}{2} =$

2.5 Dividing Integer Fractions

Two or more integer fractions with or without a common denominator are divided using the steps given as in each case below:

Case I - Dividing Two Integer Fractions with or Without a Common Denominator

Divide two integer fractions using the following steps:

Step 1 Change the integer number a to an integer fraction of the form $\frac{a}{1}$, e.g., change 9 to $\frac{9}{1}$.

Step 2 a. Change the division sign to a multiplication sign.

b. Replace the numerator of the second fraction with its denominator.

c. Replace the denominator of the second fraction with its numerator.

d. Multiply the numerator of the first fraction with the numerator of the second fraction to obtain the new numerator.

e. Multiply the denominator of the first fraction with the denominator of the second fraction to obtain the new denominator.

Step 3 Simplify the fraction to its lowest term (see Section 2.1).

Step 4 Change the improper fraction to a mixed fraction if the fraction obtained from Step 3 is an improper fraction (see Section 2.1 Appendix).

Examples with Steps

The following examples show the steps as to how two integer fractions with or without a common denominator are divided:

Example 2.5-1

$$\boxed{\frac{3}{5} \div \frac{8}{15}} =$$

Solution:

Step 1 $\boxed{\textit{Not Applicable}}$

Step 2 $$\boxed{\frac{3}{5} \div \frac{8}{15}} = \boxed{\frac{3}{5} \times \frac{15}{8}} = \boxed{\frac{3 \times 15}{5 \times 8}} = \boxed{\frac{45}{40}}$$

Step 3 $$\boxed{\frac{45}{40}} = \boxed{\frac{45 \div 5}{40 \div 5}} = \boxed{\frac{9}{8}}$$

Step 4 $$\boxed{\frac{9}{8}} = \boxed{\mathbf{1\frac{1}{8}}}$$

Example 2.5-2

$$\boxed{9 \div \frac{6}{12}} =$$

Solution:

Step 1 $\boxed{9 \div \frac{6}{12}} = \boxed{\frac{9}{1} \div \frac{6}{12}}$

Step 2 $\boxed{\frac{9}{1} \div \frac{6}{12}} = \boxed{\frac{9}{1} \times \frac{12}{6}} = \boxed{\frac{9 \times 12}{1 \times 6}} = \boxed{\frac{108}{6}}$

Step 3 $\boxed{\frac{108}{6}} = \boxed{\frac{108 \div 6}{6 \div 6}} = \boxed{\frac{18}{1}} = \boxed{\mathbf{18}}$

Step 4 *Not Applicable*

Example 2.5-3

$\boxed{\frac{320}{465} \div \frac{75}{100}} =$

Solution:

Step 1 *Not Applicable*

Step 2 $\boxed{\frac{320}{465} \div \frac{75}{100}} = \boxed{\frac{320}{465} \times \frac{100}{75}} = \boxed{\frac{320 \times 100}{465 \times 75}} = \boxed{\frac{32000}{34875}}$

Step 3 $\boxed{\frac{32000}{34875}} = \boxed{\frac{32000 \div 25}{34875 \div 25}} = \boxed{\frac{1280}{1395}} = \boxed{\frac{1280 \div 5}{1395 \div 5}} = \boxed{\mathbf{\frac{256}{279}}}$

Step 4 *Not Applicable*

Example 2.5-4

$\boxed{\frac{125}{65} \div 230} =$

Solution:

Step 1 $\boxed{\frac{125}{65} \div 230} = \boxed{\frac{125}{65} \div \frac{230}{1}}$

Step 2 $\boxed{\frac{125}{65} \div \frac{230}{1}} = \boxed{\frac{125}{65} \times \frac{1}{230}} = \boxed{\frac{125 \times 1}{65 \times 230}} = \boxed{\frac{125}{14950}}$

Step 3 $\boxed{\frac{125}{14950}} = \boxed{\frac{125 \div 25}{14950 \div 25}} = \boxed{\mathbf{\frac{5}{598}}}$

Step 4 *Not Applicable*

Example 2.5-5

$\boxed{\frac{32}{18} \div \frac{50}{12}} =$

Solution:

Step 1 *Not Applicable*

Step 2 $\frac{32}{18} \div \frac{50}{12} = \frac{32}{18} \times \frac{12}{50} = \frac{32 \times 12}{18 \times 50} = \frac{384}{900}$

Step 3 $\frac{384}{900} = \frac{384 \div 4}{900 \div 4} = \frac{96}{225} = \frac{96 \div 3}{225 \div 3} = \mathbf{\frac{32}{75}}$

Step 4 *Not Applicable*

In general, two integer fractions with or without a common denominator are divided in the following way:

$$\frac{a}{b} \div \frac{c}{d} = \frac{a}{b} \times \frac{d}{c} = \frac{a \times d}{b \times c} = \frac{ad}{bc}$$

Example 2.5-6

$$\frac{3}{5} \div \frac{2}{15} = \frac{3}{5} \times \frac{15}{2} = \frac{3 \times \overset{3}{\cancel{15}}}{\underset{1}{\cancel{5}} \times 2} = \frac{3 \times 3}{1 \times 2} = \frac{9}{2} = \mathbf{4\frac{1}{2}}$$

Case II - Dividing Three Integer Fractions with or Without a Common Denominator

Divide three integer fractions using the following steps:

Step 1 Change the integer number a to an integer fraction of the form $\frac{a}{1}$, e.g., change 58 to $\frac{58}{1}$.

Step 2 a. Select the two fractions grouped by parentheses.

b. Divide the grouped fractions following Steps 2a through 2e, outlined in Section 2.5, Case I above, to obtain a new integer fraction.

c. Divide the new integer fraction by the third fraction by repeating Steps 2a through 2e, outlined in Section 2.5, Case I above.

Step 3 Simplify the fraction to its lowest term (see Section 2.1).

Step 4 Change the improper fraction to a mixed fraction if the fraction obtained from Step 3 is an improper fraction (see Section 2.1 Appendix).

Examples with Steps

The following examples show the steps as to how three integer fractions are divided:

Example 2.5-7

$$\left(\frac{3}{5} \div 4\right) \div \frac{9}{25} =$$

Solution:

Step 1 $\boxed{\left(\frac{3}{5}\div 4\right)\div\frac{9}{25}} = \boxed{\left(\frac{3}{5}\div\frac{4}{1}\right)\div\frac{9}{25}}$

Step 2 $\boxed{\left(\frac{3}{5}\div\frac{4}{1}\right)\div\frac{9}{25}} = \boxed{\left(\frac{3}{5}\times\frac{1}{4}\right)\div\frac{9}{25}} = \boxed{\left(\frac{3\times 1}{5\times 4}\right)\div\frac{9}{25}} = \boxed{\left(\frac{3}{20}\right)\div\frac{9}{25}} = \boxed{\frac{3}{20}\div\frac{9}{25}}$

$= \boxed{\frac{3}{20}\times\frac{25}{9}} = \boxed{\frac{3\times 25}{20\times 9}} = \boxed{\frac{75}{180}}$

Step 3 $\boxed{\frac{75}{180}} = \boxed{\frac{75\div 15}{180\div 15}} = \boxed{\mathbf{\frac{5}{12}}}$

Step 4 *Not Applicable*

Example 2.5-8

$\boxed{235\div\left(\frac{68}{15}\div\frac{33}{12}\right)} =$

Solution:

Step 1 $\boxed{235\div\left(\frac{68}{15}\div\frac{33}{12}\right)} = \boxed{\frac{235}{1}\div\left(\frac{68}{15}\div\frac{33}{12}\right)}$

Step 2 $\boxed{\frac{235}{1}\div\left(\frac{68}{15}\div\frac{33}{12}\right)} = \boxed{\frac{235}{1}\div\left(\frac{68}{15}\times\frac{12}{33}\right)} = \boxed{\frac{235}{1}\div\left(\frac{68\times 12}{15\times 33}\right)} = \boxed{\frac{235}{1}\div\left(\frac{816}{495}\right)}$

$= \boxed{\frac{235}{1}\div\frac{816}{495}} = \boxed{\frac{235}{1}\times\frac{495}{816}} = \boxed{\frac{235\times 495}{1\times 816}} = \boxed{\frac{116325}{816}}$

Step 3 $\boxed{\frac{116325}{816}} = \boxed{\frac{116325\div 3}{816\div 3}} = \boxed{\frac{38775}{272}}$

Step 4 $\boxed{\frac{38775}{272}} = \boxed{\mathbf{142\frac{151}{272}}}$

Example 2.5-9

$\boxed{\left(\frac{4}{5}\div\frac{2}{3}\right)\div\frac{1}{5}} =$

Solution:

Step 1 *Not Applicable*

Step 2 $\boxed{\left(\frac{4}{5}\div\frac{2}{3}\right)\div\frac{1}{5}} = \boxed{\left(\frac{4}{5}\times\frac{3}{2}\right)\div\frac{1}{5}} = \boxed{\left(\frac{4\times 3}{5\times 2}\right)\div\frac{1}{5}} = \boxed{\left(\frac{12}{10}\right)\div\frac{1}{5}} = \boxed{\frac{12}{10}\div\frac{1}{5}} = \boxed{\frac{12}{10}\times\frac{5}{1}}$

$$= \boxed{\frac{12\times 5}{10\times 1}} = \boxed{\frac{60}{10}}$$

Step 3 $\boxed{\frac{60}{10}} = \boxed{\frac{60\div 10}{10\div 10}} = \boxed{\frac{6}{1}} = \boxed{\mathbf{6}}$

Step 4 *Not Applicable*

Example 2.5-10

$$\boxed{\frac{12}{30}\div\left(\frac{15}{6}\div\frac{12}{5}\right)} =$$

Solution:

Step 1 *Not Applicable*

Step 2 $\boxed{\frac{12}{30}\div\left(\frac{15}{6}\div\frac{12}{5}\right)} = \boxed{\frac{12}{30}\div\left(\frac{15}{6}\times\frac{5}{12}\right)} = \boxed{\frac{12}{30}\div\left(\frac{15\times 5}{6\times 12}\right)} = \boxed{\frac{12}{30}\div\left(\frac{75}{72}\right)} = \boxed{\frac{12}{30}\div\frac{75}{72}}$

$$= \boxed{\frac{12}{30}\times\frac{72}{75}} = \boxed{\frac{12\times 72}{30\times 75}} = \boxed{\frac{864}{2250}}$$

Step 3 $\boxed{\frac{864}{2250}} = \boxed{\frac{864\div 2}{2250\div 2}} = \boxed{\frac{432}{1125}} = \boxed{\frac{432\div 9}{1125\div 9}} = \boxed{\mathbf{\frac{48}{125}}}$

Step 4 *Not Applicable*

Example 2.5-11

$$\boxed{\frac{9}{6}\div\left(\frac{7}{6}\div\frac{5}{6}\right)} =$$

Solution:

Step 1 *Not Applicable*

Step 2 $\boxed{\frac{9}{6}\div\left(\frac{7}{6}\div\frac{5}{6}\right)} = \boxed{\frac{9}{6}\div\left(\frac{7}{6}\times\frac{6}{5}\right)} = \boxed{\frac{9}{6}\div\left(\frac{7\times 6}{6\times 5}\right)} = \boxed{\frac{9}{6}\div\left(\frac{42}{30}\right)} = \boxed{\frac{9}{6}\div\frac{42}{30}} = \boxed{\frac{9}{6}\times\frac{30}{42}}$

$$= \boxed{\frac{9\times 30}{6\times 42}} = \boxed{\frac{270}{252}}$$

Step 3 $\boxed{\frac{270}{252}} = \boxed{\frac{270\div 6}{252\div 6}} = \boxed{\frac{45}{42}}$

Step 4 $\boxed{\frac{45}{42}} = \boxed{\mathbf{1\frac{3}{42}}}$

In general, three integer fractions with or without a common denominator are divided as in the following cases:

Case I.

$$\left(\frac{a}{b}\div\frac{c}{d}\right)\div\frac{e}{f}=\left(\frac{a}{b}\times\frac{d}{c}\right)\div\frac{e}{f}=\left(\frac{a\times d}{b\times c}\right)\div\frac{e}{f}=\left(\frac{ad}{bc}\right)\div\frac{e}{f}=\frac{ad}{bc}\div\frac{e}{f}=\frac{ad}{bc}\times\frac{f}{e}=\frac{ad\times f}{bc\times e}=\frac{adf}{bce}$$

Example 2.5-12

$$\left(\frac{2}{5}\div\frac{3}{6}\right)\div\frac{8}{5}=\left(\frac{2}{5}\times\frac{6}{3}\right)\div\frac{8}{5}=\left(\frac{2\times\overset{2}{\cancel{6}}}{5\times\underset{1}{\cancel{3}}}\right)\div\frac{8}{5}=\left(\frac{2\times 2}{5\times 1}\right)\div\frac{8}{5}=\left(\frac{2\times 2}{5\times 1}\right)\div\frac{8}{5}=\left(\frac{4}{5}\right)\div\frac{8}{5}=\frac{4}{5}\div\frac{8}{5}$$

$$=\frac{4}{5}\times\frac{5}{8}=\frac{\overset{1}{\cancel{4}}\times\overset{1}{\cancel{5}}}{\underset{1}{\cancel{5}}\times\underset{2}{\cancel{8}}}=\frac{1\times 1}{1\times 2}=\frac{1}{2}$$

Case II.

$$\frac{a}{b}\div\left(\frac{c}{d}\div\frac{e}{f}\right)=\frac{a}{b}\div\left(\frac{c}{d}\times\frac{f}{e}\right)=\frac{a}{b}\div\left(\frac{c\times f}{d\times e}\right)=\frac{a}{b}\div\left(\frac{cf}{de}\right)=\frac{a}{b}\div\frac{cf}{de}=\frac{a}{b}\times\frac{de}{cf}=\frac{a\times de}{b\times cf}=\frac{ade}{bcf}$$

Example 2.5-13

$$\frac{2}{5}\div\left(\frac{3}{6}\div\frac{8}{5}\right)=\frac{2}{5}\div\left(\frac{3}{6}\times\frac{5}{8}\right)=\frac{2}{5}\div\left(\frac{\overset{1}{\cancel{3}}\times 5}{\underset{2}{\cancel{6}}\times 8}\right)=\frac{2}{5}\div\left(\frac{1\times 5}{2\times 8}\right)=\frac{2}{5}\div\left(\frac{5}{16}\right)=\frac{2}{5}\div\frac{5}{16}=\frac{2}{5}\times\frac{16}{5}=\frac{2\times 16}{5\times 5}$$

$$=\frac{32}{25}=1\frac{7}{25}$$

Additional Examples - Dividing Integer Fractions

The following examples further illustrate how to divide integer fractions:

Example 2.5-14

$$\frac{4}{5}\div\frac{2}{15}=\frac{4}{5}\times\frac{15}{2}=\frac{\overset{2}{\cancel{4}}\times\overset{3}{\cancel{15}}}{\underset{1}{\cancel{5}}\times\underset{1}{\cancel{2}}}=\frac{2\times 3}{1\times 1}=\frac{6}{1}=6$$

Example 2.5-15

$$\frac{3}{5}\div 24=\frac{3}{5}\div\frac{24}{1}=\frac{3}{5}\times\frac{1}{24}=\frac{\overset{1}{\cancel{3}}\times 1}{5\times\underset{8}{\cancel{24}}}=\frac{1\times 1}{5\times 8}=\frac{1}{40}$$

Example 2.5-16

$$\left(\frac{3}{5}\div\frac{1}{5}\right)\div\frac{4}{15}=\left(\frac{3}{5}\times\frac{5}{1}\right)\div\frac{4}{15}=\left(\frac{3\times 5}{5\times 1}\right)\div\frac{4}{15}=\left(\frac{15}{5}\right)\div\frac{4}{15}=\frac{15}{5}\div\frac{4}{15}=\frac{15}{5}\times\frac{15}{4}=\frac{15\times\overset{3}{\cancel{15}}}{\underset{1}{\cancel{5}}\times 4}=\frac{15\times 3}{1\times 4}$$

$$= \boxed{\frac{45}{4}} = \boxed{\mathbf{11\frac{1}{4}}}$$

Example 2.5-17

$$\boxed{25 \div \left(\frac{2}{8} \div \frac{4}{3}\right)} = \boxed{\frac{25}{1} \div \left(\frac{2}{8} \times \frac{3}{4}\right)} = \boxed{\frac{25}{1} \div \left(\frac{\overset{1}{\cancel{2}} \times 3}{8 \times \underset{2}{\cancel{4}}}\right)} = \boxed{\frac{25}{1} \div \left(\frac{1 \times 3}{8 \times 2}\right)} = \boxed{\frac{25}{1} \div \left(\frac{3}{16}\right)} = \boxed{\frac{25}{1} \div \frac{3}{16}} = \boxed{\frac{25}{1} \times \frac{16}{3}}$$

$$= \boxed{\frac{25 \times 16}{1 \times 3}} = \boxed{\frac{400}{3}} = \boxed{\mathbf{133\frac{1}{3}}}$$

Example 2.5-18

$$\boxed{\left(\frac{5}{7} \div \frac{3}{49}\right) \div 12} = \boxed{\left(\frac{5}{7} \times \frac{49}{3}\right) \div \frac{12}{1}} = \boxed{\left(\frac{5 \times \overset{7}{\cancel{49}}}{\underset{1}{\cancel{7}} \times 3}\right) \div \frac{12}{1}} = \boxed{\left(\frac{5 \times 7}{1 \times 3}\right) \div \frac{12}{1}} = \boxed{\left(\frac{35}{3}\right) \div \frac{12}{1}} = \boxed{\frac{35}{3} \div \frac{12}{1}} = \boxed{\frac{35}{3} \times \frac{1}{12}}$$

$$= \boxed{\frac{35 \times 1}{3 \times 12}} = \boxed{\mathbf{\frac{35}{36}}}$$

Example 2.5-19

$$\boxed{\left(\frac{9}{16} \div \frac{3}{32}\right) \div \left(\frac{4}{8} \div \frac{8}{1}\right)} = \boxed{\left(\frac{9}{16} \times \frac{32}{3}\right) \div \left(\frac{4}{8} \times \frac{1}{8}\right)} = \boxed{\left(\frac{\overset{3}{\cancel{9}} \times \overset{2}{\cancel{32}}}{\underset{1}{\cancel{16}} \times \underset{1}{\cancel{3}}}\right) \div \left(\frac{\overset{1}{\cancel{4}} \times 1}{8 \times \underset{2}{\cancel{8}}}\right)} = \boxed{\left(\frac{3 \times 2}{1 \times 1}\right) \div \left(\frac{1 \times 1}{8 \times 2}\right)} = \boxed{\left(\frac{6}{1}\right) \div \left(\frac{1}{16}\right)}$$

$$= \boxed{\frac{6}{1} \div \frac{1}{16}} = \boxed{\frac{6}{1} \times \frac{16}{1}} = \boxed{\frac{6 \times 16}{1 \times 1}} = \boxed{\frac{96}{1}} = \boxed{\mathbf{96}}$$

Example 2.5-20

$$\boxed{\left(\frac{1}{4} \div \frac{3}{8}\right) \div \left(5 \div \frac{2}{3}\right)} = \boxed{\left(\frac{1}{4} \times \frac{8}{3}\right) \div \left(\frac{5}{1} \div \frac{2}{3}\right)} = \boxed{\left(\frac{1 \times \overset{2}{\cancel{8}}}{\underset{1}{\cancel{4}} \times 3}\right) \div \left(\frac{5}{1} \times \frac{3}{2}\right)} = \boxed{\left(\frac{1 \times 2}{1 \times 3}\right) \div \left(\frac{5 \times 3}{1 \times 2}\right)} = \boxed{\left(\frac{2}{3}\right) \div \left(\frac{15}{2}\right)} = \boxed{\frac{2}{3} \div \frac{15}{2}}$$

$$= \boxed{\frac{2}{3} \times \frac{2}{15}} = \boxed{\frac{2 \times 2}{3 \times 15}} = \boxed{\mathbf{\frac{4}{45}}}$$

Example 2.5-21

$$\boxed{\left(15 \div \frac{3}{4}\right) \div \left(\frac{2}{5} \div 12\right)} = \boxed{\left(\frac{15}{1} \div \frac{3}{4}\right) \div \left(\frac{2}{5} \div \frac{12}{1}\right)} = \boxed{\left(\frac{15}{1} \times \frac{4}{3}\right) \div \left(\frac{2}{5} \times \frac{1}{12}\right)} = \boxed{\left(\frac{\overset{5}{\cancel{15}} \times 4}{1 \times \underset{1}{\cancel{3}}}\right) \div \left(\frac{\overset{1}{\cancel{2}} \times 1}{5 \times \underset{6}{\cancel{12}}}\right)}$$

$$= \boxed{\left(\frac{5 \times 4}{1 \times 1}\right) \div \left(\frac{1 \times 1}{5 \times 6}\right)} = \boxed{\left(\frac{20}{1}\right) \div \left(\frac{1}{30}\right)} = \boxed{\frac{20}{1} \div \frac{1}{30}} = \boxed{\frac{20}{1} \times \frac{30}{1}} = \boxed{\frac{20 \times 30}{1 \times 1}} = \boxed{\frac{600}{1}} = \boxed{\mathbf{600}}$$

Example 2.5-22

$$\boxed{\left(\frac{3}{25} \div \frac{6}{10}\right) \div \left(\frac{2}{3} \div 4\right)} = \boxed{\left(\frac{3}{25} \times \frac{10}{6}\right) \div \left(\frac{2}{3} \div \frac{4}{1}\right)} = \boxed{\left(\frac{3}{25} \times \frac{\overset{5}{\cancel{10}}}{\underset{3}{\cancel{6}}}\right) \div \left(\frac{2}{3} \times \frac{1}{4}\right)} = \boxed{\left(\frac{3}{25} \times \frac{5}{3}\right) \div \left(\frac{\overset{1}{\cancel{2}} \times 1}{3 \times \underset{2}{\cancel{4}}}\right)}$$

$$= \boxed{\left(\frac{\overset{1}{\cancel{3}}\times\overset{1}{\cancel{5}}}{\underset{5}{\cancel{25}}\times\underset{1}{\cancel{3}}}\right)\div\left(\frac{1\times1}{3\times2}\right)} = \boxed{\left(\frac{1\times1}{5\times1}\right)\div\left(\frac{1}{6}\right)} = \boxed{\left(\frac{1}{5}\right)\div\frac{1}{6}} = \boxed{\frac{1}{5}\div\frac{1}{6}} = \boxed{\frac{1}{5}\times\frac{6}{1}} = \boxed{\frac{6}{5}} = \boxed{\mathbf{1\frac{1}{5}}}$$

Example 2.5-23

$$\boxed{\left[\frac{1}{8}\div\left(\frac{3}{8}\div\frac{6}{8}\right)\right]\div\frac{4}{16}} = \boxed{\left[\frac{1}{8}\div\left(\frac{3}{8}\times\frac{8}{6}\right)\right]\div\frac{4}{16}} = \boxed{\left[\frac{1}{8}\div\left(\frac{\overset{1}{\cancel{3}}\times\overset{1}{\cancel{8}}}{\underset{1}{\cancel{8}}\times\underset{2}{\cancel{6}}}\right)\right]\div\frac{4}{16}} = \boxed{\left[\frac{1}{8}\div\left(\frac{1\times1}{1\times2}\right)\right]\div\frac{4}{16}} = \boxed{\left[\frac{1}{8}\div\left(\frac{1}{2}\right)\right]\div\frac{4}{16}}$$

$$= \boxed{\left[\frac{1}{8}\div\frac{1}{2}\right]\div\frac{4}{16}} = \boxed{\left[\frac{1}{8}\times\frac{2}{1}\right]\div\frac{4}{16}} = \boxed{\left[\frac{1\times\overset{1}{\cancel{2}}}{\underset{4}{\cancel{8}}\times1}\right]\div\frac{4}{16}} = \boxed{\left[\frac{1\times1}{4\times1}\right]\div\frac{4}{16}} = \boxed{\left[\frac{1}{4}\right]\div\frac{4}{16}} = \boxed{\frac{1}{4}\div\frac{4}{16}} = \boxed{\frac{1}{4}\times\frac{16}{4}}$$

$$= \boxed{\frac{1\times16}{4\times4}} = \boxed{\frac{\overset{1}{\cancel{16}}}{\underset{1}{\cancel{16}}}} = \boxed{\frac{1}{1}} = \boxed{1}$$

Example 2.5-24

$$\boxed{\left[\left(\frac{8}{4}\div\frac{16}{2}\right)\div\left(\frac{1}{4}\div\frac{8}{16}\right)\right]\div\frac{2}{4}} = \boxed{\left[\left(\frac{8}{4}\times\frac{2}{16}\right)\div\left(\frac{1}{4}\times\frac{16}{8}\right)\right]\div\frac{2}{4}} = \boxed{\left[\left(\frac{\overset{1}{\cancel{8}}\times\overset{1}{\cancel{2}}}{\underset{2}{\cancel{4}}\times\underset{2}{\cancel{16}}}\right)\div\left(\frac{1\times\overset{2}{\cancel{16}}}{4\times\underset{1}{\cancel{8}}}\right)\right]\div\frac{2}{4}}$$

$$= \boxed{\left[\left(\frac{1\times1}{2\times2}\right)\div\left(\frac{1\times\overset{1}{\cancel{2}}}{\underset{2}{\cancel{4}}\times1}\right)\right]\div\frac{2}{4}} = \boxed{\left[\left(\frac{1}{4}\right)\div\left(\frac{1\times1}{2\times1}\right)\right]\div\frac{2}{4}} = \boxed{\left[\left(\frac{1}{4}\right)\div\left(\frac{1}{2}\right)\right]\div\frac{2}{4}} = \boxed{\left[\frac{1}{4}\div\frac{1}{2}\right]\div\frac{2}{4}} = \boxed{\left[\frac{1}{4}\times\frac{2}{1}\right]\div\frac{2}{4}}$$

$$= \boxed{\left[\frac{1\times\overset{1}{\cancel{2}}}{\underset{2}{\cancel{4}}\times1}\right]\div\frac{2}{4}} = \boxed{\left[\frac{1\times1}{2\times1}\right]\div\frac{2}{4}} = \boxed{\left[\frac{1}{2}\right]\div\frac{2}{4}} = \boxed{\frac{1}{2}\div\frac{2}{4}} = \boxed{\frac{1}{2}\times\frac{4}{2}} = \boxed{\frac{1\times4}{2\times2}} = \boxed{\frac{\overset{1}{\cancel{4}}}{\underset{1}{\cancel{4}}}} = \boxed{\frac{1}{1}} = \boxed{1}$$

Practice Problems - Dividing Integer Fractions

Section 2.5 Practice Problems - Divide the following integer fractions:

1. $\frac{8}{10}\div\frac{4}{30}=$
2. $\left(\frac{3}{8}\div\frac{12}{16}\right)\div\frac{4}{8}=$
3. $\left(\frac{4}{16}\div\frac{1}{32}\right)\div 8=$
4. $12\div\left(\frac{9}{8}\div\frac{27}{16}\right)=$
5. $\left(\frac{2}{20}\div\frac{4}{5}\right)\div 2=$
6. $\left(\frac{4}{15}\div\frac{8}{30}\right)\div\left(\frac{1}{5}\div\frac{4}{35}\right)=$
7. $\left(\frac{2}{5}\div\frac{4}{10}\right)\div\left(\frac{9}{1}\div\frac{18}{4}\right)=$
8. $\left(\frac{4}{5}\div\frac{2}{5}\right)\div\left(\frac{8}{5}\div 4\right)=$
9. $\left(\frac{6}{10}\div 1\right)\div\left(\frac{4}{6}\div\frac{1}{3}\right)=$
10. $\left[\left(\frac{9}{8}\div\frac{18}{16}\right)\div\frac{4}{2}\right]\div\frac{1}{8}=$

Chapter 3
Exponents

Quick Reference to Chapter 3 Case Problems

Case III - Adding and Subtracting Positive Integer Exponents, *p. 144*

Case III a - Addition and Subtraction of Variables and Numbers Raised to Positive Exponential Terms, p. 144

$\boxed{x^3+3y^2+2x^3-y^2+5}=$; $\boxed{\left(a^3+2a^2+4^3\right)-\left(4a^3+20\right)}=$; $\boxed{a^{3b}+2a^{2b}-4a^{3b}+5+3a^{2b}}=$

Case III b - Addition and Subtraction of Positive Exponential Terms in Fraction Form, p. 148

$\boxed{a^2-\dfrac{1}{3-a^2}}=$; $\boxed{\dfrac{a^2-b^2}{2}+3^2b^2}=$; $\boxed{\dfrac{x^2-y^2}{3}+\dfrac{x^2}{2}}=$

3.4 Operations with Negative Integer Exponents 153

Case I - Multiplying Negative Integer Exponents, *p. 153*

Case I a - Multiplying Negative Integer Exponents (Simple Cases), p. 154

$\boxed{5^{-2}\cdot 5\cdot a^{-3}\cdot b^{-3}\cdot a^{-1}\cdot b}=$; $\boxed{\left(x^{-2}y^{-2}z^2\right)\cdot\left(x^{-1}y^3z^{-4}\right)}=$; $\boxed{\left(3^{-2}\cdot 2^{-1}\right)\cdot\left(3^{-1}\cdot 2^{-2}\right)\cdot 2}=$

Case I b - Multiplying Negative Integer Exponents (More Difficult Cases), p. 160

$\boxed{\left(a^2\right)^{-2}\cdot\left(a^3\cdot b^{-3}\right)^2}=$; $\boxed{\left[(-2)^{-2}\right]^{-3}\cdot\left(x^{-a}\cdot x^{3a}\right)^{-2}}=$; $\boxed{y^{-2}\cdot\left(2\cdot y^{-3}\right)^0\cdot x^{-2}\cdot y}=$

Case II - Dividing Negative Integer Exponents, *p. 165*

Case II a - Dividing Negative Integer Exponents (Simple Cases), p. 165

$\boxed{\dfrac{2^{-3}u^{-1}v^{-3}}{2^{-1}u^{-2}v}}=$; $\boxed{\dfrac{e^{-5}f^{-2}e^0}{2^{-3}f^5e^{-4}}}=$; $\boxed{\dfrac{2^{-3}\cdot a^{-1}}{(-2)^{-2}a^{-3}}}=$

Case II b - Dividing Negative Integer Exponents (More Difficult Cases), p. 170

$\boxed{\left(\dfrac{2^2\cdot b^3}{b^{-4}}\right)^{-2}}=$; $\boxed{\dfrac{z^{-7}\cdot(a\cdot b)^{-3}}{z^2\cdot a^{-2}}}=$; $\boxed{\dfrac{a^2\cdot b^{-4}\cdot c^5}{a^{-2}\cdot b^2\cdot c^{-3}}}=$

Case III - Adding and Subtracting Negative Integer Exponents, *p. 175*

Case III a - Addition and Subtraction of Variables and Numbers Raised to Negative Exponential Terms, p. 175

$\boxed{x^{-1}+x^{-2}+2x^{-1}-4x^{-2}+5^{-2}}=$; $\boxed{a^{-1}-b^{-1}+2a^{-1}+3b^{-1}}=$; $\boxed{k^{-2n}+k^{-3n}-3k^{-2n}+2^{-2}}=$

Case III b - Addition and Subtraction of Negative Exponential Terms in Fraction Form, p. 181

$\boxed{x^{-2}-\dfrac{2}{2-x^{-2}}}=$; $\boxed{\dfrac{2^{-a}+4^{-a}}{2^{-a}}}=$; $\boxed{\dfrac{a^{-2}}{b^{-1}}-\dfrac{4}{3-a^{-2}}}=$

Chapter 3 - Exponents

In this chapter the student will learn how to solve and simplify expressions involving positive and negative integer exponents. Positive integer exponents are addressed in Section 3.1 and negative integer exponents are addressed in Section 3.2. Simplifying positive and negative exponential expressions in mathematical operations involving multiplication, division, addition, and subtraction are addressed in Sections 3.3 and 3.4, respectively. Cases presented in each section are concluded by solving additional examples with practice problems to further enhance the student's ability on the subject.

3.1 Positive Integer Exponents

Integer exponents are defined as a^n where a is referred to as the **base**, and n is the **integer exponent**. Note that the base a can be a real number or a variable. The integer exponent n can be a positive or a negative integer. In this section, real numbers raised to positive integer exponents (Case I) and variables raised to positive integer exponents (Case II) are addressed.

Case I Real Numbers Raised to Positive Integer Exponents

In general, real numbers raised to positive integer exponents are shown as:

$$a^{+n} = a^n = a \cdot a \cdot a \cdot a \ldots a \qquad \textit{where n is a positive integer and } a \neq 0$$

For example,

$$8^{+4} = 8^4 = 8 \cdot 8 \cdot 8 \cdot 8 = 4096$$

Real numbers raised to a positive integer exponent are solved using the following steps:

Step 1 Multiply the base a by itself as many times as the number specified in the exponent. For example, 2^5 implies that multiply 2 by itself 5 times, i.e., $2^5 = 2 \cdot 2 \cdot 2 \cdot 2 \cdot 2$.

Step 2 Multiply the real numbers to obtain the product, i.e., $2 \cdot 2 \cdot 2 \cdot 2 \cdot 2 = 32$.

Examples with Steps

The following examples show the steps as to how real numbers raised to positive integer exponents are solved:

Example 3.1-1

$$\boxed{2^3} =$$

Solution:

Step 1 $\boxed{2^3} = \boxed{2 \cdot 2 \cdot 2}$

Step 2 $\boxed{2 \cdot 2 \cdot 2} = \boxed{\mathbf{8}}$

Example 3.1-2

$$\boxed{1.2^4} =$$

Solution:

Step 1 $\boxed{1.2^4} = \boxed{(1.2)\cdot(1.2)\cdot(1.2)\cdot(1.2)}$

Step 2 $\boxed{(1.2)\cdot(1.2)\cdot(1.2)\cdot(1.2)} = \boxed{\mathbf{2.074}}$

Example 3.1-3

$\boxed{(-3)^5} =$

Solution:

Step 1 $\boxed{(-3)^5} = \boxed{(-3)\cdot(-3)\cdot(-3)\cdot(-3)\cdot(-3)}$

Step 2 $\boxed{(-3)\cdot(-3)\cdot(-3)\cdot(-3)\cdot(-3)} = \boxed{\mathbf{-243}}$

Example 3.1-4

$\boxed{100^3} =$

Solution:

Step 1 $\boxed{100^3} = \boxed{100\cdot 100\cdot 100}$

Step 2 $\boxed{100\cdot 100\cdot 100} = \boxed{\mathbf{1000000}}$

Example 3.1-5

$\boxed{-(-5)^6} =$

Solution:

Step 1 $\boxed{-(-5)^6} = \boxed{-[(-5)\cdot(-5)\cdot(-5)\cdot(-5)\cdot(-5)\cdot(-5)]}$

Step 2 $\boxed{-[(-5)\cdot(-5)\cdot(-5)\cdot(-5)\cdot(-5)\cdot(-5)]} = \boxed{-(+15625)} = \boxed{\mathbf{-15625}}$

Note that:

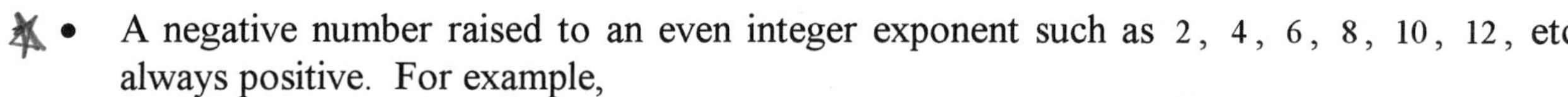

- A negative number raised to an even integer exponent such as 2, 4, 6, 8, 10, 12, etc. is always positive. For example,

  $(-3)^6 = (+3)^6 = +729 = 729 \qquad (-2)^2 = (+2)^2 = +4 = 4 \qquad (-5)^4 = (+5)^4 = +625 = 625$

- A negative number raised to an odd integer exponent such as 1, 3, 5, 7, 9, 11, etc. is always negative. For example,

  $(-3)^5 = -243 \qquad (-2)^3 = -8 \qquad (-3)^7 = -2187$

Additional Examples - Real Numbers Raised to Positive Integer Exponents

The following examples further illustrate how to solve real numbers raised to positive integer exponents:

Example 3.1-6

$\boxed{(-10)^0} = \boxed{1}$ (See the note on page 120 on numbers raised to the zero power.)

Example 3.1-7

$\boxed{-(6)^5} = \boxed{-(6 \cdot 6 \cdot 6 \cdot 6 \cdot 6)} = \boxed{-(7776)} = \boxed{\mathbf{-7776}}$

Example 3.1-8

$\boxed{18^3} = \boxed{18 \cdot 18 \cdot 18} = \boxed{\mathbf{5832}}$

Example 3.1-9

$\boxed{5.8^0} = \boxed{\mathbf{1}}$

Example 3.1-10

$\boxed{0.34^4} = \boxed{0.34 \cdot 0.34 \cdot 0.34 \cdot 0.34} = \boxed{\mathbf{0.0134}}$

Example 3.1-11

$\boxed{0^5} = \boxed{0 \cdot 0 \cdot 0 \cdot 0 \cdot 0} = \boxed{\mathbf{0}}$

Example 3.1-12

$\boxed{(-3)^4} = \boxed{(-3) \cdot (-3) \cdot (-3) \cdot (-3)} = \boxed{+81} = \boxed{\mathbf{81}}$

Example 3.1-13

$\boxed{(-4.25)^3} = \boxed{(-4.25) \cdot (-4.25) \cdot (-4.25)} = \boxed{\mathbf{-76.77}}$

Example 3.1-14

$\boxed{(10.45)^4} = \boxed{(10.45) \cdot (10.45) \cdot (10.45) \cdot (10.45)} = \boxed{\mathbf{11925.19}}$

Example 3.1-15

$\boxed{-(-20)^3} = \boxed{-[(-20) \cdot (-20) \cdot (-20)]} = \boxed{-[-8000]} = \boxed{+8000} = \boxed{\mathbf{8000}}$

Practice Problems - Real Numbers Raised to Positive Integer Exponents

Section 3.1 Case I Practice Problems - Solve the following exponential expressions with real numbers raised to positive integer exponents:

1. $4^3 =$
2. $(-10)^4 =$
3. $0.25^3 =$
4. $12^5 =$
5. $-(3)^5 =$
6. $489^0 =$
7. $100^3 =$
8. $3.6^3 =$
9. $6^4 =$
10. $(-2.4)^4 =$

Case II Variables Raised to Positive Integer Exponents

In the exponential expression a^n the base a can be a variable such as x, y, z, m, n, and b. Variables raised to a positive integer exponent are solved using the following step:

Step Multiply the base a by itself as many times as the number specified in the exponent. For example, x^4 implies that multiply x by itself 4 times, i.e., $x^4 = x \cdot x \cdot x \cdot x$.

Examples with Steps

The following examples show the step as to how variables raised to positive integer exponents are solved:

Example 3.1-16

$$\boxed{b^4} =$$

Solution:

Step $\boxed{b^4} = \boxed{\mathbf{b \cdot b \cdot b \cdot b}}$

Example 3.1-17

$$\boxed{w^5(xy)^3} =$$

Solution:

Step $\boxed{w^5(xy)^3} = \boxed{(w \cdot w \cdot w \cdot w \cdot w) \cdot (xy) \cdot (xy) \cdot (xy)} = \boxed{\mathbf{(w \cdot w \cdot w \cdot w \cdot w) \cdot (xy \cdot xy \cdot xy)}}$

Example 3.1-18

$$\boxed{a^6 b^4} =$$

Solution:

Step $\boxed{a^6 b^4} = \boxed{\mathbf{(a \cdot a \cdot a \cdot a \cdot a \cdot a) \cdot (b \cdot b \cdot b \cdot b)}}$

Example 3.1-19

$$\boxed{(a\,b)^4} =$$

Solution:

Step $\boxed{(a\,b)^4} = \boxed{(ab) \cdot (ab) \cdot (ab) \cdot (ab)} = \boxed{\mathbf{ab \cdot ab \cdot ab \cdot ab}}$

Example 3.1-20

$$\boxed{z^7 \cdot (ab)^3} =$$

Solution:

Step $\boxed{z^7 \cdot (ab)^3} = \boxed{(z \cdot z \cdot z \cdot z \cdot z \cdot z \cdot z) \cdot (ab) \cdot (ab) \cdot (ab)} = \boxed{\mathbf{(z \cdot z \cdot z \cdot z \cdot z \cdot z \cdot z) \cdot ab \cdot ab \cdot ab}}$

Additional Examples - Variables Raised to Positive Integer Exponents

The following examples further illustrate how to solve variables raised to positive integer exponents:

Example 3.1-21

$$\boxed{(cd)^3} = \boxed{(cd) \cdot (cd) \cdot (cd)} = \boxed{\mathbf{cd \cdot cd \cdot cd}}$$

Example 3.1-22

$\boxed{x^3 \cdot y^2 \cdot z^4} = \boxed{(x \cdot x \cdot x) \cdot (y \cdot y) \cdot (z \cdot z \cdot z \cdot z)}$

Example 3.1-23

$\boxed{a^4 \cdot b^2 \cdot (cd)^3} = \boxed{(a \cdot a \cdot a \cdot a) \cdot (b \cdot b) \cdot (cd) \cdot (cd) \cdot (cd)} = \boxed{\mathbf{(a \cdot a \cdot a \cdot a) \cdot (b \cdot b) \cdot (cd \cdot cd \cdot cd)}}$

Example 3.1-24

$\boxed{(xy)^4 \cdot (ab)^0 \cdot w^2} = \boxed{(xy)^4 \cdot 1 \cdot w^2} = \boxed{(xy)^4 \cdot w^2} = \boxed{(xy) \cdot (xy) \cdot (xy) \cdot (xy) \cdot (w \cdot w)} = \boxed{\mathbf{(xy \cdot xy \cdot xy \cdot xy) \cdot (w \cdot w)}}$

Example 3.1-25

$\boxed{z^0 \cdot a^4 \cdot (bxy)^3} = \boxed{1 \cdot a^4 \cdot (bxy)^3} = \boxed{a^4 \cdot (bxy)^3} = \boxed{(a \cdot a \cdot a \cdot a) \cdot (bxy) \cdot (bxy) \cdot (bxy)} = \boxed{\mathbf{(a \cdot a \cdot a \cdot a) \cdot (bxy \cdot bxy \cdot bxy)}}$

Example 3.1-26

$\boxed{a^3 \cdot \left(b^2 c^2\right) \cdot (xy)^2} = \boxed{(a \cdot a \cdot a) \cdot (b \cdot b \cdot c \cdot c) \cdot (xy) \cdot (xy)} = \boxed{\mathbf{(a \cdot a \cdot a) \cdot (b \cdot b) \cdot (c \cdot c) \cdot (xy \cdot xy)}}$

Example 3.1-27

$\boxed{\left(x^2 y^2 z^2\right)^0 \cdot (xy)^3 \cdot z^2} = \boxed{1 \cdot (xy)^3 \cdot z^2} = \boxed{(xy)^3 \cdot z^2} = \boxed{(xy) \cdot (xy) \cdot (xy) \cdot (z \cdot z)} = \boxed{\mathbf{(xy \cdot xy \cdot xy) \cdot (z \cdot z)}}$

Example 3.1-28

$\boxed{m^2 \cdot n^2 \cdot \left(a^2 b^2\right) \cdot z} = \boxed{(m \cdot m) \cdot (n \cdot n) \cdot (ab) \cdot (ab) \cdot z} = \boxed{\mathbf{(m \cdot m) \cdot (n \cdot n) \cdot (ab \cdot ab) \cdot z}}$

Example 3.1-29

$\boxed{\left(ab^2 c\right)^0 \cdot a^0 \cdot c^2 \cdot d^2} = \boxed{1 \cdot 1 \cdot c^2 \cdot d^2} = \boxed{c^2 d^2} = \boxed{\mathbf{(c \cdot c) \cdot (d \cdot d)}}$

Example 3.1-30

$\boxed{(ab)^2 \cdot (cd)^3 \cdot x^4} = \boxed{(ab) \cdot (ab) \cdot (cd) \cdot (cd) \cdot (cd) \cdot (x \cdot x \cdot x \cdot x)} = \boxed{\mathbf{(ab \cdot ab) \cdot (cd \cdot cd \cdot cd) \cdot (x \cdot x \cdot x \cdot x)}}$

Practice Problems - Variables Raised to Positive Integer Exponents

Section 3.1 Case II Practice Problems - Solve the following exponential expressions with variables raised to positive integer exponents:

1. $c^5 =$
2. $w^4 z^2 =$
3. $a^3 \cdot b^6 \cdot c^2 =$
4. $y^3 \cdot (zw)^2 =$
5. $(ab)^4 \cdot (xy)^2 =$
6. $(xyz)^5 =$
7. $a^3 b^2 =$
8. $z^4 \cdot w^3 \cdot (ab)^2 =$
9. $(xyzw)^4 \cdot b^3 =$
10. $a^3 \cdot b^2 \cdot (cd)^4 =$

3.2 Negative Integer Exponents

Negative integer exponents are defined as a^{-n} where a is referred to as the **base**, and n is the **integer exponent**. Again, note that the base a can be a real number or a variable. The integer exponent n can be a positive or a negative integer. In this section, real numbers raised to negative integer exponents (Case I) and variables raised to negative integer exponents (Case II) are addressed.

Case I Real Numbers Raised to Negative Integer Exponents

In general, real numbers raised to negative integer exponents are shown as:

$$a^{-n} = \frac{1}{a^{+n}} = \frac{1}{a^n} = \frac{1}{a \cdot a \cdot a \cdot a \ldots \cdot a} \qquad \textit{where } n \textit{ is a positive integer and } a \neq 0$$

For example,

$$5^{-4} = \frac{1}{5^4} = \frac{1}{5 \cdot 5 \cdot 5 \cdot 5} = \frac{1}{625}$$

Real numbers raised to a negative integer exponent are solved using the following steps:

Step 1 Change the negative integer exponent a^{-n} to a positive integer exponent of the form $\frac{1}{a^n}$. For example, change 3^{-4} to $\frac{1}{3^4}$.

Step 2 Multiply the base a in the denominator by itself as many times as the number specified in the exponent. For example, rewrite $\frac{1}{3^4}$ as $\frac{1}{3 \cdot 3 \cdot 3 \cdot 3}$.

Step 3 Multiply the real numbers in the denominator to obtain the answer, i.e., $\frac{1}{3 \cdot 3 \cdot 3 \cdot 3} = \frac{1}{81}$.

Examples with Steps

The following examples show the steps as to how real numbers raised to negative integer exponents are solved:

Example 3.2-1

$$\boxed{4^{-3}} =$$

Solution:

Step 1 $$\boxed{4^{-3}} = \boxed{\frac{1}{4^3}}$$

Step 2 $$\boxed{\frac{1}{4^3}} = \boxed{\frac{1}{4 \cdot 4 \cdot 4}}$$

Step 3 $$\boxed{\frac{1}{4 \cdot 4 \cdot 4}} = \boxed{\mathbf{\frac{1}{64}}}$$

Example 3.2-2

$$\boxed{3.2^{-4}} =$$

Solution:

Step 1 $\boxed{3.2^{-4}} = \boxed{\frac{1}{3.2^4}}$

Step 2 $\boxed{\frac{1}{3.2^4}} = \boxed{\frac{1}{(3.2)\cdot(3.2)\cdot(3.2)\cdot(3.2)}}$

Step 3 $\boxed{\frac{1}{(3.2)\cdot(3.2)\cdot(3.2)\cdot(3.2)}} = \boxed{\mathbf{\frac{1}{104.86}}}$

Example 3.2-3

$\boxed{(-8)^{-3}} =$

Solution:

Step 1 $\boxed{(-8)^{-3}} = \boxed{\frac{1}{(-8)^3}}$

Step 2 $\boxed{\frac{1}{(-8)^3}} = \boxed{\frac{1}{(-8)\cdot(-8)\cdot(-8)}}$

Step 3 $\boxed{\frac{1}{(-8)\cdot(-8)\cdot(-8)}} = \boxed{\frac{1}{-512}} = \boxed{\mathbf{-\frac{1}{512}}}$

Example 3.2-4

$\boxed{-(6)^{-5}} =$

Solution:

Step 1 $\boxed{-(6)^{-5}} = \boxed{-\frac{1}{(6)^5}} = \boxed{-\frac{1}{6^5}}$

Step 2 $\boxed{-\frac{1}{6^5}} = \boxed{-\frac{1}{6\cdot6\cdot6\cdot6\cdot6}}$

Step 3 $\boxed{-\frac{1}{6\cdot6\cdot6\cdot6\cdot6}} = \boxed{\mathbf{-\frac{1}{7776}}}$

Example 3.2-5

$\boxed{(-3.4)^{-6}} =$

Solution:

Step 1 $\boxed{(-3.4)^{-6}} = \boxed{\frac{1}{(-3.4)^6}}$

Step 2

$$\frac{1}{(-3.4)^6} = \frac{1}{(-3.4)\cdot(-3.4)\cdot(-3.4)\cdot(-3.4)\cdot(-3.4)\cdot(-3.4)}$$

Step 3

$$\frac{1}{(-3.4)\cdot(-3.4)\cdot(-3.4)\cdot(-3.4)\cdot(-3.4)\cdot(-3.4)} = \frac{1}{+1544.80} = \mathbf{\frac{1}{1544.80}}$$

Additional Examples - Real Numbers Raised to Negative Integer Exponents

The following examples further illustrate how to solve real numbers raised to negative integer exponents:

Example 3.2-6

$$2^{-3} = \frac{1}{2^3} = \frac{1}{2\cdot2\cdot2} = \mathbf{\frac{1}{8}}$$

Example 3.2-7

$$-(6)^{-4} = -\frac{1}{(6)^4} = -\frac{1}{6^4} = -\frac{1}{6\cdot6\cdot6\cdot6} = \mathbf{-\frac{1}{1296}}$$

Example 3.2-8

$$(-25)^{-1} = \frac{1}{(-25)^1} = \frac{1}{-25} = \mathbf{-\frac{1}{25}}$$

Example 3.2-9

$$(25)^{-3} = \frac{1}{(25)^3} = \frac{1}{25^3} = \frac{1}{25\cdot25\cdot25} = \mathbf{\frac{1}{15625}}$$

Example 3.2-10

$$(5.2)^{-4} = \frac{1}{(5.2)^4} = \frac{1}{5.2^4} = \frac{1}{(5.2)\cdot(5.2)\cdot(5.2)\cdot(5.2)} = \mathbf{\frac{1}{731.16}}$$

Example 3.2-11

$$0.32^{-5} = \frac{1}{0.32^5} = \frac{1}{(0.32)\cdot(0.32)\cdot(0.32)\cdot(0.32)\cdot(0.32)} = \mathbf{\frac{1}{0.00335}}$$

Example 3.2-12

$$234^{-1} = \frac{1}{234^1} = \mathbf{\frac{1}{234}}$$

Example 3.2-13

$$(-9)^{-4} = \frac{1}{(-9)^4} = \frac{1}{(-9)\cdot(-9)\cdot(-9)\cdot(-9)} = \frac{1}{+6561} = \mathbf{\frac{1}{6561}}$$

Example 3.2-14

$$\boxed{-4.5^{-3}} = \boxed{-\frac{1}{4.5^3}} = \boxed{-\frac{1}{(4.5)\cdot(4.5)\cdot(4.5)}} = \boxed{-\frac{1}{91.125}}$$

Example 3.2-15

$$\boxed{-(-4.5)^{-3}} = \boxed{-\frac{1}{(-4.5)^3}} = \boxed{-\frac{1}{(-4.5)\cdot(-4.5)\cdot(-4.5)}} = \boxed{-\frac{1}{-91.125}} = \boxed{+\frac{1}{91.125}} = \boxed{\frac{1}{91.125}}$$

Note 1: Any number or variable raised to the zero power is always equal to 1. For example, $55^0 = 1$, $(-15)^0 = 1$, $(5{,}689{,}763)^0 = 1$, $[(5x+2)-8]^0 = 1$, $\left[(axy)^3\right]^0 = 1$, $\left(\sqrt{x^3y^2z}\right)^0 = 1$, $(x+y+z)^0 = 1$, $\left(\frac{5y^3+2y^2+3}{3y}\right)^0 = 1$, $\left[\left(z^4-3z^2+2z+4\right)+3z\right]^0 = 1$.

Note 2: Zero raised to the zero power is not defined, i.e., 0^0 is undefined.

Note 3: Any number or variable divided by zero is not defined, i.e., $\frac{1}{0}$, $\frac{x}{0}$, $\frac{x^3y^2z}{0}$, $\frac{3x^3+2x^2+3}{0}$, $\frac{355}{0}$, $\frac{3\sqrt{5x^3}+4\sqrt{x}-6}{0}$, etc. are undefined.

Note 4: Zero divided by any number or variable is always equal to zero, i.e., $\frac{0}{1} = 0$, $\frac{0}{2{,}560} = 0$, $\frac{0}{\sqrt{10}} = 0$, $\frac{0}{x^2} = 0$, $\frac{0}{\sqrt[3]{a^2}} = 0$.

Practice Problems - Real Numbers Raised to Negative Integer Exponents

Section 3.2 Case I Practice Problems - Solve the following exponential expressions with real numbers raised to negative integer exponents:

1. $4^{-3} =$
2. $(-5)^{-4} =$
3. $0.25^{-3} =$
4. $12^{-5} =$
5. $-(3)^{-4} =$
6. $48^{-2} =$
7. $(-10)^{-3} =$
8. $3.2^{-1} =$
9. $6^{-3} =$
10. $(-4.5)^{-2} =$

Case II Variables Raised to Negative Integer Exponents

In the exponential expression a^{-n} the base a can be a variable such as p, q, r, s, k, and l. Variables raised to a negative integer exponent are solved using the following steps:

Step 1 Change the negative integer exponent a^{-n} to a positive integer exponent of the form $\frac{1}{a^n}$. For example, change p^{-5} to $\frac{1}{p^5}$.

Step 2 Multiply the base a in the denominator by itself as many times as the number specified in the exponent. For example, rewrite $\frac{1}{p^5}$ as $\frac{1}{p \cdot p \cdot p \cdot p \cdot p}$.

Examples with Steps

The following examples show the step as to how variables raised to negative integer exponents are solved:

Example 3.2-16

$$\boxed{b^{-3}} =$$

Solution:

Step 1

$$\boxed{b^{-3}} = \boxed{\frac{1}{b^3}}$$

Step 2

$$\boxed{\frac{1}{b^3}} = \boxed{\mathbf{\frac{1}{b \cdot b \cdot b}}}$$

Example 3.2-17

$$\boxed{w^{-4}(x\,y)^{-2}} =$$

Solution:

Step 1

$$\boxed{w^{-4}(x\,y)^{-2}} = \boxed{\frac{1}{w^4} \cdot \frac{1}{(x\,y)^2}} = \boxed{\frac{1 \cdot 1}{w^4 \cdot (x\,y)^2}} = \boxed{\frac{1}{w^4 (x\,y)^2}}$$

Step 2

$$\boxed{\frac{1}{w^4 (x\,y)^2}} = \boxed{\mathbf{\frac{1}{(w \cdot w \cdot w \cdot w) \cdot (xy \cdot xy)}}}$$

Example 3.2-18

$$\boxed{c^{-5} d^{-4}} =$$

Solution:

Step 1

$$\boxed{c^{-5} d^{-4}} = \boxed{\frac{1}{c^5} \cdot \frac{1}{d^4}} = \boxed{\frac{1 \cdot 1}{c^5 \cdot d^4}} = \boxed{\frac{1}{c^5 d^4}}$$

Step 2

$$\boxed{\frac{1}{c^5 d^4}} = \boxed{\mathbf{\frac{1}{(c \cdot c \cdot c \cdot c \cdot c) \cdot (d \cdot d \cdot d \cdot d)}}}$$

Example 3.2-19

$$\boxed{(a\,b)^{-4}x^{-2}} =$$

Solution:

Step 1

$$\boxed{(a\,b)^{-4}x^{-2}} = \boxed{\frac{1}{(a\,b)^4}\cdot\frac{1}{x^2}} = \boxed{\frac{1\cdot 1}{(a\,b)^4\cdot x^2}} = \boxed{\frac{1}{(a\,b)^4 x^2}}$$

Step 2

$$\boxed{\frac{1}{(a\,b)^4 x^2}} = \boxed{\frac{\mathbf{1}}{(\boldsymbol{ab\cdot ab\cdot ab\cdot ab})\cdot(\boldsymbol{x\cdot x})}}$$

Example 3.2-20

$$\boxed{z^{-5}\cdot(a\,b)^{-3}w^{-1}} =$$

Solution:

Step 1

$$\boxed{z^{-5}\cdot(a\,b)^{-3}\cdot w^{-1}} = \boxed{\frac{1}{z^5}\cdot\frac{1}{(ab)^3}\cdot\frac{1}{w^1}} = \boxed{\frac{1\cdot 1\cdot 1}{z^5\cdot(a\,b)^3\cdot w^1}} = \boxed{\frac{1}{z^5(a\,b)^3 w}}$$

Step 2

$$\boxed{\frac{1}{z^5(a\,b)^3 w}} = \boxed{\frac{\mathbf{1}}{(\boldsymbol{z\cdot z\cdot z\cdot z\cdot z})\cdot(\boldsymbol{ab\cdot ab\cdot ab})\cdot \boldsymbol{w}}}$$

Additional Examples - Variables Raised to Negative Integer Exponents

The following examples further illustrate how to solve variables raised to negative integer exponents:

Example 3.2-21

$$\boxed{(b\,c\,d)^{-3}} = \boxed{\frac{1}{(b\,c\,d)^3}} = \boxed{\frac{\mathbf{1}}{\boldsymbol{bcd\cdot bcd\cdot bcd}}}$$

Example 3.2-22

$$\boxed{x^{-4}\cdot y^{-3}\cdot z^{-4}} = \boxed{\frac{1}{x^4}\cdot\frac{1}{y^3}\cdot\frac{1}{z^4}} = \boxed{\frac{1\cdot 1\cdot 1}{x^4\cdot y^3\cdot z^4}} = \boxed{\frac{1}{x^4y^3z^4}} = \boxed{\frac{\mathbf{1}}{(\boldsymbol{x\cdot x\cdot x\cdot x})\cdot(\boldsymbol{y\cdot y\cdot y})\cdot(\boldsymbol{z\cdot z\cdot z\cdot z})}}$$

Example 3.2-23

$$\boxed{a^{-3}\cdot b^{-2}\cdot(c\,d)^{-2}} = \boxed{\frac{1}{a^3}\cdot\frac{1}{b^2}\cdot\frac{1}{(c\,d)^2}} = \boxed{\frac{1\cdot 1\cdot 1}{a^3\cdot b^2\cdot(c\,d)^2}} = \boxed{\frac{1}{a^3b^2(c\,d)^2}} = \boxed{\frac{\mathbf{1}}{(\boldsymbol{a\cdot a\cdot a})\cdot(\boldsymbol{b\cdot b})\cdot(\boldsymbol{cd\cdot cd})}}$$

Example 3.2-24

$$\boxed{(x\,y)^{-3}\cdot(a\,b)^0\cdot w^{-3}\cdot z^{-2}} = \boxed{(x\,y)^{-3}\cdot 1\cdot w^{-3}\cdot z^{-2}} = \boxed{(x\,y)^{-3}\cdot w^{-3}\cdot z^{-2}} = \boxed{\frac{1}{(x\,y)^3}\cdot\frac{1}{w^3}\cdot\frac{1}{z^2}} = \boxed{\frac{1\cdot 1\cdot 1}{(x\,y)^3\cdot w^3\cdot z^2}}$$

$$= \boxed{\frac{1}{(x\,y)^3w^3z^2}} = \boxed{\frac{\mathbf{1}}{(\boldsymbol{xy\cdot xy\cdot xy})\cdot(\boldsymbol{w\cdot w\cdot w})\cdot(\boldsymbol{z\cdot z})}}$$

Example 3.2-25

$$\boxed{z^{-3}\cdot a^{-3}\cdot(xyz)^{-2}} = \boxed{\frac{1}{z^3}\cdot\frac{1}{a^3}\cdot\frac{1}{(xyz)^2}} = \boxed{\frac{1\cdot1\cdot1}{z^3\cdot a^3\cdot(xyz)^2}} = \boxed{\frac{1}{z^3a^3(xyz)^2}} = \boxed{\frac{\mathbf{1}}{(z\cdot z\cdot z)\cdot(a\cdot a\cdot a)\cdot(xyz\cdot xyz)}}$$

Example 3.2-26

$$\boxed{(bcd)^0\cdot(ab)^{-2}\cdot x^{-1}\cdot y^{-3}} = \boxed{1\cdot(ab)^{-2}\cdot x^{-1}\cdot y^{-3}} = \boxed{(ab)^{-2}\cdot x^{-1}\cdot y^{-3}} = \boxed{\frac{1}{(ab)^2}\cdot\frac{1}{x^1}\cdot\frac{1}{y^3}} = \boxed{\frac{1\cdot1\cdot1}{(ab)^2\cdot x^1\cdot y^3}}$$

$$= \boxed{\frac{1}{(ab)^2xy^3}} = \boxed{\frac{\mathbf{1}}{(ab\cdot ab)\cdot x\cdot(y\cdot y\cdot y)}}$$

Example 3.2-27

$$\boxed{2a\cdot(xy)^{-3}\cdot z^{-2}\cdot w^0} = \boxed{2a\cdot(xy)^{-3}\cdot z^{-2}\cdot1} = \boxed{2a\cdot(xy)^{-3}\cdot z^{-2}} = \boxed{\frac{2a}{1}\cdot\frac{1}{(xy)^3}\cdot\frac{1}{z^2}} = \boxed{\frac{2a\cdot1\cdot1}{1\cdot(xy)^3\cdot z^2}}$$

$$= \boxed{\frac{2a}{(xy)^3z^2}} = \boxed{\frac{\mathbf{2a}}{(xy\cdot xy\cdot xy)\cdot(z\cdot z)}}$$

Example 3.2-28

$$\boxed{12\cdot a^{-2}\cdot(cd)^{-3}\cdot k^{-1}} = \boxed{\frac{12}{1}\cdot\frac{1}{a^2}\cdot\frac{1}{(cd)^3}\cdot\frac{1}{k^1}} = \boxed{\frac{12\cdot1\cdot1\cdot1}{1\cdot a^2\cdot(cd)^3\cdot k}} = \boxed{\frac{12}{a^2(cd)^3k}} = \boxed{\frac{\mathbf{12}}{(a\cdot a)\cdot(cd\cdot cd\cdot cd)\cdot k}}$$

Example 3.2-29

$$\boxed{5a\cdot x^{-3}\cdot y^{-2}\cdot z^{-2}} = \boxed{\frac{5a}{1}\cdot\frac{1}{x^3}\cdot\frac{1}{y^2}\cdot\frac{1}{z^2}} = \boxed{\frac{5a\cdot1\cdot1\cdot1}{1\cdot x^3\cdot y^2\cdot z^2}} = \boxed{\frac{5a}{x^3y^2z^2}} = \boxed{\frac{\mathbf{5a}}{(x\cdot x\cdot x)\cdot(y\cdot y)\cdot(z\cdot z)}}$$

Example 3.2-30

$$\boxed{8\cdot(xyz)^{-3}\cdot(xy)^0\cdot a^{-1}} = \boxed{8\cdot(xyz)^{-3}\cdot1\cdot a^{-1}} = \boxed{8\cdot(xyz)^{-3}\cdot a^{-1}} = \boxed{\frac{8}{(xyz)^3}\cdot\frac{1}{a^1}} = \boxed{\frac{8\cdot1}{1\cdot(xyz)^3\cdot a}}$$

$$= \boxed{\frac{8}{(xyz)^3a}} = \boxed{\frac{\mathbf{8}}{(xyz\cdot xyz\cdot xyz)\cdot a}}$$

Practice Problems - Variables Raised to Negative Integer Exponents

Section 3.2 Case II Practice Problems - Solve the following exponential expressions with variables raised to negative integer exponents:

1. $c^{-6} =$
2. $a^{-1}w^{-3} =$
3. $a^{-3}\cdot b^{-4}\cdot c^0 =$
4. $y^{-3}\cdot(zw)^{-4} =$
5. $(ab)^{-3}\cdot(xy)^{-1}\cdot z^{-2} =$
6. $c^{-2}\cdot(xyz)^{-4} =$
7. $a^{-2}b^{-1} =$
8. $z^{-4}\cdot w^{-2}\cdot(abc)^2 =$
9. $(xyzw)^{-1}\cdot b^{-2}\cdot(ab)^0 =$
10. $(ad)^3\cdot b^{-2}\cdot(xy)^{-4} =$

3.3 Operations with Positive Integer Exponents

To multiply, divide, add, and subtract integer exponents, we need to know the following laws of exponents (shown in Table 3.3-1). These laws are used to simplify the work in solving problems with exponential expressions and should be memorized.

Table 3.3-1: Exponent Laws 1 through 7 (Positive Integer Exponents)

Law	Formula	Description
I. **Multiplication**	$a^m \cdot a^n = a^{m+n}$	When multiplying positive exponential terms, if bases a are the same, add the exponents m and n.
II. **Power of a Power**	$\left(a^m\right)^n = a^{mn}$	When raising an exponential term to a power, multiply the powers (exponents) m and n.
III. **Power of a Product**	$(a \cdot b)^m = a^m \cdot b^m$	When raising a product to a power, raise each factor a and b to the power m.
IV. **Power of a Fraction**	$\left(\frac{a}{b}\right)^m = \frac{a^m}{b^m}$	When raising a fraction to a power, raise the numerator and the denominator to the power m.
V. **Division**	$\frac{a^m}{a^n} = a^m \cdot a^{-n} = a^{m-n}$	When dividing exponential terms, if the bases a are the same, subtract exponents m and n.
VI. **Negative Power**	$a^{-n} = \frac{1}{a^n}$	A non-zero base a raised to the $-n$ power equals 1 divided by the base a to the n power.
VII. **Zero Power**	$a^0 = 1$	A non-zero base a raised to the zero power is always equal to 1.

In this section students learn how to multiply (Case I), divide (Case II), and add or subtract (Case III) positive integer exponents by one another.

Case I Multiplying Positive Integer Exponents

Positive integer exponents are multiplied by each other using the exponent laws I through III shown in Table 3.3-2.

Table 3.3-2: Exponent Laws 1 through 3 (Positive Integer Exponents)

Law	Formula	Description
I. **Multiplication**	$a^m \cdot a^n = a^{m+n}$	When multiplying positive exponential terms, if bases a are the same, add the exponents m and n.
II. **Power of a Power**	$\left(a^m\right)^n = a^{mn}$	When raising an exponential term to a power, multiply the powers (exponents) m and n.
III. **Power of a Product**	$(a \cdot b)^m = a^m \cdot b^m$	When raising a product to a power, raise each factor a and b to the power m.

Multiplication of expressions by positive integer exponents is divided to two cases: Case I a addressing simple cases and Case I b addressing more difficult cases.

Case I a Multiplying Positive Integer Exponents (Simple Cases)

Positive integer exponents are multiplied by each other using the following steps:

Step 1 Group the exponential terms with similar bases.

Step 2 Apply the Multiplication Law (Law I) from Table 3.3-2 and simplify the exponential expressions by adding the exponents with similar bases.

Examples with Steps

The following examples show the steps as to how positive integer exponents are multiplied by one another:

Example 3.3-1

$$\boxed{\left(x^3 y^2\right)\cdot\left(x^2 y\right)\cdot y^3} =$$

Solution:

Step 1 $$\boxed{\left(x^3 y^2\right)\cdot\left(x^2 y\right)\cdot y^3} = \boxed{\left(x^3 x^2\right)\cdot\left(y^3 y^2 y\right)} = \boxed{\left(x^3 x^2\right)\cdot\left(y^3 y^2 y^1\right)}$$

Step 2 $$\boxed{\left(x^3 x^2\right)\cdot\left(y^3 y^2 y^1\right)} = \boxed{\left(x^{3+2}\right)\cdot\left(y^{3+2+1}\right)} = \boxed{x^5\cdot y^6} = \boxed{\mathbf{x^5 y^6}}$$

Example 3.3-2

$$\boxed{\left(-\frac{1}{5}a^2\right)\cdot(10ab)\cdot\left(-\frac{1}{4}ab^2\right)} =$$

Solution:

Step 1 $$\boxed{\left(-\frac{1}{5}a^2\right)\cdot(10ab)\cdot\left(-\frac{1}{4}ab^2\right)} = \boxed{\left(-\frac{1}{5}\times-\frac{1}{4}\times 10\right)\cdot\left(a^2 aa\right)\cdot\left(b^2 b\right)} = \boxed{\left(\frac{10}{20}\right)\cdot\left(a^2 a^1 a^1\right)\cdot\left(b^2 b^1\right)}$$

Step 2 $$\boxed{\left(\frac{10}{20}\right)\cdot\left(a^2 a^1 a^1\right)\cdot\left(b^2 b^1\right)} = \boxed{\frac{\overset{1}{\cancel{10}}}{\underset{2}{\cancel{20}}}\cdot\left(a^{2+1+1}\right)\cdot\left(b^{2+1}\right)} = \boxed{\frac{1}{2}\cdot a^4\cdot b^3} = \boxed{\mathbf{\frac{1}{2}a^4 b^3}}$$

Example 3.3-3

$$\boxed{\frac{1}{2}x\cdot w^2\cdot z^2\cdot x^0\cdot w\cdot z\cdot\left(\frac{3}{4}z^3\right)^0} =$$

Solution:

Step 1 $$\boxed{\frac{1}{2}x\cdot w^2\cdot z^2\cdot x^0\cdot w\cdot z\cdot\left(\frac{3}{4}z^3\right)^0} = \boxed{\frac{1}{2}x\cdot w^2\cdot z^2\cdot 1\cdot w\cdot z\cdot 1} = \boxed{\frac{1}{2}x\cdot w^2\cdot z^2\cdot w\cdot z}$$

$$= \boxed{\frac{1}{2}x\cdot\left(w^2\cdot w\right)\cdot\left(z^2\cdot z\right)} = \boxed{\frac{1}{2}x\cdot\left(w^2\cdot w^1\right)\cdot\left(z^2\cdot z^1\right)}$$

Step 2 $\boxed{\frac{1}{2}x\cdot\left(w^2\cdot w^1\right)\cdot\left(z^2\cdot z^1\right)}=\boxed{\frac{1}{2}x\cdot\left(w^{2+1}\right)\cdot\left(z^{2+1}\right)}=\boxed{\frac{1}{2}x\cdot w^3\cdot z^3}=\boxed{\mathbf{\frac{1}{2}xw^3z^3}}$

Note - Non zero numbers or variables raised to the zero power are always equal to 1, i.e., $10^0=1$, $(23456)^0=1$, $a^0=1$ *for* $a\neq 0$, $(a\cdot b)^0=1$ *for* $a\cdot b\neq 0$, $(x\cdot y\cdot z)^0=1$ *for* $x\cdot y\cdot z\neq 0$, etc.

Example 3.3-4

$\boxed{\left(e^3e^5e\right)\cdot\left(-\frac{4}{32}e^2\right)}=$

Solution:

Step 1 $\boxed{\left(e^3e^5e\right)\cdot\left(-\frac{4}{32}e^2\right)}=\boxed{-\frac{4}{32}\left(e^5e^3e^2e\right)}=\boxed{-\frac{\overset{1}{\cancel{4}}}{\underset{8}{\cancel{32}}}\left(e^5e^3e^2e^1\right)}$

Step 2 $\boxed{-\frac{\overset{1}{\cancel{4}}}{\underset{8}{\cancel{32}}}\left(e^5e^3e^2e^1\right)}=\boxed{-\frac{1}{8}\left(e^{5+3+2+1}\right)}=\boxed{\mathbf{-\frac{1}{8}e^{11}}}$

Example 3.3-5

$\boxed{\left(r^3s^3s\right)\cdot\left(r^2s^2s^0\right)\cdot rs^4}=$

Solution:

Step 1 $\boxed{\left(r^3s^3s\right)\cdot\left(r^2s^2s^0\right)\cdot rs^4}=\boxed{\left(r^3r^2r\right)\cdot\left(s^4s^3s^2ss^0\right)}=\boxed{\left(r^3r^2r^1\right)\cdot\left(s^4s^3s^2s^1s^0\right)}$

Step 2 $\boxed{\left(r^3r^2r^1\right)\cdot\left(s^4s^3s^2s^1s^0\right)}=\boxed{\left(r^{3+2+1}\right)\cdot\left(s^{4+3+2+1+0}\right)}=\boxed{r^6\cdot s^{10}}=\boxed{\mathbf{r^6s^{10}}}$

Additional Examples - Multiplying Positive Integer Exponents (Simple Cases)

The following examples further illustrate how to multiply positive exponential terms by one another:

Example 3.3-6

$\boxed{2^3\cdot 2^5}=\boxed{2^{3+5}}=\boxed{2^8}=\boxed{\mathbf{256}}$

Example 3.3-7

$\boxed{3^0\cdot 3\cdot 3^2\cdot 3^3}=\boxed{3^0\cdot 3^1\cdot 3^2\cdot 3^3}=\boxed{3^{0+1+2+3}}=\boxed{3^6}=\boxed{\mathbf{729}}$

Example 3.3-8

$\boxed{a\cdot a^2\cdot a^5}=\boxed{a^1\cdot a^2\cdot a^5}=\boxed{a^{1+2+5}}=\boxed{\mathbf{a^8}}$

Example 3.3-9

$\boxed{w^2\cdot w^3\cdot w^5}=\boxed{w^{2+3+5}}=\boxed{\mathbf{w^{10}}}$

Example 3.3-10

$$w\cdot y\cdot z\cdot w\cdot y\cdot z\cdot y\cdot z\cdot z^2 = (w\cdot w)\cdot(y\cdot y\cdot y)\cdot\left(z\cdot z\cdot z\cdot z^2\right) = \left(w^1\cdot w^1\right)\cdot\left(y^1\cdot y^1\cdot y^1\right)\cdot\left(z^1\cdot z^1\cdot z^1\cdot z^2\right)$$

$$= \left(w^{1+1}\right)\cdot\left(y^{1+1+1}\right)\cdot\left(z^{1+1+1+2}\right) = \mathbf{w^2y^3z^5}$$

Example 3.3-11

$$x^2\cdot y^2\cdot z^3\cdot x^2\cdot y^2\cdot z^4\cdot x = \left(x^2\cdot x^2\cdot x\right)\cdot\left(y^2\cdot y^2\right)\cdot\left(z^3\cdot z^4\right) = \left(x^{2+2+1}\right)\cdot\left(y^{2+2}\right)\cdot\left(z^{3+4}\right) = \mathbf{x^5y^4z^7}$$

Example 3.3-12

$$(2-3)^2\cdot\left(5x^3y^2\right)\cdot(-2xy) = (-1)^2\cdot(5\times-2)\cdot\left(x^3x^1\right)\cdot\left(y^2y^1\right) = (+1)\cdot(-10)\cdot\left(x^{3+1}\right)\cdot\left(y^{2+1}\right) = \mathbf{-10x^4y^3}$$

Example 3.3-13

$$(-3-2)^2\left(-4k^2p^2\right)\cdot(-5kp) = \left[(-5)^2\cdot(-4\times-5)\right]\cdot\left(k^2k\right)\cdot\left(p^2p\right) = [(+25)\cdot(+20)]\cdot\left(k^2k^1\right)\cdot\left(p^2p^1\right)$$

$$= 500\cdot\left(k^{2+1}\right)\cdot\left(p^{2+1}\right) = \mathbf{500k^3p^3}$$

Example 3.3-14

$$(3-5)^2\cdot\left(x^2y^2z^4\right)\cdot\left(-xyz^2\right) = (-2)^2\cdot\left(-x^2x\right)\cdot\left(y^2y\right)\cdot\left(z^4z^2\right) = (+4)\cdot\left(-x^2x^1\right)\cdot\left(y^2y^1\right)\cdot\left(z^4z^2\right)$$

$$= 4\cdot\left(-x^{2+1}\right)\cdot\left(y^{2+1}\right)\cdot\left(z^{4+2}\right) = \mathbf{-4x^3y^3z^6}$$

Example 3.3-15

$$3x\cdot4y\cdot-2z\cdot x^2y^3 = -(3\cdot4\cdot2)\cdot\left(x\cdot x^2\right)\cdot\left(y\cdot y^3\right)\cdot z = (-24)\cdot\left(x^1\cdot x^2\right)\cdot\left(y^1\cdot y^3\right)\cdot z = -24\cdot\left(x^{1+2}\right)\cdot\left(y^{1+3}\right)\cdot z$$

$$= \mathbf{-24x^3y^4z}$$

Example 3.3-16

$$(-5-1)^3\cdot\left(a^2b^2c^2\right)\cdot\left(a^3b\right)\cdot a = (-6)^3\cdot\left(a^2a^3a\right)\cdot\left(b^2b\right)\cdot c^2 = (-6\cdot-6\cdot-6)\cdot\left(a^2a^3a^1\right)\cdot\left(b^2b^1\right)\cdot c^2$$

$$= -216\cdot\left(a^{2+3+1}\right)\cdot\left(b^{2+1}\right)\cdot c^2 = \mathbf{-216a^6b^3c^2}$$

Example 3.3-17

$$\left(5h^5k^5\right)\cdot\left(-3h^0k^2\right) = (5\times-3)\cdot\left(h^5h^0\right)\cdot\left(k^5k^2\right) = (-15)\cdot\left(h^{5+0}\right)\cdot\left(k^{5+2}\right) = \mathbf{-15h^5k^7}$$

Example 3.3-18

$$-(2-4)^2\cdot\left(2m^5\right)\cdot\left(-3m^2\right) = -(-2)^2\cdot(2\times-3)\cdot\left(m^5\cdot m^2\right) = -[(-2\cdot-2)\cdot(-6)]\cdot\left(m^{5+2}\right) = -[(+4)\cdot(-6)]\cdot m^7$$

$$= \boxed{-(-24)\cdot m^7} = \boxed{+24\cdot m^7} = \boxed{\mathbf{24m^7}}$$

Example 3.3-19

$$\boxed{\left(c^2d^2e^2\right)\cdot\left(-3d^3\right)\cdot\left(c^2e^2\right)} = \boxed{-3\left(c^2c^2\right)\cdot\left(d^2d^3\right)\cdot\left(e^2e^2\right)} = \boxed{-3\left(c^{2+2}\right)\cdot\left(d^{2+3}\right)\cdot\left(e^{2+2}\right)} = \boxed{\mathbf{-3\,c^4d^5e^4}}$$

Example 3.3-20

$$\boxed{\frac{1}{2}a^2\cdot b\cdot a^3\cdot a} = \boxed{\frac{1}{2}\left(a^2a^3a\right)\cdot b} = \boxed{\frac{1}{2}\left(a^2a^3a^1\right)\cdot b} = \boxed{\frac{1}{2}\left(a^{2+3+1}\right)\cdot b} = \boxed{\frac{1}{2}\,a^6b} = \boxed{\mathbf{\frac{a^6b}{2}}}$$

Example 3.3-21

$$\boxed{\left(-x^2y^2\right)\cdot\left(x^2y^2\right)\cdot 3x} = \boxed{-3\left(xx^2x^2\right)\cdot\left(y^2y^2\right)} = \boxed{-3\left(x^1x^2x^2\right)\cdot\left(y^2y^2\right)} = \boxed{-3\left(x^{1+2+2}\right)\cdot\left(y^{2+2}\right)} = \boxed{\mathbf{-3\,x^5y^4}}$$

Example 3.3-22

$$\boxed{\left(\frac{1}{4}r^2s^2\right)\cdot\left(-4r^3s\right)} = \boxed{\frac{-4}{4}\left(r^2r^3\right)\cdot\left(s^2s\right)} = \boxed{-\frac{\overset{1}{\cancel{4}}}{\underset{1}{\cancel{4}}}\left(r^2r^3\right)\cdot\left(s^2s^1\right)} = \boxed{-\frac{1}{1}\left(r^{2+3}\right)\cdot\left(s^{2+1}\right)} = \boxed{\mathbf{-r^5s^3}}$$

Example 3.3-23

$$\boxed{\left(\frac{2}{7}y^3z^3\right)\cdot\left(9y\,z^2\right)\cdot\left(-2z^2z\right)} = \boxed{\left(-\frac{2}{7}\times 2\times 9\right)\cdot\left(y^3y\right)\cdot\left(z^3z^2z^2z\right)} = \boxed{\left(-\frac{36}{7}\right)\cdot\left(y^3y^1\right)\cdot\left(z^3z^2z^2z^1\right)}$$

$$= \boxed{-\frac{36}{7}\left(y^{3+1}\right)\cdot\left(z^{3+2+2+1}\right)} = \boxed{-\frac{36}{7}y^4z^8} = \boxed{\mathbf{-5\frac{1}{7}\left(y^4z^8\right)}}$$

Example 3.3-24

$$\boxed{\left(-2m^3n^3\right)\cdot\left(n^2m\right)\cdot\left(-3m^2n\right)} = \boxed{(-2\times -3)\cdot\left(m^3m^2m\right)\cdot\left(n^3n^2n\right)} = \boxed{(+6)\cdot\left(m^3m^2m^1\right)\cdot\left(n^3n^2n^1\right)}$$

$$= \boxed{6\left(m^{3+2+1}\right)\cdot\left(n^{3+2+1}\right)} = \boxed{6\,m^6\cdot n^6} = \boxed{\mathbf{6\,m^6n^6}}$$

Example 3.3-25

$$\boxed{\left(x^2y^2z^3\right)\cdot(-2\,x\,y)\cdot\left(-\frac{1}{4}x\,y\,z^4\right)} = \boxed{\left(-2\times -\frac{1}{4}\right)\cdot\left(x^2x\,x\right)\cdot\left(y^2y\,y\right)\cdot\left(z^4z^3\right)} = \boxed{\left(+\frac{2}{4}\right)\cdot\left(x^2x^1x^1\right)\cdot\left(y^2y^1y^1\right)\cdot\left(z^4z^3\right)}$$

$$= \boxed{\frac{\overset{1}{\cancel{2}}}{\underset{2}{\cancel{4}}}\cdot\left(x^{2+1+1}\right)\cdot\left(y^{2+1+1}\right)\cdot\left(z^{4+3}\right)} = \boxed{\frac{1}{2}\cdot x^4\cdot y^4\cdot z^7} = \boxed{\mathbf{\frac{1}{2}x^4y^4z^7}}$$

Practice Problems - Multiplying Positive Integer Exponents (Simple Cases)

Section 3.3 Case I a Practice Problems - Multiply the following positive integer exponents:

1. $x^2 \cdot x^3 \cdot x =$

2. $2 \cdot a^2 \cdot b^0 \cdot a^3 \cdot b^2 =$

3. $\frac{4}{-6} a^2 b^3 a b^4 b^5 =$

4. $2^3 \cdot 2^2 \cdot x^{2a} \cdot x^{3a} \cdot x^a =$

5. $\left(x \cdot y^2 \cdot z^3\right)^0 \cdot w^2 z^3 z w^4 z^2 =$

6. $2^0 \cdot 4^2 \cdot 4^2 \cdot 2^2 \cdot 4^1 =$

7. $\left(x^2 x^3\right) \cdot \left(\frac{2}{3} x y^2\right) \cdot \left(-2x^2 y\right) =$

8. $\left(p^3 \cdot q^2 \cdot r\right) \cdot \left(p \cdot q^2 \cdot r^3\right) =$

9. $\frac{-2}{-8} \cdot r^2 \cdot s \cdot 2^4 \cdot r \cdot s^3 =$

10. $-2 \cdot k^2 \cdot l \cdot \frac{3}{-4} \cdot k \cdot l^2 \cdot k^3 =$

Case I b Multiplying Positive Integer Exponents (More Difficult Cases)

Positive integer exponents are multiplied by each other using the following steps:

Step 1 Apply the Power of a Power Law (Law II) and/or the Power of a Product Law (Law III) from Table 3.3-2.

Step 2 Apply the Multiplication Law (Law I) from Table 3.3-2 and simplify the exponential expressions by adding the exponents with similar bases.

Examples with Steps

The following examples show the steps as to how positive integer exponents are multiplied by one another:

Example 3.3-26

$$\left(y^2\right)^3 \cdot \left(z^3\right)^2 =$$

Solution:

Step 1 $\left(y^2\right)^3 \cdot \left(z^3\right)^2 = \left(y^{2\times3}\right) \cdot \left(z^{3\times2}\right) = y^6 \cdot z^6 = \mathbf{y^6 z^6}$

Step 2 *Not Applicable*

Example 3.3-27

$$(-2)^3 \cdot \left(3 \cdot a^2\right)^3 =$$

Solution:

Step 1 $(-2)^3 \cdot \left(3 \cdot a^2\right)^3 = \left(-2^{1\times3}\right) \cdot \left(3^{1\times3} \cdot a^{2\times3}\right) = \left(-2^3\right) \cdot \left(3^3 \cdot a^6\right)$

Step 2 $\left(-2^3\right) \cdot \left(3^3 \cdot a^6\right) = (-2 \cdot -2 \cdot -2) \cdot \left(27 \cdot a^6\right) = (-8 \cdot 27) \cdot a^6 = \mathbf{-216\, a^6}$

Example 3.3-28

$$\left(-2^2\right)^4 \cdot \left(x^a \cdot x^{3a}\right)^2 =$$

Solution:

Step 1 $\left(-2^2\right)^4 \cdot \left(x^a \cdot x^{3a}\right)^2 = \left(+2^{2\times4}\right) \cdot \left(x^{a\times2} \cdot x^{3a\times2}\right) = 2^8 \cdot \left(x^{2a} \cdot x^{6a}\right)$

Step 2 $2^8 \cdot \left(x^{2a} \cdot x^{6a}\right) = 256 \cdot \left(x^{2a} \cdot x^{6a}\right) = 256 \cdot \left(x^{2a+6a}\right) = \mathbf{256\, x^{8a}}$

Example 3.3-29

$$\left(3 \cdot b^2 \cdot c^3\right)^4 \cdot \left(b^2\right)^2 =$$

Solution:

Step 1 $\boxed{\left(3\cdot b^{2}\cdot c^{3}\right)^{4}\cdot\left(b^{2}\right)^{2}}=\boxed{\left(3^{1\times 4}\cdot b^{2\times 4}\cdot c^{3\times 4}\right)\cdot\left(b^{2\times 2}\right)}=\boxed{\left(3^{4}\cdot b^{8}\cdot c^{12}\right)\cdot b^{4}}$

Step 2 $\boxed{\left(3^{4}\cdot b^{8}\cdot c^{12}\right)\cdot b^{4}}=\boxed{81\cdot b^{8}\cdot c^{12}\cdot b^{4}}=\boxed{81\cdot\left(b^{8}\cdot b^{4}\right)\cdot c^{12}}=\boxed{81\cdot\left(b^{8+4}\right)\cdot c^{12}}$

$=\boxed{81\cdot b^{12}\cdot c^{12}}=\boxed{\mathbf{81\,b^{12}c^{12}}}$

Example 3.3-30

$\boxed{\left[\left(2\cdot x^{2}\cdot y^{3}\right)^{2}\cdot x^{3}\right]^{2}}=$

Solution:

Step 1 $\boxed{\left[\left(2\cdot x^{2}\cdot y^{3}\right)^{2}\cdot x^{3}\right]^{2}}=\boxed{\left[\left(2^{1\times 2}\cdot x^{2\times 2}\cdot y^{3\times 2}\right)\cdot x^{3}\right]^{2}}=\boxed{\left[\left(2^{2}\cdot x^{4}\cdot y^{6}\right)\cdot x^{3}\right]^{2}}$

$=\boxed{\left(2^{2\times 2}\cdot x^{4\times 2}\cdot y^{6\times 2}\right)\cdot x^{3\times 2}}=\boxed{\left(2^{4}\cdot x^{8}\cdot y^{12}\right)\cdot x^{6}}$

Step 2 $\boxed{\left(2^{4}\cdot x^{8}\cdot y^{12}\right)\cdot x^{6}}=\boxed{16\cdot x^{8}\cdot y^{12}\cdot x^{6}}=\boxed{16\cdot\left(x^{8}\cdot x^{6}\right)\cdot y^{12}}=\boxed{16\cdot\left(x^{8+6}\right)\cdot y^{12}}$

$=\boxed{\mathbf{16\,x^{14}y^{12}}}$

Additional Examples - Multiplying Positive Integer Exponents (More Difficult Cases)

The following examples further illustrate how to multiply positive exponential terms by one another:

Example 3.3-31

$\boxed{\left(y^{2}\cdot y^{3}\right)^{2}\cdot y^{0}}=\boxed{\left(y^{2\times 2}\cdot y^{3\times 2}\right)\cdot 1}=\boxed{y^{4}\cdot y^{6}}=\boxed{y^{4+6}}=\boxed{\mathbf{y^{10}}}$

Example 3.3-32

$\boxed{\left(2^{3}\cdot 2^{2}\right)^{2}\cdot\left(a^{5}\right)^{2}}=\boxed{\left(2^{3\times 2}\cdot 2^{2\times 2}\right)\cdot\left(a^{5\times 2}\right)}=\boxed{\left(2^{6}\cdot 2^{4}\right)\cdot a^{10}}=\boxed{\left(2^{6+4}\right)\cdot a^{10}}=\boxed{2^{10}\cdot a^{10}}=\boxed{\mathbf{1024\,a^{10}}}$

Example 3.3-33

$\boxed{\left(-3\times 5^{2a}\right)^{3}\cdot\left(2\times 5^{3a}\right)^{4}}=\boxed{\left(-3^{1\times 3}\times 5^{2a\times 3}\right)\cdot\left(2^{1\times 4}\times 5^{3a\times 4}\right)}=\boxed{\left(-3^{3}\times 5^{6a}\right)\cdot\left(2^{4}\times 5^{12a}\right)}=\boxed{\left(-27\times 5^{6a}\right)\cdot\left(16\times 5^{12a}\right)}$

$=\boxed{-(27\cdot 16)\cdot\left(5^{6a}\cdot 5^{12a}\right)}=\boxed{-432\cdot\left(5^{6a+12a}\right)}=\boxed{\mathbf{-432\left(5^{18a}\right)}}$

Example 3.3-34

$$\boxed{w^3\cdot\left(5\cdot w^2\right)^5}=\boxed{w^3\cdot\left(5^{1\times5}\cdot w^{2\times5}\right)}=\boxed{w^3\cdot\left(5^5\cdot w^{10}\right)}=\boxed{5^5\cdot\left(w^3\cdot w^{10}\right)}=\boxed{3125\cdot\left(w^{3+10}\right)}=\boxed{\mathbf{3125\,w^{13}}}$$

Example 3.3-35

$$\boxed{y^2\cdot\left(2\cdot y^3\right)^2}=\boxed{y^2\cdot\left(2^{1\times2}\cdot y^{3\times2}\right)}=\boxed{y^2\cdot\left(2^2\cdot y^6\right)}=\boxed{y^2\cdot\left(4\cdot y^6\right)}=\boxed{4\cdot\left(y^2\cdot y^6\right)}=\boxed{4\cdot\left(y^{2+6}\right)}=\boxed{\mathbf{4\,y^8}}$$

Example 3.3-36

$$\boxed{a^2\cdot\left(a^2\cdot b^2\right)\cdot\left(a^3\cdot b\right)\cdot\left(a^0\cdot b^3\right)}=\boxed{a^2\cdot a^2\cdot b^2\cdot a^3\cdot b\cdot a^0\cdot b^3}=\boxed{\left(a^2\cdot a^2\cdot a^3\cdot a^0\right)\cdot\left(b^2\cdot b^1\cdot b^3\right)}$$

$$=\boxed{\left(a^{2+2+3+0}\right)\cdot\left(b^{2+1+3}\right)}=\boxed{a^7\cdot b^6}=\boxed{\mathbf{a^7b^6}}$$

Example 3.3-37

$$\boxed{\left(x^3\cdot y\right)^2\cdot\left(x^2\cdot y\cdot z\right)^4\cdot\left(x\cdot z^3\right)^2}=\boxed{\left(x^{3\times2}\cdot y^{1\times2}\right)\cdot\left(x^{2\times4}\cdot y^{1\times4}\cdot z^{1\times4}\right)\cdot\left(x^{1\times2}\cdot z^{3\times2}\right)}$$

$$=\boxed{\left(x^6\cdot y^2\right)\cdot\left(x^8\cdot y^4\cdot z^4\right)\cdot\left(x^2\cdot z^6\right)}=\boxed{x^6\cdot y^2\cdot x^8\cdot y^4\cdot z^4\cdot x^2\cdot z^6}=\boxed{\left(x^6\cdot x^8\cdot x^2\right)\cdot\left(y^2\cdot y^4\right)\cdot\left(z^4\cdot z^6\right)}$$

$$=\boxed{\left(x^{6+8+2}\right)\cdot\left(y^{2+4}\right)\cdot\left(z^{4+6}\right)}=\boxed{x^{16}\cdot y^6\cdot z^{10}}=\boxed{\mathbf{x^{16}y^6z^{10}}}$$

Example 3.3-38

$$\boxed{12\cdot\left(x^3\cdot y\right)^2\cdot\left(y\cdot x^3\right)^2\cdot y}=\boxed{12\cdot\left(x^{3\times2}\cdot y^{1\times2}\right)\cdot\left(y^{1\times2}\cdot x^{3\times2}\right)\cdot y^1}=\boxed{12\cdot\left(x^6\cdot y^2\right)\cdot\left(y^2\cdot x^6\right)\cdot y^1}$$

$$=\boxed{12\cdot x^6\cdot y^2\cdot y^2\cdot x^6\cdot y^1}=\boxed{12\cdot\left(x^6\cdot x^6\right)\cdot\left(y^2\cdot y^2\cdot y^1\right)}=\boxed{12\cdot\left(x^{6+6}\right)\cdot\left(y^{2+2+1}\right)}=\boxed{12\cdot x^{12}\cdot y^5}=\boxed{\mathbf{12\,x^{12}y^5}}$$

Example 3.3-39

$$\boxed{\left(2^3\cdot a^2\cdot x\right)^3\cdot\left(3\cdot a\cdot x^2\right)^2\cdot\left(a^3\right)^0}=\boxed{\left(2^{3\times3}\cdot a^{2\times3}\cdot x^{1\times3}\right)\cdot\left(3^{1\times2}\cdot a^{1\times2}\cdot x^{2\times2}\right)\cdot1}=\boxed{\left(2^9\cdot a^6\cdot x^3\right)\cdot\left(3^2\cdot a^2\cdot x^4\right)}$$

$$=\boxed{512\cdot a^6\cdot x^3\cdot9\cdot a^2\cdot x^4}=\boxed{(512\cdot9)\cdot\left(a^6\cdot a^2\right)\cdot\left(x^3\cdot x^4\right)}=\boxed{4608\cdot\left(a^{6+2}\right)\cdot\left(x^{3+4}\right)}=\boxed{\mathbf{4608\,a^8x^7}}$$

Example 3.3-40

$$\boxed{\left[\left(5^2\cdot y^3\cdot z^2\right)^2\cdot\left(y^3\cdot x^2\cdot z\right)^0\right]\cdot\left[\left(y^2\cdot z^3\right)^3\cdot x^2\right]^2}=\boxed{\left[\left(5^{2\times2}\cdot y^{3\times2}\cdot z^{2\times2}\right)\cdot1\right]\cdot\left[\left(y^{2\times3}\cdot z^{3\times3}\right)\cdot x^2\right]^2}$$

$$=\boxed{\left(5^4\cdot y^6\cdot z^4\right)\cdot\left[\left(y^6\cdot z^9\right)\cdot x^2\right]^2}=\boxed{\left(625\cdot y^6\cdot z^4\right)\cdot\left[y^{6\times2}\cdot z^{9\times2}\cdot x^{2\times2}\right]}=\boxed{\left(625\cdot y^6\cdot z^4\right)\cdot\left(y^{12}\cdot z^{18}\cdot x^4\right)}$$

$$=\boxed{625\cdot y^6\cdot z^4\cdot y^{12}\cdot z^{18}\cdot x^4}=\boxed{625\cdot x^4\cdot\left(y^6\cdot y^{12}\right)\cdot\left(z^4\cdot z^{18}\right)}=\boxed{625\cdot x^4\cdot\left(y^{6+12}\right)\cdot\left(z^{4+18}\right)}$$

$= \boxed{625 \cdot x^4 \cdot y^{18} \cdot z^{22}} = \boxed{\mathbf{625\, x^4 y^{18} z^{22}}}$

Practice Problems - Multiplying Positive Integer Exponents (More Difficult Cases)

Section 3.3 Case I b Practice Problems - Multiply the following positive integer exponents:

1. $\left(x^2 \cdot x^3\right)^2 \cdot x =$
2. $2 \cdot \left(p^2 \cdot q^0\right)^3 \cdot p^2 q =$
3. $\left(a^2 \cdot b^3\right)^2 \cdot \left(a \cdot b^4\right)^2 =$
4. $2^3 \cdot 2^2 \cdot \left(x^{2a} \cdot x^{3a}\right)^2 =$
5. $\left(h \cdot k^2\right)^0 \cdot \left(h^2\right)^{3a} \cdot h^a \cdot k =$
6. $2^0 \cdot 3^2 \cdot 3^3 \cdot 2^2 \cdot 2 =$
7. $u^2 \cdot \left(u^3 \cdot v\right)^4 \cdot \left(u \cdot v^2\right)^2 =$
8. $\left(x^3 \cdot y^2 \cdot z\right)^2 \cdot \left(x \cdot y^2 \cdot z^3\right) =$
9. $5^0 \cdot \left(r^2 \cdot s\right)^2 \cdot \left(3^2 \cdot r \cdot s^3\right)^3 =$
10. $\left(-3 \cdot x^2\right)^3 \cdot \left[\left(2 \cdot x \cdot y^2\right)^3 \cdot x\right]^2 =$

Case II Dividing Positive Integer Exponents

Positive integer exponents are divided by one another using the exponent laws I through VI shown in Table 3.3-1. These laws are used in order to simplify division of positive integer exponents by each other. Division of expressions by positive integer exponents is divided to two cases: Case II a addressing simple cases and Case II b addressing more difficult cases.

Case II a Dividing Positive Integer Exponents (Simple Cases)

Positive integer exponents are divided by one another using the following steps:

Step 1 a. Apply the Division and/or the Negative Power Laws (Laws V and VI) from Table 3.3-1.

b. Group the exponential terms with similar bases.

Step 2 Apply the Multiplication Law (Law I) from Table 3.3-1 and simplify the exponential expressions by adding the exponents with similar bases.

Examples with Steps

The following examples show the steps as to how positive integer exponents are divided by one another:

Example 3.3-41

$$\boxed{\frac{2ab}{-4a^3b^4}} =$$

Solution:

Step 1

$$\boxed{\frac{2ab}{-4a^3b^4}} = \boxed{-\frac{2}{4}\frac{a^1b^1}{a^3b^4}} = \boxed{-\frac{2}{4}\frac{1}{\left(a^3a^{-1}\right)\cdot\left(b^4b^{-1}\right)}}$$

Step 2

$$\boxed{-\frac{2}{4}\frac{1}{\left(a^3a^{-1}\right)\cdot\left(b^4b^{-1}\right)}} = \boxed{-\frac{\overset{1}{\cancel{2}}}{\underset{2}{\cancel{4}}}\frac{1}{\left(a^{3-1}\right)\cdot\left(b^{4-1}\right)}} = \boxed{-\frac{1}{2}\frac{1}{a^2\cdot b^3}} = \boxed{-\frac{1}{2}\left(\frac{1}{a^2b^3}\right)}$$

Example 3.3-42

$$\boxed{\frac{x^3y^3}{x^2y}} =$$

Solution:

Step 1

$$\boxed{\frac{x^3y^3}{x^2y}} = \boxed{\frac{x^3y^3}{x^2y^1}} = \boxed{\frac{\left(x^3x^{-2}\right)\cdot\left(y^3y^{-1}\right)}{1}}$$

Step 2

$$\boxed{\frac{\left(x^3x^{-2}\right)\cdot\left(y^3y^{-1}\right)}{1}} = \boxed{\frac{\left(x^{3-2}\right)\cdot\left(y^{3-1}\right)}{1}} = \boxed{\frac{x^1\cdot y^2}{1}} = \boxed{\frac{xy^2}{1}} = \boxed{xy^2}$$

Example 3.3-43

$$\frac{-3xy^4z^3y}{-15x^2z^2} =$$

Solution:

Step 1

$$\frac{-3xy^4z^3y}{-15x^2z^2} = \frac{-3x^1y^4z^3y^1}{-15x^2z^2} = +\frac{3}{15}\frac{x^1y^4z^3y^1}{x^2z^2} = \frac{3}{15}\frac{\left(y^4y^1\right)\cdot\left(z^3z^{-2}\right)}{x^2x^{-1}}$$

Step 2

$$\frac{3}{15}\frac{\left(y^4y^1\right)\cdot\left(z^3z^{-2}\right)}{x^2x^{-1}} = \frac{\overset{1}{\cancel{3}}}{\underset{5}{\cancel{15}}}\frac{\left(y^{4+1}\right)\cdot\left(z^{3-2}\right)}{x^{2-1}} = \frac{1}{5}\frac{y^5\cdot z^1}{x^1} = \mathbf{\frac{1}{5}\left(\frac{y^5z}{x}\right)}$$

Example 3.3-44

$$\frac{\left(x^5y^2z\right)\cdot\left(z^3\right)}{(xy)\cdot\left(z^2y^2\right)} =$$

Solution:

Step 1

$$\frac{\left(x^5y^2z\right)\cdot\left(z^3\right)}{(xy)\cdot\left(z^2y^2\right)} = \frac{x^5y^2z\,z^3}{x\,y\,z^2y^2} = \frac{x^5y^2z^1z^3}{x^1y^1z^2y^2} = \frac{\left(x^5x^{-1}\right)\cdot\left(z^1z^3z^{-2}\right)}{y^1y^2y^{-2}}$$

Step 2

$$\frac{\left(x^5x^{-1}\right)\cdot\left(z^1z^3z^{-2}\right)}{y^1y^2y^{-2}} = \frac{\left(x^{5-1}\right)\cdot\left(z^{1+3-2}\right)}{y^{1+2-2}} = \frac{x^4\cdot z^2}{y^1} = \mathbf{\frac{x^4z^2}{y}}$$

Example 3.3-45

$$\left(\frac{u^2v^3w^2}{8u^7v^5}\right)\cdot\left(\frac{u}{v^2}\right) =$$

Solution:

Step 1

$$\left(\frac{u^2v^3w^2}{8u^7v^5}\right)\cdot\left(\frac{u}{v^2}\right) = \frac{1}{8}\frac{u^2v^3w^2u^1}{u^7v^5v^2} = \frac{1}{8}\frac{w^2}{\left(u^7u^{-2}u^{-1}\right)\cdot\left(v^5v^2v^{-3}\right)}$$

Step 2

$$\frac{1}{8}\frac{w^2}{\left(u^7u^{-2}u^{-1}\right)\cdot\left(v^5v^2v^{-3}\right)} = \frac{1}{8}\frac{w^2}{\left(u^{7-2-1}\right)\cdot\left(v^{5+2-3}\right)} = \frac{1}{8}\frac{w^2}{u^4\cdot v^4} = \mathbf{\frac{1}{8}\left(\frac{w^2}{u^4v^4}\right)}$$

Additional Examples - Dividing Positive Integer Exponents (Simple Cases)

The following examples further illustrate how to divide positive integer exponential terms by one another:

Example 3.3-46

$$\frac{3a^2b^3}{6a^6} = \frac{3}{6}\frac{b^3}{\left(a^6a^{-2}\right)} = \frac{\overset{1}{\cancel{3}}}{\underset{2}{\cancel{6}}}\frac{b^3}{\left(a^{6-2}\right)} = \mathbf{\frac{1}{2}\left(\frac{b^3}{a^4}\right)}$$

Example 3.3-47

$$\boxed{\frac{x^4y^3z}{x^2y^2}} = \boxed{\frac{\left(x^4x^{-2}\right)\cdot\left(y^3y^{-2}\right)\cdot z}{1}} = \boxed{\frac{\left(x^{4-2}\right)\cdot\left(y^{3-2}\right)\cdot z}{1}} = \boxed{\frac{x^2\cdot y^1\cdot z}{1}} = \boxed{\frac{x^2yz}{1}} = \boxed{\mathbf{x^2yz}}$$

Example 3.3-48

$$\boxed{-\frac{11m^2n^2}{44m}} = \boxed{-\frac{11}{44}\frac{m^2n^2}{m^1}} = \boxed{-\frac{\overset{1}{\cancel{11}}}{\underset{4}{\cancel{44}}}\frac{\left(m^2m^{-1}\right)\cdot n^2}{1}} = \boxed{-\frac{1}{4}\frac{\left(m^{2-1}\right)\cdot n^2}{1}} = \boxed{-\frac{1}{4}\frac{m^1\cdot n^2}{1}} = \boxed{-\frac{1}{4}\frac{mn^2}{1}} = \boxed{-\frac{\mathbf{1}}{\mathbf{4}}\mathbf{mn^2}}$$

Example 3.3-49

$$\boxed{\frac{5a^3b^5}{15a^2b^0}} = \boxed{\frac{5}{15}\frac{a^3b^5}{a^2\cdot 1}} = \boxed{\frac{\overset{1}{\cancel{5}}}{\underset{3}{\cancel{15}}}\frac{\left(a^3a^{-2}\right)\cdot b^5}{1}} = \boxed{\frac{1}{3}\frac{\left(a^{3-2}\right)\cdot b^5}{1}} = \boxed{\frac{1}{3}\frac{a^1\cdot b^5}{1}} = \boxed{\frac{\mathbf{1}}{\mathbf{3}}\mathbf{ab^5}}$$

Example 3.3-50

$$\boxed{\frac{8u^3w^3z^2}{2u^3w^2z}} = \boxed{\frac{8u^3w^3z^2}{2u^3w^2z^1}} = \boxed{\frac{\overset{4}{\cancel{8}}\cdot\left(w^3w^{-2}\right)\cdot\left(z^2\cdot z^{-1}\right)}{\underset{1}{\cancel{2}}\cdot\left(u^3u^{-3}\right)}} = \boxed{\frac{4\cdot\left(w^{3-2}\right)\cdot\left(z^{2-1}\right)}{1\cdot\left(u^{3-3}\right)}} = \boxed{\frac{4\cdot w^1\cdot z^1}{u^0}} = \boxed{\frac{4wz}{1}} = \boxed{\mathbf{4wz}}$$

Example 3.3-51

$$\boxed{\frac{f^6g^4}{f^5g^5}} = \boxed{\frac{f^6f^{-5}}{g^5g^{-4}}} = \boxed{\frac{f^{6-5}}{g^{5-4}}} = \boxed{\frac{f^1}{g^1}} = \boxed{\frac{\mathbf{f}}{\mathbf{g}}}$$

Example 3.3-52

$$\boxed{\frac{3a^5b^6c^2}{-6abc}} = \boxed{-\frac{3}{6}\frac{a^5b^6c^2}{a^1b^1c^1}} = \boxed{-\frac{\overset{1}{\cancel{3}}}{\underset{2}{\cancel{6}}}\frac{\left(a^5a^{-1}\right)\cdot\left(b^6b^{-1}\right)\cdot\left(c^2c^{-1}\right)}{1}} = \boxed{-\frac{1}{2}\frac{\left(a^{5-1}\right)\cdot\left(b^{6-1}\right)\cdot\left(c^{2-1}\right)}{1}} = \boxed{-\frac{1}{2}\frac{a^4\cdot b^5\cdot c^1}{1}}$$

$$= \boxed{-\frac{\mathbf{1}}{\mathbf{2}}\mathbf{a^4b^5c}}$$

Example 3.3-53

$$\boxed{\frac{100\,p^2t^2u}{5\,pt^4u^5}} = \boxed{\frac{100}{5}\frac{p^2t^2u^1}{p^1t^4u^5}} = \boxed{\frac{\overset{20}{\cancel{100}}}{\underset{1}{\cancel{5}}}\frac{p^2p^{-1}}{\left(t^4t^{-2}\right)\cdot\left(u^5u^{-1}\right)}} = \boxed{\frac{20}{1}\frac{p^{2-1}}{\left(t^{4-2}\right)\cdot\left(u^{5-1}\right)}} = \boxed{\frac{20}{1}\left(\frac{p^1}{t^2\cdot u^4}\right)} = \boxed{\frac{\mathbf{20\,p}}{\mathbf{t^2u^4}}}$$

Example 3.3-54

$$\boxed{\frac{4k^3lm^2}{kl^2m^3}} = \boxed{\frac{4k^3l^1m^2}{k^1l^2m^3}} = \boxed{\frac{4\left(k^3k^{-1}\right)}{\left(l^2l^{-1}\right)\cdot\left(m^3m^{-2}\right)}} = \boxed{\frac{4\left(k^{3-1}\right)}{\left(l^{2-1}\right)\cdot\left(m^{3-2}\right)}} = \boxed{\frac{4k^2}{l^1\cdot m^1}} = \boxed{\frac{\mathbf{4k^2}}{\mathbf{lm}}}$$

Example 3.3-55

$$\boxed{\left(\frac{w^2z^2}{z}\right)\cdot\left(\frac{w}{z^3}\right)} = \boxed{\left(\frac{w^2z^2}{z^1}\right)\cdot\left(\frac{w^1}{z^3}\right)} = \boxed{\frac{w^2z^2\cdot w^1}{z^1\cdot z^3}} = \boxed{\frac{w^2w^1}{z^1z^3z^{-2}}} = \boxed{\frac{w^{2+1}}{z^{1+3-2}}} = \boxed{\frac{\mathbf{w^3}}{\mathbf{z^2}}}$$

Example 3.3-56

$$\frac{-4a^2b^3c^4}{-3ab^2c} = \frac{-4}{-3}\frac{a^2b^3c^4}{a^1b^2c^1} = +\frac{4}{3}\frac{\left(a^2a^{-1}\right)\cdot\left(b^3b^{-2}\right)\cdot\left(c^4c^{-1}\right)}{1} = \frac{4}{3}\frac{\left(a^{2-1}\right)\cdot\left(b^{3-2}\right)\cdot\left(c^{4-1}\right)}{1} = \frac{4}{3}\frac{a^1\cdot b^1\cdot c^3}{1}$$

$$= \frac{4}{3}\frac{abc^3}{1} = \frac{4}{3}abc^3 = \mathbf{1\frac{1}{3}\left(abc^3\right)}$$

Example 3.3-57

$$\frac{5m^3n^3l^5}{-10n^2nl^3} = \frac{5}{-10}\frac{m^3n^3l^5}{n^2n^1l^3} = -\frac{\overset{1}{\cancel{5}}}{\underset{2}{\cancel{10}}}\frac{m^3\cdot\left(l^5l^{-3}\right)}{\left(n^2n^1n^{-3}\right)} = -\frac{1}{2}\frac{m^3\cdot\left(l^{5-3}\right)}{\left(n^{2+1-3}\right)} = -\frac{1}{2}\frac{m^3\cdot l^2}{n^0} = -\frac{1}{2}\frac{m^3\cdot l^2}{1} = \mathbf{-\frac{1}{2}m^3l^2}$$

Example 3.3-58

$$\frac{3pqt^2}{-15p^2q^4} = \frac{3}{-15}\frac{p^1q^1t^2}{p^2q^4} = \frac{3}{-15}\frac{t^2}{\left(p^2p^{-1}\right)\cdot\left(q^4q^{-1}\right)} = -\frac{\overset{1}{\cancel{3}}}{\underset{5}{\cancel{15}}}\frac{t^2}{\left(p^{2-1}\right)\cdot\left(q^{4-1}\right)} = -\frac{1}{5}\frac{t^2}{p^1\cdot q^3} = \mathbf{-\frac{1}{5}\left(\frac{t^2}{pq^3}\right)}$$

Example 3.3-59

$$\frac{c^2d^3e^{11}}{8c^8d^9e^3} = \frac{1}{8}\frac{c^2d^3e^{11}}{c^8d^9e^3} = \frac{1}{8}\frac{e^{11}e^{-3}}{\left(c^8c^{-2}\right)\cdot\left(d^9d^{-3}\right)} = \frac{1}{8}\frac{e^{11-3}}{\left(c^{8-2}\right)\cdot\left(d^{9-3}\right)} = \frac{1}{8}\frac{e^8}{c^6\cdot d^6} = \mathbf{\frac{1}{8}\left(\frac{e^8}{c^6d^6}\right)}$$

Example 3.3-60

$$\frac{-3u^2v^3}{-6uv^5} = \frac{-3}{-6}\frac{u^2v^3}{u^1v^5} = \frac{3}{6}\frac{u^2u^{-1}}{v^5v^{-3}} = \frac{\overset{1}{\cancel{3}}}{\underset{2}{\cancel{6}}}\frac{u^{2-1}}{v^{5-3}} = \frac{1}{2}\frac{u^1}{v^2} = \mathbf{\frac{1}{2}\left(\frac{u}{v^2}\right)}$$

Example 3.3-61

$$\frac{\left(x^2y^5z^3\right)\cdot\left(xyz\right)}{\left(-4x^2y^2z\right)} = \frac{x^2y^5z^3xyz}{-4x^2y^2z} = \frac{1}{-4}\frac{x^2y^5z^3x^1y^1z^1}{x^2y^2z^1} = -\frac{1}{4}\frac{\left(x^2x^{-2}x^1\right)\cdot\left(y^5y^{-2}y^1\right)\cdot\left(z^3z^{-1}z^1\right)}{1}$$

$$= -\frac{1}{4}\frac{\left(x^{2-2+1}\right)\cdot\left(y^{5-2+1}\right)\cdot\left(z^{3-1+1}\right)}{1} = -\frac{1}{4}\frac{x^1\cdot y^4\cdot z^3}{1} = -\frac{1}{4}\frac{xy^4z^3}{1} = \mathbf{-\frac{1}{4}xy^4z^3}$$

Example 3.3-62

$$\frac{\left(w^2x^3\right)\cdot\left(wxy\right)}{w^3x^3y^3} = \frac{w^2x^3wxy}{w^3x^3y^3} = \frac{w^2x^3w^1x^1y^1}{w^3x^3y^3} = \frac{x^3x^{-3}x^1}{\left(w^3w^{-2}w^{-1}\right)\cdot\left(y^3y^{-1}\right)} = \frac{x^{3-3+1}}{\left(w^{3-2-1}\right)\cdot\left(y^{3-1}\right)} = \frac{x^1}{w^0\cdot y^2}$$

$$= \frac{x}{1\cdot y^2} = \mathbf{\frac{x}{y^2}}$$

Example 3.3-63

$$\boxed{\frac{(abc)\cdot\left(a^2b^2c^3\right)\cdot b^3}{\left(-3^3b^3c^7\right)\cdot(ab)}} = \boxed{\frac{abca^2b^2c^3b^3}{-27b^3c^7ab}} = \boxed{\frac{1}{-27}\,\frac{a^1b^1c^1a^2b^2c^3b^3}{b^3c^7a^1b^1}} = \boxed{-\frac{1}{27}\,\frac{\left(a^1a^2a^{-1}\right)\cdot\left(b^1b^2b^3b^{-3}b^{-1}\right)}{\left(c^7c^{-1}c^{-3}\right)}}$$

$$= \boxed{-\frac{1}{27}\,\frac{\left(a^{1+2-1}\right)\cdot\left(b^{1+2+3-3-1}\right)}{\left(c^{7-1-3}\right)}} = \boxed{-\frac{1}{27}\,\frac{a^2\cdot b^2}{c^3}} = \boxed{-\frac{\mathbf{1}}{\mathbf{27}}\left(\frac{\mathbf{a^2b^2}}{\mathbf{c^3}}\right)}$$

Example 3.3-64

$$\boxed{\frac{-5^2\cdot\left(bc^2\right)}{(2ab)\cdot a^3}} = \boxed{\frac{-25\cdot\left(bc^2\right)}{(2ab)\cdot a^3}} = \boxed{-\frac{25}{2}\,\frac{b^1c^2}{a^1b^1a^3}} = \boxed{\frac{-25}{2}\,\frac{\left(b^1b^{-1}\right)\cdot c^2}{a^1a^3}} = \boxed{-\frac{25}{2}\,\frac{\left(b^{1-1}\right)\cdot c^2}{a^{1+3}}} = \boxed{-\frac{25}{2}\,\frac{b^0\cdot c^2}{a^4}}$$

$$= \boxed{-\frac{25}{2}\,\frac{1\cdot c^2}{a^4}} = \boxed{-\frac{25}{2}\left(\frac{c^2}{a^4}\right)} = \boxed{\mathbf{-12\frac{1}{2}\left(\frac{c^2}{a^4}\right)}}$$

Example 3.3-65

$$\boxed{\left(\frac{kl^2m^2}{l^3m^3}\right)\cdot\left(\frac{k^2l}{m^2l^2}\right)\cdot\left(\frac{m}{-2}\right)} = \boxed{\frac{1}{-2}\,\frac{kl^2m^2k^2lm}{l^3m^3m^2l^2}} = \boxed{-\frac{1}{2}\,\frac{k^1l^2m^2k^2l^1m^1}{l^3m^3m^2l^2}} = \boxed{-\frac{1}{2}\,\frac{k^1k^2}{\left(l^3l^2l^{-2}l^{-1}\right)\cdot\left(m^3m^2m^{-2}m^{-1}\right)}}$$

$$= \boxed{-\frac{1}{2}\,\frac{k^{1+2}}{\left(l^{3+2-2-1}\right)\cdot\left(m^{3+2-2-1}\right)}} = \boxed{-\frac{1}{2}\,\frac{k^3}{l^2\cdot m^2}} = \boxed{-\frac{\mathbf{1}}{\mathbf{2}}\left(\frac{\mathbf{k^3}}{\mathbf{l^2m^2}}\right)}$$

Practice Problems - Dividing Positive Integer Exponents (Simple Cases)

Section 3.3 Case II a Practice Problems - Divide the following positive integer exponents:

1. $\frac{x^5}{x^3} =$
2. $\frac{a^2b^3}{a} =$
3. $\frac{a^3b^3c^2}{a^2b^6c} =$
4. $\frac{3^2\cdot\left(rs^2\right)}{(2rs)\cdot r^3} =$
5. $\frac{2p^2q^3pr^4}{-6p^4q^2r} =$
6. $\frac{\left(k^2l^3\right)\cdot\left(kl^2m^0\right)}{k^4l^3m^5} =$
7. $\frac{2\cdot a^5\cdot b^2\cdot c}{-a\cdot b\cdot c^3} =$
8. $\frac{-c^3d^6e^5}{8c^6d^2e^3e^2}\cdot\frac{-2c}{3d} =$
9. $\frac{-2\left(m^3n^3l^5\right)\cdot 3m^2}{\left(10n^2n\right)\cdot\left(l^2l^4\right)} =$
10. $\frac{-5\left(x^2y^5z^4\right)\cdot(xyz)}{-2x^4y^2z^3} =$

Case II b Dividing Positive Integer Exponents (More Difficult Cases)

Positive integer exponents are divided by one another using the following steps:

Step 1 Apply exponent laws such as the Power of a Power Law, the Power of a Product Law, and the Power of a Fraction Law (Laws II, III, and IV) from Table 3.3-1.

Step 2 a. Apply the Division and/or the Negative Power Law (Laws V and VI) from Table 3.3-1.

b. Group the exponential terms with similar bases.

c. Apply the Multiplication Law (Law I) from Table 3.3-1 and simplify the exponential expressions by adding the exponents with similar bases.

Note that the objective is to write the final answer without a negative exponent.

Examples with Steps

The following examples show the steps as to how positive integer exponents are divided by one another:

Example 3.3-66

$$\left(\frac{x^3}{x^2}\right)^2 =$$

Solution:

Step 1 $$\left(\frac{x^3}{x^2}\right)^2 = \left(\frac{x^{3\times 2}}{x^{2\times 2}}\right) = \frac{x^6}{x^4}$$

Step 2 $$\frac{x^6}{x^4} = \frac{x^6 \cdot x^{-4}}{1} = \frac{x^{6-4}}{1} = \frac{x^2}{1} = \boldsymbol{x^2}$$

Example 3.3-67

$$\left(\frac{2^2 \cdot a^3}{a^4}\right)^3 =$$

Solution:

Step 1 $$\left(\frac{2^2 \cdot a^3}{a^4}\right)^3 = \frac{2^{2\times 3} \cdot a^{3\times 3}}{a^{4\times 3}} = \frac{2^6 \cdot a^9}{a^{12}}$$

Step 2 $$\frac{2^6 \cdot a^9}{a^{12}} = \frac{64}{a^{12} \cdot a^{-9}} = \frac{64}{a^{12-9}} = \boldsymbol{\frac{64}{a^3}}$$

Example 3.3-68

$$\left(\frac{4\cdot c^2}{3\cdot b^3}\right)^2 =$$

Solution:

Step 1

$$\left(\frac{4\cdot c^2}{3\cdot b^3}\right)^2 = \left(\frac{4^{1\times 2}\cdot c^{2\times 2}}{3^{1\times 2}\cdot b^{3\times 2}}\right) = \frac{4^2\cdot c^4}{3^2\cdot b^6}$$

Step 2

$$\frac{4^2\cdot c^4}{3^2\cdot b^6} = \frac{16\cdot c^4}{9\cdot b^6} = \frac{16}{9}\left(\frac{c^4}{b^6}\right) = \mathbf{1\frac{7}{9}\left(\frac{c^4}{b^6}\right)}$$

Example 3.3-69

$$\frac{z\cdot(a\cdot b)^3}{z^2\cdot a^2} =$$

Solution:

Step 1

$$\frac{z\cdot(a\cdot b)^3}{z^2\cdot a^2} = \frac{z\cdot\left(a^{1\times 3}\cdot b^{1\times 3}\right)}{z^2\cdot a^2} = \frac{z\cdot\left(a^3\cdot b^3\right)}{z^2\cdot a^2}$$

Step 2

$$\frac{z\cdot\left(a^3\cdot b^3\right)}{z^2\cdot a^2} = \frac{z^1\cdot a^3\cdot b^3}{z^2\cdot a^2} = \frac{\left(a^3\cdot a^{-2}\right)\cdot b^3}{z^2\cdot z^{-1}} = \frac{\left(a^{3-2}\right)\cdot b^3}{z^{2-1}} = \frac{a^1\cdot b^3}{z^1} = \mathbf{\frac{a\,b^3}{z}}$$

Example 3.3-70

$$\left(\frac{\left(a^2\cdot b^3\right)^4}{\left(a^3\cdot b\cdot c\right)^2}\right)^3 =$$

Solution:

Step 1

$$\left(\frac{\left(a^2\cdot b^3\right)^4}{\left(a^3\cdot b\cdot c\right)^2}\right)^3 = \left(\frac{a^{2\times 4}\cdot b^{3\times 4}}{a^{3\times 2}\cdot b^{1\times 2}\cdot c^{1\times 2}}\right)^3 = \left(\frac{a^8\cdot b^{12}}{a^6\cdot b^2\cdot c^2}\right)^3 = \left(\frac{a^{8\times 3}\cdot b^{12\times 3}}{a^{6\times 3}\cdot b^{2\times 3}\cdot c^{2\times 3}}\right)$$

$$= \frac{a^{24}\cdot b^{36}}{a^{18}\cdot b^6\cdot c^6}$$

Step 2

$$\frac{a^{24}\cdot b^{36}}{a^{18}\cdot b^6\cdot c^6} = \frac{\left(a^{24}\cdot a^{-18}\right)\cdot\left(b^{36}\cdot b^{-6}\right)}{c^6} = \frac{\left(a^{24-18}\right)\cdot\left(b^{36-6}\right)}{c^6} = \frac{a^6\cdot b^{30}}{c^6} = \mathbf{\frac{a^6 b^{30}}{c^6}}$$

Additional Examples - Dividing Positive Integer Exponents (More Difficult Cases)

The following examples further illustrate how to divide exponential expressions by one another:

Example 3.3-71

$$\boxed{\left(\frac{3}{5}\right)^2\cdot\left(\frac{a^2}{a^3}\right)}=\boxed{\left(\frac{3^{1\times2}}{5^{1\times2}}\right)\cdot\left(\frac{1}{a^3\cdot a^{-2}}\right)}=\boxed{\left(\frac{3^2}{5^2}\right)\cdot\left(\frac{1}{a^{3-2}}\right)}=\boxed{\frac{9}{25}\cdot\frac{1}{a^1}}=\boxed{\mathbf{\frac{9}{25\,a}}}$$

Example 3.3-72

$$\boxed{\left(\frac{a^0\cdot x^2}{a^2\cdot y^3}\right)^2}=\boxed{\frac{a^{0\times2}\cdot x^{2\times2}}{a^{2\times2}\cdot y^{3\times2}}}=\boxed{\frac{a^0\cdot x^4}{a^4\cdot y^6}}=\boxed{\frac{1\cdot x^4}{a^4\cdot y^6}}=\boxed{\frac{x^4}{a^4\cdot y^6}}=\boxed{\mathbf{\frac{x^4}{a^4\,y^6}}}$$

Example 3.3-73

$$\boxed{\frac{a^2\cdot b^5\cdot c^5}{a^3\cdot b^2\cdot c^3}}=\boxed{\frac{\left(b^5\cdot b^{-2}\right)\cdot\left(c^5\cdot c^3\right)}{a^3a^{-2}}}=\boxed{\frac{\left(b^{5-2}\right)\cdot\left(c^{5-3}\right)}{a^{3-2}}}=\boxed{\frac{b^3\cdot c^2}{a^1}}=\boxed{\mathbf{\frac{b^3c^2}{a}}}$$

Example 3.3-74

$$\boxed{\left(\frac{x^2\cdot y^0}{x^5}\right)^3}=\boxed{\frac{x^{2\times3}\cdot y^{0\times3}}{x^{5\times3}}}=\boxed{\frac{x^6\cdot y^0}{x^{15}}}=\boxed{\frac{y^0}{x^{15}\cdot x^{-6}}}=\boxed{\frac{1}{x^{15-6}}}=\boxed{\mathbf{\frac{1}{x^9}}}$$

Example 3.3-75

$$\boxed{\left(\frac{x\cdot y^3}{x^3\cdot y^2}\right)^2}=\boxed{\frac{x^{1\times2}\cdot y^{3\times2}}{x^{3\times2}\cdot y^{2\times2}}}=\boxed{\frac{x^2\cdot y^6}{x^6\cdot y^4}}=\boxed{\frac{y^6\cdot y^{-4}}{x^6\cdot x^{-2}}}=\boxed{\frac{y^{6-4}}{x^{6-2}}}=\boxed{\mathbf{\frac{y^2}{x^4}}}$$

Example 3.3-76

$$\boxed{\left(\frac{5\cdot a^3\cdot b^3}{a^2\cdot b^4}\right)^3}=\boxed{\frac{5^{1\times3}\cdot a^{3\times3}\cdot b^{3\times3}}{a^{2\times3}\cdot b^{4\times3}}}=\boxed{\frac{5^3\cdot a^9\cdot b^9}{a^6\cdot b^{12}}}=\boxed{\frac{125\cdot\left(a^9\cdot a^{-6}\right)}{b^{12}\cdot b^{-9}}}=\boxed{\frac{125\cdot\left(a^{9-6}\right)}{b^{12-9}}}=\boxed{\frac{125\cdot a^3}{b^3}}=\boxed{\mathbf{125\,\frac{a^3}{b^3}}}$$

Example 3.3-77

$$\boxed{\left(\frac{3^2\cdot 3^3\cdot x^5}{3^4\cdot x\cdot x^3}\right)^3\cdot\left(3^2\cdot x^3\right)^2}=\boxed{\left(\frac{3^{2+3}\cdot x^5}{3^4\cdot x^{1+3}}\right)^3\cdot\left(3^{2\times2}\cdot x^{3\times2}\right)}=\boxed{\left(\frac{3^5\cdot x^5}{3^4\cdot x^4}\right)^3\cdot\left(3^4\cdot x^6\right)}=\boxed{\left(\frac{3^{5\times3}\cdot x^{5\times3}}{3^{4\times3}\cdot x^{4\times3}}\right)\cdot\left(3^4\cdot x^6\right)}$$

$$=\boxed{\left(\frac{3^{15}\cdot x^{15}}{3^{12}\cdot x^{12}}\right)\cdot\left(\frac{3^4\cdot x^6}{1}\right)}=\boxed{\frac{\left(3^{15}\cdot 3^4\right)\cdot\left(x^{15}\cdot x^6\right)}{3^{12}\cdot x^{12}\cdot 1}}=\boxed{\frac{\left(3^{15+4}\right)\cdot\left(x^{15+6}\right)}{3^{12}\cdot x^{12}}}=\boxed{\frac{3^{19}\cdot x^{21}}{3^{12}\cdot x^{12}}}=\boxed{\frac{\left(3^{19}\cdot 3^{-12}\right)\cdot\left(x^{21}\cdot x^{-12}\right)}{1}}$$

$$=\boxed{\frac{\left(3^{19-12}\right)\cdot\left(x^{21-12}\right)}{1}}=\boxed{\frac{3^7\cdot x^9}{1}}=\boxed{3^7\cdot x^9}=\boxed{\mathbf{2187\,x^9}}$$

Example 3.3-78

$$\frac{\left(a\cdot a^2\right)^4\cdot\left(x^3\right)^2}{\left(a^3\cdot y^2\right)^0}=\frac{\left(a^{1+2}\right)^4\cdot\left(x^{3\times 2}\right)}{\left(a^{3\times 0}\cdot y^{2\times 0}\right)}=\frac{\left(a^3\right)^4\cdot\left(x^{3\times 2}\right)}{\left(a^0\cdot y^0\right)}=\frac{\left(a^{3\times 4}\right)\cdot x^6}{(1\cdot 1)}=\frac{a^{12}\cdot x^6}{1}=\mathbf{a^{12}x^6}$$

Example 3.3-79

$$\frac{4\cdot x^0\cdot\left(-2\cdot x^2\right)^3}{3^2\cdot\left(-x^3\cdot y\right)^2}=\frac{4\cdot 1\cdot\left(-2^{1\times 3}\cdot x^{2\times 3}\right)}{9\cdot\left(+x^{3\times 2}\cdot y^{1\times 2}\right)}=\frac{4\cdot\left(-2^3\cdot x^6\right)}{9\cdot x^6\cdot y^2}=\frac{4\cdot\left(-8\cdot x^6\right)}{9\cdot x^6\cdot y^2}=\frac{-32\cdot\left(x^6\cdot x^{-6}\right)}{9\cdot y^2}$$

$$=-\frac{32}{9}\frac{x^{6-6}}{y^2}=-\frac{32}{9}\frac{x^0}{y^2}=-\frac{32}{9}\left(\frac{1}{y^2}\right)=\mathbf{-3\frac{5}{9}\left(\frac{1}{y^2}\right)}$$

Example 3.3-80

$$\left[\frac{\left(a^2\cdot b\right)^3\cdot(x\cdot y)^2}{a^3\cdot x\cdot y^3}\right]^4=\left[\frac{\left(a^{2\times 3}\cdot b^{1\times 3}\right)\cdot\left(x^{1\times 2}\cdot y^{1\times 2}\right)}{a^3\cdot x\cdot y^3}\right]^4=\left[\frac{\left(a^6\cdot b^3\right)\cdot\left(x^2\cdot y^2\right)}{a^3\cdot x^1\cdot y^3}\right]^4=\left[\frac{\left(a^6\cdot a^{-3}\right)\cdot b^3\cdot\left(x^2\cdot x^{-1}\right)}{y^3\cdot y^{-2}}\right]^4$$

$$=\left[\frac{\left(a^{6-3}\right)\cdot b^3\cdot\left(x^{2-1}\right)}{y^{3-2}}\right]^4=\left[\frac{a^3\cdot b^3\cdot x^1}{y^1}\right]^4=\frac{a^{3\times 4}\cdot b^{3\times 4}\cdot x^{1\times 4}}{y^{1\times 4}}=\frac{a^{12}\cdot b^{12}\cdot x^4}{y^4}=\mathbf{\frac{a^{12}\,b^{12}\,x^4}{y^4}}$$

Note: The following is another way of solving this problem:

$$\left[\frac{\left(a^2\cdot b\right)^3\cdot(x\cdot y)^2}{a^3\cdot x\cdot y^3}\right]^4=\left[\frac{\left(a^{2\times 3}\cdot b^{1\times 3}\right)\cdot\left(x^{1\times 2}\cdot y^{1\times 2}\right)}{a^3\cdot x\cdot y^3}\right]^4=\left[\frac{\left(a^6\cdot b^3\right)\cdot\left(x^2\cdot y^2\right)}{a^3\cdot x^1\cdot y^3}\right]^4$$

$$=\left[\left(a^6\cdot b^3\right)\cdot\left(x^2\cdot y^2\right)\cdot a^{-3}\cdot x^{-1}\cdot y^{-3}\right]^4=\left[\left(a^6\cdot a^{-3}\right)\cdot b^3\cdot\left(x^2\cdot x^{-1}\right)\cdot\left(y^2\cdot y^{-3}\right)\right]^4=\left[\left(a^{6-3}\right)\cdot b^3\cdot\left(x^{2-1}\right)\cdot\left(y^{2-3}\right)\right]^4$$

$$=\left[a^3\cdot b^3\cdot x^1\cdot y^{-1}\right]^4=a^{3\times 4}\cdot b^{3\times 4}\cdot x^{1\times 4}\cdot y^{-1\times 4}=a^{12}\cdot b^{12}\cdot x^4\cdot y^{-4}=\frac{a^{12}\cdot b^{12}\cdot x^4}{1}\cdot\frac{1}{y^4}=\frac{a^{12}\cdot b^{12}\cdot x^4\cdot 1}{y^4\cdot 1}$$

$$=\mathbf{\frac{a^{12}\,b^{12}\,x^4}{y^4}}$$

Practice Problems - Dividing Positive Integer Exponents (More Difficult Cases)

Section 3.3 Case II b Practice Problems - Divide the following positive integer exponents:

1. $\left(\dfrac{x^5}{x^3}\right)^3 =$

2. $\dfrac{\left(a^2 \cdot b^3\right)^2}{a^3} =$

3. $\left(\dfrac{a^3 \cdot b^6 \cdot c^2}{a^2 \cdot b^3}\right)^0 =$

4. $\dfrac{y^3 \cdot (z \cdot w)^2}{-y^2 \cdot w^3 \cdot z^4} =$

5. $\dfrac{(a \cdot b)^4 \cdot (x \cdot y)^2}{a^3 \cdot y^2 \cdot x} =$

6. $\dfrac{-3(x \cdot y \cdot z)^5 \cdot x^2}{6(x \cdot y)^3 \cdot z^7} =$

7. $\dfrac{2^2 \cdot a^3 \cdot b^2 \cdot c}{-2^3 \cdot a \cdot \left(b \cdot c^2\right)^2} =$

8. $\left(\dfrac{z^4 \cdot w^3 \cdot (a \cdot b)^2}{a^2 \cdot w^0 \cdot z}\right)^2 =$

9. $\dfrac{3^3 \cdot 3^2 \cdot (y \cdot z)^4 \cdot b^3}{3^6 \cdot b^2 \cdot y^3} =$

10. $\left(\dfrac{a \cdot b^2 \cdot (c \cdot d)^4}{c^5 \cdot d^6}\right)^3 \cdot a^2 b^2 =$

Case III Adding and Subtracting Positive Integer Exponents

A common source of mistakes among students is in dealing with addition and subtraction of exponential expressions. In this section two classes of positive integer exponents are addressed. The first class deals with addition and subtraction of variables and numbers that are raised to positive exponents (Case III a). The second class deals with addition and subtraction of positive integer exponents that are in fraction form (Case III b). These two cases are addressed below:

Case III a Addition and Subtraction of Variables and Numbers Raised to Positive Exponential Terms

Positive exponential expressions are added and subtracted using the following steps:

Step 1 Group the exponential terms with similar bases.

Step 2 Simplify the exponential expressions by adding or subtracting the like terms.

Note that **like terms** are defined as terms having the same variables raised to the same power. For example, x^3 *and* $2x^3$; y^2 *and* $4y^2$ are like terms of one another.

Examples with Steps

The following examples show the steps as to how exponential expressions having positive integer exponents are added or subtracted:

Example 3.3-81

$$\boxed{x^3+3y^2+2x^3-y^2+5}=$$

Solution:

Step 1 $\boxed{x^3+3y^2+2x^3-y^2+5}=\boxed{\left(x^3+2x^3\right)+\left(3y^2-y^2\right)+5}$

Step 2 $\boxed{\left(x^3+2x^3\right)+\left(3y^2-y^2\right)+5}=\boxed{(1+2)x^3+(3-1)y^2+5}=\boxed{\mathbf{3x^3+2y^2+5}}$

Example 3.3-82

$$\boxed{\left(2^3+x^2+4y\right)-\left(3x^2+y\right)+2x^2}=$$

Solution:

Step 1 $\boxed{\left(2^3+x^2+4y\right)-\left(3x^2+y\right)+2x^2}=\boxed{8+x^2+4y-3x^2-y+2x^2}$

$=\boxed{\left(x^2+2x^2-3x^2\right)+(4y-y)+8}$

Step 2 $\boxed{\left(x^2+2x^2-3x^2\right)+(4y-y)+8}=\boxed{(1+2-3)x^2+(4-1)y+8}=\boxed{0x^2+3y+8}=\boxed{\mathbf{3y+8}}$

Example 3.3-83

$$\boxed{a^{3b}+2a^{2b}-4a^{3b}+5+3a^{2b}}=$$

Solution:

Step 1 $\boxed{a^{3b}+2a^{2b}-4a^{3b}+5+3a^{2b}}=\boxed{\left(a^{3b}-4a^{3b}\right)+\left(2a^{2b}+3a^{2b}\right)+5}$

Step 2 $\boxed{\left(a^{3b}-4a^{3b}\right)+\left(2a^{2b}+3a^{2b}\right)+5}=\boxed{(1-4)a^{3b}+(2+3)a^{2b}+5}=\boxed{\mathbf{-3a^{3b}+5a^{2b}+5}}$

Example 3.3-84

$\boxed{\left(a^2\right)^2+\left(a^3b^3\right)^2-3a^4+2a^6b^6}=$

Solution:

Step 1 $\boxed{\left(a^2\right)^2+\left(a^3b^3\right)^2-3a^4+2a^6b^6}=\boxed{\left(a^{2\times 2}\right)+\left(a^{3\times 2}b^{3\times 2}\right)-3a^4+2a^6b^6}$

$=\boxed{a^4+a^6b^6-3a^4+2a^6b^6}=\boxed{\left(a^4-3a^4\right)+\left(a^6b^6+2a^6b^6\right)}$

Step 2 $\boxed{\left(a^4-3a^4\right)+\left(a^6b^6+2a^6b^6\right)}=\boxed{(1-3)a^4+(1+2)a^6b^6}=\boxed{\mathbf{3a^6b^6-2a^4}}$

Example 3.3-85

$\boxed{\left(3z^4+2z^3-4z^2+5z\right)-\left(z^4+3z^3-7z\right)}=$

Solution:

Step 1 $\boxed{\left(3z^4+2z^3-4z^2+5z\right)-\left(z^4+3z^3-7z\right)}=\boxed{3z^4+2z^3-4z^2+5z-z^4-3z^3+7z}$

$=\boxed{\left(3z^4-z^4\right)+\left(2z^3-3z^3\right)-4z^2+(5z+7z)}$

Step 2 $\boxed{\left(3z^4-z^4\right)+\left(2z^3-3z^3\right)-4z^2+(5z+7z)}=\boxed{(3-1)z^4+(2-3)z^3-4z^2+(5+7)z}$

$=\boxed{\mathbf{2z^4-z^3-4z^2+12z}}$

Additional Examples - Addition and Subtraction of Variables and Numbers Raised to Positive Exponential Terms

The following examples further illustrate addition and subtraction of exponential terms:

Example 3.3-86

$\boxed{5x^3+3x^2+2x^3-x^2+5}=\boxed{\left(5x^3+2x^3\right)+\left(3x^2-x^2\right)+5}=\boxed{(5+2)x^3+(3-1)x^2+5}=\boxed{\mathbf{7x^3+2x^2+5}}$

Example 3.3-87

$\boxed{\left(-2m^4-3m^2+2m^4+3m-10\right)-\left(5m^2+2m+3\right)}=\boxed{-2m^4-3m^2+2m^4+3m-10-5m^2-2m-3}$

$= \boxed{\left(-2m^4 + 2m^4\right) + \left(-3m^2 - 5m^2\right) + (3m - 2m) + (-10 - 3)} = \boxed{(-2+2)m^4 + (-3-5)m^2 + (3-2)m - 13}$

$= \boxed{0m^4 - 8m^2 + m - 13} = \boxed{\mathbf{8m^2 + m - 13}}$

Example 3.3-88

$\boxed{\left(x^5 - 3x^3 + x\right) - \left(-3x^5 - 2x^3 + 3\right)} = \boxed{\left(x^5 - 3x^3 + x\right) + \left(3x^5 + 2x^3 - 3\right)} = \boxed{\left(x^5 + 3x^5\right) + \left(-3x^3 + 2x^3\right) + x - 3}$

$= \boxed{(1+3)x^5 + (-3+2)x^3 + x - 3} = \boxed{\mathbf{4x^5 - x^3 + x - 3}}$

Example 3.3-89

$\boxed{x^2 + 3x^2 + y^2 + x - 4y^2 - 5^2 + 2x^2 + 6x} = \boxed{\left(x^2 + 3x^2 + 2x^2\right) + \left(y^2 - 4y^2\right) + (x + 6x) - 25}$

$= \boxed{(1+3+2)x^2 + (1-4)y^2 + (1+6)x - 25} = \boxed{\mathbf{6x^2 - 3y^2 + 7x - 25}}$

Example 3.3-90

$\boxed{\left(k^4 - 3k^2 - 4\right) - \left(-3k^4 + 5k^2\right) - 3} = \boxed{\left(k^4 - 3k^2 - 4\right) + \left(3k^4 - 5k^2\right) - 3} = \boxed{\left(k^4 + 3k^4\right) + \left(-3k^2 - 5k^2\right) + (-4 - 3)}$

$= \boxed{(1+3)k^4 + (-3-5)k^2 - 7} = \boxed{\mathbf{4k^4 - 8k^2 - 7}}$

Example 3.3-91

$\boxed{\left(-5w^3 - 3w - 5\right) - \left(3w^3 - w - 4\right) + 5w + 2} = \boxed{\left(-5w^3 - 3w - 5\right) + \left(-3w^3 + w + 4\right) + 5w + 2}$

$= \boxed{\left(-5w^3 - 3w^3\right) + (-3w + 5w + w) + (-5 + 4 + 2)} = \boxed{(-5-3)w^3 + (-3+5+1)w + 1} = \boxed{\mathbf{-8w^3 + 3w + 1}}$

Example 3.3-92

$\boxed{\left(5x^5 - 4x^4 + 3x^2 - 2x^5 + 6\right) - \left(2x^4 - 3x^3 - 4x^2 + 2\right)} = \boxed{\left(5x^5 - 4x^4 + 3x^2 - 2x^5 + 6\right) + \left(-2x^4 + 3x^3 + 4x^2 - 2\right)}$

$= \boxed{\left(5x^5 - 2x^5\right) + \left(-4x^4 - 2x^4\right) + 3x^3 + \left(3x^2 + 4x^2\right) + (6 - 2)} = \boxed{(5-2)x^5 + (-4-2)x^4 + 3x^3 + (3+4)x^2 + 4}$

$= \boxed{\mathbf{3x^5 - 6x^4 + 3x^3 + 7x^2 + 4}}$

Example 3.3-93

$\boxed{\left(a^2b^3 + 3a^2 - b + 2^4\right) + \left(2a^2b^3 + 2a^2 + 3^0\right) - 3^3} = \boxed{a^2b^3 + 3a^2 - b + 16 + 2a^2b^3 + 2a^2 + 1 - 27}$

$= \boxed{\left(a^2b^3 + 2a^2b^3\right) + \left(3a^2 + 2a^2\right) - b + (16 + 1 - 27)} = \boxed{(1+2)a^2b^3 + (3+2)a^2 - b - 10} = \boxed{\mathbf{3a^2b^3 + 5a^2 - b - 10}}$

Example 3.3-94

$$\boxed{-\left(-3n^2+4m^3-4m^2\right)-\left(3m^3+m^2-6m^3\right)-2n^2} = \boxed{\left(3n^2-4m^3+4m^2\right)+\left(-3m^3-m^2+6m^3\right)-2n^2}$$

$$= \boxed{\left(3n^2-2n^2\right)+\left(-4m^3-3m^3+6m^3\right)+\left(4m^2-m^2\right)} = \boxed{(3-2)n^2+(-4-3+6)m^3+(4-1)m^2}$$

$$= \boxed{\mathbf{n^2-m^3+3m^2}}$$

Example 3.3-95

$$\boxed{\left(c^3+12c^2+8c+8\right)+\left(-2c^3-5c^2+4c\right)-4c^3-2c} = \boxed{c^3+12c^2+8c+8-2c^3-5c^2+4c-4c^3-2c}$$

$$= \boxed{\left(c^3-2c^3-4c^3\right)+\left(12c^2-5c^2\right)+(8c+4c-2c)+8} = \boxed{(1-2-4)c^3+(12-5)c^2+(8+4-2)c+8}$$

$$= \boxed{\mathbf{-5c^3+7c^2+10c+8}}$$

Practice Problems - Addition and Subtraction of Variables and Numbers Raised to Positive Exponential Terms

Section 3.3 Case III a Practice Problems - Add or subtract the following positive integer exponential expressions:

1. $x^2+4xy-2x^2-2xy+z^3=$
2. $\left(a^3+2a^2+4^3\right)-\left(4a^3+20\right)=$
3. $3x^4+2x^2+2x^4-\left(x^4-2x^2+3\right)=$
4. $-\left(-2l^3a^3+2l^2a^2-5^3\right)-\left(4l^3a^3-20\right)=$
5. $\left(m^{3n}-4m^{2n}\right)-\left(2m^{3n}+3m^{2n}\right)+5m=$
6. $\left(-7z^3+3z-5\right)-\left(-3z^3+z-4\right)+5z+20=$
7. $\left(a^3\right)^2+\left(a^2\cdot b^2\right)^2-5a^6+3a^4b^4+2a^6=$
8. $\left(k^5+10k^2+5\right)+\left(-2k^5-5k^2+5k\right)-4k^3-k=$
9. $\left(3x^2+xy-x^2+3x^3\right)-\left(2x^3-y^3-4y^3+x^3\right)=$
10. $\left(xy^2+20x^2+5x\right)-\left(3xy^2+20x\right)+2^4=$

Case III b Addition and Subtraction of Positive Exponential Terms in Fraction Form

A special class of positive integer exponents are in the form of fractions. This class of positive integer exponents are added and subtracted and further simplified by applying the fraction techniques, discussed in Chapter 2, and the exponent laws (see Table 3.3-1). It is recommended that students review addition and subtraction of integer fractions (see Sections 2.2 and 2.3) before proceeding with this section. Positive integer exponents in fraction form are added and subtracted using the following steps:

Step 1 Change the exponential expression x^a to $\frac{x^a}{1}$.

Step 2 Simplify the exponential expression by:

a. Using the fraction techniques learned in Chapter 2, and

b. Using appropriate exponent laws such as the Multiplication Law (Law I) from Table 3.3-1.

Examples with Steps

The following examples show the steps as to how exponential expressions in fraction form are added and subtracted:

Example 3.3-96

$$3^3 + \frac{1}{2^3} =$$

Solution:

Step 1

$$3^3 + \frac{1}{2^3} = \frac{3^3}{1} + \frac{1}{2^3} = \frac{27}{1} + \frac{1}{8}$$

Step 2

$$\frac{27}{1} + \frac{1}{8} = \frac{(27 \cdot 8) + (1 \cdot 1)}{1 \cdot 8} = \frac{216 + 1}{8} = \mathbf{\frac{217}{8}}$$

Example 3.3-97

$$a^2 - \frac{1}{3 - a^2} =$$

Solution:

Step 1

$$a^2 - \frac{1}{3 - a^2} = \frac{a^2}{1} - \frac{1}{3 - a^2}$$

Step 2

$$\frac{a^2}{1} - \frac{1}{3 - a^2} = \frac{\left[a^2 \cdot \left(3 - a^2\right)\right] - (1 \cdot 1)}{1 \cdot \left(3 - a^2\right)} = \frac{\left(3 \cdot a^2 - a^2 \cdot a^2\right) - 1}{3 - a^2} = \frac{3a^2 - a^{2+2} - 1}{3 - a^2}$$

$$= \frac{3a^2 - a^4 - 1}{3 - a^2} = \mathbf{\frac{-a^4 + 3a^2 - 1}{-a^2 + 3}}$$

Example 3.3-98

$$\boxed{\frac{1}{x}+x^2}=$$

Solution:

Step 1

$$\boxed{\frac{1}{x}+x^2}=\boxed{\frac{1}{x}+\frac{x^2}{1}}$$

Step 2

$$\boxed{\frac{1}{x}+\frac{x^2}{1}}=\boxed{\frac{(1\cdot 1)+\left(x^2\cdot x\right)}{(x\cdot 1)}}=\boxed{\frac{1+\left(x^2\cdot x^1\right)}{x}}=\boxed{\frac{1+\left(x^{2+1}\right)}{x}}=\boxed{\frac{1+x^3}{x}}=\boxed{\mathbf{\frac{x^3+1}{x}}}$$

Example 3.3-99

$$\boxed{(x+1)-\frac{(1+3x)}{2x}}=$$

Solution:

Step 1

$$\boxed{(x+1)-\frac{(1+3x)}{2x}}=\boxed{\frac{(x+1)}{1}-\frac{(1+3x)}{2x}}$$

Step 2

$$\boxed{\frac{(x+1)}{1}-\frac{(1+3x)}{2x}}=\boxed{\frac{[(x+1)\cdot 2x]-[1\cdot(1+3x)]}{(1\cdot 2x)}}=\boxed{\frac{\left(2x^2+2x\right)-(1+3x)}{2x}}$$

$$=\boxed{\frac{2x^2+2x-1-3x}{2x}}=\boxed{\frac{2x^2+(2x-3x)-1}{2x}}=\boxed{\frac{2x^2+(2-3)x-1}{2x}}=\boxed{\mathbf{\frac{2x^2-x-1}{2x}}}$$

Example 3.3-100

$$\boxed{\frac{a^2-b^2}{2}+3^2b^2}=$$

Solution:

Step 1

$$\boxed{\frac{a^2-b^2}{2}+3^2b^2}=\boxed{\frac{a^2-b^2}{2}+9\,b^2}=\boxed{\frac{a^2-b^2}{2}+\frac{9\,b^2}{1}}$$

Step 2

$$\boxed{\frac{a^2-b^2}{2}+\frac{9\,b^2}{1}}=\boxed{\frac{\left[\left(a^2-b^2\right)\cdot 1\right]+\left(9b^2\cdot 2\right)}{2\cdot 1}}=\boxed{\frac{\left(a^2-b^2\right)+18b^2}{2}}=\boxed{\frac{a^2+\left(18\,b^2-b^2\right)}{2}}$$

$$=\boxed{\frac{a^2+(18-1)b^2}{2}}=\boxed{\mathbf{\frac{a^2+17\,b^2}{2}}}$$

Additional Examples - Addition and Subtraction of Positive Exponential Terms in Fraction Form

The following examples further illustrate addition and subtraction of exponential terms in fraction form:

Example 3.3-101

$$\frac{3x^2+2xy-x^2}{x^2}-\frac{2x^3+y^3-2y^3-x^3+xy}{x^2}=\frac{(3x^2+2xy-x^2)-(2x^3+y^3-2y^3-x^3+xy)}{x^2}$$

$$=\frac{3x^2+2xy-x^2-2x^3-y^3+2y^3+x^3-xy}{x^2}=\frac{(-2x^3+x^3)+(3x^2-x^2)+(2y^3-y^3)+(2xy-xy)}{x^2}$$

$$=\frac{(-2+1)x^3+(3-1)x^2+(2-1)y^3+(2-1)xy}{x^2}=\boldsymbol{\frac{-x^3+2x^2+y^3+xy}{x^2}}$$

Example 3.3-102

$$\frac{a^2-b^2}{2}+\frac{b^2}{2^2}=\frac{a^2-b^2}{2}+\frac{b^2}{4}=\frac{[(a^2-b^2)\cdot 4]+(2\cdot b^2)}{2\cdot 4}=\frac{(4a^2-4b^2)+2b^2}{8}=\frac{4a^2+(-4b^2+2b^2)}{8}$$

$$=\frac{4a^2+(-4+2)b^2}{8}=\frac{4a^2-2b^2}{8}=\frac{\overset{1}{\cancel{2}}(2a^2-b^2)}{\underset{4}{\cancel{8}}}=\boldsymbol{\frac{2a^2-b^2}{4}}$$

Example 3.3-103

$$\frac{x+3^2}{x^2}+\frac{1}{x}=\frac{x+9}{x^2}+\frac{1}{x}=\frac{[(x+9)\cdot x]+(1\cdot x^2)}{x^2\cdot x}=\frac{(x\cdot x+9\cdot x)+x^2}{x^3}=\frac{x^2+9x+x^2}{x^3}=\frac{(x^2+x^2)+9x}{x^3}$$

$$=\frac{(1+1)x^2+9x}{x^3}=\frac{2x^2+9x}{x^3}=\frac{x(2x+9)}{x^3}=\frac{2x+9}{x^3\cdot x^{-1}}=\frac{2x+9}{x^{3-1}}=\boldsymbol{\frac{2x+9}{x^2}}$$

Example 3.3-104

$$\frac{5^2}{x^2-y}-x^3=\frac{25}{x^2-y}-\frac{x^3}{1}=\frac{(25\cdot 1)-[x^3\cdot(x^2-y)]}{(x^2-y)\cdot 1}=\frac{25-(x^3\cdot x^2-x^3\cdot y)}{x^2-y}=\frac{25-(x^{3+2}-x^3y)}{x^2-y}$$

$$=\frac{25-x^5+x^3y}{x^2-y}=\boldsymbol{\frac{-x^5+x^3y+25}{x^2-y}}$$

Example 3.3-105

$$\frac{x^2+3x}{5}-\frac{2x^2-5x}{6}=\frac{[(x^2+3x)\cdot 6]-[(2x^2-5x)\cdot 5]}{5\cdot 6}=\frac{(6x^2+18x)-(10x^2-25x)}{30}$$

$$= \boxed{\frac{6x^2 + 18x - 10x^2 + 25x}{30}} = \boxed{\frac{\left(6x^2 - 10x^2\right) + (18x + 25x)}{30}} = \boxed{\frac{(6-10)x^2 + (18+25)x}{30}} = \boxed{\mathbf{\frac{-4x^2 + 43x}{30}}}$$

Example 3.3-106

$$\boxed{\frac{x^2 - y^2}{3} + \frac{x^2}{2}} = \boxed{\frac{\left[\left(x^2 - y^2\right)\cdot 2\right] + \left(3 \cdot x^2\right)}{3 \cdot 2}} = \boxed{\frac{2x^2 - 2y^2 + 3x^2}{6}} = \boxed{\frac{\left(2x^2 + 3x^2\right) - 2y^2}{6}} = \boxed{\frac{(2+3)x^2 - 2y^2}{6}}$$

$$= \boxed{\mathbf{\frac{5x^2 - 2y^2}{6}}}$$

Example 3.3-107

$$\boxed{\frac{3x^2 + 4}{xy} - \frac{2x}{y}} = \boxed{\frac{\left[\left(3x^2 + 4\right)\cdot y\right] - (2x \cdot xy)}{xy \cdot y}} = \boxed{\frac{3x^2 y + 4y - 2x^2 y}{xy^2}} = \boxed{\frac{\left(3x^2 y - 2x^2 y\right) + 4\,y}{xy^2}} = \boxed{\frac{(3-2)x^2 y + 4\,y}{xy^2}}$$

$$= \boxed{\frac{x^2 y + 4\,y}{xy^2}} = \boxed{\frac{\left(x^2 + 4\right)y}{xy^2}} = \boxed{\frac{x^2 + 4}{x\left(y^2 \cdot y^{-1}\right)}} = \boxed{\frac{x^2 + 4}{x\left(y^{2-1}\right)}} = \boxed{\frac{x^2 + 4}{xy^1}} = \boxed{\mathbf{\frac{x^2 + 4}{xy}}}$$

Example 3.3-108

$$\boxed{\frac{2x^3 + 3x^2 + 1}{2} + \frac{5x^3 - 3x^2}{3}} = \boxed{\frac{\left[\left(2x^3 + 3x^2 + 1\right)\cdot 3\right] + \left[\left(5x^3 - 3x^2\right)\cdot 2\right]}{2 \cdot 3}} = \boxed{\frac{\left(6x^3 + 9x^2 + 3\right) + \left(10x^3 - 6x^2\right)}{6}}$$

$$= \boxed{\frac{\left(6x^3 + 10x^3\right) + \left(9x^2 - 6x^2\right) + 3}{6}} = \boxed{\frac{(6+10)x^3 + (9-6)x^2 + 3}{6}} = \boxed{\mathbf{\frac{16x^3 + 3x^2 + 3}{6}}}$$

Example 3.3-109

$$\boxed{\frac{a^2 + b^2}{a^2 - b^2} + 3^2} = \boxed{\frac{a^2 + b^2}{a^2 - b^2} + 9} = \boxed{\frac{a^2 + b^2}{a^2 - b^2} + \frac{9}{1}} = \boxed{\frac{\left[\left(a^2 + b^2\right)\cdot 1\right] + \left[9 \cdot \left(a^2 - b^2\right)\right]}{\left(a^2 - b^2\right)\cdot 1}} = \boxed{\frac{\left(a^2 + b^2\right) + \left(9a^2 - 9b^2\right)}{a^2 - b^2}}$$

$$= \boxed{\frac{\left(a^2 + 9\,a^2\right) + \left(b^2 - 9b^2\right)}{a^2 - b^2}} = \boxed{\frac{(1+9)a^2 + (1-9)b^2}{a^2 - b^2}} = \boxed{\mathbf{\frac{10\,a^2 - 8\,b^2}{a^2 - b^2}}}$$

Example 3.3-110

$$\boxed{\frac{x^3 + x^2 + x}{2} - \frac{2x^3 - x^2 + 2}{3}} = \boxed{\frac{\left[\left(x^3 + x^2 + x\right)\cdot 3\right] - \left[\left(2x^3 - x^2 + 2\right)\cdot 2\right]}{2 \cdot 3}} = \boxed{\frac{\left(3x^3 + 3x^2 + 3x\right) - \left(4x^3 - 2x^2 + 4\right)}{6}}$$

$$= \boxed{\frac{3x^3 + 3x^2 + 3x - 4x^3 + 2x^2 - 4}{6}} = \boxed{\frac{\left(-4x^3 + 3x^3\right) + \left(3x^2 + 2x^2\right) + 3x - 4}{6}} = \boxed{\frac{(-4+3)x^3 + (3+2)x^2 + 3x - 4}{6}}$$

$$= \boxed{\frac{-x^3 + 5x^2 + 3x - 4}{6}}$$

Practice Problems - Addition and Subtraction of Positive Exponential Terms in Fraction Form

Section 3.3 Case III b Practice Problems - Simplify the following positive integer exponential expressions shown in fraction form:

1. $\dfrac{2-2^3}{7} + \dfrac{4^2}{3} =$
2. $\dfrac{b^2 + 3b - 4b^2}{c} - \dfrac{4b + b^2}{c} =$
3. $\dfrac{2a^3 - 3b^3}{a^3 + b^3} - 2^2 =$
4. $\dfrac{3x^2 + 3x}{5} - \dfrac{x^2 + x}{5} =$
5. $\dfrac{y^2}{y - y^3} + \dfrac{y}{2} =$
6. $\dfrac{b+2}{1+b} + \dfrac{1}{b} =$
7. $\dfrac{x + y^2}{x - y^2} + \dfrac{2^2}{3} =$
8. $\dfrac{4l^2 + 5}{lm} - \dfrac{5l}{m} =$
9. $\dfrac{4xy^2}{2x - 3} + y^2 =$
10. $\dfrac{2a^3}{a+1} - \dfrac{a^2}{3} =$

3.4 Operations with Negative Integer Exponents

To proceed with simplification of negative exponents, we need to know the Negative Power Law in addition to the other exponent laws (shown in Table 3.4-1). The Negative Power Law states that a base raised to a negative exponent is equal to one divided by the same base raised to the positive exponent, or vice versa, i.e.,

$$a^{-n} = \frac{1}{a^n}$$

and

$$a^n = \frac{1}{a^{-n}} \qquad \text{since} \qquad a^n = \frac{1}{a^{-n}} = \frac{1}{\frac{1}{a^n}} = \frac{\frac{1}{1}}{\frac{1}{a^n}} = \frac{1 \times a^n}{1 \times 1} = \frac{a^n}{1} = a^n.$$

Note that the objective is to write the final answer without a negative exponent. To achieve that the exponent laws are used when simplifying expressions having negative integer exponents. These laws are used to simplify the work in solving exponential expressions and should be memorized.

Table 3.4-1: Exponent Laws 1 through 6 (Negative Integer Exponents)

I. **Multiplication**	$a^{-m} \cdot a^{-n} = a^{-m-n}$	When multiplying negative exponential terms, if bases a are the same, add the negative exponents $-m$ and $-n$.
II. **Power of a Power**	$\left(a^{-m}\right)^{-n} = a^{-m \times -n}$	When raising a negative exponential term to a negative power, multiply the negative powers (exponents) $-m$ and $-n$.
III. **Power of a Product**	$(a \cdot b)^{-m} = a^{-m} \cdot b^{-m}$	When raising a product to a negative power, raise each factor a and b to the negative power $-m$.
IV. **Power of a Fraction**	$\left(\frac{a}{b}\right)^{-m} = \frac{a^{-m}}{b^{-m}}$	When raising a fraction to a negative power, raise the numerator and the denominator to the negative power $-m$.
V. **Division**	$\frac{a^{-m}}{a^{-n}} = a^{-m} \cdot a^n = a^{-m+n}$	When dividing negative exponential terms, if the bases a are the same, add exponents $-m$ and n.
VI. **Negative Power**	$a^{-n} = \frac{1}{a^n}$	A non-zero base a raised to the $-n$ power equals 1 divided by the base a to the n power

In this section students learn how to multiply (Case I), divide (Case II), and add or subtract (Case III) negative integer exponents by one another.

Case I Multiplying Negative Integer Exponents

Negative integer exponents are multiplied by each other using the exponent laws I through III shown in Table 3.4-2.

Table 3.4-2: Exponent Laws 1 through 3 (Negative Integer Exponents)

I. **Multiplication**	$a^{-m} \cdot a^{-n} = a^{-m-n}$	When multiplying negative exponential terms, if bases a are the same, add the negative exponents $-m$ and $-n$.
II. **Power of a Power**	$\left(a^{-m}\right)^{-n} = a^{-m \times -n}$	When raising a negative exponential term to a negative power, multiply the negative powers (exponents) $-m$ and $-n$.
III. **Power of a Product**	$(a \cdot b)^{-m} = a^{-m} \cdot b^{-m}$	When raising a product to a negative power, raise each factor a and b to the negative power $-m$.

Multiplication of expressions by negative exponents is divided to two cases: Case I a addressing simple cases and Case I b addressing more difficult cases.

Case I a Multiplying Negative Integer Exponents (Simple Cases)

Negative integer exponents are multiplied by each other using the following steps:

Step 1 Group the exponential terms with similar bases.

Step 2 Apply the Multiplication Law (Law I) from Table 3.4-2 and simplify the exponential expressions by adding the exponents with similar bases.

Step 3 Change the negative integer exponents to positive integer exponents.

Examples With Steps

The following examples show the steps as to how negative integer exponents are multiplied by one another:

Example 3.4-1

$$\boxed{3^{-2} \cdot 2 \cdot 3^{-1} \cdot 2^{-3}} =$$

Solution:

Step 1

$$\boxed{3^{-2} \cdot 2 \cdot 3^{-1} \cdot 2^{-3}} = \boxed{3^{-2} \cdot 2^{1} \cdot 3^{-1} \cdot 2^{-3}} = \boxed{\left(3^{-2} \cdot 3^{-1}\right) \cdot \left(2^{-3} \cdot 2^{1}\right)}$$

Step 2

$$\boxed{\left(3^{-2} \cdot 3^{-1}\right) \cdot \left(2^{-3} \cdot 2^{1}\right)} = \boxed{\left(3^{-2-1}\right) \cdot \left(2^{-3+1}\right)} = \boxed{3^{-3} \cdot 2^{-2}}$$

Step 3

$$\boxed{3^{-3} \cdot 2^{-2}} = \boxed{\frac{1}{3^{3} \cdot 2^{2}}} = \boxed{\frac{1}{(3 \cdot 3 \cdot 3) \cdot (2 \cdot 2)}} = \boxed{\mathbf{\frac{1}{108}}}$$

Example 3.4-2

$$\boxed{5^{-2} \cdot 5 \cdot a^{-3} \cdot b^{-3} \cdot a^{-1} \cdot b} =$$

Solution:

Step 1 $\boxed{5^{-2}\cdot 5\cdot a^{-3}\cdot b^{-3}\cdot a^{-1}\cdot b} = \boxed{5^{-2}\cdot 5^{1}\cdot a^{-3}\cdot b^{-3}\cdot a^{-1}\cdot b^{1}} = \boxed{\left(5^{-2}5^{1}\right)\cdot\left(a^{-3}a^{-1}\right)\cdot\left(b^{1}b^{-3}\right)}$

Step 2 $\boxed{\left(5^{-2}5^{1}\right)\cdot\left(a^{-3}a^{-1}\right)\cdot\left(b^{1}b^{-3}\right)} = \boxed{\left(5^{-2+1}\right)\cdot\left(a^{-3-1}\right)\cdot\left(b^{1-3}\right)} = \boxed{5^{-1}\cdot a^{-4}\cdot b^{-2}} = \boxed{5^{-1}a^{-4}b^{-2}}$

Step 3 $\boxed{5^{-1}a^{-4}b^{-2}} = \boxed{\frac{1}{5^{1}}\cdot\frac{1}{a^{4}}\cdot\frac{1}{b^{2}}} = \boxed{\frac{1\cdot 1\cdot 1}{5\cdot a^{4}\cdot b^{2}}} = \boxed{\mathbf{\frac{1}{5a^{4}b^{2}}}}$

Example 3.4-3

$\boxed{\left(x^{-2}y^{-2}z^{2}\right)\cdot\left(x^{-1}y^{3}z^{-4}\right)} =$

Solution:

Step 1 $\boxed{\left(x^{-2}y^{-2}z^{2}\right)\cdot\left(x^{-1}y^{3}z^{-4}\right)} = \boxed{x^{-2}y^{-2}z^{2}\cdot x^{-1}y^{3}z^{-4}} = \boxed{\left(x^{-2}x^{-1}\right)\cdot\left(y^{3}y^{-2}\right)\cdot\left(z^{2}z^{-4}\right)}$

Step 2 $\boxed{\left(x^{-2}x^{-1}\right)\cdot\left(y^{3}y^{-2}\right)\cdot\left(z^{2}z^{-4}\right)} = \boxed{\left(x^{-2-1}\right)\cdot\left(y^{3-2}\right)\cdot\left(z^{2-4}\right)} = \boxed{x^{-3}\cdot y^{1}\cdot z^{-2}} = \boxed{x^{-3}\cdot y\cdot z^{-2}}$

Step 3 $\boxed{x^{-3}\cdot y\cdot z^{-2}} = \boxed{\frac{1}{x^{3}}\cdot\frac{y}{1}\cdot\frac{1}{z^{2}}} = \boxed{\frac{1\cdot y\cdot 1}{x^{3}\cdot 1\cdot z^{2}}} = \boxed{\mathbf{\frac{y}{x^{3}z^{2}}}}$

Example 3.4-4

$\boxed{\left(r^{-2}s^{-1}t^{0}\right)\cdot\left(r^{3}s^{-3}t^{-4}\right)} =$

Solution:

Step 1 $\boxed{\left(r^{-2}s^{-1}t^{0}\right)\cdot\left(r^{3}s^{-3}t^{-4}\right)} = \boxed{r^{-2}s^{-1}t^{0}\cdot r^{3}s^{-3}t^{-4}} = \boxed{\left(r^{-2}r^{3}\right)\cdot\left(s^{-1}s^{-3}\right)\cdot\left(t^{-4}t^{0}\right)}$

Step 2 $\boxed{\left(r^{-2}r^{3}\right)\cdot\left(s^{-1}s^{-3}\right)\cdot\left(t^{-4}t^{0}\right)} = \boxed{\left(r^{-2+3}\right)\cdot\left(s^{-1-3}\right)\cdot\left(t^{-4+0}\right)} = \boxed{r^{1}\cdot s^{-4}\cdot t^{-4}} = \boxed{r\cdot s^{-4}\cdot t^{-4}}$

Step 3 $\boxed{r\cdot s^{-4}\cdot t^{-4}} = \boxed{\frac{r}{1}\cdot\frac{1}{s^{4}}\cdot\frac{1}{t^{4}}} = \boxed{\frac{r\cdot 1\cdot 1}{1\cdot s^{4}\cdot t^{4}}} = \boxed{\mathbf{\frac{r}{s^{4}t^{4}}}}$

Example 3.4-5

$\boxed{-\left(2^{-3}\right)\cdot\left(w^{-2}y^{-3}z^{-1}\right)\cdot\left(w^{-4}y^{-1}z^{3}\right)} =$

Solution:

Step 1 $\boxed{-\left(2^{-3}\right)\cdot\left(w^{-2}y^{-3}z^{-1}\right)\cdot\left(w^{-4}y^{-1}z^{3}\right)} = \boxed{-\left(2^{-3}\right)\cdot w^{-2}y^{-3}z^{-1}\cdot w^{-4}y^{-1}z^{3}}$

$= \boxed{-\left(2^{-3}\right)\cdot\left(w^{-2}w^{-4}\right)\cdot\left(y^{-3}y^{-1}\right)\cdot\left(z^{-1}z^{3}\right)}$

Step 2 $\boxed{-\left(2^{-3}\right)\cdot\left(w^{-2}w^{-4}\right)\cdot\left(y^{-3}y^{-1}\right)\cdot\left(z^{-1}z^{3}\right)} = \boxed{-\left(2^{-3}\right)\cdot\left(w^{-2-4}\right)\cdot\left(y^{-3-1}\right)\cdot\left(z^{-1+3}\right)}$

$= \boxed{-2^{-3}\cdot w^{-6}\cdot y^{-4}\cdot z^{2}}$

Step 3 $\boxed{-2^{-3}\cdot w^{-6}\cdot y^{-4}\cdot z^{2}} = \boxed{-\frac{1}{2^3}\cdot\frac{1}{w^6}\cdot\frac{1}{y^4}\cdot z^2} = \boxed{-\frac{1}{(2\cdot 2\cdot 2)}\cdot\frac{1}{w^6}\cdot\frac{1}{y^4}\cdot\frac{z^2}{1}}$

$= \boxed{-\frac{1\cdot 1\cdot 1\cdot z^2}{8\cdot w^6\cdot y^4\cdot 1}} = \boxed{-\frac{z^2}{8w^6y^4}}$

Additional Examples - Multiplying Negative Integer Exponents (Simple Cases)

The following examples further illustrate how to multiply negative exponential terms by one another:

Example 3.4-6

$\boxed{a^{-2}\cdot a^{3}\cdot a^{-1}} = \boxed{a^{-2+3-1}} = \boxed{a^{0}} = \boxed{1}$

Example 3.4-7

$\boxed{5^{-1}\cdot 5^{-2}} = \boxed{5^{-1-2}} = \boxed{5^{-3}} = \boxed{\frac{1}{5^3}} = \boxed{\frac{1}{5\cdot 5\cdot 5}} = \boxed{\frac{1}{125}}$

Example 3.4-8

$\boxed{3^{-2}\cdot 3^{-1}\cdot 2^{-2}\cdot 3^{-4}\cdot 2} = \boxed{3^{-2}\cdot 3^{-1}\cdot 2^{-2}\cdot 3^{-4}\cdot 2^{1}} = \boxed{\left(3^{-2}\cdot 3^{-1}\cdot 3^{-4}\right)\cdot\left(2^{1}\cdot 2^{-2}\right)} = \boxed{\left(3^{-2-1-4}\right)\cdot\left(2^{1-2}\right)}$

$= \boxed{3^{-7}\cdot 2^{-1}} = \boxed{\frac{1}{3^7}\cdot\frac{1}{2^1}} = \boxed{\frac{1}{3\cdot 3\cdot 3\cdot 3\cdot 3\cdot 3\cdot 3}\cdot\frac{1}{2}} = \boxed{\frac{1}{2187}\cdot\frac{1}{2}} = \boxed{\frac{1\cdot 1}{2187\cdot 2}} = \boxed{\frac{1}{4374}}$

Example 3.4-9

$\boxed{a^{-5}\cdot b^{-3}\cdot a^{-1}\cdot b^{-2}} = \boxed{\left(a^{-5}\cdot a^{-1}\right)\cdot\left(b^{-3}\cdot b^{-2}\right)} = \boxed{\left(a^{-5-1}\right)\cdot\left(b^{-3-2}\right)} = \boxed{a^{-6}\cdot b^{-5}} = \boxed{a^{-6}b^{-5}} = \boxed{\frac{1}{a^6b^5}}$

Example 3.4-10

$\boxed{\left(3^{-2}\cdot 2^{-1}\right)\cdot\left(3^{-1}\cdot 2^{-2}\right)\cdot 2} = \boxed{3^{-2}\cdot 2^{-1}\cdot 3^{-1}\cdot 2^{-2}\cdot 2^{1}} = \boxed{\left(3^{-2}\cdot 3^{-1}\right)\cdot\left(2^{-2}\cdot 2^{-1}\cdot 2^{1}\right)} = \boxed{\left(3^{-2-1}\right)\cdot\left(2^{-2-1+1}\right)}$

$= \boxed{3^{-3}\cdot 2^{-2}} = \boxed{\frac{1}{3^3}\cdot\frac{1}{2^2}} = \boxed{\frac{1}{3\cdot 3\cdot 3}\cdot\frac{1}{2\cdot 2}} = \boxed{\frac{1}{27}\cdot\frac{1}{4}} = \boxed{\frac{1\cdot 1}{27\cdot 4}} = \boxed{\frac{1}{108}}$

Example 3.4-11

$\boxed{\left(5^{-3}\cdot 5^{-1}\right)\cdot\left(2^{-3}\cdot 5^{-2}\right)} = \boxed{5^{-3}\cdot 5^{-1}\cdot 2^{-3}\cdot 5^{-2}} = \boxed{\left(5^{-3}\cdot 5^{-1}\cdot 5^{-2}\right)\cdot 2^{-3}} = \boxed{\left(5^{-3-1-2}\right)\cdot 2^{-3}} = \boxed{5^{-6}\cdot 2^{-3}} = \boxed{\frac{1}{5^6}\cdot\frac{1}{2^3}}$

$$=\boxed{\frac{1}{(5\cdot5\cdot5\cdot5\cdot5\cdot5)}\cdot\frac{1}{(2\cdot2\cdot2)}}=\boxed{\frac{1}{15625}\cdot\frac{1}{8}}=\boxed{\frac{1\cdot1}{(15625\cdot8)}}=\boxed{\mathbf{\frac{1}{125000}}}$$

Example 3.4-12

$$\boxed{\left(a^{-2}b^{-3}c\right)\cdot\left(a^{-1}b\right)\cdot c^{-2}}=\boxed{a^{-2}\cdot b^{-3}\cdot c^{1}\cdot a^{-1}\cdot b^{1}\cdot c^{-2}}=\boxed{\left(a^{-2}\cdot a^{-1}\right)\cdot\left(b^{-3}\cdot b^{1}\right)\cdot\left(c^{-2}\cdot c^{1}\right)}$$

$$=\boxed{\left(a^{-2-1}\right)\cdot\left(b^{-3+1}\right)\cdot\left(c^{-2+1}\right)}=\boxed{a^{-3}\cdot b^{-2}\cdot c^{-1}}=\boxed{\frac{1}{a^{3}}\cdot\frac{1}{b^{2}}\cdot\frac{1}{c^{1}}}=\boxed{\frac{1\cdot1\cdot1}{a^{3}\cdot b^{2}\cdot c}}=\boxed{\mathbf{\frac{1}{a^{3}b^{2}c}}}$$

Example 3.4-13

$$\boxed{-(2-3)^{-1}\cdot\left(5^{-3}\cdot2^{-2}\cdot3\right)\cdot\left(5^{2}\cdot2^{3}\cdot3^{-2}\right)}=\boxed{-(-1)^{-1}\cdot5^{-3}\cdot2^{-2}\cdot3^{1}\cdot5^{2}\cdot2^{3}\cdot3^{-2}}$$

$$=\boxed{-(-1)^{-1}\cdot\left(5^{-3}\cdot5^{2}\right)\cdot\left(3^{1}\cdot3^{-2}\right)\cdot\left(2^{3}\cdot2^{-2}\right)}=\boxed{-(-1)^{-1}\cdot\left(5^{-3+2}\right)\cdot\left(3^{1-2}\right)\cdot\left(2^{3-2}\right)}=\boxed{-(-1)^{-1}\cdot5^{-1}\cdot3^{-1}\cdot2}$$

$$=\boxed{-\frac{1}{(-1)^{1}}\cdot\frac{1}{5^{1}}\cdot\frac{1}{3^{1}}\cdot\frac{2}{1}}=\boxed{-\frac{1}{-1}\cdot\frac{1}{5}\cdot\frac{1}{3}\cdot\frac{2}{1}}=\boxed{+\frac{1}{1}\cdot\frac{1}{5}\cdot\frac{1}{3}\cdot\frac{2}{1}}=\boxed{\frac{1\cdot1\cdot1\cdot2}{1\cdot5\cdot3\cdot1}}=\boxed{\mathbf{\frac{2}{15}}}$$

Example 3.4-14

$$\boxed{(-1+3)^{-2}\left(r^{-2}s^{2}t\right)\cdot\left(r^{3}s^{-2}t^{-3}s\right)}=\boxed{(2)^{-2}r^{-2}s^{2}t^{1}\cdot r^{3}s^{-2}t^{-3}s^{1}}=\boxed{2^{-2}\left(r^{-2}r^{3}\right)\cdot\left(s^{2}s^{-2}s^{1}\right)\cdot\left(t^{1}t^{-3}\right)}$$

$$=\boxed{2^{-2}\left(r^{-2+3}\right)\cdot\left(s^{2-2+1}\right)\cdot\left(t^{1-3}\right)}=\boxed{2^{-2}r^{1}\cdot s^{1}\cdot t^{-2}}=\boxed{2^{-2}\cdot r\cdot s\cdot t^{-2}}=\boxed{\frac{1}{2^{2}}\cdot\frac{r}{1}\cdot\frac{s}{1}\cdot\frac{1}{t^{2}}}=\boxed{\frac{1\cdot r\cdot s\cdot1}{2\cdot2\cdot1\cdot1\cdot t^{2}}}=\boxed{\mathbf{\frac{rs}{4t^{2}}}}$$

Example 3.4-15

$$\boxed{r\frac{m^{-3}}{-2}\cdot m^{-2}\cdot m^{-2}\cdot-4mr^{2}}=\boxed{\frac{-4}{-2}r^{1}\cdot m^{-3}\cdot m^{-2}\cdot m^{-2}\cdot m^{1}\cdot r^{2}}=\boxed{+\frac{\overset{2}{\cancel{4}}}{\underset{1}{\cancel{2}}}\left(m^{-3}\cdot m^{-2}\cdot m^{-2}\cdot m^{1}\right)\cdot\left(r^{1}\cdot r^{2}\right)}$$

$$=\boxed{\frac{2}{1}\left(m^{-3-2-2+1}\right)\cdot\left(r^{1+2}\right)}=\boxed{\frac{2}{1}m^{-6}\cdot r^{3}}=\boxed{\frac{2}{1}\cdot\frac{1}{m^{6}}\cdot\frac{r^{3}}{1}}=\boxed{\frac{2\cdot1\cdot r^{3}}{1\cdot m^{6}\cdot1}}=\boxed{\mathbf{\frac{2r^{3}}{m^{6}}}}$$

Example 3.4-16

$$\boxed{\left(-\frac{1}{3}a^{-5}b^{-2}\right)\cdot\left(-ba^{-2}b^{4}\right)}=\boxed{+\frac{1}{3}a^{-5}b^{-2}\cdot b^{1}a^{-2}b^{4}}=\boxed{\frac{1}{3}\left(a^{-5}a^{-2}\right)\cdot\left(b^{-2}b^{1}b^{4}\right)}=\boxed{\frac{1}{3}\left(a^{-5-2}\right)\cdot\left(b^{-2+1+4}\right)}$$

$$=\boxed{\frac{1}{3}\cdot a^{-7}\cdot b^{3}}=\boxed{\frac{1}{3}\cdot\frac{1}{a^{7}}\cdot\frac{b^{3}}{1}}=\boxed{\frac{1\cdot1\cdot b^{3}}{3\cdot a^{7}\cdot1}}=\boxed{\mathbf{\frac{b^{3}}{3a^{7}}}}$$

Example 3.4-17

$$\boxed{4^{-1}\cdot5^{-4}\cdot2^{-3}\cdot2^{2}\cdot5^{4}\cdot5^{0}\cdot4^{2}}=\boxed{\left(5^{-4}\cdot5^{4}\cdot5^{0}\right)\cdot\left(4^{2}\cdot4^{-1}\right)\cdot\left(2^{-3}\cdot2^{2}\right)}=\boxed{\left(5^{-4+4+0}\right)\cdot\left(4^{2-1}\right)\cdot\left(2^{-3+2}\right)}$$

$$= \boxed{5^0 \cdot 4^1 \cdot 2^{-1}} = \boxed{1 \cdot 4 \cdot 2^{-1}} = \boxed{4 \cdot \frac{1}{2^1}} = \boxed{\frac{4}{1} \cdot \frac{1}{2}} = \boxed{\frac{4 \cdot 1}{1 \cdot 2}} = \boxed{\frac{\frac{2}{\cancel{4}}}{\frac{\cancel{2}}{1}}} = \boxed{\frac{2}{1}} = \boxed{\mathbf{2}}$$

Example 3.4-18

$$\boxed{(-3-1)^{-3} \cdot \left(a^{-3}b^2c^4\right) \cdot \left(a^2b^{-3}c^{-1}\right)} = \boxed{(-4)^{-3} \cdot a^{-3}b^2c^4 \cdot a^2b^{-3}c^{-1}} = \boxed{(-4)^{-3} \cdot \left(a^{-3}a^2\right) \cdot \left(b^2b^{-3}\right) \cdot \left(c^4c^{-1}\right)}$$

$$= \boxed{(-4)^{-3} \cdot \left(a^{-3+2}\right) \cdot \left(b^{2-3}\right) \cdot \left(c^{4-1}\right)} = \boxed{(-4)^{-3} \cdot a^{-1} \cdot b^{-1} \cdot c^3} = \boxed{\frac{1}{(-4)^3} \cdot \frac{1}{a^1} \cdot \frac{1}{b^1} \cdot \frac{c^3}{1}} = \boxed{\frac{1}{(-4 \cdot -4 \cdot -4)} \cdot \frac{1}{a} \cdot \frac{1}{b} \cdot \frac{c^3}{1}}$$

$$\boxed{\frac{1}{-64} \cdot \frac{1}{a} \cdot \frac{1}{b} \cdot \frac{c^3}{1}} = \boxed{-\frac{1 \cdot 1 \cdot 1 \cdot c^3}{64 \cdot a \cdot b \cdot 1}} = \boxed{-\frac{\mathbf{c^3}}{\mathbf{64ab}}}$$

Example 3.4-19

$$\boxed{(2-5)^{-2} \cdot \left(x^3y^2z^2\right) \cdot \left(x^{-4}y^{-3}z^{-2}\right)} = \boxed{(-3)^{-2} \cdot x^3y^2z^2 \cdot x^{-4}y^{-3}z^{-2}} = \boxed{(-3)^{-2} \cdot \left(x^3x^{-4}\right) \cdot \left(y^2y^{-3}\right) \cdot \left(z^2z^{-2}\right)}$$

$$= \boxed{(-3)^{-2} \cdot \left(x^{3-4}\right) \cdot \left(y^{2-3}\right) \cdot \left(z^{2-2}\right)} = \boxed{(-3)^{-2} \cdot x^{-1} \cdot y^{-1} \cdot z^0} = \boxed{(-3)^{-2} \cdot x^{-1} \cdot y^{-1} \cdot 1} = \boxed{(-3)^{-2} \cdot x^{-1} \cdot y^{-1}}$$

$$= \boxed{\frac{1}{(-3)^2} \cdot \frac{1}{x^1} \cdot \frac{1}{y^1}} = \boxed{\frac{1}{(-3 \cdot -3)} \cdot \frac{1}{x} \cdot \frac{1}{y}} = \boxed{+\frac{1}{9} \cdot \frac{1}{x} \cdot \frac{1}{y}} = \boxed{\frac{1 \cdot 1 \cdot 1}{9 \cdot x \cdot y}} = \boxed{\frac{\mathbf{1}}{\mathbf{9xy}}}$$

Example 3.4-20

$$\boxed{\left(3^{-5} \cdot a^{-3} \cdot b^2\right) \cdot \left(3^2 \cdot 3 \cdot a^2 \cdot b^{-4}\right)} = \boxed{3^{-5} \cdot a^{-3} \cdot b^2 \cdot 3^2 \cdot 3^1 \cdot a^2 \cdot b^{-4}} = \boxed{\left(3^{-5} \cdot 3^2 \cdot 3^1\right) \cdot \left(a^{-3} \cdot a^2\right) \cdot \left(b^2 \cdot b^{-4}\right)}$$

$$= \boxed{\left(3^{-5+2+1}\right) \cdot \left(a^{-3+2}\right) \cdot \left(b^{2-4}\right)} = \boxed{3^{-2} \cdot a^{-1} \cdot b^{-2}} = \boxed{\frac{1}{3^2} \cdot \frac{1}{a^1} \cdot \frac{1}{b^2}} = \boxed{\frac{1}{3 \cdot 3} \cdot \frac{1}{a} \cdot \frac{1}{b^2}} = \boxed{\frac{1 \cdot 1 \cdot 1}{9 \cdot a \cdot b^2}} = \boxed{\frac{\mathbf{1}}{\mathbf{9ab^2}}}$$

Example 3.4-21

$$\boxed{(-2)^{-3}\left(w^{-2}y^{-3}\right) \cdot \left(w^{-1}y^2\right)} = \boxed{(-2)^{-3}w^{-2}y^{-3} \cdot w^{-1}y^2} = \boxed{(-2)^{-3}\left(w^{-2-1}\right) \cdot \left(y^{-3+2}\right)} = \boxed{(-2)^{-3}w^{-3} \cdot y^{-1}}$$

$$= \boxed{\frac{1}{(-2)^3} \frac{1}{w^3} \cdot \frac{1}{y^1}} = \boxed{\frac{1}{(-2 \cdot -2 \cdot -2)} \frac{1}{w^3} \cdot \frac{1}{y}} = \boxed{\frac{1}{-8} \frac{1}{w^3} \cdot \frac{1}{y}} = \boxed{-\frac{1 \cdot 1 \cdot 1}{8 \cdot w^3 \cdot y}} = \boxed{-\frac{\mathbf{1}}{\mathbf{8w^3y}}}$$

Example 3.4-22

$$\boxed{(-4)^{-2} \cdot \left(h^{-2}l^{-5}m^{-6}\right) \cdot \left(h^{-3}l^{-1}m\right)} = \boxed{(-4)^{-2} \cdot h^{-2}l^{-5}m^{-6} \cdot h^{-3}l^{-1}m^1} = \boxed{(-4)^{-2} \cdot \left(h^{-3}h^{-2}\right) \cdot \left(l^{-5}l^{-1}\right) \cdot \left(m^{-6}m^1\right)}$$

$$= \boxed{(-4)^{-2} \cdot \left(h^{-3-2}\right) \cdot \left(l^{-5-1}\right) \cdot \left(m^{-6+1}\right)} = \boxed{(-4)^{-2} \cdot h^{-5} \cdot l^{-6} \cdot m^{-5}} = \boxed{\frac{1}{(-4)^2} \cdot \frac{1}{h^5} \cdot \frac{1}{l^6} \cdot \frac{1}{m^5}} = \boxed{\frac{1}{(-4 \cdot -4)} \cdot \frac{1}{h^5} \cdot \frac{1}{l^6} \cdot \frac{1}{m^5}}$$

$$= \boxed{\frac{1}{16}\cdot\frac{1}{h^5}\cdot\frac{1}{l^6}\cdot\frac{1}{m^5}} = \boxed{\frac{1\cdot1\cdot1\cdot1}{16\cdot h^5\cdot l^6\cdot m^5}} = \boxed{\frac{\mathbf{1}}{\mathbf{16h^5l^6m^5}}}$$

Example 3.4-23

$$\boxed{-\left(4^{-2}\right)\cdot\left(k^{-3}m^{-8}n^4k^{-2}m^2\right)} = \boxed{-\left(4^{-2}\right)\cdot\left(k^{-3}k^{-2}\right)\cdot\left(m^2m^{-8}\right)\cdot n^4} = \boxed{-\left(4^{-2}\right)\cdot\left(k^{-3-2}\right)\cdot\left(m^{2-8}\right)\cdot n^4}$$

$$= \boxed{-\left(4^{-2}\right)\cdot k^{-5}\cdot m^{-6}\cdot n^4} = \boxed{-\frac{1}{4^2}\cdot\frac{1}{k^5}\cdot\frac{1}{m^6}\cdot n^4} = \boxed{-\frac{1}{(4\cdot4)}\cdot\frac{1}{k^5}\cdot\frac{1}{m^6}\cdot\frac{n^4}{1}} = \boxed{-\frac{1\cdot1\cdot1\cdot n^4}{16\cdot k^5\cdot m^6\cdot1}} = \boxed{-\frac{\mathbf{n^4}}{\mathbf{16k^5m^6}}}$$

Example 3.4-24

$$\boxed{\left(\frac{2}{-5}\right)^{-2}x^{-3}y^{-2}y^3x^2zz^{-4}} = \boxed{\frac{2^{-2}}{(-5)^{-2}}\cdot\left(x^{-3}x^2\right)\cdot\left(y^{-2}y^3\right)\cdot\left(z^1z^{-4}\right)} = \boxed{\frac{(-5)^2}{2^2}\cdot\left(x^{-3+2}\right)\cdot\left(y^{-2+3}\right)\cdot\left(z^{1-4}\right)}$$

$$= \boxed{\frac{(-5\cdot-5)}{2\cdot2}\cdot x^{-1}\cdot y^1\cdot z^{-3}} = \boxed{\frac{25}{4}\cdot x^{-1}\cdot y\cdot z^{-3}} = \boxed{\frac{25}{4}\cdot\frac{1}{x^1}\cdot\frac{y}{1}\cdot\frac{1}{z^3}} = \boxed{\frac{25\cdot1\cdot y\cdot1}{4\cdot x\cdot1\cdot z^3}} = \boxed{\frac{\mathbf{25y}}{\mathbf{4xz^3}}}$$

Example 3.4-25

$$\boxed{\left(\frac{-4}{-3}\right)^{-4}4^2v^{-3}4^{-3}v^2v^{-1}} = \boxed{\frac{(-4)^{-4}}{(-3)^{-4}}\cdot\left(4^24^{-3}\right)\cdot\left(v^2v^{-1}v^{-3}\right)} = \boxed{\frac{(-3)^4}{(-4)^4}\cdot\left(4^{2-3}\right)\cdot\left(v^{2-1-3}\right)} = \boxed{\frac{(-3)^4}{(-4)^4}\cdot4^{-1}\cdot v^{-2}}$$

$$= \boxed{\frac{(-3\cdot-3\cdot-3\cdot-3)}{(-4\cdot-4\cdot-4\cdot-4)}\cdot4^{-1}\cdot v^{-2}} = \boxed{\frac{81}{256}\cdot\frac{1}{4^1}\cdot\frac{1}{v^2}} = \boxed{\frac{81\cdot1\cdot1}{(256\cdot4)\cdot v^2}} = \boxed{\frac{\mathbf{81}}{\mathbf{1024v^2}}}$$

Practice Problems - Multiplying Negative Integer Exponents (Simple Cases)

Section 3.4 Case I a Practice Problems - Multiply the following exponential expressions by one another:

1. $\left(3^{-3}\cdot2^{-1}\right)\cdot\left(2^{-3}\cdot3^{-2}\cdot2\right) =$
2. $a^{-6}\cdot b^{-4}\cdot a^{-1}\cdot b^{-2}\cdot a^0 =$
3. $\left(a^{-2}\cdot b^{-3}\right)^2\cdot\left(a\cdot b^{-2}\right) =$
4. $(-2)^{-4}\left(r^{-2}s^2t\right)\cdot\left(r^3st^{-2}s^{-1}\right) =$
5. $\left(\frac{4}{5}\right)^{-4}2^2v^{-5}2^{-4}v^3v^{-2} =$
6. $2^{-1}\cdot3^2\cdot3^{-5}\cdot2^2\cdot2^0 =$
7. $(-3)^{-3}\left(k^{-2}l^{-4}\right)\cdot\left(k^{-3}l^2\right) =$
8. $-\left(2^{-2}\right)\cdot\left(h^{-3}m^{-3}n^{-4}hm^{-2}n\right) =$
9. $\left(-\frac{1}{5}a^{-3}b^{-5}\right)\cdot\left(-ba^{-2}b^3\right) =$
10. $\frac{-2}{-5}\cdot m^3\cdot m^{-5}\cdot r^{-2}\cdot m\cdot r^3 =$

Case I b Multiplying Negative Integer Exponents (More Difficult Cases)

Negative integer exponents are multiplied by one another using the following steps:

Step 1 Apply the Power of a Power Law (Law II) and/or the Power of a Product Law (Law III) from Table 3.4-2.

Step 2 Apply the Multiplication Law (Law I) from Table 3.4-2 and simplify the exponential expressions by adding the exponents with similar bases.

Step 3 Change the negative integer exponents to positive integer exponents.

Examples with Steps

The following examples show the steps as to how negative integer exponents are multiplied by one another:

Example 3.4-26

$$\boxed{\left(a^2\right)^{-2}\cdot\left(a^3\cdot b^{-3}\right)^2} =$$

Solution:

Step 1 $$\boxed{\left(a^2\right)^{-2}\cdot\left(a^3\cdot b^{-3}\right)^2} = \boxed{\left(a^{2\times -2}\right)\cdot\left(a^{3\times 2}\cdot b^{-3\times 2}\right)} = \boxed{a^{-4}\cdot a^6\cdot b^{-6}}$$

Step 2 $$\boxed{a^{-4}\cdot a^6\cdot b^{-6}} = \boxed{\left(a^{-4}\cdot a^6\right)\cdot b^{-6}} = \boxed{\left(a^{-4+6}\right)\cdot b^{-6}} = \boxed{a^2\cdot b^{-6}}$$

Step 3 $$\boxed{a^2\cdot b^{-6}} = \boxed{\frac{a^2}{1}\cdot\frac{1}{b^6}} = \boxed{\frac{a^2\cdot 1}{1\cdot b^6}} = \boxed{\mathbf{\frac{a^2}{b^6}}}$$

Example 3.4-27

$$\boxed{(2)^{-3}\cdot\left(2\cdot b^{-2}\right)^3\cdot b^4} =$$

Solution:

Step 1 $$\boxed{(2)^{-3}\cdot\left(2\cdot b^{-2}\right)^3\cdot b^4} = \boxed{\left(2^{1\times -3}\right)\cdot\left(2^{1\times 3}\cdot b^{-2\times 3}\right)\cdot b^4} = \boxed{2^{-3}\cdot 2^3\cdot b^{-6}\cdot b^4}$$

Step 2 $$\boxed{2^{-3}\cdot 2^3\cdot b^{-6}\cdot b^4} = \boxed{\left(2^{-3}\cdot 2^3\right)\cdot\left(b^{-6}\cdot b^4\right)} = \boxed{\left(2^{-3+3}\right)\cdot\left(b^{-6+4}\right)} = \boxed{2^0\cdot b^{-2}} = \boxed{1\cdot b^{-2}} = \boxed{b^{-2}}$$

Step 3 $$\boxed{b^{-2}} = \boxed{\mathbf{\frac{1}{b^2}}}$$

Example 3.4-28

$$\boxed{\left[(-2)^{-2}\right]^{-3}\cdot\left(x^{-a}\cdot x^{3a}\right)^{-2}} =$$

Solution:

Step 1 $\left[(-2)^{-2}\right]^{-3}\cdot\left(x^{-a}\cdot x^{3a}\right)^{-2}=(-2)^{-2\times-3}\cdot\left(x^{-a\times-2}\cdot x^{3a\times-2}\right)=(-2)^{6}\cdot\left(x^{2a}\cdot x^{-6a}\right)$

Step 2 $(-2)^{6}\cdot\left(x^{2a}\cdot x^{-6a}\right)=+64\cdot\left(x^{2a}\cdot x^{-6a}\right)=64\cdot\left(x^{2a-6a}\right)=64\cdot x^{-4a}$

Step 3 $64\cdot x^{-4a}=\frac{64}{1}\cdot\frac{1}{x^{4a}}=\frac{64\cdot 1}{1\cdot x^{4a}}=\mathbf{\frac{64}{x^{4a}}}$

Example 3.4-29

$\left(2\cdot a^{-1}\cdot b^{2}\cdot c^{-3}\right)^{4}\cdot\left(a^{-2}\cdot c^{-5}\right)^{2}=$

Solution:

Step 1 $\left(2\cdot a^{-1}\cdot b^{2}\cdot c^{-3}\right)^{4}\cdot\left(a^{-2}\cdot c^{-5}\right)^{2}=\left(2^{1\times 4}\cdot a^{-1\times 4}\cdot b^{2\times 4}\cdot c^{-3\times 4}\right)\cdot\left(a^{-2\times 2}\cdot c^{-5\times 2}\right)$

$=\left(2^{4}\cdot a^{-4}\cdot b^{8}\cdot c^{-12}\right)\cdot\left(a^{-4}\cdot c^{-10}\right)$

Step 2 $\left(2^{4}\cdot a^{-4}\cdot b^{8}\cdot c^{-12}\right)\cdot\left(a^{-4}\cdot c^{-10}\right)=16\cdot a^{-4}\cdot b^{8}\cdot c^{-12}\cdot a^{-4}\cdot c^{-10}$

$=16\cdot\left(a^{-4}\cdot a^{-4}\right)\cdot b^{8}\cdot\left(c^{-12}\cdot c^{-10}\right)=16\cdot\left(a^{-4-4}\right)\cdot b^{8}\cdot\left(c^{-12-10}\right)=16\cdot a^{-8}\cdot b^{8}\cdot c^{-22}$

Step 3 $16\cdot a^{-8}\cdot b^{8}\cdot c^{-22}=\frac{16}{1}\cdot\frac{1}{a^{8}}\cdot\frac{b^{8}}{1}\cdot\frac{1}{c^{22}}=\frac{16\cdot 1\cdot b^{8}\cdot 1}{1\cdot a^{8}\cdot 1\cdot c^{22}}=\mathbf{\frac{16\,b^{8}}{a^{8}c^{22}}}$

Example 3.4-30

$\left[\left(4^{-1}\cdot x^{2}\cdot w^{3}\right)^{-2}\cdot w\right]^{-3}=$

Solution:

Step 1 $\left[\left(4^{-1}\cdot x^{2}\cdot w^{3}\right)^{-2}\cdot w\right]^{-3}=\left[\left(4^{-1\times-2}\cdot x^{2\times-2}\cdot w^{3\times-2}\right)\cdot w\right]^{-3}=\left[\left(4^{2}\cdot x^{-4}\cdot w^{-6}\right)\cdot w\right]^{-3}$

$=\left(4^{2\times-3}\cdot x^{-4\times-3}\cdot w^{-6\times-3}\right)\cdot w^{1\times-3}=\left(4^{-6}\cdot x^{12}\cdot w^{18}\right)\cdot w^{-3}=4^{-6}\cdot x^{12}\cdot\left(w^{18}\cdot w^{-3}\right)$

Step 2 $4^{-6}\cdot x^{12}\cdot\left(w^{18}\cdot w^{-3}\right)=4^{-6}\cdot x^{12}\cdot\left(w^{18-3}\right)=4^{-6}\cdot x^{12}\cdot w^{15}$

Step 3 $4^{-6}\cdot x^{12}\cdot w^{15}=\frac{1}{4^{6}}\cdot\frac{x^{12}}{1}\cdot\frac{w^{15}}{1}=\frac{1}{4096}\cdot\frac{x^{12}}{1}\cdot\frac{w^{15}}{1}=\frac{1\cdot x^{12}\cdot w^{15}}{4096\cdot 1\cdot 1}=\mathbf{\frac{x^{12}w^{15}}{4096}}$

Additional Examples - Multiplying Negative Integer Exponents (More Difficult Cases)

The following examples further illustrate how to multiply negative integer exponents by one another:

Example 3.4-31

$$\boxed{a^{-2}\cdot a^{3}\cdot a^{-1}}=\boxed{a^{-2+3-1}}=\boxed{a^{0}}=\boxed{\mathbf{1}}$$

Example 3.4-32

$$\boxed{\left(2^{-3}\cdot 2^{2}\right)^{-2}\cdot\left(a^{-4}\right)^{2}}=\boxed{\left(2^{-3+2}\right)^{-2}\cdot\left(a^{-4\times 2}\right)}=\boxed{\left(2^{-1}\right)^{-2}\cdot a^{-8}}=\boxed{\left(2^{-1\times -2}\right)\cdot a^{-8}}=\boxed{2^{2}\cdot a^{-8}}=\boxed{4\cdot\frac{1}{a^{8}}}=\boxed{\mathbf{\frac{4}{a^{8}}}}$$

Example 3.4-33

$$\boxed{(-4)^{-2}\cdot\left(3^{2a}\right)^{-2}\cdot\left(2^{-1}\cdot 3^{-3a}\right)^{4}}=\boxed{(-4)^{-2}\cdot\left(3^{2a\times -2}\right)\cdot\left(2^{-1\times 4}\cdot 3^{-3a\times 4}\right)}=\boxed{(-4)^{-2}\cdot 3^{-4a}\cdot 2^{-4}\cdot 3^{-12a}}$$

$$=\boxed{(-4)^{-2}\cdot 2^{-4}\cdot\left(3^{-4a}\cdot 3^{-12a}\right)}=\boxed{(-4)^{-2}\cdot 2^{-4}\cdot\left(3^{-4a-12a}\right)}=\boxed{\frac{1}{(-4)^{2}}\cdot\frac{1}{2^{4}}\cdot 3^{-16a}}=\boxed{\frac{1}{(-4\cdot -4)}\cdot\frac{1}{16}\cdot\frac{1}{3^{16a}}}$$

$$=\boxed{\frac{1}{+16}\cdot\frac{1}{16}\cdot\frac{1}{3^{16a}}}=\boxed{\frac{1\cdot 1\cdot 1}{(16\cdot 16)\cdot 3^{16a}}}=\boxed{\mathbf{\frac{1}{(256)\cdot 3^{16a}}}}$$

Example 3.4-34

$$\boxed{r^{3}\cdot\left(2^{2}\cdot r^{2}\right)^{-5}}=\boxed{r^{3}\cdot\left(2^{2\times -5}\cdot r^{2\times -5}\right)}=\boxed{r^{3}\cdot\left(2^{-10}\cdot r^{-10}\right)}=\boxed{r^{3}\cdot 2^{-10}\cdot r^{-10}}=\boxed{2^{-10}\cdot\left(r^{3}\cdot r^{-10}\right)}$$

$$=\boxed{2^{-10}\cdot\left(r^{3-10}\right)}=\boxed{2^{-10}\cdot r^{-7}}=\boxed{\frac{1}{2^{10}}\cdot\frac{1}{r^{7}}}=\boxed{\frac{1\cdot 1}{2^{10}\cdot r^{7}}}=\boxed{\mathbf{\frac{1}{1024\,r^{7}}}}$$

Example 3.4-35

$$\boxed{y^{-2}\cdot\left(2\cdot y^{-3}\right)^{0}\cdot x^{-2}\cdot y}=\boxed{y^{-2}\cdot 1\cdot x^{-2}\cdot y}=\boxed{y^{-2}\cdot x^{-2}\cdot y^{1}}=\boxed{\left(y^{-2}\cdot y^{1}\right)\cdot x^{-2}}=\boxed{\left(y^{-2+1}\right)\cdot x^{-2}}=\boxed{y^{-1}\cdot x^{-2}}$$

$$=\boxed{\frac{1}{y^{1}}\cdot\frac{1}{x^{2}}}=\boxed{\mathbf{\frac{1}{x^{2}y}}}$$

Example 3.4-36

$$\boxed{\left(a^{-2}\cdot b^{2}\right)^{-2}\cdot\left(a^{-4}\cdot b\right)\cdot\left(a^{0}\cdot b^{3}\right)^{-1}}=\boxed{\left(a^{-2\times -2}\cdot b^{2\times -2}\right)\cdot\left(a^{-4}\cdot b\right)\cdot\left(a^{0\times -1}\cdot b^{3\times -1}\right)}=\boxed{\left(a^{4}\cdot b^{-4}\right)\cdot\left(a^{-4}\cdot b\right)\cdot\left(a^{0}\cdot b^{-3}\right)}$$

$$=\boxed{a^{4}\cdot b^{-4}\cdot a^{-4}\cdot b\cdot 1\cdot b^{-3}}=\boxed{\left(a^{4}\cdot a^{-4}\right)\cdot\left(b^{-4}\cdot b^{1}\cdot b^{-3}\right)}=\boxed{\left(a^{4-4}\right)\cdot\left(b^{-4+1-3}\right)}=\boxed{a^{0}\cdot b^{-6}}=\boxed{1\cdot b^{-6}}=\boxed{b^{-6}}=\boxed{\mathbf{\frac{1}{b^{6}}}}$$

Example 3.4-37

$$\left(x^3\cdot y\right)^{-2}\cdot\left(x^2\cdot y^{-1}\cdot z^{-2}\right)^4\cdot\left(x\cdot z^3\right)^{-1}=\left(x^{3\times-2}\cdot y^{1\times-2}\right)\cdot\left(x^{2\times4}\cdot y^{-1\times4}\cdot z^{-2\times4}\right)\cdot\left(x^{1\times-1}\cdot z^{3\times-1}\right)$$

$$=\left(x^{-6}\cdot y^{-2}\right)\cdot\left(x^8\cdot y^{-4}\cdot z^{-8}\right)\cdot\left(x^{-1}\cdot z^{-3}\right)=x^{-6}\cdot y^{-2}\cdot x^8\cdot y^{-4}\cdot z^{-8}\cdot x^{-1}\cdot z^{-3}$$

$$=\left(x^{-6}\cdot x^8\cdot x^{-1}\right)\cdot\left(y^{-2}\cdot y^{-4}\right)\cdot\left(z^{-8}\cdot z^{-3}\right)=\left(x^{-6+8-1}\right)\cdot\left(y^{-2-4}\right)\cdot\left(z^{-8-3}\right)=x^1\cdot y^{-6}\cdot z^{-11}=\frac{x}{1}\cdot\frac{1}{y^6}\cdot\frac{1}{z^{11}}$$

$$=\frac{x\cdot1\cdot1}{1\cdot y^6\cdot z^{11}}=\mathbf{\frac{x}{y^6z^{11}}}$$

Example 3.4-38

$$y^0\cdot\left(x^{-3}\cdot y\right)^2\cdot\left(y\cdot x^{-3}\right)^{-2}\cdot y^{-1}=1\cdot\left(x^{-3\times2}\cdot y^{1\times2}\right)\cdot\left(y^{1\times-2}\cdot x^{-3\times-2}\right)\cdot y^{-1}=\left(x^{-6}\cdot y^2\right)\cdot\left(y^{-2}\cdot x^6\right)\cdot y^{-1}$$

$$=x^{-6}\cdot y^2\cdot y^{-2}\cdot x^6\cdot y^{-1}=\left(x^{-6}\cdot x^6\right)\cdot\left(y^2\cdot y^{-2}\cdot y^{-1}\right)=\left(x^{-6+6}\right)\cdot\left(y^{2-2-1}\right)=x^0\cdot y^{-1}=1\cdot y^{-1}=y^{-1}=\mathbf{\frac{1}{y}}$$

Example 3.4-39

$$\left(5\cdot a^2\cdot w\right)^{-3}\cdot\left(3\cdot a\cdot w^2\right)^0\cdot\left(a^{-3}\right)^2=\left(5\cdot a^2\cdot w\right)^{-3}\cdot1\cdot\left(a^{-3}\right)^2=\left(5^{1\times-3}\cdot a^{2\times-3}\cdot w^{1\times-3}\right)\cdot\left(a^{-3\times2}\right)$$

$$=5^{-3}\cdot a^{-6}\cdot w^{-3}\cdot a^{-6}=5^{-3}\cdot\left(a^{-6}\cdot a^{-6}\right)\cdot w^{-3}=5^{-3}\cdot\left(a^{-6-6}\right)\cdot w^{-3}=5^{-3}\cdot a^{-12}\cdot w^{-3}=\frac{1}{5^3}\cdot\frac{1}{a^{12}}\cdot\frac{1}{w^3}$$

$$=\frac{1\cdot1\cdot1}{5^3\cdot a^{12}\cdot w^3}=\mathbf{\frac{1}{125\,a^{12}w^3}}$$

Example 3.4-40

$$\left(2^2\cdot y^{-2}\cdot z^4\right)^3\cdot\left[\left(y^{-2}\cdot z^3\right)^{-1}\cdot x^2\right]^{-2}=\left(2^{2\times3}\cdot y^{-2\times3}\cdot z^{4\times3}\right)\cdot\left[\left(y^{-2\times-1}\cdot z^{3\times-1}\right)\cdot x^2\right]^{-2}$$

$$=\left(2^6\cdot y^{-6}\cdot z^{12}\right)\cdot\left[\left(y^2\cdot z^{-3}\right)\cdot x^2\right]^{-2}=\left(2^6\cdot y^{-6}\cdot z^{12}\right)\cdot\left[\left(y^{2\times-2}\cdot z^{-3\times-2}\right)\cdot x^{2\times-2}\right]=2^6\cdot y^{-6}\cdot z^{12}\cdot y^{-4}\cdot z^6\cdot x^{-4}$$

$$=2^6\cdot x^{-4}\cdot\left(y^{-6}\cdot y^{-4}\right)\cdot\left(z^{12}\cdot z^6\right)=2^6\cdot x^{-4}\cdot\left(y^{-6-4}\right)\cdot\left(z^{12+6}\right)=64\cdot x^{-4}\cdot y^{-10}\cdot z^{18}=\frac{64}{1}\cdot\frac{1}{x^4}\cdot\frac{1}{y^{10}}\cdot\frac{z^{18}}{1}$$

$$=\frac{64\cdot1\cdot1\cdot z^{18}}{1\cdot x^4\cdot y^{10}\cdot1}=\mathbf{\frac{64\,z^{18}}{x^4y^{10}}}$$

Practice Problems - Multiplying Negative Integer Exponents (More Difficult Cases)

Section 3.4 Case I b Practice Problems - Multiply the following exponential expressions by one another:

1. $\left(a^2 \cdot a^3\right)^{-2} =$
2. $2 \cdot \left(a^2 \cdot y^{-1}\right)^{-2} \cdot y^{-3} =$
3. $\left(a^{-2} \cdot b^{-3}\right)^2 \cdot \left(a \cdot b^{-2}\right) =$
4. $2^{-3} \cdot \left(x^{-2a} \cdot x^{3a}\right)^{-2} =$
5. $\left(x \cdot y^{-1} \cdot z^3\right)^{-3a} \cdot \left(x^{-2}\right)^{-2a} =$
6. $2^{-1} \cdot 5^2 \cdot 5^{-5} \cdot 2^0 =$
7. $y^2 \cdot \left(x^{-2}\right)^4 \cdot \left(-x \cdot y^{-4}\right)^2 =$
8. $\left(x^3 \cdot y^2\right)^{-3} \cdot \left(x^5 \cdot y^2 \cdot z^{-3}\right) =$
9. $\left(x^2 \cdot y\right)^{-1} \cdot \left(3^{-1} \cdot x \cdot y^3\right)^{-4} =$
10. $\left(5 \cdot x^2\right)^{-1} \cdot \left[\left(x \cdot y^2\right)^{-2} \cdot x\right]^{-2} =$

Case II Dividing Negative Integer Exponents

Negative integer exponents are divided by one another using the exponent laws I through VI shown in Table 3.4-1. These laws are used in order to simplify division of negative fractional exponents by each other. Division of expressions with negative integer exponents is divided to two cases: Case II a addressing simple cases and Case II b addressing more difficult cases.

Case II a Dividing Negative Integer Exponents (Simple Cases)

Negative integer exponents are divided by one another using the following steps:

Step 1 a. Apply the Division and/or the Negative Power Laws (Laws V and VI) from Table 3.4-1.
b. Group the exponential terms with similar bases.

Step 2 Apply the Multiplication Law (Law I) from Table 3.4-1 and simplify the exponential expressions by adding the exponents with similar bases.

Examples with Steps

The following examples show the steps as to how negative integer exponents are divided by one another:

Example 3.4-41

$$\frac{5^{-2}}{5^{-3}} =$$

Solution:

Step 1

$$\frac{5^{-2}}{5^{-3}} = \frac{5^{3}\cdot 5^{-2}}{1} =$$

Step 2

$$\frac{5^{3}\cdot 5^{-2}}{1} = \frac{5^{3-2}}{1} = \frac{5^{1}}{1} = \boxed{5}$$

Example 3.4-42

$$\frac{(2-4)^{-3}}{-(-3-1)^{-4}} =$$

Solution:

Step 1

$$\frac{(2-4)^{-3}}{-(-3-1)^{-4}} = \frac{(-2)^{-3}}{-(-4)^{-4}} = \frac{(-4)^{4}}{-(-2)^{3}} = -\frac{(-4)^{4}}{(-2)^{3}}$$

Step 2

$$-\frac{(-4)^{4}}{(-2)^{3}} = -\frac{-4\cdot -4\cdot -4\cdot -4}{-2\cdot -2\cdot -2} = -\frac{+256}{-8} = +\frac{\overset{32}{\cancel{256}}}{\underset{1}{\cancel{8}}} = \frac{32}{1} = \boxed{32}$$

Example 3.4-43

$$\frac{a^{-2}b^{-3}}{a^{-1}b^{-2}} =$$

Solution:

Step 1 $\boxed{\frac{a^{-2}b^{-3}}{a^{-1}b^{-2}}} = \boxed{\frac{1}{\left(a^{2}a^{-1}\right)\cdot\left(b^{3}b^{-2}\right)}}$

Step 2 $\boxed{\frac{1}{\left(a^{2}a^{-1}\right)\cdot\left(b^{3}b^{-2}\right)}} = \boxed{\frac{1}{\left(a^{2-1}\right)\cdot\left(b^{3-2}\right)}} = \boxed{\frac{1}{a^{1}\cdot b^{1}}} = \boxed{\frac{\mathbf{1}}{\mathbf{ab}}}$

Example 3.4-44

$\boxed{\frac{2^{-3}u^{-1}v^{-3}}{2^{-1}u^{-2}v}} =$

Solution:

Step 1 $\boxed{\frac{2^{-3}u^{-1}v^{-3}}{2^{-1}u^{-2}v}} = \boxed{\frac{2^{-3}u^{-1}v^{-3}}{2^{-1}u^{-2}v^{1}}} = \boxed{\frac{u^{2}u^{-1}}{\left(2^{3}2^{-1}\right)\cdot\left(v^{3}v^{1}\right)}}$

Step 2 $\boxed{\frac{u^{2}u^{-1}}{\left(2^{3}2^{-1}\right)\cdot\left(v^{3}v^{1}\right)}} = \boxed{\frac{u^{2-1}}{\left(2^{3-1}\right)\cdot\left(v^{3+1}\right)}} = \boxed{\frac{u^{1}}{2^{2}\cdot v^{4}}} = \boxed{\frac{u}{4\cdot v^{4}}} = \boxed{\frac{\mathbf{u}}{\mathbf{4v^{4}}}}$

Example 3.4-45

$\boxed{\frac{e^{-5}f^{-2}e^{0}}{2^{-3}f^{5}e^{-4}}} =$

Solution:

Step 1 $\boxed{\frac{e^{-5}f^{-2}e^{0}}{2^{-3}f^{5}e^{-4}}} = \boxed{\frac{2^{3}\cdot e^{0}}{\left(f^{5}f^{2}\right)\cdot\left(e^{5}e^{-4}\right)}}$

Step 2 $\boxed{\frac{2^{3}\cdot e^{0}}{\left(f^{5}f^{2}\right)\cdot\left(e^{5}e^{-4}\right)}} = \boxed{\frac{8\cdot 1}{\left(f^{5+2}\right)\cdot\left(e^{5-4}\right)}} = \boxed{\frac{8}{f^{7}\cdot e^{1}}} = \boxed{\frac{\mathbf{8}}{\mathbf{e\,f^{7}}}}$

Additional Examples - Dividing Negative Integer Exponents (Simple Cases)

The following examples further illustrate how to divide negative integer exponents by one another:

Example 3.4-46

$\boxed{\frac{2^{-3}}{2^{-5}}} = \boxed{\frac{2^{5}\cdot 2^{-3}}{1}} = \boxed{\frac{2^{5-3}}{1}} = \boxed{\frac{2^{2}}{1}} = \boxed{\frac{4}{1}} = \boxed{\mathbf{4}}$

Example 3.4-47

$\boxed{\frac{a^{-2}c^{-3}}{-ac^{4}}} = \boxed{-\frac{a^{-2}c^{-3}}{a^{1}c^{4}}} = \boxed{-\frac{1}{\left(a^{1}a^{2}\right)\cdot\left(c^{4}c^{3}\right)}} = \boxed{-\frac{1}{\left(a^{1+2}\right)\cdot\left(c^{4+3}\right)}} = \boxed{-\frac{\mathbf{1}}{\mathbf{a^{3}c^{7}}}}$

Example 3.4-48

$$\boxed{\frac{(-3)^{-3}}{-(3)^{-3}}} = \boxed{-\frac{(-3)^{-3}}{(3)^{-3}}} = \boxed{-\frac{(3)^{3}}{(-3)^{3}}} = \boxed{-\frac{3\cdot 3\cdot 3}{-3\cdot -3\cdot -3}} = \boxed{-\frac{27}{-27}} = \boxed{+\frac{\overset{1}{\cancel{27}}}{\underset{1}{\cancel{27}}}} = \boxed{\frac{1}{1}} = \boxed{1}$$

Example 3.4-49

$$\boxed{\frac{(-2)^{-4}}{-(-2)^{-3}}} = \boxed{-\frac{(-2)^{-4}}{(-2)^{-3}}} = \boxed{-\frac{(-2)^{3}}{(-2)^{4}}} = \boxed{-\frac{-2\cdot -2\cdot -2}{-2\cdot -2\cdot -2\cdot -2}} = \boxed{-\frac{-8}{16}} = \boxed{+\frac{\overset{1}{\cancel{8}}}{\underset{2}{\cancel{16}}}} = \boxed{\frac{1}{2}}$$

Example 3.4-50

$$\boxed{\frac{-2^{-3}\cdot a^{-1}}{(-2)^{-2}a^{-3}}} = \boxed{-\frac{2^{-3}\cdot a^{-1}}{(-2)^{-2}a^{-3}}} = \boxed{-\frac{(-2)^{2}\cdot\left(a^{3}\cdot a^{-1}\right)}{2^{3}}} = \boxed{-\frac{(-2\cdot -2)\cdot\left(a^{3-1}\right)}{2\cdot 2\cdot 2}} = \boxed{-\frac{4\cdot a^{2}}{8}} = \boxed{-\frac{\overset{1}{\cancel{4}}}{\underset{2}{\cancel{8}}}a^{2}} = \boxed{-\frac{1}{2}a^{2}}$$

Example 3.4-51

$$\boxed{\frac{a^{2}}{2^{-3}\cdot a^{-2}}} = \boxed{\frac{2^{3}\cdot a^{2}\cdot a^{2}}{1}} = \boxed{\frac{8\cdot a^{2+2}}{1}} = \boxed{\frac{8a^{4}}{1}} = \boxed{8a^{4}}$$

Example 3.4-52

$$\boxed{\frac{(-3)^{-3}\cdot a^{-3}}{-a^{-3}}} = \boxed{-\frac{(-3)^{-3}\cdot a^{-3}}{a^{-3}}} = \boxed{-\frac{a^{3}\cdot a^{-3}}{(-3)^{3}}} = \boxed{-\frac{a^{3-3}}{-3\cdot -3\cdot -3}} = \boxed{-\frac{a^{0}}{-27}} = \boxed{+\frac{1}{27}} = \boxed{\frac{1}{27}}$$

Example 3.4-53

$$\boxed{\frac{a^{-3}b^{-2}}{a^{-3}b^{-1}}} = \boxed{\frac{a^{3}a^{-3}}{b^{2}b^{-1}}} = \boxed{\frac{a^{3-3}}{b^{2-1}}} = \boxed{\frac{a^{0}}{b^{1}}} = \boxed{\frac{1}{b}}$$

Example 3.4-54

$$\boxed{\frac{\left(x^{-3}y^{-3}\right)\cdot z^{3}}{\left(x^{-4}y^{-2}\right)\cdot z^{-3}}} = \boxed{\frac{\left(x^{4}x^{-3}\right)\cdot\left(z^{3}z^{3}\right)}{y^{3}y^{-2}}} = \boxed{\frac{\left(x^{4-3}\right)\cdot\left(z^{3+3}\right)}{y^{3-2}}} = \boxed{\frac{x^{1}\cdot z^{6}}{y^{1}}} = \boxed{\frac{xz^{6}}{y}}$$

Example 3.4-55

$$\boxed{\frac{2^{-2}x^{-2}w^{-2}}{4^{-3}x^{-1}w^{-3}}} = \boxed{\frac{4^{3}\left(w^{3}w^{-2}\right)}{2^{2}\left(x^{2}x^{-1}\right)}} = \boxed{\frac{(4\cdot 4\cdot 4)\cdot\left(w^{3-2}\right)}{(2\cdot 2)\cdot\left(x^{2-1}\right)}} = \boxed{\frac{64\cdot w^{1}}{4\cdot x^{1}}} = \boxed{\frac{\overset{16}{\cancel{64}}\cdot w}{\underset{1}{\cancel{4}}\cdot x}} = \boxed{\frac{16\cdot w}{1\cdot x}} = \boxed{\frac{16w}{x}}$$

Example 3.4-56

$$\boxed{\frac{(-2)^{-3}\cdot a^{-3}b^{-3}c^{-1}}{-(4)^{-2}\cdot a^{-3}b^{3}c^{-1}}} = \boxed{-\frac{(-2)^{-3}\cdot a^{-3}b^{-3}c^{-1}}{(4)^{-2}\cdot a^{-3}b^{3}c^{-1}}} = \boxed{-\frac{(4)^{2}\cdot\left(a^{3}a^{-3}\right)\cdot\left(c^{1}c^{-1}\right)}{(-2)^{3}\cdot\left(b^{3}b^{3}\right)}} = \boxed{-\frac{(4\cdot 4)\cdot\left(a^{3-3}\right)\cdot\left(c^{1-1}\right)}{(-2\cdot -2\cdot -2)\cdot\left(b^{3+3}\right)}}$$

$$= -\frac{16 \cdot a^0 \cdot c^0}{-8 \cdot b^6} = +\frac{16 \cdot 1 \cdot 1}{8 \cdot b^6} = \frac{\overset{2}{\cancel{16}}}{\underset{1}{\cancel{8}} \cdot b^6} = \frac{2}{1 \cdot b^6} = \boldsymbol{\frac{2}{b^6}}$$

Example 3.4-57

$$\frac{a^{-1}b^0c^{-2}d^5d^{-3}}{-8a^{-1}b^{-2}c^2d} = -\frac{a^{-1}b^0c^{-2}d^5d^{-3}}{8a^{-1}b^{-2}c^2d^1} = -\frac{\left(a^1a^{-1}\right)\cdot\left(b^0b^2\right)\cdot\left(d^5d^{-3}d^{-1}\right)}{8\left(c^2c^2\right)} = -\frac{\left(a^{1-1}\right)\cdot\left(b^{0+2}\right)\cdot\left(d^{5-3-1}\right)}{8\left(c^{2+2}\right)}$$

$$= -\frac{a^0 \cdot b^2 \cdot d^1}{8 \cdot c^4} = -\frac{1 \cdot b^2 \cdot d}{8c^4} = -\boldsymbol{\frac{b^2d}{8c^4}}$$

Example 3.4-58

$$\frac{3^{-2} \cdot 4^{-2}v^{-3}w^3}{(-1)^{-2} \cdot 4^{-1}v^{-1}w^2} = \frac{(-1)^2 \cdot \left(w^3w^{-2}\right)}{3^2 \cdot \left(4^24^{-1}\right) \cdot \left(v^3v^{-1}\right)} = \frac{(-1 \cdot -1) \cdot \left(w^{3-2}\right)}{(3 \cdot 3) \cdot \left(4^{2-1}\right) \cdot \left(v^{3-1}\right)} = \frac{+1 \cdot w^1}{(9 \cdot 4) \cdot v^2} = \frac{1 \cdot w}{36 \cdot v^2} = \boldsymbol{\frac{w}{36v^2}}$$

Example 3.4-59

$$\frac{a^{-2}b^{-3}}{a^2b^{-1}c^{-3}} = \frac{c^3}{\left(a^2a^2\right)\cdot\left(b^3b^{-1}\right)} = \frac{c^3}{\left(a^{2+2}\right)\cdot\left(b^{3-1}\right)} = \frac{c^3}{a^4 \cdot b^2} = \boldsymbol{\frac{c^3}{a^4b^2}}$$

Example 3.4-60

$$\frac{(-3)^{-2}c^3d^3}{-(3)^{-4}c^{-3}d^{-2}} = -\frac{(3)^4 \cdot \left(c^3c^3\right)\cdot\left(d^3d^2\right)}{(-3)^2} = -\frac{(3 \cdot 3 \cdot 3 \cdot 3)\cdot\left(c^{3+3}\right)\cdot\left(d^{3+2}\right)}{-3 \cdot -3} = -\frac{81 \cdot c^6 \cdot d^5}{9} = -\frac{\overset{9}{\cancel{81}}c^6d^5}{\underset{1}{\cancel{9}}}$$

$$= -\frac{9c^6d^5}{1} = \boldsymbol{-9c^6d^5}$$

Example 3.4-61

$$\frac{4^{-1}c^{-2}e^{-3}e^2}{2^2c^{-3}c^2} = \frac{e^{-3}e^2}{2^2 \cdot 4^1 \cdot \left(c^{-3}c^2c^2\right)} = \frac{e^{-3+2}}{(4 \cdot 4)\cdot\left(c^{-3+2+2}\right)} = \frac{e^{-1}}{16 \cdot c^1} = \frac{1}{16 \cdot c \cdot e^1} = \boldsymbol{\frac{1}{16ce}}$$

Example 3.4-62

$$\frac{u^{-3}v^8w^{-3}u^2v^{-2}}{u^{-2}u^3} = \frac{v^8v^{-2}}{\left(u^{-2}u^3u^3u^{-2}\right)\cdot w^3} = \frac{v^{8-2}}{\left(u^{-2+3+3-2}\right)\cdot w^3} = \frac{v^6}{u^2 \cdot w^3} = \boldsymbol{\frac{v^6}{u^2w^3}}$$

Example 3.4-63

$$\frac{\left(p^{-2}q^{-3}r^{-4}\right)\cdot\left(p^{-1}q^{-2}r^2\right)}{p^3q^{-2}} = \frac{p^{-2}q^{-3}r^{-4}p^{-1}q^{-2}r^2}{p^3q^{-2}} = \frac{r^{-4}r^2}{\left(p^3p^2p^1\right)\cdot\left(q^3q^2q^{-2}\right)} = \frac{r^{-4+2}}{\left(p^{3+2+1}\right)\cdot\left(q^{3+2-2}\right)}$$

$$= \boxed{\frac{r^{-2}}{p^6 \cdot q^3}} = \boxed{\frac{\mathbf{1}}{\mathbf{p^6 q^3 r^2}}}$$

Example 3.4-64

$$\boxed{\frac{(2-3)^{-2}(xy)^{-2}(wz)^{-3}}{(-3-4)^{-1}(xy)^{-2}}} = \boxed{\frac{(-1)^{-2}(xy)^{-2}(wz)^{-3}}{(-7)^{-1}(xy)^{-2}}} = \boxed{\frac{(-7)^{1}(xy)^{-2}(xy)^{2}}{(-1)^{2}(wz)^{3}}} = \boxed{\frac{-7\cdot(xy)^{-2+2}}{(-1\cdot-1)\cdot(wz)^{3}}} = \boxed{-\frac{7\cdot(xy)^{0}}{1\cdot(wz)^{3}}}$$

$$= \boxed{-\frac{7\cdot 1}{(wz)^3}} = \boxed{-\frac{\mathbf{7}}{\mathbf{(wz)^3}}}$$

Example 3.4-65

$$\boxed{\frac{-(-5+3)^{-3}\cdot x^{-3}y^{-2}z}{-(-3-1)^{2}\cdot x^{-2}y^{-1}z^{-3}}} = \boxed{+\frac{(-2)^{-3}\cdot x^{-3}y^{-2}z}{(-4)^{2}\cdot x^{-2}y^{-1}z^{-3}}} = \boxed{\frac{z^3\cdot z^1}{(-4)^2\cdot(-2)^3\cdot\left(x^3x^{-2}\right)\cdot\left(y^2y^{-1}\right)}}$$

$$= \boxed{\frac{z^{3+1}}{(-4\cdot-4)\cdot(-2\cdot-2\cdot-2)\cdot\left(x^{3-2}\right)\cdot\left(y^{2-1}\right)}} = \boxed{\frac{z^4}{(16)\cdot(-8)\cdot x^1\cdot y^1}} = \boxed{\frac{z^4}{-128\cdot x\cdot y}} = \boxed{-\frac{\mathbf{z^4}}{\mathbf{128\,x\,y}}}$$

Practice Problems - Dividing Negative Integer Exponents (Simple Cases)

Section 3.4 Case II a Practice Problems - Divide the following negative integer exponents:

1. $\dfrac{x^{-2}x}{x^3x^0} =$
2. $\dfrac{-2a^{-2}b^3}{-6a^{-1}b^{-2}} =$
3. $\dfrac{-(-3)^{-4}}{3\cdot(-3)^{-3}} =$
4. $\dfrac{-3^3y^{-3}yw}{(-3)^{-2}y^2w^{-3}} =$
5. $\dfrac{a^{-2}b^2a^{-5}y^{-2}}{a^{-3}y} =$
6. $\dfrac{(x\cdot y\cdot z)^0\cdot yx^{-2}}{x^{-4}y^{-1}} =$
7. $\dfrac{2^{-1}a^3b^{-2}c}{8ab^{-1}c^{-2}} =$
8. $\dfrac{-4^{-2}z^4w^2a}{a^2w^{-2}z^0} =$
9. $\dfrac{2^{-3}2^2y^{-3}yb^3}{2^2b^{-1}y^3} =$
10. $\dfrac{2^{-3}a^{-3}b^2c^{-1}d^{-3}}{(-2)^{-3}b^{-1}c^3d} =$

Case II b Dividing Negative Integer Exponents (More Difficult Cases)

Negative integer exponents are divided by one another using the following steps:

Step 1 Apply exponent laws such as the Power of a Power Law, the Power of a Product Law, and the Power of a Fraction Law (Laws II and III, and IV) from Table 3.4-1.

Step 2 a. Apply the Division and/or the Negative Power Law (Laws V and VI) from Table 3.4-1.

b. Group the exponential terms with similar bases.

c. Apply the Multiplication Law (Law I) from Table 3.4-1 and simplify the exponential expressions by adding the exponents with similar bases.

Step 3 Change the negative integer exponents to positive integer exponents.

Examples with Steps

The following examples show the steps as to how negative integer exponents are divide by one another:

Example 3.4-66

$$\left(\frac{x^{-3}}{x^2}\right)^{-2} =$$

Solution:

Step 1 $\left(\frac{x^{-3}}{x^2}\right)^{-2} = \frac{x^{-3\times-2}}{x^{2\times-2}} = \frac{x^6}{x^{-4}}$

Step 2 $\frac{x^6}{x^{-4}} = \frac{x^6 \cdot x^4}{1} = x^{6+4} = \mathbf{x^{10}}$

Step 3 *Not Applicable*

Example 3.4-67

$$\left(\frac{2^2 \cdot b^3}{b^{-4}}\right)^{-2} =$$

Solution:

Step 1 $\left(\frac{2^2 \cdot b^3}{b^{-4}}\right)^{-2} = \frac{2^{2\times-2} \cdot b^{3\times-2}}{b^{-4\times-2}} = \frac{2^{-4} \cdot b^{-6}}{b^8}$

Step 2 $\frac{2^{-4} \cdot b^{-6}}{b^8} = \frac{2^{-4}}{b^6 \cdot b^8} = \frac{2^{-4}}{b^{6+8}} = \frac{2^{-4}}{b^{14}}$

Step 3
$$\boxed{\frac{2^{-4}}{b^{14}}} = \boxed{\frac{1}{2^4 b^{14}}} = \boxed{\frac{1}{(2\cdot2\cdot2\cdot2)\cdot b^{14}}} = \boxed{\mathbf{\frac{1}{16\,b^{14}}}}$$

Example 3.4-68

$$\boxed{\left(\frac{4\cdot c}{3\cdot c^{-3}}\right)^{-3}} =$$

Solution:

Step 1
$$\boxed{\left(\frac{4\cdot c}{3\cdot c^{-3}}\right)^{-3}} = \boxed{\frac{4^{1\times-3}\cdot c^{1\times-3}}{3^{1\times-3}\cdot c^{-3\times-3}}} = \boxed{\frac{4^{-3}\cdot c^{-3}}{3^{-3}\cdot c^{9}}}$$

Step 2
$$\boxed{\frac{4^{-3}\cdot c^{-3}}{3^{-3}\cdot c^{9}}} = \boxed{\frac{4^{-3}\cdot\left(c^{-3}\cdot c^{-9}\right)}{3^{-3}}} = \boxed{\frac{4^{-3}\cdot\left(c^{-3-9}\right)}{3^{-3}}} = \boxed{\frac{4^{-3}\cdot c^{-12}}{3^{-3}}}$$

Step 3
$$\boxed{\frac{4^{-3}\cdot c^{-12}}{3^{-3}}} = \boxed{\frac{3^3}{4^3\cdot c^{12}}} = \boxed{\frac{3\cdot3\cdot3}{(4\cdot4\cdot4)\cdot c^{12}}} = \boxed{\mathbf{\frac{27}{64}\,\frac{1}{c^{12}}}}$$

Example 3.4-69

$$\boxed{\frac{z^{-7}\cdot(a\cdot b)^{-3}}{z^2\cdot a^{-2}}} =$$

Solution:

Step 1
$$\boxed{\frac{z^{-7}\cdot(a\cdot b)^{-3}}{z^2\cdot a^{-2}}} = \boxed{\frac{z^{-7}\cdot\left(a^{1\times-3}\cdot b^{1\times-3}\right)}{z^2\cdot a^{-2}}} = \boxed{\frac{z^{-7}\cdot a^{-3}\cdot b^{-3}}{z^2\cdot a^{-2}}}$$

Step 2
$$\boxed{\frac{z^{-7}\cdot a^{-3}\cdot b^{-3}}{z^2\cdot a^{-2}}} = \boxed{\frac{\left(a^2\cdot a^{-3}\right)\cdot b^{-3}}{z^7\cdot z^2}} = \boxed{\frac{\left(a^{2-3}\right)\cdot b^{-3}}{z^{7+2}}} = \boxed{\frac{a^{-1}\cdot b^{-3}}{z^9}}$$

Step 3
$$\boxed{\frac{a^{-1}\cdot b^{-3}}{z^9}} = \boxed{\frac{1}{a^1\cdot b^3\cdot z^9}} = \boxed{\mathbf{\frac{1}{a b^3 z^9}}}$$

Example 3.4-70

$$\boxed{\left(\frac{\left(a^2\cdot b^3\right)^{-4}}{\left(a^{-3}\cdot b\cdot c\right)^2}\right)^{-2}} =$$

Solution:

Step 1
$$\boxed{\left(\frac{\left(a^2\cdot b^3\right)^{-4}}{\left(a^{-3}\cdot b\cdot c\right)^2}\right)^{-2}} = \boxed{\left(\frac{a^{2\times-4}\cdot b^{3\times-4}}{a^{-3\times2}\cdot b^{1\times2}\cdot c^{1\times2}}\right)^{-2}} = \boxed{\left(\frac{a^{-8}\cdot b^{-12}}{a^{-6}\cdot b^2\cdot c^2}\right)^{-2}}$$

$$= \frac{a^{-8\times-2}\cdot b^{-12\times-2}}{a^{-6\times-2}\cdot b^{2\times-2}\cdot c^{2\times-2}} = \frac{a^{16}\cdot b^{24}}{a^{12}\cdot b^{-4}\cdot c^{-4}}$$

Step 2

$$\frac{a^{16}\cdot b^{24}}{a^{12}\cdot b^{-4}\cdot c^{-4}} = \frac{\left(a^{16}\cdot a^{-12}\right)\cdot\left(b^{24}\cdot b^{4}\right)}{c^{-4}} = \frac{\left(a^{16-12}\right)\cdot\left(b^{24+4}\right)}{c^{-4}} = \frac{a^{4}\cdot b^{28}}{c^{-4}}$$

Step 3

$$\frac{a^{4}\cdot b^{28}}{c^{-4}} = \frac{a^{4}\cdot b^{28}\cdot c^{4}}{1} = \mathbf{a^4 b^{28} c^4}$$

Additional Examples - Dividing Negative Integer Exponents (More Difficult Cases)

The following examples further illustrate how to divide negative integer exponential expressions by one another: (Again note that in solving this class of problems one needs to ensure that the final answer is without a negative exponent.)

Example 3.4-71

$$\frac{2^{-1}}{5^{-2}}\cdot\left(\frac{a^{2}}{a^{3}}\right)^{-3} = \frac{2^{-1}}{5^{-2}}\cdot\left(\frac{a^{2\times-3}}{a^{3\times-3}}\right) = \frac{2^{-1}}{5^{-2}}\cdot\frac{a^{-6}}{a^{-9}} = \frac{5^{2}}{2^{1}}\cdot\frac{\left(a^{-6}\cdot a^{9}\right)}{1} = \frac{5\cdot5}{2}\cdot\frac{a^{-6+9}}{1} = \frac{25\cdot a^{3}}{2\cdot1} = \frac{25}{2}a^{3} = \mathbf{12\frac{1}{2}a^{3}}$$

Example 3.4-72

$$\left(\frac{3^{-2}\cdot a^{0}\cdot w^{2}}{a^{2}\cdot w^{-3}}\right)^{-2} = \frac{3^{-2\times-2}\cdot a^{0\times-2}\cdot w^{2\times-2}}{a^{2\times-2}\cdot w^{-3\times-2}} = \frac{3^{4}\cdot a^{0}\cdot w^{-4}}{a^{-4}\cdot w^{6}} = \frac{3^{4}\cdot1}{a^{-4}\cdot\left(w^{6}\cdot w^{4}\right)} = \frac{3\cdot3\cdot3\cdot3}{a^{-4}\cdot\left(w^{6+4}\right)} = \frac{81}{a^{-4}\cdot w^{10}}$$

$$= \frac{81\cdot a^{4}}{w^{10}} = \mathbf{\frac{81\,a^{4}}{w^{10}}}$$

Example 3.4-73

$$\frac{a^{2}\cdot b^{-4}\cdot c^{5}}{a^{-2}\cdot b^{2}\cdot c^{-3}} = \frac{\left(a^{2}\cdot a^{2}\right)\cdot\left(c^{5}\cdot c^{3}\right)}{b^{2}\cdot b^{4}} = \frac{\left(a^{2+2}\right)\cdot\left(c^{5+3}\right)}{b^{2+4}} = \frac{a^{4}\cdot c^{8}}{b^{6}} = \mathbf{\frac{a^{4}c^{8}}{b^{6}}}$$

Example 3.4-74

$$\left(\frac{x^{-3}\cdot y^{-2}}{x^{5}}\right)^{3} = \frac{x^{-3\times3}\cdot y^{-2\times3}}{x^{5\times3}} = \frac{x^{-9}\cdot y^{-6}}{x^{15}} = \frac{y^{-6}}{x^{9}\cdot x^{15}} = \frac{y^{-6}}{x^{9+15}} = \frac{y^{-6}}{x^{24}} = \frac{1}{x^{24}\cdot y^{6}} = \mathbf{\frac{1}{x^{24}y^{6}}}$$

Example 3.4-75

$$\left(\frac{x^{-1}\cdot y^{-4}}{x^{-2}\cdot y^{2}}\right)^{-2} = \frac{x^{-1\times-2}\cdot y^{-4\times-2}}{x^{-2\times-2}\cdot y^{2\times-2}} = \frac{x^{2}\cdot y^{8}}{x^{4}\cdot y^{-4}} = \frac{y^{4}\cdot y^{8}}{x^{4}\cdot x^{-2}} = \frac{y^{4+8}}{x^{4-2}} = \mathbf{\frac{y^{12}}{x^{2}}}$$

Example 3.4-76

$$\left(\frac{5\cdot a^{3}\cdot b^{-3}}{a^{-2}\cdot b^{4}}\right)^{-3} = \frac{5^{1\times-3}\cdot a^{3\times-3}\cdot b^{-3\times-3}}{a^{-2\times-3}\cdot b^{4\times-3}} = \frac{5^{-3}\cdot a^{-9}\cdot b^{9}}{a^{6}\cdot b^{-12}} = \frac{b^{12}\cdot b^{9}}{5^{3}\cdot\left(a^{9}\cdot a^{6}\right)} = \frac{b^{12+9}}{125\cdot\left(a^{9+6}\right)} = \mathbf{\frac{b^{21}}{125\,a^{15}}}$$

Example 3.4-77

$$\left(\frac{a^{5}\cdot(b\,x\,y)^{3}}{x^{-2}}\right)\cdot\frac{x^{-3}}{y^{-2}} = \left(\frac{a^{5}\cdot\left(b^{1\times3}\cdot x^{1\times3}\cdot y^{1\times3}\right)}{x^{-2}}\right)\cdot\frac{x^{-3}}{y^{-2}} = \frac{a^{5}\cdot b^{3}\cdot x^{3}\cdot y^{3}}{x^{-2}}\cdot\frac{x^{-3}}{y^{-2}} = \frac{a^{5}\cdot b^{3}\cdot x^{3}\cdot y^{3}\cdot x^{-3}}{x^{-2}\cdot y^{-2}}$$

$$= a^{5}\cdot b^{3}\cdot\left(x^{3}\cdot x^{2}\cdot x^{-3}\right)\cdot\left(y^{3}\cdot y^{2}\right) = a^{5}\cdot b^{3}\cdot\left(x^{3+2-3}\right)\cdot\left(y^{3+2}\right) = a^{5}\cdot b^{3}\cdot x^{2}\cdot y^{5} = \mathbf{a^{5}\,b^{3}\,x^{2}\,y^{5}}$$

Example 3.4-78

$$\frac{\left(x^{-2}\right)^{-3}\cdot y^{-4}}{\left(x^{2}\cdot a^{3}\right)^{-1}} = \frac{\left(x^{-2\times-3}\right)\cdot y^{-4}}{x^{2\times-1}\cdot a^{3\times-1}} = \frac{x^{6}\cdot y^{-4}}{x^{-2}\cdot a^{-3}} = \frac{\left(x^{6}\cdot x^{2}\right)\cdot y^{-4}}{a^{-3}} = \frac{\left(x^{6+2}\right)\cdot y^{-4}}{a^{-3}} = \frac{x^{8}\cdot a^{3}}{y^{4}} = \mathbf{\frac{a^{3}x^{8}}{y^{4}}}$$

Example 3.4-79

$$\left(\frac{2^{-2}\cdot 2^{-4}\cdot a^{3}}{\left(a^{-2}\cdot 3^{2}\right)^{0}}\right)\cdot\left(\frac{a^{-2}}{b^{2}}\right)^{-4} = \left(\frac{2^{-2-4}\cdot a^{3}}{1}\right)\cdot\left(\frac{a^{-2\times-4}}{b^{2\times-4}}\right) = \left(\frac{2^{-6}\cdot a^{3}}{1}\right)\cdot\left(\frac{a^{8}}{b^{-8}}\right) = \frac{2^{-6}\cdot a^{3}}{1}\cdot\frac{a^{8}}{b^{-8}}$$

$$= \frac{2^{-6}\cdot\left(a^{3}\cdot a^{8}\right)}{1\cdot b^{-8}} = \frac{2^{-6}\cdot\left(a^{3+8}\right)}{b^{-8}} = \frac{2^{-6}\cdot a^{11}}{b^{-8}} = \frac{a^{11}}{2^{6}\cdot b^{-8}} = \frac{a^{11}\cdot b^{8}}{2\cdot2\cdot2\cdot2\cdot2\cdot2} = \mathbf{\frac{a^{11}b^{8}}{64}}$$

Example 3.4-80

$$\left(w^{2}\cdot z^{-2}\cdot a^{3}\right)^{-1}\cdot\left(\frac{a^{-2}}{w^{-1}\cdot z^{2}}\right)\cdot\frac{1}{a^{4}} = \left(w^{2\times-1}\cdot z^{-2\times-1}\cdot a^{3\times-1}\right)\cdot\left(\frac{a^{-2}\cdot w}{z^{2}}\right)\cdot\frac{1}{a^{4}} = \frac{\left(w^{-2}\cdot z^{2}\cdot a^{-3}\right)}{1}\cdot\left(\frac{a^{-2}\cdot w}{z^{2}}\right)\cdot\frac{1}{a^{4}}$$

$$= \frac{\left(w^{-2}\cdot z^{2}\cdot a^{-3}\right)\cdot\left(a^{-2}\cdot w\right)\cdot 1}{1\cdot z^{2}\cdot a^{4}} = \frac{w^{-2}\cdot z^{2}\cdot a^{-3}\cdot a^{-2}\cdot w^{1}}{z^{2}\cdot a^{4}} = \frac{\left(w^{-2}\cdot w^{1}\right)\cdot\left(a^{-3}\cdot a^{-2}\cdot a^{-4}\right)}{z^{2}\cdot z^{-2}} = \frac{\left(w^{-2+1}\right)\cdot\left(a^{-3-2-4}\right)}{z^{2-2}}$$

$$= \frac{w^{-1}\cdot a^{-9}}{z^{0}} = \frac{w^{-1}\cdot a^{-9}}{1} = \frac{1}{w^{1}\cdot a^{9}} = \mathbf{\frac{1}{a^{9}w}}$$

Practice Problems - Dividing Negative Integer Exponents (More Difficult Cases)

Section 3.4 Case II b Practice Problems - Divide the following negative integer exponents:

1. $\dfrac{\left(x^{-2}\cdot x\right)^{-3}}{x^{-3}} =$

2. $\dfrac{-(2)^{-2}a^{-2}b^{3}}{\left(a^{-1}b\right)^{-1}} =$

3. $\left(\dfrac{a^{3}\cdot b^{6}\cdot c^{2}}{a^{2}\cdot b^{3}}\right)^{-2} =$

4. $\dfrac{y^{3}\cdot(yzw)^{-1}}{z\cdot y^{-2}\cdot w^{-3}} =$

5. $\dfrac{(a\cdot b)^{-2}\cdot(xy)^{-1}}{a^{-3}\cdot y} =$

6. $\dfrac{(xyz)^{0}\cdot x^{-2}}{\left(x\cdot y^{-1}\right)^{-3}} =$

7. $\dfrac{2^{-1}\cdot a^{3}\cdot b^{-2}\cdot c}{\left(b\cdot c^{-2}\right)^{2}} =$

8. $\left(\dfrac{z^{4}\cdot w^{-2}\cdot(a\cdot b)^{-1}}{a^{2}\cdot w^{-2}\cdot z^{0}}\right)^{2} =$

9. $\dfrac{2^{-3}\cdot 2^{2}\cdot(y\cdot z)^{-3}\cdot b^{3}}{b^{-1}\cdot y^{3}} =$

10. $\left(\dfrac{a^{-3}\cdot b^{2}\cdot\left(c^{-1}\cdot d\right)^{4}}{c^{3}\cdot d}\right)^{-2} =$

Case III Adding and Subtracting Negative Integer Exponents

In this section two classes of negative integer exponents are addressed. The first class deals with addition and subtraction of variables and numbers that are raised to negative exponents (Case III a). The second class deals with addition and subtraction of negative integer exponents that are in fraction form (Case III b). These two cases are addressed below:

Case III a Addition and Subtraction of Variables and Numbers Raised to Negative Exponential Terms

Negative exponential expressions are added and subtracted using the following steps:

Step 1 Group the exponential terms with similar bases.

Step 2 Apply the Negative Power Law (Law VI) from Table 3.4-1, i.e., change a^{-n} to $\frac{1}{a^n}$.

Step 3 Simplify the exponential expression by:

a. Using the fraction techniques learned in Chapter 2, and

b. Using appropriate exponent laws such as the Multiplication Law (Law I) from Table 3.4-1.

Examples with Steps

The following examples show the steps as to how exponential expressions having negative integer exponents are added or subtracted:

Example 3.4-81

$$\boxed{3^{-3}+3^{-2}} =$$

Solution:

Step 1 $\boxed{\textit{Not Applicable}}$

Step 2 $\boxed{3^{-3}+3^{-2}} = \boxed{\frac{1}{3^3}+\frac{1}{3^2}} = \boxed{\frac{1}{27}+\frac{1}{9}}$

Step 3 $\boxed{\frac{1}{27}+\frac{1}{9}} = \boxed{\frac{(1\cdot 9)+(1\cdot 27)}{27\cdot 9}} = \boxed{\frac{9+27}{243}} = \boxed{\frac{\overset{4}{\cancel{36}}}{\underset{27}{\cancel{243}}}} = \boxed{\frac{4}{27}}$

Example 3.4-82

$$\boxed{x^{-1}+x^{-2}+2x^{-1}-4x^{-2}+5^{-2}} =$$

Solution:

Step 1 $\boxed{x^{-1}+x^{-2}+2x^{-1}-4x^{-2}+5^{-2}} = \boxed{\left(x^{-1}+2x^{-1}\right)+\left(x^{-2}-4x^{-2}\right)+5^{-2}}$

$= \boxed{(1+2)x^{-1}+(1-4)x^{-2}+5^{-2}} = \boxed{3x^{-1}-3\,x^{-2}+5^{-2}}$

Step 2

$$\boxed{3x^{-1}-3x^{-2}+5^{-2}}=\boxed{\frac{3}{x^1}-\frac{3}{x^2}+\frac{1}{5^2}}=\boxed{\frac{3}{x}-\frac{3}{x^2}+\frac{1}{5^2}}$$

Step 3

$$\boxed{\frac{3}{x}-\frac{3}{x^2}+\frac{1}{5^2}}=\boxed{\left(\frac{3}{x}-\frac{3}{x^2}\right)+\frac{1}{25}}=\boxed{\left(\frac{(3\cdot x^2)-(3\cdot x)}{x\cdot x^2}\right)+\frac{1}{25}}=\boxed{\left(\frac{3x^2-3x}{x^{1+2}}\right)+\frac{1}{25}}$$

$$=\boxed{\frac{3x^2-3x}{x^3}+\frac{1}{25}}=\boxed{\frac{25\cdot\left(3x^2-3x\right)+1\cdot x^3}{x^3\cdot 25}}=\boxed{\frac{75x^2-75x+x^3}{25x^3}}=\boxed{\frac{x^3+75x^2-75x}{25x^3}}$$

$$=\boxed{\frac{x\left(x^2+75x-75\right)}{25x^3}}=\boxed{\frac{x^2+75x-75}{25\left(x^3x^{-1}\right)}}=\boxed{\frac{x^2+75x-75}{25x^{3-1}}}=\boxed{\mathbf{\frac{x^2+75x-75}{25x^2}}}$$

Example 3.4-83

$$\boxed{5m^{-2}-7m^{-3}+mn^{-4}-2m^{-2}+3m^{-3}}=$$

Solution:

Step 1

$$\boxed{5m^{-2}-7m^{-3}+mn^{-4}-2m^{-2}+3m^{-3}}=\boxed{\left(5m^{-2}-2m^{-2}\right)+\left(-7m^{-3}+3m^{-3}\right)+mn^{-4}}$$

$$=\boxed{(5-2)m^{-2}+(-7+3)m^{-3}+mn^{-4}}=\boxed{3m^{-2}-4m^{-3}+mn^{-4}}$$

Step 2

$$\boxed{3m^{-2}-4m^{-3}+mn^{-4}}=\boxed{\left(3m^{-2}-4m^{-3}\right)+mn^{-4}}=\boxed{\left(\frac{3}{m^2}-\frac{4}{m^3}\right)+\frac{m}{n^4}}$$

Step 3

$$\boxed{\left(\frac{3}{m^2}-\frac{4}{m^3}\right)+\frac{m}{n^4}}=\boxed{\left(\frac{3\cdot m^3-4\cdot m^2}{m^2\cdot m^3}\right)+\frac{m}{n^4}}=\boxed{\frac{3m^3-4m^2}{m^2m^3}+\frac{m}{n^4}}$$

$$=\boxed{\frac{3m^3-4m^2}{m^{2+3}}+\frac{m}{n^4}}=\boxed{\frac{3m^3-4m^2}{m^5}+\frac{m}{n^4}}=\boxed{\frac{\left[\left(3m^3-4m^2\right)\cdot n^4\right]+\left(m\cdot m^5\right)}{m^5\cdot n^4}}$$

$$=\boxed{\frac{\left[3m^3\cdot n^4-4m^2\cdot n^4\right]+m^1\cdot m^5}{m^5n^4}}=\boxed{\frac{3m^3n^4-4m^2n^4+m^{1+5}}{m^5n^4}}$$

$$=\boxed{\frac{3m^3n^4-4m^2n^4+m^6}{m^5n^4}}=\boxed{\frac{m^2\left(3mn^4-4n^4+m^4\right)}{m^5n^4}}=\boxed{\frac{3mn^4-4n^4+m^4}{\left(m^5m^{-2}\right)n^4}}$$

$$=\boxed{\frac{3mn^4-4n^4+m^4}{m^{5-2}n^4}}=\boxed{\mathbf{\frac{3mn^4-4n^4+m^4}{m^3n^4}}}$$

Example 3.4-84

$$\boxed{x^{-3}+y^{-2}-3x^{-3}+2y^{-2}}=$$

Solution:

Step 1

$$\boxed{x^{-3}+y^{-2}-3x^{-3}+2y^{-2}}=\boxed{\left(x^{-3}-3x^{-3}\right)+\left(y^{-2}+2y^{-2}\right)}=\boxed{(1-3)x^{-3}+(1+2)y^{-2}}$$

$$=\boxed{-2x^{-3}+3y^{-2}}$$

Step 2

$$\boxed{-2x^{-3}+3y^{-2}}=\boxed{\frac{-2}{x^3}+\frac{3}{y^2}}$$

Step 3

$$\boxed{\frac{-2}{x^3}+\frac{3}{y^2}}=\boxed{\frac{\left(-2\cdot y^2\right)+\left(3\cdot x^3\right)}{x^3\cdot y^2}}=\boxed{\mathbf{\frac{-2y^2+3x^3}{x^3y^2}}}$$

Example 3.4-85

$$\boxed{a^{-3b}+2a^{-2b}-4a^{-3b}+5+3a^{-2b}}=$$

Solution:

Step 1

$$\boxed{a^{-3b}+2a^{-2b}-4a^{-3b}+5+3a^{-2b}}=\boxed{\left(a^{-3b}-4a^{-3b}\right)+\left(2a^{-2b}+3a^{-2b}\right)+5}$$

$$=\boxed{(1-4)a^{-3b}+(2+3)a^{-2b}+5}=\boxed{-3a^{-3b}+5a^{-2b}+5}$$

Step 2

$$\boxed{-3a^{-3b}+5a^{-2b}+5}=\boxed{-\frac{3}{a^{3b}}+\frac{5}{a^{2b}}+\frac{5}{1}}$$

Step 3

$$\boxed{-\frac{3}{a^{3b}}+\frac{5}{a^{2b}}+\frac{5}{1}}=\boxed{\left(-\frac{3}{a^{3b}}+\frac{5}{a^{2b}}\right)+\frac{5}{1}}=\boxed{\left(\frac{\left(-3\cdot a^{2b}\right)+\left(5\cdot a^{3b}\right)}{a^{3b}\cdot a^{2b}}\right)+\frac{5}{1}}$$

$$=\boxed{\frac{-3a^{2b}+5a^{3b}}{a^{3b+2b}}+\frac{5}{1}}=\boxed{\frac{-3a^{2b}+5a^{3b}}{a^{5b}}+\frac{5}{1}}=\boxed{\frac{\left[\left(-3a^{2b}+5a^{3b}\right)\cdot 1\right]+\left(5\cdot a^{5b}\right)}{a^{5b}\cdot 1}}$$

$$=\boxed{\frac{-3a^{2b}+5a^{3b}+5a^{5b}}{a^{5b}}}=\boxed{\frac{a^{2b}\left(-3+5a^{b}+5a^{3b}\right)}{a^{5b}}}=\boxed{\frac{-3+5a^{b}+5a^{3b}}{a^{5b}a^{-2b}}}$$

$$=\boxed{\frac{-3+5a^{b}+5a^{3b}}{a^{5b-2b}}}=\boxed{\mathbf{\frac{5a^{3b}+5a^{b}-3}{a^{3b}}}}$$

Additional Examples - Addition and Subtraction of Variables and Numbers Raised to Negative Exponential Terms

The following examples further illustrate addition and subtraction of negative exponential terms:

Example 3.4-86

$$(x+y)^{-5} = \frac{1}{(x+y)^5}$$

Note: $(x \cdot y)^{-5} = \frac{1}{(x \cdot y)^5} = \frac{1}{x^5 y^5}$

Example 3.4-87

$$a^{-1} - b^{-1} + 2a^{-1} + 3b^{-1} = \left(a^{-1} + 2a^{-1}\right) + \left(3b^{-1} - b^{-1}\right) = (1+2)a^{-1} + (3-1)b^{-1} = 3a^{-1} + 2b^{-1} = \frac{3}{a} + \frac{2}{b}$$

$$= \frac{(3 \cdot b) + (2 \cdot a)}{a \cdot b} = \frac{2a + 3b}{ab}$$

Example 3.4-88

$$x^{-1} + y^{-2} + 5y^{-2} = x^{-1} + \left(y^{-2} + 5y^{-2}\right) = x^{-1} + 6y^{-2} = \frac{1}{x} + \frac{6}{y^2} = \frac{\left(1 \cdot y^2\right) + (6 \cdot x)}{x \cdot y^2} = \frac{6x + y^2}{xy^2}$$

Example 3.4-89

$$a^{-1} + b^{-2} + c^{-3} + 3b^{-2} = a^{-1} + \left(b^{-2} + 3b^{-2}\right) + c^{-3} = a^{-1} + (1+3)b^{-2} + c^{-3} = a^{-1} + 4b^{-2} + c^{-3}$$

$$= \left(a^{-1} + 4b^{-2}\right) + c^{-3} = \left(\frac{1}{a} + \frac{4}{b^2}\right) + \frac{1}{c^3} = \left(\frac{\left(1 \cdot b^2\right) + (4 \cdot a)}{a \cdot b^2}\right) + \frac{1}{c^3} = \left(\frac{b^2 + 4a}{a\,b^2}\right) + \frac{1}{c^3}$$

$$= \frac{\left[\left(b^2 + 4a\right) \cdot c^3\right] + \left(1 \cdot ab^2\right)}{a\,b^2 \cdot c^3} = \frac{b^2 \cdot c^3 + 4a \cdot c^3 + a\,b^2}{a\,b^2\,c^3} = \frac{b^2c^3 + 4ac^3 + ab^2}{ab^2c^3}$$

Example 3.4-90

$$5x^{-3} + 3x^{-2} + 2x^{-3} - x^{-2} + 5^{-2} = \left(5x^{-3} + 2x^{-3}\right) + \left(3x^{-2} - x^{-2}\right) + 5^{-2} = (5+2)x^{-3} + (3-1)x^{-2} + 5^{-2}$$

$$= 7x^{-3} + 2x^{-2} + 5^{-2} = \frac{7}{x^3} + \frac{2}{x^2} + \frac{1}{5^2} = \left(\frac{7}{x^3} + \frac{2}{x^2}\right) + \frac{1}{25} = \left(\frac{7 \cdot x^2 + 2 \cdot x^3}{x^3 \cdot x^2}\right) + \frac{1}{25} = \frac{7x^2 + 2x^3}{x^{3+2}} + \frac{1}{25}$$

$$= \frac{x^2(7+2x)}{x^5} + \frac{1}{25} = \frac{7+2x}{x^5 \cdot x^{-2}} + \frac{1}{25} = \frac{7+2x}{x^{5-2}} + \frac{1}{25} = \frac{7+2x}{x^3} + \frac{1}{25} = \frac{[(7+2x) \cdot 25] + \left(1 \cdot x^3\right)}{25 \cdot x^3}$$

$$= \frac{175 + 50x + x^3}{25x^3} = \frac{x^3 + 50x + 175}{25x^3}$$

Example 3.4-91

$$-2m^{-4}-3m^{-2}+2m^{-4}+3m^{0}-m^{-2}-10=\left(-2m^{-4}+2m^{-4}\right)+\left(-3m^{-2}-m^{-2}\right)+(3-10)$$

$$=(-2+2)m^{-4}+(-3-1)m^{-2}-7=0m^{-4}-4m^{-2}-7=-4m^{-2}-7=-\frac{4}{m^2}-\frac{7}{1}=\frac{-(4\cdot 1)-\left(7\cdot m^2\right)}{m^2\cdot 1}$$

$$=\frac{-4-7m^2}{m^2}=\mathbf{-\frac{7m^2+4}{m^2}}$$

Example 3.4-92

$$\left(x^{-5}-3x^{-3}+x\right)-\left(-3x^{-5}+2x^{-3}\right)=x^{-5}-3x^{-3}+x+3x^{-5}-2x^{-3}=\left(x^{-5}+3x^{-5}\right)+\left(-3x^{-3}-2x^{-3}\right)+x$$

$$=(1+3)x^{-5}+(-3-2)x^{-3}+x=4x^{-5}-5x^{-3}+x=\frac{4}{x^5}-\frac{5}{x^3}+\frac{x}{1}=\left(\frac{4}{x^5}-\frac{5}{x^3}\right)+\frac{x}{1}$$

$$=\left(\frac{4\cdot x^3-5\cdot x^5}{x^5\cdot x^3}\right)+\frac{x}{1}=\left(\frac{4x^3-5x^5}{x^{5+3}}\right)+\frac{x}{1}=\left(\frac{4x^3-5x^5}{x^8}\right)+\frac{x}{1}=\frac{\left[\left(4x^3-5x^5\right)\cdot 1\right]+\left(x\cdot x^8\right)}{x^8\cdot 1}$$

$$=\frac{4x^3-5x^5+x^9}{x^8}=\frac{x^3\left(4-5x^2+x^6\right)}{x^8}=\frac{4-5x^2+x^6}{x^8x^{-3}}=\frac{4-5x^2+x^6}{x^{8-3}}=\mathbf{\frac{x^6-5x^2+4}{x^5}}$$

Example 3.4-93

$$x^{-2}+3x^{-2}+y^{-2}+x^2-4y^{-2}-5^2+2x^2=\left(x^{-2}+3x^{-2}\right)+\left(y^{-2}-4y^{-2}\right)+\left(x^2+2x^2\right)-25$$

$$=(1+3)x^{-2}+(1-4)y^{-2}+(1+2)x^2-25=4x^{-2}-3y^{-2}+3x^2-25=\left(\frac{4}{x^2}-\frac{3}{y^2}\right)+3x^2-25$$

$$=\left(\frac{4\cdot y^2-3\cdot x^2}{x^2\cdot y^2}\right)+\frac{3x^2-25}{1}=\frac{4y^2-3x^2}{x^2y^2}+\frac{3x^2-25}{1}=\frac{\left[\left(4y^2-3x^2\right)\cdot 1\right]+\left[x^2y^2\left(3x^2-25\right)\right]}{x^2y^2}$$

$$=\mathbf{\frac{4y^2-3x^2+3x^4y^2-25x^2y^2}{x^2y^2}}$$

Example 3.4-94

$$\left(k^{-4}-3k^{-2}-5\right)-\left(-3k^{-4}+5k^{-2}\right)-4=k^{-4}-3k^{-2}-5+3k^{-4}-5k^{-2}-4$$

$$=\left(k^{-4}+3k^{-4}\right)+\left(-3k^{-2}-5k^{-2}\right)+(-4-5)=4k^{-4}-8k^{-2}-9=\frac{4}{k^4}-\frac{8}{k^2}-\frac{9}{1}=\left(\frac{4}{k^4}-\frac{8}{k^2}\right)-\frac{9}{1}$$

$$= \boxed{\left(\frac{4\cdot k^2 - 8\cdot k^4}{k^4\cdot k^2}\right) - \frac{9}{1}} = \boxed{\frac{4k^2 - 8k^4}{k^{4+2}} - \frac{9}{1}} = \boxed{\frac{4k^2 - 8k^4}{k^6} - \frac{9}{1}} = \boxed{\frac{\left[\left(4k^2 - 8k^4\right)\cdot 1\right] - \left(9\cdot k^6\right)}{k^6}}$$

$$= \boxed{\frac{4k^2 - 8k^4 - 9k^6}{k^6}} = \boxed{\frac{k^2\left(4 - 8k^2 - 9k^4\right)}{k^6}} = \boxed{\frac{4 - 8k^2 - 9k^4}{k^6\cdot k^{-2}}} = \boxed{\frac{4 - 8k^2 - 9k^4}{k^{6-2}}} = \boxed{\frac{\mathbf{-9k^4 - 8k^2 + 4}}{\mathbf{k^4}}}$$

Example 3.4-95

$$\boxed{\left(-5h^{-3} - 3h^{-1} - 5\right) - \left(3h^{-3} - h - 4\right) + 5h^{-1} + 2 + h^{-3}} = \boxed{-5h^{-3} - 3h^{-1} - 5 - 3h^{-3} + h + 4 + 5h^{-1} + 2 + h^{-3}}$$

$$= \boxed{\left(-5h^{-3} - 3h^{-3} + h^{-3}\right) + \left(-3h^{-1} + 5h^{-1}\right) + h + (-5 + 4 + 2)} = \boxed{(-5 - 3 + 1)h^{-3} + (-3 + 5)h^{-1} + h + 1}$$

$$= \boxed{-7h^{-3} + 2h^{-1} + h + 1} = \boxed{-\frac{7}{h^3} + \frac{2}{h^1} + \frac{h+1}{1}} = \boxed{\left(-\frac{7}{h^3} + \frac{2}{h}\right) + \frac{h+1}{1}} = \boxed{\left(\frac{-7\cdot h + 2\cdot h^3}{h^3\cdot h}\right) + \frac{h+1}{1}}$$

$$= \boxed{\frac{-7h + 2h^3}{h^{3+1}} + \frac{h+1}{1}} = \boxed{\frac{-7h + 2h^3}{h^4} + \frac{h+1}{1}} = \boxed{\frac{\left[\left(-7h + 2h^3\right)\cdot 1\right] + \left[(h+1)\cdot h^4\right]}{h^4\cdot 1}} = \boxed{\frac{-7h + 2h^3 + h^5 + h^4}{h^4}}$$

$$= \boxed{\frac{h^5 + h^4 + 2h^3 - 7h}{h^4}} = \boxed{\frac{h\cdot\left(h^4 + h^3 + 2h^2 - 7\right)}{h^4}} = \boxed{\frac{h^4 + h^3 + 2h^2 - 7}{h^4\cdot h^{-1}}} = \boxed{\frac{h^4 + h^3 + 2h^2 - 7}{h^{4-1}}}$$

$$= \boxed{\frac{\mathbf{h^4 + h^3 + 2h^2 - 7}}{\mathbf{h^3}}}$$

Practice Problems - Addition and Subtraction of Variables and Numbers Raised to Negative Exponential Terms

Section 3.4 Case III a Practice Problems - Simplify the following negative integer exponential expressions:

1. $x^{-1} + 2x^{-2} + 3x^{-1} - 6x^{-2} =$
2. $\left(3a^{-4} - b^{-2}\right) + \left(-2a^{-4} + 3b^{-2}\right) =$
3. $(xy)^{-1} + y^{-2} + 4(xy)^{-1} - 3y^{-2} + 2^{-3} =$
4. $4x^{-1} + y^{-3} + 5y^{-3} =$
5. $m^{-5} - \left(m^{-2} - 3m^{-5} + m^0\right) + 3m^{-2} =$
6. $\left(a^3\right)^{-2} + \left(a^{-2}b\right)^2 - 6a^{-6} + 3a^{-4}b^2 =$
7. $3x^{-4} + 3x^{-2} + 2x^{-4} - \left(x^{-4} + 3x^{-2}\right) =$
8. $k^{-2n} + k^{-3n} - 3k^{-2n} + 2^{-2} =$
9. $w^{-2} + 3w^{-4} + 2w^{-2} - \left(w^0 - 4w^{-2}\right) =$
10. $\left(-7z^{-3} + 3z - 5\right) - \left(-3z^{-3} + z - 2\right) + 4z =$

Case III b Addition and Subtraction of Negative Exponential Terms in Fraction Form

A special class of negative integer exponents are in the form of fractions. This class of negative integer exponents are added and subtracted by applying the fraction techniques, discussed in Chapter 2, and the exponent laws (see Table 3.4-1). It is recommended that students review addition and subtraction of integer fractions (see Sections 2.2 and 2.3) before proceeding with this section. Negative integer exponents in fraction form are added and subtracted using the following steps:

Step 1 Apply the Negative Power Law (Law VI) from Table 3.4-1, i.e., change a^{-n} to $\frac{1}{a^n}$.

Step 2 Simplify the exponential expression by:

a. Using the fraction techniques learned in Chapter 2, and

b. Using appropriate exponent laws such as the Multiplication Law (Law I) from Table 3.4-1.

Examples with Steps

The following examples show the steps as to how negative exponential expressions in fraction form are added and subtracted:

Example 3.4-96

$$x^{-2} - \frac{2}{2-x^{-2}} =$$

Solution:

Step 1

$$x^{-2} - \frac{2}{2-x^{-2}} = \frac{1}{x^2} - \frac{2}{2-\frac{1}{x^2}}$$

Step 2

$$\frac{1}{x^2} - \frac{2}{2-\frac{1}{x^2}} = \frac{1}{x^2} - \frac{2}{\frac{2}{1}-\frac{1}{x^2}} = \frac{1}{x^2} - \frac{2}{\frac{2\cdot x^2 - 1\cdot 1}{1\cdot x^2}} = \frac{1}{x^2} - \frac{2}{\frac{2x^2-1}{x^2}}$$

$$= \frac{1}{x^2} - \frac{\frac{2}{1}}{\frac{2x^2-1}{x^2}} = \frac{1}{x^2} - \frac{2\cdot x^2}{1\cdot\left(2x^2-1\right)} = \frac{1}{x^2} - \frac{2x^2}{2x^2-1} = \frac{\left[1\cdot\left(2x^2-1\right)\right]-\left(2x^2\cdot x^2\right)}{x^2\cdot\left(2x^2-1\right)}$$

$$= \frac{2x^2-1-2x^{2+2}}{2\left(x^2\cdot x^2\right)-1\cdot x^2} = \frac{2x^2-1-2x^4}{2\left(x^{2+2}\right)-x^2} = \mathbf{\frac{-2x^4+2x^2-1}{2x^4-x^2}}$$

Example 3.4-97

$$\frac{x-y^{-3}}{x} =$$

Solution:

Step 1

$$\frac{x-y^{-3}}{x} = \frac{x-\frac{1}{y^3}}{x}$$

Step 2

$$\frac{x-\frac{1}{y^3}}{x} = \frac{\frac{x}{1}-\frac{1}{y^3}}{x} = \frac{\frac{(x\cdot y^3)-(1\cdot 1)}{1\cdot y^3}}{x} = \frac{\frac{xy^3-1}{y^3}}{x} = \frac{\frac{x\,y^3-1}{y^3}}{\frac{x}{1}} = \frac{(x\,y^3-1)\cdot 1}{y^3\cdot x}$$

$$= \boldsymbol{\frac{x\,y^3-1}{x\,y^3}}$$

Example 3.4-98

$$\frac{2^{-a}+4^{-a}}{2^{-a}} =$$

Solution:

Step 1

$$\frac{2^{-a}+4^{-a}}{2^{-a}} = \frac{\frac{1}{2^a}+\frac{1}{4^a}}{\frac{1}{2^a}}$$

Step 2

$$\frac{\frac{1}{2^a}+\frac{1}{4^a}}{\frac{1}{2^a}} = \frac{\frac{(1\cdot 4^a)+(1\cdot 2^a)}{2^a\cdot 4^a}}{\frac{1}{2^a}} = \frac{\frac{4^a+2^a}{2^a\cdot 4^a}}{\frac{1}{2^a}} = \frac{(4^a+2^a)\cdot 2^a}{(2^a\cdot 4^a)\cdot 1} = \frac{(4^a+2^a)\cdot 2^a}{2^a\cdot 4^a}$$

$$= \frac{(4^a+2^a)\cdot(2^a\cdot 2^{-a})}{4^a} = \frac{(4^a+2^a)\cdot(2^{a-a})}{4^a} = \frac{(4^a+2^a)\cdot 2^0}{4^a} = \frac{(4^a+2^a)\cdot 1}{4^a}$$

$$= \boldsymbol{\frac{4^a+2^a}{4^a}}$$

Example 3.4-99

$$x^{-3}+\frac{8}{x^{-2}} =$$

Solution:

Step 1

$$x^{-3}+\frac{8}{x^{-2}} = \frac{1}{x^3}+\frac{8}{\frac{1}{x^2}}$$

Step 2

$$\boxed{\frac{1}{x^3}+\frac{8}{\frac{1}{x^2}}}=\boxed{\frac{1}{x^3}+\frac{\frac{8}{1}}{\frac{1}{x^2}}}=\boxed{\frac{1}{x^3}+\frac{8\cdot x^2}{1\cdot 1}}=\boxed{\frac{1}{x^3}+\frac{8x^2}{1}}=\boxed{\frac{(1\cdot 1)+8\left(x^2\cdot x^3\right)}{x^3\cdot 1}}$$

$$=\boxed{\frac{1+8x^{2+3}}{x^3}}=\boxed{\frac{1+8x^5}{x^3}}=\boxed{\mathbf{\frac{8x^5+1}{x^3}}}$$

Example 3.4-100

$$\boxed{\frac{a^{-2}}{b^{-1}}-\frac{4}{3-a^{-2}}}=$$

Solution:

Step 1

$$\boxed{\frac{a^{-2}}{b^{-1}}-\frac{4}{3-a^{-2}}}=\boxed{\frac{\frac{1}{a^2}}{\frac{1}{b}}-\frac{4}{3-\frac{1}{a^2}}}$$

Step 2

$$\boxed{\frac{\frac{1}{a^2}}{\frac{1}{b}}-\frac{4}{3-\frac{1}{a^2}}}=\boxed{\frac{1\cdot b}{a^2\cdot 1}-\frac{4}{\frac{3}{1}-\frac{1}{a^2}}}=\boxed{\frac{b}{a^2}-\frac{4}{\frac{3\cdot a^2-1\cdot 1}{1\cdot a^2}}}=\boxed{\frac{b}{a^2}-\frac{\frac{4}{1}}{\frac{3a^2-1}{a^2}}}$$

$$=\boxed{\frac{b}{a^2}-\frac{4\cdot a^2}{1\cdot\left(3a^2-1\right)}}=\boxed{\frac{b}{a^2}-\frac{4a^2}{3a^2-1}}=\boxed{\frac{\left[b\cdot\left(3a^2-1\right)\right]-\left(4a^2\cdot a^2\right)}{a^2\cdot\left(3a^2-1\right)}}$$

$$=\boxed{\frac{3a^2b-b-4a^{2+2}}{3a^{2+2}-a^2}}=\boxed{\frac{3a^2b-b-4a^4}{3a^4-a^2}}=\boxed{\mathbf{\frac{-4a^4+3a^2b-b}{3a^4-a^2}}}$$

Additional Examples - Addition and Subtraction of Negative Exponential Terms in Fraction Form

The following examples further illustrate addition and subtraction of negative exponential expressions in fraction form:

Example 3.4-101

$$\boxed{\frac{5}{x^{-1}-y^{-1}}}=\boxed{\frac{5}{\frac{1}{x}-\frac{1}{y}}}=\boxed{\frac{5}{\frac{(1\cdot y)-(1\cdot x)}{x\cdot y}}}=\boxed{\frac{5}{\frac{y-x}{x\,y}}}=\boxed{\frac{\frac{5}{1}}{\frac{y-x}{x\,y}}}=\boxed{\frac{5\cdot(x\,y)}{1\cdot(y-x)}}=\boxed{\mathbf{\frac{5\,x\,y}{y-x}}}$$

Example 3.4-102

$$\boxed{\frac{x-y^{-1}}{3}}=\boxed{\frac{x-\frac{1}{y}}{3}}=\boxed{\frac{\frac{x}{1}-\frac{1}{y}}{3}}=\boxed{\frac{\frac{(x\cdot y)-(1\cdot 1)}{1\cdot y}}{3}}=\boxed{\frac{\frac{x\,y-1}{y}}{3}}=\boxed{\frac{\frac{x\,y-1}{y}}{\frac{3}{1}}}=\boxed{\frac{(x\,y-1)\cdot 1}{y\cdot 3}}=\boxed{\mathbf{\frac{x\,y-1}{3\,y}}}$$

Example 3.4-103

$$\frac{x^{-2}}{2^{-1}+2^{-2}} = \frac{\frac{1}{x^2}}{\frac{1}{2}+\frac{1}{2^2}} = \frac{\frac{1}{x^2}}{\frac{1}{2}+\frac{1}{4}} = \frac{\frac{1}{x^2}}{\frac{(1\cdot 4)+(1\cdot 2)}{2\cdot 4}} = \frac{\frac{1}{x^2}}{\frac{4+2}{8}} = \frac{\frac{1}{x^2}}{\frac{6}{8}} = \frac{1\cdot 8}{x^2\cdot 6} = \frac{\overset{4}{\cancel{8}}}{\underset{3}{\cancel{6}}\,x^2} = \frac{4}{3x^2} = \mathbf{1\frac{1}{3}\left(\frac{1}{x^2}\right)}$$

Example 3.4-104

$$\frac{x^{-1}}{x^{-2}-y} = \frac{\frac{1}{x}}{\frac{1}{x^2}-y} = \frac{\frac{1}{x}}{\frac{1}{x^2}-\frac{y}{1}} = \frac{\frac{1}{x}}{\frac{(1\cdot 1)-(x^2\cdot y)}{x^2\cdot 1}} = \frac{\frac{1}{x}}{\frac{1-x^2y}{x^2}} = \frac{1\cdot x^2}{x\cdot\left(1-x^2y\right)} = \frac{x^2}{x^1\cdot\left(1-x^2y\right)} = \frac{x^2\cdot x^{-1}}{\left(1-x^2y\right)}$$

$$= \frac{x^{2-1}}{1-x^2y} = \mathbf{\frac{x}{1-x^2y}}$$

Example 3.4-105

$$\frac{a^{-1}+b^{-2}}{a^{-1}-b^{-2}} = \frac{\frac{1}{a}+\frac{1}{b^2}}{\frac{1}{a}-\frac{1}{b^2}} = \frac{\frac{\left(1\cdot b^2\right)+(1\cdot a)}{a\cdot b^2}}{\frac{\left(1\cdot b^2\right)-(1\cdot a)}{a\cdot b^2}} = \frac{\frac{b^2+a}{a\cdot b^2}}{\frac{b^2-a}{a\cdot b^2}} = \frac{\left(b^2+a\right)\cdot a\cdot b^2}{\left(b^2-a\right)\cdot a\cdot b^2} = \frac{\left(b^2+a\right)\cdot a^1\cdot b^2}{\left(b^2-a\right)\cdot a^1\cdot b^2}$$

$$= \frac{\left(b^2+a\right)\cdot\left(a^1\cdot a^{-1}\right)\cdot\left(b^2\cdot b^{-2}\right)}{b^2-a} = \frac{\left(b^2+a\right)\cdot a^{1-1}\cdot b^{2-2}}{b^2-a} = \frac{\left(b^2+a\right)\cdot a^0\cdot b^0}{b^2-a} = \frac{\left(b^2+a\right)\cdot 1\cdot 1}{b^2-a} = \mathbf{\frac{b^2+a}{b^2-a}}$$

Example 3.4-106

$$x^{-1}-\left(x^{-1}+1\right)^{-1} = \frac{1}{x}-\frac{1}{\left(x^{-1}+1\right)} = \frac{1}{x}-\frac{1}{\frac{1}{x}+1} = \frac{1}{x}-\frac{1}{\frac{1}{x}+\frac{1}{1}} = \frac{1}{x}-\frac{1}{\frac{(1\cdot 1)+(1\cdot x)}{x}} = \frac{1}{x}-\frac{1}{\frac{1+x}{x}}$$

$$= \frac{1}{x}-\frac{\frac{1}{1}}{\frac{1+x}{x}} = \frac{1}{x}-\frac{1\cdot x}{1\cdot(1+x)} = \frac{1}{x}-\frac{x}{1+x} = \frac{[1\cdot(1+x)]-(x\cdot x)}{x\cdot(1+x)} = \frac{1+x-x^2}{x+x^2} = \mathbf{\frac{-x^2+x+1}{x^2+x}}$$

Example 3.4-107

$$\left(m^{-3}-m^{-2}\right)^3+3m^{-9}-4m^{-6}+2m^{-9} = \left(m^{-3\times 3}-m^{-2\times 3}\right)+3m^{-9}-4m^{-6}+2m^{-9}$$

$$= m^{-9}-m^{-6}+3m^{-9}-4m^{-6}+2m^{-9} = \left(m^{-9}+3m^{-9}+2m^{-9}\right)-\left(m^{-6}+4m^{-6}\right) = (1+3+2)m^{-9}-(1+4)m^{-6}$$

$$= 6m^{-9}-5m^{-6} = \frac{6}{m^9}-\frac{5}{m^6} = \frac{\left(6\cdot m^6\right)-\left(5\cdot m^9\right)}{m^9\cdot m^6} = \frac{6m^6-5m^9}{m^9m^6} = \frac{\cancel{m^6}\left(6-5m^3\right)}{m^9\cancel{m^6}} = \frac{6-5m^3}{m^9} = \mathbf{\frac{-5m^3+6}{m^9}}$$

Example 3.4-108

$$\boxed{\frac{k^{-2}}{k-k^{-2}}}=\boxed{\frac{\frac{1}{k^2}}{k-\frac{1}{k^2}}}=\boxed{\frac{\frac{1}{k^2}}{\frac{k}{1}-\frac{1}{k^2}}}=\boxed{\frac{\frac{1}{k^2}}{\frac{\left(k\cdot k^2\right)-(1\cdot 1)}{1\cdot k^2}}}=\boxed{\frac{\frac{1}{k^2}}{\frac{k^{1+2}-1}{k^2}}}=\boxed{\frac{\frac{1}{k^2}}{\frac{k^3-1}{k^2}}}=\boxed{\frac{1\cdot k^2}{k^2\cdot\left(k^3-1\right)}}=\boxed{\frac{k^2}{k^2\left(k^3-1\right)}}$$

$$=\boxed{\frac{k^2\cdot k^{-2}}{k^3-1}}=\boxed{\frac{k^{2-2}}{k^3-1}}=\boxed{\frac{k^0}{k^3-1}}=\boxed{\mathbf{\frac{1}{k^3-1}}}$$

Example 3.4-109

$$\boxed{\frac{\left(x^3\right)^{-2}}{x^0+y^{-2}}}=\boxed{\frac{x^{3\times -2}}{1+\frac{1}{y^2}}}=\boxed{\frac{x^{-6}}{1+\frac{1}{y^2}}}=\boxed{\frac{\frac{1}{x^6}}{\frac{1}{1}+\frac{1}{y^2}}}=\boxed{\frac{\frac{1}{x^6}}{\frac{\left(1\cdot y^2\right)+(1\cdot 1)}{1\cdot y^2}}}=\boxed{\frac{\frac{1}{x^6}}{\frac{y^2+1}{y^2}}}=\boxed{\frac{1\cdot y^2}{x^6\cdot\left(y^2+1\right)}}=\boxed{\mathbf{\frac{y^2}{x^6y^2+x^6}}}$$

Example 3.4-110

$$\boxed{\frac{a^{-3}+c^{-3}}{a^{-3}-c^{-3}}-3a^{-4}}=\boxed{\frac{\frac{1}{a^3}+\frac{1}{c^3}}{\frac{1}{a^3}-\frac{1}{c^3}}-\frac{3}{a^4}}=\boxed{\frac{\frac{\left(1\cdot c^3\right)+\left(1\cdot a^3\right)}{a^3\cdot c^3}}{\frac{\left(1\cdot c^3\right)-\left(1\cdot a^3\right)}{a^3\cdot c^3}}-\frac{3}{a^4}}=\boxed{\frac{\frac{c^3+a^3}{a^3c^3}}{\frac{c^3-a^3}{a^3c^3}}-\frac{3}{a^4}}=\boxed{\frac{\left(c^3+a^3\right)\cdot a^3c^3}{\left(c^3-a^3\right)\cdot a^3c^3}-\frac{3}{a^4}}$$

$$=\boxed{\frac{\left(c^3+a^3\right)\cdot\left(a^3\cdot a^{-3}\right)}{\left(c^3-a^3\right)\cdot\left(c^3\cdot c^{-3}\right)}-\frac{3}{a^4}}=\boxed{\frac{\left(c^3+a^3\right)\cdot\left(a^{3-3}\right)}{\left(c^3-a^3\right)\cdot\left(c^{3-3}\right)}-\frac{3}{a^4}}=\boxed{\frac{\left(c^3+a^3\right)\cdot a^0}{\left(c^3-a^3\right)\cdot c^0}-\frac{3}{a^4}}=\boxed{\frac{\left(c^3+a^3\right)\cdot 1}{\left(c^3-a^3\right)\cdot 1}-\frac{3}{a^4}}$$

$$=\boxed{\frac{c^3+a^3}{c^3-a^3}-\frac{3}{a^4}}=\boxed{\frac{\left[\left(c^3+a^3\right)\cdot a^4\right]-\left[3\cdot\left(c^3-a^3\right)\right]}{\left(c^3-a^3\right)\cdot a^4}}=\boxed{\frac{c^3a^4+\left(a^3\cdot a^4\right)-3c^3+3a^3}{c^3\cdot a^4-\left(a^3\cdot a^4\right)}}$$

$$=\boxed{\frac{c^3a^4+a^{3+4}-3c^3+3a^3}{c^3a^4-a^{3+4}}}=\boxed{\frac{c^3a^4+a^7-3c^3+3a^3}{c^3a^4-a^7}}=\boxed{\mathbf{\frac{a^7+c^3a^4+3a^3-3c^3}{-a^7+c^3a^4}}}$$

Practice Problems - Addition and Subtraction of Negative Exponential Terms in Fraction Form

Section 3.4 Case III b Practice Problems - Simplify the following negative integer exponential expressions shown in fraction form:

1. $\dfrac{2^{-1}-2^{-3}}{2} =$

2. $\left(a^{-3}+\dfrac{a^{-2}}{2}\right)^{2} =$

3. $\dfrac{2}{x^{-1}}+x^{-1} =$

4. $\dfrac{5}{x+x^{-1}} =$

5. $\dfrac{y^{-1}}{y-y^{-1}} =$

6. $\dfrac{b^{-2}}{b^{-1}+b^{-2}} =$

7. $\dfrac{x^{-1}+y^{-2}}{x^{-1}-y^{-2}} =$

8. $\dfrac{3a}{a^{-1}-b^{-1}} =$

9. $\dfrac{x^{-1}}{(x-3)^{-1}}+y^{-1} =$

10. $\dfrac{2a^{-1}\cdot b^{-1}}{a^{-1}-b} =$

Chapter 4
Radicals

Quick Reference to Chapter 4 Case Problems

Case III b - Multiplying Monomial and Binomial Expressions in Radical Form, with Variables, p. 234

$\sqrt{x^3y^2}\cdot\left(5+\sqrt{x^3y^2}\right)=$; $\sqrt[4]{x^5}\cdot\left(\sqrt[4]{162x^6}-2\sqrt[4]{16x^7}\right)=$; $\sqrt[3]{x}\cdot\left(\sqrt[3]{125x^2}-\sqrt[3]{128x^4}\right)=$

Case I - Rationalizing Radical Expressions with Monomial Denominators, *p. 240*

Case I a - Rationalizing Radical Expressions - Monomial Denominators with Real Numbers, p. 240

$\dfrac{-8\sqrt{3}}{32\sqrt{45}}=$; $\dfrac{5\sqrt{(-2)^2\cdot 3}}{-\sqrt{7}}=$; $\dfrac{3\sqrt[5]{8}}{\sqrt[5]{81}}=$

Case I b - Rationalizing Radical Expressions - Monomial Denominators with Variables, p. 246

$\dfrac{\sqrt{25a^4b^5}}{\sqrt{36a^7b^2}}=$; $\sqrt{\dfrac{48m^5n^3}{150m^7n^2}}=$; $\dfrac{\sqrt[4]{256x^6y^2z^2}}{\sqrt[4]{32x^9y^7z}}=$

Case II - Rationalizing Radical Expressions with Binomial Denominators, *p. 254*

Case II a - Rationalizing Radical Expressions - Binomial Denominators with Real Numbers, p. 254

$\dfrac{8}{2-\sqrt{2}}=$; $\dfrac{\sqrt{125}}{\sqrt{3}-\sqrt{5}}=$; $\dfrac{3+\sqrt{5}}{\sqrt{3}+\sqrt{5}}=$

Case II b - Rationalizing Radical Expressions - Binomial Denominators with Variables, p. 263

$\dfrac{\sqrt{p^3}}{\sqrt{p^5}-\sqrt{q}}=$; $\dfrac{\sqrt{r^2}+\sqrt{s^3}}{\sqrt{r^4}-\sqrt{s}}=$; $\dfrac{-3x+\sqrt{y^3}}{x+\sqrt{y^5}}=$

Case I - Adding and Subtracting Radical Terms (Simple Cases), *p. 271*

$6\sqrt{2}+4\sqrt{2}=$; $20\sqrt[5]{3}-8\sqrt[5]{3}+5\sqrt[5]{3}=$; $a\sqrt{xy}+b\sqrt[3]{xy}-c^2\sqrt{xy}-d=$

Case II - Adding and Subtracting Radical Terms (More Difficult Cases), *p. 274*

$\sqrt[5]{3a}+\sqrt[5]{96a^6}+\sqrt[5]{729a}=$; $2\sqrt{300x^3}+5\sqrt{12x}+8\sqrt{3x^7}=$; $2\sqrt{32a^3}+5\sqrt{50a^5}-2\sqrt{2a}=$

Chapter 4 - Radicals

The objective of this chapter is to introduce the student to methods for solving problems involving radical expressions. In Section 4.1 the student is introduced to the concept of roots and radicals and learns about rational, irrational, real, and imaginary numbers. Simplification of radical expressions with real and variable radicands is also discussed in section 4.1. Multiplication of two monomial and two binomial radical terms, including how monomial and binomial radical expressions are multiplied by one another is addressed in Section 4.2. Section 4.3 covers division of radical expressions and rationalization of radical terms with monomial and binomial denominators. Addition and subtraction of radical terms are addressed in Section 4.4. Cases presented in each section are concluded by solving additional examples with practice problems to further enhance the student's ability on the subject.

4.1 Introduction to Radicals

In this section a description of roots and radicals (Case I), classification of numbers (Case II), and simplification of radical expressions with real numbers and variables (Cases III and IV) are discussed.

Case I Roots and Radical Expressions

In the general radical expression $\sqrt[a]{b} = c$, the symbol $\sqrt{\ }$ is called a **radical sign**. The expression under the radical b is called the **radicand**, a is called the **index**, and the positive square root of the number c is called the **principal square root**.
Exponents are a kind of shorthand for multiplication. For example, $5 \times 5 = 25$ can be expressed in exponential form as $5^2 = 25$. Radical signs are used to reverse this process. For example, to write the reverse of $5^2 = 25$ we take the square root of the terms on both sides of the equal sign, i.e., we write $\sqrt{25} = \sqrt{5^2} = 5$. Note that since $5^2 = 25$ and $(-5)^2 = 25$, we use $\sqrt{25}$ to indicate the positive square root of 25 is equal to 5 and $-\sqrt{25}$ to indicate the negative square root of 25 is equal to -5. Table 4-1 provides square roots, cube roots, fourth roots, and fifth roots of some common numbers used in solving radical expressions. This table should be used as a reference when simplifying radical terms. The students **are not** encouraged to memorize this table. Following are a few examples on simplifying radical expressions using Table 4-1:

a. $\sqrt{64} = \sqrt{8^2} = 8$

b. $-2\sqrt{25} = -2\sqrt{5^2} = -(2 \cdot 5) = -10$

c. $5\sqrt[5]{32} = 5\sqrt[5]{2^5} = (5 \cdot 2) = 10$

d. $\sqrt{125} = \sqrt{25 \cdot 5} = \sqrt{5^2 \cdot 5} = 5\sqrt{5}$

e. $\sqrt[2]{147} = \sqrt{49 \cdot 3} = \sqrt{7^2 \cdot 3} = 7\sqrt{3}$

f. $2\sqrt{32} = 2\sqrt{16 \cdot 2} = 2\sqrt{4^2 \cdot 2} = (2 \cdot 4)\sqrt{2} = 8\sqrt{2}$

g. $\sqrt[5]{2048} = \sqrt[5]{1024 \cdot 2} = \sqrt[5]{4^5 \cdot 2} = 4\sqrt[5]{2}$

h. $\sqrt[3]{375} = \sqrt[3]{125 \cdot 3} = \sqrt[3]{5^3 \cdot 3} = 5\sqrt[3]{3}$

i. $2\sqrt{250} = 2\sqrt{25 \cdot 10} = 2\sqrt{5^2 \cdot 10} = (2 \cdot 5)\sqrt{10} = 10\sqrt{10}$

j. $\sqrt[4]{324} = \sqrt[4]{81 \cdot 4} = \sqrt[4]{3^4 \cdot 4} = 3\sqrt[4]{4}$

k. $\sqrt[3]{648} = \sqrt[3]{216 \cdot 3} = \sqrt[3]{6^3 \cdot 3} = 6\sqrt[3]{3}$

l. $-\sqrt[2]{324} = -\sqrt[2]{81 \cdot 4} = -\sqrt{9^2 \cdot 2^2} = -(9 \cdot 2) = -18$

Table 4-1: Square roots, cube roots, fourth roots, and fifth roots

Square Roots	***Cube Roots***
$\sqrt{1}=\sqrt{1^2}=(1)^{\frac{1}{2}}=\left(1^2\right)^{\frac{1}{2}}=1$ *Note:* $\sqrt[2]{a}=\sqrt{a}$	$\sqrt[3]{1}=\sqrt[3]{1^3}=(1)^{\frac{1}{3}}=\left(1^3\right)^{\frac{1}{3}}=1$
$\sqrt{4}=\sqrt{2^2}=(4)^{\frac{1}{2}}=\left(2^2\right)^{\frac{1}{2}}=2$	$\sqrt[3]{8}=\sqrt[3]{2^3}=(8)^{\frac{1}{3}}=\left(2^3\right)^{\frac{1}{3}}=2$
$\sqrt{9}=\sqrt{3^2}=(9)^{\frac{1}{2}}=\left(3^2\right)^{\frac{1}{2}}=3$	$\sqrt[3]{27}=\sqrt[3]{3^3}=(27)^{\frac{1}{3}}=\left(3^3\right)^{\frac{1}{3}}=3$
$\sqrt{16}=\sqrt{4^2}=(16)^{\frac{1}{2}}=\left(4^2\right)^{\frac{1}{2}}=4$	$\sqrt[3]{64}=\sqrt[3]{4^3}=(64)^{\frac{1}{3}}=\left(4^3\right)^{\frac{1}{3}}=4$
$\sqrt{25}=\sqrt{5^2}=(25)^{\frac{1}{2}}=\left(5^2\right)^{\frac{1}{2}}=5$	$\sqrt[3]{125}=\sqrt[3]{5^3}=(125)^{\frac{1}{3}}=\left(5^3\right)^{\frac{1}{3}}=5$
$\sqrt{36}=\sqrt{6^2}=(36)^{\frac{1}{2}}=\left(6^2\right)^{\frac{1}{2}}=6$	$\sqrt[3]{216}=\sqrt[3]{6^3}=(216)^{\frac{1}{3}}=\left(6^3\right)^{\frac{1}{3}}=6$
$\sqrt{49}=\sqrt{7^2}=(49)^{\frac{1}{2}}=\left(7^2\right)^{\frac{1}{2}}=7$	$\sqrt[3]{343}=\sqrt[3]{7^3}=(343)^{\frac{1}{3}}=\left(7^3\right)^{\frac{1}{3}}=7$
$\sqrt{64}=\sqrt{8^2}=(64)^{\frac{1}{2}}=\left(8^2\right)^{\frac{1}{2}}=8$	$\sqrt[3]{512}=\sqrt[3]{8^3}=(512)^{\frac{1}{3}}=\left(8^3\right)^{\frac{1}{3}}=8$
$\sqrt{81}=\sqrt{9^2}=(81)^{\frac{1}{2}}=\left(9^2\right)^{\frac{1}{2}}=9$	$\sqrt[3]{729}=\sqrt[3]{9^3}=(729)^{\frac{1}{3}}=\left(9^3\right)^{\frac{1}{3}}=9$
$\sqrt{100}=\sqrt{10^2}=(100)^{\frac{1}{2}}=\left(10^2\right)^{\frac{1}{2}}=10$	$\sqrt[3]{1000}=\sqrt[3]{10^3}=(1000)^{\frac{1}{3}}=\left(10^3\right)^{\frac{1}{3}}=10$
Fourth Roots	***Fifth Roots***
$\sqrt[4]{1}=\sqrt[4]{1^4}=(1)^{\frac{1}{4}}=\left(1^4\right)^{\frac{1}{4}}=1$	$\sqrt[5]{1}=\sqrt[5]{1^5}=(1)^{\frac{1}{5}}=\left(1^5\right)^{\frac{1}{5}}=1$
$\sqrt[4]{16}=\sqrt[4]{2^4}=(16)^{\frac{1}{4}}=\left(2^4\right)^{\frac{1}{4}}=2$	$\sqrt[5]{32}=\sqrt[5]{2^5}=(32)^{\frac{1}{5}}=\left(2^5\right)^{\frac{1}{5}}=2$
$\sqrt[4]{81}=\sqrt[4]{3^4}=(81)^{\frac{1}{4}}=\left(3^4\right)^{\frac{1}{4}}=3$	$\sqrt[5]{243}=\sqrt[5]{3^5}=(243)^{\frac{1}{5}}=\left(3^5\right)^{\frac{1}{5}}=3$
$\sqrt[4]{256}=\sqrt[4]{4^4}=(256)^{\frac{1}{4}}=\left(4^4\right)^{\frac{1}{4}}=4$	$\sqrt[5]{1024}=\sqrt[5]{4^5}=(1024)^{\frac{1}{5}}=\left(4^5\right)^{\frac{1}{5}}=4$
$\sqrt[4]{625}=\sqrt[4]{5^4}=(625)^{\frac{1}{4}}=\left(5^4\right)^{\frac{1}{4}}=5$	$\sqrt[5]{3125}=\sqrt[5]{5^5}=(3125)^{\frac{1}{5}}=\left(5^5\right)^{\frac{1}{5}}=5$
$\sqrt[4]{1296}=\sqrt[4]{6^4}=(1296)^{\frac{1}{4}}=\left(6^4\right)^{\frac{1}{4}}=6$	$\sqrt[5]{7776}=\sqrt[5]{6^5}=(7776)^{\frac{1}{5}}=\left(6^5\right)^{\frac{1}{5}}=6$
$\sqrt[4]{2401}=\sqrt[4]{7^4}=(2401)^{\frac{1}{4}}=\left(7^4\right)^{\frac{1}{4}}=7$	$\sqrt[5]{16807}=\sqrt[5]{7^5}=(16807)^{\frac{1}{5}}=\left(7^5\right)^{\frac{1}{5}}=7$
$\sqrt[4]{4096}=\sqrt[4]{8^4}=(4096)^{\frac{1}{4}}=\left(8^4\right)^{\frac{1}{4}}=8$	$\sqrt[5]{32768}=\sqrt[5]{8^5}=(32768)^{\frac{1}{5}}=\left(8^5\right)^{\frac{1}{5}}=8$
$\sqrt[4]{6561}=\sqrt[4]{9^4}=(6561)^{\frac{1}{4}}=\left(9^4\right)^{\frac{1}{4}}=9$	$\sqrt[5]{59049}=\sqrt[5]{9^5}=(59049)^{\frac{1}{5}}=\left(9^5\right)^{\frac{1}{5}}=9$
$\sqrt[4]{10000}=\sqrt[4]{10^4}=(10000)^{\frac{1}{4}}=\left(10^4\right)^{\frac{1}{4}}=10$	$\sqrt[5]{100000}=\sqrt[5]{10^5}=(100000)^{\frac{1}{5}}=\left(10^5\right)^{\frac{1}{5}}=10$

Practice Problems - Roots and Radical Expressions

Section 4.1 Case I Practice Problems - Simplify the following radical expressions by using Table 4-1:

1. $\sqrt[2]{98} =$
2. $3\sqrt{75} =$
3. $\sqrt[3]{125} =$
4. $\sqrt[5]{3125} =$
5. $\sqrt[4]{162} =$
6. $\sqrt[2]{192} =$
7. $-\sqrt[3]{64} =$
8. $\sqrt{250} =$
9. $\sqrt[3]{54} =$
10. $\sqrt[5]{486} =$

Case II Rational, Irrational, Real, and Imaginary Numbers

A **rational number** is a number that **can** be expressed as:

1. An integer fraction $\frac{a}{b}$, where a and b are integer numbers and $b \neq 0$. For example: $\frac{3}{8}$, $-\frac{4}{5}$, $\frac{25}{100}$, and $\frac{2}{7}$ are rational numbers.
2. The square root of a perfect square, the cube root of a perfect cube, etc. For example: $\sqrt{36} = \sqrt{6^2} = 6$, $\sqrt{49} = \sqrt{7^2} = 7$, $-\sqrt[3]{125} = -\sqrt[3]{5^3} = -5$, $\sqrt[4]{81} = \sqrt[4]{3^4} = 3$, and $-\sqrt[5]{1024} = -\sqrt[5]{4^5} = -4$ are rational numbers.
3. An integer (a whole number). For example: $5 = \frac{5}{1}$, 0, $-25 = -\frac{25}{1}$, $\frac{0}{28} = 0$, and 125 are rational numbers.
4. A terminating decimal. For example: $0.25 = \frac{25}{100}$, -0.75, $5.5 = -5\frac{1}{2}$, and $-3.8 = -3\frac{4}{5}$ are rational numbers.
5. A repeating decimal. For example: $-0.3333333... = -\frac{1}{3}$, $0.45454545...$, and $1.666666... = \frac{5}{3}$ are rational numbers.

An **irrational number** is a number that:

1. Can not be expressed as an integer fraction $\frac{a}{b}$, where a and b are integer numbers and $b \neq 0$. For example: π, $\frac{2}{\sqrt{2}}$, and $-\frac{5}{\sqrt{3}}$ are irrational numbers.
2. Can not be expressed as the square root of a perfect square, the cube root of a perfect cube, etc. For example: $\sqrt{5}$, $-\sqrt{7}$, $\sqrt{12}$, $\sqrt[3]{4}$, $-\sqrt[5]{6}$, and $\sqrt{3}$ are irrational numbers.
3. Is not a terminating or repeating decimal. For example: $0.432643...$, $-8.346723...$, and $3.14159...$ are irrational numbers.

The **real numbers** consist of all the rational and irrational numbers. For example: π, $\frac{2}{\sqrt{2}}$, $-\sqrt{7}$, $\sqrt[3]{4}$, $-\sqrt[5]{6}$, $\sqrt{3}$, $\sqrt{36}=\sqrt{6^2}=6$, $0.25=\frac{25}{100}$, -0.75, $-5.5=-5\frac{1}{2}$, $-3.8=-3\frac{4}{5}$, $5=\frac{5}{1}$, 0, and $-25=-\frac{25}{1}$ are real numbers.

The **not real numbers** or **imaginary numbers** are square root of any negative real number. For example: $\sqrt{-15}$, $\sqrt{-9}$, $\sqrt{-45}$, and $\sqrt{-36}$ are imaginary numbers.

Table 4-2 provides a sample of rational, irrational, real, and imaginary numbers:

Table 4-2: Rational, Irrational, Real, and Imaginary Numbers

Number	Rational	Irrational	Real	Not Real
25	rational		real	
$\sqrt{25}$	rational		real	
$\frac{12}{5}$	rational		real	
$\frac{\sqrt{6}}{3}$		irrational	real	
$\sqrt{-13}$				not real
$\frac{3}{\sqrt{3}}$		irrational	real	
$3\sqrt{64}$	rational		real	
$0.3333333...$	rational		real	
$12\frac{3}{5}$	rational		real	
$-23\frac{3}{\sqrt{5}}$		irrational	real	
$\sqrt{-49}$				not real
$-\sqrt{49}$	rational		real	
$3.13425638...$		irrational	real	
$\sqrt{7}$		irrational	real	
$\frac{3}{8}$	rational		real	
$\sqrt[3]{216}$	rational		real	
-0.85	rational		real	
$0.27272727...$	rational		real	
$-\sqrt{-3}$				not real

Note that $\sqrt{a^2} = \pm a$. However, since we are only interested in the positive value of a, we express $\sqrt{a^2}$ as the absolute value of a, i.e., $\sqrt{a^2} = |a|$.

Examples:

1. $\sqrt{7^2} = |7| = 7$
2. $-\sqrt{7^2} = -|7| = -7$
3. $\sqrt{(-7)^2} = |-7| = 7$
4. $\sqrt{-(7)^2} = \sqrt{-49} = \textit{not real}$

In this book, and in the remainder of this chapter, all real and variable terms under radical sign represent positive numbers. Therefore, it is not necessary to show the answers in absolute value form.

Practice Problems - Rational, Irrational, Real, and Imaginary Numbers

Section 4.1 Case II Practice Problems - Identify which one of the following numbers are rational, irrational, real, or not real:

1. $\frac{5}{8} =$
2. $\sqrt{45} =$
3. $450 =$
4. $-\frac{2}{\sqrt{10}} =$
5. $-\sqrt{-5} =$
6. $\frac{\sqrt{5}}{-2} =$
7. $0.1111111\ldots =$
8. $-0.2367432\ldots =$
9. $\sqrt[5]{7776} =$
10. $-0.35 =$

Case III Simplifying Radical Expressions with Real Numbers as Radicand

Radical expressions with a real number as radicand are simplified using the following general rule:

$$\sqrt[n]{a^n} = a^{\frac{n}{n}} = a \qquad \text{The } n^{th} \text{ root of } a^n \text{ is } a$$

Where a is a positive real number and n is an integer.

Radicals of the form $\sqrt[n]{a^n} = a$ are simplified using the following steps:

Step 1 Factor out the radicand a^n to a perfect square, cube, fourth, fifth, etc. term (use Table 4-1). Write any term under the radical that exceeds the index n as multiple sum of the index.

Step 2 Use the Multiplication Law for exponents (see Section 3.3) by writing a^{m+n} in the form of $a^m \cdot a^n$.

Step 3 Simplify the radical expression by using the general rule $\sqrt[n]{a^n} = a$. Note that any term under the radical which is less than the index n stays inside the radical.

Examples with Steps

The following examples show the steps as to how radical expressions with real terms are simplified:

Example 4.1-1

$$\boxed{\sqrt[2]{64}} =$$

Solution:

Step 1 $\boxed{\sqrt[2]{64}} = \boxed{\sqrt{64}} = \boxed{\sqrt{8 \cdot 8}} = \boxed{\sqrt{8^1 \cdot 8^1}} = \boxed{\sqrt{8^{1+1}}} = \boxed{\sqrt{8^2}}$

Step 2 *Not Applicable*

Step 3 $\boxed{\sqrt{8^2}} = \boxed{8}$

Example 4.1-2

$$\boxed{\frac{1}{-8}\sqrt[2]{72}} =$$

Solution:

Step 1 $\boxed{\frac{1}{-8}\sqrt[2]{72}} = \boxed{-\frac{1}{8}\sqrt{72}} = \boxed{-\frac{1}{8}\sqrt{36 \cdot 2}} = \boxed{-\frac{1}{8}\sqrt{(6 \cdot 6) \cdot 2}} = \boxed{-\frac{1}{8}\sqrt{\left(6^1 \cdot 6^1\right) \cdot 2}}$

$= \boxed{-\frac{1}{8}\sqrt{\left(6^{1+1}\right) \cdot 2}} = \boxed{-\frac{1}{8}\sqrt{6^2 \cdot 2}}$

Step 2 *Not Applicable*

Step 3 $-\frac{1}{8}\sqrt{6^2\cdot 2} = -\frac{1}{8}\cdot 6\sqrt{2} = -\frac{\overset{3}{\cancel{6}}}{\underset{4}{\cancel{8}}}\sqrt{2} = \boxed{-\frac{3}{4}\sqrt{2}}$

Example 4.1-3

$\frac{-3}{-2}\sqrt{400} =$

Solution:

Step 1 $\frac{-3}{-2}\sqrt{400} = +\frac{3}{2}\sqrt{20\cdot 20} = \frac{3}{2}\sqrt{20\cdot 20} = \frac{3}{2}\sqrt{20\cdot 20} = \frac{3}{2}\sqrt{20^1\cdot 20^1} = \frac{3}{2}\sqrt{20^{1+1}}$

$= \frac{3}{2}\sqrt{20^2}$

Step 2 *Not Applicable*

Step 3 $\frac{3}{2}\sqrt{20^2} = \frac{3}{2}\cdot 20 = \frac{3}{2}\cdot\frac{20}{1} = \frac{3\cdot 20}{2\cdot 1} = \frac{\overset{30}{\cancel{60}}}{\underset{1}{\cancel{2}}} = \frac{30}{1} = \boxed{30}$

Example 4.1-4

$\frac{-3}{10}\sqrt[4]{5^7\cdot 4} =$

Solution:

Step 1 $\frac{-3}{10}\sqrt[4]{5^7\cdot 4} = -\frac{3}{10}\sqrt[4]{5^{4+3}\cdot 4}$

Step 2 $-\frac{3}{10}\sqrt[4]{5^{4+3}\cdot 4} = -\frac{3}{10}\sqrt[4]{5^4\cdot 5^3\cdot 4}$

Step 3 $-\frac{3}{10}\sqrt[4]{5^4\cdot 5^3\cdot 4} = -\frac{3}{10}\cdot 5\sqrt[4]{5^3\cdot 4} = -\frac{\overset{3}{\cancel{15}}}{\underset{2}{\cancel{10}}}\sqrt[4]{5^3\cdot 4} = -\frac{3}{2}\sqrt[4]{125\cdot 4} = \boxed{-\frac{3}{2}\sqrt[4]{500}}$

Example 4.1-5

$\frac{\sqrt[3]{162}}{9} =$

Solution:

Step 1 $\frac{\sqrt[3]{162}}{9} = \frac{\sqrt[3]{81\cdot 2}}{9} = \frac{\sqrt[3]{3^4\cdot 2}}{9} = \frac{\sqrt[3]{3^{3+1}\cdot 2}}{9}$

Step 2 $\frac{\sqrt[3]{3^{3+1}\cdot 2}}{9} = \frac{\sqrt[3]{3^3\cdot 3^1\cdot 2}}{9}$

Step 3 $\frac{\sqrt[3]{3^3\cdot3^1\cdot2}}{9} = \frac{\overset{1}{\cancel{3}}\sqrt[3]{3\cdot2}}{\underset{3}{\cancel{9}}} = \mathbf{\frac{\sqrt[3]{6}}{3}}$

Additional Examples - Simplifying Radical Expressions with Real Numbers as Radicand

The following examples further illustrate how to solve radical expressions with real numbers as radicand:

Example 4.1-6

$$\sqrt[2]{25} = \sqrt{25} = \sqrt{5\cdot5} = \sqrt{5^1\cdot5^1} = \sqrt{5^{1+1}} = \sqrt{5^2} = \mathbf{5}$$

Example 4.1-7

$$\sqrt[2]{18} = \sqrt{18} = \sqrt{9\cdot2} = \sqrt{(3\cdot3)\cdot2} = \sqrt{\left(3^1\cdot3^1\right)\cdot2} = \sqrt{\left(3^{1+1}\right)\cdot2} = \sqrt{3^2\cdot2} = \mathbf{3\sqrt{2}}$$

Example 4.1-8

$$\sqrt[2]{125} = \sqrt{125} = \sqrt{25\cdot5} = \sqrt{(5\cdot5)\cdot5} = \sqrt{\left(5^1\cdot5^1\right)\cdot5} = \sqrt{\left(5^{1+1}\right)\cdot5} = \sqrt{5^2\cdot5} = \mathbf{5\sqrt{5}}$$

Example 4.1-9

$$\frac{-2}{-9}\sqrt[2]{9} = +\frac{2}{9}\sqrt{9} = \frac{2}{9}\sqrt{3\cdot3} = \frac{2}{9}\sqrt{3^1\cdot3^1} = \frac{2}{9}\sqrt{3^{1+1}} = \frac{2}{9}\sqrt{3^2} = \frac{2}{9}\cdot3 = \frac{2}{9}\cdot\frac{3}{1} = \frac{2\cdot\overset{1}{\cancel{3}}}{\underset{3}{\cancel{9}}\cdot1} = \frac{2\cdot1}{3\cdot1} = \mathbf{\frac{2}{3}}$$

Example 4.1-10

$$\sqrt[2]{2400} = \sqrt{2400} = \sqrt{400\cdot6} = \sqrt{(20\cdot20)\cdot6} = \sqrt{\left(20^1\cdot20^1\right)\cdot6} = \sqrt{\left(20^{1+1}\right)\cdot6} = \sqrt{20^2\cdot6} = \mathbf{20\sqrt{6}}$$

Example 4.1-11

$$\frac{1}{-5}\sqrt[2]{1000} = -\frac{1}{5}\sqrt{1000} = -\frac{1}{5}\sqrt{100\cdot10} = -\frac{1}{5}\sqrt{(10\cdot10)\cdot10} = -\frac{1}{5}\sqrt{\left(10^1\cdot10^1\right)\cdot10} = -\frac{1}{5}\sqrt{\left(10^{1+1}\right)\cdot10}$$

$$= -\frac{1}{5}\sqrt{10^2\cdot10} = -\frac{1}{5}\cdot10\sqrt{10} = -\frac{\overset{2}{\cancel{10}}}{\underset{1}{\cancel{5}}}\sqrt{10} = -\frac{2}{1}\sqrt{10} = \mathbf{-2\sqrt{10}}$$

Example 4.1-12

$$\sqrt[2]{120} = \sqrt{120} = \sqrt{4\cdot30} = \sqrt{(2\cdot2)\cdot30} = \sqrt{\left(2^1\cdot2^1\right)\cdot30} = \sqrt{\left(2^{1+1}\right)\cdot30} = \sqrt{2^2\cdot30} = \mathbf{2\sqrt{30}}$$

Example 4.1-13

$$\frac{-2}{16}\sqrt[2]{28} = -\frac{2}{16}\sqrt{28} = -\frac{\overset{1}{\cancel{2}}}{\underset{8}{\cancel{16}}}\sqrt{4\cdot7} = -\frac{1}{8}\sqrt{(2\cdot2)\cdot7} = -\frac{1}{8}\sqrt{\left(2^1\cdot2^1\right)\cdot7} = -\frac{1}{8}\sqrt{\left(2^{1+1}\right)\cdot7} = -\frac{1}{8}\sqrt{2^2\cdot7}$$

$$= \boxed{-\frac{1}{8}\cdot 2\sqrt{7}} = \boxed{-\frac{\overset{1}{\cancel{2}}}{\underset{4}{\cancel{8}}}\cdot\sqrt{7}} = \boxed{-\frac{1}{4}\sqrt{7}} = \boxed{-\frac{\sqrt{7}}{4}}$$

Example 4.1-14

$$\boxed{\sqrt[2]{75}} = \boxed{\sqrt{75}} = \boxed{\sqrt{25\cdot 3}} = \boxed{\sqrt{(5\cdot 5)\cdot 3}} = \boxed{\sqrt{\left(5^1\cdot 5^1\right)\cdot 3}} = \boxed{\sqrt{\left(5^{1+1}\right)\cdot 3}} = \boxed{\sqrt{5^2\cdot 3}} = \boxed{5\sqrt{3}}$$

Example 4.1-15

$$\boxed{\sqrt[2]{81}} = \boxed{\sqrt{81}} = \boxed{\sqrt{9\cdot 9}} = \boxed{\sqrt{9^1\cdot 9^1}} = \boxed{\sqrt{9^{1+1}}} = \boxed{\sqrt{9^2}} = \boxed{9}$$

Example 4.1-16

$$\boxed{\frac{2}{15}\sqrt[2]{50}} = \boxed{\frac{2}{15}\sqrt{50}} = \boxed{\frac{2}{15}\sqrt{25\cdot 2}} = \boxed{\frac{2}{15}\sqrt{(5\cdot 5)\cdot 2}} = \boxed{\frac{2}{15}\sqrt{\left(5^1\cdot 5^1\right)\cdot 2}} = \boxed{\frac{2}{15}\sqrt{\left(5^{1+1}\right)\cdot 2}} = \boxed{\frac{2}{15}\sqrt{5^2\cdot 2}}$$

$$= \boxed{\frac{2}{15}\cdot 5\sqrt{2}} = \boxed{\frac{\overset{2}{\cancel{10}}}{\underset{3}{\cancel{15}}}\sqrt{2}} = \boxed{\frac{2}{3}\sqrt{2}}$$

Example 4.1-17

$$\boxed{\frac{\sqrt[2]{48}}{16}} = \boxed{\frac{\sqrt{48}}{16}} = \boxed{\frac{\sqrt{16\cdot 3}}{16}} = \boxed{\frac{\sqrt{(4\cdot 4)\cdot 3}}{16}} = \boxed{\frac{\sqrt{\left(4^1\cdot 4^1\right)\cdot 3}}{16}} = \boxed{\frac{\sqrt{\left(4^{1+1}\right)\cdot 3}}{16}} = \boxed{\frac{\sqrt{4^2\cdot 3}}{16}} = \boxed{\frac{\overset{1}{\cancel{4}}\sqrt{3}}{\underset{4}{\cancel{16}}}} = \boxed{\frac{\sqrt{3}}{4}}$$

Example 4.1-18

$$\boxed{\frac{1}{-20}\sqrt[3]{375}} = \boxed{-\frac{1}{20}\sqrt[3]{125\cdot 3}} = \boxed{-\frac{1}{20}\sqrt[3]{5^3\cdot 3}} = \boxed{-\frac{1}{20}\cdot 5\sqrt[3]{3}} = \boxed{-\frac{\overset{1}{\cancel{5}}}{\underset{4}{\cancel{20}}}\cdot\sqrt[3]{3}} = \boxed{-\frac{1}{4}\cdot\sqrt[3]{3}} = \boxed{-\frac{\sqrt[3]{3}}{4}}$$

Example 4.1-19

$$\boxed{-\frac{1}{32}\sqrt[4]{32}} = \boxed{-\frac{1}{32}\sqrt[4]{16\cdot 2}} = \boxed{-\frac{1}{32}\sqrt[4]{2^4\cdot 2}} = \boxed{-\frac{1}{32}\cdot 2\sqrt[4]{2}} = \boxed{-\frac{\overset{1}{\cancel{2}}}{\underset{16}{\cancel{32}}}\cdot\sqrt[4]{2}} = \boxed{-\frac{1}{16}\cdot\sqrt[4]{2}} = \boxed{-\frac{\sqrt[4]{2}}{16}}$$

Example 4.1-20

$$\boxed{\frac{-3}{27}\sqrt[5]{486}} = \boxed{-\frac{\overset{1}{\cancel{3}}}{\underset{9}{\cancel{27}}}\sqrt[5]{243\cdot 2}} = \boxed{-\frac{1}{9}\sqrt[5]{3^5\cdot 2}} = \boxed{-\frac{1}{9}\cdot 3\sqrt[5]{2}} = \boxed{-\frac{\overset{1}{\cancel{3}}}{\underset{3}{\cancel{9}}}\cdot\sqrt[5]{2}} = \boxed{-\frac{1}{3}\cdot\sqrt[5]{2}} = \boxed{-\frac{\sqrt[5]{2}}{3}}$$

Example 4.1-21

$$\boxed{3\sqrt[4]{1875}} = \boxed{3\sqrt[4]{625\cdot 3}} = \boxed{3\sqrt[4]{5^4\cdot 3}} = \boxed{(3\cdot 5)\sqrt[4]{3}} = \boxed{15\sqrt[4]{3}}$$

Example 4.1-22

$$\boxed{\frac{-3}{-2}\sqrt[5]{4096}} = \boxed{+\frac{3}{2}\sqrt[5]{1024\cdot 4}} = \boxed{\frac{3}{2}\sqrt[5]{4^5\cdot 4}} = \boxed{\frac{3}{2}\cdot 4\sqrt[5]{4}} = \boxed{\frac{12}{2}\sqrt[5]{4}} = \boxed{\frac{\overset{6}{\cancel{12}}}{\underset{1}{\cancel{2}}}\sqrt[5]{4}} = \boxed{\frac{6}{1}\sqrt[5]{4}} = \boxed{6\sqrt[5]{4}}$$

Example 4.1-23

$\sqrt[3]{2^{11}} = \sqrt[3]{2^{3+3+3+2}} = \sqrt[3]{\left(2^3 \cdot 2^3 \cdot 2^3\right) \cdot 2^2} = \left(2^1 \cdot 2^1 \cdot 2^1\right) \cdot \sqrt[3]{2^2} = \left(2^{1+1+1}\right) \cdot \sqrt[3]{4} = 2^3 \cdot \sqrt[3]{4} = \mathbf{8\sqrt[3]{4}}$

Example 4.1-24

$\sqrt[5]{2 \cdot 3^7} = \sqrt[5]{2 \cdot 3^{5+2}} = \sqrt[5]{2 \cdot 3^5 \cdot 3^2} = 3\sqrt[5]{2 \cdot 3^2} = 3\sqrt[5]{2 \cdot 9} = \mathbf{3\sqrt[5]{18}}$

Example 4.1-25

$\sqrt[4]{4 \cdot 3^{10}} = \sqrt[4]{4 \cdot 3^{4+4+2}} = \sqrt[4]{4 \cdot \left(3^4 \cdot 3^4\right) \cdot 3^2} = \left(3^1 \cdot 3^1\right) \cdot \sqrt[4]{4 \cdot 3^2} = \left(3^{1+1}\right) \cdot \sqrt[4]{4 \cdot 9} = 3^2 \cdot \sqrt[4]{36} = \mathbf{9\sqrt[4]{36}}$

Practice Problems - Simplifying Radical Expressions with Real Numbers as Radicand

Section 4.1 Case III Practice Problems - Simplify the following radical expressions:

1. $-\sqrt{49} =$
2. $\sqrt{54} =$
3. $-\sqrt{500} =$
4. $\sqrt[5]{3^5 \cdot 5} =$
5. $\sqrt[2]{216} =$
6. $-\frac{1}{4}\sqrt[4]{4^5 \cdot 2} =$
7. $\sqrt[4]{162} =$
8. $\frac{-2}{-9}\sqrt[4]{3^9} =$
9. $-4\sqrt[2]{1800} =$
10. $\frac{1}{-6}\sqrt{100000} =$

Case IV Simplifying Radical Expressions with Variables as Radicand

Radical expressions with variable radicands are simplified using the following general rule:

$$\sqrt[n]{x^n} = x^{\frac{n}{n}} = x \qquad \text{The } n^{th} \text{ root of } x^n \text{ is } x$$

Where x is a variable and n is an integer.

Radicals of the form $\sqrt[n]{x^n} = x$ are simplified using the following steps:

Step 1 Factor out the radicand x^n to a perfect square, cube, fourth, fifth, etc. term (use Table 4-1). Write any term under the radical that exceeds the index n as multiple sum of the index.

Step 2 Use the Multiplication Law for exponents (see Section 3.3) by writing x^{m+n} in the form of $x^m \cdot x^n$.

Step 3 Simplify the radical expression by using the general rule $\sqrt[n]{x^n} = x$. Note that any term under the radical which is less than the index n stays inside the radical.

Examples with Steps

The following examples show the steps as to how radical expressions with variable terms are simplified:

Example 4.1-26

$$\boxed{\sqrt[2]{64y^3}} =$$

Solution:

Step 1 $\boxed{\sqrt[2]{64y^3}} = \boxed{\sqrt{64y^3}} = \boxed{\sqrt{(8\cdot 8)y^{2+1}}} = \boxed{\sqrt{8^2 y^{2+1}}}$

Step 2 $\boxed{\sqrt{8^2 y^{2+1}}} = \boxed{\sqrt{8^2 y^2 \cdot y^1}}$

Step 3 $\boxed{\sqrt{8^2 y^2 \cdot y^1}} = \boxed{8y\sqrt{y^1}} = \boxed{\mathbf{8y\sqrt{y}}}$

Example 4.1-27

$$\boxed{x\sqrt[2]{x^4}} =$$

Solution:

Step 1 $\boxed{x\sqrt[2]{x^4}} = \boxed{x\sqrt{x^4}} = \boxed{x\sqrt{x^{2+2}}}$

Step 2 $\boxed{x\sqrt{x^{2+2}}} = \boxed{x\sqrt{x^2 \cdot x^2}}$

Step 3 $\boxed{x\sqrt{x^2 \cdot x^2}} = \boxed{x \cdot x^1 \cdot x^1} = \boxed{x^1 \cdot x^1 \cdot x^1} = \boxed{x^{1+1+1}} = \boxed{\mathbf{x^3}}$

Example 4.1-28

$$-2x^3y^2\sqrt{27x^3y^5} =$$

Solution:

Step 1

$$-2x^3y^2\sqrt{27x^3y^5} = -2x^3y^2\sqrt{(9\cdot 3)x^{2+1}y^{2+2+1}} = -2x^3y^2\sqrt{\left(3^2\cdot 3\right)x^{2+1}y^{2+2+1}}$$

Step 2

$$-2x^3y^2\sqrt{\left(3^2\cdot 3\right)x^{2+1}y^{2+2+1}} = -(2\cdot 3)x^3y^2\sqrt{3\left(x^2\cdot x^1\right)\cdot\left(y^2\cdot y^2\cdot y^1\right)}$$

Step 3

$$-(2\cdot 3)x^3y^2\sqrt{3\left(x^2\cdot x^1\right)\cdot\left(y^2\cdot y^2\cdot y^1\right)} = -6\left(x^3\cdot x^1\right)\cdot\left(y^2\cdot y^1\cdot y^1\right)\sqrt{3x^1\cdot y^1}$$

$$= -6\left(x^{3+1}\right)\cdot\left(y^{2+1+1}\right)\sqrt{3x^1\cdot y^1} = \mathbf{-6x^4y^4\sqrt{3xy}}$$

Example 4.1-29

$$\sqrt[3]{56p^4q^3r^5} =$$

Solution:

Step 1

$$\sqrt[3]{56p^4q^3r^5} = \sqrt[3]{(8\cdot 7)p^{3+1}q^3r^{3+2}} = \sqrt[3]{\left(2^3\cdot 7\right)p^{3+1}q^3r^{3+2}}$$

Step 2

$$\sqrt[3]{\left(2^3\cdot 7\right)p^{3+1}q^3r^{3+2}} = \sqrt[3]{\left(2^3\cdot 7\right)\cdot\left(p^3\cdot p^1\right)\cdot q^3\cdot\left(r^3\cdot r^2\right)}$$

Step 3

$$\sqrt[3]{\left(2^3\cdot 7\right)\cdot\left(p^3\cdot p^1\right)\cdot q^3\cdot\left(r^3\cdot r^2\right)} = 2pqr\sqrt[3]{7\cdot p^1\cdot r^2} = \mathbf{2pqr\sqrt[3]{7pr^2}}$$

Example 4.1-30

$$-\sqrt[2]{48x^7y^2z^6} =$$

Solution:

Step 1

$$-\sqrt[2]{48x^7y^2z^6} = -\sqrt{48x^7y^2z^6} = -\sqrt{(16\cdot 3)x^{2+2+2+1}y^2z^{2+2+2}}$$

$$= -\sqrt{\left(4^2\cdot 3\right)x^{2+2+2+1}y^2z^{2+2+2}}$$

Step 2

$$-\sqrt{\left(4^2\cdot 3\right)x^{2+2+2+1}y^2z^{2+2+2}} = -\sqrt{\left(4^2\cdot 3\right)\cdot\left(x^2\cdot x^2\cdot x^2\cdot x^1\right)\cdot y^2\cdot\left(z^2\cdot z^2\cdot z^2\right)}$$

Step 3

$$-\sqrt{\left(4^2\cdot 3\right)\cdot\left(x^2\cdot x^2\cdot x^2\cdot x^1\right)\cdot y^2\cdot\left(z^2\cdot z^2\cdot z^2\right)} = -4\left(x^1\cdot x^1\cdot x^1\right)\cdot y^1\cdot\left(z^1\cdot z^1\cdot z^1\right)\sqrt{3\cdot x^1}$$

$$= -4\left(x^{1+1+1}\right)\cdot y\cdot\left(z^{1+1+1}\right)\sqrt{3\cdot x} = \mathbf{-4x^3yz^3\sqrt{3x}}$$

Additional Examples - Simplifying Radical Expressions with Variables as Radicand

The following examples further illustrate how to solve radical expressions with variables as radicand:

Example 4.1-31

$$\boxed{\sqrt{125x^3}} = \boxed{\sqrt{(25\cdot 5)x^{2+1}}} = \boxed{\sqrt{\left(5^2\cdot 5\right)x^2\cdot x^1}} = \boxed{5x\sqrt{5\cdot x^1}} = \boxed{\mathbf{5x\sqrt{5x}}}$$

Example 4.1-32

$$\boxed{\sqrt{x^5y^7}} = \boxed{\sqrt{x^{2+2+1}y^{2+2+2+1}}} = \boxed{\sqrt{\left(x^2\cdot x^2\cdot x^1\right)\cdot\left(y^2\cdot y^2\cdot y^2\cdot y^1\right)}} = \boxed{\left(x^1\cdot x^1\right)\cdot\left(y^1\cdot y^1\cdot y^1\right)\sqrt{x^1\cdot y^1}}$$

$$= \boxed{x^{1+1}\cdot y^{1+1+1}\sqrt{xy}} = \boxed{\mathbf{x^2y^3\sqrt{xy}}}$$

Example 4.1-33

$$\boxed{m\sqrt{m^4n^5}} = \boxed{m\sqrt{m^{2+2}n^{2+2+1}}} = \boxed{m\sqrt{\left(m^2\cdot m^2\right)\cdot\left(n^2\cdot n^2\cdot n^1\right)}} = \boxed{\left(m^1\cdot m^1\cdot m^1\right)\cdot\left(n^1\cdot n^1\right)\sqrt{n^1}}$$

$$= \boxed{m^{1+1+1}\cdot n^{1+1}\sqrt{n}} = \boxed{\mathbf{m^3n^2\sqrt{n}}}$$

Example 4.1-34

$$\boxed{\sqrt{16a^3b^5}} = \boxed{\sqrt{4^2a^{2+1}b^{2+2+1}}} = \boxed{4\sqrt{\left(a^2\cdot a^1\right)\cdot\left(b^2\cdot b^2\cdot b^1\right)}} = \boxed{4\cdot a^1\cdot\left(b^1\cdot b^1\right)\sqrt{a^1\cdot b^1}} = \boxed{4\cdot a\cdot b^{1+1}\sqrt{ab}}$$

$$= \boxed{\mathbf{4ab^2\sqrt{ab}}}$$

Example 4.1-35

$$\boxed{\sqrt{30r^4s^3t^6}} = \boxed{\sqrt{30r^{2+2}s^{2+1}t^{2+2+2}}} = \boxed{\sqrt{30\left(r^2\cdot r^2\right)\cdot\left(s^2\cdot s^1\right)\cdot\left(t^2\cdot t^2\cdot t^2\right)}} = \boxed{\left(r^1\cdot r^1\right)\cdot s^1\cdot\left(t^1\cdot t^1\cdot t^1\right)\sqrt{30s^1}}$$

$$= \boxed{r^{1+1}st^{1+1+1}\sqrt{30s}} = \boxed{\mathbf{r^2st^3\sqrt{30s}}}$$

Example 4.1-36

$$\boxed{\frac{1}{-30}\sqrt{75x^3y^2z^4}} = \boxed{-\frac{1}{30}y\sqrt{(25\cdot 3)\cdot x^{2+1}z^{2+2}}} = \boxed{-\frac{y}{30}\sqrt{\left(5^2\cdot 3\right)\cdot\left(x^2\cdot x^1\right)\cdot\left(z^2\cdot z^2\right)}} = \boxed{-\frac{y}{30}\cdot 5\cdot x^1\cdot\left(z^1\cdot z^1\right)\sqrt{3\cdot x^1}}$$

$$= \boxed{-\frac{y}{30}\cdot 5\cdot x\cdot z^{1+1}\sqrt{3x}} = \boxed{-\frac{\overset{1}{\cancel{5}}xyz^2\sqrt{3x}}{\underset{6}{\cancel{30}}}} = \boxed{-\frac{\mathbf{xyz^2\sqrt{3x}}}{\mathbf{6}}}$$

Example 4.1-37

$$\boxed{-8\sqrt{8u^3v^7}} = \boxed{-8\sqrt{(4\cdot 2)u^{2+1}v^{2+2+2+1}}} = \boxed{-8\sqrt{\left(2^2\cdot 2\right)\cdot\left(u^2\cdot u^1\right)\cdot\left(v^2\cdot v^2\cdot v^2\cdot v^1\right)}}$$

$$= \boxed{-(8\cdot 2)\cdot u^1\cdot\left(v^1\cdot v^1\cdot v^1\right)\sqrt{2\cdot u^1\cdot v^1}} = \boxed{-16u\cdot\left(v^{1+1+1}\right)\sqrt{2uv}} = \boxed{\mathbf{-16uv^3\sqrt{2uv}}}$$

Example 4.1-38

$$\sqrt[3]{x^5 y^7} = \sqrt[3]{x^{3+2} y^{3+3+1}} = \sqrt[3]{\left(x^3 \cdot x^2\right)\cdot\left(y^3 \cdot y^3 \cdot y^1\right)} = x \cdot \left(y^1 \cdot y^1\right)\sqrt[3]{x^2 \cdot y^1} = x \cdot y^{1+1}\sqrt[3]{x^2 y} = \mathbf{xy^2\sqrt[3]{x^2 y}}$$

Example 4.1-39

$$\frac{-1}{-4}\sqrt[5]{\left(2^5 \cdot 3\right)x^7 y^8 z^3} = +\frac{1}{4}\sqrt[5]{\left(2^5 \cdot 3\right)x^{5+2} y^{5+3} z^3} = 2 \cdot \frac{1}{4}\sqrt[5]{3\left(x^5 \cdot x^2\right)\cdot\left(y^5 \cdot y^3\right)\cdot z^3} = 2 \cdot \frac{1}{4} xy\sqrt[5]{3x^2 y^3 z^3}$$

$$= \frac{\overset{1}{\cancel{2}}\,xy}{\underset{2}{\cancel{4}}} \cdot \sqrt[5]{3x^2 y^3 z^3} = \mathbf{\frac{xy}{2}\sqrt[5]{3x^2 y^3 z^3}}$$

Example 4.1-40

$$\sqrt[4]{256a^5 b^6 c^4} = \sqrt[4]{4^4 a^{4+1} b^{4+2} c^4} = 4c\sqrt[4]{\left(a^4 \cdot a^1\right)\cdot\left(b^4 \cdot b^2\right)} = 4cab\sqrt[4]{a^1 \cdot b^2} = \mathbf{4abc\sqrt[4]{ab^2}}$$

Example 4.1-41

$$\frac{1}{-50}\sqrt{625x^5 y^6 z^3} = -\frac{1}{50}\sqrt{25^2 x^{2+2+1} y^{2+2+2} z^{2+1}} = -\frac{25}{50}\sqrt{\left(x^2 \cdot x^2 \cdot x^1\right)\cdot\left(y^2 \cdot y^2 \cdot y^2\right)\cdot\left(z^2 \cdot z^1\right)}$$

$$= -\frac{\overset{1}{\cancel{25}}}{\underset{2}{\cancel{50}}}\left(x^1 \cdot x^1\right)\cdot\left(y^1 \cdot y^1 \cdot y^1\right)\cdot z\sqrt{x^1 \cdot z^1} = -\frac{1}{2}\left(x^{1+1}\right)\cdot\left(y^{1+1+1}\right)\cdot z\sqrt{x \cdot z} = \mathbf{-\frac{1}{2}\left(x^2 y^3 z\sqrt{xz}\right)}$$

Example 4.1-42

$$r^2 t\sqrt{27r^3 t^6} = r^2 t\sqrt{(9 \cdot 3)r^{2+1} t^{2+2+2}} = r^2 t\sqrt{\left(3^2 \cdot 3\right)\cdot\left(r^2 \cdot r^1\right)\cdot\left(t^2 \cdot t^2 \cdot t^2\right)} = 3r^2 t \cdot r^1 \cdot t^1 \cdot t^1 \cdot t^1\sqrt{3 \cdot r}$$

$$= 3\left(r^2 \cdot r^1\right)\cdot\left(t^1 \cdot t^1 \cdot t^1 \cdot t^1\right)\sqrt{3r} = 3\left(r^{2+1}\right)\cdot\left(t^{1+1+1+1}\right)\sqrt{3r} = \mathbf{3r^3 t^4\sqrt{3r}}$$

Example 4.1-43

$$\sqrt[4]{32u^6 v^5 w^9} = \sqrt[4]{(16 \cdot 2)u^{4+2} v^{4+1} w^{4+4+1}} = \sqrt[4]{\left(2^4 \cdot 2\right)\cdot\left(u^4 \cdot u^2\right)\cdot\left(v^4 \cdot v^1\right)\cdot\left(w^4 \cdot w^4 \cdot w^1\right)} = 2uv \cdot \left(w^1 \cdot w^1\right)\sqrt[4]{2u^2 vw}$$

$$= 2uv \cdot \left(w^{1+1}\right)\sqrt[4]{2u^2 vw} = \mathbf{2uvw^2\sqrt[4]{2u^2 vw}}$$

Example 4.1-44

$$x^2 y^2\sqrt{99x^5 y^3} = x^2 y^2\sqrt{(9 \cdot 11)x^{2+2+1} y^{2+1}} = x^2 y^2\sqrt{\left(3^2 \cdot 11\right)\cdot\left(x^2 \cdot x^2 \cdot x^1\right)\cdot\left(y^2 \cdot y^1\right)}$$

$$= 3x^2 y^2 \cdot x^1 \cdot x^1 \cdot y^1\sqrt{11 \cdot x^1 \cdot y^1} = 3\left(x^2 \cdot x^1 \cdot x^1\right)\cdot\left(y^2 \cdot y^1\right)\sqrt{11xy} = 3\left(x^{2+1+1}\right)\cdot\left(y^{2+1}\right)\sqrt{11xy} = \mathbf{3x^4 y^3\sqrt{11xy}}$$

Example 4.1-45

$$\boxed{\frac{-2}{-6}x\sqrt{8x^9}} = \boxed{+\frac{\overset{1}{2}}{\underset{3}{\cancel{6}}}x\sqrt{(4\cdot 2)x^{2+2+2+2+1}}} = \boxed{\frac{1}{3}x\sqrt{\left(2^2\cdot 2\right)\cdot\left(x^2\cdot x^2\cdot x^2\cdot x^2\cdot x^1\right)}} = \boxed{\frac{1}{3}\cdot 2x\cdot\left(x^1\cdot x^1\cdot x^1\cdot x^1\right)\sqrt{2\cdot x^1}}$$

$$= \boxed{\frac{2x}{3}\cdot\left(x^{1+1+1+1}\right)\sqrt{2x}} = \boxed{\frac{2x}{3}\cdot x^4\sqrt{2x}} = \boxed{\frac{2x^1\cdot x^4}{3}\cdot\sqrt{2x}} = \boxed{\frac{2x^{1+4}}{3}\cdot\sqrt{2x}} = \boxed{\frac{2x^5}{3}\cdot\frac{\sqrt{2x}}{1}} = \boxed{\frac{2x^5\cdot\sqrt{2x}}{3\cdot 1}}$$

$$= \boxed{\frac{2x^5\sqrt{2x}}{3}} = \boxed{\mathbf{\frac{2}{3}\left(x^5\sqrt{2x}\right)}}$$

Example 4.1-46

$$\boxed{-m^3\sqrt[2]{50k^2m^4n^3}} = \boxed{-m^3\sqrt{50k^2m^4n^3}} = \boxed{-m^3k\sqrt{(25\cdot 2)m^{2+2}n^{2+1}}} = \boxed{-m^3k\sqrt{\left(5^2\cdot 2\right)\cdot\left(m^2\cdot m^2\right)\cdot\left(n^2\cdot n^1\right)}}$$

$$= \boxed{-5\left(m^3\cdot m^1\cdot m^1\right)nk\sqrt{2\cdot n}} = \boxed{-5\left(m^{3+1+1}\right)nk\sqrt{2\cdot n}} = \boxed{\mathbf{-5km^5n\sqrt{2n}}}$$

Example 4.1-47

$$\boxed{\sqrt[4]{81a^4b^4c^5}} = \boxed{\sqrt[4]{3^4a^4b^4c^{4+1}}} = \boxed{3ab\sqrt[4]{c^4\cdot c^1}} = \boxed{3abc\sqrt[4]{c^1}} = \boxed{\mathbf{3abc\sqrt[4]{c}}}$$

Example 4.1-48

$$\boxed{xy\sqrt{72x^2y^4z}} = \boxed{(x\cdot x)y\sqrt{(36\cdot 2)y^{2+2}z}} = \boxed{\left(x^1\cdot x^1\right)y\sqrt{\left(6^2\cdot 2\right)\cdot\left(y^2\cdot y^2\right)\cdot z}} = \boxed{6x^{1+1}\cdot\left(y^1\cdot y^1\cdot y^1\right)\sqrt{2z}}$$

$$= \boxed{6x^2\cdot\left(y^{1+1+1}\right)\sqrt{2z}} = \boxed{\mathbf{6x^2y^3\sqrt{2z}}}$$

Example 4.1-49

$$\boxed{\frac{1}{-2}x^2y^3\sqrt[3]{8x^4y^4}} = \boxed{-\frac{1}{2}x^2y^3\sqrt[3]{2^3x^{3+1}y^{3+1}}} = \boxed{-\frac{\overset{1}{\cancel{2}}}{\underset{1}{\cancel{2}}}x^2y^3\sqrt[3]{\left(x^3\cdot x^1\right)\cdot\left(y^3\cdot y^1\right)}} = \boxed{-\frac{1}{1}\left(x^2\cdot x^1\right)\cdot\left(y^3\cdot y^1\right)\sqrt[3]{x^1y^1}}$$

$$= \boxed{-\left(x^{2+1}\right)\cdot\left(y^{3+1}\right)\sqrt[3]{xy}} = \boxed{\mathbf{-x^3y^4\sqrt[3]{xy}}}$$

Example 4.1-50

$$\boxed{-5\sqrt[3]{4^3w^8x^5y^6z^4}} = \boxed{-(5\cdot 4)\sqrt[3]{w^{3+3+2}x^{3+2}y^{3+3}z^{3+1}}} = \boxed{-(5\cdot 4)\sqrt[3]{\left(w^3\cdot w^3\cdot w^2\right)\cdot\left(x^3\cdot x^2\right)\cdot\left(y^3\cdot y^3\right)\cdot\left(z^3\cdot z^1\right)}}$$

$$= \boxed{-20\left(w^1\cdot w^1\right)\cdot x\cdot\left(y^1\cdot y^1\right)\cdot z\sqrt[3]{w^2x^2z^1}} = \boxed{-20\cdot\left(w^{1+1}\right)\cdot x\cdot\left(y^{1+1}\right)\cdot z\sqrt[3]{w^2x^2z}} = \boxed{\mathbf{-20xy^2w^2z\sqrt[3]{x^2w^2z}}}$$

Practice Problems - Simplifying Radical Expressions with Variables as Radicand

Section 4.1 Case IV Practice Problems - Simplify the following radical expressions:

1. $\sqrt{y^3} =$
2. $x\sqrt[2]{x^4} =$
3. $x^3y^2\sqrt[2]{x^3y^5} =$
4. $-2x\sqrt{8xy^3} =$
5. $\frac{1}{-12}\sqrt[3]{216a^5b^6c^7} =$
6. $uv^2\sqrt[5]{u^6v^8w^4} =$
7. $\frac{-3}{-10}\sqrt[4]{1250l^8m^7n^6} =$
8. $-x^2\sqrt[3]{729x^3y^5z^6} =$
9. $\frac{-5}{-6}\sqrt{72r^3s^5t^7} =$
10. $\frac{-3x}{4}\sqrt[2]{100x^5y^6z^3} =$

4.2 Multiplying Radical Expressions

Radicals are multiplied by each other by using the following general product rule:

$$a\sqrt[n]{x}\cdot b\sqrt[n]{y}\cdot c\sqrt[n]{z}=(a\cdot b\cdot c)\sqrt[n]{x\cdot y\cdot z}=abc\sqrt[n]{xyz}$$

Note that radicals can only be multiplied by each other if they have the same index n. In this section, students learn how radical expressions in monomial (Case I) and binomial form (Case II) are multiplied by each other.

Note: A monomial expression in radical form is defined as:

$$\sqrt{8x^5}\,,\ \sqrt{y}\,,\ \sqrt{27}\,,\ -3\sqrt[5]{x^6y^7}\,,\ \sqrt[3]{x^2y^5}\,,\ 2\sqrt{125}\,,\ etc.$$

A binomial expression in radical form is defined as:

$$\sqrt{x}+\sqrt{y}\,,\ 1+\sqrt{8x}\,,\ xy+\sqrt{y^3}\,,\ x^3y^3-\sqrt{x^2y}\,,\ 9-\sqrt[5]{x^5y^6}\,,\ \sqrt[3]{64}+\sqrt[3]{x^2y^5}\,,\ etc.$$

Case I Multiplying Monomial Expressions in Radical Form

Monomial expressions in radical form are multiplied by each other using the above general product rule. Multiplication of monomial expressions is divided into two cases. Case I a - multiplication of monomial terms in radical form, with real numbers and Case I b - multiplication of monomial terms in radical form, with variables.

Case I a Multiplying Monomial Expressions in Radical Form, with Real Numbers

Radical expressions with real numbers as radicands are multiplied by each other using the following steps:

Step 1 Simplify the radical terms (see Section 4.1, Case III).

Step 2 Multiply the radical terms by using the product rule. Repeat Step 1, if necessary.

$$k_1\sqrt[n]{a}\cdot k_2\sqrt[n]{b}\cdot k_3\sqrt[n]{c}=(k_1\cdot k_2\cdot k_3)\sqrt[n]{a\cdot b\cdot c}=k_1k_2k_3\sqrt[n]{abc} \qquad a,\ b,\ and\ c\geq 0$$

Examples with Steps

The following examples show the steps as to how radical expressions in monomial form are multiplied by one another:

Example 4.2-1

$$\boxed{\sqrt{5}\cdot\sqrt{15}}=$$

Solution:

Step 1 *Not Applicable*

Step 2 $\boxed{\sqrt{5}\cdot\sqrt{15}}=\boxed{\sqrt{5\cdot 15}}=\boxed{\sqrt{75}}=\boxed{\sqrt{25\cdot 3}}=\boxed{\sqrt{5^2\cdot 3}}=\boxed{5\sqrt{3}}$

Example 4.2-2

$$\boxed{\sqrt[2]{12}\cdot\sqrt[2]{32}}=$$

Solution:

Step 1 $\boxed{\sqrt[2]{12}\cdot\sqrt[2]{32}}=\boxed{\sqrt{12}\cdot\sqrt{32}}=\boxed{\sqrt{4\cdot 3}\cdot\sqrt{16\cdot 2}}=\boxed{\sqrt{2^2\cdot 3}\cdot\sqrt{4^2\cdot 2}}=\boxed{2\sqrt{3}\cdot 4\sqrt{2}}$

Step 2 $\boxed{2\sqrt{3}\cdot 4\sqrt{2}} = \boxed{(2\cdot 4)\cdot\left(\sqrt{3\cdot 2}\right)} = \boxed{\mathbf{8\sqrt{6}}}$

Example 4.2-3

$\boxed{\sqrt{98}\cdot\sqrt{48}\cdot\sqrt{108}} =$

Solution:

Step 1 $\boxed{\sqrt{98}\cdot\sqrt{48}\cdot\sqrt{108}} = \boxed{\sqrt{49\cdot 2}\cdot\sqrt{16\cdot 3}\cdot\sqrt{36\cdot 3}} = \boxed{\sqrt{7^2\cdot 2}\cdot\sqrt{4^2\cdot 3}\cdot\sqrt{6^2\cdot 3}}$

$= \boxed{7\sqrt{2}\cdot 4\sqrt{3}\cdot 6\sqrt{3}}$

Step 2 $\boxed{7\sqrt{2}\cdot 4\sqrt{3}\cdot 6\sqrt{3}} = \boxed{(7\cdot 4\cdot 6)\cdot\left(\sqrt{2}\cdot\sqrt{3}\cdot\sqrt{3}\right)} = \boxed{168\left(\sqrt{2\cdot 3\cdot 3}\right)} = \boxed{168\sqrt{2\cdot 3^2}}$

$= \boxed{(168\cdot 3)\sqrt{2}} = \boxed{\mathbf{504\sqrt{2}}}$

Example 4.2-4

$\boxed{\left(-2\sqrt[3]{512}\right)\cdot\left(-5\sqrt[3]{108}\right)} =$

Solution:

Step 1 $\boxed{\left(-2\sqrt[3]{512}\right)\cdot\left(-5\sqrt[3]{108}\right)} = \boxed{\left(-2\sqrt[3]{8^3}\right)\cdot\left(-5\sqrt[3]{27\cdot 4}\right)} = \boxed{(-2\cdot 8)\cdot\left(-5\sqrt[3]{3^3\cdot 4}\right)}$

$= \boxed{(-16)\cdot(-5\cdot 3)\sqrt[3]{4}} = \boxed{(-16)\cdot(-15)\sqrt[3]{4}} = \boxed{+240\sqrt[3]{4}} = \boxed{\mathbf{240\sqrt[3]{4}}}$

Step 2 $\boxed{\textit{Not Applicable}}$

Example 4.2-5

$\boxed{\left(\sqrt{50}\cdot\sqrt{54}\right)\cdot\left(\sqrt[4]{243}\cdot\sqrt[4]{64}\right)} =$

Solution:

Step 1 $\boxed{\left(\sqrt{50}\cdot\sqrt{54}\right)\cdot\left(\sqrt[4]{243}\cdot\sqrt[4]{64}\right)} = \boxed{\left(\sqrt{25\cdot 2}\cdot\sqrt{9\cdot 6}\right)\cdot\left(\sqrt[4]{81\cdot 3}\cdot\sqrt[4]{16\cdot 4}\right)}$

$= \boxed{\left(\sqrt{5^2\cdot 2}\cdot\sqrt{3^2\cdot 6}\right)\cdot\left(\sqrt[4]{3^4\cdot 3}\cdot\sqrt[4]{2^4\cdot 4}\right)} = \boxed{\left(5\sqrt{2}\cdot 3\sqrt{6}\right)\cdot\left(3\sqrt[4]{3}\cdot 2\sqrt[4]{4}\right)}$

Step 2 $\boxed{\left(5\sqrt{2}\cdot 3\sqrt{6}\right)\cdot\left(3\sqrt[4]{3}\cdot 2\sqrt[4]{4}\right)} = \boxed{\left(5\cdot 3\sqrt{2\cdot 6}\right)\cdot\left(3\cdot 2\sqrt[4]{3\cdot 4}\right)} = \boxed{\left(15\sqrt{12}\right)\cdot\left(6\sqrt[4]{12}\right)}$

$= \boxed{(15\cdot 6)\sqrt{4\cdot 3}\cdot\sqrt[4]{12}} = \boxed{90\sqrt{2^2\cdot 3}\cdot\sqrt[4]{12}} = \boxed{(90\cdot 2)\sqrt{3}\cdot\sqrt[4]{12}} = \boxed{\mathbf{180\sqrt{3}\cdot\sqrt[4]{12}}}$

Additional Examples - Multiplying Monomial Expressions in Radical Form, with Real Numbers

The following examples further illustrate how to multiply radical terms by one another:

Example 4.2-6

$$\boxed{\sqrt{12}\cdot\sqrt{9}} = \boxed{\left(\sqrt{4\cdot 3}\right)\cdot\sqrt{3^2}} = \boxed{\left(\sqrt{2^2\cdot 3}\right)\cdot 3} = \boxed{2\sqrt{3}\cdot 3} = \boxed{(2\cdot 3)\sqrt{3}} = \boxed{\mathbf{6\sqrt{3}}}$$

Example 4.2-7

$$\boxed{\sqrt{9}\cdot\sqrt{8}} = \boxed{\sqrt{3^2}\cdot\sqrt{4\cdot 2}} = \boxed{3\cdot\sqrt{2^2\cdot 2}} = \boxed{3\cdot 2\sqrt{2}} = \boxed{\mathbf{6\sqrt{2}}}$$

Example 4.2-8

$$\boxed{\sqrt{20}\cdot\sqrt{50}} = \boxed{\sqrt{4\cdot 5}\cdot\sqrt{25\cdot 2}} = \boxed{\sqrt{2^2\cdot 5}\cdot\sqrt{5^2\cdot 2}} = \boxed{2\sqrt{5}\cdot 5\sqrt{2}} = \boxed{(2\cdot 5)\cdot\left(\sqrt{5\cdot 2}\right)} = \boxed{\mathbf{10\sqrt{10}}}$$

Note that we can also simplify the radical terms in the following way:

$$\boxed{\sqrt{20}\cdot\sqrt{50}} = \boxed{\sqrt{20\cdot 50}} = \boxed{\sqrt{1000}} = \boxed{\sqrt{100\cdot 10}} = \boxed{\sqrt{10^2\cdot 10}} = \boxed{\mathbf{10\sqrt{10}}}$$

Example 4.2-9

$$\boxed{\sqrt{12}\cdot\sqrt{90}} = \boxed{\sqrt{4\cdot 3}\cdot\sqrt{9\cdot 10}} = \boxed{\sqrt{2^2\cdot 3}\cdot\sqrt{3^2\cdot 10}} = \boxed{2\sqrt{3}\cdot 3\sqrt{10}} = \boxed{(2\cdot 3)\cdot\left(\sqrt{3\cdot 10}\right)} = \boxed{\mathbf{6\sqrt{30}}}$$

Example 4.2-10

$$\boxed{\sqrt{50}\cdot\sqrt{32}\cdot\sqrt{3}} = \boxed{\sqrt{25\cdot 2}\cdot\sqrt{16\cdot 2}\cdot\sqrt{3}} = \boxed{\sqrt{5^2\cdot 2}\cdot\sqrt{4^2\cdot 2}\cdot\sqrt{3}} = \boxed{5\sqrt{2}\cdot 4\sqrt{2}\cdot\sqrt{3}} = \boxed{(5\cdot 4)\cdot\left(\sqrt{2}\cdot\sqrt{2}\cdot\sqrt{3}\right)}$$

$$= \boxed{20\cdot\left(\sqrt{2\cdot 2\cdot 3}\right)} = \boxed{20\cdot\left(\sqrt{2^1\cdot 2^1\cdot 3}\right)} = \boxed{20\cdot\left(\sqrt{2^{1+1}\cdot 3}\right)} = \boxed{20\sqrt{2^2\cdot 3}} = \boxed{(20\cdot 2)\sqrt{3}} = \boxed{\mathbf{40\sqrt{3}}}$$

Example 4.2-11

$$\boxed{\sqrt{6}\cdot\sqrt{48}\cdot\sqrt{45}} = \boxed{\sqrt{6}\cdot\sqrt{16\cdot 3}\cdot\sqrt{9\cdot 5}} = \boxed{\sqrt{6}\cdot\sqrt{4^2\cdot 3}\cdot\sqrt{3^2\cdot 5}} = \boxed{\sqrt{6}\cdot 4\sqrt{3}\cdot 3\sqrt{5}} = \boxed{(4\cdot 3)\cdot\left(\sqrt{6}\cdot\sqrt{3}\cdot\sqrt{5}\right)}$$

$$= \boxed{12\cdot\left(\sqrt{6\cdot 3\cdot 5}\right)} = \boxed{12\cdot\left(\sqrt{90}\right)} = \boxed{12\cdot\sqrt{9\cdot 10}} = \boxed{12\cdot\sqrt{3^2\cdot 10}} = \boxed{(12\cdot 3)\sqrt{10}} = \boxed{\mathbf{36\sqrt{10}}}$$

Example 4.2-12

$$\boxed{\sqrt{10}\cdot\sqrt{5}\cdot\sqrt{5}} = \boxed{\sqrt{10\cdot 5\cdot 5}} = \boxed{\sqrt{250}} = \boxed{\sqrt{25\cdot 10}} = \boxed{\sqrt{5^2\cdot 10}} = \boxed{\mathbf{5\sqrt{10}}}$$

Example 4.2-13

$$\boxed{\sqrt{1000}\cdot\sqrt{20}\cdot\sqrt{75}} = \boxed{\sqrt{100\cdot 10}\cdot\sqrt{4\cdot 5}\cdot\sqrt{25\cdot 3}} = \boxed{\sqrt{10^2\cdot 10}\cdot\sqrt{2^2\cdot 5}\cdot\sqrt{5^2\cdot 3}} = \boxed{10\sqrt{10}\cdot 2\sqrt{5}\cdot 5\sqrt{3}}$$

$$= \boxed{(10\cdot 2\cdot 5)\cdot\left(\sqrt{10}\cdot\sqrt{5}\cdot\sqrt{3}\right)} = \boxed{100\cdot\left(\sqrt{10\cdot 5\cdot 3}\right)} = \boxed{100\cdot\sqrt{150}} = \boxed{100\cdot\sqrt{25\cdot 6}} = \boxed{100\cdot\sqrt{5^2\cdot 6}} = \boxed{100\cdot 5\sqrt{6}}$$

$$= \boxed{\mathbf{500\sqrt{6}}}$$

Example 4.2-14

$$\sqrt[3]{128}\cdot\sqrt[3]{500} = \sqrt[3]{2\cdot 64}\cdot\sqrt[3]{4\cdot 125} = \sqrt[3]{2\cdot 4^3}\cdot\sqrt[3]{4\cdot 5^3} = 4\sqrt[3]{2}\cdot 5\sqrt[3]{4} = (4\cdot 5)\cdot\left(\sqrt[3]{2}\cdot\sqrt[3]{4}\right) = 20\cdot\left(\sqrt[3]{2\cdot 4}\right)$$

$$= 20\cdot\sqrt[3]{8} = 20\cdot\sqrt[3]{2^3} = 20\cdot 2 = \boxed{\mathbf{40}}$$

Example 4.2-15

$$\sqrt[4]{54}\cdot\sqrt[4]{48} = \sqrt[4]{2\cdot 27}\cdot\sqrt[4]{3\cdot 16} = \sqrt[4]{2\cdot 3^3}\cdot\sqrt[4]{3\cdot 2^4} = \sqrt[4]{2\cdot 3^3}\cdot 2\sqrt[4]{3^1} = 2\sqrt[4]{2\cdot 3^3\cdot 3^1} = 2\sqrt[4]{2\cdot 3^{3+1}} = 2\sqrt[4]{2\cdot 3^4}$$

$$= (2\cdot 3)\cdot\sqrt[4]{2} = \boxed{\mathbf{6\sqrt[4]{2}}}$$

Example 4.2-16

$$\sqrt[5]{5^4\cdot 3^2}\cdot\sqrt[5]{5\cdot 3^6} = \sqrt[5]{\left(5^4\cdot 5^1\right)\cdot\left(3^2\cdot 3^6\right)} = \sqrt[5]{\left(5^{4+1}\right)\cdot\left(3^{2+6}\right)} = \sqrt[5]{5^5\cdot 3^8} = 5\sqrt[5]{3^{5+3}} = 5\sqrt[5]{3^5\cdot 3^3}$$

$$= (5\cdot 3)\cdot\sqrt[5]{3^3} = \boxed{\mathbf{15\sqrt[5]{27}}}$$

Example 4.2-17

$$\sqrt[3]{6^2}\cdot\sqrt[3]{5^5}\cdot\sqrt[3]{6^4}\cdot\sqrt[3]{5^2} = \sqrt[3]{\left(6^2\cdot 6^4\right)\cdot\left(5^5\cdot 5^2\right)} = \sqrt[3]{\left(6^{2+4}\right)\cdot\left(5^{5+2}\right)} = \sqrt[3]{6^6\cdot 5^7} = \sqrt[3]{6^{3+3}\cdot 5^{3+3+1}}$$

$$= \sqrt[3]{\left(6^3\cdot 6^3\right)\cdot\left(5^3\cdot 5^3\right)\cdot 5^1} = \left(6^1\cdot 6^1\right)\cdot\left(5^1\cdot 5^1\right)\cdot\sqrt[3]{5^1} = 6^{1+1}\cdot 5^{1+1}\cdot\sqrt[3]{5} = 6^2\cdot 5^2\cdot\sqrt[3]{5} = (36\cdot 25)\cdot\sqrt[3]{5} = \boxed{\mathbf{900\sqrt[3]{5}}}$$

Example 4.2-18

$$\sqrt[4]{16}\cdot\sqrt[3]{27}\cdot\sqrt[4]{32}\cdot\sqrt[3]{192} = \sqrt[4]{2^4}\cdot\sqrt[3]{3^3}\cdot\sqrt[4]{16\cdot 2}\cdot\sqrt[3]{64\cdot 3} = 2\cdot 3\cdot\left(\sqrt[4]{2^4\cdot 2}\right)\cdot\left(\sqrt[3]{4^3\cdot 3}\right) = 6\cdot\left(2\sqrt[4]{2}\right)\cdot\left(4\sqrt[3]{3}\right)$$

$$= (6\cdot 2\cdot 4)\cdot\sqrt[4]{2}\cdot\sqrt[3]{3} = \boxed{\mathbf{48\sqrt[4]{2}\,\sqrt[3]{3}}}$$

Example 4.2-19

$$\sqrt[5]{1024}\cdot\sqrt{36}\cdot\sqrt{45}\cdot\sqrt[5]{243}\cdot\sqrt{3} = \sqrt[5]{4^5}\cdot\sqrt{6^2}\cdot\sqrt{9\cdot 5}\cdot\sqrt[5]{3^5}\cdot\sqrt{3} = 4\cdot 6\cdot\sqrt{3^2\cdot 5}\cdot 3\cdot\sqrt{3} = 4\cdot 6\cdot 3\sqrt{5}\cdot 3\cdot\sqrt{3}$$

$$= (4\cdot 6\cdot 3\cdot 3)\cdot\left(\sqrt{5}\cdot\sqrt{3}\right) = 216\cdot\sqrt{5\cdot 3} = \boxed{\mathbf{216\sqrt{15}}}$$

Example 4.2-20

$$\sqrt{45}\cdot\sqrt[3]{27}\cdot\sqrt[3]{3^2}\cdot\sqrt{20}\cdot\sqrt[3]{6} = \sqrt{5\cdot 9}\cdot\sqrt[3]{3^3}\cdot\sqrt[3]{3^2}\cdot\sqrt{4\cdot 5}\cdot\sqrt[3]{6} = \sqrt{5\cdot 3^2}\cdot 3\cdot\sqrt[3]{3^2}\cdot\sqrt{2^2\cdot 5}\cdot\sqrt[3]{6}$$

$$= 3\sqrt{5}\cdot 3\sqrt[3]{3^2}\cdot 2\sqrt{5}\cdot\sqrt[3]{6} = (3\cdot 3\cdot 2)\cdot\left(\sqrt{5}\cdot\sqrt{5}\right)\cdot\left(\sqrt[3]{3^2}\cdot\sqrt[3]{6}\right) = 18\cdot\left(\sqrt{5\cdot 5}\right)\cdot\left(\sqrt[3]{9\cdot 6}\right) = 18\cdot\left(\sqrt{5^{1+1}}\right)\cdot\left(\sqrt[3]{54}\right)$$

$$= 18\cdot\sqrt{5^2}\cdot\sqrt[3]{27\cdot 2} = (18\cdot 5)\cdot\sqrt[3]{3^3\cdot 2} = (90\cdot 3)\sqrt[3]{2} = \boxed{\mathbf{270\sqrt[3]{2}}}$$

Practice Problems - Multiplying Monomial Expressions in Radical Form, with Real Numbers

Section 4.2 Case I a Practice Problems - Multiply the following radical expressions:

1. $\sqrt{72} \cdot \sqrt{75} =$
2. $-3\sqrt{20} \cdot 2\sqrt{32} =$
3. $\sqrt[2]{16} \cdot \sqrt[2]{27} =$
4. $\sqrt{64} \cdot \sqrt{100} \cdot \sqrt{54} =$
5. $-\sqrt{125} \cdot -2\sqrt{98} =$
6. $\sqrt[4]{625} \cdot \sqrt[4]{324} \cdot \sqrt[4]{48} =$
7. $\sqrt[2]{192} \cdot \sqrt[2]{48} \cdot \sqrt[2]{300} =$
8. $\sqrt{75} \cdot \sqrt{150} =$
9. $\sqrt[3]{343} \cdot \sqrt[3]{128} \cdot \sqrt[3]{108} =$
10. $\sqrt{225} \cdot \sqrt{800} \cdot \sqrt{18} =$

Case I b Multiplying Monomial Expressions in Radical Form, with Variables

Monomial expressions in radical form with variables as radicands are multiplied by each other using the following steps:

Step 1 Simplify the radical terms (see Section 4.1, Case IV).

Step 2 Multiply the radical terms by using the product rule. Repeat Step 1, if necessary.

$$a\sqrt[n]{x}\cdot b\sqrt[n]{y}\cdot c\sqrt[n]{z}=(a\cdot b\cdot c)\sqrt[n]{x\cdot y\cdot z}=abc\sqrt[n]{xyz}$$

Examples with Steps

The following examples show the steps as to how monomial radical expressions are multiplied by one another:

Example 4.2-21

$$\boxed{\sqrt{x^3y^5}\cdot x\sqrt{xy^2}}=$$

Solution:

Step 1
$$\boxed{\sqrt{x^3y^5}\cdot x\sqrt{xy^2}}=\boxed{\sqrt{x^{2+1}y^{2+2+1}}\cdot xy\sqrt{x}}=\boxed{\sqrt{\left(x^2\cdot x^1\right)\cdot\left(y^2\cdot y^2\cdot y^1\right)}\cdot xy\sqrt{x}}$$
$$=\boxed{x\left(y^1\cdot y^1\right)\sqrt{xy}\cdot xy\sqrt{x}}=\boxed{xy^{1+1}\sqrt{xy}\cdot xy\sqrt{x}}=\boxed{xy^2\sqrt{xy}\cdot xy\sqrt{x}}$$

Step 2
$$\boxed{xy^2\sqrt{xy}\cdot xy\sqrt{x}}=\boxed{xy^2\cdot xy\sqrt{xy\cdot x}}=\boxed{(x\cdot x)\cdot\left(y^2\cdot y\right)\sqrt{(x\cdot x)\cdot y}}=\boxed{x^2\cdot y^3\sqrt{x^2\cdot y}}$$
$$=\boxed{\left(x^2\cdot x\right)y^3\sqrt{y}}=\boxed{x^3y^3\sqrt{y}}$$

Example 4.2-22

$$\boxed{\left(\frac{-3}{-12}\sqrt{r^2s^5t^3}\right)\cdot\left(\frac{1}{-5}\sqrt{r^2s^3t}\right)}=$$

Solution:

Step 1
$$\boxed{\left(\frac{-3}{-12}\sqrt{r^2s^5t^3}\right)\cdot\left(\frac{1}{-5}\sqrt{r^2s^3t}\right)}=\boxed{\left(+\frac{\overset{1}{\cancel{3}}}{\underset{4}{\cancel{12}}}r\sqrt{s^{2+2+1}t^{2+1}}\right)\cdot\left(-\frac{1}{5}r\sqrt{s^{2+1}t}\right)}$$
$$=\boxed{\left(\frac{1}{4}r\sqrt{\left(s^2\cdot s^2\cdot s^1\right)\cdot\left(t^2\cdot t^1\right)}\right)\cdot\left(-\frac{1}{5}r\sqrt{s^2\cdot s^1t}\right)}=\boxed{\left(\frac{1}{4}r\left(s^1\cdot s^1\right)t\sqrt{st}\right)\cdot\left(-\frac{1}{5}rs\sqrt{st}\right)}$$
$$=\boxed{\left(\frac{1}{4}r\left(s^{1+1}\right)t\sqrt{st}\right)\cdot\left(-\frac{1}{5}rs\sqrt{st}\right)}=\boxed{\left(\frac{1}{4}rs^2t\sqrt{st}\right)\cdot\left(-\frac{1}{5}rs\sqrt{st}\right)}$$

Step 2
$$\boxed{\left(\frac{1}{4}rs^2t\sqrt{st}\right)\cdot\left(-\frac{1}{5}rs\sqrt{st}\right)}=\boxed{\left(-\frac{1}{5}\cdot\frac{1}{4}\right)\cdot\left(rs^2t\cdot rs\right)\cdot\left(\sqrt{st}\cdot\sqrt{st}\right)}$$

$$= \boxed{\left(-\frac{1\times 1}{5\times 4}\right)\cdot(r\cdot r)\cdot\left(s^2\cdot s\right)\cdot t\left(\sqrt{st\cdot st}\right)} = \boxed{-\frac{1}{20}\cdot r^2 s^3 t\sqrt{s^2t^2}} = \boxed{-\frac{1}{20}\cdot r^2 s^3 t\cdot st}$$

$$= \boxed{-\frac{1}{20}\cdot r^2\left(s^3\cdot s\right)\cdot(t\cdot t)} = \boxed{-\frac{1}{20}\cdot r^2 s^4 t^2} = \boxed{-\frac{\mathbf{r^2s^4t^2}}{\mathbf{20}}}$$

Example 4.2-23

$$\boxed{\sqrt[5]{32u^6v^7w^3}\cdot\sqrt[5]{u^2v^3w^2}} =$$

Solution:

Step 1

$$\boxed{\sqrt[5]{32u^6v^7w^3}\cdot\sqrt[5]{u^2v^3w^2}} = \boxed{\sqrt[5]{2^5u^{5+1}v^{5+2}w^3}\cdot\sqrt[5]{u^2v^3w^2}}$$

$$= \boxed{\left[2\sqrt[5]{\left(u^5\cdot u^1\right)\cdot\left(v^5\cdot v^2\right)\cdot w^3}\right]\cdot\left(\sqrt[5]{u^2v^3w^2}\right)} = \boxed{\left(2uv\sqrt[5]{uv^2w^3}\right)\cdot\left(\sqrt[5]{u^2v^3w^2}\right)}$$

Step 2

$$\boxed{\left(2uv\sqrt[5]{uv^2w^3}\right)\cdot\left(\sqrt[5]{u^2v^3w^2}\right)} = \boxed{2uv\sqrt[5]{\left(uv^2w^3\right)\cdot\left(u^2v^3w^2\right)}}$$

$$= \boxed{2uv\sqrt[5]{\left(u^1\cdot u^2\right)\cdot\left(v^2\cdot v^3\right)\cdot\left(w^3\cdot w^2\right)}} = \boxed{2uv\sqrt[5]{u^{1+2}\cdot v^{2+3}\cdot w^{3+2}}} = \boxed{2uv\sqrt[5]{u^3v^5w^5}}$$

$$= \boxed{2u\cdot(v\cdot v)\cdot w\sqrt[5]{u^3}} = \boxed{\mathbf{2uv^2w\sqrt[5]{u^3}}}$$

Example 4.2-24

$$\boxed{\sqrt{40p^3q^5}\cdot\sqrt{125p^2q}\cdot\sqrt{36p^5q^2}}$$

Solution:

Step 1

$$\boxed{\sqrt{40p^3q^5}\cdot\sqrt{125p^2q}\cdot\sqrt{36p^5q^2}} = \boxed{\sqrt{(4\cdot 10)p^{2+1}q^{2+2+1}}\cdot p\sqrt{(25\cdot 5)q}\cdot q\sqrt{(6\cdot 6)p^{2+2+1}}}$$

$$= \boxed{\left[\sqrt{\left(2^2\cdot 10\right)\cdot\left(p^2\cdot p^1\right)\cdot\left(q^2\cdot q^2\cdot q^1\right)}\right]\cdot\left[p\sqrt{\left(5^2\cdot 5\right)q}\right]\cdot\left[q\sqrt{6^2\cdot p^2\cdot p^2\cdot p^1}\right]}$$

$$= \boxed{\left[2p\left(q^1\cdot q^1\right)\sqrt{10pq}\right]\cdot\left[5p\sqrt{5q}\right]\cdot\left[6\left(p^1\cdot p^1\right)q\sqrt{p}\right]}$$

$$= \boxed{\left[2p\left(q^{1+1}\right)\sqrt{10pq}\right]\cdot\left[5p\sqrt{5q}\right]\cdot\left[6p^{1+1}q\sqrt{p}\right]} = \boxed{\left[2pq^2\sqrt{10pq}\right]\cdot\left[5p\sqrt{5q}\right]\cdot\left[6p^2q\sqrt{p}\right]}$$

Step 2

$$\boxed{\left[2pq^2\sqrt{10pq}\right]\cdot\left[5p\sqrt{5q}\right]\cdot\left[6p^2q\sqrt{p}\right]} = \boxed{(2\cdot 5\cdot 6)\cdot\left(p\cdot p\cdot p^2\right)\cdot\left(q^2\cdot q\right)\cdot\left(\sqrt{10pq}\cdot\sqrt{5q}\cdot\sqrt{p}\right)}$$

$$= \boxed{60p^4q^3\sqrt{(10\cdot 5)\cdot(p\cdot p)\cdot(q\cdot q)}} = \boxed{60p^4q^3\sqrt{50p^2q^2}} = \boxed{60\left(p^4\cdot p\right)\left(q^3\cdot q\right)\sqrt{50}}$$

$$= \boxed{\left(60\cdot\sqrt{25\cdot 2}\right)p^5q^4} = \boxed{\left(60\cdot\sqrt{5^2\cdot 2}\right)p^5q^4} = \boxed{(60\cdot 5)\sqrt{2}\,p^5q^4} = \boxed{\mathbf{300\sqrt{2}\,p^5q^4}}$$

Example 4.2-25

$$\boxed{\left(-\sqrt[4]{625m^6n}\right)\cdot\left(\sqrt[4]{16m^4n^3}\right)\cdot\left(-\sqrt[4]{mn^4}\right)} =$$

Solution:

Step 1

$$\boxed{\left(-\sqrt[4]{625m^6n}\right)\cdot\left(\sqrt[4]{16m^4n^3}\right)\cdot\left(-\sqrt[4]{mn^4}\right)} = \boxed{\left(-\sqrt[4]{5^4m^{4+2}n}\right)\cdot\left(m\sqrt[4]{2^4n^3}\right)\cdot\left(-n\sqrt[4]{m}\right)}$$

$$= \boxed{\left(-5\sqrt[4]{m^4\cdot m^2\cdot n}\right)\cdot\left(2m\sqrt[4]{n^3}\right)\cdot\left(-n\sqrt[4]{m}\right)} = \boxed{\left(-5m\sqrt[4]{m^2n}\right)\cdot\left(2m\sqrt[4]{n^3}\right)\cdot\left(-n\sqrt[4]{m}\right)}$$

Step 2

$$\boxed{\left(-5m\sqrt[4]{m^2n}\right)\cdot\left(2m\sqrt[4]{n^3}\right)\cdot\left(-n\sqrt[4]{m}\right)} = \boxed{(-5\cdot 2\cdot -1)\cdot(m\cdot m\cdot n)\cdot\left(\sqrt[4]{m^2n}\cdot\sqrt[4]{n^3}\cdot\sqrt[4]{m}\right)}$$

$$= \boxed{10m^2n\sqrt[4]{\left(m^2\cdot m\right)\cdot\left(n^3\cdot n\right)}} = \boxed{10m^2n\sqrt[4]{m^3n^4}} = \boxed{10m^2(n\cdot n)\sqrt[4]{m^3}} = \boxed{\mathbf{10m^2n^2\sqrt[4]{m^3}}}$$

Additional Examples - Multiplying Monomial Expressions in Radical Form, with Variables

The following examples further illustrate how to multiply radical terms by one another:

Example 4.2-26

$$\boxed{\sqrt{x^3}\cdot\sqrt{x}} = \boxed{\sqrt{x^{2+1}}\cdot\sqrt{x}} = \boxed{\sqrt{x^2\cdot x^1}\cdot\sqrt{x}} = \boxed{x\sqrt{x}\cdot\sqrt{x}} = \boxed{x\sqrt{x\cdot x}} = \boxed{x\sqrt{x^1\cdot x^1}} = \boxed{x\sqrt{x^{1+1}}} = \boxed{x\sqrt{x^2}} = \boxed{x\cdot x}$$

$$= \boxed{x^1\cdot x^1} = \boxed{x^{1+1}} = \boxed{\mathbf{x^2}}$$

Example 4.2-27

$$\boxed{\sqrt{x^4y^2z}\cdot\sqrt{xy^3z}} = \boxed{y\sqrt{x^{2+2}z}\cdot\sqrt{xy^{2+1}z}} = \boxed{\left[y\sqrt{\left(x^2\cdot x^2\right)\cdot z}\right]\cdot\left[\sqrt{x\cdot\left(y^2\cdot y^1\right)\cdot z}\right]} = \boxed{\left(y\cdot x^1\cdot x^1\sqrt{z}\right)\cdot\left(y\sqrt{xyz}\right)}$$

$$= \boxed{\left(yx^{1+1}\sqrt{z}\right)\cdot\left(y\sqrt{xyz}\right)} = \boxed{\left(y\cdot yx^2\right)\cdot\left(\sqrt{z}\cdot\sqrt{xyz}\right)} = \boxed{\left(y^1y^1x^2\right)\cdot\left(\sqrt{xyz\cdot z}\right)} = \boxed{\left(y^{1+1}x^2\right)\cdot\left(\sqrt{xyz^1\cdot z^1}\right)}$$

$$= \boxed{\left(y^2x^2\right)\cdot\left(\sqrt{xyz^{1+1}}\right)} = \boxed{y^2x^2\cdot\left(\sqrt{xyz^2}\right)} = \boxed{y^2x^2\cdot\left(z\sqrt{xy}\right)} = \boxed{\mathbf{x^2y^2z\sqrt{xy}}}$$

Example 4.2-28

$$\boxed{-\frac{r}{2}\sqrt{r^6s^4t^5}\cdot\frac{1}{3}\sqrt{rst}} = \boxed{-\frac{r}{2}\sqrt{r^{2+2+2}s^{2+2}t^{2+2+1}}\cdot\frac{1}{3}\sqrt{rst}} = \boxed{\left[-\frac{1}{2}r\sqrt{\left(r^2\cdot r^2\cdot r^2\right)\cdot\left(s^2\cdot s^2\right)\cdot\left(t^2\cdot t^2\cdot t^1\right)}\right]\cdot\frac{1}{3}\sqrt{rst}}$$

$$= \boxed{\left[-\frac{1}{2}\left(r^1\cdot r^1\cdot r^1\cdot r^1\right)\cdot\left(s^1\cdot s^1\right)\cdot\left(t^1\cdot t^1\right)\sqrt{t^1}\right]\cdot\frac{1}{3}\sqrt{rst}} = \boxed{-\left(\frac{1}{2}\cdot\frac{1}{3}\right)\cdot\left(r^{1+1+1+1}\right)\cdot\left(s^{1+1}\right)\cdot\left(t^{1+1}\right)\sqrt{t}\cdot\sqrt{rst}}$$

$$= \boxed{-\left(\frac{1\times1}{2\times3}\right)\cdot r^4s^2t^2\sqrt{t\cdot rst}} = \boxed{-\frac{1}{6}\cdot r^4s^2t^2\sqrt{rs\cdot(t\cdot t)}} = \boxed{-\frac{1}{6}\cdot r^4s^2t^2\sqrt{rs\cdot\left(t^1\cdot t^1\right)}} = \boxed{-\frac{1}{6}\cdot r^4s^2t^2\sqrt{rs\cdot t^{1+1}}}$$

$$= \boxed{-\frac{1}{6}\cdot r^4s^2t^2\sqrt{rst^2}} = \boxed{-\frac{1}{6}\cdot r^4s^2t^2t^1\sqrt{rs}} = \boxed{-\frac{1}{6}\cdot r^4s^2t^{2+1}\sqrt{rs}} = \boxed{\mathbf{-\frac{r^4s^2t^3}{6}\sqrt{rs}}}$$

Example 4.2-29

$$\boxed{\sqrt[5]{u^9v^6w^8}\cdot\sqrt[5]{u^2v^2w}} = \boxed{\sqrt[5]{u^{5+4}v^{5+1}w^{5+3}}\cdot\sqrt[5]{u^2v^2w}} = \boxed{\sqrt[5]{\left(u^5\cdot u^4\right)\cdot\left(v^5\cdot v^1\right)\cdot\left(w^5\cdot w^3\right)}\cdot\sqrt[5]{u^2v^2w}}$$

$$= \boxed{u\cdot v\cdot w\cdot\sqrt[5]{u^4v^1w^3}\cdot\sqrt[5]{u^2v^2w}} = \boxed{uvw\cdot\sqrt[5]{\left(u^4v^1w^3\right)\cdot\left(u^2v^2w^1\right)}} = \boxed{uvw\cdot\sqrt[5]{\left(u^4u^2\right)\cdot\left(v^1v^2\right)\cdot\left(w^3w^1\right)}}$$

$$= \boxed{uvw\cdot\sqrt[5]{\left(u^{4+2}\right)\cdot\left(v^{1+2}\right)\cdot\left(w^{3+1}\right)}} = \boxed{uvw\cdot\sqrt[5]{u^6v^3w^4}} = \boxed{uvw\cdot\sqrt[5]{u^{5+1}v^3w^4}} = \boxed{uvw\cdot\sqrt[5]{\left(u^5\cdot u^1\right)v^3w^4}}$$

$$= \boxed{uvw\cdot u\sqrt[5]{u^1v^3w^4}} = \boxed{(u\cdot u)vw\cdot\sqrt[5]{uv^3w^4}} = \boxed{\left(u^1\cdot u^1\right)vw\cdot\sqrt[5]{uv^3w^4}} = \boxed{u^{1+1}vw\cdot\sqrt[5]{uv^3w^4}} = \boxed{\mathbf{u^2vw\sqrt[5]{uv^3w^4}}}$$

Example 4.2-30

$$\boxed{\sqrt[3]{\frac{8}{125}p^5q^7}\cdot\sqrt[3]{\frac{2}{27}p^3q^4}} = \boxed{\sqrt[3]{\frac{2^3}{5^3}p^{3+2}q^{3+3+1}}\cdot p\sqrt[3]{\frac{2}{3^3}q^{3+1}}} = \boxed{\frac{2}{5}\sqrt[3]{\left(p^3\cdot p^2\right)\cdot\left(q^3\cdot q^3\cdot q^1\right)}\cdot\frac{p}{3}\sqrt[3]{\frac{2}{1}q^3\cdot q^1}}$$

$$= \boxed{\left(\frac{2}{5}pq^1\cdot q^1\sqrt[3]{p^2q}\right)\cdot\left(\frac{pq}{3}\sqrt[3]{2q}\right)} = \boxed{\left(\frac{2}{5}pq^{1+1}\sqrt[3]{p^2q}\right)\cdot\left(\frac{pq}{3}\sqrt[3]{2q}\right)} = \boxed{\left(\frac{2}{5}pq^2\sqrt[3]{p^2q}\right)\cdot\left(\frac{pq}{3}\sqrt[3]{2q}\right)}$$

$$= \boxed{\left(\frac{2}{5}\cdot\frac{1}{3}\right)\cdot(p\cdot p)\cdot\left(q^2\cdot q\right)\cdot\sqrt[3]{2p^2\cdot(qq)}} = \boxed{\left(\frac{2\cdot1}{5\cdot3}\right)\cdot\left(p^1\cdot p^1\right)\cdot\left(q^2\cdot q^1\right)\cdot\sqrt[3]{2p^2\cdot\left(q^1q^1\right)}}$$

$$= \boxed{\frac{2}{15}\cdot p^{1+1}\cdot q^{2+1}\cdot\sqrt[3]{2p^2\cdot q^{1+1}}} = \boxed{\mathbf{\frac{2}{15}p^2q^3\sqrt[3]{2p^2q^2}}}$$

Example 4.2-31

$$\boxed{\sqrt{x^2y^2z^2}\cdot\sqrt{xz^3}} = \boxed{xyz\cdot\sqrt{xz^{2+1}}} = \boxed{xyz\cdot\sqrt{xz^2\cdot z^1}} = \boxed{xy(z\cdot z)\sqrt{x\cdot z^1}} = \boxed{xy\cdot\left(z^1z^1\right)\cdot\sqrt{xz}} = \boxed{xy\cdot z^{1+1}\cdot\sqrt{xz}}$$

$$= \boxed{xy\cdot z^2\cdot\sqrt{xz}} = \boxed{\mathbf{xyz^2\sqrt{xz}}}$$

Example 4.2-32

$$\boxed{\sqrt{100p^2q^5}\cdot\sqrt{8pq}\cdot\sqrt{36p^3q}} = \boxed{p\sqrt{10^2q^{2+2+1}}\cdot\sqrt{4\cdot2pq}\cdot\sqrt{6^2p^{2+1}q}} = \boxed{10p\sqrt{q^2\cdot q^2\cdot q^1}\cdot\sqrt{2^2\cdot2pq}\cdot6\sqrt{p^2\cdot p^1q}}$$

$$= \boxed{\left(10pq^1\cdot q^1\sqrt{q}\right)\cdot\left(2\sqrt{2pq}\right)\cdot\left(6p\sqrt{pq}\right)} = \boxed{\left(10pq^{1+1}\sqrt{q}\right)\cdot\left(2\sqrt{2pq}\right)\cdot\left(6p\sqrt{pq}\right)}$$

$$= \boxed{(10\cdot2\cdot6)\cdot\left(p^1p^1q^2\right)\cdot\left(\sqrt{q}\cdot\sqrt{2pq}\cdot\sqrt{pq}\right)} = \boxed{120\cdot\left(p^{1+1}q^2\right)\cdot\left(\sqrt{q\cdot2pq\cdot pq}\right)} = \boxed{120p^2q^2\cdot\sqrt{2\left(p^1p^1\right)\cdot\left(q^1q^1q^1\right)}}$$

$$= \boxed{120p^2q^2\cdot\sqrt{2p^{1+1}\cdot q^{1+1+1}}} = \boxed{120p^2q^2\cdot\sqrt{2p^2\cdot q^{2+1}}} = \boxed{120p^2q^2\cdot p\sqrt{2q^2\cdot q^1}} = \boxed{120p^2q^2\cdot pq\sqrt{2q}}$$

$$= \boxed{120\left(p^2\cdot p\right)\cdot\left(q^2\cdot q\right)\cdot\sqrt{2q}} = \boxed{120\left(p^2\cdot p^1\right)\cdot\left(q^2\cdot q^1\right)\cdot\sqrt{2q}} = \boxed{120p^{2+1}\cdot q^{2+1}\cdot\sqrt{2q}} = \boxed{\mathbf{120p^3q^3\sqrt{2q}}}$$

Example 4.2-33

$$\boxed{\sqrt[4]{m^5n^5}\cdot\sqrt[4]{16\cdot mn^3}\cdot\sqrt[4]{81m}} = \boxed{\sqrt[4]{m^{4+1}n^{4+1}}\cdot\sqrt[4]{2^4\cdot mn^3}\cdot\sqrt[4]{3^4m}} = \boxed{\left[\sqrt[4]{\left(m^4\cdot m^1\right)\cdot\left(n^4\cdot n^1\right)}\right]\cdot\left(\sqrt[4]{2^4\cdot mn^3}\right)\cdot\left(\sqrt[4]{3^4m}\right)}$$

$$= \boxed{mn\sqrt[4]{m^1n^1}\cdot 2\sqrt[4]{mn^3}\cdot 3\sqrt[4]{m}} = \boxed{(2\cdot 3)\cdot mn\sqrt[4]{mn}\cdot\sqrt[4]{mn^3}\cdot\sqrt[4]{m}} = \boxed{6\cdot mn\sqrt[4]{mn\cdot mn^3\cdot m}} = \boxed{6mn\sqrt[4]{(mmm)\cdot\left(nn^3\right)}}$$

$$= \boxed{6mn\sqrt[4]{\left(m^1m^1m^1\right)\cdot\left(n^1n^3\right)}} = \boxed{6mn\sqrt[4]{\left(m^{1+1+1}\right)\cdot\left(n^{1+3}\right)}} = \boxed{6mn\sqrt[4]{m^3n^4}} = \boxed{6mn\cdot n\sqrt[4]{m^3}} = \boxed{6m\cdot\left(n^1\cdot n^1\right)\cdot\sqrt[4]{m^3}}$$

$$= \boxed{6m\cdot n^{1+1}\cdot\sqrt[4]{m^3}} = \boxed{6m\cdot n^2\cdot\sqrt[4]{m^3}} = \boxed{\mathbf{6mn^2\sqrt[4]{m^3}}}$$

Example 4.2-34

$$\boxed{x\sqrt{x^2y^2z}\cdot-\sqrt{x^3y^3z}\cdot y^2\sqrt{xyz^3}} = \boxed{\left(xxy\sqrt{z}\right)\cdot\left(-\sqrt{x^{2+1}y^{2+1}z}\right)\cdot\left(y^2\sqrt{xyz^{2+1}}\right)}$$

$$= \boxed{\left(xxy\sqrt{z}\right)\cdot\left(-\sqrt{\left(x^2\cdot x^1\right)\cdot\left(y^2\cdot y^1\right)\cdot z}\right)\cdot\left(y^2\sqrt{xy\cdot\left(z^2\cdot z^1\right)}\right)} = \boxed{\left(xxy\sqrt{z}\right)\cdot\left(-xy\sqrt{xyz}\right)\cdot\left(y^2z\sqrt{xyz}\right)}$$

$$= \boxed{-\left(xxy\cdot xy\cdot y^2z\right)\cdot\left(\sqrt{z}\cdot\sqrt{xyz}\cdot\sqrt{xyz}\right)} = \boxed{-\left[(xxx)\cdot\left(yyy^2\right)\cdot z\right]\cdot\left(\sqrt{z\cdot xyz\cdot xyz}\right)} = \boxed{-\left(x^3\cdot y^4\cdot z\right)\cdot\left(\sqrt{(xx)\cdot(yy)\cdot(zzz)}\right)}$$

$$= \boxed{-\left(x^3y^4z\right)\cdot\left(\sqrt{x^2y^2z^2z}\right)} = \boxed{-\left(x^3y^4z\right)\cdot xyz\sqrt{z^1}} = \boxed{-\left(x^3x\right)\cdot\left(y^4y\right)\cdot(zz)\sqrt{z}} = \boxed{-\left(x^3x^1\right)\cdot\left(y^4y^1\right)\cdot\left(z^1z^1\right)\sqrt{z}}$$

$$= \boxed{-\left(x^{3+1}\right)\cdot\left(y^{4+1}\right)\cdot\left(z^{1+1}\right)\sqrt{z}} = \boxed{\mathbf{-x^4y^5z^2\sqrt{z}}}$$

Example 4.2-35

$$\boxed{\sqrt{3ab}\cdot\sqrt{2a^3}\cdot\sqrt{12b^5}} = \boxed{\sqrt{3ab}\cdot\sqrt{2a^{2+1}}\cdot\sqrt{4\cdot 3b^{2+2+1}}} = \boxed{\sqrt{3ab}\cdot\sqrt{2a^2a^1}\cdot\sqrt{2^2\cdot 3b^2\cdot b^2\cdot b^1}}$$

$$= \boxed{\sqrt{3ab}\cdot a\sqrt{2a^1}\cdot 2b^1b^1\sqrt{3b^1}} = \boxed{\sqrt{3ab}\cdot a\sqrt{2a^1}\cdot 2b^{1+1}\sqrt{3b^1}} = \boxed{\sqrt{3ab}\cdot a\sqrt{2a^1}\cdot 2b^2\sqrt{3b^1}}$$

$$= \boxed{2ab^2\sqrt{3a^1b^1}\cdot\sqrt{2a^1}\cdot\sqrt{3b^1}} = \boxed{2ab^2\sqrt{\left(3a^1b^1\right)\cdot\left(2a^1\right)\cdot\left(3b^1\right)}} = \boxed{2ab^2\sqrt{(3\cdot 2\cdot 3)\cdot\left(a^1a^1\right)\cdot\left(b^1b^1\right)}}$$

$$= \boxed{2ab^2\sqrt{\left(3^2\cdot 2\right)\cdot a^{1+1}\cdot b^{1+1}}} = \boxed{2ab^2\sqrt{\left(3^2\cdot 2\right)\cdot a^2\cdot b^2}} = \boxed{2ab^2\cdot 3ab\sqrt{2}} = \boxed{(2\cdot 3)\cdot(a\cdot a)\cdot\left(b^2\cdot b\right)\sqrt{2}}$$

$$= \boxed{6\cdot\left(a^1\cdot a^1\right)\cdot\left(b^2\cdot b^1\right)\sqrt{2}} = \boxed{6\cdot\left(a^{1+1}\right)\cdot\left(b^{2+1}\right)\sqrt{2}} = \boxed{\mathbf{6a^2b^3\sqrt{2}}}$$

Example 4.2-36

$$\sqrt{14u^4v^6z}\cdot\sqrt{24\cdot u^5v} = \sqrt{14u^{2+2}v^{2+2+2}z}\cdot\sqrt{(4\cdot 6)\cdot u^{2+2+1}v} = \sqrt{14\left(u^2u^2\right)\cdot\left(v^2v^2v^2\right)z}\cdot\sqrt{\left(2^2\cdot 6\right)\cdot\left(u^2u^2u^1\right)v}$$

$$= \left(u^1u^1v^1v^1v^1\sqrt{14z}\right)\cdot\left(2u^1u^1\sqrt{6uv}\right) = \left(u^{1+1}v^{1+1+1}\sqrt{14z}\right)\cdot\left(2u^{1+1}\sqrt{6uv}\right) = \left(u^2v^3\sqrt{14z}\right)\cdot\left(2u^2\sqrt{6uv}\right)$$

$$= \left(2u^2u^2v^3\right)\cdot\left(\sqrt{14z}\cdot\sqrt{6uv}\right) = \left(2u^{2+2}v^3\right)\cdot\left(\sqrt{(14\cdot 6)uvz}\right) = 2u^4v^3\sqrt{84uvz} = 2u^4v^3\sqrt{(4\cdot 21)uvz}$$

$$= 2u^4v^3\sqrt{\left(2^2\cdot 21\right)uvz} = 2u^4v^3\cdot 2\sqrt{21uvz} = (2\cdot 2)u^4v^3\sqrt{21uvz} = \mathbf{4u^4v^3\sqrt{21uvz}}$$

Example 4.2-37

$$\sqrt{t^8u^5}\cdot\sqrt{t^2u^7}\cdot u\sqrt{t^6} = \left(\sqrt{t^{2+2+2+2}u^{2+2+1}}\right)\cdot\left(t\sqrt{u^{2+2+2+1}}\right)\cdot\left(u\sqrt{t^{2+2+2}}\right)$$

$$= \left(\sqrt{t^2t^2t^2t^2u^2u^2u^1}\right)\cdot\left(t\sqrt{u^2u^2u^2u^1}\right)\cdot\left(u\sqrt{t^2t^2t^2}\right) = \left(t^1t^1t^1t^1u^1u^1\sqrt{u^1}\right)\cdot\left(tu^1u^1u^1\sqrt{u^1}\right)\cdot\left(ut^1t^1t^1\right)$$

$$= \left(t^{1+1+1+1}u^{1+1}\sqrt{u}\right)\cdot\left(tu^{1+1+1}\sqrt{u}\right)\cdot\left(ut^{1+1+1}\right) = \left(t^4u^2\sqrt{u}\right)\cdot\left(tu^3\sqrt{u}\right)\cdot\left(ut^3\right) = \left(t^4\cdot t\cdot t^3\right)\cdot\left(u^2\cdot u^3\cdot u\right)\cdot\left(\sqrt{u}\cdot\sqrt{u}\right)$$

$$= \left(t^4\cdot t^1\cdot t^3\right)\cdot\left(u^2\cdot u^3\cdot u^1\right)\cdot\left(\sqrt{u\cdot u}\right) = \left(t^{4+1+3}\right)\cdot\left(u^{2+3+1}\right)\cdot\left(\sqrt{u^1\cdot u^1}\right) = t^8u^6\sqrt{u^{1+1}} = t^8u^6\sqrt{u^2} = t^8u^6u^1$$

$$= t^8u^{6+1} = \mathbf{t^8u^7}$$

Example 4.2-38

$$\sqrt{x^3y^3}\cdot\sqrt{x^2y^5}\cdot y\sqrt{x^6} = \left(\sqrt{x^{2+1}y^{2+1}}\right)\cdot\left(\sqrt{x^2y^{2+2+1}}\right)\cdot\left(y\sqrt{x^{2+2+2}}\right)$$

$$= \left(\sqrt{\left(x^2\cdot x^1\right)\cdot\left(y^2\cdot y^1\right)}\right)\cdot\left(\sqrt{x^2\cdot\left(y^2\cdot y^2\cdot y^1\right)}\right)\cdot\left(y\sqrt{x^2\cdot x^2\cdot x^2}\right) = \left(xy\sqrt{x^1\cdot y^1}\right)\cdot\left(xy^1y^1\sqrt{y^1}\right)\cdot\left(yx^1x^1x^1\right)$$

$$= \left(xy\sqrt{xy}\right)\cdot\left(xy^{1+1}\sqrt{y}\right)\cdot\left(yx^{1+1+1}\right) = \left(xy\sqrt{xy}\right)\cdot\left(xy^2\sqrt{y}\right)\cdot\left(yx^3\right) = \left(xy\cdot xy^2\cdot yx^3\right)\cdot\left(\sqrt{xy}\cdot\sqrt{y}\right)$$

$$= \left(x\cdot x\cdot x^3\right)\cdot\left(y\cdot y^2\cdot y\right)\cdot\left(\sqrt{xy\cdot y}\right) = \left(x^1\cdot x^1\cdot x^3\right)\cdot\left(y^1\cdot y^2\cdot y^1\right)\cdot\left(\sqrt{xy^1\cdot y^1}\right) = \left(x^{1+1+3}\right)\cdot\left(y^{1+2+1}\right)\cdot\left(\sqrt{xy^{1+1}}\right)$$

$$= x^5\cdot y^4\cdot\sqrt{xy^2} = x^5\cdot y^4\cdot y^1\sqrt{x} = x^5\cdot y^{4+1}\sqrt{x} = \mathbf{x^5y^5\sqrt{x}}$$

Example 4.2-39

$$\sqrt{20a^3b}\cdot\sqrt{5a^2b}\cdot\sqrt{18a^5b^5} = \left[\sqrt{(4\cdot 5)a^{2+1}b}\right]\cdot\left(a\sqrt{5b}\right)\cdot\left[\sqrt{(9\cdot 2)a^{2+2+1}b^{2+2+1}}\right]$$

$$= \left[\left[\sqrt{\left(2^2\cdot 5\right)\cdot\left(a^2a^1\right)b}\right]\cdot\left(a\sqrt{5b}\right)\cdot\left[\sqrt{\left(3^2\cdot 2\right)\cdot\left(a^2a^2a^1\right)\cdot\left(b^2b^2b^1\right)}\right]\right] = \left(2a\sqrt{5ab}\right)\cdot\left(a\sqrt{5b}\right)\cdot\left(3a^1a^1b^1b^1\sqrt{2ab}\right)$$

$= \boxed{\left(2a\sqrt{5ab}\right)\cdot\left(a\sqrt{5b}\right)\cdot\left(3a^{1+1}b^{1+1}\sqrt{2ab}\right)} = \boxed{\left(2a\sqrt{5ab}\right)\cdot\left(a\sqrt{5b}\right)\cdot\left(3a^2b^2\sqrt{2ab}\right)}$

$= \boxed{(2\cdot 3)\cdot\left(a\cdot a\cdot a^2\right)\cdot b^2\cdot\left(\sqrt{5ab}\cdot\sqrt{5b}\cdot\sqrt{2ab}\right)} = \boxed{6\cdot\left(a^1\cdot a^1\cdot a^2\right)\cdot b^2\cdot\left(\sqrt{5ab\cdot 5b\cdot 2ab}\right)}$

$= \boxed{6\cdot\left(a^{1+1+2}\right)\cdot b^2\cdot\sqrt{(5\cdot 5\cdot 2)\cdot(aa)\cdot(bbb)}} = \boxed{6a^4b^2\sqrt{\left(5^2\cdot 2\right)\cdot a^2\cdot b^2b}} = \boxed{6a^4b^2\cdot(5\cdot ab)\sqrt{2b}}$

$= \boxed{(6\cdot 5)\cdot\left(a^4a\right)\cdot\left(b^2b\right)\cdot\sqrt{2b}} = \boxed{30\cdot\left(a^4a^1\right)\cdot\left(b^2b^1\right)\cdot\sqrt{2b}} = \boxed{30\cdot\left(a^{4+1}\right)\cdot\left(b^{2+1}\right)\cdot\sqrt{2b}} = \boxed{\mathbf{30a^5b^3\sqrt{2b}}}$

Example 4.2-40

$\boxed{\sqrt{40x^4y^5z^3}\cdot\sqrt{x^2y^3z}\cdot\sqrt{25xy}} = \boxed{\sqrt{(4\cdot 10)x^{2+2}y^{2+2+1}z^{2+1}}\cdot\sqrt{x^2y^{2+1}z}\cdot\sqrt{5^2xy}}$

$= \boxed{\left[\sqrt{\left(2^2\cdot 10\right)\cdot\left(x^2x^2\right)\cdot\left(y^2y^2y^1\right)\cdot\left(z^2z^1\right)}\right]\cdot\left(x\sqrt{y^2y^1z}\right)\cdot\left(5\sqrt{xy}\right)} = \boxed{\left[2\left(x^1x^1\right)\cdot\left(y^1y^1\right)z\sqrt{10yz}\right]\cdot\left(xy\sqrt{yz}\right)\cdot\left(5\sqrt{xy}\right)}$

$= \boxed{\left(2x^{1+1}y^{1+1}z\sqrt{10yz}\right)\cdot\left(xy\sqrt{yz}\right)\cdot\left(5\sqrt{xy}\right)} = \boxed{\left[(2\cdot 5)x^2y^2z\sqrt{10yz}\right]\cdot\left(xy\sqrt{yz}\right)\cdot\left(\sqrt{xy}\right)}$

$= \boxed{\left[10\left(x^2\cdot x\right)\cdot\left(y^2\cdot y\right)\cdot z\right]\cdot\sqrt{10yz\cdot yz\cdot xy}} = \boxed{\left[10\left(x^2\cdot x^1\right)\cdot\left(y^2\cdot y^1\right)\cdot z\right]\cdot\sqrt{10x\cdot(yyy)\cdot(zz)}}$

$= \boxed{\left[10x^{2+1}\cdot y^{2+1}\cdot z\right]\cdot\sqrt{10x\cdot\left(y^1y^1y^1\right)\cdot\left(z^1z^1\right)}} = \boxed{10x^3y^3z\cdot\sqrt{10x\cdot\left(y^{1+1+1}\right)\cdot\left(z^{1+1}\right)}} = \boxed{10x^3y^3z\cdot\sqrt{10xy^3z^2}}$

$= \boxed{10x^3y^3z\cdot yz\sqrt{10xy}} = \boxed{10x^3\left(y^3\cdot y\right)\cdot(z\cdot z)\cdot\sqrt{10xy}} = \boxed{10x^3\left(y^3\cdot y^1\right)\cdot\left(z^1\cdot z^1\right)\cdot\sqrt{10xy}}$

$= \boxed{10x^3\left(y^{3+1}\right)\cdot\left(z^{1+1}\right)\cdot\sqrt{10xy}} = \boxed{\mathbf{10x^3y^4z^2\sqrt{10xy}}}$

Practice Problems - Multiplying Monomial Expressions in Radical Form, with Variables

Section 4.2 Case I b Practice Problems - Multiply the following radical expressions:

1. $\sqrt{x^5y^6}\cdot\sqrt{x^2y^3} =$
2. $\sqrt{a^5b^5}\cdot\sqrt{a^2b^3}\cdot b^2\sqrt{a^4} =$
3. $\sqrt[5]{x^8y^4z^7}\cdot z^2\sqrt[5]{x^3y^2} =$
4. $\sqrt{x^5y^3z^2}\cdot\sqrt{x^3y^2z} =$
5. $\sqrt[3]{x^5y^6z}\cdot\sqrt[3]{xyz}\cdot\sqrt[3]{x^4yz^4} =$
6. $\sqrt[4]{u^5v^6}\cdot\sqrt[4]{uv^8}\cdot\sqrt[4]{u^2v^2} =$
7. $\sqrt{40r^3s}\cdot\sqrt{36r^2s}\cdot\sqrt{4r^5s^5} =$
8. $\sqrt[3]{125u^6v^8w}\cdot\sqrt[3]{54uv^2w^5} =$
9. $\sqrt[3]{m^4n^3l}\cdot\sqrt{m^2l}\cdot\sqrt[3]{m^4n^5l^4}\cdot\sqrt{mn^5l^3} =$
10. $\sqrt[5]{a^{10}b^4c^2}\cdot\sqrt[5]{abc^3}\cdot\sqrt[5]{a^2b^{10}c^5} =$

Case II Multiplying Binomial Expressions in Radical Form

To multiply two binomial radical expressions the following multiplication method known as the **FOIL** method need to be memorized:

$$(a+b)\cdot(c+d)=(a\cdot c)+(a\cdot d)+(b\cdot c)+(b\cdot d)$$

Multiply the **F**irst two terms, i.e., $(a\cdot c)$.

Multiply the **O**uter two terms, i.e., $(a\cdot d)$.

Multiply the **I**nner two terms, i.e., $(b\cdot c)$.

Multiply the **L**ast two terms, i.e., $(b\cdot d)$.

Examples:

1. $\boxed{\left(\sqrt{u}+\sqrt{v}\right)\cdot\left(\sqrt{u}-\sqrt{v}\right)}=\boxed{\left(\sqrt{u}\cdot\sqrt{u}\right)-\left(\sqrt{u}\sqrt{v}\right)+\left(\sqrt{v}\cdot\sqrt{u}\right)-\left(\sqrt{v}\cdot\sqrt{v}\right)}=\boxed{\left(\sqrt{u\cdot u}\right)-\left(\sqrt{u\cdot v}\right)+\left(\sqrt{u\cdot v}\right)-\left(\sqrt{v\cdot v}\right)}$

   $=\boxed{\sqrt{u^2}-\sqrt{v^2}}=\boxed{\mathbf{u-v}}$

2. $\boxed{\left(3-\sqrt{5}\right)\cdot\left(5+\sqrt{7}\right)}=\boxed{(3\cdot5)+\left(3\cdot\sqrt{7}\right)-\left(5\cdot\sqrt{5}\right)-\left(\sqrt{5}\cdot\sqrt{7}\right)}=\boxed{15+3\sqrt{7}-5\sqrt{5}-\sqrt{5\cdot7}}=\boxed{\mathbf{15+3\sqrt{7}-5\sqrt{5}-\sqrt{35}}}$

Multiplication of binomial expressions in radical form is divided into two cases. Case II a - multiplication of binomial expressions in radical form, with real numbers and Case II b - multiplication of binomial expressions in radical form, with variables.

Case II a Multiplying Binomial Expressions in Radical Form, with Real Numbers

Binomial radical expressions are multiplied by each other using the following steps:

Step 1 Simplify the radical terms (see Section 4.1, Case III).

Step 2 Use the FOIL method to multiply each term. Repeat Step 1, if necessary.

$$(a+b)\cdot(c+d)=(a\cdot c)+(a\cdot d)+(b\cdot c)+(b\cdot d)$$

Examples with Steps

The following examples show the steps as to how binomial radical expressions with real numbers as radicands are multiplied by one another:

Example 4.2-41

$$\boxed{\left(2+\sqrt{2}\right)\cdot\left(5-\sqrt{8}\right)}=$$

Solution:

Step 1 $\boxed{\left(2+\sqrt{2}\right)\cdot\left(5-\sqrt{8}\right)}=\boxed{\left(2+\sqrt{2}\right)\cdot\left(5-\sqrt{4\cdot2}\right)}=\boxed{\left(2+\sqrt{2}\right)\cdot\left(5-\sqrt{2^2\cdot2}\right)}$

$=\boxed{\left(2+\sqrt{2}\right)\cdot\left(5-2\sqrt{2}\right)}$

Step 2 $\boxed{\left(2+\sqrt{2}\right)\cdot\left(5-2\sqrt{2}\right)}=\boxed{(2\cdot5)-\left(2\cdot2\sqrt{2}\right)+\left(5\cdot\sqrt{2}\right)-\left(2\sqrt{2}\cdot\sqrt{2}\right)}$

$= \boxed{10 - 4\sqrt{2} + 5\sqrt{2} - 2\sqrt{2 \cdot 2}} = \boxed{10 - 4\sqrt{2} + 5\sqrt{2} - 2\sqrt{2^2}} = \boxed{10 + (-4+5)\sqrt{2} - (2 \cdot 2)}$

$= \boxed{10 + \sqrt{2} - 4} = \boxed{(10-4) + \sqrt{2}} = \boxed{\mathbf{6 + \sqrt{2}}}$

Example 4.2-42

$\boxed{\left(2\sqrt{5} - 3\right) \cdot \left(2 + \sqrt{3}\right)} =$

Solution:

Step 1 $\boxed{\textit{Not Applicable}}$ The binomial terms are in their simplified form.

Step 2 $\boxed{\left(2\sqrt{5} - 3\right) \cdot \left(2 + \sqrt{3}\right)} = \boxed{\left(2 \cdot 2\sqrt{5}\right) + \left(2\sqrt{5} \cdot \sqrt{3}\right) - (3 \cdot 2) - \left(3 \cdot \sqrt{3}\right)}$

$= \boxed{4\sqrt{5} + 2\sqrt{5 \cdot 3} - 6 - 3\sqrt{3}} = \boxed{\mathbf{4\sqrt{5} + 2\sqrt{15} - 3\sqrt{3} - 6}}$

Example 4.2-43

$\boxed{\left(2\sqrt[4]{162} + 3\right) \cdot \left(3\sqrt[4]{2} + 5\right)} =$

Solution:

Step 1 $\boxed{\left(2\sqrt[4]{162} + 3\right) \cdot \left(3\sqrt[4]{2} + 5\right)} = \boxed{\left(2\sqrt[4]{81 \cdot 2} + 3\right) \cdot \left(3\sqrt[4]{2} + 5\right)} = \boxed{\left(2\sqrt[4]{3^4 \cdot 2} + 3\right) \cdot \left(3\sqrt[4]{2} + 5\right)}$

$= \boxed{\left(2 \cdot 3\sqrt[4]{2} + 3\right) \cdot \left(3\sqrt[4]{2} + 5\right)} = \boxed{\left(6\sqrt[4]{2} + 3\right) \cdot \left(3\sqrt[4]{2} + 5\right)}$

Step 2 $\boxed{\left(6\sqrt[4]{2} + 3\right) \cdot \left(3\sqrt[4]{2} + 5\right)} = \boxed{\left(6\sqrt[4]{2} \cdot 3\sqrt[4]{2}\right) + \left(5 \cdot 6\sqrt[4]{2}\right) + \left(3 \cdot 3\sqrt[4]{2}\right) + (3 \cdot 5)}$

$= \boxed{(6 \cdot 3)\sqrt[4]{2 \cdot 2} + 30\sqrt[4]{2} + 9\sqrt[4]{2} + 15} = \boxed{18\sqrt[4]{4} + (30+9)\sqrt[4]{2} + 15} = \boxed{\mathbf{18\sqrt[4]{4} + 39\sqrt[4]{2} + 15}}$

Example 4.2-44

$\boxed{\left(\sqrt{24} + 3\sqrt{60}\right) \cdot \left(\sqrt{25} - \sqrt{72}\right)} =$

Solution:

Step 1 $\boxed{\left(\sqrt{24} + 3\sqrt{60}\right) \cdot \left(\sqrt{25} - \sqrt{72}\right)} = \boxed{\left(\sqrt{4 \cdot 6} + 3\sqrt{4 \cdot 15}\right) \cdot \left(\sqrt{5 \cdot 5} - \sqrt{36 \cdot 2}\right)}$

$= \boxed{\left(\sqrt{2^2 \cdot 6} + 3\sqrt{2^2 \cdot 15}\right) \cdot \left(\sqrt{5^2} - \sqrt{6^2 \cdot 2}\right)} = \boxed{\left(2\sqrt{6} + (3 \cdot 2)\sqrt{15}\right) \cdot \left(5 - 6\sqrt{2}\right)}$

$= \boxed{\left(2\sqrt{6} + 6\sqrt{15}\right) \cdot \left(5 - 6\sqrt{2}\right)}$

Step 2 $\boxed{\left(2\sqrt{6} + 6\sqrt{15}\right) \cdot \left(5 - 6\sqrt{2}\right)} = \boxed{\left(5 \cdot 2\sqrt{6}\right) - \left(2\sqrt{6} \cdot 6\sqrt{2}\right) + \left(5 \cdot 6\sqrt{15}\right) - \left(6\sqrt{15} \cdot 6\sqrt{2}\right)}$

$= \boxed{10\sqrt{6} - (2 \cdot 6)\sqrt{6 \cdot 2} + 30\sqrt{15} - (6 \cdot 6)\sqrt{15 \cdot 2}} = \boxed{10\sqrt{6} - 12\sqrt{12} + 30\sqrt{15} - 36\sqrt{30}}$

$$= \boxed{10\sqrt{6} - 12\sqrt{4 \cdot 3} + 30\sqrt{15} - 36\sqrt{30}} = \boxed{10\sqrt{6} - 12\sqrt{2^2 \cdot 3} + 30\sqrt{15} - 36\sqrt{30}}$$

$$= \boxed{10\sqrt{6} - (12 \cdot 2)\sqrt{3} + 30\sqrt{15} - 36\sqrt{30}} = \boxed{\mathbf{10\sqrt{6} - 24\sqrt{3} + 30\sqrt{15} - 36\sqrt{30}}}$$

Example 4.2-45

$$\boxed{\left(\sqrt{64} + \sqrt{5}\right) \cdot \left(\sqrt{25} - \sqrt{5}\right)} =$$

Solution:

Step 1

$$\boxed{\left(\sqrt{64} + \sqrt{5}\right) \cdot \left(\sqrt{25} - \sqrt{5}\right)} = \boxed{\left(\sqrt{8^2} + \sqrt{5}\right) \cdot \left(\sqrt{5^2} - \sqrt{5}\right)} = \boxed{\left(8 + \sqrt{5}\right) \cdot \left(5 - \sqrt{5}\right)}$$

Step 2

$$\boxed{\left(8 + \sqrt{5}\right) \cdot \left(5 - \sqrt{5}\right)} = \boxed{(8 \cdot 5) - \left(8 \cdot \sqrt{5}\right) + \left(5 \cdot \sqrt{5}\right) - \left(\sqrt{5} \cdot \sqrt{5}\right)} = \boxed{40 - 8\sqrt{5} + 5\sqrt{5} - \sqrt{5 \cdot 5}}$$

$$= \boxed{40 + (-8 + 5)\sqrt{5} - \sqrt{5^2}} = \boxed{40 - 3\sqrt{5} - 5} = \boxed{(40 - 5) - 3\sqrt{5}} = \boxed{\mathbf{35 - 3\sqrt{5}}}$$

Additional Examples - Multiplying Binomial Expressions in Radical Form, with Real Numbers

The following examples further illustrate how to multiply radical expressions by one another:

Example 4.2-46

$$\boxed{\left(3 + \sqrt{300}\right) \cdot \left(8 - \sqrt{50}\right)} = \boxed{\left(3 + \sqrt{100 \cdot 3}\right) \cdot \left(8 - \sqrt{25 \cdot 2}\right)} = \boxed{\left(3 + \sqrt{10^2 \cdot 3}\right) \cdot \left(8 - \sqrt{5^2 \cdot 2}\right)} = \boxed{\left(3 + 10\sqrt{3}\right) \cdot \left(8 - 5\sqrt{2}\right)}$$

$$= \boxed{(3 \cdot 8) - (3 \cdot 5)\sqrt{2} + (8 \cdot 10)\sqrt{3} - (10 \cdot 5)\sqrt{3} \cdot \sqrt{2}} = \boxed{24 - 15\sqrt{2} + 80\sqrt{3} - 50\sqrt{3 \cdot 2}} = \boxed{\mathbf{-15\sqrt{2} + 80\sqrt{3} - 50\sqrt{6} + 24}}$$

Example 4.2-47

$$\boxed{\left(\sqrt{98} - 3\sqrt{3}\right) \cdot \left(\sqrt{8} - \sqrt{108}\right)} = \boxed{\left(\sqrt{49 \cdot 2} - 3\sqrt{3}\right) \cdot \left(\sqrt{4 \cdot 2} - \sqrt{36 \cdot 3}\right)} = \boxed{\left(\sqrt{7^2 \cdot 2} - 3\sqrt{3}\right) \cdot \left(\sqrt{2^2 \cdot 2} - \sqrt{6^2 \cdot 3}\right)}$$

$$= \boxed{\left(7\sqrt{2} - 3\sqrt{3}\right) \cdot \left(2\sqrt{2} - 6\sqrt{3}\right)} = \boxed{(7 \cdot 2)\left(\sqrt{2} \cdot \sqrt{2}\right) - (7 \cdot 6)\left(\sqrt{2} \cdot \sqrt{3}\right) - (3 \cdot 2)\left(\sqrt{3} \cdot \sqrt{2}\right) + (3 \cdot 6)\left(\sqrt{3} \cdot \sqrt{3}\right)}$$

$$= \boxed{14\sqrt{2 \cdot 2} - 42\sqrt{2 \cdot 3} - 6\sqrt{3 \cdot 2} + 18\sqrt{3 \cdot 3}} = \boxed{14\sqrt{2^2} - 42\sqrt{6} - 6\sqrt{6} + 18\sqrt{3^2}} = \boxed{(14 \cdot 2) - (42 + 6)\sqrt{6} + (18 \cdot 3)}$$

$$= \boxed{28 - 48\sqrt{6} + 54} = \boxed{(28 + 54) - 48\sqrt{6}} = \boxed{\mathbf{82 - 48\sqrt{6}}}$$

Example 4.2-48

$$\boxed{\left(3 + \sqrt{12}\right) \cdot \left(\sqrt{75} - \sqrt{2}\right)} = \boxed{\left(3 + \sqrt{4 \cdot 3}\right) \cdot \left(\sqrt{25 \cdot 3} - \sqrt{2}\right)} = \boxed{\left(3 + \sqrt{2^2 \cdot 3}\right) \cdot \left(\sqrt{5^2 \cdot 3} - \sqrt{2}\right)}$$

$$= \boxed{\left(3 + 2\sqrt{3}\right) \cdot \left(5\sqrt{3} - \sqrt{2}\right)} = \boxed{(3 \cdot 5)\sqrt{3} - \left(3 \cdot \sqrt{2}\right) + (2 \cdot 5)\left(\sqrt{3} \cdot \sqrt{3}\right) - 2\left(\sqrt{3} \cdot \sqrt{2}\right)} = \boxed{15\sqrt{3} - 3\sqrt{2} + 10\sqrt{3 \cdot 3} - 2\sqrt{3 \cdot 2}}$$

$$= \boxed{15\sqrt{3} - 3\sqrt{2} + 10\sqrt{3^2} - 2\sqrt{6}} = \boxed{15\sqrt{3} - 3\sqrt{2} + (10 \cdot 3) - 2\sqrt{6}} = \boxed{\mathbf{15\sqrt{3} - 3\sqrt{2} - 2\sqrt{6} + 30}}$$

Example 4.2-49

$$\left(\sqrt[3]{3\cdot5^3}-\sqrt[3]{4\cdot2^3}\right)\cdot\left(\sqrt[3]{3}-\sqrt[3]{2\cdot2^3}\right)=\left(5\sqrt[3]{3}-2\sqrt[3]{4}\right)\cdot\left(\sqrt[3]{3}-2\sqrt[3]{2}\right)$$

$$=\left(5\sqrt[3]{3}\cdot\sqrt[3]{3}\right)-(5\cdot2)\left(\sqrt[3]{3}\cdot\sqrt[3]{2}\right)-\left(2\sqrt[3]{4}\cdot\sqrt[3]{3}\right)+(2\cdot2)\left(\sqrt[3]{4}\cdot\sqrt[3]{2}\right)=\left(5\sqrt[3]{3\cdot3}\right)-10\left(\sqrt[3]{3\cdot2}\right)-\left(2\sqrt[3]{4\cdot3}\right)+4\left(\sqrt[3]{4\cdot2}\right)$$

$$=5\sqrt[3]{9}-10\sqrt[3]{6}-2\sqrt[3]{12}+4\sqrt[3]{8}=5\sqrt[3]{9}-10\sqrt[3]{6}-2\sqrt[3]{12}+4\sqrt[3]{2^3}=5\sqrt[3]{9}-10\sqrt[3]{6}-2\sqrt[3]{12}+(4\cdot2)$$

$$=\mathbf{5\sqrt[3]{9}-10\sqrt[3]{6}-2\sqrt[3]{12}+8}$$

Example 4.2-50

$$\left(6\sqrt{48}+2\right)\cdot\left(2\sqrt{18}-4\right)=\left(6\sqrt{3\cdot16}+2\right)\cdot\left(2\sqrt{2\cdot9}-4\right)=\left(6\sqrt{3\cdot4^2}+2\right)\cdot\left(2\sqrt{2\cdot3^2}-4\right)$$

$$=\left(6\cdot4\sqrt{3}+2\right)\cdot\left(2\cdot3\sqrt{2}-4\right)=\left(24\sqrt{3}+2\right)\cdot\left(6\sqrt{2}-4\right)=\left(24\sqrt{3}\cdot6\sqrt{2}\right)-\left(4\cdot24\sqrt{3}\right)+\left(2\cdot6\sqrt{2}\right)-(2\cdot4)$$

$$=\left(24\cdot6\sqrt{3\cdot2}\right)-\left(96\sqrt{3}\right)+\left(12\sqrt{2}\right)-8=\mathbf{144\sqrt{6}-96\sqrt{3}+12\sqrt{2}-8}$$

Example 4.2-51

$$3\sqrt{2}\cdot\left(\sqrt{10}+4\sqrt{20}\right)=\left(3\sqrt{2}\cdot\sqrt{10}\right)+\left(3\sqrt{2}\cdot4\sqrt{20}\right)=\left(3\sqrt{2\cdot10}\right)+\left(3\cdot4\sqrt{2\cdot20}\right)=\left(3\sqrt{20}\right)+\left(12\sqrt{40}\right)$$

$$=3\sqrt{4\cdot5}+12\sqrt{4\cdot10}=3\sqrt{2^2\cdot5}+12\sqrt{2^2\cdot10}=3\cdot2\sqrt{5}+12\cdot2\sqrt{10}=\mathbf{6\sqrt{5}+24\sqrt{10}}$$

Example 4.2-52

$$\left[\left(-\sqrt{3}+2\right)\cdot\left(3-\sqrt{3}\right)\right]\cdot\left(\sqrt{3}-4\right)=\left[\left(-3\cdot\sqrt{3}\right)+\left(\sqrt{3}\cdot\sqrt{3}\right)+(2\cdot3)-\left(2\cdot\sqrt{3}\right)\right]\cdot\left(\sqrt{3}-4\right)$$

$$=\left[-3\sqrt{3}+\sqrt{3\cdot3}+6-2\sqrt{3}\right]\cdot\left(\sqrt{3}-4\right)=\left[\left(-3\sqrt{3}-2\sqrt{3}\right)+\sqrt{3^2}+6\right]\cdot\left(\sqrt{3}-4\right)=\left[-5\sqrt{3}+3+6\right]\cdot\left(\sqrt{3}-4\right)$$

$$=\left[-5\sqrt{3}+9\right]\cdot\left(\sqrt{3}-4\right)=\left(-5\sqrt{3}\cdot\sqrt{3}\right)+(5\cdot4)\sqrt{3}+9\cdot\sqrt{3}-(9\cdot4)=-5\sqrt{3\cdot3}+20\sqrt{3}+9\sqrt{3}-36$$

$$=-5\sqrt{3^2}+(20+9)\sqrt{3}-36=-(5\cdot3)+29\sqrt{3}-36=-15-36+29\sqrt{3}=\mathbf{-51+29\sqrt{3}}$$

Example 4.2-53

$$\left(\sqrt[5]{243}+\sqrt{2}\right)\cdot\left(\sqrt{5}-\sqrt{72}\right)=\left(\sqrt[5]{3^5}+\sqrt{2}\right)\cdot\left(\sqrt{5}-\sqrt{36\cdot2}\right)=\left(3+\sqrt{2}\right)\cdot\left(\sqrt{5}-\sqrt{6^2\cdot2}\right)=\left(3+\sqrt{2}\right)\cdot\left(\sqrt{5}-6\sqrt{2}\right)$$

$$=3\sqrt{5}-3\cdot6\sqrt{2}+\sqrt{2}\cdot\sqrt{5}-6\sqrt{2}\cdot\sqrt{2}=3\sqrt{5}-18\sqrt{2}+\sqrt{2\cdot5}-6\sqrt{2\cdot2}=3\sqrt{5}-18\sqrt{2}+\sqrt{10}-6\sqrt{2^2}$$

$$=3\sqrt{5}-18\sqrt{2}+\sqrt{10}-6\cdot2=\mathbf{3\sqrt{5}-18\sqrt{2}+\sqrt{10}-12}$$

Example 4.2-54

$$\left(\sqrt[3]{216}+\sqrt{3}\right)\cdot\left(\sqrt{2}+2\right)=\left(\sqrt[3]{6^3}+\sqrt{3}\right)\cdot\left(\sqrt{2}+2\right)=\left(6+\sqrt{3}\right)\cdot\left(\sqrt{2}+2\right)=\left(6\cdot\sqrt{2}\right)+\left(6\cdot 2\right)+\left(\sqrt{3}\cdot\sqrt{2}\right)+\left(2\cdot\sqrt{3}\right)$$

$$=6\sqrt{2}+12+\sqrt{3\cdot 2}+2\sqrt{3}=6\sqrt{2}+12+\sqrt{6}+2\sqrt{3}=\mathbf{6\sqrt{2}+\sqrt{6}+2\sqrt{3}+12}$$

Example 4.2-55

$$\left(\sqrt[5]{243}\cdot\sqrt{5}+2\right)\cdot\left(\sqrt[3]{64}\cdot\sqrt{5}-\sqrt{2}\right)=\left(\sqrt[5]{3^5}\cdot\sqrt{5}+2\right)\cdot\left(\sqrt[3]{4^3}\cdot\sqrt{5}-\sqrt{2}\right)=\left(3\sqrt{5}+2\right)\cdot\left(4\sqrt{5}-\sqrt{2}\right)$$

$$=\left(3\cdot 4\right)\cdot\left(\sqrt{5}\cdot\sqrt{5}\right)-\left(3\sqrt{5}\cdot\sqrt{2}\right)+\left(2\cdot 4\sqrt{5}\right)-\left(2\cdot\sqrt{2}\right)=12\sqrt{5\cdot 5}-3\sqrt{5\cdot 2}+8\sqrt{5}-2\sqrt{2}$$

$$=12\sqrt{5^2}-3\sqrt{10}+8\sqrt{5}-2\sqrt{2}=\left(12\cdot 5\right)-3\sqrt{10}+8\sqrt{5}-2\sqrt{2}=\mathbf{60-3\sqrt{10}+8\sqrt{5}-2\sqrt{2}}$$

Example 4.2-56

$$\left(4+\sqrt{2}\right)\cdot\left[\left(\sqrt{2}-3\right)\cdot\left(-5-\sqrt{2}\right)\right]=\left(4+\sqrt{2}\right)\cdot\left[\left(-5\cdot\sqrt{2}\right)-\left(\sqrt{2}\cdot\sqrt{2}\right)+\left(3\cdot 5\right)+\left(3\cdot\sqrt{2}\right)\right]$$

$$=\left(4+\sqrt{2}\right)\cdot\left[-5\sqrt{2}-\sqrt{2\cdot 2}+15+3\sqrt{2}\right]=\left(4+\sqrt{2}\right)\cdot\left[-5\sqrt{2}+3\sqrt{2}-\sqrt{2^2}+15\right]$$

$$=\left(4+\sqrt{2}\right)\cdot\left[\left(-5+3\right)\sqrt{2}-2+15\right]=\left(4+\sqrt{2}\right)\cdot\left[-2\sqrt{2}+13\right]=-\left(4\cdot 2\sqrt{2}\right)+\left(4\cdot 13\right)-\left(2\sqrt{2}\cdot\sqrt{2}\right)+\left(13\cdot\sqrt{2}\right)$$

$$=-8\sqrt{2}+52-2\sqrt{2\cdot 2}+13\sqrt{2}=-8\sqrt{2}+13\sqrt{2}+52-2\sqrt{2^2}=\left(-8+13\right)\sqrt{2}+52-2\cdot 2=5\sqrt{2}+52-4$$

$$=\mathbf{5\sqrt{2}+48}$$

Example 4.2-57

$$\left(\sqrt[3]{16}-1\right)\cdot\left(5-\sqrt[3]{54}\right)=\left(\sqrt[3]{8\cdot 2}-1\right)\cdot\left(5-\sqrt[3]{27\cdot 2}\right)=\left(\sqrt[3]{2^3\cdot 2}-1\right)\cdot\left(5-\sqrt[3]{3^3\cdot 2}\right)=\left(2\sqrt[3]{2}-1\right)\cdot\left(5-3\sqrt[3]{2}\right)$$

$$=\left(2\cdot 5\right)\sqrt[3]{2}-\left(2\cdot 3\right)\cdot\left(\sqrt[3]{2}\cdot\sqrt[3]{2}\right)-\left(1\cdot 5\right)+\left(1\cdot 3\right)\sqrt[3]{2}=10\sqrt[3]{2}-6\sqrt[3]{2\cdot 2}-5+3\sqrt[3]{2}=10\sqrt[3]{2}+3\sqrt[3]{2}-6\sqrt[3]{2^2}-5$$

$$=\left(10+3\right)\sqrt[3]{2}-6\sqrt[3]{4}-5=\mathbf{13\sqrt[3]{2}-6\sqrt[3]{4}-5}$$

Example 4.2-58

$$\left(\sqrt{9}-2\sqrt{125}\right)\cdot\left(3\sqrt{18}-4\sqrt{90}\right)=\left(\sqrt{3^2}-2\sqrt{25\cdot 5}\right)\cdot\left(3\sqrt{9\cdot 2}-4\sqrt{9\cdot 10}\right)=\left(3-2\sqrt{5^2\cdot 5}\right)\cdot\left(3\sqrt{3^2\cdot 2}-4\sqrt{3^2\cdot 10}\right)$$

$$=\left(3-\left(2\cdot 5\right)\sqrt{5}\right)\cdot\left(\left(3\cdot 3\right)\sqrt{2}-\left(4\cdot 3\right)\sqrt{10}\right)=\left(3-10\sqrt{5}\right)\cdot\left(9\sqrt{2}-12\sqrt{10}\right)$$

$$=\left(3\cdot 9\right)\sqrt{2}-\left(3\cdot 12\right)\sqrt{10}-\left(10\cdot 9\right)\cdot\left(\sqrt{5}\cdot\sqrt{2}\right)+\left(10\cdot 12\right)\cdot\left(\sqrt{5}\sqrt{10}\right)=27\sqrt{2}-36\sqrt{10}-90\sqrt{5\cdot 2}+120\sqrt{5\cdot 10}$$

$= \boxed{27\sqrt{2} - 36\sqrt{10} - 90\sqrt{10} + 120\sqrt{50}} = \boxed{27\sqrt{2} + (-36 - 90)\sqrt{10} + 120\sqrt{50}} = \boxed{27\sqrt{2} - 126\sqrt{10} + 120\sqrt{25 \cdot 2}}$

$= \boxed{27\sqrt{2} - 126\sqrt{10} + 120\sqrt{5^2 \cdot 2}} = \boxed{27\sqrt{2} - 126\sqrt{10} + (120 \cdot 5)\sqrt{2}} = \boxed{27\sqrt{2} - 126\sqrt{10} + 600\sqrt{2}}$

$= \boxed{\left(27\sqrt{2} + 600\sqrt{2}\right) - 126\sqrt{10}} = \boxed{(27 + 600)\sqrt{2} - 126\sqrt{10}} = \boxed{\mathbf{627\sqrt{2} - 126\sqrt{10}}}$

Example 4.2-59

$\boxed{\left(\sqrt{225} - \sqrt{2}\right) \cdot \left(\sqrt{200} + \sqrt{8}\right)} = \boxed{\left(\sqrt{15 \cdot 15} - \sqrt{2}\right) \cdot \left(\sqrt{100 \cdot 2} + \sqrt{4 \cdot 2}\right)} = \boxed{\left(\sqrt{15^2} - \sqrt{2}\right) \cdot \left(\sqrt{10^2 \cdot 2} + \sqrt{2^2 \cdot 2}\right)}$

$= \boxed{\left(15 - \sqrt{2}\right) \cdot \left(10\sqrt{2} + 2\sqrt{2}\right)} = \boxed{\left(15 - \sqrt{2}\right) \cdot (10 + 2)\sqrt{2}} = \boxed{\left(15 - \sqrt{2}\right) \cdot \left(12\sqrt{2}\right)} = \boxed{(15 \cdot 12)\sqrt{2} - 12\left(\sqrt{2} \cdot \sqrt{2}\right)}$

$= \boxed{180\sqrt{2} - 12\sqrt{2 \cdot 2}} = \boxed{180\sqrt{2} - 12\sqrt{2^2}} = \boxed{180\sqrt{2} - 12 \cdot 2} = \boxed{\mathbf{180\sqrt{2} - 24}}$

Example 4.2-60

$\boxed{\left(\sqrt[4]{16} - \sqrt[4]{2^5}\right) \cdot \left(1 - \sqrt[4]{2 \cdot 3^5}\right)} = \boxed{\left(\sqrt[4]{2^4} - \sqrt[4]{2^{4+1}}\right) \cdot \left(1 - \sqrt[4]{2 \cdot 3^{4+1}}\right)} = \boxed{\left(2 - \sqrt[4]{2^4 \cdot 2^1}\right) \cdot \left(1 - \sqrt[4]{2 \cdot 3^4 \cdot 3^1}\right)}$

$= \boxed{\left(2 - 2\sqrt[4]{2}\right) \cdot \left(1 - 3\sqrt[4]{2 \cdot 3}\right)} = \boxed{\left(2 - 2\sqrt[4]{2}\right) \cdot \left(1 - 3\sqrt[4]{6}\right)} = \boxed{(1 \cdot 2) - (2 \cdot 3)\sqrt[4]{6} - (1 \cdot 2)\sqrt[4]{2} + (2 \cdot 3)\left(\sqrt[4]{2} \cdot \sqrt[4]{6}\right)}$

$= \boxed{2 - 6\sqrt[4]{6} - 2\sqrt[4]{2} + 6\sqrt[4]{2 \cdot 6}} = \boxed{\mathbf{2 - 6\sqrt[4]{6} - 2\sqrt[4]{2} + 6\sqrt[4]{12}}}$

Practice Problems - Multiplying Binomial Expressions in Radical Form, with Real Numbers

Section 4.2 Case II a Practice Problems - Multiply the following radical expressions:

1. $\left(2\sqrt{3} + 1\right) \cdot \left(2 + \sqrt{2}\right) =$
2. $\left(1 + \sqrt{5}\right) \cdot \left(\sqrt{8} + \sqrt{5}\right) =$
3. $\left(2 - \sqrt{2}\right) \cdot \left(3 + \sqrt{2}\right) =$
4. $\left(5 + \sqrt{5}\right) \cdot \left(5 - \sqrt{5^3}\right) =$
5. $\left(2 + \sqrt{6}\right) \cdot \left(\sqrt[4]{16} - \sqrt{18}\right) =$
6. $\left(2 - \sqrt{5}\right) \cdot \left(\sqrt{45} + \sqrt[4]{81}\right) =$
7. $\left(2 - \sqrt{m}\right) \cdot \left(\sqrt{8} + \sqrt{m^3}\right) =$
8. $\left(\sqrt{32} - \sqrt{3}\right) \cdot \left(2 + \sqrt{3}\right) =$
9. $\left(a\sqrt{x} + \sqrt{x^2}\right) \cdot \left(\sqrt{a} - \sqrt{x}\right) =$
10. $\left(\sqrt{2} + \sqrt{3}\right) \cdot \left(\sqrt{32} - \sqrt{75}\right) =$

Case II b Multiplying Binomial Expressions in Radical Form, with Variables

Binomial radical expressions with variables as radicands are multiplied by each other using the following steps:

Step 1 Simplify the radical terms (see Section 4.1, Case IV).

Step 2 Use the FOIL method to multiply each term. Repeat Step 1, if necessary.

$$(a+b)\cdot(c+d)=(a\cdot c)+(a\cdot d)+(b\cdot c)+(b\cdot d)$$

Examples with Steps

The following examples show the steps as to how binomial radical expressions are multiplied by one another:

Example 4.2-61

$$\boxed{\left(x+\sqrt{x}\right)\cdot\left(x-x\sqrt{x}\right)}=$$

Solution:

Step 1 *Not Applicable*

Step 2

$$\boxed{\left(x+\sqrt{x}\right)\cdot\left(x-x\sqrt{x}\right)}=\boxed{(x\cdot x)-(x\cdot x)\sqrt{x}+x\sqrt{x}-x\left(\sqrt{x}\cdot\sqrt{x}\right)}$$

$$=\boxed{x^2-x^2\sqrt{x}+x\sqrt{x}-x\sqrt{x\cdot x}}=\boxed{x^2-x^2\sqrt{x}+x\sqrt{x}-x\sqrt{x^2}}$$

$$=\boxed{\left(x\sqrt{x}-x^2\sqrt{x}\right)+x^2-x\cdot x}=\boxed{x\sqrt{x}(1-x)+x^2-x^2}=\boxed{\mathbf{x\sqrt{x}(1-x)}}$$

Example 4.2-62

$$\boxed{\left(\sqrt{u^5v^6}-uv\right)\cdot\left(\sqrt{u^3v^2}+uv\right)}=$$

Solution:

Step 1

$$\boxed{\left(\sqrt{u^5v^6}-uv\right)\cdot\left(\sqrt{u^3v^2}+uv\right)}=\boxed{\left(\sqrt{u^{2+2+1}v^{2+2+2}}-uv\right)\cdot\left(v\sqrt{u^{2+1}}+uv\right)}$$

$$=\boxed{\left(\sqrt{\left(u^2\cdot u^2\cdot u^1\right)\cdot\left(v^2\cdot v^2\cdot v^2\right)}-uv\right)\cdot\left(v\sqrt{u^2\cdot u^1}+uv\right)}$$

$$=\boxed{\left(\left(u^1\cdot u^1\right)\cdot\left(v^1\cdot v^1\cdot v^1\right)\sqrt{u}-uv\right)\cdot\left(uv\sqrt{u}+uv\right)}=\boxed{\left(\left(u^{1+1}\right)\cdot\left(v^{1+1+1}\right)\sqrt{u}-uv\right)\cdot\left(uv\sqrt{u}+uv\right)}$$

$$=\boxed{\left(u^2v^3\sqrt{u}-uv\right)\cdot\left(uv\sqrt{u}+uv\right)}$$

Step 2

$$\boxed{\left(u^2v^3\sqrt{u}-uv\right)\cdot\left(uv\sqrt{u}+uv\right)}$$

$$= \boxed{\left(u^2v^3 \cdot uv\right)\cdot\left(\sqrt{u}\cdot\sqrt{u}\right)+\left(u^2v^3 \cdot uv\right)\sqrt{u}-\left(uv \cdot uv\right)\sqrt{u}-\left(uv \cdot uv\right)}$$

$$= \boxed{\left(u^2 \cdot u^1\right)\cdot\left(v^3 \cdot v^1\right)\sqrt{u \cdot u}+\left(u^2 \cdot u^1\right)\cdot\left(v^3 \cdot v^1\right)\sqrt{u}-(u \cdot u)(v \cdot v)\sqrt{u}-(u \cdot u)\cdot(v \cdot v)}$$

$$= \boxed{\left(u^{2+1}\right)\cdot\left(v^{3+1}\right)\sqrt{u^2}+\left(u^{2+1}\right)\cdot\left(v^{3+1}\right)\sqrt{u}-u^2v^2\sqrt{u}-u^2v^2}$$

$$= \boxed{u^3v^4 \cdot u+u^3v^4\sqrt{u}-u^2v^2\sqrt{u}-u^2v^2} = \boxed{\left(u^3 \cdot u^1\right)v^4-u^2v^2+u^3v^4\sqrt{u}-u^2v^2\sqrt{u}}$$

$$= \boxed{u^{3+1}v^4-u^2v^2+u^2v^2\sqrt{u}\left(uv^2-1\right)} = \boxed{u^4v^4-u^2v^2+u^2v^2\sqrt{u}\left(uv^2-1\right)}$$

$$= \boxed{\mathbf{u^2v^2\left(u^2v^2-1\right)+u^2v^2\sqrt{u}\left(uv^2-1\right)}}$$

Example 4.2-63

$$\boxed{\left(x-\sqrt[3]{x^5y^4}\right)\cdot\left(x+\sqrt[3]{xy^3}\right)} =$$

Solution:

Step 1

$$\boxed{\left(x-\sqrt[3]{x^5y^4}\right)\cdot\left(x+\sqrt[3]{xy^3}\right)} = \boxed{\left(x-\sqrt[3]{x^{3+2}y^{3+1}}\right)\cdot\left(x+y\sqrt[3]{x}\right)}$$

$$= \boxed{\left(x-\sqrt[3]{x^3 \cdot x^2y^3 \cdot y^1}\right)\cdot\left(x+y\sqrt[3]{x}\right)} = \boxed{\left(x-xy\sqrt[3]{x^2y}\right)\cdot\left(x+y\sqrt[3]{x}\right)}$$

Step 2

$$\boxed{\left(x-xy\sqrt[3]{x^2y}\right)\cdot\left(x+y\sqrt[3]{x}\right)} = \boxed{(x \cdot x)+\left(x \cdot y\sqrt[3]{x}\right)-\left(x \cdot xy\sqrt[3]{x^2y}\right)-(xy \cdot y)\left(\sqrt[3]{x^2y}\cdot\sqrt[3]{x}\right)}$$

$$= \boxed{x^2+xy\sqrt[3]{x}-x^2y\sqrt[3]{x^2y}-xy^2\sqrt[3]{\left(x^2 \cdot x^1\right)y}} = \boxed{x^2+xy\sqrt[3]{x}-x^2y\sqrt[3]{x^2y}-xy^2\sqrt[3]{\left(x^{2+1}\right)y}}$$

$$= \boxed{x^2+xy\sqrt[3]{x}-x^2y\sqrt[3]{x^2y}-xy^2\sqrt[3]{x^3y}} = \boxed{x^2+xy\sqrt[3]{x}-x^2y\sqrt[3]{x^2y}-(x \cdot x)y^2\sqrt[3]{y}}$$

$$= \boxed{\mathbf{x^2+xy\sqrt[3]{x}-x^2y\sqrt[3]{x^2y}-x^2y^2\sqrt[3]{y}}}$$

Example 4.2-64

$$\boxed{\left(a+2\sqrt{b}\right)\cdot\left(a-5\sqrt{b}\right)} =$$

Solution:

Step 1 *Not Applicable*

Step 2

$$\boxed{\left(a+2\sqrt{b}\right)\cdot\left(a-5\sqrt{b}\right)} = \boxed{(a \cdot a)-\left(a \cdot 5\sqrt{b}\right)+\left(a \cdot 2\sqrt{b}\right)-\left(2\sqrt{b}\cdot 5\sqrt{b}\right)}$$

$$= \boxed{a^2 - 5a\sqrt{b} + 2a\sqrt{b} - (2\cdot 5)\sqrt{b\cdot b}} = \boxed{a^2 + (-5a + 2a)\sqrt{b} - 10\sqrt{b^2}}$$

$$= \boxed{\mathbf{a^2 - 3a\sqrt{b} - 10b}}$$

Example 4.2-65

$$\boxed{\left(a^2b^2 + \sqrt{ab}\right)\cdot\left(a\sqrt{b^3} - \sqrt{a^5b^4}\right)} =$$

Solution:

Step 1

$$\boxed{\left(a^2b^2 + \sqrt{ab}\right)\cdot\left(a\sqrt{b^3} - \sqrt{a^5b^4}\right)} = \boxed{\left(a^2b^2 + \sqrt{ab}\right)\cdot\left(a\sqrt{b^{2+1}} - \sqrt{a^{2+2+1}b^{2+2}}\right)}$$

$$= \boxed{\left(a^2b^2 + \sqrt{ab}\right)\cdot\left(a\sqrt{b^2b^1} - \sqrt{a^2a^2a^1b^2b^2}\right)} = \boxed{\left(a^2b^2 + \sqrt{ab}\right)\cdot\left(ab\sqrt{b} - a^1a^1b^1b^1\sqrt{a}\right)}$$

$$= \boxed{\left(a^2b^2 + \sqrt{ab}\right)\cdot\left(ab\sqrt{b} - a^{1+1}b^{1+1}\sqrt{a}\right)} = \boxed{\left(a^2b^2 + \sqrt{ab}\right)\cdot\left(ab\sqrt{b} - a^2b^2\sqrt{a}\right)}$$

Step 2

$$\boxed{\left(a^2b^2 + \sqrt{ab}\right)\cdot\left(ab\sqrt{b} - a^2b^2\sqrt{a}\right)}$$

$$= \boxed{\left(a^2b^2\cdot ab\sqrt{b}\right) - \left(a^2b^2\cdot a^2b^2\sqrt{a}\right) + \left(ab\sqrt{ab}\cdot\sqrt{b}\right) - \left(\sqrt{ab}\cdot a^2b^2\sqrt{a}\right)}$$

$$= \boxed{\left(a^2\cdot a^1\right)\cdot\left(b^2\cdot b^1\right)\cdot\sqrt{b} - \left(a^2\cdot a^2\right)\cdot\left(b^2\cdot b^2\right)\cdot\sqrt{a} + \left(ab\sqrt{ab\cdot b}\right) - \left(a^2b^2\sqrt{a\cdot ab}\right)}$$

$$= \boxed{a^{2+1}\cdot b^{2+1}\cdot\sqrt{b} - a^{2+2}\cdot b^{2+2}\cdot\sqrt{a} + ab\sqrt{ab^2} - a^2b^2\sqrt{a^2b}}$$

$$= \boxed{a^3b^3\sqrt{b} - a^4b^4\sqrt{a} + ab\cdot b\sqrt{a} - a^2\cdot ab^2\sqrt{b}} = \boxed{a^3b^3\sqrt{b} - a^4b^4\sqrt{a} + ab^2\sqrt{a} - a^3b^2\sqrt{b}}$$

$$= \boxed{\left(a^3b^3\sqrt{b} - a^3b^2\sqrt{b}\right) + \left(ab^2\sqrt{a} - a^4b^4\sqrt{a}\right)} = \boxed{\mathbf{a^3b^2\sqrt{b}(b-1) + ab^2\sqrt{a}\left(1 - a^3b^2\right)}}$$

Additional Examples - Multiplying Binomial Expressions in Radical Form, with Variables

Note: To minimize the number of steps, details such as:

$\boxed{abc\cdot c = abc^1\cdot c^1 = abc^{1+1} = abc^2}$, $\boxed{\sqrt{abc^3} = \sqrt{abc^{2+1}} = \sqrt{abc^2\cdot c^1} = c\sqrt{abc}}$, or $\boxed{x\cdot x^2 = x^1\cdot x^2 = x^{1+2} = x^3}$

are not shown. By now, it should be clear to the student that:

$\boxed{abc\cdot c = abc^2}$, $\boxed{\sqrt{abc^3} = c\sqrt{abc}}$, $\boxed{x\cdot x^2 = x^3}$, $\boxed{\sqrt{x^3} = x\sqrt{x}}$, $\boxed{\sqrt{a^3b^3} = ab\sqrt{ab}}$, etc.

The following examples further illustrate how to multiply binomial expressions in radical form:

Example 4.2-66

$$\boxed{\left(5 + \sqrt{x^3}\right)\cdot\left(5 - \sqrt{x^3}\right)} = \boxed{\left(5 + x\sqrt{x}\right)\cdot\left(5 - x\sqrt{x}\right)} = \boxed{(5\cdot 5) - \left(5\cdot x\sqrt{x}\right) + \left(5\cdot x\sqrt{x}\right) - (x\cdot x)\cdot\left(\sqrt{x}\cdot\sqrt{x}\right)}$$

$$= \boxed{25 - 5x\sqrt{x} + 5x\sqrt{x} - x^2\sqrt{x \cdot x}} = \boxed{25 - x^2\sqrt{x^2}} = \boxed{25 - x^2 \cdot x} = \boxed{\mathbf{25 - x^3}}$$

Example 4.2-67

$$\boxed{\left(a + b\sqrt{w}\right) \cdot \left(a - b\sqrt{w}\right)} = \boxed{(a \cdot a) - (a \cdot b)\sqrt{w} + (a \cdot b)\sqrt{w} - (b \cdot b) \cdot \left(\sqrt{w} \cdot \sqrt{w}\right)} = \boxed{a^2 - ab\sqrt{w} + ab\sqrt{w} - b^2\sqrt{w \cdot w}}$$

$$= \boxed{a^2 - b^2\sqrt{w^2}} = \boxed{\mathbf{a^2 - b^2 w}}$$

Example 4.2-68

$$\boxed{\left(r + \sqrt{s^3}\right) \cdot \left(r^2 - \sqrt{s}\right)} = \boxed{\left(r + s\sqrt{s}\right) \cdot \left(r^2 - \sqrt{s}\right)} = \boxed{\left(r \cdot r^2\right) - \left(r \cdot \sqrt{s}\right) + \left(r^2 \cdot s\sqrt{s}\right) - \left(s\sqrt{s} \cdot \sqrt{s}\right)}$$

$$= \boxed{r^3 - r\sqrt{s} + r^2 s\sqrt{s} - s\sqrt{s \cdot s}} = \boxed{r^3 - r\sqrt{s} + r^2 s\sqrt{s} - s\sqrt{s^2}} = \boxed{r^3 - r\sqrt{s} + r^2 s\sqrt{s} - s \cdot s} = \boxed{r^3 - r\sqrt{s} + r^2 s\sqrt{s} - s^2}$$

$$= \boxed{\left(r^3 - s^2\right) + \left(r^2 s\sqrt{s} - r\sqrt{s}\right)} = \boxed{\mathbf{\left(r^3 - s^2\right) + r\sqrt{s}(rs - 1)}}$$

Example 4.2-69

$$\boxed{\left(\sqrt{x} + \sqrt{y}\right) \cdot \left(\sqrt{x^3} - \sqrt{y^3}\right)} = \boxed{\left(\sqrt{x} + \sqrt{y}\right) \cdot \left(x\sqrt{x} - y\sqrt{y}\right)} = \boxed{\left(x\sqrt{x} \cdot \sqrt{x}\right) - \left(y\sqrt{x} \cdot \sqrt{y}\right) + \left(x\sqrt{x} \cdot \sqrt{y}\right) - \left(y\sqrt{y} \cdot \sqrt{y}\right)}$$

$$= \boxed{x\sqrt{x \cdot x} - y\sqrt{x \cdot y} + x\sqrt{x \cdot y} - y\sqrt{y \cdot y}} = \boxed{x\sqrt{x^2} - y\sqrt{xy} + x\sqrt{xy} - y\sqrt{y^2}} = \boxed{x \cdot x - y\sqrt{xy} + x\sqrt{xy} - y \cdot y}$$

$$= \boxed{x^2 - y\sqrt{xy} + x\sqrt{xy} - y^2} = \boxed{\left(x^2 - y^2\right) + \left(x\sqrt{xy} - y\sqrt{xy}\right)} = \boxed{\left(x^2 - y^2\right) + \sqrt{xy}(x - y)}$$

$$= \boxed{[(x - y)(x + y)] + \sqrt{xy}(x - y)} = \boxed{\mathbf{(x - y)\left[(x + y) + \sqrt{xy}\right]}}$$

Note: $\boxed{x^2 - y^2 = (x - y)(x + y)}$

Example 4.2-70

$$\boxed{\left(\sqrt[4]{x^5 y^4} + 3\right) \cdot \left(\sqrt[4]{x^4 y^5} - 3\right)} = \boxed{\left(y\sqrt[4]{x^{4+1}} + 3\right) \cdot \left(x\sqrt[4]{y^{4+1}} - 3\right)} = \boxed{\left(y\sqrt[4]{x^4 \cdot x^1} + 3\right) \cdot \left(x\sqrt[4]{y^4 \cdot y^1} - 3\right)}$$

$$= \boxed{\left(xy\sqrt[4]{x} + 3\right) \cdot \left(xy\sqrt[4]{y} - 3\right)} = \boxed{(xy \cdot xy) \cdot \left(\sqrt[4]{x} \cdot \sqrt[4]{y}\right) - \left(3 \cdot xy\sqrt[4]{x}\right) + \left(3 \cdot xy\sqrt[4]{y}\right) - (3 \cdot 3)}$$

$$= \boxed{(x \cdot x)(y \cdot y)\sqrt[4]{xy} - 3xy\sqrt[4]{x} + 3xy\sqrt[4]{y} - 9} = \boxed{\mathbf{x^2 y^2 \sqrt[4]{xy} - 3xy\sqrt[4]{x} + 3xy\sqrt[4]{y} - 9}}$$

Example 4.2-71

$$\boxed{\left(8 + x^2\sqrt{x}\right) \cdot \left(8 - \sqrt{x^3}\right)} = \boxed{\left(8 + x^2\sqrt{x}\right) \cdot \left(8 - x\sqrt{x}\right)} = \boxed{(8 \cdot 8) - \left(8 \cdot x\sqrt{x}\right) + \left(8 \cdot x^2\sqrt{x}\right) - \left(x \cdot x^2\right)\left(\sqrt{x} \cdot \sqrt{x}\right)}$$

$$= \boxed{64 - 8x\sqrt{x} + 8x^2\sqrt{x} - x^3\sqrt{x \cdot x}} = \boxed{64 - 8x\sqrt{x} + 8x^2\sqrt{x} - x^3\sqrt{x^2}} = \boxed{64 - 8x\sqrt{x} + 8x^2\sqrt{x} - x^3 \cdot x}$$

$$= \boxed{64 - 8x\sqrt{x} + 8x^2\sqrt{x} - x^4} = \boxed{\left(64 - x^4\right) + \left(8x^2\sqrt{x} - 8x\sqrt{x}\right)} = \boxed{\left(64 - x^4\right) + 8x\sqrt{x}(x - 1)}$$

$$= \boxed{\left[\left(8-x^2\right)\left(8+x^2\right)\right]+8x\sqrt{x}(x-1)}$$ Note: $\boxed{64-x^4=\left(8-x^2\right)\left(8+x^2\right)}$

Example 4.2-72

$$\boxed{\left(5a\sqrt{x}+\sqrt{x^3}\right)\cdot\left(2a\sqrt{y}-b\sqrt{x}\right)}=\boxed{\left(5a\sqrt{x}+x\sqrt{x}\right)\cdot\left(2a\sqrt{y}-b\sqrt{x}\right)}$$

$$=\boxed{(5a\cdot 2a)\cdot\left(\sqrt{x}\cdot\sqrt{y}\right)-(5a\cdot b)\cdot\left(\sqrt{x}\cdot\sqrt{x}\right)+\left(2ax\sqrt{x}\cdot\sqrt{y}\right)-\left(bx\sqrt{x}\cdot\sqrt{x}\right)}$$

$$=\boxed{10a^2\left(\sqrt{x\cdot y}\right)-5ab\left(\sqrt{x\cdot x}\right)+\left(2ax\sqrt{x\cdot y}\right)-\left(bx\sqrt{x\cdot x}\right)}=\boxed{10a^2\sqrt{xy}-5ab\sqrt{x^2}+2ax\sqrt{xy}-bx\sqrt{x^2}}$$

$$=\boxed{10a^2\sqrt{xy}-5abx+2ax\sqrt{xy}-bx\cdot x}=\boxed{10a^2\sqrt{xy}-5abx+2ax\sqrt{xy}-bx^2}=\boxed{\mathbf{2a\sqrt{xy}(5a+x)-bx(5a+x)}}$$

Example 4.2-73

$$\boxed{\left(\sqrt[4]{81}\cdot\sqrt{a}+2\right)\cdot\left(\sqrt[3]{8}\cdot\sqrt{b}+\sqrt{16}\right)}=\boxed{\left(\sqrt[4]{3^4}\cdot\sqrt{a}+2\right)\cdot\left(\sqrt[3]{2^3}\cdot\sqrt{b}+\sqrt{4^2}\right)}=\boxed{\left(3\sqrt{a}+2\right)\cdot\left(2\sqrt{b}+4\right)}$$

$$=\boxed{(3\cdot 2)\cdot\left(\sqrt{a}\cdot\sqrt{b}\right)+\left(3\cdot 4\sqrt{a}\right)+\left(2\cdot 2\sqrt{b}\right)+(2\cdot 4)}=\boxed{6\cdot\left(\sqrt{a\cdot b}\right)+\left(12\sqrt{a}\right)+\left(4\sqrt{b}\right)+(8)}$$

$$=\boxed{\mathbf{6\sqrt{ab}+12\sqrt{a}+4\sqrt{b}+8}}$$

Example 4.2-74

$$\boxed{\left(x+\sqrt[3]{x}\right)\cdot\left(x^2+\sqrt[3]{x}\right)}=\boxed{\left(x\cdot x^2\right)+\left(x\cdot\sqrt[3]{x}\right)+\left(x^2\cdot\sqrt[3]{x}\right)+\left(\sqrt[3]{x}\cdot\sqrt[3]{x}\right)}=\boxed{x^3+x\sqrt[3]{x}+x^2\sqrt[3]{x}+\sqrt[3]{x\cdot x}}$$

$$=\boxed{x^3+x\sqrt[3]{x}+x^2\sqrt[3]{x}+\sqrt[3]{x^2}}=\boxed{\left(x^3+\sqrt[3]{x^2}\right)+\left(x\sqrt[3]{x}+x^2\sqrt[3]{x}\right)}=\boxed{\mathbf{x^3+\sqrt[3]{x^2}+x\sqrt[3]{x}(1+x)}}$$

Example 4.2-75

$$\boxed{\left(\sqrt{a^2bc^3}+ab\right)\cdot\left(\sqrt{ab^2c}-ab\right)}=\boxed{\left(a\sqrt{bc^{2+1}}+ab\right)\cdot\left(b\sqrt{ac}-ab\right)}=\boxed{\left(a\sqrt{bc^2c^1}+ab\right)\cdot\left(b\sqrt{ac}-ab\right)}$$

$$=\boxed{\left(ac\sqrt{bc}+ab\right)\cdot\left(b\sqrt{ac}-ab\right)}=\boxed{\left(ac\sqrt{bc}\cdot b\sqrt{ac}\right)-\left(ac\sqrt{bc}\cdot ab\right)+\left(ab\cdot b\sqrt{ac}\right)-(ab\cdot ab)}$$

$$=\boxed{\left(abc\sqrt{bc\cdot ac}\right)-\left(ac\cdot ab\sqrt{bc}\right)+\left(ab^2\sqrt{ac}\right)-\left(a^2b^2\right)}=\boxed{abc\sqrt{abc^2}-a^2bc\sqrt{bc}+ab^2\sqrt{ac}-a^2b^2}$$

$$=\boxed{abc\cdot c\sqrt{ab}-a^2bc\sqrt{bc}+ab^2\sqrt{ac}-a^2b^2}=\boxed{abc^2\sqrt{ab}-a^2bc\sqrt{bc}+ab^2\sqrt{ac}-a^2b^2}$$

$$=\boxed{\mathbf{abc\sqrt{b}\left(c\sqrt{a}-a\sqrt{c}\right)+ab^2\left(\sqrt{ac}-a\right)}}$$

Example 4.2-76

$$\boxed{\left(5+\sqrt[4]{m^5n^6}\right)\cdot\left(5-\sqrt[4]{m^3n^2}\right)}=\boxed{\left(5+\sqrt[4]{m^{4+1}n^{4+2}}\right)\cdot\left(5-\sqrt[4]{m^3n^2}\right)}=\boxed{\left(5+\sqrt[4]{m^4\cdot m^1\cdot n^4\cdot n^2}\right)\cdot\left(5-\sqrt[4]{m^3n^2}\right)}$$

$$= \boxed{\left(5+mn\sqrt[4]{mn^2}\right)\cdot\left(5-\sqrt[4]{m^3n^2}\right)} = \boxed{(5\cdot 5)-\left(5\cdot\sqrt[4]{m^3n^2}\right)+\left(5\cdot mn\sqrt[4]{mn^2}\right)-\left(mn\sqrt[4]{mn^2}\cdot\sqrt[4]{m^3n^2}\right)}$$

$$= \boxed{25-5\sqrt[4]{m^3n^2}+5mn\sqrt[4]{mn^2}-mn\sqrt[4]{mn^2\cdot m^3n^2}} = \boxed{25-5\sqrt[4]{m^3n^2}+5mn\sqrt[4]{mn^2}-mn\sqrt[4]{\left(m\cdot m^3\right)\left(n^2\cdot n^2\right)}}$$

$$= \boxed{25-5\sqrt[4]{m^3n^2}+5mn\sqrt[4]{mn^2}-mn\sqrt[4]{m^4n^4}} = \boxed{25-5\sqrt[4]{m^3n^2}+5mn\sqrt[4]{mn^2}-mn\cdot mn}$$

$$= \boxed{25-5\sqrt[4]{m^3n^2}+5mn\sqrt[4]{mn^2}-(m\cdot m)(n\cdot n)} = \boxed{\mathbf{\left(25-m^2n^2\right)+5\sqrt[4]{mn^2}\left(mn-\sqrt[4]{m^2}\right)}}$$

Example 4.2-77

$$\boxed{\left(a+\sqrt[3]{a^5}\right)\cdot\left(a-\sqrt[3]{a^2}\right)} = \boxed{(a\cdot a)-\left(a\cdot\sqrt[3]{a^2}\right)+\left(a\cdot\sqrt[3]{a^5}\right)-\left(\sqrt[3]{a^5}\cdot\sqrt[3]{a^2}\right)} = \boxed{a^2-a\sqrt[3]{a^2}+a\sqrt[3]{a^5}-\sqrt[3]{a^5\cdot a^2}}$$

$$= \boxed{a^2-a\sqrt[3]{a^2}+a\sqrt[3]{a^{3+2}}-\sqrt[3]{a^{5+2}}} = \boxed{a^2-a\sqrt[3]{a^2}+a\sqrt[3]{a^3\cdot a^2}-\sqrt[3]{a^7}} = \boxed{a^2-a\sqrt[3]{a^2}+(a\cdot a)\sqrt[3]{a^2}-\sqrt[3]{a^{3+3+1}}}$$

$$= \boxed{a^2-a\sqrt[3]{a^2}+a^2\sqrt[3]{a^2}-\sqrt[3]{a^3\cdot a^3\cdot a^1}} = \boxed{a^2-a\sqrt[3]{a^2}+a^2\sqrt[3]{a^2}-a^1\cdot a^1\sqrt[3]{a}} = \boxed{a^2-a\sqrt[3]{a^2}+a^2\sqrt[3]{a^2}-a^{1+1}\sqrt[3]{a}}$$

$$= \boxed{a^2-a\sqrt[3]{a^2}+a^2\sqrt[3]{a^2}-a^2\sqrt[3]{a}} = \boxed{\left(a^2-a^2\sqrt[3]{a}\right)+\left(a^2\sqrt[3]{a^2}-a\sqrt[3]{a^2}\right)} = \boxed{\mathbf{a^2\left(1-\sqrt[3]{a}\right)+a\sqrt[3]{a^2}\,(a-1)}}$$

Example 4.2-78

$$\boxed{\left[\left(a-\sqrt{ab}\right)\cdot\left(a+\sqrt{ab}\right)\right]\cdot\left(-a-\sqrt{ab}\right)} = \boxed{\left[(a\cdot a)+\left(a\cdot\sqrt{ab}\right)-\left(a\cdot\sqrt{ab}\right)-\left(\sqrt{ab}\cdot\sqrt{ab}\right)\right]\cdot\left(-a-\sqrt{ab}\right)}$$

$$= \boxed{\left[a^2+a\sqrt{ab}-a\sqrt{ab}-\sqrt{ab\cdot ab}\right]\cdot\left(-a-\sqrt{ab}\right)} = \boxed{\left[a^2-\sqrt{a^2b^2}\right]\cdot\left(-a-\sqrt{ab}\right)} = \boxed{\left[a^2-ab\right]\cdot\left(-a-\sqrt{ab}\right)}$$

$$= \boxed{-\left(a^2\cdot a\right)-\left(a^2\cdot\sqrt{ab}\right)+(a\cdot ab)+\left(ab\cdot\sqrt{ab}\right)} = \boxed{-a^3-a^2\sqrt{ab}+a^2b+ab\sqrt{ab}}$$

$$= \boxed{\left(-a^3+a^2b\right)+\left(-a^2\sqrt{ab}+ab\sqrt{ab}\right)} = \boxed{a^2(-a+b)+a\sqrt{ab}(-a+b)} = \boxed{\mathbf{a(b-a)\left[a+\sqrt{ab}\right]}}$$

Example 4.2-79

$$\boxed{\left(\sqrt{x^3y^3}+\sqrt{xy}\right)\cdot\left(\sqrt{xy^2}-\sqrt{xy}\right)} = \boxed{\left(xy\sqrt{xy}+\sqrt{xy}\right)\cdot\left(y\sqrt{x}-\sqrt{xy}\right)}$$

$$= \boxed{(xy\cdot y)\left(\sqrt{xy}\cdot\sqrt{x}\right)-xy\left(\sqrt{xy}\cdot\sqrt{xy}\right)+y\left(\sqrt{xy}\cdot\sqrt{x}\right)-\left(\sqrt{xy}\cdot\sqrt{xy}\right)} = \boxed{xy^2\sqrt{xy\cdot x}-xy\sqrt{xy\cdot xy}+y\sqrt{xy\cdot x}-\sqrt{xy\cdot xy}}$$

$$= \boxed{xy^2\sqrt{(x\cdot x)y}-xy\sqrt{(x\cdot x)\cdot(y\cdot y)}+y\sqrt{(x\cdot x)\cdot y}-\sqrt{(x\cdot x)\cdot(y\cdot y)}} = \boxed{xy^2\sqrt{x^2y}-xy\sqrt{x^2y^2}+y\sqrt{x^2y}-\sqrt{x^2y^2}}$$

$$= \boxed{xy^2\cdot x\sqrt{y}-xy\cdot xy+y\cdot x\sqrt{y}-xy} = \boxed{(x\cdot x)y^2\sqrt{y}-(x\cdot x)(y\cdot y)+xy\sqrt{y}-xy} = \boxed{x^2y^2\sqrt{y}-x^2y^2+xy\sqrt{y}-xy}$$

$$= \boxed{x^2y^2\left(\sqrt{y}-1\right)+xy\left(\sqrt{y}-1\right)} = \boxed{\mathbf{xy\left(\sqrt{y}-1\right)\left[xy+1\right]}}$$

Example 4.2-80

$$\boxed{\left(\sqrt{w^5}-\sqrt{u^2v^2w^2}\right)\cdot\left(\sqrt{w}+\sqrt{u^4v^2w^6}\right)} = \boxed{\left(\sqrt{w^{2+2+1}}-uvw\right)\cdot\left(\sqrt{w}+\sqrt{u^{2+2}v^2w^{2+2+2}}\right)}$$

$$= \boxed{\left(\sqrt{w^2\cdot w^2\cdot w^1}-uvw\right)\cdot\left(\sqrt{w}+\sqrt{u^2\cdot u^2\cdot v^2\cdot w^2\cdot w^2\cdot w^2}\right)}$$

$$= \boxed{\left(w^1\cdot w^1\sqrt{w}-uvw\right)\cdot\left[\sqrt{w}+\left(u^1\cdot u^1\right)\cdot v\cdot\left(w^1\cdot w^1\cdot w^1\right)\right]} = \boxed{\left(w^{1+1}\sqrt{w}-uvw\right)\cdot\left[\sqrt{w}+\left(u^{1+1}\right)\cdot v\cdot\left(w^{1+1+1}\right)\right]}$$

$$= \boxed{\left(w^2\sqrt{w}-uvw\right)\cdot\left(\sqrt{w}+u^2vw^3\right)} = \boxed{\left(w^2\sqrt{w}\cdot\sqrt{w}\right)+\left(w^2\sqrt{w}\cdot u^2vw^3\right)-\left(uvw\cdot\sqrt{w}\right)-\left(uvw\cdot u^2vw^3\right)}$$

$$= \boxed{\left(w^2\sqrt{w\cdot w}\right)+u^2v\left(w^3w^2\right)\sqrt{w}-uvw\sqrt{w}-\left[\left(u\cdot u^2\right)\left(v\cdot v\right)\left(w\cdot w^3\right)\right]} = \boxed{w^2\sqrt{w^2}+u^2vw^5\sqrt{w}-uvw\sqrt{w}-u^3v^2w^4}$$

$$= \boxed{w^2\cdot w-u^3v^2w^4+u^2vw^5\sqrt{w}-uvw\sqrt{w}} = \boxed{\left(w^3-u^3v^2w^4\right)+\left(u^2vw^5\sqrt{w}-uvw\sqrt{w}\right)}$$

$$= \boxed{\mathbf{w^3\left(1-u^3v^2w\right)+uvw\sqrt{w}\left(uw^4-1\right)}}$$

Practice Problems - Multiplying Binomial Expressions in Radical Form, with Variables

Section 4.2 Case II b Practice Problems - Multiply the following binomial expressions in radical form:

1. $\left(a+\sqrt{b}\right)\cdot\left(a-\sqrt{b^3}\right) =$
2. $\left(a+x\sqrt{x}\right)\cdot\left(a-\sqrt{x}\right) =$
3. $\left(5a+\sqrt{x^5}\right)\cdot\left(2a-\sqrt{x}\right) =$
4. $\left(4+\sqrt{r}\right)\cdot\left(7-\sqrt{r}\right) =$
5. $\left(2+\sqrt{x^3y^3}\right)\cdot\left(2-y\sqrt{x^3}\right) =$
6. $\left(m+\sqrt[3]{m^4}\right)\cdot\left(m-\sqrt[3]{m^5}\right) =$
7. $\left(\sqrt{r}+2\right)\cdot\left(4\sqrt{r^3}-2\right) =$
8. $\left(4+\sqrt{a}\right)\cdot\left(4-\sqrt{a}\right) =$
9. $\left(3+\sqrt[4]{x^5}\right)\cdot\left(3-\sqrt[4]{x^7}\right) =$
10. $\left(1+\sqrt{xy}\right)\cdot\left(1-\sqrt{x^3y^3}\right) =$

Case III Multiplying Monomial and Binomial Expressions in Radical Form

To multiply monomial and binomial expressions in radical form the following general multiplication rule is used:

$$a \cdot (b+c) = a \cdot b + a \cdot c$$

Multiplication of monomials by binomial expressions in radical form is divided into two cases. Case III a - multiplication of monomial and binomial expressions in radical form, with real numbers and Case III b - multiplication of monomial and binomial expressions in radical form, with variables.

Case III a Multiplying Monomial and Binomial Expressions in Radical Form, with Real Numbers

Monomial and binomial expressions in radical form are multiplied by each other using the following steps:

Step 1 Simplify the radical terms (see Section 4.1, Case III).

Step 2 Multiply each term using the general multiplication rule, i.e., $a \cdot (b+c) = a \cdot b + a \cdot c$. Repeat Step 1, if necessary.

Examples with Steps

The following examples show the steps as to how monomial and binomial expressions in radical form are multiplied by one another:

Example 4.2-81

$$\boxed{\sqrt{5} \cdot \left(\sqrt{50} + 2\sqrt{27}\right)} =$$

Solution:

Step 1

$$\boxed{\sqrt{5} \cdot \left(\sqrt{50} + 2\sqrt{27}\right)} = \boxed{\sqrt{5} \cdot \left(\sqrt{25 \cdot 2} + 2\sqrt{9 \cdot 3}\right)} = \boxed{\sqrt{5} \cdot \left(\sqrt{5^2 \cdot 2} + 2\sqrt{3^2 \cdot 3}\right)}$$

$$= \boxed{\sqrt{5} \cdot \left[5\sqrt{2} + (2 \cdot 3)\sqrt{3}\right]} = \boxed{\sqrt{5} \cdot \left[5\sqrt{2} + 6\sqrt{3}\right]}$$

Step 2

$$\boxed{\sqrt{5} \cdot \left[5\sqrt{2} + 6\sqrt{3}\right]} = \boxed{5\left(\sqrt{2} \cdot \sqrt{5}\right) + 6\left(\sqrt{5} \cdot \sqrt{3}\right)} = \boxed{5\sqrt{2 \cdot 5} + 6\sqrt{5 \cdot 3}} = \boxed{\mathbf{5\sqrt{10} + 6\sqrt{15}}}$$

Example 4.2-82

$$\boxed{2\sqrt{2} \cdot \left(5 + \sqrt{20}\right)} =$$

Solution:

Step 1

$$\boxed{2\sqrt{2} \cdot \left(5 + \sqrt{20}\right)} = \boxed{2\sqrt{2} \cdot \left(5 + \sqrt{4 \cdot 5}\right)} = \boxed{2\sqrt{2} \cdot \left(5 + \sqrt{2^2 \cdot 5}\right)} = \boxed{2\sqrt{2} \cdot \left(5 + 2\sqrt{5}\right)}$$

Step 2

$$\boxed{2\sqrt{2} \cdot \left(5 + 2\sqrt{5}\right)} = \boxed{\left[(2 \cdot 5)\sqrt{2}\right] + \left[(2 \cdot 2) \cdot \left(\sqrt{2} \cdot \sqrt{5}\right)\right]} = \boxed{10\sqrt{2} + 4\sqrt{2 \cdot 5}} = \boxed{10\sqrt{2} + 4\sqrt{10}}$$

$$= \boxed{\mathbf{2\left(5\sqrt{2} + 2\sqrt{10}\right)}}$$

Example 4.2-83

$$-2\sqrt{24}\cdot\left(\sqrt{36}-\sqrt{125}\right)=$$

Solution:

Step 1

$$-2\sqrt{24}\cdot\left(\sqrt{36}-\sqrt{125}\right)=-2\sqrt{4\cdot6}\cdot\left(\sqrt{6^2}-\sqrt{25\cdot5}\right)=-2\sqrt{2^2\cdot6}\cdot\left(6-\sqrt{5^2\cdot5}\right)$$

$$=-(2\cdot2)\sqrt{6}\cdot\left(6-5\sqrt{5}\right)=-4\sqrt{6}\cdot\left(6-5\sqrt{5}\right)$$

Step 2

$$-4\sqrt{6}\cdot\left(6-5\sqrt{5}\right)=-(4\cdot6)\sqrt{6}+(4\cdot5)\cdot\left(\sqrt{6}\cdot\sqrt{5}\right)=-24\sqrt{6}+20\sqrt{6\cdot5}$$

$$=-24\sqrt{6}+20\sqrt{30}=\mathbf{4\left(5\sqrt{30}-6\sqrt{6}\right)}$$

Example 4.2-84

$$-2\sqrt[4]{4}\cdot\left(\sqrt[4]{64}-\sqrt[4]{162}\right)=$$

Solution:

Step 1

$$-2\sqrt[4]{4}\cdot\left(\sqrt[4]{64}-\sqrt[4]{162}\right)=-2\sqrt[4]{4}\cdot\left(\sqrt[4]{16\cdot4}-\sqrt[4]{81\cdot2}\right)=-2\sqrt[4]{4}\cdot\left(\sqrt[4]{2^4\cdot4}-\sqrt[4]{3^4\cdot2}\right)$$

$$=-2\sqrt[4]{4}\cdot\left(2\sqrt[4]{4}-3\sqrt[4]{2}\right)$$

Step 2

$$-2\sqrt[4]{4}\cdot\left(2\sqrt[4]{4}-3\sqrt[4]{2}\right)=-(2\cdot2)\cdot\left(\sqrt[4]{4}\cdot\sqrt[4]{4}\right)+(2\cdot3)\cdot\left(\sqrt[4]{4}\cdot\sqrt[4]{2}\right)=-4\left(\sqrt[4]{4\cdot4}\right)+6\left(\sqrt[4]{4\cdot2}\right)$$

$$=-4\sqrt[4]{16}+6\sqrt[4]{8}=-4\sqrt[4]{2^4}+6\sqrt[4]{8}=-(4\cdot2)+6\sqrt[4]{8}=-8+6\sqrt[4]{8}=\mathbf{2\left(3\sqrt[4]{8}-4\right)}$$

Example 4.2-85

$$\sqrt[3]{4}\cdot\left(\sqrt[3]{16}-\sqrt[3]{500}\right)=$$

Solution:

Step 1

$$\sqrt[3]{4}\cdot\left(\sqrt[3]{16}-\sqrt[3]{500}\right)=\sqrt[3]{4}\cdot\left(\sqrt[3]{8\cdot2}-\sqrt[3]{125\cdot4}\right)=\sqrt[3]{4}\cdot\left(\sqrt[3]{2^3\cdot2}-\sqrt[3]{5^3\cdot4}\right)$$

$$=\sqrt[3]{4}\cdot\left(2\sqrt[3]{2}-5\sqrt[3]{4}\right)$$

Step 2

$$\sqrt[3]{4}\cdot\left(2\sqrt[3]{2}-5\sqrt[3]{4}\right)=2\left(\sqrt[3]{4}\cdot\sqrt[3]{2}\right)-5\left(\sqrt[3]{4}\cdot\sqrt[3]{4}\right)=2\left(\sqrt[3]{4\cdot2}\right)-5\left(\sqrt[3]{4\cdot4}\right)=2\sqrt[3]{8}-5\sqrt[3]{16}$$

$$=2\sqrt[3]{2^3}-5\sqrt[3]{8\cdot2}=(2\cdot2)-5\sqrt[3]{2^3\cdot2}=4-(5\cdot2)\sqrt[3]{2}=4-10\sqrt[3]{2}=\mathbf{2\left(2-5\sqrt[3]{2}\right)}$$

Additional Examples - Multiplying Monomial and Binomial Expressions in Radical Form, with Real Numbers

The following examples further illustrate how to multiply radical terms by one another:

Example 4.2-86

$$\boxed{\left(2\sqrt{3}\right)\cdot\left(3\sqrt{6}+2\right)}=\boxed{(2\cdot3)\cdot\left(\sqrt{3}\cdot\sqrt{6}\right)+(2\cdot2)\sqrt{3}}=\boxed{6\left(\sqrt{3\cdot6}\right)+4\sqrt{3}}=\boxed{6\sqrt{18}+4\sqrt{3}}=\boxed{6\sqrt{9\cdot2}+4\sqrt{3}}$$

$$=\boxed{6\sqrt{3^2\cdot2}+4\sqrt{3}}=\boxed{(6\cdot3)\sqrt{2}+4\sqrt{3}}=\boxed{18\sqrt{2}+4\sqrt{3}}=\boxed{\mathbf{2\left(9\sqrt{2}+2\sqrt{3}\right)}}$$

Example 4.2-87

$$\boxed{3\sqrt{5}\cdot\left(\sqrt{5}+6\sqrt{10}\right)}=\boxed{\left(3\sqrt{5}\cdot\sqrt{5}\right)+\left(3\sqrt{5}\cdot6\sqrt{10}\right)}=\boxed{\left(3\sqrt{5\cdot5}\right)+(3\cdot6)\sqrt{5\cdot10}}=\boxed{3\sqrt{5^2}+18\sqrt{50}}$$

$$=\boxed{(3\cdot5)+18\sqrt{25\cdot2}}=\boxed{15+18\sqrt{5^2\cdot2}}=\boxed{15+(18\cdot5)\sqrt{2}}=\boxed{15+90\sqrt{2}}=\boxed{\mathbf{15\left(1+6\sqrt{2}\right)}}$$

Example 4.2-88

$$\boxed{-2\sqrt{6}\cdot\left(-\sqrt{5}+\sqrt{50}\right)}=\boxed{\left(+2\sqrt{6}\cdot\sqrt{5}\right)-\left(2\sqrt{6}\cdot\sqrt{50}\right)}=\boxed{2\sqrt{6\cdot5}-2\sqrt{6\cdot50}}=\boxed{2\sqrt{30}-2\sqrt{300}}=\boxed{2\sqrt{30}-2\sqrt{3\cdot100}}$$

$$=\boxed{2\sqrt{30}-2\sqrt{10^2\cdot3}}=\boxed{2\sqrt{30}-(2\cdot10)\sqrt{3}}=\boxed{2\sqrt{30}-20\sqrt{3}}=\boxed{\mathbf{2\left(\sqrt{30}-10\sqrt{3}\right)}}$$

Example 4.2-89

$$\boxed{\sqrt{3}\cdot\left(2\sqrt{3}+\sqrt{6}\right)}=\boxed{\left(2\sqrt{3}\cdot\sqrt{3}\right)+\left(\sqrt{3}\cdot\sqrt{6}\right)}=\boxed{\left(2\sqrt{3\cdot3}\right)+\left(\sqrt{3\cdot6}\right)}=\boxed{2\sqrt{3^1\cdot3^1}+\sqrt{18}}=\boxed{2\sqrt{3^{1+1}}+\sqrt{9\cdot2}}$$

$$=\boxed{2\sqrt{3^2}+\sqrt{3^2\cdot2}}=\boxed{(2\cdot3)+3\sqrt{2}}=\boxed{6+3\sqrt{2}}=\boxed{\mathbf{3\left(2+\sqrt{2}\right)}}$$

Example 4.2-90

$$\boxed{\sqrt{5}\cdot\left(2\sqrt{5}+3\right)}=\boxed{\left(2\sqrt{5}\cdot\sqrt{5}\right)+\left(3\cdot\sqrt{5}\right)}=\boxed{2\sqrt{5\cdot5}+3\sqrt{5}}=\boxed{2\sqrt{5^1\cdot5^1}+3\sqrt{5}}=\boxed{2\sqrt{5^{1+1}}+3\sqrt{5}}=\boxed{2\sqrt{5^2}+3\sqrt{5}}$$

$$=\boxed{(2\cdot5)+3\sqrt{5}}=\boxed{\mathbf{10+3\sqrt{5}}}$$

Example 4.2-91

$$\boxed{3\sqrt{2}\cdot\left(\sqrt{10}+4\sqrt{20}\right)}=\boxed{\left(3\sqrt{2}\cdot\sqrt{10}\right)+\left(3\sqrt{2}\cdot4\sqrt{20}\right)}=\boxed{\left(3\sqrt{2\cdot10}\right)+(3\cdot4)\sqrt{2\cdot20}}=\boxed{3\sqrt{20}+12\sqrt{40}}$$

$$=\boxed{3\sqrt{4\cdot5}+12\sqrt{4\cdot10}}=\boxed{3\sqrt{2^2\cdot5}+12\sqrt{2^2\cdot10}}=\boxed{(3\cdot2)\sqrt{5}+(12\cdot2)\sqrt{10}}=\boxed{6\sqrt{5}+24\sqrt{10}}=\boxed{\mathbf{6\left(\sqrt{5}+4\sqrt{10}\right)}}$$

Example 4.2-92

$$\boxed{-\sqrt{7}\cdot\left(-\sqrt{7}+2\sqrt{3}\right)}=\boxed{+\left(\sqrt{7}\cdot\sqrt{7}\right)-\left(2\sqrt{7}\cdot\sqrt{3}\right)}=\boxed{\sqrt{7\cdot7}-2\sqrt{7\cdot3}}=\boxed{\sqrt{7^1\cdot7^1}-2\sqrt{21}}=\boxed{\sqrt{7^{1+1}}-2\sqrt{21}}$$

$$=\boxed{\sqrt{7^2}-2\sqrt{21}}=\boxed{\mathbf{7-2\sqrt{21}}}$$

Example 4.2-93

$$\sqrt[3]{5}\cdot\left(\sqrt[3]{25}-\sqrt[3]{216}\right)=\sqrt[3]{5}\cdot\left(\sqrt[3]{5^2}-\sqrt[3]{6^3}\right)=\left(\sqrt[3]{5}\cdot\sqrt[3]{5^2}\right)-\left(\sqrt[3]{5}\cdot\sqrt[3]{6^3}\right)=\left(\sqrt[3]{5\cdot 5^2}\right)-\left(\sqrt[3]{5}\cdot 6\right)=\sqrt[3]{5^1\cdot 5^2}-6\sqrt[3]{5}$$

$$=\sqrt[3]{5^{1+2}}-6\sqrt[3]{5}=\sqrt[3]{5^3}-6\sqrt[3]{5}=\mathbf{5-6\sqrt[3]{5}}$$

Example 4.2-94

$$\sqrt[4]{4^2}\cdot\left(3+\sqrt[4]{4^3}\right)=\left(3\cdot\sqrt[4]{4^2}\right)+\left(\sqrt[4]{4^2}\cdot\sqrt[4]{4^3}\right)=3\sqrt[4]{16}+\sqrt[4]{4^2\cdot 4^3}=3\sqrt[4]{2^4}+\sqrt[4]{4^{2+3}}=(3\cdot 2)+\sqrt[4]{4^5}$$

$$=6+\sqrt[4]{4^{4+1}}=6+\sqrt[4]{4^4\cdot 4^1}=6+4\sqrt[4]{4}=\mathbf{2\left(3+2\sqrt[4]{4}\right)}$$

Example 4.2-95

$$3\sqrt{3}\cdot\left(\sqrt{6}+2\sqrt{12}\right)=\left(3\sqrt{3}\cdot\sqrt{6}\right)+\left[(2\cdot 3)\sqrt{3}\cdot\sqrt{12}\right]=3\sqrt{3\cdot 6}+6\sqrt{3\cdot 12}=3\sqrt{18}+6\sqrt{36}=3\sqrt{9\cdot 2}+6\sqrt{6^2}$$

$$=3\sqrt{3^2\cdot 2}+(6\cdot 6)=(3\cdot 3)\sqrt{2}+36=9\sqrt{2}+36=\mathbf{9\left(4+\sqrt{2}\right)}$$

Practice Problems - Multiplying Monomial and Binomial Expressions in Radical Form, with Real Numbers

Section 4.2 Case III a Practice Problems - Multiply the following radical expressions:

1. $2\sqrt{3}\cdot\left(2+\sqrt{2}\right)=$
2. $\sqrt{5}\cdot\left(\sqrt{8}+\sqrt{5}\right)=$
3. $-\sqrt{8}\cdot\left(3-\sqrt{3}\right)=$
4. $4\sqrt{98}\cdot\left(3-\sqrt{2^3}\right)=$
5. $\sqrt[4]{48}\cdot\left(\sqrt[4]{324}+\sqrt[4]{32}\right)=$
6. $2\sqrt{5}\cdot\left(\sqrt{45}+\sqrt[4]{81}\right)=$
7. $-\sqrt[5]{64}\cdot\left(-\sqrt[5]{486}+4\right)=$
8. $\sqrt{32}\cdot\left(2+\sqrt{3}\right)=$
9. $\left(3\sqrt{44}+\sqrt{27}\right)\cdot\sqrt{8}=$
10. $-\sqrt{2}\cdot\left(\sqrt{32}-2\sqrt{75}\right)=$

Case III b Multiplying Monomial and Binomial Expressions in Radical Form, with Variables

Monomial and binomial expressions in radical form are multiplied by each other using the following steps:

Step 1 Simplify the radical terms (see Section 4.1, Case IV).

Step 2 Multiply each term using the general multiplication rule, i.e., $x\cdot(y+z)=x\cdot y+x\cdot z$. Repeat Step 1, if necessary.

Examples with Steps

The following examples show the steps as to how monomial and binomial expressions in radical form are multiplied by one another:

Example 4.2-96

$$\boxed{2\sqrt{x}\cdot\left(3\sqrt{xy}-\sqrt{x^2y}\right)}=$$

Solution:

Step 1

$$\boxed{2\sqrt{x}\cdot\left(3\sqrt{xy}-\sqrt{x^2y}\right)}=\boxed{2\sqrt{x}\cdot\left(3\sqrt{xy}-x\sqrt{y}\right)}$$

Step 2

$$\boxed{2\sqrt{x}\cdot\left(3\sqrt{xy}-x\sqrt{y}\right)}=\boxed{(2\cdot3)\left(\sqrt{x}\cdot\sqrt{xy}\right)-2x\left(\sqrt{x}\cdot\sqrt{y}\right)}=\boxed{6\sqrt{x\cdot xy}-2x\sqrt{x\cdot y}}$$

$$=\boxed{6\sqrt{x^2y}-2x\sqrt{xy}}=\boxed{6x\sqrt{y}-2x\sqrt{xy}}=\boxed{\mathbf{2x\sqrt{y}\left(3-\sqrt{x}\right)}}$$

Example 4.2-97

$$\boxed{\sqrt{x^3y^2}\cdot\left(5+\sqrt{x^3y^2}\right)}=$$

Solution:

Step 1

$$\boxed{\sqrt{x^3y^2}\cdot\left(5+\sqrt{x^3y^2}\right)}=\boxed{y\sqrt{x^{2+1}}\cdot\left(5+y\sqrt{x^{2+1}}\right)}=\boxed{y\sqrt{x^2\cdot x^1}\cdot\left(5+y\sqrt{x^2\cdot x^1}\right)}$$

$$=\boxed{xy\sqrt{x}\cdot\left(5+xy\sqrt{x}\right)}$$

Step 2

$$\boxed{xy\sqrt{x}\cdot\left(5+xy\sqrt{x}\right)}=\boxed{\left(5\cdot xy\sqrt{x}\right)+\left[(xy\cdot xy)\cdot\left(\sqrt{x}\cdot\sqrt{x}\right)\right]}=\boxed{5xy\sqrt{x}+\left[x^2y^2\cdot\left(\sqrt{x\cdot x}\right)\right]}$$

$$=\boxed{5xy\sqrt{x}+\left(x^2y^2\cdot\sqrt{x^2}\right)}=\boxed{5xy\sqrt{x}+\left(x^2y^2\cdot x\right)}=\boxed{5xy\sqrt{x}+\left(x^2\cdot x\right)y^2}$$

$$=\boxed{5xy\sqrt{x}+\left(x^2\cdot x^1\right)y^2}=\boxed{5xy\sqrt{x}+\left(x^{2+1}\right)y^2}=\boxed{5xy\sqrt{x}+x^3y^2}=\boxed{\mathbf{xy\left(5\sqrt{x}+x^2y\right)}}$$

Example 4.2-98

$$-2\sqrt{x^5}\cdot\left(\sqrt{x^3}-4\sqrt{x^4}\right)=$$

Solution:

Step 1

$$-2\sqrt{x^5}\cdot\left(\sqrt{x^3}-4\sqrt{x^4}\right)=-2\sqrt{x^{2+2+1}}\cdot\left(\sqrt{x^{2+1}}-4\sqrt{x^{2+2}}\right)$$

$$=-2\sqrt{\left(x^2\cdot x^2\right)\cdot x^1}\cdot\left(\sqrt{x^2\cdot x^1}-4\sqrt{x^2\cdot x^2}\right)=-2\left(x^1\cdot x^1\right)\sqrt{x}\cdot\left[x\sqrt{x}-4\left(x^1\cdot x^1\right)\right]$$

$$=-2x^{1+1}\sqrt{x}\cdot\left[x\sqrt{x}-4\left(x^{1+1}\right)\right]=-2x^2\sqrt{x}\cdot\left(x\sqrt{x}-4x^2\right)$$

Step 2

$$-2x^2\sqrt{x}\cdot\left(x\sqrt{x}-4x^2\right)=\left[-2\left(x^2\cdot x\right)\cdot\left(\sqrt{x}\cdot\sqrt{x}\right)\right]+\left[(2\cdot4)\cdot\left(x^2\cdot x^2\right)\sqrt{x}\right]$$

$$=-2x^3\left(\sqrt{x\cdot x}\right)+8x^4\sqrt{x}=-2x^3\sqrt{x^2}+8x^4\sqrt{x}=-2\left(x^3\cdot x\right)+8x^4\sqrt{x}$$

$$=-2x^4+8x^4\sqrt{x}=\mathbf{2x^4\left(4\sqrt{x}-1\right)}$$

Example 4.2-99

$$\sqrt[4]{x^5}\cdot\left(\sqrt[4]{162x^6}-2\sqrt[4]{16x^7}\right)=$$

Solution:

Step 1

$$\sqrt[4]{x^5}\cdot\left(\sqrt[4]{162x^6}-2\sqrt[4]{16x^7}\right)=\sqrt[4]{x^{4+1}}\cdot\left(\sqrt[4]{(81\cdot2)x^{4+2}}-2\sqrt[4]{2^4x^{4+3}}\right)$$

$$=\sqrt[4]{x^4\cdot x^1}\cdot\left[\sqrt[4]{\left(3^4\cdot2\right)\cdot\left(x^4\cdot x^2\right)}-(2\cdot2)\sqrt[4]{x^4\cdot x^3}\right]=x\sqrt[4]{x}\cdot\left(3x\sqrt[4]{2x^2}-4x\sqrt[4]{x^3}\right)$$

Step 2

$$x\sqrt[4]{x}\cdot\left(3x\sqrt[4]{2x^2}-4x\sqrt[4]{x^3}\right)=\left[(3x\cdot x)\cdot\left(\sqrt[4]{x}\cdot\sqrt[4]{2x^2}\right)\right]-\left[(4x\cdot x)\cdot\left(\sqrt[4]{x}\cdot\sqrt[4]{x^3}\right)\right]$$

$$=\left[3x^2\cdot\left(\sqrt[4]{2x^2\cdot x}\right)\right]-\left[4x^2\cdot\left(\sqrt[4]{x^3\cdot x}\right)\right]=3x^2\sqrt[4]{2x^3}-4x^2\sqrt[4]{x^4}=3x^2\sqrt[4]{2x^3}-4\left(x^2\cdot x\right)$$

$$=3x^2\sqrt[4]{2x^3}-4x^3=\mathbf{x^2\left(3\sqrt[4]{2x^3}-4x\right)}$$

Example 4.2-100

$$\sqrt[3]{x}\cdot\left(\sqrt[3]{125x^2}-\sqrt[3]{128x^4}\right)=$$

Solution:

Step 1

$$\sqrt[3]{x}\cdot\left(\sqrt[3]{125x^2}-\sqrt[3]{128x^4}\right)=\sqrt[3]{x}\cdot\left[\sqrt[3]{5^3x^2}-\sqrt[3]{(64\cdot2)x^{3+1}}\right]$$

$$=\sqrt[3]{x}\cdot\left[5\sqrt[3]{x^2}-\sqrt[3]{\left(4^3\cdot2\right)\cdot\left(x^3\cdot x^1\right)}\right]=\sqrt[3]{x}\cdot\left(5\sqrt[3]{x^2}-4x\sqrt[3]{2x}\right)$$

Step 2

$$\sqrt[3]{x}\cdot\left(5\sqrt[3]{x^2}-4x\sqrt[3]{2x}\right)=5\left(\sqrt[3]{x}\cdot\sqrt[3]{x^2}\right)-4x\left(\sqrt[3]{x}\cdot\sqrt[3]{2x}\right)=5\left(\sqrt[3]{x^2\cdot x}\right)-4x\left(\sqrt[3]{2x\cdot x}\right)$$

$$=5\sqrt[3]{x^3}-4x\sqrt[3]{2x^2}=5x-4x\sqrt[3]{2x^2}=\mathbf{x\left(5-4\sqrt[3]{2x^2}\right)}$$

Additional Examples - Multiplying Monomial and Binomial Expressions in Radical Form, with Variables

The following examples further illustrate how to multiply radical terms by one another:

Example 4.2-101

$$5a\sqrt{x}\cdot\left(2a\sqrt{y}-3b\sqrt{x}\right)=\left(5a\sqrt{x}\cdot2a\sqrt{y}\right)-\left(5a\sqrt{x}\cdot3b\sqrt{x}\right)=(5a\cdot2a)\sqrt{x\cdot y}-(5\cdot3)\cdot(a\cdot b)\sqrt{x\cdot x}$$

$$=\left(10a^2\sqrt{xy}\right)-\left(15ab\sqrt{x^2}\right)=10a^2\sqrt{xy}-15abx=\mathbf{5a\left(2a\sqrt{xy}-3bx\right)}$$

Example 4.2-102

$$\sqrt{xy}\cdot\left(\sqrt{x}-\sqrt{y}\right)=\left(\sqrt{xy}\cdot\sqrt{x}\right)-\left(\sqrt{xy}\cdot\sqrt{y}\right)=\sqrt{(x\cdot x)y}-\sqrt{x(y\cdot y)}=\sqrt{\left(x^1x^1\right)y}-\sqrt{x\left(y^1y^1\right)}$$

$$=\sqrt{x^{1+1}y}-\sqrt{xy^{1+1}}=\sqrt{x^2y}-\sqrt{xy^2}=\mathbf{x\sqrt{y}-y\sqrt{x}}$$

Example 4.2-103

$$3\sqrt{x^3}\cdot\left(\sqrt{x^5}-2\sqrt{x^6}\right)=3\sqrt{x^{2+1}}\cdot\left(\sqrt{x^{2+2+1}}-2\sqrt{x^{2+2+2}}\right)=3\sqrt{x^2\cdot x^1}\cdot\left(\sqrt{x^2\cdot x^2\cdot x^1}-2\sqrt{x^2\cdot x^2\cdot x^2}\right)$$

$$=3x\sqrt{x}\cdot\left(x^1x^1\sqrt{x}-2x^1x^1x^1\right)=3x\sqrt{x}\cdot\left(x^{1+1}\sqrt{x}-2x^{1+1+1}\right)=3x\sqrt{x}\cdot\left(x^2\sqrt{x}-2x^3\right)$$

$$=\left[\left(3x\sqrt{x}\right)\cdot\left(x^2\sqrt{x}\right)\right]-\left[2x^3\cdot\left(3x\sqrt{x}\right)\right]=\left[\left(3x\cdot x^2\right)\cdot\left(\sqrt{x}\cdot\sqrt{x}\right)\right]-\left[(2\cdot3)\cdot\left(x^3\cdot x\sqrt{x}\right)\right]$$

$$=\left[\left(3x^1\cdot x^2\right)\cdot\left(\sqrt{x\cdot x}\right)\right]-\left[6\left(x^3\cdot x^1\sqrt{x}\right)\right]=\left[3x^{1+2}\cdot\left(\sqrt{x^1\cdot x^1}\right)\right]-\left[6\left(x^{3+1}\sqrt{x}\right)\right]=3x^3\sqrt{x^{1+1}}-6\left(x^4\sqrt{x}\right)$$

$$=3x^3\sqrt{x^2}-6x^4\sqrt{x}=\left(3x^3\cdot x\right)-6x^4\sqrt{x}=\left(3x^3\cdot x^1\right)-6x^4\sqrt{x}=3x^{3+1}-6x^4\sqrt{x}=3x^4-6x^4\sqrt{x}$$

$= \boxed{\mathbf{3x^4\left(1-2\sqrt{x}\right)}}$

Note that we can also solve the above problem without simplifying each radical term first as is shown below:

$$\boxed{3\sqrt{x^3}\cdot\left(\sqrt{x^5}-2\sqrt{x^6}\right)} = \boxed{3\left(\sqrt{x^3}\cdot\sqrt{x^5}\right)-\left[(2\cdot 3)\left(\sqrt{x^3}\cdot\sqrt{x^6}\right)\right]} = \boxed{3\left(\sqrt{x^3\cdot x^5}\right)-6\left(\sqrt{x^3\cdot x^6}\right)}$$

$$= \boxed{3\sqrt{x^{3+5}}-6\sqrt{x^{3+6}}} = \boxed{3\sqrt{x^8}-6\sqrt{x^9}} = \boxed{3\sqrt{x^{2+2+2+2}}-6\sqrt{x^{2+2+2+2+1}}}$$

$$= \boxed{3\sqrt{\left(x^2\cdot x^2\cdot x^2\cdot x^2\right)}-6\sqrt{\left(x^2\cdot x^2\cdot x^2\cdot x^2\right)\cdot x^1}} = \boxed{3\left(x^1\cdot x^1\cdot x^1\cdot x^1\right)-6\left(x^1\cdot x^1\cdot x^1\cdot x^1\right)\sqrt{x}}$$

$$= \boxed{3\left(x^{1+1+1+1}\right)-6\left(x^{1+1+1+1}\right)\sqrt{x}} = \boxed{3x^4-6x^4\sqrt{x}} = \boxed{\mathbf{3x^4\left(1-2\sqrt{x}\right)}}$$

Example 4.2-104

$$\boxed{\sqrt{xy^2}\cdot\left(5x-\sqrt{x^5y^3}\right)} = \boxed{y\sqrt{x}\cdot\left(5x-\sqrt{x^{2+2+1}y^{2+1}}\right)} = \boxed{y\sqrt{x}\cdot\left(5x-\sqrt{x^2\cdot x^2\cdot x^1\cdot y^2\cdot y^1}\right)}$$

$$= \boxed{y\sqrt{x}\cdot\left(5x-x^1\cdot x^1\cdot y^1\sqrt{x\cdot y}\right)} = \boxed{y\sqrt{x}\cdot\left(5x-x^{1+1}y\sqrt{xy}\right)} = \boxed{y\sqrt{x}\cdot\left(5x-x^2y\sqrt{xy}\right)}$$

$$= \boxed{\left(5x\cdot y\sqrt{x}\right)-\left(y\sqrt{x}\cdot x^2y\sqrt{xy}\right)} = \boxed{\left(5xy\sqrt{x}\right)-\left(x^2y\cdot y\sqrt{xy\cdot x}\right)} = \boxed{5xy\sqrt{x}-x^2y^2\sqrt{x^2y}} = \boxed{5xy\sqrt{x}-x^2\cdot xy^2\sqrt{y}}$$

$$= \boxed{5xy\sqrt{x}-x^3y^2\sqrt{y}} = \boxed{\mathbf{xy\left(5\sqrt{x}-x^2y\sqrt{y}\right)}}$$

Example 4.2-105

$$\boxed{-2\sqrt{a^3b^3}\cdot\left(ab-\sqrt{a^2b}\right)} = \boxed{-2\sqrt{a^{2+1}b^{2+1}}\cdot\left(ab-a\sqrt{b}\right)} = \boxed{-2\sqrt{\left(a^2\cdot a^1\right)\cdot\left(b^2\cdot b^1\right)}\cdot\left(ab-a\sqrt{b}\right)}$$

$$= \boxed{-2ab\sqrt{ab}\cdot\left(ab-a\sqrt{b}\right)} = \boxed{\left(-2ab\sqrt{ab}\right)\cdot(ab)+\left(2ab\sqrt{ab}\right)\cdot\left(a\sqrt{b}\right)} = \boxed{\left[-2(a\cdot a)\cdot(b\cdot b)\sqrt{ab}\right]+\left[2(a\cdot a)b\sqrt{ab\cdot b}\right]}$$

$$= \boxed{\left(-2a^2b^2\sqrt{ab}\right)+\left(2a^2b\sqrt{ab^2}\right)} = \boxed{-2a^2b^2\sqrt{ab}+2a^2(b\cdot b)\sqrt{a}} = \boxed{-2a^2b^2\sqrt{ab}+2a^2b^2\sqrt{a}}$$

$$= \boxed{\mathbf{2a^2b^2\sqrt{a}\left(1-\sqrt{b}\right)}}$$

Example 4.2-106

$$\boxed{\sqrt[3]{u^5v^4}\cdot\left(uv-\sqrt[3]{uv}\right)} = \boxed{\sqrt[3]{u^{3+2}v^{3+1}}\cdot\left(uv-\sqrt[3]{uv}\right)} = \boxed{\sqrt[3]{u^3\cdot u^2\cdot v^3\cdot v^1}\cdot\left(uv-\sqrt[3]{uv}\right)} = \boxed{uv\sqrt[3]{u^2v}\cdot\left(uv-\sqrt[3]{uv}\right)}$$

$$= \boxed{\left(uv\sqrt[3]{u^2v}\right)\cdot(uv)-\left(uv\sqrt[3]{u^2v}\right)\left(\sqrt[3]{uv}\right)} = \boxed{\left[(u\cdot u)\cdot(v\cdot v)\sqrt[3]{u^2v}\right]-\left[uv\sqrt[3]{\left(u^2u^1\right)\cdot\left(v^1v^1\right)}\right]}$$

$$= \boxed{\left[u^2v^2\sqrt[3]{u^2v}\right] - \left[uv\sqrt[3]{u^{2+1}\cdot v^{1+1}}\right]} = \boxed{u^2v^2\sqrt[3]{u^2v} - uv\sqrt[3]{u^3v^2}} = \boxed{u^2v^2\sqrt[3]{u^2v} - (u\cdot u)v\sqrt[3]{v^2}}$$

$$= \boxed{u^2v^2\sqrt[3]{u^2v} - u^2v\sqrt[3]{v^2}} = \boxed{\mathbf{u^2v\left(v\sqrt[3]{u^2v} - \sqrt[3]{v^2}\right)}}$$

Example 4.2-107

$$\boxed{\sqrt{mn}\cdot\left(\sqrt{mn^2} - \sqrt{m^5n^6}\right)} = \boxed{\sqrt{mn}\cdot\left(n\sqrt{m} - \sqrt{m^{2+2+1}n^{2+2+2}}\right)} = \boxed{\sqrt{mn}\cdot\left[n\sqrt{m} - \sqrt{\left(m^2\cdot m^2\cdot m^1\right)\cdot\left(n^2\cdot n^2\cdot n^2\right)}\right]}$$

$$= \boxed{\sqrt{mn}\cdot\left[n\sqrt{m} - \left(m^1\cdot m^1\right)\cdot\left(n^1\cdot n^1\cdot n^1\right)\sqrt{m}\right]} = \boxed{\sqrt{mn}\cdot\left[n\sqrt{m} - \left(m^{1+1}\right)\cdot\left(n^{1+1+1}\right)\sqrt{m}\right]} = \boxed{\sqrt{mn}\cdot\left(n\sqrt{m} - m^2n^3\sqrt{m}\right)}$$

$$= \boxed{\left(n\sqrt{mn}\cdot\sqrt{m}\right) - m^2n^3\left(\sqrt{mn}\cdot\sqrt{m}\right)} = \boxed{\left[n\sqrt{(m\cdot m)n}\right] - m^2n^3\left[\sqrt{(m\cdot m)n}\right]} = \boxed{\left(n\sqrt{m^2n}\right) - m^2n^3\left(\sqrt{m^2n}\right)}$$

$$= \boxed{mn\sqrt{n} - \left(m^2\cdot m\right)n^3\sqrt{n}} = \boxed{mn\sqrt{n} - m^3n^3\sqrt{n}} = \boxed{\mathbf{mn\sqrt{n}\left(1 - m^2n^2\right)}}$$

Example 4.2-108

$$\boxed{\sqrt[4]{x^5y^4}\cdot\left(\sqrt[4]{x^2y^5} - \sqrt[4]{x^6y^4}\right)} = \boxed{y\sqrt[4]{x^{4+1}}\cdot\left(\sqrt[4]{x^2y^{4+1}} - y\sqrt[4]{x^{4+2}}\right)} = \boxed{y\sqrt[4]{x^4\cdot x^1}\cdot\left(\sqrt[4]{x^2\cdot y^4\cdot y^1} - y\sqrt[4]{x^4\cdot x^2}\right)}$$

$$= \boxed{xy\sqrt[4]{x}\left(y\sqrt[4]{x^2y} - xy\sqrt[4]{x^2}\right)} = \boxed{x(y\cdot y)\left(\sqrt[4]{x}\cdot\sqrt[4]{x^2y}\right) - (x\cdot x)(y\cdot y)\left(\sqrt[4]{x}\cdot\sqrt[4]{x^2}\right)}$$

$$= \boxed{xy^2\left(\sqrt[4]{x\cdot x^2y}\right) - x^2y^2\left(\sqrt[4]{x\cdot x^2}\right)} = \boxed{xy^2\left(\sqrt[4]{x^3y}\right) - x^2y^2\left(\sqrt[4]{x^3}\right)} = \boxed{\mathbf{xy^2\left(\sqrt[4]{x^3y} - x\sqrt[4]{x^3}\right)}}$$

Example 4.2-109

$$\boxed{\sqrt[3]{a^5b^4c^3}\cdot\left(\sqrt[3]{a^2b^5c^6} - \sqrt[3]{a^6b^3}\right)} = \boxed{c\sqrt[3]{a^{3+2}b^{3+1}}\cdot\left(\sqrt[3]{a^2b^{3+2}c^{3+3}} - b\sqrt[3]{a^{3+3}}\right)}$$

$$= \boxed{c\sqrt[3]{\left(a^3\cdot a^2\right)\cdot\left(b^3\cdot b^1\right)}\cdot\left[\sqrt[3]{a^2\cdot\left(b^3\cdot b^2\right)\cdot\left(c^3\cdot c^3\right)} - b\sqrt[3]{a^3\cdot a^3}\right]} = \boxed{abc\sqrt[3]{a^2b}\cdot\left[b\cdot\left(c^1\cdot c^1\right)\sqrt[3]{a^2b^2} - \left(a^1\cdot a^1\right)\cdot b\right]}$$

$$= \boxed{abc\sqrt[3]{a^2b}\cdot\left[b\cdot c^{1+1}\sqrt[3]{a^2b^2} - a^{1+1}\cdot b\right]} = \boxed{abc\sqrt[3]{a^2b}\cdot\left[bc^2\sqrt[3]{a^2b^2} - a^2b\right]}$$

$$= \boxed{\left(abc\cdot bc^2\right)\cdot\left(\sqrt[3]{a^2b}\cdot\sqrt[3]{a^2b^2}\right) - \left(abc\cdot a^2b\right)\sqrt[3]{a^2b}} = \boxed{ab^2c^3\cdot\left(\sqrt[3]{a^2b\cdot a^2b^2}\right) - a^3b^2c\sqrt[3]{a^2b}}$$

$$= \boxed{ab^2c^3\cdot\left(\sqrt[3]{a^4b^3}\right) - a^3b^2c\sqrt[3]{a^2b}} = \boxed{ab^2c^3\cdot\left(b\sqrt[3]{a^{3+1}}\right) - a^3b^2c\sqrt[3]{a^2b}} = \boxed{ab^2c^3\cdot\left(b\sqrt[3]{a^3\cdot a^1}\right) - a^3b^2c\sqrt[3]{a^2b}}$$

$$= \boxed{ab^2c^3\cdot\left(ab\sqrt[3]{a}\right) - a^3b^2c\sqrt[3]{a^2b}} = \boxed{a^2b^3c^3\sqrt[3]{a} - a^3b^2c\sqrt[3]{a^2b}} = \boxed{\mathbf{a^2b^2c\left(bc^2\sqrt[3]{a} - a\sqrt[3]{a^2b}\right)}}$$

Example 4.2-110

$$5r\sqrt{s}\cdot\left(r^2s^2+s\sqrt{r^3s^5}\right)=\left(5r\cdot r^2s^2\sqrt{s}\right)+\left(5r\cdot s\sqrt{r^3s^5s}\right)=\left(5r^1\cdot r^2s^2\sqrt{s}\right)+\left(5rs\sqrt{r^3s^5s^1}\right)$$

$$=\left(5r^{1+2}s^2\sqrt{s}\right)+\left(5rs\sqrt{r^3s^{5+1}}\right)=\left(5r^3s^2\sqrt{s}\right)+\left(5rs\sqrt{r^3s^6}\right)=\left(5r^3s^2\sqrt{s}\right)+\left(5rs\sqrt{r^{2+1}s^{2+2+2}}\right)$$

$$=\left(5r^3s^2\sqrt{s}\right)+\left(5rs\sqrt{r^2\cdot r^1\cdot s^2\cdot s^2\cdot s^2}\right)=\left(5r^3s^2\sqrt{s}\right)+\left(5rs\cdot r^1\cdot s^1\cdot s^1\cdot s^1\sqrt{r}\right)$$

$$=\left(5r^3s^2\sqrt{s}\right)+\left[5\left(r^1\cdot r^1\right)\cdot\left(s^1\cdot s^1\cdot s^1\cdot s^1\right)\cdot\sqrt{r}\right]=\left(5r^3s^2\sqrt{s}\right)+\left(5r^{1+1}\cdot s^{1+1+1+1}\cdot\sqrt{r}\right)=5r^3s^2\sqrt{s}+5r^2s^4\sqrt{r}$$

$$=\mathbf{5r^2s^2\left(r\sqrt{s}+s^2\sqrt{r}\right)}$$

Practice Problems - Multiplying Monomial and Binomial Expressions in Radical Form, with Variables

Section 4.2 Case III b Practice Problems - Multiply the following radical expressions:

1. $x\sqrt{x^3}\cdot\left(x^2+\sqrt{x}\right)=$
2. $\sqrt{a^5b^4}\cdot\left(\sqrt{a^2b}+\sqrt{ab^3}\right)=$
3. $\left(w^3-\sqrt{u^3w^2}\right)\cdot\sqrt{u^4w^5}=$
4. $-\sqrt{m}\cdot\left(-\sqrt{m}+\sqrt{m^3}\right)=$
5. $\left(\sqrt[5]{x^4y^6}+\sqrt[5]{x}\right)\cdot\sqrt[5]{x^{10}y^7}=$
6. $\sqrt[3]{p^5q^2}\cdot\left(p^2+\sqrt[3]{p^4q}\right)=$
7. $\sqrt[3]{xy^4}\cdot\left(\sqrt[3]{x^5y^3}+\sqrt[3]{x^2y^2}\right)=$
8. $2\sqrt[4]{r}\cdot\left(\sqrt[4]{r^5s^6}-3\sqrt[4]{r^2s^7}\right)=$
9. $3a\sqrt{b}\cdot\left(a-\sqrt{b^3}\right)=$
10. $-3\sqrt{m^5}\cdot\left(m-4\sqrt{m^3}\right)=$

4.3 Dividing Radical Expressions

Radicals are divided by each other using the following general rule:

$$\sqrt[n]{\frac{x}{y}} = \frac{\sqrt[n]{x}}{\sqrt[n]{y}} \qquad x \geq 0,\ y \rangle 0$$

In section 4.1 the difference between rational and irrational numbers was discussed. We learned that the square root of non perfect squares, the cube root of non perfect cubes, etc. are irrational numbers. For example, $\sqrt{3}, \sqrt{7}, \sqrt{10}, \sqrt[3]{4}, \sqrt[5]{7}, etc.$ are classified as irrational numbers. In division of radicals, if the denominator of a fractional radical expression is not a rational number, we rationalize the denominator by changing the radicand of the denominator to a perfect square, a perfect cube, etc. In this section, students learn how to rationalize radical expressions with monomial (Case I) and binomial (Case II) denominators.

Case I Rationalizing Radical Expressions with Monomial Denominators

Simplification of radical expressions being divided requires rationalization of the denominator. A monomial and irrational denominator is rationalized by multiplying the numerator and the denominator by the irrational denominator. This change the radicand of the denominator to a perfect square.

Examples:

1. $\frac{\sqrt{1}}{\sqrt{7}} = \frac{\sqrt{1}}{\sqrt{7}} \times \frac{\sqrt{7}}{\sqrt{7}} = \frac{\sqrt{1 \cdot 7}}{\sqrt{7 \cdot 7}} = \frac{\sqrt{7}}{\sqrt{7^2}} = \frac{\sqrt{7}}{7}$

   Note that $\sqrt{7}$ is an irrational number. By multiplying $\sqrt{7}$ by itself the denominator is changed to a rational number, i.e., 7.

2. $\sqrt{\frac{20}{3}} = \frac{\sqrt{20}}{\sqrt{3}} = \frac{\sqrt{4 \cdot 5}}{\sqrt{3}} = \frac{\sqrt{2^2 \cdot 5}}{\sqrt{3}} = \frac{2\sqrt{5}}{\sqrt{3}} = \frac{2\sqrt{5}}{\sqrt{3}} \times \frac{\sqrt{3}}{\sqrt{3}} = \frac{2\sqrt{5 \cdot 3}}{\sqrt{3} \cdot \sqrt{3}} = \frac{2\sqrt{15}}{\sqrt{3^2}} = \mathbf{\frac{2\sqrt{15}}{3}}$

   Again, note that $\sqrt{3}$ is an irrational number. By multiplying $\sqrt{3}$ by itself the denominator is changed to a rational number, i.e., 3.

Division of monomial terms in radical form is divided into two cases. Case I a - rationalization of radical expressions, with real numbers and Case I b - rationalization of radical expressions , with variables.

Case I a Rationalizing Radical Expressions - Monomial Denominators with Real Numbers

Radical expressions with monomial denominators are simplified using the following steps:

Step 1 Change the radical expression $\sqrt{\frac{a}{b}}$ to $\frac{\sqrt{a}}{\sqrt{b}}$ and simplify (see Section 4.1, Case III).

Step 2 Rationalize the denominator by multiplying the numerator and the denominator of the radical expression $\frac{\sqrt{a}}{\sqrt{b}}$ by $\sqrt{b}$.

Step 3 Simplify the radical expression (see Section 4.1, Case III).

Examples with Steps

The following examples show the steps as to how radical expressions with monomial denominators are simplified:

Example 4.3-1

$$\boxed{\frac{-8\sqrt{3}}{32\sqrt{45}}} =$$

Solution:

Step 1

$$\boxed{\frac{-8\sqrt{3}}{32\sqrt{45}}} = \boxed{-\frac{\overset{1}{\cancel{8}}\sqrt{3}}{\underset{4}{\cancel{32}}\sqrt{45}}} = \boxed{-\frac{\sqrt{3}}{4\sqrt{9\cdot 5}}} = \boxed{-\frac{\sqrt{3}}{4\sqrt{3^2\cdot 5}}} = \boxed{-\frac{\sqrt{3}}{4\cdot 3\sqrt{5}}} = \boxed{-\frac{\sqrt{3}}{12\sqrt{5}}}$$

Step 2

$$\boxed{-\frac{\sqrt{3}}{12\sqrt{5}}} = \boxed{-\frac{\sqrt{3}}{12\sqrt{5}} \times \frac{\sqrt{5}}{\sqrt{5}}}$$

Step 3

$$\boxed{-\frac{\sqrt{3}}{12\sqrt{5}} \times \frac{\sqrt{5}}{\sqrt{5}}} = \boxed{-\frac{\sqrt{3}\times\sqrt{5}}{12\sqrt{5}\times\sqrt{5}}} = \boxed{-\frac{\sqrt{3\cdot 5}}{12\sqrt{5\cdot 5}}} = \boxed{-\frac{\sqrt{15}}{12\sqrt{5^1\cdot 5^1}}} = \boxed{-\frac{\sqrt{15}}{12\sqrt{5^{1+1}}}}$$

$$= \boxed{-\frac{\sqrt{15}}{12\sqrt{5^2}}} = \boxed{-\frac{\sqrt{15}}{12\cdot 5}} = \boxed{-\frac{\sqrt{15}}{60}}$$

Example 4.3-2

$$\boxed{\frac{5\sqrt{(-2)^2\cdot 3}}{-\sqrt{7}}} =$$

Solution:

Step 1

$$\boxed{\frac{5\sqrt{(-2)^2\cdot 3}}{-\sqrt{7}}} = \boxed{\frac{5\sqrt{(-2)^2\cdot 3}}{-\sqrt{7}}} = \boxed{-\frac{5\sqrt{4\cdot 3}}{\sqrt{7}}} = \boxed{-\frac{5\sqrt{2^2\cdot 3}}{\sqrt{7}}} = \boxed{-\frac{(5\cdot 2)\sqrt{3}}{\sqrt{7}}} = \boxed{-\frac{10\sqrt{3}}{\sqrt{7}}}$$

Step 2

$$\boxed{-\frac{10\sqrt{3}}{\sqrt{7}}} = \boxed{-\frac{10\sqrt{3}}{\sqrt{7}} \times \frac{\sqrt{7}}{\sqrt{7}}}$$

Step 3

$$\boxed{-\frac{10\sqrt{3}}{\sqrt{7}} \times \frac{\sqrt{7}}{\sqrt{7}}} = \boxed{-\frac{10\sqrt{3}\times\sqrt{7}}{\sqrt{7}\times\sqrt{7}}} = \boxed{-\frac{10\sqrt{3\cdot 7}}{\sqrt{7\cdot 7}}} = \boxed{-\frac{10\sqrt{21}}{\sqrt{7^1\cdot 7^1}}} = \boxed{-\frac{10\sqrt{21}}{\sqrt{7^{1+1}}}} = \boxed{-\frac{10\sqrt{21}}{\sqrt{7^2}}}$$

$$= \boxed{-\frac{10\sqrt{21}}{7}}$$

Example 4.3-3

$$\boxed{\frac{100}{\sqrt[2]{1000}}} =$$

Solution:

Step 1 $\boxed{\frac{100}{\sqrt[2]{1000}}} = \boxed{\frac{100}{\sqrt{1000}}} = \boxed{\frac{100}{\sqrt{100\cdot 10}}} = \boxed{\frac{100}{\sqrt{10^2\cdot 10}}} = \boxed{\frac{\overset{10}{\cancel{100}}}{\underset{1}{\cancel{10}}\cdot\sqrt{10}}} = \boxed{\frac{10}{1\cdot\sqrt{10}}} = \boxed{\frac{10}{\sqrt{10}}}$

Step 2 $\boxed{\frac{10}{\sqrt{10}}} = \boxed{\frac{10}{\sqrt{10}} \times \frac{\sqrt{10}}{\sqrt{10}}}$

Step 3 $\boxed{\frac{10}{\sqrt{10}} \times \frac{\sqrt{10}}{\sqrt{10}}} = \boxed{\frac{10\times\sqrt{10}}{\sqrt{10}\times\sqrt{10}}} = \boxed{\frac{10\cdot\sqrt{10}}{\sqrt{10\cdot 10}}} = \boxed{\frac{10\cdot\sqrt{10}}{\sqrt{10^{1+1}}}} = \boxed{\frac{10\cdot\sqrt{10}}{\sqrt{10^2}}} = \boxed{\frac{\overset{1}{\cancel{10}}\cdot\sqrt{10}}{\underset{1}{\cancel{10}}}} = \boxed{\frac{1\cdot\sqrt{10}}{1}}$

$= \boxed{\mathbf{\sqrt{10}}}$

Example 4.3-4

$\boxed{\frac{3\sqrt[5]{8}}{\sqrt[5]{81}}} =$

Solution:

Step 1 $\boxed{\frac{3\sqrt[5]{8}}{\sqrt[5]{81}}} = \boxed{\frac{3\sqrt[5]{8}}{\sqrt[5]{3^4}}}$

Step 2 $\boxed{\frac{3\sqrt[5]{8}}{\sqrt[5]{3^4}}} = \boxed{\frac{3\sqrt[5]{8}}{\sqrt[5]{3^4}} \times \frac{\sqrt[5]{3^1}}{\sqrt[5]{3^1}}}$

Note that radical expressions with third, fourth, or higher root in the denominator can also be rationalized by changing the denominator to a perfect third, fourth, or higher power.

Step 3 $\boxed{\frac{3\sqrt[5]{8}}{\sqrt[5]{3^4}} \times \frac{\sqrt[5]{3^1}}{\sqrt[5]{3^1}}} = \boxed{\frac{3\sqrt[5]{8}\times\sqrt[5]{3^1}}{\sqrt[5]{3^4}\times\sqrt[5]{3^1}}} = \boxed{\frac{3\sqrt[5]{8\cdot 3}}{\sqrt[5]{3^4\cdot 3^1}}} = \boxed{\frac{3\sqrt[5]{24}}{\sqrt[5]{3^{4+1}}}} = \boxed{\frac{3\sqrt[5]{24}}{\sqrt[5]{3^5}}} = \boxed{\frac{\overset{1}{\cancel{3}}\cdot\sqrt[5]{24}}{\underset{1}{\cancel{3}}}} = \boxed{\frac{1\cdot\sqrt[5]{24}}{1}}$

$= \boxed{\mathbf{\sqrt[5]{24}}}$

Example 4.3-5

$\boxed{\sqrt[4]{\frac{2}{25}}} =$

Solution:

Step 1 $\boxed{\sqrt[4]{\frac{2}{25}}} = \boxed{\sqrt[4]{\frac{2}{5^2}}} = \boxed{\frac{\sqrt[4]{2}}{\sqrt[4]{5^2}}}$

Step 2 $\boxed{\frac{\sqrt[4]{2}}{\sqrt[4]{5^2}}} = \boxed{\frac{\sqrt[4]{2}}{\sqrt[4]{5^2}} \times \frac{\sqrt[4]{5^2}}{\sqrt[4]{5^2}}}$

Step 3 $\frac{\sqrt[4]{2}}{\sqrt[4]{5^2}} \times \frac{\sqrt[4]{5^2}}{\sqrt[4]{5^2}} = \frac{\sqrt[4]{2 \cdot 5^2}}{\sqrt[4]{5^2 \cdot 5^2}} = \frac{\sqrt[4]{2 \cdot 25}}{\sqrt[4]{5^{2+2}}} = \frac{\sqrt[4]{50}}{\sqrt[4]{5^4}} = \boxed{\frac{\sqrt[4]{50}}{5}}$

Additional Examples: Rationalizing Radical Expressions - Monomial Denominators with Real Numbers

The following examples further illustrate how to solve radical expressions with monomial denominators:

Example 4.3-6

$$\sqrt[2]{\frac{1}{4}} = \sqrt{\frac{1}{4}} = \frac{\sqrt{1}}{\sqrt{4}} = \frac{\sqrt{1}}{\sqrt{2^2}} = \boxed{\frac{1}{2}}$$

Example 4.3-7

$$\frac{8\sqrt{3}}{\sqrt{2}} = \frac{8\sqrt{3}}{\sqrt{2}} \times \frac{\sqrt{2}}{\sqrt{2}} = \frac{8\sqrt{3} \times \sqrt{2}}{\sqrt{2} \times \sqrt{2}} = \frac{8\sqrt{3 \cdot 2}}{\sqrt{2 \cdot 2}} = \frac{8\sqrt{6}}{\sqrt{2^1 \cdot 2^1}} = \frac{8\sqrt{6}}{\sqrt{2^{1+1}}} = \frac{8\sqrt{6}}{\sqrt{2^2}} = \frac{\overset{4}{\cancel{8}}\sqrt{6}}{\underset{1}{\cancel{2}}} = \frac{4\sqrt{6}}{1} = \boxed{4\sqrt{6}}$$

Example 4.3-8

$$\sqrt[2]{\frac{5}{7}} = \sqrt{\frac{5}{7}} = \frac{\sqrt{5}}{\sqrt{7}} = \frac{\sqrt{5}}{\sqrt{7}} \times \frac{\sqrt{7}}{\sqrt{7}} = \frac{\sqrt{5} \times \sqrt{7}}{\sqrt{7} \times \sqrt{7}} = \frac{\sqrt{5 \cdot 7}}{\sqrt{7 \cdot 7}} = \frac{\sqrt{35}}{\sqrt{7^1 \cdot 7^1}} = \frac{\sqrt{35}}{\sqrt{7^{1+1}}} = \frac{\sqrt{35}}{\sqrt{7^2}} = \boxed{\frac{\sqrt{35}}{7}}$$

Example 4.3-9

$$\frac{1}{\sqrt[2]{5}} = \frac{1}{\sqrt{5}} = \frac{1}{\sqrt{5}} \times \frac{\sqrt{5}}{\sqrt{5}} = \frac{1 \times \sqrt{5}}{\sqrt{5} \times \sqrt{5}} = \frac{\sqrt{5}}{\sqrt{5 \cdot 5}} = \frac{\sqrt{5}}{\sqrt{5^1 \cdot 5^1}} = \frac{\sqrt{5}}{\sqrt{5^{1+1}}} = \frac{\sqrt{5}}{\sqrt{5^2}} = \boxed{\frac{\sqrt{5}}{5}}$$

Example 4.3-10

$$\frac{40\sqrt{12}}{5\sqrt{6}} = \frac{\overset{8}{\cancel{40}}\sqrt{12}}{\underset{1}{\cancel{5}}\sqrt{6}} = \frac{8\sqrt{12}}{\sqrt{6}} = \frac{8\sqrt{4 \cdot 3}}{\sqrt{6}} = \frac{8\sqrt{2^2 \cdot 3}}{\sqrt{6}} = \frac{(8 \cdot 2)\sqrt{3}}{\sqrt{6}} = \frac{16\sqrt{3}}{\sqrt{6}} = \frac{16\sqrt{3}}{\sqrt{6}} \times \frac{\sqrt{6}}{\sqrt{6}} = \frac{16\sqrt{3} \times \sqrt{6}}{\sqrt{6} \times \sqrt{6}}$$

$$= \frac{16\sqrt{3 \cdot 6}}{\sqrt{6 \cdot 6}} = \frac{16\sqrt{18}}{\sqrt{6^1 \cdot 6^1}} = \frac{16\sqrt{9 \cdot 2}}{\sqrt{6^{1+1}}} = \frac{16\sqrt{3^2 \cdot 2}}{\sqrt{6^2}} = \frac{(16 \cdot 3)\sqrt{2}}{6} = \frac{\overset{8}{\cancel{48}}\sqrt{2}}{\underset{1}{\cancel{6}}} = \frac{8\sqrt{2}}{1} = \boxed{8\sqrt{2}}$$

Example 4.3-11

$$\sqrt{\frac{1000}{36}} = \sqrt{\frac{\overset{250}{\cancel{1000}}}{\underset{9}{\cancel{36}}}} = \sqrt{\frac{250}{9}} = \frac{\sqrt{250}}{\sqrt{9}} = \frac{\sqrt{25 \cdot 10}}{\sqrt{3^2}} = \frac{\sqrt{5^2 \times 10}}{3} = \boxed{\frac{5\sqrt{10}}{3}}$$

Example 4.3-12

$$\frac{-2\sqrt{8}}{-5} = +\frac{2\sqrt{8}}{5} = \frac{2\sqrt{4 \cdot 2}}{5} = \frac{2\sqrt{2^2 \cdot 2}}{5} = \frac{(2 \cdot 2)\sqrt{2}}{5} = \boxed{\frac{4\sqrt{2}}{5}}$$

Example 4.3-13

$$\frac{\sqrt[2]{40}}{\sqrt[2]{5}} = \frac{\sqrt{40}}{\sqrt{5}} = \sqrt{\frac{\overset{8}{\cancel{40}}}{\underset{1}{\cancel{5}}}} = \sqrt{\frac{8}{1}} = \sqrt{8} = \sqrt{4 \cdot 2} = \sqrt{2^2 \cdot 2} = \mathbf{2\sqrt{2}}$$

The above problem can also be solved in the following way:

$$\frac{\sqrt[2]{40}}{\sqrt[2]{5}} = \frac{\sqrt{40}}{\sqrt{5}} = \frac{\sqrt{4 \cdot 10}}{\sqrt{5}} = \frac{\sqrt{2^2 \cdot 10}}{\sqrt{5}} = \frac{2\sqrt{10}}{\sqrt{5}} \times \frac{\sqrt{5}}{\sqrt{5}} = \frac{2\sqrt{10} \times \sqrt{5}}{\sqrt{5} \times \sqrt{5}} = \frac{2\sqrt{10 \cdot 5}}{\sqrt{5 \cdot 5}} = \frac{2\sqrt{50}}{\sqrt{5^1 \cdot 5^1}} = \frac{2\sqrt{25 \cdot 2}}{\sqrt{5^{1+1}}}$$

$$= \frac{2\sqrt{5^2 \cdot 2}}{\sqrt{5^2}} = \frac{(2 \cdot 5)\sqrt{2}}{5} = \frac{\overset{2}{\cancel{10}}\sqrt{2}}{\underset{1}{\cancel{5}}} = \frac{2\sqrt{2}}{1} = \mathbf{2\sqrt{2}}$$

Example 4.3-14

$$\frac{8\sqrt{3}}{\sqrt{2}} = \frac{8\sqrt{3}}{\sqrt{2}} = \frac{8\sqrt{3}}{\sqrt{2}} \times \frac{\sqrt{2}}{\sqrt{2}} = \frac{8\sqrt{3} \times \sqrt{2}}{\sqrt{2} \times \sqrt{2}} = \frac{8\sqrt{3 \cdot 2}}{\sqrt{2 \cdot 2}} = \frac{8\sqrt{6}}{\sqrt{2^1 \cdot 2^1}} = \frac{8\sqrt{6}}{\sqrt{2^{1+1}}} = \frac{8\sqrt{6}}{\sqrt{2^2}} = \frac{\overset{4}{\cancel{8}}\sqrt{6}}{\underset{1}{\cancel{2}}} = \frac{4\sqrt{6}}{1} = \mathbf{4\sqrt{6}}$$

Example 4.3-15

$$\frac{-2}{\sqrt[2]{5}} = \frac{-2}{\sqrt{5}} = -\frac{2}{\sqrt{5}} \times \frac{\sqrt{5}}{\sqrt{5}} = -\frac{2 \times \sqrt{5}}{\sqrt{5} \times \sqrt{5}} = -\frac{2\sqrt{5}}{\sqrt{5 \cdot 5}} = -\frac{2\sqrt{5}}{\sqrt{5^1 \cdot 5^1}} = -\frac{2\sqrt{5}}{\sqrt{5^{1+1}}} = -\frac{2\sqrt{5}}{\sqrt{5^2}} = \mathbf{-\frac{2\sqrt{5}}{5}}$$

Example 4.3-16

$$\sqrt[3]{\frac{1}{5}} = \frac{\sqrt[3]{1}}{\sqrt[3]{5^1}} = \frac{\sqrt[3]{1}}{\sqrt[3]{5^1}} \times \frac{\sqrt[3]{5^2}}{\sqrt[3]{5^2}} = \frac{\sqrt[3]{1} \times \sqrt[3]{5^2}}{\sqrt[3]{5^1} \times \sqrt[3]{5^2}} = \frac{\sqrt[3]{1 \cdot 5^2}}{\sqrt[3]{5^1 \cdot 5^2}} = \frac{\sqrt[3]{5^2}}{\sqrt[3]{5^{1+2}}} = \frac{\sqrt[3]{25}}{\sqrt[3]{5^3}} = \mathbf{\frac{\sqrt[3]{25}}{5}}$$

Example 4.3-17

$$\sqrt[5]{\frac{32}{8^3}} = \frac{\sqrt[5]{2^5}}{\sqrt[5]{8^3}} = \frac{2}{\sqrt[5]{8^3}} \times \frac{\sqrt[5]{8^2}}{\sqrt[5]{8^2}} = \frac{2 \times \sqrt[5]{8^2}}{\sqrt[5]{8^3} \times \sqrt[5]{8^2}} = \frac{2 \cdot \sqrt[5]{8^2}}{\sqrt[5]{8^3 \cdot 8^2}} = \frac{2 \cdot \sqrt[5]{64}}{\sqrt[5]{8^{3+2}}} = \frac{2 \cdot \sqrt[5]{64}}{\sqrt[5]{8^5}} = \frac{\overset{1}{\cancel{2}} \cdot \sqrt[5]{64}}{\underset{4}{\cancel{8}}} = \frac{1 \cdot \sqrt[5]{64}}{4} = \mathbf{\frac{\sqrt[5]{64}}{4}}$$

Example 4.3-18

$$\sqrt[4]{\frac{64}{256}} = \frac{\sqrt[4]{16 \cdot 4}}{\sqrt[4]{4^4}} = \frac{\sqrt[4]{2^4 \cdot 4}}{4} = \frac{\overset{1}{\cancel{2}} \cdot \sqrt[4]{4}}{\underset{2}{\cancel{4}}} = \frac{1 \cdot \sqrt[4]{4}}{2} = \mathbf{\frac{\sqrt[4]{4}}{2}}$$

Example 4.3-19

$$\sqrt[3]{\frac{4}{7}} = \frac{\sqrt[3]{4}}{\sqrt[3]{7^1}} = \frac{\sqrt[3]{4}}{\sqrt[3]{7^1}} \times \frac{\sqrt[3]{7^2}}{\sqrt[3]{7^2}} = \frac{\sqrt[3]{4} \times \sqrt[3]{7^2}}{\sqrt[3]{7^1} \times \sqrt[3]{7^2}} = \frac{\sqrt[3]{4 \cdot 7^2}}{\sqrt[3]{7^1 \cdot 7^2}} = \frac{\sqrt[3]{4 \cdot 49}}{\sqrt[3]{7^{1+2}}} = \frac{\sqrt[3]{196}}{\sqrt[3]{7^3}} = \mathbf{\frac{\sqrt[3]{196}}{7}}$$

Example 4.3-20

$$\boxed{\frac{7}{\sqrt[4]{49}}} = \boxed{\frac{7}{\sqrt[4]{7^2}}} = \boxed{\frac{7}{\sqrt[4]{7^2}} \times \frac{\sqrt[4]{7^2}}{\sqrt[4]{7^2}}} = \boxed{\frac{7 \times \sqrt[4]{7^2}}{\sqrt[4]{7^2} \times \sqrt[4]{7^2}}} = \boxed{\frac{7 \cdot \sqrt[4]{7^2}}{\sqrt[4]{7^2 \cdot 7^2}}} = \boxed{\frac{7 \cdot \sqrt[4]{49}}{\sqrt[4]{7^{2+2}}}} = \boxed{\frac{7 \cdot \sqrt[4]{49}}{\sqrt[4]{7^4}}} = \boxed{\frac{\overset{1}{\cancel{7}} \cdot \sqrt[4]{49}}{\underset{1}{\cancel{7}}}} = \boxed{\frac{1 \cdot \sqrt[4]{49}}{1}}$$

$$= \boxed{\sqrt[4]{49}}$$

Practice Problems: Rationalizing Radical Expressions - Monomial Denominators with Real Numbers

Section 4.3 Case I a Practice Problems - Solve the following radical expressions:

1. $\sqrt{\frac{1}{8}} =$
2. $\sqrt[2]{\frac{50}{7}} =$
3. $\frac{\sqrt{75}}{-5} =$
4. $\sqrt[3]{\frac{25}{16}} =$
5. $\sqrt[5]{\frac{32}{8}} =$
6. $\frac{-3\sqrt{100}}{-5\sqrt{3000}} =$
7. $\frac{-20}{\sqrt[2]{45}} =$
8. $\frac{1}{\sqrt[4]{324}} =$
9. $\sqrt[2]{\frac{3}{48}} =$
10. $\sqrt[5]{\frac{243}{256}} =$

Case I b Rationalizing Radical Expressions - Monomial Denominators with Variables

Radical expressions with monomial denominators are simplified using the following steps:

Step 1 Change the radical expression $\sqrt{\frac{x}{y}}$ to $\frac{\sqrt{x}}{\sqrt{y}}$ and simplify (see Section 4.1, Case IV).

Step 2 Rationalize the denominator by multiplying the numerator and the denominator of the radical expression $\frac{\sqrt{x}}{\sqrt{y}}$ by $\sqrt{y}$.

Step 3 Simplify the radical expression (see Section 4.1, Case IV).

Examples with Steps

The following examples show the steps as to how radical expressions with monomial denominators are simplified:

Example 4.3-21

$$\frac{\sqrt{25a^4b^5}}{\sqrt{36a^7b^2}} =$$

Solution:

Step 1

$$\frac{\sqrt{25a^4b^5}}{\sqrt{36a^7b^2}} = \sqrt{\frac{25a^4b^5}{36a^7b^2}} = \sqrt{\frac{5^2b^5b^{-2}}{6^2a^7a^{-4}}} = \frac{5}{6}\sqrt{\frac{b^{5-2}}{a^{7-4}}} = \frac{5}{6}\sqrt{\frac{b^3}{a^3}} = \frac{5}{6}\frac{\sqrt{b^3}}{\sqrt{a^3}}$$

$$= \frac{5}{6}\frac{\sqrt{b^{2+1}}}{\sqrt{a^{2+1}}} = \frac{5}{6}\frac{\sqrt{b^2b^1}}{\sqrt{a^2a^1}} = \frac{5b}{6a}\frac{\sqrt{b}}{\sqrt{a}}$$

Step 2

$$\frac{5b}{6a}\frac{\sqrt{b}}{\sqrt{a}} = \frac{5b}{6a}\frac{\sqrt{b}}{\sqrt{a}} \times \frac{\sqrt{a}}{\sqrt{a}}$$

Step 3

$$\frac{5b}{6a}\frac{\sqrt{b}}{\sqrt{a}} \times \frac{\sqrt{a}}{\sqrt{a}} = \frac{5b}{6a}\frac{\sqrt{b} \times \sqrt{a}}{\sqrt{a} \times \sqrt{a}} = \frac{5b}{6a}\frac{\sqrt{b \cdot a}}{\sqrt{a^1 \cdot a^1}} = \frac{5b}{6a}\frac{\sqrt{ab}}{\sqrt{a^{1+1}}} = \frac{5b}{6a}\frac{\sqrt{ab}}{\sqrt{a^2}} = \frac{5b}{6a}\frac{\sqrt{ab}}{a}$$

$$= \frac{5b \cdot \sqrt{ab}}{6a \cdot a} = \frac{5b\sqrt{ab}}{6a^1a^1} = \frac{5b\sqrt{ab}}{6a^{1+1}} = \mathbf{\frac{5b\sqrt{ab}}{6a^2}}$$

Example 4.3-22

$$\sqrt{\frac{48m^5n^3}{150m^7n^2}} =$$

Solution:

Step 1

$$\sqrt{\frac{48m^5n^3}{150m^7n^2}} = \sqrt{\frac{(16 \cdot 3)n^3n^{-2}}{(25 \cdot 6)m^7m^{-5}}} = \sqrt{\frac{(4^2 \cdot 3)n^{3-2}}{(5^2 \cdot 6)m^{7-5}}} = \frac{4}{5}\sqrt{\frac{3n^1}{6m^2}} = \frac{4}{5}\sqrt{\frac{\overset{1}{\cancel{3}}n}{\underset{2}{\cancel{6}}m^2}}$$

$$= \boxed{\frac{4}{5}\sqrt{\frac{n}{2m^2}}} = \boxed{\frac{4}{5m}\sqrt{\frac{n}{2}}} = \boxed{\frac{4}{5m}\frac{\sqrt{n}}{\sqrt{2}}}$$

Step 2 $\boxed{\frac{4}{5m}\frac{\sqrt{n}}{\sqrt{2}}} = \boxed{\frac{4}{5m}\frac{\sqrt{n}}{\sqrt{2}} \times \frac{\sqrt{2}}{\sqrt{2}}} =$

Step 3 $\boxed{\frac{4}{5m}\frac{\sqrt{n}}{\sqrt{2}} \times \frac{\sqrt{2}}{\sqrt{2}}} = \boxed{\frac{4}{5m}\frac{\sqrt{n}\times\sqrt{2}}{\sqrt{2}\times\sqrt{2}}} = \boxed{\frac{4}{5m}\frac{\sqrt{n\cdot 2}}{\sqrt{2\cdot 2}}} = \boxed{\frac{4}{5m}\frac{\sqrt{2n}}{\sqrt{2^1\cdot 2^1}}} = \boxed{\frac{4}{5m}\frac{\sqrt{2n}}{\sqrt{2^{1+1}}}}$

$$= \boxed{\frac{4}{5m}\frac{\sqrt{2n}}{\sqrt{2^2}}} = \boxed{\frac{4}{5m}\frac{\sqrt{2n}}{2}} = \boxed{\frac{4\cdot\sqrt{2n}}{(5\cdot 2)m}} = \boxed{\frac{\overset{2}{\cancel{4}}\sqrt{2n}}{\underset{5}{\cancel{10}}m}} = \boxed{\mathbf{\frac{2\sqrt{2n}}{5m}}}$$

Example 4.3-23

$$\boxed{\frac{\sqrt[4]{256x^6y^2z^2}}{\sqrt[4]{32x^9y^7z}}} =$$

Solution:

Step 1 $\boxed{\frac{\sqrt[4]{256x^6y^2z^2}}{\sqrt[4]{32x^9y^7z}}} = \boxed{\sqrt[4]{\frac{256x^6y^2z^2}{32x^9y^7z^1}}} = \boxed{\sqrt[4]{\frac{4^4z^2z^{-1}}{(16\cdot 2)x^9x^{-6}y^7y^{-2}}}} = \boxed{\sqrt[4]{\frac{4^4z^{2-1}}{(2^4\cdot 2)x^{9-6}y^{7-2}}}}$

$$= \boxed{\frac{4}{2}\sqrt[4]{\frac{z^1}{2x^3y^5}}} = \boxed{\frac{\overset{2}{\cancel{4}}}{\underset{1}{\cancel{2}}}\sqrt[4]{\frac{z}{2x^3y^{4+1}}}} = \boxed{\frac{2}{1}\sqrt[4]{\frac{z}{2x^3y^4y^1}}} = \boxed{\frac{2}{1\cdot y}\sqrt[4]{\frac{z}{2x^3y}}} = \boxed{\frac{2}{y}\frac{\sqrt[4]{z}}{\sqrt[4]{2x^3y}}}$$

Step 2 $\boxed{\frac{2}{y}\frac{\sqrt[4]{z}}{\sqrt[4]{2x^3y}}} = \boxed{\frac{2}{y}\frac{\sqrt[4]{z}}{\sqrt[4]{2^1x^3y^1}}} = \boxed{\frac{2}{y}\frac{\sqrt[4]{z}}{\sqrt[4]{2^1x^3y^1}} \times \frac{\sqrt[4]{2^3x^1y^3}}{\sqrt[4]{2^3x^1y^3}}}$

Step 3 $\boxed{\frac{2}{y}\frac{\sqrt[4]{z}}{\sqrt[4]{2^1x^3y^1}} \times \frac{\sqrt[4]{2^3x^1y^3}}{\sqrt[4]{2^3x^1y^3}}} = \boxed{\frac{2}{y}\frac{\sqrt[4]{z}\times\sqrt[4]{2^3x^1y^3}}{\sqrt[4]{2^1x^3y^1}\times\sqrt[4]{2^3x^1y^3}}} = \boxed{\frac{2}{y}\frac{\sqrt[4]{z\cdot 2^3x^1y^3}}{\sqrt[4]{2^1x^3y^1\cdot 2^3x^1y^3}}}$

$$= \boxed{\frac{2}{y}\frac{\sqrt[4]{8xy^3z}}{\sqrt[4]{(2^1\cdot 2^3)\cdot(x^3\cdot x^1)\cdot(y^1\cdot y^3)}}} = \boxed{\frac{2}{y}\frac{\sqrt[4]{8xy^3z}}{\sqrt[4]{(2^{1+3})\cdot(x^{3+1})\cdot(y^{1+3})}}} = \boxed{\frac{2}{y}\frac{\sqrt[4]{8xy^3z}}{\sqrt[4]{2^4\cdot x^4\cdot y^4}}}$$

$$= \boxed{\frac{2}{y}\frac{\sqrt[4]{8xy^3z}}{2xy}} = \boxed{\frac{2\cdot\sqrt[4]{8xy^3z}}{2x(y\cdot y)}} = \boxed{\frac{\overset{1}{\cancel{2}}\cdot\sqrt[4]{8xy^3z}}{\underset{1}{\cancel{2}}x(y^1\cdot y^1)}} = \boxed{\frac{\sqrt[4]{8xy^3z}}{x(y^{1+1})}} = \boxed{\mathbf{\frac{\sqrt[4]{8xy^3z}}{xy^2}}}$$

Example 4.3-24

$$\boxed{\sqrt{\frac{44u^5v^2w^3}{11u^2v^3w^8}}} =$$

Solution:

Step 1

$$\sqrt{\frac{44u^5v^2w^3}{11u^2v^3w^8}} = \sqrt{\frac{(4\cdot 11)\cdot\left(u^5u^{-2}\right)}{11\left(v^3v^{-2}\right)\cdot\left(w^8w^{-3}\right)}} = \sqrt{\frac{2^2\cdot \overset{1}{\cancel{11}}\left(u^{5-2}\right)}{\underset{1}{\cancel{11}}\left(v^{3-2}\right)\cdot\left(w^{8-3}\right)}} = \sqrt{\frac{2^2u^3}{v^1w^5}} = \sqrt{\frac{2^2u^3}{v^1w^5}}$$

$$= \frac{\sqrt{2^2u^3}}{\sqrt{vw^5}} = \frac{2\sqrt{u^{2+1}}}{\sqrt{vw^{2+2+1}}} = \frac{2\sqrt{u^2\cdot u^1}}{\sqrt{vw^2\cdot w^2\cdot w^1}} = \frac{2u\sqrt{u}}{w^1w^1\sqrt{vw}} = \frac{2u\sqrt{u}}{w^{1+1}\sqrt{vw}} = \frac{2u\sqrt{u}}{w^2\sqrt{vw}}$$

Step 2

$$\frac{2u\sqrt{u}}{w^2\sqrt{vw}} = \frac{2u\sqrt{u}}{w^2\sqrt{v^1w^1}} = \frac{2u\sqrt{u}}{w^2\sqrt{v^1w^1}}\times\frac{\sqrt{v^1w^1}}{\sqrt{v^1w^1}}$$

Step 3

$$\frac{2u\sqrt{u}}{w^2\sqrt{v^1w^1}}\times\frac{\sqrt{v^1w^1}}{\sqrt{v^1w^1}} = \frac{2u\sqrt{u}\times\sqrt{v^1w^1}}{w^2\sqrt{v^1w^1}\times\sqrt{v^1w^1}} = \frac{2u\sqrt{u\cdot vw}}{w^2\sqrt{v^1w^1\cdot v^1w^1}}$$

$$= \frac{2u\sqrt{uvw}}{w^2\sqrt{\left(v^1\cdot v^1\right)\cdot\left(w^1\cdot w^1\right)}} = \frac{2u\sqrt{uvw}}{w^2\sqrt{\left(v^{1+1}\right)\cdot\left(w^{1+1}\right)}} = \frac{2u\sqrt{uvw}}{w^2\sqrt{v^2w^2}} = \frac{2u\sqrt{uvw}}{w^2v^1w^1}$$

$$= \frac{2u\sqrt{uvw}}{vw^{2+1}} = \boldsymbol{\frac{2u\sqrt{uvw}}{vw^3}}$$

Example 4.3-25

$$\frac{\sqrt[3]{8r^7st}}{\sqrt[3]{9r^9s^2t^4}} =$$

Solution:

Step 1

$$\frac{\sqrt[3]{8r^7st}}{\sqrt[3]{9r^9s^2t^4}} = \sqrt[3]{\frac{8r^7s^1t^1}{9r^9s^2t^4}} = \sqrt[3]{\frac{8}{9\left(r^9r^{-7}\right)\cdot\left(s^2s^{-1}\right)\cdot\left(t^4t^{-1}\right)}} = \sqrt[3]{\frac{8}{9\left(r^{9-7}\right)\cdot\left(s^{2-1}\right)\cdot\left(t^{4-1}\right)}}$$

$$= \sqrt[3]{\frac{2^3}{3^2r^2s^1t^3}} = \frac{\sqrt[3]{2^3}}{\sqrt[3]{3^2r^2s^1t^3}} = \frac{2}{t\sqrt[3]{3^2r^2s^1}}$$

Step 2

$$\frac{2}{t\sqrt[3]{3^2r^2s^1}} = \frac{2}{t\sqrt[3]{3^2r^2s^1}}\times\frac{\sqrt[3]{3^1r^1s^2}}{\sqrt[3]{3^1r^1s^2}}$$

Step 3

$$\frac{2}{t\sqrt[3]{3^2r^2s^1}}\times\frac{\sqrt[3]{3^1r^1s^2}}{\sqrt[3]{3^1r^1s^2}} = \frac{2\times\sqrt[3]{3^1r^1s^2}}{t\sqrt[3]{3^2r^2s^1}\times\sqrt[3]{3^1r^1s^2}} = \frac{2\sqrt[3]{3rs^2}}{t\sqrt[3]{\left(3^2r^2s^1\right)\cdot\left(3^1r^1s^2\right)}}$$

$$= \frac{2\sqrt[3]{3rs^2}}{t\sqrt[3]{\left(3^2\cdot 3^1\right)\cdot\left(r^2\cdot r^1\right)\cdot\left(s^1\cdot s^2\right)}} = \frac{2\sqrt[3]{3rs^2}}{t\sqrt[3]{\left(3^{2+1}\right)\cdot\left(r^{2+1}\right)\cdot\left(s^{1+2}\right)}} = \frac{2\sqrt[3]{3rs^2}}{t\sqrt[3]{3^3r^3s^3}} = \mathbf{\frac{2\sqrt[3]{3rs^2}}{3rst}}$$

Additional Examples: Rationalizing Radical Expressions - Monomial Denominators with Variables

The following examples further illustrate how to rationalize radical expressions:

Example 4.3-26

$$\frac{\sqrt{20w^5}}{\sqrt{5w^2}} = \sqrt{\frac{20w^5}{5w^2}} = \sqrt{\frac{\overset{4}{\cancel{20}}\left(w^5\cdot w^{-2}\right)}{\underset{1}{\cancel{5}}}} = \sqrt{\frac{4w^{5-2}}{1}} = \frac{\sqrt{4w^3}}{1} = \sqrt{4w^3} = \sqrt{2^2w^{2+1}} = 2\sqrt{w^2\cdot w^1} = \mathbf{2w\sqrt{w}}$$

Example 4.3-27

$$\frac{\sqrt{75k^3l^2}}{\sqrt{15k^5l}} = \sqrt{\frac{75k^3l^2}{15k^5l}} = \sqrt{\frac{\overset{5}{\cancel{75}}\left(l^2\cdot l^{-1}\right)}{\underset{1}{\cancel{15}}\left(k^5\cdot k^{-3}\right)}} = \sqrt{\frac{5l^{2-1}}{k^{5-3}}} = \sqrt{\frac{5l^1}{k^2}} = \sqrt{\frac{5l}{k^2}} = \frac{\sqrt{5l}}{\sqrt{k^2}} = \mathbf{\frac{\sqrt{5l}}{k}}$$

Example 4.3-28

$$\frac{\sqrt{18u^5v^2}}{\sqrt{3uv^5}} = \sqrt{\frac{18u^5v^2}{3uv^5}} = \sqrt{\frac{\overset{6}{\cancel{18}}\left(u^5\cdot u^{-1}\right)}{\underset{1}{\cancel{3}}\left(v^5\cdot v^{-2}\right)}} = \sqrt{\frac{6u^{5-1}}{v^{5-2}}} = \sqrt{\frac{6u^4}{v^3}} = \frac{\sqrt{6u^4}}{\sqrt{v^3}} = \frac{\sqrt{6u^{2+2}}}{\sqrt{v^{2+1}}} = \frac{\sqrt{6u^2\cdot u^2}}{\sqrt{v^2\cdot v^1}}$$

$$= \frac{u^1\cdot u^1\sqrt{6}}{v\sqrt{v}} = \frac{u^{1+1}\sqrt{6}}{v\sqrt{v}} = \frac{u^2\sqrt{6}}{v\sqrt{v}} = \frac{u^2\sqrt{6}}{v\sqrt{v}}\times\frac{\sqrt{v}}{\sqrt{v}} = \frac{u^2\sqrt{6}\cdot\sqrt{v}}{v\sqrt{v}\cdot\sqrt{v}} = \frac{u^2\sqrt{6v}}{v\sqrt{v^2}} = \frac{u^2\sqrt{6v}}{vv} = \mathbf{\frac{u^2\sqrt{6v}}{v^2}}$$

Example 4.3-29

$$\frac{\sqrt{60x^3y^7}}{\sqrt{15xy}} = \sqrt{\frac{60x^3y^7}{15xy}} = \sqrt{\frac{\overset{4}{\cancel{60}}x^3y^7}{\underset{1}{\cancel{15}}x^1y^1}} = \sqrt{\frac{4\left(x^3\cdot x^{-1}\right)\cdot\left(y^7\cdot y^{-1}\right)}{1}} = \frac{\sqrt{4x^{3-1}\cdot y^{7-1}}}{1} = \frac{\sqrt{2^2\cdot x^2\cdot y^6}}{1}$$

$$= 2x\sqrt{y^{2+2+2}} = 2x\sqrt{y^2\cdot y^2\cdot y^2} = 2x\cdot y^1\cdot y^1\cdot y^1 = 2x\cdot y^{1+1+1} = 2x\cdot y^3 = \mathbf{2xy^3}$$

Example 4.3-30

$$\frac{\sqrt{15a^8b^3}}{\sqrt{3a^5b}} = \sqrt{\frac{15a^8b^3}{3a^5b}} = \sqrt{\frac{\overset{5}{\cancel{15}}\left(a^8a^{-5}\right)\cdot\left(b^3b^{-1}\right)}{\underset{1}{\cancel{3}}}} = \sqrt{\frac{5\left(a^{8-5}\right)\cdot\left(b^{3-1}\right)}{1}} = \sqrt{5a^3b^2} = b\sqrt{5a^{2+1}} = b\sqrt{5a^2\cdot a^1}$$

$$= ab\sqrt{5\cdot a} = \mathbf{ab\sqrt{5a}}$$

Example 4.3-31

$$\frac{\sqrt[4]{625x^5y^3}}{-2\sqrt[4]{16x^7y^5}} = -\frac{1}{2}\sqrt[4]{\frac{625x^5y^3}{16x^7y^5}} = -\frac{1}{2}\sqrt[4]{\frac{5^4}{2^4\left(x^7x^{-5}\right)\cdot\left(y^5y^{-3}\right)}} = -\frac{1}{2}\cdot\frac{5}{2}\sqrt[4]{\frac{1}{\left(x^{7-5}\right)\cdot\left(y^{5-3}\right)}}$$

$$= -\frac{1}{2}\cdot\frac{5}{2}\sqrt[4]{\frac{1}{\left(x^2\right)\cdot\left(y^2\right)}} = -\frac{1\cdot5}{2\cdot2}\sqrt[4]{\frac{1}{x^2y^2}} = -\frac{5}{4}\frac{\sqrt[4]{1}}{\sqrt[4]{x^2y^2}} = -\frac{5}{4}\frac{1}{\sqrt[4]{x^2y^2}} = -\frac{5}{4}\frac{1}{\sqrt[4]{x^2y^2}}\times\frac{\sqrt[4]{x^2y^2}}{\sqrt[4]{x^2y^2}}$$

$$= -\frac{5}{4}\frac{1\cdot\sqrt[4]{x^2y^2}}{\sqrt[4]{x^2y^2}\cdot\sqrt[4]{x^2y^2}} = -\frac{5}{4}\frac{\sqrt[4]{x^2y^2}}{\sqrt[4]{x^2y^2\cdot x^2y^2}} = -\frac{5}{4}\frac{\sqrt[4]{x^2y^2}}{\sqrt[4]{x^{2+2}y^{2+2}}} = -\frac{5}{4}\frac{\sqrt[4]{x^2y^2}}{\sqrt[4]{x^4y^4}} = \mathbf{-1\frac{1}{4}\left(\frac{\sqrt[4]{x^2y^2}}{xy}\right)}$$

Example 4.3-32

$$\frac{\sqrt[5]{m^8n^6}}{\sqrt[5]{8^3m^6n^2}} = \sqrt[5]{\frac{m^8n^6}{8^3m^6n^2}} = \sqrt[5]{\frac{m^8m^{-6}n^6n^{-2}}{8^3}} = \sqrt[5]{\frac{m^{8-6}n^{6-2}}{8^3}} = \sqrt[5]{\frac{m^2n^4}{8^3}} = \frac{\sqrt[5]{m^2n^4}}{\sqrt[5]{8^3}} = \frac{\sqrt[5]{m^2n^4}}{\sqrt[5]{8^3}}\times\frac{\sqrt[5]{8^2}}{\sqrt[5]{8^2}}$$

$$= \frac{\sqrt[5]{8^2\cdot m^2n^4}}{\sqrt[5]{8^3\cdot8^2}} = \frac{\sqrt[5]{8^2m^2n^4}}{\sqrt[5]{8^{3+2}}} = \frac{\sqrt[5]{8^2m^2n^4}}{\sqrt[5]{8^5}} = \mathbf{\frac{\sqrt[5]{64m^2n^4}}{8}}$$

Example 4.3-33

$$\sqrt[3]{\frac{x^3y^5z^6}{2}} = \frac{\sqrt[3]{x^3y^5z^6}}{\sqrt[3]{2}} = \frac{\sqrt[3]{x^3y^{3+2}z^{3+3}}}{\sqrt[3]{2}} = \frac{x\sqrt[3]{y^3\cdot y^2\cdot z^3\cdot z^3}}{\sqrt[3]{2}} = \frac{xyz^2\sqrt[3]{y^2}}{\sqrt[3]{2}} = \frac{xyz^2\sqrt[3]{y^2}}{\sqrt[3]{2}}\times\frac{\sqrt[3]{2^2}}{\sqrt[3]{2^2}}$$

$$= \frac{xyz^2\sqrt[3]{y^2}\cdot\sqrt[3]{2^2}}{\sqrt[3]{2}\cdot\sqrt[3]{2^2}} = \frac{xyz^2\sqrt[3]{2^2y^2}}{\sqrt[3]{2^1\cdot2^2}} = \frac{xyz^2\sqrt[3]{4y^2}}{\sqrt[3]{2^{1+2}}} = \frac{xyz^2\sqrt[3]{4y^2}}{\sqrt[3]{2^3}} = \mathbf{\frac{xyz^2\sqrt[3]{4y^2}}{2}}$$

Example 4.3-34

$$\frac{\sqrt[3]{2r^5s^4t}}{\sqrt[3]{5^2r^4t^2}} = \sqrt[3]{\frac{2r^5s^4t^1}{5^2r^4t^2}} = \sqrt[3]{\frac{2\left(r^5r^{-4}\right)\cdot s^4}{5^2\cdot\left(t^2t^{-1}\right)}} = \sqrt[3]{\frac{2\left(r^{5-4}\right)\cdot s^{3+1}}{5^2\cdot\left(t^{2-1}\right)}} = \sqrt[3]{\frac{2r^1\cdot\left(s^3\cdot s^1\right)}{5^2\cdot t^1}} = s\cdot\sqrt[3]{\frac{2r\cdot s}{5^2\cdot t}} = \frac{s\sqrt[3]{2rs}}{\sqrt[3]{5^2t}}$$

$$= \frac{s\sqrt[3]{2rs}}{\sqrt[3]{5^2t}}\times\frac{\sqrt[3]{5^1t^2}}{\sqrt[3]{5^1t^2}} = \frac{s\sqrt[3]{2rs}\cdot\sqrt[3]{5^1t^2}}{\sqrt[3]{5^2t}\cdot\sqrt[3]{5^1t^2}} = \frac{s\sqrt[3]{(2\cdot5)rst^2}}{\sqrt[3]{\left(5^2\cdot5^1\right)\cdot\left(t^1\cdot t^2\right)}} = \frac{s\sqrt[3]{10rst^2}}{\sqrt[3]{5^3t^3}} = \mathbf{\frac{s\sqrt[3]{10rst^2}}{5t}}$$

Radical expressions can be simplified in different ways. For example, we can also simplify the above radical expression the following way:

$$\frac{\sqrt[3]{2r^5s^4t}}{\sqrt[3]{5^2r^4t^2}} = \frac{\sqrt[3]{2r^{3+2}s^{3+1}t}}{\sqrt[3]{5^2r^{3+1}t^2}} = \frac{\sqrt[3]{2r^3\cdot r^2\cdot s^3\cdot s^1t}}{\sqrt[3]{5^2r^3\cdot r^1\cdot t^2}} = \frac{rs\sqrt[3]{2r^2s^1t}}{r\sqrt[3]{5^2r^1t^2}} = \frac{\overset{1}{\cancel{r}}s\sqrt[3]{2r^2st}}{\underset{1}{\cancel{r}}\sqrt[3]{5^2rt^2}} = \frac{s\sqrt[3]{2r^2st}}{\sqrt[3]{5^2rt^2}}$$

$$= \boxed{\frac{s\sqrt[3]{2r^2st}}{\sqrt[3]{5^2rt^2}} \times \frac{\sqrt[3]{5^1r^2t^1}}{\sqrt[3]{5^1r^2t^1}}} = \boxed{\frac{s\sqrt[3]{2r^2st}\cdot\sqrt[3]{5^1r^2t^1}}{\sqrt[3]{5^2rt^2}\cdot\sqrt[3]{5^1r^2t^1}}} = \boxed{\frac{s\sqrt[3]{(2\cdot5)\cdot\left(r^2\cdot r^2\right)\cdot s\left(t^1\cdot t^1\right)}}{\sqrt[3]{\left(5^2\cdot5^1\right)\cdot\left(r^1\cdot r^2\right)\cdot\left(t^2\cdot t^1\right)}}} = \boxed{\frac{s\sqrt[3]{10\cdot\left(r^{2+2}\right)\cdot s\left(t^{1+1}\right)}}{\sqrt[3]{\left(5^{2+1}\right)\cdot\left(r^{1+2}\right)\cdot\left(t^{2+1}\right)}}}$$

$$= \boxed{\frac{s\sqrt[3]{10r^4st^2}}{\sqrt[3]{5^3r^3t^3}}} = \boxed{\frac{s\sqrt[3]{10r^{3+1}st^2}}{5rt}} = \boxed{\frac{s\sqrt[3]{10r^3\cdot r^1st^2}}{5rt}} = \boxed{\frac{\overset{1}{\not r}\,s\sqrt[3]{10rst^2}}{5\underset{1}{\not r}t}} = \boxed{\frac{\mathbf{s\sqrt[3]{10rst^2}}}{\mathbf{5t}}}$$

Example 4.3-35

$$\boxed{\frac{\sqrt{40k^3l^2}}{\sqrt{5kl}}} = \boxed{\sqrt{\frac{40k^3l^2}{5kl}}} = \boxed{\sqrt{\frac{\overset{8}{\not 40}k^3l^2}{\underset{1}{\not 5}k^1l^1}}} = \boxed{\sqrt{\frac{8\left(k^3k^{-1}\right)\cdot\left(l^2l^{-1}\right)}{1}}} = \boxed{\sqrt{\frac{8\left(k^{3-1}\right)\cdot\left(l^{2-1}\right)}{1}}} = \boxed{\sqrt{\frac{(4\cdot2)k^2l}{1}}}$$

$$= \boxed{\frac{\sqrt{\left(2^2\cdot2\right)k^2l}}{1}} = \boxed{\frac{2k\sqrt{2l}}{1}} = \boxed{\mathbf{2k\sqrt{2l}}}$$

Example 4.3-36

$$\boxed{\sqrt{\frac{21u^5v^2}{3u^2v^3}}} = \boxed{\sqrt{\frac{\overset{7}{\not 21}\left(u^5u^{-2}\right)}{\underset{1}{\not 3}\left(v^3v^{-2}\right)}}} = \boxed{\sqrt{\frac{7u^{5-2}}{v^{3-2}}}} = \boxed{\sqrt{\frac{7u^3}{1v^1}}} = \boxed{\frac{\sqrt{7u^3}}{\sqrt{v}}} = \boxed{\frac{\sqrt{7u^{2+1}}}{\sqrt{v}} \times \frac{\sqrt{v}}{\sqrt{v}}} = \boxed{\frac{\sqrt{7u^2u^1}\cdot\sqrt{v}}{\sqrt{v}\cdot\sqrt{v}}} = \boxed{\frac{u\sqrt{7u\cdot v}}{\sqrt{v^1\cdot v^1}}}$$

$$= \boxed{\frac{u\sqrt{7uv}}{\sqrt{v^{1+1}}}} = \boxed{\frac{u\sqrt{7uv}}{\sqrt{v^2}}} = \boxed{\frac{\mathbf{u\sqrt{7uv}}}{\mathbf{v}}}$$

Example 4.3-37

$$\boxed{\frac{\sqrt[5]{a^7b^8c^3}}{\sqrt[5]{243a^2b^2c^9}}} = \boxed{\sqrt[5]{\frac{a^7b^8c^3}{243a^2b^2c^9}}} = \boxed{\sqrt[5]{\frac{\left(a^7a^{-2}\right)\cdot\left(b^8b^{-2}\right)}{3^5\cdot\left(c^9c^{-3}\right)}}} = \boxed{\frac{1}{3}\sqrt[5]{\frac{\left(a^{7-2}\right)\cdot\left(b^{8-2}\right)}{c^{9-3}}}} = \boxed{\frac{1}{3}\sqrt[5]{\frac{a^5b^6}{c^6}}} = \boxed{\frac{1}{3}\frac{\sqrt[5]{a^5b^6}}{\sqrt[5]{c^6}}}$$

$$= \boxed{\frac{1}{3}\frac{a\sqrt[5]{b^{5+1}}}{\sqrt[5]{c^{5+1}}}} = \boxed{\frac{1}{3}\frac{a\sqrt[5]{b^5\cdot b^1}}{\sqrt[5]{c^5\cdot c^1}}} = \boxed{\frac{1}{3}\frac{ab\sqrt[5]{b}}{c\sqrt[5]{c}}} = \boxed{\frac{ab}{3c}\frac{\sqrt[5]{b}}{\sqrt[5]{c}}} = \boxed{\frac{ab}{3c}\frac{\sqrt[5]{b}}{\sqrt[5]{c^1}} \times \frac{\sqrt[5]{c^4}}{\sqrt[5]{c^4}}} = \boxed{\frac{ab}{3c}\frac{\sqrt[5]{b}\times\sqrt[5]{c^4}}{\sqrt[5]{c^1}\times\sqrt[5]{c^4}}} = \boxed{\frac{ab}{3c}\frac{\sqrt[5]{b\cdot c^4}}{\sqrt[5]{c^1\cdot c^4}}}$$

$$= \boxed{\frac{ab}{3c}\frac{\sqrt[5]{bc^4}}{\sqrt[5]{c^{1+4}}}} = \boxed{\frac{ab}{3c}\frac{\sqrt[5]{bc^4}}{\sqrt[5]{c^5}}} = \boxed{\frac{ab}{3c}\frac{\sqrt[5]{bc^4}}{c}} = \boxed{\frac{ab\cdot\sqrt[5]{bc^4}}{3(c\cdot c)}} = \boxed{\frac{ab\sqrt[5]{bc^4}}{3\left(c^1\cdot c^1\right)}} = \boxed{\frac{ab\sqrt[5]{bc^4}}{3c^{1+1}}} = \boxed{\frac{\mathbf{ab\sqrt[5]{bc^4}}}{\mathbf{3c^2}}}$$

Example 4.3-38

$$\boxed{\sqrt{\frac{72x^5y^2z}{81x^3y^6z^4}}} = \boxed{\sqrt{\frac{(36\cdot2)x^5y^2z^1}{9^2x^3y^6z^4}}} = \boxed{\sqrt{\frac{\left(6^2\cdot2\right)\cdot\left(x^5x^{-3}\right)}{9^2\cdot\left(y^6y^{-2}\right)\cdot\left(z^4\cdot z^{-1}\right)}}} = \boxed{\frac{6}{9}\sqrt{\frac{2\cdot\left(x^{5-3}\right)}{\left(y^{6-2}\right)\cdot\left(z^{4-1}\right)}}} = \boxed{\frac{\overset{2}{\not 6}}{\underset{3}{\not 9}}\sqrt{\frac{2x^2}{y^4z^3}}}$$

$$= \boxed{\frac{2}{3}\sqrt{\frac{2x^2}{y^{2+2}z^{2+1}}}} = \boxed{\frac{2}{3}\sqrt{\frac{2x^2}{y^2\cdot y^2\cdot z^2\cdot z^1}}} = \boxed{\frac{2x}{3y^1y^1z^1}\sqrt{\frac{2}{z}}} = \boxed{\frac{2x}{3y^{1+1}z}\sqrt{\frac{2}{z}}} = \boxed{\frac{2x}{3y^2z}\frac{\sqrt{2}}{\sqrt{z}}} = \boxed{\frac{2x}{3y^2z}\frac{\sqrt{2}}{\sqrt{z}}\times\frac{\sqrt{z}}{\sqrt{z}}}$$

$$= \boxed{\frac{2x}{3y^2z}\frac{\sqrt{2}\times\sqrt{z}}{\sqrt{z}\times\sqrt{z}}} = \boxed{\frac{2x}{3y^2z}\frac{\sqrt{2\cdot z}}{\sqrt{z\cdot z}}} = \boxed{\frac{2x}{3y^2z}\frac{\sqrt{2z}}{\sqrt{z^1z^1}}} = \boxed{\frac{2x}{3y^2z}\frac{\sqrt{2z}}{\sqrt{z^{1+1}}}} = \boxed{\frac{2x}{3y^2z}\frac{\sqrt{2z}}{\sqrt{z^2}}} = \boxed{\frac{2x}{3y^2z}\frac{\sqrt{2z}}{z}} = \boxed{\frac{2x\cdot\sqrt{2z}}{3y^2\cdot z\cdot z}}$$

$$= \boxed{\frac{2x\sqrt{2z}}{3y^2\left(z^1\cdot z^1\right)}} = \boxed{\frac{2x\sqrt{2z}}{3y^2z^{1+1}}} = \boxed{\mathbf{\frac{2x\sqrt{2z}}{3y^2z^2}}}$$

Example 4.3-39

$$\boxed{\frac{\sqrt[4]{256x^5y^3z^6}}{\sqrt[4]{x^4y^2z}}} = \boxed{\sqrt[4]{\frac{256x^5y^3z^6}{x^4y^2z}}} = \boxed{\sqrt[4]{\frac{4^4\left(x^5x^{-4}\right)\cdot\left(y^3y^{-2}\right)\cdot\left(z^6z^{-1}\right)}{1}}} = \boxed{4\sqrt[4]{\frac{\left(x^{5-4}\right)\cdot\left(y^{3-2}\right)\cdot\left(z^{6-1}\right)}{1}}}$$

$$= \boxed{4\sqrt[4]{\frac{x^1y^1z^5}{1}}} = \boxed{4\sqrt[4]{xyz^5}} = \boxed{4\sqrt[4]{xyz^{4+1}}} = \boxed{4\sqrt[4]{xyz^4\cdot z^1}} = \boxed{\mathbf{4z\sqrt[4]{xyz}}}$$

Example 4.3-40

$$\boxed{\frac{\sqrt{1000p^5q^3r^4}}{\sqrt{400pq^2r^7}}} = \boxed{\sqrt{\frac{1000p^5q^3r^4}{400p^1q^2r^7}}} = \boxed{\sqrt{\frac{\overset{5}{\cancel{1000}}\cdot\left(p^5p^{-1}\right)\cdot\left(q^3q^{-2}\right)}{\underset{2}{\cancel{400}}\cdot\left(r^7r^{-4}\right)}}} = \boxed{\sqrt{\frac{5\cdot\left(p^{5-1}\right)\cdot\left(q^{3-2}\right)}{2\cdot\left(r^{7-4}\right)}}} = \boxed{\sqrt{\frac{5p^4q^1}{2r^3}}}$$

$$= \boxed{\sqrt{\frac{5p^{2+2}q}{2r^{2+1}}}} = \boxed{\sqrt{\frac{5p^2\cdot p^2\cdot q}{2r^2\cdot r^1}}} = \boxed{\frac{p^1p^1}{r}\sqrt{\frac{5q}{2r}}} = \boxed{\frac{p^{1+1}}{r}\frac{\sqrt{5q}}{\sqrt{2r}}} = \boxed{\frac{p^2}{r}\frac{\sqrt{5q}}{\sqrt{2r}}\times\frac{\sqrt{2r}}{\sqrt{2r}}} = \boxed{\frac{p^2}{r}\frac{\sqrt{5q}\times\sqrt{2r}}{\sqrt{2r}\times\sqrt{2r}}}$$

$$= \boxed{\frac{p^2}{r}\frac{\sqrt{(5\cdot 2)\cdot(q\cdot r)}}{\sqrt{(2\cdot 2)\cdot(r\cdot r)}}} = \boxed{\frac{p^2}{r}\frac{\sqrt{10\cdot qr}}{\sqrt{\left(2^1\cdot 2^1\right)\cdot\left(r^1\cdot r^1\right)}}} = \boxed{\frac{p^2}{r}\frac{\sqrt{10qr}}{\sqrt{2^{1+1}r^{1+1}}}} = \boxed{\frac{p^2}{r}\frac{\sqrt{10qr}}{\sqrt{2^2r^2}}} = \boxed{\frac{p^2}{r}\frac{\sqrt{10qr}}{2r}}$$

$$= \boxed{\frac{p^2\cdot\sqrt{10qr}}{2r\cdot r}} = \boxed{\mathbf{\frac{p^2\sqrt{10qr}}{2r^2}}}$$

Practice Problems: Rationalizing Radical Expressions - Monomial Denominators with Variables

Section 4.3 Case I b Practice Problems - Solve the following radical expressions:

1. $\dfrac{\sqrt{80x}}{\sqrt{4x^4}} =$

2. $\dfrac{\sqrt{48u^3v^2}}{\sqrt{16uv^3}} =$

3. $\sqrt{\dfrac{100x^3y^5}{25x^5y}} =$

4. $\dfrac{\sqrt[4]{81x^3y^2z^2}}{\sqrt[4]{2x^6y^5z^3}} =$

5. $\sqrt[5]{\dfrac{32a^2b^3}{256a^6b}} =$

6. $\dfrac{\sqrt{18u^3v^2}}{\sqrt{4uv^3}} =$

7. $\dfrac{\sqrt{5k^5l^2}}{\sqrt{40k^3l}} =$

8. $\dfrac{\sqrt[4]{625x^4y^2}}{-3\sqrt[4]{81x^6y^6}} =$

9. $\sqrt{\dfrac{x^3y^5z}{81xy^6z^3}} =$

10. $\dfrac{\sqrt[3]{m^3n^2}}{\sqrt[3]{5^3m^7n^4}} =$

Case II Rationalizing Radical Expressions with Binomial Denominators

Simplification of fractional radical expressions with binomial denominators requires rationalization of the denominator. A binomial denominator is rationalized by multiplying the numerator and the denominator by its conjugate. Two binomials that differ only by the sign between them are called **conjugates** of each other. Note that whenever conjugates are multiplied by each other, the two similar but opposite in sign middle terms drop out.

Examples:

1. The conjugate of $2+\sqrt{3}$ is $2-\sqrt{3}$.
2. The conjugate of $\sqrt{6}-10$ is $\sqrt{6}+10$.
3. The conjugate of $\sqrt{3}-\sqrt{5}$ is $\sqrt{3}+\sqrt{5}$.
4. The conjugate of $\sqrt{7}+\sqrt{2}$ is $\sqrt{7}-\sqrt{2}$.

Division of binomial terms in radical form is divided into two cases. Case I a - rationalization of radical expressions, with real numbers and Case II b - rationalization of radical expressions, with variables.

Case II a Rationalizing Radical Expressions - Binomial Denominators with Real Numbers

Radical expressions with binomial denominators are simplified using the following steps:

Step 1 Simplify the radical terms in the numerator and the denominator (see Section 4.1, Case III).

Step 2 Rationalize the denominator by multiplying the numerator and the denominator by its conjugate.

Step 3 Simplify the radical expression using the **FOIL** method (see Section 4.2, Case IIa).

Examples with Steps

The following examples show the steps as to how radical expressions with two terms in the denominator are simplified:

Example 4.3-41

$$\frac{8}{2-\sqrt{2}}$$

Solution:

Step 1 *Not Applicable*

Step 2

$$\frac{8}{2-\sqrt{2}} = \frac{8}{2-\sqrt{2}} \times \frac{2+\sqrt{2}}{2+\sqrt{2}}$$

Step 3

$$\frac{8}{2-\sqrt{2}} \times \frac{2+\sqrt{2}}{2+\sqrt{2}} = \frac{8 \times \left(2+\sqrt{2}\right)}{\left(2-\sqrt{2}\right) \times \left(2+\sqrt{2}\right)} = \frac{8 \cdot \left(2+\sqrt{2}\right)}{(2 \cdot 2)+\left(2 \cdot \sqrt{2}\right)-\left(2 \cdot \sqrt{2}\right)-\left(\sqrt{2} \cdot \sqrt{2}\right)}$$

$$= \boxed{\frac{8\left(2+\sqrt{2}\right)}{4+2\sqrt{2}-2\sqrt{2}-\sqrt{2\cdot 2}}} = \boxed{\frac{8\left(2+\sqrt{2}\right)}{4-\sqrt{2^1\cdot 2^1}}} = \boxed{\frac{8\left(2+\sqrt{2}\right)}{4-\sqrt{2^{1+1}}}} = \boxed{\frac{8\left(2+\sqrt{2}\right)}{4-\sqrt{2^2}}} = \boxed{\frac{8\left(2+\sqrt{2}\right)}{4-2}}$$

$$= \boxed{\frac{\overset{4}{\cancel{8}}\left(2+\sqrt{2}\right)}{\underset{1}{\cancel{2}}}} = \boxed{\frac{4\left(2+\sqrt{2}\right)}{1}} = \boxed{\mathbf{4\left(2+\sqrt{2}\right)}}$$

Example 4.3-42

$$\boxed{\frac{\sqrt{125}}{\sqrt{3}-\sqrt{5}}} =$$

Solution:

Step 1

$$\boxed{\frac{\sqrt{125}}{\sqrt{3}-\sqrt{5}}} = \boxed{\frac{\sqrt{25\cdot 5}}{\sqrt{3}-\sqrt{5}}} = \boxed{\frac{\sqrt{5^2\cdot 5}}{\sqrt{3}-\sqrt{5}}} = \boxed{\frac{5\sqrt{5}}{\sqrt{3}-\sqrt{5}}}$$

Step 2

$$\boxed{\frac{5\sqrt{5}}{\sqrt{3}-\sqrt{5}}} = \boxed{\frac{5\sqrt{5}}{\sqrt{3}-\sqrt{5}}\times\frac{\sqrt{3}+\sqrt{5}}{\sqrt{3}+\sqrt{5}}}$$

Step 3

$$\boxed{\frac{5\sqrt{5}}{\sqrt{3}-\sqrt{5}}\times\frac{\sqrt{3}+\sqrt{5}}{\sqrt{3}+\sqrt{5}}} = \boxed{\frac{5\sqrt{5}\times\left(\sqrt{3}+\sqrt{5}\right)}{\left(\sqrt{3}-\sqrt{5}\right)\times\left(\sqrt{3}+\sqrt{5}\right)}}$$

$$= \boxed{\frac{5\left(\sqrt{5}\cdot\sqrt{3}\right)+5\left(\sqrt{5}\cdot\sqrt{5}\right)}{\left(\sqrt{3}\cdot\sqrt{3}\right)+\left(\sqrt{3}\cdot\sqrt{5}\right)-\left(\sqrt{5}\cdot\sqrt{3}\right)-\left(\sqrt{5}\cdot\sqrt{5}\right)}} = \boxed{\frac{5\sqrt{5\cdot 3}+5\sqrt{5\cdot 5}}{\sqrt{3\cdot 3}+\sqrt{3\cdot 5}-\sqrt{5\cdot 3}-\sqrt{5\cdot 5}}}$$

$$= \boxed{\frac{5\sqrt{15}+5\sqrt{5^2}}{\sqrt{3^2}+\sqrt{15}-\sqrt{15}-\sqrt{5^2}}} = \boxed{\frac{5\sqrt{15}+5\cdot 5}{3-5}} = \boxed{\frac{5\sqrt{15}+25}{-2}} = \boxed{-\frac{\mathbf{5\left(\sqrt{15}+5\right)}}{\mathbf{2}}}$$

Example 4.2-43

$$\boxed{\frac{3+\sqrt{5}}{\sqrt{3}+\sqrt{5}}} =$$

Solution:

Step 1 *Not Applicable*

Step 2

$$\boxed{\frac{3+\sqrt{5}}{\sqrt{3}+\sqrt{5}}} = \boxed{\frac{3+\sqrt{5}}{\sqrt{3}+\sqrt{5}}\times\frac{\sqrt{3}-\sqrt{5}}{\sqrt{3}-\sqrt{5}}}$$

Step 3

$$\boxed{\frac{3+\sqrt{5}}{\sqrt{3}+\sqrt{5}}\times\frac{\sqrt{3}-\sqrt{5}}{\sqrt{3}-\sqrt{5}}} = \boxed{\frac{\left(3+\sqrt{5}\right)\times\left(\sqrt{3}-\sqrt{5}\right)}{\left(\sqrt{3}+\sqrt{5}\right)\times\left(\sqrt{3}-\sqrt{5}\right)}} = \boxed{\frac{\left(3\cdot\sqrt{3}\right)-\left(3\cdot\sqrt{5}\right)+\left(\sqrt{5}\cdot\sqrt{3}\right)-\left(\sqrt{5}\cdot\sqrt{5}\right)}{\left(\sqrt{3}\cdot\sqrt{3}\right)-\left(\sqrt{3}\cdot\sqrt{5}\right)+\left(\sqrt{3}\cdot\sqrt{5}\right)-\left(\sqrt{5}\cdot\sqrt{5}\right)}}$$

$$= \boxed{\frac{3\cdot\sqrt{3} - 3\cdot\sqrt{5} + \sqrt{5\cdot 3} - \sqrt{5\cdot 5}}{\sqrt{3\cdot 3} - \sqrt{3\cdot 5} + \sqrt{3\cdot 5} - \sqrt{5\cdot 5}}} = \boxed{\frac{3\sqrt{3} - 3\sqrt{5} + \sqrt{15} - \sqrt{5^2}}{\sqrt{3^2} - \sqrt{15} + \sqrt{15} - \sqrt{5^2}}} = \boxed{\frac{3\sqrt{3} - 3\sqrt{5} + \sqrt{15} - 5}{\sqrt{3^2} - \sqrt{5^2}}}$$

$$= \boxed{\frac{3\sqrt{3} - 3\sqrt{5} + \sqrt{15} - 5}{3-5}} = \boxed{\frac{3\sqrt{3} - 3\sqrt{5} + \sqrt{15} - 5}{-2}} = \boxed{-\frac{3\sqrt{3} - 3\sqrt{5} + \sqrt{15} - 5}{2}}$$

$$= \boxed{\frac{-3\sqrt{3} + 3\sqrt{5} - \sqrt{15} + 5}{2}} = \boxed{\frac{3(\sqrt{5} - \sqrt{3}) - \sqrt{15} + 5}{2}}$$

Example 4.3-44

$$\boxed{\frac{\sqrt{8} + \sqrt{4}}{4 - \sqrt{2}}} =$$

Solution:

Step 1

$$\boxed{\frac{\sqrt{8} + \sqrt{4}}{4 - \sqrt{2}}} = \boxed{\frac{\sqrt{2^2\cdot 2} + \sqrt{2^2}}{4 - \sqrt{2}}} = \boxed{\frac{2\sqrt{2} + 2}{4 - \sqrt{2}}}$$

Step 2

$$\boxed{\frac{2\sqrt{2} + 2}{4 - \sqrt{2}}} = \boxed{\frac{2\sqrt{2} + 2}{4 - \sqrt{2}} \times \frac{4 + \sqrt{2}}{4 + \sqrt{2}}}$$

Step 3

$$\boxed{\frac{2\sqrt{2} + 2}{4 - \sqrt{2}} \times \frac{4 + \sqrt{2}}{4 + \sqrt{2}}} = \boxed{\frac{(2\sqrt{2} + 2) \times (4 + \sqrt{2})}{(4 - \sqrt{2}) \times (4 + \sqrt{2})}} = \boxed{\frac{(4\cdot 2\sqrt{2}) + (2\sqrt{2}\cdot\sqrt{2}) + (2\cdot 4) + (2\cdot\sqrt{2})}{(4\cdot 4) + (4\cdot\sqrt{2}) - (4\cdot\sqrt{2}) - (\sqrt{2}\cdot\sqrt{2})}}$$

$$= \boxed{\frac{8\sqrt{2} + 2\sqrt{2\cdot 2} + 8 + 2\sqrt{2}}{16 + 4\sqrt{2} - 4\sqrt{2} - \sqrt{2\cdot 2}}} = \boxed{\frac{8\sqrt{2} + 2\sqrt{2^2} + 8 + 2\sqrt{2}}{16 - \sqrt{2^2}}} = \boxed{\frac{8\sqrt{2} + (2\cdot 2) + 8 + 2\sqrt{2}}{16 - 2}}$$

$$= \boxed{\frac{8\sqrt{2} + 4 + 8 + 2\sqrt{2}}{14}} = \boxed{\frac{(8+2)\sqrt{2} + 12}{14}} = \boxed{\frac{10\sqrt{2} + 12}{14}} = \boxed{\frac{\overset{1}{\cancel{2}}\cdot(5\sqrt{2} + 6)}{\underset{7}{\cancel{14}}}} = \boxed{\frac{1\cdot(5\sqrt{2} + 6)}{7}}$$

$$= \boxed{\frac{5\sqrt{2} + 6}{7}}$$

Example 4.3-45

$$\boxed{\frac{\sqrt{5} - \sqrt{98}}{\sqrt{5} + \sqrt{7}}} =$$

Solution:

Step 1

$$\boxed{\frac{\sqrt{5} - \sqrt{98}}{\sqrt{5} + \sqrt{7}}} = \boxed{\frac{\sqrt{5} - \sqrt{49\cdot 2}}{\sqrt{5} + \sqrt{7}}} = \boxed{\frac{\sqrt{5} - \sqrt{7^2\cdot 2}}{\sqrt{5} + \sqrt{7}}} = \boxed{\frac{\sqrt{5} - 7\sqrt{2}}{\sqrt{5} + \sqrt{7}}}$$

Step 2

$$\boxed{\frac{\sqrt{5}-7\sqrt{2}}{\sqrt{5}+\sqrt{7}}} = \boxed{\frac{\sqrt{5}-7\sqrt{2}}{\sqrt{5}+\sqrt{7}} \times \frac{\sqrt{5}-\sqrt{7}}{\sqrt{5}-\sqrt{7}}}$$

Step 3

$$\boxed{\frac{\sqrt{5}-7\sqrt{2}}{\sqrt{5}+\sqrt{7}} \times \frac{\sqrt{5}-\sqrt{7}}{\sqrt{5}-\sqrt{7}}} = \boxed{\frac{\left(\sqrt{5}-7\sqrt{2}\right)\times\left(\sqrt{5}-\sqrt{7}\right)}{\left(\sqrt{5}+\sqrt{7}\right)\times\left(\sqrt{5}-\sqrt{7}\right)}}$$

$$= \boxed{\frac{\left(\sqrt{5}\cdot\sqrt{5}\right)-\left(\sqrt{5}\cdot\sqrt{7}\right)-\left(7\sqrt{2}\cdot\sqrt{5}\right)+\left(7\sqrt{2}\cdot\sqrt{7}\right)}{\left(\sqrt{5}\cdot\sqrt{5}\right)-\left(\sqrt{5}\cdot\sqrt{7}\right)+\left(\sqrt{7}\cdot\sqrt{5}\right)-\left(\sqrt{7}\cdot\sqrt{7}\right)}} = \boxed{\frac{\sqrt{5\cdot5}-\sqrt{5\cdot7}-7\sqrt{2\cdot5}+7\sqrt{2\cdot7}}{\sqrt{5\cdot5}-\sqrt{5\cdot7}+\sqrt{7\cdot5}-\sqrt{7\cdot7}}}$$

$$= \boxed{\frac{\sqrt{5^2}-\sqrt{35}-7\sqrt{10}+7\sqrt{14}}{\sqrt{5^2}-\sqrt{35}+\sqrt{35}-\sqrt{7^2}}} = \boxed{\frac{5-\sqrt{35}-7\sqrt{10}+7\sqrt{14}}{5-7}} = \boxed{\frac{5-\sqrt{35}-7\sqrt{10}+7\sqrt{14}}{-2}}$$

$$= \boxed{-\frac{5-\sqrt{35}-7\sqrt{10}+7\sqrt{14}}{2}} = \boxed{\mathbf{\frac{\sqrt{35}+7\sqrt{10}-7\sqrt{14}-5}{2}}}$$

Additional Examples: Rationalizing Radical Expressions - Binomial Denominators with Real Numbers

The following examples further illustrate how to rationalize radical expressions with binomial denominators:

Example 4.3-46

$$\boxed{\frac{\sqrt{5}}{3+\sqrt{5}}} = \boxed{\frac{\sqrt{5}}{3+\sqrt{5}} \times \frac{3-\sqrt{5}}{3-\sqrt{5}}} = \boxed{\frac{\sqrt{5}\times\left(3-\sqrt{5}\right)}{\left(3+\sqrt{5}\right)\times\left(3-\sqrt{5}\right)}} = \boxed{\frac{\left(3\cdot\sqrt{5}\right)-\left(\sqrt{5\cdot5}\right)}{\left(3\cdot3\right)-\left(3\cdot\sqrt{5}\right)+\left(3\cdot\sqrt{5}\right)-\left(\sqrt{5}\cdot\sqrt{5}\right)}}$$

$$= \boxed{\frac{3\sqrt{5}-\sqrt{5^2}}{9-3\sqrt{5}+3\sqrt{5}-\sqrt{5^2}}} = \boxed{\frac{3\sqrt{5}-5}{9-\sqrt{5^2}}} = \boxed{\frac{3\sqrt{5}-5}{9-5}} = \boxed{\mathbf{\frac{3\sqrt{5}-5}{4}}}$$

Example 4.3-47

$$\boxed{\frac{2+\sqrt{2}}{2-\sqrt{2}}} = \boxed{\frac{2+\sqrt{2}}{2-\sqrt{2}} \times \frac{2+\sqrt{2}}{2+\sqrt{2}}} = \boxed{\frac{\left(2+\sqrt{2}\right)\times\left(2+\sqrt{2}\right)}{\left(2-\sqrt{2}\right)\times\left(2+\sqrt{2}\right)}} = \boxed{\frac{\left(2\cdot2\right)+\left(2\cdot\sqrt{2}\right)+\left(2\cdot\sqrt{2}\right)+\left(\sqrt{2}\cdot\sqrt{2}\right)}{\left(2-\sqrt{2}\right)\times\left(2+\sqrt{2}\right)}}$$

$$= \boxed{\frac{4+2\sqrt{2}+2\sqrt{2}+\sqrt{2\cdot2}}{\left(2\cdot2\right)+\left(2\cdot\sqrt{2}\right)-\left(2\cdot\sqrt{2}\right)-\left(\sqrt{2}\cdot\sqrt{2}\right)}} = \boxed{\frac{4+4\sqrt{2}+\sqrt{2^2}}{4+2\sqrt{2}-2\sqrt{2}-\sqrt{2\cdot2}}} = \boxed{\frac{4+4\sqrt{2}+2}{4-\sqrt{2^2}}} = \boxed{\frac{(4+2)+4\sqrt{2}}{4-2}}$$

$$= \boxed{\frac{6+4\sqrt{2}}{2}} = \boxed{\frac{\overset{1}{\cancel{2}}\cdot\left(3+2\sqrt{2}\right)}{\underset{1}{\cancel{2}}}} = \boxed{\frac{1\cdot\left(3+2\sqrt{2}\right)}{1}} = \boxed{\frac{3+2\sqrt{2}}{1}} = \boxed{\mathbf{3+2\sqrt{2}}}$$

Example 4.3-48

$$\boxed{\frac{5}{3+\sqrt{3}}} = \boxed{\frac{5}{3+\sqrt{3}} \times \frac{3-\sqrt{3}}{3-\sqrt{3}}} = \boxed{\frac{5\times(3-\sqrt{3})}{(3+\sqrt{3})\times(3-\sqrt{3})}} = \boxed{\frac{5(3-\sqrt{3})}{(3\cdot 3)-(3\cdot\sqrt{3})+(3\cdot\sqrt{3})-(\sqrt{3}\cdot\sqrt{3})}}$$

$$= \boxed{\frac{5(3-\sqrt{3})}{9-3\sqrt{3}+3\sqrt{3}-\sqrt{3^2}}} = \boxed{\frac{5(3-\sqrt{3})}{9-\sqrt{3^2}}} = \boxed{\frac{5(3-\sqrt{3})}{9-3}} = \boxed{\frac{\mathbf{5(3-\sqrt{3})}}{\mathbf{6}}}$$

Example 4.3-49

$$\boxed{\frac{-\sqrt{7}}{\sqrt{7}+\sqrt{5}}} = \boxed{\frac{-\sqrt{7}}{\sqrt{7}+\sqrt{5}} \times \frac{\sqrt{7}-\sqrt{5}}{\sqrt{7}-\sqrt{5}}} = \boxed{\frac{-\sqrt{7}\times(\sqrt{7}-\sqrt{5})}{(\sqrt{7}+\sqrt{5})\times(\sqrt{7}-\sqrt{5})}} = \boxed{\frac{-\sqrt{7}\cdot\sqrt{7}+\sqrt{7}\cdot\sqrt{5}}{(\sqrt{7}\cdot\sqrt{7})-(\sqrt{7}\cdot\sqrt{5})+(\sqrt{5}\cdot\sqrt{7})-(\sqrt{5}\cdot\sqrt{5})}}$$

$$= \boxed{\frac{-\sqrt{7\cdot 7}+\sqrt{7\cdot 5}}{\sqrt{7\cdot 7}-\sqrt{7\cdot 5}+\sqrt{7\cdot 5}-\sqrt{5\cdot 5}}} = \boxed{\frac{-\sqrt{7^2}+\sqrt{35}}{\sqrt{7^2}-\sqrt{35}+\sqrt{35}-\sqrt{5^2}}} = \boxed{\frac{-7+\sqrt{35}}{7-5}} = \boxed{\frac{\mathbf{-7+\sqrt{35}}}{\mathbf{2}}}$$

Example 4.3-50

$$\boxed{\frac{8-\sqrt{5}}{\sqrt{8}+\sqrt{7}}} = \boxed{\frac{8-\sqrt{5}}{\sqrt{8}+\sqrt{7}} \times \frac{\sqrt{8}-\sqrt{7}}{\sqrt{8}-\sqrt{7}}} = \boxed{\frac{(8-\sqrt{5})\times(\sqrt{8}-\sqrt{7})}{(\sqrt{8}+\sqrt{7})\times(\sqrt{8}-\sqrt{7})}} = \boxed{\frac{(8\cdot\sqrt{8})-(8\cdot\sqrt{7})-(\sqrt{5}\cdot\sqrt{8})+(\sqrt{5}\cdot\sqrt{7})}{(\sqrt{8}\cdot\sqrt{8})-(\sqrt{8}\cdot\sqrt{7})+(\sqrt{8}\cdot\sqrt{7})-(\sqrt{7}\cdot\sqrt{7})}}$$

$$= \boxed{\frac{8\sqrt{8}-8\sqrt{7}-\sqrt{5\cdot 8}+\sqrt{5\cdot 7}}{\sqrt{8\cdot 8}-\sqrt{8\cdot 7}+\sqrt{8\cdot 7}-\sqrt{7\cdot 7}}} = \boxed{\frac{8\sqrt{4\cdot 2}-8\sqrt{7}-\sqrt{40}+\sqrt{35}}{\sqrt{8^2}-\sqrt{56}+\sqrt{56}-\sqrt{7^2}}} = \boxed{\frac{8\sqrt{2^2\cdot 2}-8\sqrt{7}-\sqrt{4\cdot 10}+\sqrt{35}}{8-7}}$$

$$= \boxed{\frac{8\cdot 2\sqrt{2}-8\sqrt{7}-\sqrt{2^2\cdot 10}+\sqrt{35}}{1}} = \boxed{\mathbf{16\sqrt{2}-8\sqrt{7}-2\sqrt{10}+\sqrt{35}}}$$

Example 4.3-51

$$\boxed{\frac{\sqrt{5}+\sqrt{3}}{\sqrt{7}+\sqrt{2}}} = \boxed{\frac{\sqrt{5}+\sqrt{3}}{\sqrt{7}+\sqrt{2}} \times \frac{\sqrt{7}-\sqrt{2}}{\sqrt{7}-\sqrt{2}}} = \boxed{\frac{(\sqrt{5}+\sqrt{3})\times(\sqrt{7}-\sqrt{2})}{(\sqrt{7}+\sqrt{2})\times(\sqrt{7}-\sqrt{2})}} = \boxed{\frac{(\sqrt{5}\cdot\sqrt{7})-(\sqrt{5}\cdot\sqrt{2})+(\sqrt{3}\cdot\sqrt{7})-(\sqrt{3}\cdot\sqrt{2})}{(\sqrt{7}\cdot\sqrt{7})-(\sqrt{7}\cdot\sqrt{2})+(\sqrt{2}\cdot\sqrt{7})-(\sqrt{2}\cdot\sqrt{2})}}$$

$$= \boxed{\frac{\sqrt{5\cdot 7}-\sqrt{5\cdot 2}+\sqrt{3\cdot 7}-\sqrt{3\cdot 2}}{\sqrt{7\cdot 7}-\sqrt{7\cdot 2}+\sqrt{2\cdot 7}-\sqrt{2\cdot 2}}} = \boxed{\frac{\sqrt{35}-\sqrt{10}+\sqrt{21}-\sqrt{6}}{\sqrt{7^2}-\sqrt{14}+\sqrt{14}-\sqrt{2^2}}} = \boxed{\frac{\sqrt{35}-\sqrt{10}+\sqrt{21}-\sqrt{6}}{7-2}}$$

$$= \boxed{\frac{\mathbf{\sqrt{35}-\sqrt{10}+\sqrt{21}-\sqrt{6}}}{\mathbf{5}}}$$

Example 4.3-52

$$\boxed{\frac{10}{10+\sqrt{200}}} = \boxed{\frac{10}{10+\sqrt{100\cdot 2}}} = \boxed{\frac{10}{10+\sqrt{10^2\cdot 2}}} = \boxed{\frac{10}{10+10\sqrt{2}}} = \boxed{\frac{\overset{1}{\cancel{10}}}{\underset{1}{\cancel{10}}\cdot(1+\sqrt{2})}} = \boxed{\frac{1}{1\cdot(1+\sqrt{2})}} = \boxed{\frac{1}{1+\sqrt{2}}}$$

$$= \boxed{\frac{1}{1+\sqrt{2}} \times \frac{1-\sqrt{2}}{1-\sqrt{2}}} = \boxed{\frac{1\times\left(1-\sqrt{2}\right)}{\left(1+\sqrt{2}\right)\times\left(1-\sqrt{2}\right)}} = \boxed{\frac{1-\sqrt{2}}{(1\cdot 1)-\left(1\cdot\sqrt{2}\right)+\left(1\cdot\sqrt{2}\right)-\left(\sqrt{2}\cdot\sqrt{2}\right)}} = \boxed{\frac{1-\sqrt{2}}{1-\sqrt{2}+\sqrt{2}-\sqrt{2\cdot 2}}}$$

$$= \boxed{\frac{1-\sqrt{2}}{1-\sqrt{2^2}}} = \boxed{\frac{1-\sqrt{2}}{1-2}} = \boxed{\frac{1-\sqrt{2}}{-1}} = \boxed{-\frac{1-\sqrt{2}}{1}} = \boxed{\frac{\sqrt{2}-1}{1}} = \boxed{\mathbf{\sqrt{2}-1}}$$

Example 4.3-53

$$\boxed{\frac{8+\sqrt{72}}{\sqrt{8}-\sqrt{18}}} = \boxed{\frac{8+\sqrt{36\cdot 2}}{\sqrt{4\cdot 2}-\sqrt{9\cdot 2}}} = \boxed{\frac{8+\sqrt{6^2\cdot 2}}{\sqrt{2^2\cdot 2}-\sqrt{3^2\cdot 2}}} = \boxed{\frac{8+6\sqrt{2}}{2\sqrt{2}-3\sqrt{2}}} = \boxed{\frac{8+6\sqrt{2}}{2\sqrt{2}-3\sqrt{2}} \times \frac{2\sqrt{2}+3\sqrt{2}}{2\sqrt{2}+3\sqrt{2}}}$$

$$= \boxed{\frac{\left(8+6\sqrt{2}\right)\times\left(2\sqrt{2}+3\sqrt{2}\right)}{\left(2\sqrt{2}-3\sqrt{2}\right)\times\left(2\sqrt{2}+3\sqrt{2}\right)}} = \boxed{\frac{\left(8\cdot 2\sqrt{2}\right)+\left(8\cdot 3\sqrt{2}\right)+\left(6\sqrt{2}\cdot 2\sqrt{2}\right)+\left(6\sqrt{2}\cdot 3\sqrt{2}\right)}{\left(2\sqrt{2}\cdot 2\sqrt{2}\right)+\left(2\sqrt{2}\cdot 3\sqrt{2}\right)-\left(3\sqrt{2}\cdot 2\sqrt{2}\right)-\left(3\sqrt{2}\cdot 3\sqrt{2}\right)}}$$

$$= \boxed{\frac{16\sqrt{2}+24\sqrt{2}+(6\cdot 2)\sqrt{2\cdot 2}+(6\cdot 3)\sqrt{2\cdot 2}}{(2\cdot 2)\sqrt{2\cdot 2}+(2\cdot 3)\sqrt{2\cdot 2}-(3\cdot 2)\sqrt{2\cdot 2}-(3\cdot 3)\sqrt{2\cdot 2}}} = \boxed{\frac{16\sqrt{2}+24\sqrt{2}+12\sqrt{2^2}+18\sqrt{2^2}}{4\sqrt{2^2}+6\sqrt{2^2}-6\sqrt{2^2}-9\sqrt{2^2}}}$$

$$= \boxed{\frac{16\sqrt{2}+24\sqrt{2}+(12\cdot 2)+(18\cdot 2)}{(4\cdot 2)+(6\cdot 2)-(6\cdot 2)-(9\cdot 2)}} = \boxed{\frac{16\sqrt{2}+24\sqrt{2}+24+36}{8+12-12-18}} = \boxed{\frac{(16+24)\sqrt{2}+60}{-10}} = \boxed{\frac{40\sqrt{2}+60}{-10}}$$

$$= \boxed{-\frac{\overset{1}{\cancel{10}}\left(4\sqrt{2}+6\right)}{\underset{1}{\cancel{10}}}} = \boxed{-\frac{4\sqrt{2}+6}{1}} = \boxed{-\left(4\sqrt{2}+6\right)} = \boxed{\mathbf{-2\left(2\sqrt{2}+3\right)}}$$

Note that the above problem can also be solved in the following way by recognizing that the radical term in the denominator can further be simplified, i.e., $2\sqrt{2}-3\sqrt{2} = (2-3)\sqrt{2} = -\sqrt{2}$.

$$\boxed{\frac{8+\sqrt{72}}{\sqrt{8}-\sqrt{18}}} = \boxed{\frac{8+\sqrt{36\cdot 2}}{\sqrt{4\cdot 2}-\sqrt{9\cdot 2}}} = \boxed{\frac{8+\sqrt{6^2\cdot 2}}{\sqrt{2^2\cdot 2}-\sqrt{3^2\cdot 2}}} = \boxed{\frac{8+6\sqrt{2}}{2\sqrt{2}-3\sqrt{2}}} = \boxed{\frac{8+6\sqrt{2}}{(2-3)\sqrt{2}}} = \boxed{\frac{8+6\sqrt{2}}{-\sqrt{2}}} = \boxed{-\frac{8+6\sqrt{2}}{\sqrt{2}}}$$

$$= \boxed{-\frac{8+6\sqrt{2}}{\sqrt{2}} \times \frac{\sqrt{2}}{\sqrt{2}}} = \boxed{-\frac{\left(8+6\sqrt{2}\right)\times\sqrt{2}}{\sqrt{2}\times\sqrt{2}}} = \boxed{-\frac{8\cdot\sqrt{2}+6\sqrt{2}\cdot\sqrt{2}}{\sqrt{2\cdot 2}}} = \boxed{-\frac{8\sqrt{2}+6\sqrt{2\cdot 2}}{\sqrt{2^2}}} = \boxed{-\frac{8\sqrt{2}+6\sqrt{2^2}}{2}}$$

$$= \boxed{-\frac{8\sqrt{2}+(6\cdot 2)}{2}} = \boxed{-\frac{8\sqrt{2}+12}{2}} = \boxed{-\frac{\overset{1}{\cancel{2}}\left(4\sqrt{2}+6\right)}{\underset{1}{\cancel{2}}}} = \boxed{-\frac{4\sqrt{2}+6}{1}} = \boxed{-\left(4\sqrt{2}+6\right)} = \boxed{\mathbf{-2\left(2\sqrt{2}+3\right)}}$$

Example 4.3-54

$$\boxed{\frac{\sqrt{7}-1}{\sqrt{7}-2}} = \boxed{\frac{\sqrt{7}-1}{\sqrt{7}-2} \times \frac{\sqrt{7}+2}{\sqrt{7}+2}} = \boxed{\frac{\left(\sqrt{7}-1\right)\times\left(\sqrt{7}+2\right)}{\left(\sqrt{7}-2\right)\times\left(\sqrt{7}+2\right)}} = \boxed{\frac{\left(\sqrt{7}\cdot\sqrt{7}\right)+\left(2\cdot\sqrt{7}\right)-\left(1\cdot\sqrt{7}\right)-(1\cdot 2)}{\left(\sqrt{7}\cdot\sqrt{7}\right)+\left(2\cdot\sqrt{7}\right)-\left(2\cdot\sqrt{7}\right)-(2\cdot 2)}}$$

$$= \boxed{\frac{\sqrt{7\cdot 7}+2\sqrt{7}-\sqrt{7}-2}{\sqrt{7\cdot 7}+2\sqrt{7}-2\sqrt{7}-4}} = \boxed{\frac{\sqrt{7^2}+2\sqrt{7}-\sqrt{7}-2}{\sqrt{7^2}-4}} = \boxed{\frac{7+2\sqrt{7}-\sqrt{7}-2}{7-4}} = \boxed{\frac{(7-2)+(2-1)\sqrt{7}}{3}} = \boxed{\mathbf{\frac{5+\sqrt{7}}{3}}}$$

Example 4.3-55

$$\boxed{\frac{3+\sqrt{27}}{3-\sqrt{3}}} = \boxed{\frac{3+\sqrt{9\cdot 3}}{3-\sqrt{3}}} = \boxed{\frac{3+\sqrt{3^2\cdot 3}}{3-\sqrt{3}}} = \boxed{\frac{3+3\sqrt{3}}{3-\sqrt{3}}} = \boxed{\frac{3+3\sqrt{3}}{3-\sqrt{3}}\times\frac{3+\sqrt{3}}{3+\sqrt{3}}} = \boxed{\frac{\left(3+3\sqrt{3}\right)\times\left(3+\sqrt{3}\right)}{\left(3-\sqrt{3}\right)\times\left(3+\sqrt{3}\right)}}$$

$$= \boxed{\frac{(3\cdot 3)+\left(3\cdot\sqrt{3}\right)+(3\cdot 3)\sqrt{3}+\left(3\sqrt{3}\cdot\sqrt{3}\right)}{(3\cdot 3)+\left(3\cdot\sqrt{3}\right)-\left(3\cdot\sqrt{3}\right)-\left(\sqrt{3}\cdot\sqrt{3}\right)}} = \boxed{\frac{9+3\sqrt{3}+9\sqrt{3}+3\sqrt{3\cdot 3}}{9+3\sqrt{3}-3\sqrt{3}-\sqrt{3\cdot 3}}} = \boxed{\frac{9+3\sqrt{3}+9\sqrt{3}+3\sqrt{3^2}}{9-\sqrt{3^2}}}$$

$$= \boxed{\frac{9+3\sqrt{3}+9\sqrt{3}+3\cdot 3}{9-3}} = \boxed{\frac{(9+9)+(3+9)\sqrt{3}}{9-3}} = \boxed{\frac{18+12\sqrt{3}}{6}} = \boxed{\frac{\overset{1}{\cancel{6}}\left(3+2\sqrt{3}\right)}{\underset{1}{\cancel{6}}}} = \boxed{\frac{3+2\sqrt{3}}{1}} = \boxed{\mathbf{3+2\sqrt{3}}}$$

Example 4.3-56

$$\boxed{\frac{\sqrt{5}-1}{\sqrt{5}+1}} = \boxed{\frac{\sqrt{5}-1}{\sqrt{5}+1}\times\frac{\sqrt{5}-1}{\sqrt{5}-1}} = \boxed{\frac{\left(\sqrt{5}-1\right)\times\left(\sqrt{5}-1\right)}{\left(\sqrt{5}+1\right)\times\left(\sqrt{5}-1\right)}} = \boxed{\frac{\left(\sqrt{5}\cdot\sqrt{5}\right)-\left(1\cdot\sqrt{5}\right)-\left(1\cdot\sqrt{5}\right)+(1\cdot 1)}{\left(\sqrt{5}\cdot\sqrt{5}\right)-\left(1\cdot\sqrt{5}\right)+\left(1\cdot\sqrt{5}\right)-(1\cdot 1)}} = \boxed{\frac{\sqrt{5\cdot 5}-\sqrt{5}-\sqrt{5}+1}{\sqrt{5\cdot 5}-\sqrt{5}+\sqrt{5}-1}}$$

$$= \boxed{\frac{\sqrt{5^2}-\sqrt{5}-\sqrt{5}+1}{\sqrt{5^2}-1}} = \boxed{\frac{5-\sqrt{5}-\sqrt{5}+1}{5-1}} = \boxed{\frac{(5+1)+(-1-1)\sqrt{5}}{5-1}} = \boxed{\frac{6-2\sqrt{5}}{4}} = \boxed{\frac{\overset{1}{\cancel{2}}\left(3-\sqrt{5}\right)}{\underset{2}{\cancel{4}}}} = \boxed{\mathbf{\frac{3-\sqrt{5}}{2}}}$$

Example 4.3-57

$$\boxed{\frac{\sqrt{8}+4\sqrt{10}}{8-\sqrt{80}}} = \boxed{\frac{\sqrt{4\cdot 2}+4\sqrt{10}}{8-\sqrt{16\cdot 5}}} = \boxed{\frac{\sqrt{2^2\cdot 2}+4\sqrt{10}}{8-\sqrt{4^2\cdot 5}}} = \boxed{\frac{2\sqrt{2}+4\sqrt{10}}{8-4\sqrt{5}}} = \boxed{\frac{2\sqrt{2}+4\sqrt{10}}{8-4\sqrt{5}}\times\frac{8+4\sqrt{5}}{8+4\sqrt{5}}}$$

$$= \boxed{\frac{\left(2\sqrt{2}+4\sqrt{10}\right)\times\left(8+4\sqrt{5}\right)}{\left(8-4\sqrt{5}\right)\times\left(8+4\sqrt{5}\right)}} = \boxed{\frac{\left(8\cdot 2\sqrt{2}\right)+(2\cdot 4)\cdot\left(\sqrt{2}\cdot\sqrt{5}\right)+\left(8\cdot 4\sqrt{10}\right)+(4\cdot 4)\cdot\left(\sqrt{10}\cdot\sqrt{5}\right)}{(8\cdot 8)+(8\cdot 4)\sqrt{5}-(8\cdot 4)\sqrt{5}-(4\cdot 4)\cdot\left(\sqrt{5}\cdot\sqrt{5}\right)}}$$

$$= \boxed{\frac{16\sqrt{2}+8\sqrt{2\cdot 5}+32\sqrt{10}+16\sqrt{10\cdot 5}}{64+32\sqrt{5}-32\sqrt{5}-16\sqrt{5\cdot 5}}} = \boxed{\frac{16\sqrt{2}+8\sqrt{10}+32\sqrt{10}+16\sqrt{50}}{64-16\sqrt{5^2}}} = \boxed{\frac{16\sqrt{2}+(8+32)\sqrt{10}+16\sqrt{25\cdot 2}}{64-(16\cdot 5)}}$$

$$= \boxed{\frac{16\sqrt{2}+40\sqrt{10}+16\sqrt{5^2\cdot 2}}{64-80}} = \boxed{\frac{40\sqrt{10}+16\sqrt{2}+(16\cdot 5)\sqrt{2}}{-16}} = \boxed{-\frac{40\sqrt{10}+16\sqrt{2}+80\sqrt{2}}{16}}$$

$$= \boxed{-\frac{40\sqrt{10}+(16+80)\sqrt{2}}{16}} = \boxed{-\frac{40\sqrt{10}+96\sqrt{2}}{16}} = \boxed{-\frac{\overset{1}{\cancel{8}}\left(5\sqrt{10}+12\sqrt{2}\right)}{\underset{2}{\cancel{16}}}} = \boxed{\mathbf{-\frac{5\sqrt{10}+12\sqrt{2}}{2}}}$$

Example 4.3-58

$$\boxed{\frac{\sqrt{5}}{\sqrt{7}+\sqrt{3}}} = \boxed{\frac{\sqrt{5}}{\sqrt{7}+\sqrt{3}} \times \frac{\sqrt{7}-\sqrt{3}}{\sqrt{7}-\sqrt{3}}} = \boxed{\frac{\sqrt{5}\times\left(\sqrt{7}-\sqrt{3}\right)}{\left(\sqrt{7}+\sqrt{3}\right)\times\left(\sqrt{7}-\sqrt{3}\right)}} = \boxed{\frac{\left(\sqrt{5}\cdot\sqrt{7}\right)-\left(\sqrt{5}\cdot\sqrt{3}\right)}{\left(\sqrt{7}\cdot\sqrt{7}\right)-\left(\sqrt{7}\cdot\sqrt{3}\right)+\left(\sqrt{3}\cdot\sqrt{7}\right)-\left(\sqrt{3}\cdot\sqrt{3}\right)}}$$

$$= \boxed{\frac{\sqrt{5\cdot7}-\sqrt{5\cdot3}}{\sqrt{7\cdot7}-\sqrt{7\cdot3}+\sqrt{3\cdot7}-\sqrt{3\cdot3}}} = \boxed{\frac{\sqrt{35}-\sqrt{15}}{\sqrt{7^2}-\sqrt{21}+\sqrt{21}-\sqrt{3^2}}} = \boxed{\frac{\sqrt{35}-\sqrt{15}}{7-3}} = \boxed{\frac{\mathbf{\sqrt{35}-\sqrt{15}}}{\mathbf{4}}}$$

Example 4.3-59

$$\boxed{\frac{-2\sqrt{3}-\sqrt{7}}{7-3\sqrt{3}}} = \boxed{\frac{-2\sqrt{3}-\sqrt{7}}{7-3\sqrt{3}} \times \frac{7+3\sqrt{3}}{7+3\sqrt{3}}} = \boxed{\frac{\left(-2\sqrt{3}-\sqrt{7}\right)\times\left(7+3\sqrt{3}\right)}{\left(7-3\sqrt{3}\right)\times\left(7+3\sqrt{3}\right)}}$$

$$= \boxed{\frac{-(2\cdot7)\sqrt{3}-(2\cdot3)\cdot\left(\sqrt{3}\cdot\sqrt{3}\right)-\left(7\cdot\sqrt{7}\right)-3\left(\sqrt{7}\cdot\sqrt{3}\right)}{(7\cdot7)+\left(7\cdot3\sqrt{3}\right)-(3\cdot7)\sqrt{3}-(3\cdot3)\cdot\left(\sqrt{3}\cdot\sqrt{3}\right)}} = \boxed{\frac{-14\sqrt{3}-6\sqrt{3\cdot3}-7\sqrt{7}-3\sqrt{7\cdot3}}{49+21\sqrt{3}-21\sqrt{3}-9\sqrt{3\cdot3}}}$$

$$= \boxed{\frac{-14\sqrt{3}-6\sqrt{3^2}-7\sqrt{7}-3\sqrt{21}}{49-9\sqrt{3^2}}} = \boxed{\frac{-14\sqrt{3}-(6\cdot3)-7\sqrt{7}-3\sqrt{21}}{49-(9\cdot3)}} = \boxed{\frac{-\left(14\sqrt{3}+7\sqrt{7}+3\sqrt{21}+18\right)}{49-27}}$$

$$= \boxed{-\frac{\mathbf{14\sqrt{3}+7\sqrt{7}+3\sqrt{21}+18}}{\mathbf{22}}}$$

Example 4.3-60

$$\boxed{\frac{\sqrt{2}+\sqrt{3}}{4-2\sqrt{3}}} = \boxed{\frac{\sqrt{2}+\sqrt{3}}{4-2\sqrt{3}} \times \frac{4+2\sqrt{3}}{4+2\sqrt{3}}} = \boxed{\frac{\left(\sqrt{2}+\sqrt{3}\right)\times\left(4+2\sqrt{3}\right)}{\left(4-2\sqrt{3}\right)\times\left(4+2\sqrt{3}\right)}} = \boxed{\frac{\left(4\cdot\sqrt{2}\right)+2\cdot\left(\sqrt{2}\cdot\sqrt{3}\right)+\left(4\cdot\sqrt{3}\right)+2\cdot\left(\sqrt{3}\cdot\sqrt{3}\right)}{(4\cdot4)+(4\cdot2)\sqrt{3}-(2\cdot4)\sqrt{3}-(2\cdot2)\left(\sqrt{3}\cdot\sqrt{3}\right)}}$$

$$= \boxed{\frac{4\sqrt{2}+2\sqrt{2\cdot3}+4\sqrt{3}+2\sqrt{3\cdot3}}{16+8\sqrt{3}-8\sqrt{3}-4\sqrt{3\cdot3}}} = \boxed{\frac{4\sqrt{2}+2\sqrt{6}+4\sqrt{3}+2\sqrt{3^2}}{16-4\sqrt{3^2}}} = \boxed{\frac{4\sqrt{2}+2\sqrt{6}+4\sqrt{3}+(2\cdot3)}{16-(4\cdot3)}}$$

$$= \boxed{\frac{4\sqrt{2}+2\sqrt{6}+4\sqrt{3}+6}{16-12}} = \boxed{\frac{\overset{1}{\cancel{2}}\left(2\sqrt{2}+\sqrt{6}+2\sqrt{3}+3\right)}{\underset{2}{\cancel{4}}}} = \boxed{\frac{\mathbf{2\sqrt{2}+\sqrt{6}+2\sqrt{3}+3}}{\mathbf{2}}}$$

Practice Problems: Rationalizing Radical Expressions - Binomial Denominators with Real Numbers

Section 4.3 Case II a Practice Problems - Solve the following radical expressions:

1. $\dfrac{7}{1+\sqrt{7}} =$

2. $\dfrac{1-\sqrt{18}}{2+\sqrt{18}} =$

3. $\dfrac{\sqrt{5}}{\sqrt{5}+\sqrt{2}} =$

4. $\dfrac{3-\sqrt{5}}{\sqrt{7}-\sqrt{4}} =$

5. $\dfrac{-3+\sqrt{3}}{4+\sqrt{5}} =$

6. $\dfrac{3-\sqrt{3}}{3+\sqrt{3}} =$

7. $\dfrac{\sqrt{72}+\sqrt{20}}{\sqrt{50}+\sqrt{27}} =$

8. $\dfrac{1-\sqrt{8}}{2+\sqrt{8}} =$

9. $\dfrac{5+5\sqrt{25}}{5-\sqrt{125}} =$

10. $\dfrac{\sqrt{5}+\sqrt{3}}{\sqrt{5}-\sqrt{3}} =$

Case II b Rationalizing Radical Expressions - Binomial Denominators with Variables

Radical expressions with binomial denominators are simplified using the following steps:

Step 1 Simplify the radical terms in the numerator and the denominator (see Section 4.1, Case IV).

Step 2 Rationalize the denominator by multiplying the numerator and the denominator by its conjugate.

Step 3 Simplify the radical expression using the **FOIL** method (see Section 4.2, Case II b).

Examples with Steps

The following examples show the steps as to how radical expressions with two terms in the denominator are simplified:

Example 4.3-61

$$\frac{\sqrt{p^3}}{\sqrt{p^5}-\sqrt{q}} =$$

Solution:

Step 1

$$\frac{\sqrt{p^3}}{\sqrt{p^5}-\sqrt{q}} = \frac{\sqrt{p^{2+1}}}{\sqrt{p^{2+2+1}}-\sqrt{q}} = \frac{\sqrt{p^2 \cdot p^1}}{\sqrt{p^2 \cdot p^2 \cdot p^1}-\sqrt{q}} = \frac{p\sqrt{p}}{p^1 \cdot p^1\sqrt{p}-\sqrt{q}}$$

$$= \frac{p\sqrt{p}}{p^{1+1}\sqrt{p}-\sqrt{q}} = \frac{p\sqrt{p}}{p^2\sqrt{p}-\sqrt{q}}$$

Step 2

$$\frac{p\sqrt{p}}{p^2\sqrt{p}-\sqrt{q}} = \frac{p\sqrt{p}}{p^2\sqrt{p}-\sqrt{q}} \times \frac{p^2\sqrt{p}+\sqrt{q}}{p^2\sqrt{p}+\sqrt{q}}$$

Step 3

$$\frac{p\sqrt{p}}{p^2\sqrt{p}-\sqrt{q}} \times \frac{p^2\sqrt{p}+\sqrt{q}}{p^2\sqrt{p}+\sqrt{q}} = \frac{p\sqrt{p} \times \left(p^2\sqrt{p}+\sqrt{q}\right)}{\left(p^2\sqrt{p}-\sqrt{q}\right) \times \left(p^2\sqrt{p}+\sqrt{q}\right)}$$

$$= \frac{\left(p \cdot p^2\right) \cdot \left(\sqrt{p} \cdot \sqrt{p}\right) + \left(p\sqrt{p} \cdot \sqrt{q}\right)}{\left(p^2 \cdot p^2\right) \cdot \left(\sqrt{p} \cdot \sqrt{p}\right) + \left(p^2\sqrt{p} \cdot \sqrt{q}\right) - \left(p^2\sqrt{q} \cdot \sqrt{p}\right) - \left(\sqrt{q} \cdot \sqrt{q}\right)}$$

$$= \frac{p^{1+2}\sqrt{p \cdot p} + p\sqrt{p \cdot q}}{p^{2+2}\sqrt{p \cdot p} + p^2\sqrt{p \cdot q} - p^2\sqrt{p \cdot q} - \sqrt{q \cdot q}} = \frac{p^3\sqrt{p^2} + p\sqrt{pq}}{p^4\sqrt{p^2} + p^2\sqrt{pq} - p^2\sqrt{pq} - \sqrt{q^2}}$$

$$= \frac{p^3 \cdot p^1 + p\sqrt{pq}}{p^4 \cdot p^1 - q} = \frac{p^{3+1} + p\sqrt{pq}}{p^{4+1} - q} = \frac{p^4 + p\sqrt{pq}}{p^5 - q} = \boldsymbol{\frac{p\left(p^3 + \sqrt{pq}\right)}{p^5 - q}}$$

Example 4.3-62

$$\boxed{\frac{a}{a+\sqrt{a}}}=$$

Solution:

Step 1 $\boxed{\textit{Not Applicable}}$

Step 2 $$\boxed{\frac{a}{a+\sqrt{a}}}=\boxed{\frac{a}{a+\sqrt{a}}\times\frac{a-\sqrt{a}}{a-\sqrt{a}}}$$

Step 3 $$\boxed{\frac{a}{a+\sqrt{a}}\times\frac{a-\sqrt{a}}{a-\sqrt{a}}}=\boxed{\frac{a\times\left(a-\sqrt{a}\right)}{\left(a+\sqrt{a}\right)\times\left(a-\sqrt{a}\right)}}=\boxed{\frac{a\cdot\left(a-\sqrt{a}\right)}{(a\cdot a)-\left(a\cdot\sqrt{a}\right)+\left(a\cdot\sqrt{a}\right)-\left(\sqrt{a}\cdot\sqrt{a}\right)}}$$

$$=\boxed{\frac{a\cdot\left(a-\sqrt{a}\right)}{a^2-a\sqrt{a}+a\sqrt{a}-\sqrt{a\cdot a}}}=\boxed{\frac{a\cdot\left(a-\sqrt{a}\right)}{a^2-\sqrt{a^2}}}=\boxed{\frac{a\cdot\left(a-\sqrt{a}\right)}{a^2-a}}=\boxed{\frac{\overset{1}{\cancel{a}}\cdot\left(a-\sqrt{a}\right)}{\underset{1}{\cancel{a}}\cdot(a-1)}}$$

$$=\boxed{\frac{1\cdot\left(a-\sqrt{a}\right)}{1\cdot(a-1)}}=\boxed{\boldsymbol{\frac{a-\sqrt{a}}{a-1}}}$$

Example 4.3-63

$$\boxed{\frac{\sqrt{x^3}}{\sqrt{8}-\sqrt{x^5}}}=$$

Solution:

Step 1 $$\boxed{\frac{\sqrt{x^3}}{\sqrt{8}-\sqrt{x^5}}}=\boxed{\frac{\sqrt{x^{2+1}}}{\sqrt{4\cdot 2}-\sqrt{x^{2+2+1}}}}=\boxed{\frac{\sqrt{x^2\cdot x^1}}{\sqrt{2^2\cdot 2}-\sqrt{x^2\cdot x^2\cdot x^1}}}=\boxed{\frac{x\sqrt{x}}{2\sqrt{2}-x^1\cdot x^1\sqrt{x}}}$$

$$=\boxed{\frac{x\sqrt{x}}{2\sqrt{2}-x^{1+1}\sqrt{x}}}=\boxed{\frac{x\sqrt{x}}{2\sqrt{2}-x^2\sqrt{x}}}$$

Step 2 $$\boxed{\frac{x\sqrt{x}}{2\sqrt{2}-x^2\sqrt{x}}}=\boxed{\frac{x\sqrt{x}}{2\sqrt{2}-x^2\sqrt{x}}\times\frac{2\sqrt{2}+x^2\sqrt{x}}{2\sqrt{2}+x^2\sqrt{x}}}$$

Step 3 $$\boxed{\frac{x\sqrt{x}}{2\sqrt{2}-x^2\sqrt{x}}\times\frac{2\sqrt{2}+x^2\sqrt{x}}{2\sqrt{2}+x^2\sqrt{x}}}=\boxed{\frac{x\sqrt{x}\times\left(2\sqrt{2}+x^2\sqrt{x}\right)}{\left(2\sqrt{2}-x^2\sqrt{x}\right)\times\left(2\sqrt{2}+x^2\sqrt{x}\right)}}$$

$$=\boxed{\frac{\left(x\sqrt{x}\cdot 2\sqrt{2}\right)+\left(x\sqrt{x}\cdot x^2\sqrt{x}\right)}{\left(2\sqrt{2}\cdot 2\sqrt{2}\right)+\left(2\sqrt{2}\cdot x^2\sqrt{x}\right)-\left(2\sqrt{2}\cdot x^2\sqrt{x}\right)-\left(x^2\cdot x^2\right)\cdot\left(\sqrt{x}\cdot\sqrt{x}\right)}}$$

$$= \boxed{\frac{(2\cdot x)\cdot\left(\sqrt{2}\cdot\sqrt{x}\right)+\left(x\cdot x^2\right)\cdot\left(\sqrt{x}\cdot\sqrt{x}\right)}{\left((2\cdot 2)\sqrt{2\cdot 2}\right)+\left(2\sqrt{2}x^2\sqrt{x}\right)-\left(2\sqrt{2}x^2\sqrt{x}\right)-\left(x^{2+2}\right)\left(\sqrt{x\cdot x}\right)}} = \boxed{\frac{2x\sqrt{2\cdot x}+x^3\sqrt{x\cdot x}}{4\sqrt{2^2}-x^4\sqrt{x^2}}}$$

$$= \boxed{\frac{2x\sqrt{2x}+x^3\sqrt{x^2}}{(4\cdot 2)-x^4x^1}} = \boxed{\frac{2x\sqrt{2x}+x^3\cdot x^1}{8-x^{4+1}}} = \boxed{\frac{2x\sqrt{2x}+x^{3+1}}{8-x^5}} = \boxed{\mathbf{\frac{2x\sqrt{2x}+x^4}{8-x^5}}}$$

Example 4.3-64

$$\boxed{\frac{\sqrt{r^2}+\sqrt{s^3}}{\sqrt{r^4}-\sqrt{s}}} =$$

Solution:

Step 1

$$\boxed{\frac{\sqrt{r^2}+\sqrt{s^3}}{\sqrt{r^4}-\sqrt{s}}} = \boxed{\frac{r+\sqrt{s^{2+1}}}{\sqrt{r^{2+2}}-\sqrt{s}}} = \boxed{\frac{r+\sqrt{s^2\cdot s^1}}{\sqrt{r^2\cdot r^2}-\sqrt{s}}} = \boxed{\frac{r+s\sqrt{s}}{r^1\cdot r^1-\sqrt{s}}} = \boxed{\frac{r+s\sqrt{s}}{r^{1+1}-\sqrt{s}}}$$

$$= \boxed{\frac{r+s\sqrt{s}}{r^2-\sqrt{s}}}$$

Step 2

$$\boxed{\frac{r+s\sqrt{s}}{r^2-\sqrt{s}}} = \boxed{\frac{r+s\sqrt{s}}{r^2-\sqrt{s}}\times\frac{r^2+\sqrt{s}}{r^2+\sqrt{s}}}$$

Step 3

$$\boxed{\frac{r+s\sqrt{s}}{r^2-\sqrt{s}}\times\frac{r^2+\sqrt{s}}{r^2+\sqrt{s}}} = \boxed{\frac{\left(r+s\sqrt{s}\right)\times\left(r^2+\sqrt{s}\right)}{\left(r^2-\sqrt{s}\right)\times\left(r^2+\sqrt{s}\right)}} = \boxed{\frac{\left(r\cdot r^2\right)+\left(r\cdot\sqrt{s}\right)+\left(r^2 s\sqrt{s}\right)+\left(s\sqrt{s}\cdot\sqrt{s}\right)}{\left(r^2\cdot r^2\right)+\left(r^2\cdot\sqrt{s}\right)-\left(r^2\cdot\sqrt{s}\right)-\left(\sqrt{s}\cdot\sqrt{s}\right)}}$$

$$= \boxed{\frac{r^{1+2}+r\sqrt{s}+r^2 s\sqrt{s}+s\sqrt{s\cdot s}}{r^{2+2}+r^2\sqrt{s}-r^2\sqrt{s}-\sqrt{s\cdot s}}} = \boxed{\frac{r^3+r\sqrt{s}+r^2 s\sqrt{s}+s\sqrt{s^2}}{r^4-\sqrt{s^2}}} = \boxed{\frac{r^3+r\sqrt{s}+r^2 s\sqrt{s}+s\cdot s}{r^4-s}}$$

$$= \boxed{\frac{r^3+r\sqrt{s}+r^2 s\sqrt{s}+s^2}{r^4-s}} = \boxed{\mathbf{\frac{\left(r^3+s^2\right)+r\sqrt{s}(1+rs)}{r^4-s}}}$$

Example 4.3-65

$$\boxed{\frac{4+\sqrt{m}}{\sqrt{m}+\sqrt{n}}} =$$

Solution:

Step 1 *Not Applicable*

Step 2

$$\boxed{\frac{4+\sqrt{m}}{\sqrt{m}+\sqrt{n}}} = \boxed{\frac{4+\sqrt{m}}{\sqrt{m}+\sqrt{n}}\times\frac{\sqrt{m}-\sqrt{n}}{\sqrt{m}-\sqrt{n}}}$$

Step 3

$$\boxed{\frac{4+\sqrt{m}}{\sqrt{m}+\sqrt{n}}\times\frac{\sqrt{m}-\sqrt{n}}{\sqrt{m}-\sqrt{n}}} = \boxed{\frac{\left(4+\sqrt{m}\right)\times\left(\sqrt{m}-\sqrt{n}\right)}{\left(\sqrt{m}+\sqrt{n}\right)\times\left(\sqrt{m}-\sqrt{n}\right)}}$$

$$= \boxed{\frac{\left(4\cdot\sqrt{m}\right)-\left(4\cdot\sqrt{n}\right)+\left(\sqrt{m}\cdot\sqrt{m}\right)-\left(\sqrt{m}\cdot\sqrt{n}\right)}{\left(\sqrt{m}\cdot\sqrt{m}\right)-\left(\sqrt{m}\cdot\sqrt{n}\right)+\left(\sqrt{n}\cdot\sqrt{m}\right)-\left(\sqrt{n}\cdot\sqrt{n}\right)}} = \boxed{\frac{4\sqrt{m}-4\sqrt{n}+\sqrt{m\cdot m}-\sqrt{m\cdot n}}{\sqrt{m\cdot m}-\sqrt{m\cdot n}+\sqrt{m\cdot n}-\sqrt{n\cdot n}}}$$

$$= \boxed{\frac{4\sqrt{m}-4\sqrt{n}+\sqrt{m^2}-\sqrt{mn}}{\sqrt{m^2}-\sqrt{n^2}}} = \boxed{\mathbf{\frac{4\sqrt{m}-4\sqrt{n}+m-\sqrt{mn}}{m-n}}}$$

Additional Examples: Rationalizing Radical Expressions - Binomial Denominators with Variables

The following examples further illustrate how to rationalize radical expressions with binomial denominators:

Example 4.3-66

$$\boxed{\frac{\sqrt{a}}{a-\sqrt{b}}} = \boxed{\frac{\sqrt{a}}{a-\sqrt{b}}\times\frac{a+\sqrt{b}}{a+\sqrt{b}}} = \boxed{\frac{\sqrt{a}\times\left(a+\sqrt{b}\right)}{\left(a-\sqrt{b}\right)\times\left(a+\sqrt{b}\right)}} = \boxed{\frac{\left(a\cdot\sqrt{a}\right)+\left(\sqrt{a}\cdot\sqrt{b}\right)}{\left(a\cdot a\right)+\left(a\cdot\sqrt{b}\right)-\left(a\cdot\sqrt{b}\right)-\left(\sqrt{b}\cdot\sqrt{b}\right)}}$$

$$= \boxed{\frac{a\sqrt{a}+\sqrt{ab}}{a^2+a\sqrt{b}-a\sqrt{b}-\sqrt{b\cdot b}}} = \boxed{\frac{a\sqrt{a}+\sqrt{ab}}{a^2-\sqrt{b^2}}} = \boxed{\mathbf{\frac{a\sqrt{a}+\sqrt{ab}}{a^2-b}}}$$

Example 4.3-67

$$\boxed{\frac{1+\sqrt{x}}{1-\sqrt{x}}} = \boxed{\frac{1+\sqrt{x}}{1-\sqrt{x}}\times\frac{1+\sqrt{x}}{1+\sqrt{x}}} = \boxed{\frac{\left(1+\sqrt{x}\right)\times\left(1+\sqrt{x}\right)}{\left(1-\sqrt{x}\right)\times\left(1+\sqrt{x}\right)}} = \boxed{\frac{(1\cdot 1)+\left(1\cdot\sqrt{x}\right)+\left(1\cdot\sqrt{x}\right)+\left(\sqrt{x}\cdot\sqrt{x}\right)}{(1\cdot 1)+\left(1\cdot\sqrt{x}\right)-\left(1\cdot\sqrt{x}\right)-\left(\sqrt{x}\cdot\sqrt{x}\right)}}$$

$$= \boxed{\frac{1+\sqrt{x}+\sqrt{x}+\sqrt{x\cdot x}}{1+\sqrt{x}-\sqrt{x}-\sqrt{x\cdot x}}} = \boxed{\frac{1+(1+1)\sqrt{x}+\sqrt{x^2}}{1-\sqrt{x^2}}} = \boxed{\mathbf{\frac{1+2\sqrt{x}+x}{1-x}}}$$

Example 4.3-68

$$\boxed{\frac{y+\sqrt{y^4}}{y-\sqrt{y}}} = \boxed{\frac{y+\sqrt{y^{2+2}}}{y-\sqrt{y}}} = \boxed{\frac{y+\sqrt{y^2\cdot y^2}}{y-\sqrt{y}}} = \boxed{\frac{y+y^1\cdot y^1}{y-\sqrt{y}}} = \boxed{\frac{y+y^{1+1}}{y-\sqrt{y}}} = \boxed{\frac{y+y^2}{y-\sqrt{y}}} = \boxed{\frac{y+y^2}{y-\sqrt{y}}\times\frac{y+\sqrt{y}}{y+\sqrt{y}}}$$

$$= \boxed{\frac{\left(y+y^2\right)\times\left(y+\sqrt{y}\right)}{\left(y-\sqrt{y}\right)\times\left(y+\sqrt{y}\right)}} = \boxed{\frac{(y\cdot y)+\left(y\cdot\sqrt{y}\right)+\left(y^2\cdot y\right)+\left(y^2\cdot\sqrt{y}\right)}{(y\cdot y)+\left(y\cdot\sqrt{y}\right)-\left(y\cdot\sqrt{y}\right)-\left(\sqrt{y}\cdot\sqrt{y}\right)}} = \boxed{\frac{y^2+y\sqrt{y}+\left(y^2y^1\right)+y^2\sqrt{y}}{y^2+y\sqrt{y}-y\sqrt{y}-\sqrt{y\cdot y}}}$$

$$= \boxed{\frac{y^2+y\sqrt{y}+\left(y^{2+1}\right)+y^2\sqrt{y}}{y^2-\sqrt{y^2}}} = \boxed{\frac{y^2+y\sqrt{y}+y^3+y^2\sqrt{y}}{y^2-y}} = \boxed{\frac{\overset{1}{\cancel{y}}\left(y+\sqrt{y}+y^2+y\sqrt{y}\right)}{\underset{1}{\cancel{y}}(y-1)}}$$

$$= \boxed{\frac{y+\sqrt{y}+y^2+y\sqrt{y}}{y-1}} = \boxed{\frac{\left(y\sqrt{y}+\sqrt{y}\right)+\left(y^2+y\right)}{y-1}} = \boxed{\frac{\sqrt{y}(y+1)+y(y+1)}{y-1}} = \boxed{\frac{\mathbf{(y+1)\left[\sqrt{y}+y\right]}}{\mathbf{y-1}}}$$

Example 4.3-69

$$\boxed{\frac{2x}{2+\sqrt{x^3}}} = \boxed{\frac{2x}{2+\sqrt{x^{2+1}}}} = \boxed{\frac{2x}{2+\sqrt{x^2\cdot x^1}}} = \boxed{\frac{2x}{2+x\sqrt{x}}} = \boxed{\frac{2x}{2+x\sqrt{x}}\times\frac{2-x\sqrt{x}}{2-x\sqrt{x}}} = \boxed{\frac{2x\times\left(2-x\sqrt{x}\right)}{\left(2+x\sqrt{x}\right)\times\left(2-x\sqrt{x}\right)}}$$

$$= \boxed{\frac{2x\left(2-x\sqrt{x}\right)}{(2\cdot 2)-\left(2\cdot x\sqrt{x}\right)+\left(2\cdot x\sqrt{x}\right)-(x\cdot x)\cdot\left(\sqrt{x}\sqrt{x}\right)}} = \boxed{\frac{2x\left(2-x\sqrt{x}\right)}{4-2x\sqrt{x}+2x\sqrt{x}-x^2\sqrt{x\cdot x}}} = \boxed{\frac{2x\left(2-x\sqrt{x}\right)}{4-x^2\sqrt{x^2}}}$$

$$= \boxed{\frac{2x\left(2-x\sqrt{x}\right)}{4-x^2\cdot x^1}} = \boxed{\frac{2x\left(2-x\sqrt{x}\right)}{4-x^{2+1}}} = \boxed{\frac{\mathbf{2x\left(2-x\sqrt{x}\right)}}{\mathbf{4-x^3}}}$$

Example 4.3-70

$$\boxed{\frac{a}{a-a\sqrt{b}}} = \boxed{\frac{a}{a-a\sqrt{b}}\times\frac{a+a\sqrt{b}}{a+a\sqrt{b}}} = \boxed{\frac{a\times\left(a+a\sqrt{b}\right)}{\left(a-a\sqrt{b}\right)\times\left(a+a\sqrt{b}\right)}} = \boxed{\frac{(a\cdot a)+(a\cdot a)\sqrt{b}}{(a\cdot a)+(a\cdot a)\sqrt{b}-(a\cdot a)\sqrt{b}-(a\cdot a)\left(\sqrt{b}\cdot\sqrt{b}\right)}}$$

$$= \boxed{\frac{a^2+a^2\sqrt{b}}{a^2+a^2\sqrt{b}-a^2\sqrt{b}-a^2\sqrt{b\cdot b}}} = \boxed{\frac{a^2+a^2\sqrt{b}}{a^2-a^2\sqrt{b^2}}} = \boxed{\frac{a^2\left(1+\sqrt{b}\right)}{a^2-a^2b}} = \boxed{\frac{\not{a}^2\left(1+\sqrt{b}\right)}{\not{a}^2(1-b)}} = \boxed{\frac{\mathbf{1+\sqrt{b}}}{\mathbf{1-b}}}$$

Note that one can simplify the above expression first before rationalizing the denominator as shown below:

$$\boxed{\frac{a}{a-a\sqrt{b}}} = \boxed{\frac{\overset{1}{\not{a}}}{\underset{1}{\not{a}}\left(1-\sqrt{b}\right)}} = \boxed{\frac{1}{1-\sqrt{b}}} = \boxed{\frac{1}{1-\sqrt{b}}\times\frac{1+\sqrt{b}}{1+\sqrt{b}}} = \boxed{\frac{1\times\left(1+\sqrt{b}\right)}{\left(1-\sqrt{b}\right)\times\left(1+\sqrt{b}\right)}}$$

$$= \boxed{\frac{1+\sqrt{b}}{(1\cdot 1)+\left(1\cdot\sqrt{b}\right)-\left(1\cdot\sqrt{b}\right)-\left(\sqrt{b}\cdot\sqrt{b}\right)}} = \boxed{\frac{1+\sqrt{b}}{1+\sqrt{b}-\sqrt{b}-\sqrt{b\cdot b}}} = \boxed{\frac{1+\sqrt{b}}{1-\sqrt{b^2}}} = \boxed{\frac{\mathbf{1+\sqrt{b}}}{\mathbf{1-b}}}$$

Example 4.3-71

$$\boxed{\frac{-3x+\sqrt{y^3}}{x+\sqrt{y^5}}} = \boxed{\frac{-3x+\sqrt{y^{2+1}}}{x+\sqrt{y^{2+2+1}}}} = \boxed{\frac{-3x+\sqrt{y^2\cdot y^1}}{x+\sqrt{y^2\cdot y^2\cdot y^1}}} = \boxed{\frac{-3x+y\sqrt{y}}{x+y^1\cdot y^1\sqrt{y}}} = \boxed{\frac{-3x+y\sqrt{y}}{x+y^{1+1}\sqrt{y}}} = \boxed{\frac{-3x+y\sqrt{y}}{x+y^2\sqrt{y}}}$$

$$= \boxed{\frac{-3x+y\sqrt{y}}{x+y^2\sqrt{y}}\times\frac{x-y^2\sqrt{y}}{x-y^2\sqrt{y}}} = \boxed{\frac{\left(-3x+y\sqrt{y}\right)\times\left(x-y^2\sqrt{y}\right)}{\left(x+y^2\sqrt{y}\right)\times\left(x-y^2\sqrt{y}\right)}}$$

$$= \boxed{\frac{-3(x\cdot x)+3\left(x\cdot y^2\right)\sqrt{y}+\left(x\cdot y\sqrt{y}\right)-\left(y\cdot y^2\right)\cdot\left(\sqrt{y}\cdot\sqrt{y}\right)}{(x\cdot x)-\left(x\cdot y^2\right)\sqrt{y}+\left(x\cdot y^2\right)\sqrt{y}-\left(y^2\cdot y^2\right)\cdot\left(\sqrt{y}\cdot\sqrt{y}\right)}} = \boxed{\frac{-3x^2+3xy^2\sqrt{y}+xy\sqrt{y}-y^1y^2\left(\sqrt{y\cdot y}\right)}{x^2-xy^2\sqrt{y}+xy^2\sqrt{y}-y^{2+2}\left(\sqrt{y\cdot y}\right)}}$$

$$= \frac{-3x^2 + 3xy^2\sqrt{y} + xy\sqrt{y} - y^1y^2\sqrt{y^2}}{x^2 - y^4\sqrt{y^2}} = \frac{-3x^2 + 3xy^2\sqrt{y} + xy\sqrt{y} - y^1y^2y^1}{x^2 - y^4y^1}$$

$$= \frac{-3x^2 + 3xy^2\sqrt{y} + xy\sqrt{y} - y^{1+2+1}}{x^2 - y^{4+1}} = \frac{3xy^2\sqrt{y} + xy\sqrt{y} - 3x^2 - y^4}{x^2 - y^5} = \boxed{\frac{xy\sqrt{y}(3y+1) - 3x^2 - y^4}{x^2 - y^5}}$$

Example 4.3-72

$$\frac{5+\sqrt{r}}{1-2\sqrt{r}} = \frac{5+\sqrt{r}}{1-2\sqrt{r}} \times \frac{1+2\sqrt{r}}{1+2\sqrt{r}} = \frac{\left(5+\sqrt{r}\right)\times\left(1+2\sqrt{r}\right)}{\left(1-2\sqrt{r}\right)\times\left(1+2\sqrt{r}\right)} = \frac{(5\cdot1)+(5\cdot2)\sqrt{r}+\left(1\cdot\sqrt{r}\right)+2\left(\sqrt{r}\cdot\sqrt{r}\right)}{(1\cdot1)+\left(1\cdot2\sqrt{r}\right)-\left(1\cdot2\sqrt{r}\right)-(2\cdot2)\cdot\left(\sqrt{r}\cdot\sqrt{r}\right)}$$

$$= \frac{5+10\sqrt{r}+\sqrt{r}+2\sqrt{r\cdot r}}{1+2\sqrt{r}-2\sqrt{r}-4\sqrt{r\cdot r}} = \frac{5+(10+1)\sqrt{r}+2\sqrt{r^2}}{1-4\sqrt{r^2}} = \boxed{\frac{5+11\sqrt{r}+2r}{1-4r}}$$

Example 4.3-73

$$\frac{\sqrt{a}-\sqrt{b}}{\sqrt{a}+\sqrt{b}} = \frac{\sqrt{a}-\sqrt{b}}{\sqrt{a}+\sqrt{b}} \times \frac{\sqrt{a}-\sqrt{b}}{\sqrt{a}-\sqrt{b}} = \frac{\left(\sqrt{a}-\sqrt{b}\right)\times\left(\sqrt{a}-\sqrt{b}\right)}{\left(\sqrt{a}+\sqrt{b}\right)\times\left(\sqrt{a}-\sqrt{b}\right)} = \frac{\left(\sqrt{a}\cdot\sqrt{a}\right)-\left(\sqrt{a}\cdot\sqrt{b}\right)-\left(\sqrt{b}\cdot\sqrt{a}\right)+\left(\sqrt{b}\cdot\sqrt{b}\right)}{\left(\sqrt{a}\cdot\sqrt{a}\right)-\left(\sqrt{a}\cdot\sqrt{b}\right)+\left(\sqrt{b}\cdot\sqrt{a}\right)-\left(\sqrt{b}\cdot\sqrt{b}\right)}$$

$$= \frac{\sqrt{a\cdot a}-\sqrt{a\cdot b}-\sqrt{a\cdot b}+\sqrt{b\cdot b}}{\sqrt{a\cdot a}-\sqrt{a\cdot b}+\sqrt{a\cdot b}-\sqrt{b\cdot b}} = \frac{\sqrt{a^2}-2\sqrt{a\cdot b}+\sqrt{b^2}}{\sqrt{a^2}-\sqrt{b^2}} = \frac{a-2\sqrt{ab}+b}{a-b} = \boxed{\frac{a+b-2\sqrt{ab}}{a-b}}$$

Example 4.3-74

$$\frac{x}{5+\sqrt{x}} = \frac{x}{5+\sqrt{x}} \times \frac{5-\sqrt{x}}{5-\sqrt{x}} = \frac{x\times\left(5-\sqrt{x}\right)}{\left(5+\sqrt{x}\right)\times\left(5-\sqrt{x}\right)} = \frac{x\left(5-\sqrt{x}\right)}{(5\cdot5)-\left(5\cdot\sqrt{x}\right)+\left(5\cdot\sqrt{x}\right)-\left(\sqrt{x}\cdot\sqrt{x}\right)}$$

$$= \frac{x\left(5-\sqrt{x}\right)}{25-5\sqrt{x}+5\sqrt{x}-\sqrt{x\cdot x}} = \frac{x\left(5-\sqrt{x}\right)}{25-\sqrt{x^2}} = \boxed{\frac{x\left(5-\sqrt{x}\right)}{25-x}}$$

Example 4.3-75

$$\frac{a+\sqrt{b}}{a-\sqrt{b}} = \frac{a+\sqrt{b}}{a-\sqrt{b}} \times \frac{a+\sqrt{b}}{a+\sqrt{b}} = \frac{\left(a+\sqrt{b}\right)\times\left(a+\sqrt{b}\right)}{\left(a-\sqrt{b}\right)\times\left(a+\sqrt{b}\right)} = \frac{(a\cdot a)+\left(a\cdot\sqrt{b}\right)+\left(a\cdot\sqrt{b}\right)+\left(\sqrt{b}\cdot\sqrt{b}\right)}{(a\cdot a)+\left(a\cdot\sqrt{b}\right)-\left(a\cdot\sqrt{b}\right)-\left(\sqrt{b}\cdot\sqrt{b}\right)}$$

$$= \frac{a^2+a\sqrt{b}+a\sqrt{b}+\sqrt{b\cdot b}}{a^2+a\sqrt{b}-a\sqrt{b}-\sqrt{b\cdot b}} = \frac{a^2+(a+a)\sqrt{b}+\sqrt{b^2}}{a^2-\sqrt{b^2}} = \boxed{\frac{a^2+2a\sqrt{b}+b}{a^2-b}}$$

Example 4.3-76

$$\frac{\sqrt{x}}{5-\sqrt{x^5}} = \frac{\sqrt{x}}{5-\sqrt{x^{2+2+1}}} = \frac{\sqrt{x}}{5-\sqrt{x^2\cdot x^2\cdot x^1}} = \frac{\sqrt{x}}{5-x^1\cdot x^1\sqrt{x}} = \frac{\sqrt{x}}{5-x^{1+1}\sqrt{x}} = \frac{\sqrt{x}}{5-x^2\sqrt{x}}$$

$$= \frac{\sqrt{x}}{5 - x^2\sqrt{x}} \times \frac{5 + x^2\sqrt{x}}{5 + x^2\sqrt{x}} = \frac{\sqrt{x} \times \left(5 + x^2\sqrt{x}\right)}{\left(5 - x^2\sqrt{x}\right) \times \left(5 + x^2\sqrt{x}\right)} = \frac{\left(5 \cdot \sqrt{x}\right) + x^2\left(\sqrt{x} \cdot \sqrt{x}\right)}{(5 \cdot 5) + \left(5 \cdot x^2\sqrt{x}\right) - \left(5 \cdot x^2\sqrt{x}\right) - \left(x^2 \cdot x^2\right) \cdot \left(\sqrt{x} \cdot \sqrt{x}\right)}$$

$$= \frac{5\sqrt{x} + x^2\sqrt{x \cdot x}}{25 + 5x^2\sqrt{x} - 5x^2\sqrt{x} - x^2x^2\sqrt{x \cdot x}} = \frac{5\sqrt{x} + x^2\sqrt{x^2}}{25 - x^2x^2\sqrt{x^2}} = \frac{5\sqrt{x} + x^2x^1}{25 - x^2x^2x^1} = \frac{5\sqrt{x} + x^{2+1}}{25 - x^{2+2+1}} = \boldsymbol{\frac{5\sqrt{x} + x^3}{25 - x^5}}$$

Example 4.3-77

$$\frac{\sqrt{x^3}}{\sqrt{x^4} - \sqrt{x}} = \frac{\sqrt{x^{2+1}}}{\sqrt{x^{2+2}} - \sqrt{x}} = \frac{\sqrt{x^2 \cdot x^1}}{\sqrt{x^2 \cdot x^2} - \sqrt{x}} = \frac{x\sqrt{x}}{x^1 \cdot x^1 - \sqrt{x}} = \frac{x\sqrt{x}}{x^{1+1} - \sqrt{x}} = \frac{x\sqrt{x}}{x^2 - \sqrt{x}}$$

$$= \frac{x\sqrt{x}}{x^2 - \sqrt{x}} \times \frac{x^2 + \sqrt{x}}{x^2 + \sqrt{x}} = \frac{x\sqrt{x} \times \left(x^2 + \sqrt{x}\right)}{\left(x^2 - \sqrt{x}\right) \times \left(x^2 + \sqrt{x}\right)} = \frac{\left(x^1 \cdot x^2\right)\sqrt{x} + \left(x\sqrt{x} \cdot \sqrt{x}\right)}{\left(x^2 \cdot x^2\right) + \left(x^2 \cdot \sqrt{x}\right) - \left(x^2 \cdot \sqrt{x}\right) - \left(\sqrt{x} \cdot \sqrt{x}\right)}$$

$$= \frac{x^{1+2}\sqrt{x} + x\sqrt{x \cdot x}}{x^{2+2} + x^2\sqrt{x} - x^2\sqrt{x} - \sqrt{x \cdot x}} = \frac{x^3\sqrt{x} + x\sqrt{x^2}}{x^4 - \sqrt{x^2}} = \frac{x^3\sqrt{x} + x \cdot x}{x^4 - x} = \frac{x^3\sqrt{x} + x^2}{x^4 - x} = \frac{x^2\left(x\sqrt{x} + 1\right)}{x\left(x^3 - 1\right)}$$

$$= \frac{x^2\left(x\sqrt{x} + 1\right)}{x^1\left(x^3 - 1\right)} = \frac{\left(x^2 \cdot x^{-1}\right) \cdot \left(x\sqrt{x} + 1\right)}{x^3 - 1} = \frac{x^{2-1}\left(x\sqrt{x} + 1\right)}{x^3 - 1} = \boldsymbol{\frac{x\left(x\sqrt{x} + 1\right)}{x^3 - 1}}$$

Example 4.3-78

$$\frac{-5}{1 - \sqrt{a^3}} = \frac{-5}{1 - \sqrt{a^{2+1}}} = \frac{-5}{1 - \sqrt{a^2 \cdot a^1}} = \frac{-5}{1 - a\sqrt{a}} = \frac{-5}{1 - a\sqrt{a}} \times \frac{1 + a\sqrt{a}}{1 + a\sqrt{a}} = \frac{-5 \times \left(1 + a\sqrt{a}\right)}{\left(1 - a\sqrt{a}\right) \times \left(1 + a\sqrt{a}\right)}$$

$$= -\frac{5\left(1 + a\sqrt{a}\right)}{(1 \cdot 1) + \left(1 \cdot a\sqrt{a}\right) - \left(1 \cdot a\sqrt{a}\right) - (a \cdot a) \cdot \left(\sqrt{a} \cdot \sqrt{a}\right)} = -\frac{5\left(1 + a\sqrt{a}\right)}{1 + a\sqrt{a} - a\sqrt{a} - a^2\sqrt{a \cdot a}} = -\frac{5\left(1 + a\sqrt{a}\right)}{1 - a^2\sqrt{a^2}}$$

$$= -\frac{5\left(1 + a\sqrt{a}\right)}{1 - a^2 \cdot a^1} = -\frac{5\left(1 + a\sqrt{a}\right)}{1 - a^{2+1}} = \boldsymbol{-\frac{5\left(1 + a\sqrt{a}\right)}{1 - a^3}}$$

Example 4.3-79

$$\frac{\sqrt{m} + 3}{\sqrt{m} - 3} = \frac{\sqrt{m} + 3}{\sqrt{m} - 3} \times \frac{\sqrt{m} + 3}{\sqrt{m} + 3} = \frac{\left(\sqrt{m} + 3\right) \times \left(\sqrt{m} + 3\right)}{\left(\sqrt{m} - 3\right) \times \left(\sqrt{m} + 3\right)} = \frac{\left(\sqrt{m} \cdot \sqrt{m}\right) + \left(3 \cdot \sqrt{m}\right) + \left(3 \cdot \sqrt{m}\right) + (3 \cdot 3)}{\left(\sqrt{m} \cdot \sqrt{m}\right) + \left(3 \cdot \sqrt{m}\right) - \left(3 \cdot \sqrt{m}\right) - (3 \cdot 3)}$$

$$= \frac{\sqrt{m \cdot m} + 3\sqrt{m} + 3\sqrt{m} + 9}{\sqrt{m \cdot m} + 3\sqrt{m} - 3\sqrt{m} - 9} = \frac{\sqrt{m^2} + (3 + 3)\sqrt{m} + 9}{\sqrt{m^2} - 9} = \boldsymbol{\frac{m + 6\sqrt{m} + 9}{m - 9}}$$

Example 4.3-80

$$\boxed{\frac{x\sqrt{x}-1}{x-\sqrt{x}}} = \boxed{\frac{x\sqrt{x}-1}{x-\sqrt{x}} \times \frac{x+\sqrt{x}}{x+\sqrt{x}}} = \boxed{\frac{\left(x\sqrt{x}-1\right)\times\left(x+\sqrt{x}\right)}{\left(x-\sqrt{x}\right)\times\left(x+\sqrt{x}\right)}} = \boxed{\frac{(x\cdot x)\sqrt{x}+\left(x\sqrt{x}\cdot\sqrt{x}\right)-(1\cdot x)-\left(1\cdot\sqrt{x}\right)}{(x\cdot x)+\left(x\cdot\sqrt{x}\right)-\left(x\cdot\sqrt{x}\right)-\left(\sqrt{x}\cdot\sqrt{x}\right)}}$$

$$= \boxed{\frac{x^2\sqrt{x}+x\sqrt{x\cdot x}-x-\sqrt{x}}{x^2+x\sqrt{x}-x\sqrt{x}-\sqrt{x\cdot x}}} = \boxed{\frac{x^2\sqrt{x}+x\sqrt{x^2}-x-\sqrt{x}}{x^2-\sqrt{x^2}}} = \boxed{\frac{x^2\sqrt{x}+(x\cdot x)-x-\sqrt{x}}{x^2-x}} = \boxed{\frac{x^2\sqrt{x}+x^2-x-\sqrt{x}}{x^2-x}}$$

$$= \boxed{\frac{\left(x^2\sqrt{x}-\sqrt{x}\right)+\left(x^2-x\right)}{x(x-1)}} = \boxed{\frac{\sqrt{x}\left(x^2-1\right)+x(x-1)}{x(x-1)}} = \boxed{\frac{\sqrt{x}[(x-1)(x+1)]+x(x-1)}{x(x-1)}} = \boxed{\frac{(x-1)\left[\sqrt{x}(x+1)+x\right]}{x(x-1)}}$$

$$= \boxed{\frac{\sqrt{x}(x+1)+x}{x}}$$

Note: $\left(x^2-1\right)=(x-1)(x+1)$

Practice Problems: Rationalizing Radical Expressions - Binomial Denominators with Variables

Section 4.3 Case II b Practice Problems - Solve the following radical expressions:

1. $\frac{5x}{1+\sqrt{x}} =$
2. $\frac{\sqrt{x}}{2-\sqrt{x}} =$
3. $\frac{1+3x}{1-2\sqrt{x}} =$
4. $\frac{a-b}{\sqrt{a}-\sqrt{b}} =$
5. $\frac{-a}{a-\sqrt{a}} =$
6. $\frac{x+y}{x+\sqrt{y}} =$
7. $\frac{5+x}{2-\sqrt{x}} =$
8. $\frac{-w+\sqrt{w}}{w+\sqrt{w}} =$
9. $\frac{\sqrt{k}-3}{1+\sqrt{k}} =$
10. $\frac{m\sqrt{m}+\sqrt{n}}{\sqrt{m}-n\sqrt{n}} =$

4.4 Adding and Subtracting Radical Expressions

Radicals are added and subtracted using the following general rule:

$$k_1\sqrt[n]{a}+k_2\sqrt[n]{a}+k_3\sqrt[n]{a}=(k_1+k_2+k_3)\sqrt[n]{a}$$

Only similar radicals can be added and subtracted. **Similar radicals** are defined as radical expressions with the same index n and the same radicand a. Note that the distributive property of multiplication (see Section 1.4) is used to group the numbers in front of the similar radical terms. In this section students learn how to add and subtract radical expressions for simple (Case I) and more difficult cases (Case II).

Case I Adding and Subtracting Radical Terms (Simple Cases)

Radicals are added and subtracted using the following steps:

Step 1 Group similar radicals.

Step 2 Simplify the radical expression.

Examples with Steps

The following examples show the steps as to how radical expressions are added and subtracted:

Example 4.4-1

$\boxed{6\sqrt{2}+4\sqrt{2}}=$

Solution:

Step 1 $\boxed{6\sqrt{2}+4\sqrt{2}}=\boxed{(6+4)\sqrt{2}}$

Step 2 $\boxed{(6+4)\sqrt{2}}=\boxed{\mathbf{10\sqrt{2}}}$

Example 4.4-2

$\boxed{5\sqrt[3]{5}+8\sqrt[3]{5}}=$

Solution:

Step 1 $\boxed{5\sqrt[3]{5}+8\sqrt[3]{5}}=\boxed{(5+8)\sqrt[3]{5}}$

Step 2 $\boxed{(5+8)\sqrt[3]{5}}=\boxed{\mathbf{13\sqrt[3]{5}}}$

Example 4.4-3

$\boxed{20\sqrt[5]{3}-8\sqrt[5]{3}+5\sqrt[5]{3}}=$

Solution:

Step 1 $\boxed{20\sqrt[5]{3}-8\sqrt[5]{3}+5\sqrt[5]{3}}=\boxed{(20-8+5)\sqrt[5]{3}}$

Step 2 $\boxed{(20-8+5)\sqrt[5]{3}}=\boxed{\mathbf{17\sqrt[5]{3}}}$

Example 4.4-4

$$\boxed{\left(6\sqrt{7}+2\sqrt{7}\right)-2\sqrt[3]{7}} =$$

Solution:

Step 1 $\boxed{\left(6\sqrt{7}+2\sqrt{7}\right)-2\sqrt[3]{7}} = \boxed{(6+2)\sqrt{7}-2\sqrt[3]{7}}$

Step 2 $\boxed{(6+2)\sqrt{7}-2\sqrt[3]{7}} = \boxed{\mathbf{8\sqrt{7}-2\sqrt[3]{7}}}$

Example 4.4-5

$$\boxed{8\sqrt[3]{4}-3\sqrt[3]{4}+7\sqrt[3]{4}-\sqrt[3]{4}} =$$

Solution:

Step 1 $\boxed{8\sqrt[3]{4}-3\sqrt[3]{4}+7\sqrt[3]{4}-\sqrt[3]{4}} = \boxed{(8-3+7-1)\sqrt[3]{4}}$

Step 2 $\boxed{(8-3+7-1)\sqrt[3]{4}} = \boxed{\mathbf{11\sqrt[3]{4}}}$

Additional Examples - Adding and Subtracting Radical Terms (Simple Cases)

The following examples further illustrate how to add and subtract radical terms:

Example 4.4-6

$$\boxed{2\sqrt{5}+3\sqrt{5}+6} = \boxed{(2+3)\sqrt{5}+6} = \boxed{\mathbf{5\sqrt{5}+6}}$$

Example 4.4-7

$$\boxed{8\sqrt[3]{4}+2\sqrt[3]{4}+5} = \boxed{(8+2)\sqrt[3]{4}+5} = \boxed{10\sqrt[3]{4}+5} = \boxed{\mathbf{5\left(2\sqrt[3]{4}+1\right)}}$$

Example 4.4-8

$$\boxed{2\sqrt[4]{3}+4\sqrt[4]{3}-3\sqrt[4]{3}+\sqrt[4]{5}} = \boxed{(2+4-3)\sqrt[4]{3}+\sqrt[4]{5}} = \boxed{\mathbf{3\sqrt[4]{3}+\sqrt[4]{5}}}$$

Note that the two radical terms have the same index (4) but have different radicands (3 and 5). Therefore, they can not be combined.

Example 4.4-9

$$\boxed{\sqrt[5]{5}+3\sqrt[5]{5}+a\sqrt[5]{5}-(4+a)\sqrt{2}} = \boxed{(1+3+a)\sqrt[5]{5}-(4+a)\sqrt{2}} = \boxed{(4+a)\sqrt[5]{5}-(4+a)\sqrt{2}} = \boxed{\mathbf{(4+a)\left[\sqrt[5]{5}-\sqrt{2}\right]}}$$

Example 4.4-10

$$\boxed{5\sqrt[3]{2x}+8\sqrt[3]{2x}-2c\sqrt[3]{2x}+4\sqrt{2x}-8\sqrt{2x}} = \boxed{(5+8-2c)\sqrt[3]{2x}+(4-8)\sqrt{2x}} = \boxed{\mathbf{(13-2c)\sqrt[3]{2x}-4\sqrt{2x}}}$$

Example 4.4-11

$$\boxed{a\sqrt{xy}+b\sqrt[3]{xy}-c^2\sqrt{xy}-d} = \boxed{a\sqrt{xy}-c^2\sqrt{xy}+b\sqrt[3]{xy}-d} = \boxed{\mathbf{\left(a-c^2\right)\sqrt{xy}+b\sqrt[3]{xy}-d}}$$

Example 4.4-12

$$\boxed{2\sqrt{75}+3\sqrt{125}+\sqrt{20}+3\sqrt{10}-4\sqrt{10}} = \boxed{2\sqrt{25\cdot 3}+3\sqrt{25\cdot 5}+\sqrt{4\cdot 5}+(3-4)\sqrt{10}}$$

$= \boxed{2\sqrt{5^2 \cdot 3} + 3\sqrt{5^2 \cdot 5} + \sqrt{2^2 \cdot 5} - \sqrt{10}} = \boxed{(2 \cdot 5)\sqrt{3} + (3 \cdot 5)\sqrt{5} + 2\sqrt{5} - \sqrt{10}} = \boxed{10\sqrt{3} + 15\sqrt{5} + 2\sqrt{5} - \sqrt{10}}$

$= \boxed{10\sqrt{3} + (15+2)\sqrt{5} - \sqrt{10}} = \boxed{\mathbf{10\sqrt{3} + 17\sqrt{5} - \sqrt{10}}}$

Example 4.4-13

$\boxed{5\sqrt[4]{3} + 8\sqrt[4]{3} + 8\sqrt[5]{3} + 3\sqrt[5]{3}} = \boxed{(5+8)\sqrt[4]{3} + (8+3)\sqrt[5]{3}} = \boxed{\mathbf{13\sqrt[4]{3} + 11\sqrt[5]{3}}}$

Example 4.4-14

$\boxed{7\sqrt{5} + 8\sqrt[3]{4} - \sqrt{5} - b\sqrt[3]{4} + \sqrt{2}} = \boxed{7\sqrt{5} - \sqrt{5} + 8\sqrt[3]{4} - b\sqrt[3]{4} + \sqrt{2}} = \boxed{(7-1)\sqrt{5} + (8-b)\sqrt[3]{4} + \sqrt{2}}$

$= \boxed{\mathbf{6\sqrt{5} + (8-b)\sqrt[3]{4} + \sqrt{2}}}$

Example 4.4-15

$\boxed{8\sqrt[3]{6} + 4\sqrt[3]{6} + a\sqrt[3]{6} - \sqrt{5} - 4\sqrt{5}} = \boxed{(8+4+a)\sqrt[3]{6} + (-1-4)\sqrt{5}} = \boxed{(12+a)\sqrt[3]{6} + (-5)\sqrt{5}} = \boxed{\mathbf{(12+a)\sqrt[3]{6} - 5\sqrt{5}}}$

Practice Problems - Adding and Subtracting Radical Terms (Simple Cases)

Section 4.4 Case I Practice Problems - Simplify the following radical expressions:

1. $5\sqrt{3} + 8\sqrt{3} =$
2. $2\sqrt[3]{3} - 4\sqrt[3]{3} =$
3. $12\sqrt[4]{5} + 8\sqrt[4]{5} + 2\sqrt[4]{3} =$
4. $a\sqrt{ab} - b\sqrt{ab} + c\sqrt{ab} =$
5. $3x\sqrt[3]{x} - 2x\sqrt[3]{x} + 4x\sqrt[3]{x^2} =$
6. $5\sqrt[3]{2} + 8\sqrt[3]{5} =$
7. $2\sqrt[5]{5} + 8\sqrt[3]{5} - 5\sqrt[5]{5} + 2\sqrt{5} =$
8. $3\sqrt{a} + 3a\sqrt{a} - 4a\sqrt{a} =$
9. $2\sqrt[3]{x^2} + 4\sqrt[4]{x^2} + 3\sqrt[5]{x^2} =$
10. $3\sqrt{ac} + 4\sqrt{ac} - 2\sqrt[3]{ac} + 3\sqrt[3]{ac} =$

Case II Adding and Subtracting Radical Terms (More Difficult Cases)

Radicals are added and subtracted using the following steps:

Step 1 Simplify the radical expression (see Section 4.1, Cases III and IV).

Step 2 Group similar radicals.

Step 3 Simplify the radical expression.

Examples with Steps

The following examples show the steps as to how radical expressions are added and subtracted:

Example 4.4-16

$$\sqrt{3x^5}+2\sqrt{27x}=$$

Solution:

Step 1

$$\sqrt{3x^5}+2\sqrt{27x}=\sqrt{3x^{2+2+1}}+2\sqrt{(9\cdot 3)\cdot x}=\sqrt{3\cdot\left(x^2\cdot x^2\cdot x^1\right)}+2\sqrt{\left(3^2\cdot 3\right)\cdot x}$$

$$=\left(x^1\cdot x^1\right)\sqrt{3\cdot x}+(2\cdot 3)\sqrt{3\cdot x}=\left(x^{1+1}\right)\sqrt{3x}+6\sqrt{3x}=x^2\sqrt{3x}+6\sqrt{3x}$$

Step 2

$$x^2\sqrt{3x}+6\sqrt{3x}=\mathbf{\left(x^2+6\right)\sqrt{3x}}$$

Step 3

Not Applicable

Example 4.4-17

$$\sqrt{48}-2\sqrt{27}-5\sqrt{75}=$$

Solution:

Step 1

$$\sqrt{48}-2\sqrt{27}-5\sqrt{75}=\sqrt{16\cdot 3}-2\sqrt{9\cdot 3}-5\sqrt{25\cdot 3}=\sqrt{4^2\cdot 3}-2\sqrt{3^2\cdot 3}-5\sqrt{5^2\cdot 3}$$

$$=4\sqrt{3}-(2\cdot 3)\sqrt{3}-(5\cdot 5)\sqrt{3}=4\sqrt{3}-6\sqrt{3}-25\sqrt{3}$$

Step 2

$$4\sqrt{3}-6\sqrt{3}-25\sqrt{3}=(4-6-25)\sqrt{3}$$

Step 3

$$(4-6-25)\sqrt{3}=\mathbf{-27\sqrt{3}}$$

Example 4.4-18

$$\sqrt[5]{3a}+\sqrt[5]{96a^6}+\sqrt[5]{729a}=$$

Solution:

Step 1

$$\sqrt[5]{3a}+\sqrt[5]{96a^6}+\sqrt[5]{729a}=\sqrt[5]{3a}+\sqrt[5]{(32\cdot 3)\cdot a^{5+1}}+\sqrt[5]{(243\cdot 3)\cdot a}$$

$$= \boxed{\sqrt[5]{3a} + \sqrt[5]{(2^5 \cdot 3) \cdot (a^5 \cdot a^1)} + \sqrt[5]{(3^5 \cdot 3) \cdot a}} = \boxed{\sqrt[5]{3a} + 2a\sqrt[5]{3 \cdot a^1} + 3\sqrt[5]{3 \cdot a}}$$

$$= \boxed{\sqrt[5]{3a} + 2a\sqrt[5]{3a} + 3\sqrt[5]{3a}}$$

Step 2 $\boxed{\sqrt[5]{3a} + 2a\sqrt[5]{3a} + 3\sqrt[5]{3a}} = \boxed{(1 + 2a + 3)\sqrt[5]{3a}}$

Step 3 $\boxed{(1 + 2a + 3)\sqrt[5]{3a}} = \boxed{(2a + 4)\sqrt[5]{3a}} = \boxed{\mathbf{2\sqrt[5]{3a}(2 + a)}}$

Example 4.4-19

$$\boxed{\sqrt[3]{a^5 b^5} + \sqrt[3]{a^8 b^2} + \sqrt[3]{a^2 b^2}} =$$

Solution:

Step 1 $\boxed{\sqrt[3]{a^5 b^5} + \sqrt[3]{a^8 b^2} + \sqrt[3]{a^2 b^2}} = \boxed{\sqrt[3]{a^{3+2} b^{3+2}} + \sqrt[3]{a^{3+3+2} b^2} + \sqrt[3]{a^2 b^2}}$

$$= \boxed{\sqrt[3]{(a^3 \cdot a^2) \cdot (b^3 \cdot b^2)} + \sqrt[3]{(a^3 \cdot a^3 \cdot a^2) \cdot b^2} + \sqrt[3]{a^2 b^2}}$$

$$= \boxed{ab\sqrt[3]{a^2 b^2} + (a^1 \cdot a^1)\sqrt[3]{a^2 b^2} + \sqrt[3]{a^2 b^2}} = \boxed{ab\sqrt[3]{a^2 b^2} + (a^{1+1})\sqrt[3]{a^2 b^2} + \sqrt[3]{a^2 b^2}}$$

$$= \boxed{ab\sqrt[3]{a^2 b^2} + a^2\sqrt[3]{a^2 b^2} + \sqrt[3]{a^2 b^2}}$$

Step 2 $\boxed{ab\sqrt[3]{a^2 b^2} + a^2\sqrt[3]{a^2 b^2} + \sqrt[3]{a^2 b^2}} = \boxed{(ab + a^2 + 1)\sqrt[3]{a^2 b^2}} = \boxed{\mathbf{(a^2 + ab + 1)\sqrt[3]{a^2 b^2}}}$

Step 3 *Not Applicable*

Example 4.4-20

$$\boxed{2\sqrt{300x^3} + 5\sqrt{12x} + 8\sqrt{3x^7}} =$$

Solution:

Step 1 $\boxed{2\sqrt{300x^3} + 5\sqrt{12x} + 8\sqrt{3x^7}} = \boxed{2\sqrt{(100 \cdot 3)x^{2+1}} + 5\sqrt{(4 \cdot 3) \cdot x} + 8\sqrt{3 \cdot x^{2+2+2+1}}}$

$$= \boxed{2\sqrt{(10^2 \cdot 3) \cdot (x^2 \cdot x^1)} + 5\sqrt{(2^2 \cdot 3) \cdot x} + 8\sqrt{3 \cdot (x^2 \cdot x^2 \cdot x^2 \cdot x^1)}}$$

$$= \boxed{(2 \cdot 10)x\sqrt{3 \cdot x} + (5 \cdot 2)\sqrt{3 \cdot x} + 8(x^1 \cdot x^1 \cdot x^1)\sqrt{3 \cdot x}} = \boxed{20x\sqrt{3x} + 10\sqrt{3x} + 8(x^{1+1+1})\sqrt{3x}}$$

$$= \boxed{20x\sqrt{3x} + 10\sqrt{3x} + 8x^3\sqrt{3x}}$$

Step 2 $\boxed{20x\sqrt{3x} + 10\sqrt{3x} + 8x^3\sqrt{3x}} = \boxed{8x^3\sqrt{3x} + 20x\sqrt{3x} + 10\sqrt{3x}} = \boxed{(8x^3 + 20x + 10)\sqrt{3x}}$

Step 3 $\boxed{(8x^3 + 20x + 10)\sqrt{3x}} = \boxed{\mathbf{2\sqrt{3x}(4x^3 + 10x + 5)}}$

$2^2 + 2 = 4 + 2 = 6$

Additional Examples - Adding and Subtracting Radical Terms (More Difficult Cases)

The following examples further illustrate how to add and subtract radical expressions:

Example 4.4-21

$$\boxed{5\sqrt[3]{27x}+\sqrt[3]{x}}=\boxed{5\sqrt[3]{3^3\cdot x}+\sqrt[3]{x}}=\boxed{(5\cdot 3)\sqrt[3]{x}+\sqrt[3]{x}}=\boxed{15\sqrt[3]{x}+\sqrt[3]{x}}=\boxed{(15+1)\cdot\sqrt[3]{x}}=\boxed{\mathbf{16\sqrt[3]{x}}}$$

Example 4.4-22

$$\boxed{5\sqrt{20}-\sqrt{5}+3\sqrt{45}}=\boxed{5\sqrt{4\cdot 5}-\sqrt{5}+3\sqrt{9\cdot 5}}=\boxed{5\sqrt{2^2\cdot 5}-\sqrt{5}+3\sqrt{3^2\cdot 5}}=\boxed{(5\cdot 2)\sqrt{5}-\sqrt{5}+(3\cdot 3)\sqrt{5}}$$

$$=\boxed{10\sqrt{5}-\sqrt{5}+9\sqrt{5}}=\boxed{(10-1+9)\sqrt{5}}=\boxed{\mathbf{18\sqrt{5}}}$$

Example 4.4-23

$$\boxed{5\sqrt{2x^3}+2\sqrt{8x}-7x\sqrt{2x}}=\boxed{5\sqrt{2x^{2+1}}+2\sqrt{2^3x}-7x\sqrt{2x}}=\boxed{5\sqrt{2x^2\cdot x^1}+2\sqrt{2^{2+1}x}-7x\sqrt{2x}}$$

$$=\boxed{5x\sqrt{2\cdot x^1}+2\sqrt{2^2\cdot 2^1\cdot x}-7x\sqrt{2x}}=\boxed{5x\sqrt{2x}+(2\cdot 2)\sqrt{2x}-7x\sqrt{2x}}=\boxed{5x\sqrt{2x}+4\sqrt{2x}-7x\sqrt{2x}}$$

$$=\boxed{4\sqrt{2x}+(5x-7x)\sqrt{2x}}=\boxed{4\sqrt{2x}-2x\sqrt{2x}}=\boxed{\mathbf{2\sqrt{2x}(2-x)}}$$

Example 4.4-24

$$\boxed{3\sqrt{32x^2}-8\sqrt{50}}=\boxed{3x\sqrt{16\cdot 2}-8\sqrt{25\cdot 2}}=\boxed{3x\sqrt{4^2\cdot 2}-8\sqrt{5^2\cdot 2}}=\boxed{(3\cdot 4)x\sqrt{2}-(8\cdot 5)\sqrt{2}}$$

$$=\boxed{12x\sqrt{2}-40\sqrt{2}}=\boxed{\mathbf{4\sqrt{2}(3x-10)}}$$

Example 4.4-25

$$\boxed{8\sqrt{3}-2\sqrt{27}+5\sqrt{192}}=\boxed{8\sqrt{3}-2\sqrt{9\cdot 3}+5\sqrt{64\cdot 3}}=\boxed{8\sqrt{3}-2\sqrt{3^2\cdot 3}+5\sqrt{8^2\cdot 3}}=\boxed{8\sqrt{3}-(2\cdot 3)\sqrt{3}+(5\cdot 8)\sqrt{3}}$$

$$=\boxed{8\sqrt{3}-6\sqrt{3}+40\sqrt{3}}=\boxed{(8-6+40)\sqrt{3}}=\boxed{\mathbf{42\sqrt{3}}}$$

Example 4.4-26

$$\boxed{2\sqrt{32a^3}+5\sqrt{50a^5}-2\sqrt{2a}}=\boxed{2\sqrt{(16\cdot 2)\cdot a^{2+1}}+5\sqrt{(25\cdot 2)\cdot a^{2+2+1}}-2\sqrt{2a}}$$

$$=\boxed{2\sqrt{\left(4^2\cdot 2\right)\cdot\left(a^2\cdot a^1\right)}+5\sqrt{\left(5^2\cdot 2\right)\cdot\left(a^2\cdot a^2\cdot a^1\right)}-2\sqrt{2a}}=\boxed{(2\cdot 4)\cdot a\sqrt{2\cdot a^1}+(5\cdot 5)\cdot(a\cdot a)\sqrt{2\cdot a^1}-2\sqrt{2a}}$$

$$=\boxed{8a\sqrt{2a}+25a^2\sqrt{2a}-2\sqrt{2a}}=\boxed{\left(8a+25a^2-2\right)\sqrt{2a}}=\boxed{\mathbf{\left(25a^2+8a-2\right)\sqrt{2a}}}$$

Example 4.4-27

$$\boxed{2\sqrt[4]{a^5}+5\sqrt[4]{a^9}+3\sqrt[4]{a^4}}=\boxed{2\sqrt[4]{a^{4+1}}+5\sqrt[4]{a^{4+4+1}}+3a}=\boxed{2\sqrt[4]{a^4\cdot a^1}+5\sqrt[4]{a^4\cdot a^4\cdot a^1}+3a}$$

$$=\boxed{2a\sqrt[4]{a^1}+5\left(a^1\cdot a^1\right)\sqrt[4]{a^1}+3a}=\boxed{2a\sqrt[4]{a}+5\left(a^{1+1}\right)\sqrt[4]{a}+3a}=\boxed{2a\sqrt[4]{a}+5a^2\sqrt[4]{a}+3a}=\boxed{\mathbf{(5a+2)a\sqrt[4]{a}+3a}}$$

Example 4.4-28

$$2\sqrt[3]{40x^5}+8x\sqrt[3]{5x^2}-\sqrt[3]{135x^8} = 2\sqrt[3]{(8\cdot5)\cdot x^{3+2}}+8x\sqrt[3]{5x^2}-\sqrt[3]{(27\cdot5)\cdot x^{3+3+2}}$$

$$= 2\sqrt[3]{\left(2^3\cdot5\right)\cdot\left(x^3\cdot x^2\right)}+8x\sqrt[3]{5x^2}-\sqrt[3]{\left(3^3\cdot5\right)\cdot\left(x^3\cdot x^3\cdot x^2\right)} = (2\cdot2)x\sqrt[3]{5\cdot x^2}+8x\sqrt[3]{5x^2}-3\left(x^1\cdot x^1\right)\sqrt[3]{5\cdot x^2}$$

$$= 4x\sqrt[3]{5x^2}+8x\sqrt[3]{5x^2}-3\left(x^{1+1}\right)\sqrt[3]{5x^2} = 4x\sqrt[3]{5x^2}+8x\sqrt[3]{5x^2}-3x^2\sqrt[3]{5x^2} = (4x+8x)\sqrt[3]{5x^2}-3x^2\sqrt[3]{5x^2}$$

$$= 12x\sqrt[3]{5x^2}-3x^2\sqrt[3]{5x^2} = \mathbf{3x\sqrt[3]{5x^2}\,(4-x)}$$

Example 4.4-29

$$5\sqrt[3]{a^5b}+3\sqrt[3]{a^2b^7}-2\sqrt[3]{a^8b^4} = 5\sqrt[3]{a^{3+2}b}+3\sqrt[3]{a^2b^{3+3+1}}-2\sqrt[3]{a^{3+3+2}b^{3+1}}$$

$$= 5\sqrt[3]{\left(a^3\cdot a^2\right)\cdot b}+3\sqrt[3]{a^2\cdot\left(b^3\cdot b^3\cdot b^1\right)}-2\sqrt[3]{\left(a^3\cdot a^3\cdot a^2\right)\cdot\left(b^3\cdot b^1\right)} = 5a\sqrt[3]{a^2b}+3\left(b^1\cdot b^1\right)\sqrt[3]{a^2b}-2\left(a^1\cdot a^1\right)b\sqrt[3]{a^2b}$$

$$= 5a\sqrt[3]{a^2b}+3\left(b^{1+1}\right)\sqrt[3]{a^2b}-2\left(a^{1+1}\right)b\sqrt[3]{a^2b} = 5a\sqrt[3]{a^2b}+3b^2\sqrt[3]{a^2b}-2a^2b\sqrt[3]{a^2b} = \left(5a+3b^2-2a^2b\right)\sqrt[3]{a^2b}$$

$$= \mathbf{\left(3b^2-2a^2b+5a\right)\sqrt[3]{a^2b}}$$

Example 4.4-30

$$4\sqrt[5]{x^{11}}-2x\sqrt[5]{x^6}+8\sqrt[5]{x^5} = 4\sqrt[5]{x^{5+5+1}}-2x\sqrt[5]{x^{5+1}}+8\cdot x = 4\sqrt[5]{x^5\cdot x^5\cdot x^1}-2x\sqrt[5]{x^5\cdot x^1}+8x$$

$$= 4\left(x^1\cdot x^1\right)\sqrt[5]{x}-2\left(x^1\cdot x^1\right)\sqrt[5]{x}+8x = 4\left(x^{1+1}\right)\sqrt[5]{x}-2\left(x^{1+1}\right)\sqrt[5]{x}+8x = 4x^2\sqrt[5]{x}-2x^2\sqrt[5]{x}+8x$$

$$= \left(4x^2-2x^2\right)\sqrt[5]{x}+8x = 2x^2\sqrt[5]{x}+8x = \mathbf{2x\left(x\sqrt[5]{x}+4\right)}$$

Practice Problems - Adding and Subtracting Radical Terms (More Difficult Cases)

Section 4.4 Case II Practice Problems - Simplify the following radical expressions:

1. $5\sqrt{2a}+\sqrt{32a} =$
2. $\sqrt[3]{27x}-\sqrt[3]{375x^4}-x\sqrt[3]{24x^7} =$
3. $2a^2\sqrt[4]{x^5}+4\sqrt[4]{81\cdot x^9}+\sqrt[4]{256x} =$
4. $\sqrt[5]{w^{11}}+5\sqrt[5]{32w^6}-2a\sqrt[5]{w^{16}} =$
5. $\sqrt{4xy}-4\sqrt{(xy)^5}+2\sqrt{49\cdot(xy)^3} =$
6. $\sqrt[3]{x^5y}+5\sqrt[3]{x^2y^7}+3\sqrt[3]{x^8y^4} =$
7. $\sqrt[5]{mn+3}+\sqrt[5]{(mn+3)^6}+2a\sqrt[5]{(mn+3)^{11}} =$
8. $\sqrt{x^3}-\sqrt{125x^5}+3\sqrt{x^3} =$
9. $\sqrt[3]{x^7y^8z^3}+2\sqrt[3]{x^4y^5z^6}+8\sqrt[3]{64\cdot xy^2} =$
10. $\sqrt[4]{512\cdot x^5y^{10}}+\sqrt[4]{48\cdot x\,y^6}-2\sqrt[4]{81\cdot x^9y^2} =$

Chapter 5
Fractional Exponents

Quick Reference to Chapter 5 Case Problems

$$\boxed{27^{\frac{2}{3}}} = ; \quad \boxed{-(256)^{\frac{1}{8}}} = ; \quad \boxed{-8^{\frac{1}{3}}} =$$

$$\boxed{x^{\frac{9}{5}}} = ; \quad \boxed{\left(x^3\right)^{\frac{2}{5}} \cdot \left(w^5\right)^{\frac{1}{3}}} = ; \quad \boxed{\left(x^2\right)^{\frac{4}{5}}} =$$

$$\boxed{81^{-\frac{1}{4}}} = ; \quad \boxed{(-2)^{-\frac{4}{3}}} = ; \quad \boxed{-(8)^{-\frac{2}{3}}} =$$

$$\boxed{y^{-\frac{9}{4}}} = ; \quad \boxed{-\left(x^3\right)^{-\frac{2}{3}}} = ; \quad \boxed{(x \cdot y)^2 \cdot \left(z^{-\frac{1}{4}}\right)^{\frac{8}{3}}} =$$

$$\boxed{\left(y^{\frac{2}{3}}\right)^3 \cdot \left(z^3\right)^{\frac{2}{5}} \cdot y} = ; \quad \boxed{\left(2^3 \cdot 2^2\right)^{\frac{2}{3}} \cdot \left(a^{\frac{3}{5}}\right)^2} = ; \quad \boxed{\left(5^2 \cdot y^4 \cdot z^2\right)^{\frac{1}{4}} \cdot \left(y^2 \cdot z^3\right)^{\frac{1}{3}}} =$$

$$\boxed{\left(\frac{2^2 \cdot a^{\frac{1}{3}}}{a^{\frac{4}{5}}}\right)^3} = ; \quad \boxed{\frac{a^{\frac{2}{3}} \cdot b^{\frac{5}{2}} \cdot c^3}{a \cdot b^{\frac{2}{5}} \cdot c^{\frac{3}{4}}}} = ; \quad \boxed{\left(\frac{3^5 \cdot 2^0}{3^2}\right)^{\frac{3}{5}}} =$$

$$\boxed{\frac{2}{x^{\frac{2}{3}} + 2} + x^{\frac{2}{3}}} = ; \quad \boxed{\frac{2^{\frac{2}{3}} + 5^{\frac{2}{3}}}{2} + \frac{2^{\frac{3}{5}}}{3}} = ; \quad \boxed{\frac{a^{\frac{3}{5}} - b^{\frac{1}{4}}}{a} + \frac{1}{a^{\frac{2}{3}}}} =$$

$\boxed{\left(a^3\right)^{-\frac{2}{3}} \cdot \left(a^2 \cdot b^{-3}\right)^{\frac{1}{6}}} = ;\quad \boxed{\left(x^a\right)^{-\frac{3}{2}} \cdot \left(x^{-b} \cdot x^{3a}\right)^{-\frac{1}{3}}} = ;\quad \boxed{\left(4^{-3} \cdot 2^2\right)^{-\frac{2}{3}} \cdot \left(2^{-1}\right)^{\frac{2}{3}}} =$

$\boxed{\left(\frac{243 \cdot a}{c^{-3}}\right)^{-\frac{4}{3}} \cdot \frac{c^2}{a}} = ;\quad \boxed{\frac{z^2 \cdot (a \cdot b)^{-\frac{2}{3}}}{z^{-\frac{1}{4}} \cdot a^{-2}}} = ;\quad \boxed{\left(\frac{3^{-2} \cdot a^0 \cdot x^2}{a^2 \cdot x^{-3}}\right)^{-\frac{1}{2}}} =$

$\boxed{\frac{x^{\frac{1}{2}} - x^{-\frac{1}{2}}}{x^{-\frac{2}{3}}}} = ;\quad \boxed{x^{-\frac{1}{2}} + y^{-2} + z^{-\frac{5}{2}}} = ;\quad \boxed{\frac{x - x^{-\frac{2}{3}}}{x + x^{\frac{1}{2}}}} =$

$\boxed{250000} = ;\quad \boxed{1234.56} = ;\quad \boxed{0.0002456} =$

$\boxed{2.45 \times 10^{+3}} = ;\quad \boxed{8.6 \times 10^{-7}} = ;\quad \boxed{5.0 \times 10^{+5}} =$

$\boxed{\left(4 \times 10^{+3}\right) \cdot \left(6 \times 10^{+2}\right)} = ;\quad \boxed{\left(2 \times 10^{-2}\right) \cdot \left(6.6 \times 10^{-3}\right)} = ;\quad \boxed{\left(2.34 \times 10^{-2}\right) \cdot \left(9.4 \times 10^{+3}\right)} =$

$\boxed{\frac{2.4 \times 10^{+5}}{2 \times 10^{+2}}} = ;\quad \boxed{\frac{2.346 \times 10^{-4}}{4 \times 10^{-2}}} = ;\quad \boxed{\frac{2.857 \times 10^{+2}}{8 \times 10^{0}}} =$

Chapter 5 - Fractional Exponents

The objective of this chapter is to improve the student's ability to solve and simplify expressions involving positive and negative fractional exponents. The steps used to solve and simplify real numbers and variables raised to positive and negative fractional exponents are addressed in Sections 5.1 and 5.2. Simplifying positive and negative fractional exponential expressions in multiplication, division, addition and subtraction are addressed in Sections 5.3 and 5.4, respectively. Chapter 5 Appendix introduces the student to the concept of scientific notation. In this section the student learns how to change numbers to scientific notation form, change scientific notation numbers to expanded form, and multiply and divide scientific notation numbers. Cases presented in each section are concluded by solving additional examples with practice problems to further enhance the student's ability.

5.1 Positive Fractional Exponents

Real numbers and variables raised to positive and negative integer exponents and the steps for their simplification were discussed in sections 3.1 and 3.2. Real numbers and variables raised to positive and negative fractional exponents, which are the most difficult class of exponents, are addressed in this and the following section. A **fractional exponent** is defined as:

$$a^{\frac{n}{m}} = \sqrt[m]{a^n} \qquad \text{where } n \text{ denotes the power and } m \text{ denotes the root}$$

The fractional exponent $\frac{n}{m}$ can be a positive or a negative integer fraction. The base a can be a real number or a variable. In this section, real numbers raised to positive fractional exponents (Case I) and variables raised to positive fractional exponents (Case II) are addressed:

Case I Real Numbers Raised to Positive Fractional Exponents

Real numbers raised to a positive fractional exponent are solved using the following steps:

Step 1 Change the fractional exponent $a^{\frac{n}{m}}$, where a is a real number, to a radical expression of the form $\sqrt[m]{a^n}$. For example, change $6^{\frac{3}{7}}$ to $\sqrt[7]{6^3}$.

Step 2 Simplify the radical expression (see Section 4.1, Case III). Note that to simplify radical expressions with real numbers as radicand we need to refer to Table 4-1 in Section 4.1, Case I.

Examples With Steps

The following examples show the steps as to how real numbers raised to positive fractional exponents are solved:

Example 5.1-1

$$\boxed{27^{\frac{2}{3}}} =$$

Solution:

Step 1 $\boxed{27^{\frac{2}{3}}} = \boxed{\sqrt[3]{27^2}}$

Step 2 $\boxed{\sqrt[3]{27^2}} = \boxed{\sqrt[3]{729}} = \boxed{\sqrt[3]{9^3}} = \boxed{9}$ (From Table 4-1 $\sqrt[3]{729} = \sqrt[3]{9^3}$)

Example 5.1-2

$\boxed{64^{\frac{1}{3}}} =$

Solution:

Step 1 $\boxed{64^{\frac{1}{3}}} = \boxed{\sqrt[3]{64}}$

Step 2 $\boxed{\sqrt[3]{64}} = \boxed{\sqrt[3]{4^3}} = \boxed{4}$ (From Table 4-1 $\sqrt[3]{64} = \sqrt[3]{4^3}$)

Example 5.1-3

$\boxed{26^{\frac{5}{4}}} =$

Solution:

Step 1 $\boxed{26^{\frac{5}{4}}} = \boxed{\sqrt[4]{26^5}}$

Step 2 $\boxed{\sqrt[4]{26^5}} = \boxed{\sqrt[4]{26^{4+1}}} = \boxed{\sqrt[4]{26^4 \cdot 26^1}} = \boxed{26 \cdot \sqrt[4]{26^1}} = \boxed{\mathbf{26\sqrt[4]{26}}}$

Example 5.1-4

$\boxed{125^{\frac{2}{3}}} =$

Solution:

Step 1 $\boxed{125^{\frac{2}{3}}} = \boxed{\sqrt[3]{125^2}} = \boxed{\sqrt[3]{\left(5^3\right)^2}}$

Step 2 $\boxed{\sqrt[3]{\left(5^3\right)^2}} = \boxed{\sqrt[3]{5^3 \cdot 5^3}} = \boxed{5 \cdot 5} = \boxed{25}$ (From Table 4-1 $\sqrt[3]{125} = \sqrt[3]{5^3}$)

Example 5.1-5

$\boxed{48^{\frac{1}{2}}} =$

Solution:

Step 1 $\boxed{48^{\frac{1}{2}}} = \boxed{\sqrt[2]{48^1}} = \boxed{\sqrt{48}}$

Step 2 $\boxed{\sqrt{48}} = \boxed{\sqrt{16 \cdot 3}} = \boxed{\sqrt{4^2 \cdot 3}} = \boxed{4 \cdot \sqrt{3}} = \boxed{\mathbf{4\sqrt{3}}}$

Additional Examples - Real Numbers Raised to Positive Fractional Exponents

The following examples further illustrate how to solve real numbers raised to positive fractional exponents:

Example 5.1-6

$$64^{\frac{2}{3}} = \sqrt[3]{64^2} = \sqrt[3]{\left(4^3\right)^2} = \sqrt[3]{4^3 \cdot 4^3} = 4 \cdot 4 = \mathbf{16}$$

Example 5.1-7

$$-(256)^{\frac{1}{8}} = -\left(\sqrt[8]{256}\right) = -\left(\sqrt[8]{4^4}\right) = -\sqrt[8]{\left(2^2\right)^4} = -\left(\sqrt[8]{2^8}\right) = -(2) = \mathbf{-2}$$

Example 5.1-8

$$2^{\frac{4}{3}} = \sqrt[3]{2^4} = \sqrt[3]{2^{3+1}} = \sqrt[3]{2^3 \cdot 2^1} = 2 \cdot \sqrt[3]{2^1} = \mathbf{2\sqrt[3]{2}}$$

Example 5.1-9

$$128^{\frac{3}{2}} = \sqrt[2]{128^3} = \sqrt{128^3} = \sqrt{128^{2+1}} = \sqrt{128^2 \cdot 128^1} = 128 \cdot \sqrt{128} = 128 \cdot \sqrt{64 \cdot 2} = 128 \cdot \sqrt{8^2 \cdot 2}$$

$$= 128 \cdot 8 \cdot \sqrt{2} = 1024 \cdot \sqrt{2} = \mathbf{1024\sqrt{2}}$$

Example 5.1-10

$$405^{\frac{1}{4}} = \sqrt[4]{405^1} = \sqrt[4]{5 \times 81} = \sqrt[4]{5 \cdot 3^4} = 3 \cdot \sqrt[4]{5} = \mathbf{3\sqrt[4]{5}}$$

Example 5.1-11

$$768^{\frac{1}{4}} = \sqrt[4]{768} = \sqrt[4]{256 \cdot 3} = \sqrt[4]{4^4 \cdot 3} = 4 \cdot \sqrt[4]{3} = \mathbf{4\sqrt[4]{3}}$$

Example 5.1-12

$$100^{\frac{5}{2}} = \sqrt[2]{100^5} = \sqrt{100^5} = \sqrt{100^{2+2+1}} = \sqrt{100^2 \cdot 100^2 \cdot 100^1} = 100 \cdot 100 \cdot \sqrt{100^1} = 10000\sqrt{10^2}$$

$$= 10000 \cdot 10 = \mathbf{100000}$$

Example 5.1-13

$$1874^{\frac{1}{4}} = \sqrt[4]{1875^1} = \sqrt[4]{3 \cdot 625} = \sqrt[4]{3 \cdot 5^4} = 5 \cdot \sqrt[4]{3} = \mathbf{5\sqrt[4]{3}}$$

Example 5.1-14

$$729^{\frac{1}{5}} = \sqrt[5]{729^1} = \sqrt[5]{729} = \sqrt[5]{243 \cdot 3} = \sqrt[5]{3^5 \cdot 3} = 3 \cdot \sqrt[5]{3} = \mathbf{3\sqrt[5]{3}}$$

Example 5.1-15

$$\boxed{729^{\frac{2}{3}}} = \boxed{\sqrt[3]{729^2}} = \boxed{\sqrt[3]{\left(9^3\right)^2}} = \boxed{\sqrt[3]{9^3 \cdot 9^3}} = \boxed{9 \cdot 9} = \boxed{\mathbf{81}}$$

Practice Problems - Real Numbers Raised to Positive Fractional Exponents

Section 5.1 Case I Practice Problems - Solve the following exponential expressions with real numbers raised to positive fractional exponents:

1. $75^{\frac{1}{2}} =$
2. $-8^{\frac{1}{3}} =$
3. $36^{\frac{5}{2}} =$
4. $72^{\frac{1}{2}} =$
5. $5^{\frac{4}{3}} =$
6. $32^{\frac{1}{5}} =$
7. $64^{\frac{2}{3}} =$
8. $125^{\frac{1}{3}} =$
9. $2^{\frac{5}{4}} =$
10. $343^{\frac{1}{3}} =$

Case II Variables Raised to Positive Fractional Exponents

Variables raised to a positive fractional exponent are solved using the following steps:

Step 1 Change the fractional exponent $x^{\frac{n}{m}}$, where x is a variable, to a radical expression of the form $\sqrt[m]{x^n}$. For example, change $p^{\frac{1}{3}}$ to $\sqrt[3]{p^1} = \sqrt[3]{p}$.

Step 2 Simplify the radical expression (see Section 4.1, Case IV).

Examples with Steps

The following examples show the steps as to how variables are raised to positive fractional exponents:

Example 5.1-16

$$\boxed{x^{\frac{9}{5}}} =$$

Solution:

Step 1 $\boxed{x^{\frac{9}{5}}} = \boxed{\sqrt[5]{x^9}}$

Step 2 $\boxed{\sqrt[5]{x^9}} = \boxed{\sqrt[5]{x^{5+4}}} = \boxed{\sqrt[5]{x^5 \cdot x^4}} = \boxed{\mathbf{x\sqrt[5]{x^4}}}$

Example 5.1-17

$$\boxed{\left(y^{\frac{3}{5}}\right)^{\frac{2}{3}}} =$$

Solution:

Step 1 $\boxed{\left(y^{\frac{3}{5}}\right)^{\frac{2}{3}}} = \boxed{y^{\frac{3}{5}\times\frac{2}{3}}} = \boxed{y^{\frac{6}{15}=\frac{2}{5}}} = \boxed{y^{\frac{2}{5}}} = \boxed{\mathbf{\sqrt[5]{y^2}}}$

Step 2 $\boxed{\textit{Not Applicable}}$

Example 5.1-18

$$\boxed{a^{\frac{8}{3}}} =$$

Solution:

Step 1 $\boxed{a^{\frac{8}{3}}} = \boxed{\sqrt[3]{a^8}}$

Step 2 $\boxed{\sqrt[3]{a^8}} = \boxed{\sqrt[3]{a^{3+3+2}}} = \boxed{\sqrt[3]{\left(a^3 \cdot a^3\right) \cdot a^2}} = \boxed{(a \cdot a) \cdot \sqrt[3]{a^2}} = \boxed{a^{1+1} \cdot \sqrt[3]{a^2}} = \boxed{\mathbf{a^2\sqrt[3]{a^2}}}$

Example 5.1-19

$$-\left(w^2\right)^{\frac{2}{3}} =$$

Solution:

Step 1

$$-\left(w^2\right)^{\frac{2}{3}} = -w^{2\times\frac{2}{3}} = -w^{\frac{2}{1}\times\frac{2}{3}} = -w^{\frac{4}{3}} = -\sqrt[3]{w^4}$$

Step 2

$$-\sqrt[3]{w^4} = -\sqrt[3]{w^{3+1}} = -\sqrt[3]{w^3 \cdot w^1} = -w \cdot \sqrt[3]{w^1} = \mathbf{-w\sqrt[3]{w}}$$

Example 5.1-20

$$(-x)^{\frac{11}{7}} =$$

Solution:

Step 1

$$(-x)^{\frac{11}{7}} = \sqrt[7]{(-x)^{11}}$$

Step 2

$$\sqrt[7]{(-x)^{11}} = \sqrt[7]{(-x)^{7+4}} = \sqrt[7]{(-x)^7 \cdot (-x)^4} = (-x) \cdot \sqrt[7]{(-x)^4}$$

$$= -x \cdot \sqrt[7]{(-x)\cdot(-x)\cdot(-x)\cdot(-x)} = -x \cdot \sqrt[7]{+x^4} = \mathbf{-x\sqrt[7]{x^4}}$$

Additional Examples - Variables Raised to Positive Fractional Exponents

The following examples further illustrate how to solve variables raised to positive fractional exponents:

Example 5.1-21

$$\left(a^4\right)^{\frac{1}{3}} = a^{4\times\frac{1}{3}} = a^{\frac{4}{1}\times\frac{1}{3}} = a^{\frac{4}{3}} = \sqrt[3]{a^4} = \sqrt[3]{a^{3+1}} = \sqrt[3]{a^3 \cdot a^1} = a \cdot \sqrt[3]{a^1} = \mathbf{a\sqrt[3]{a}}$$

Example 5.1-22

$$\left(x^2\right)^{\frac{4}{5}} = x^{2\times\frac{4}{5}} = x^{\frac{2}{1}\times\frac{4}{5}} = x^{\frac{8}{5}} = \sqrt[5]{x^8} = \sqrt[5]{x^{5+3}} = \sqrt[5]{x^5 \cdot x^3} = x \cdot \sqrt[5]{x^3} = \mathbf{x\sqrt[5]{x^3}}$$

Example 5.1-23

$$\left(x^{\frac{1}{6}}\right)^3 = x^{\frac{1}{6}\times 3} = x^{\frac{1}{6}\times\frac{3}{1}} = x^{\frac{3}{6}=\frac{1}{2}} = x^{\frac{1}{2}} = \mathbf{\sqrt{x}}$$

Example 5.1-24

$$a^{\frac{2}{5}} \cdot b^{\frac{8}{3}} = \sqrt[5]{a^2} \cdot \sqrt[3]{b^8} = \sqrt[5]{a^2} \cdot \sqrt[3]{b^{3+3+2}} = \sqrt[5]{a^2} \cdot \sqrt[3]{\left(b^3 \cdot b^3\right) \cdot b^2} = \sqrt[5]{a^2} \cdot (b \cdot b) \cdot \sqrt[3]{b^2} = \sqrt[5]{a^2} \cdot b^{1+1} \cdot \sqrt[3]{b^2}$$

$$= \sqrt[5]{a^2}\cdot b^2\cdot\sqrt[3]{b^2} = \mathbf{b^2\sqrt[5]{a^2}\sqrt[3]{b^2}}$$

Example 5.1-25

$$\left(y^{\frac{2}{3}}\right)^5 = y^{\frac{2}{3}\times 5} = y^{\frac{2}{3}\times\frac{5}{1}} = y^{\frac{2\times 5}{3\times 1}} = y^{\frac{10}{3}} = \sqrt[3]{y^{10}} = \sqrt[3]{y^{3+3+3+1}} = \sqrt[3]{\left(y^3\cdot y^3\cdot y^3\right)\cdot y^1} = (y\cdot y\cdot y)\cdot\sqrt[3]{y^1}$$

$$= y^{1+1+1}\cdot\sqrt[3]{y} = \mathbf{y^3\sqrt[3]{y}}$$

Example 5.1-26

$$2\cdot x^{\frac{2}{3}}\cdot\left(z^{\frac{1}{2}}\right)^{\frac{8}{3}} = 2\cdot x^{\frac{2}{3}}\cdot z^{\frac{1}{2}\times\frac{8}{3}} = 2\cdot x^{\frac{2}{3}}\cdot z^{\frac{8}{6}=\frac{4}{3}} = 2\cdot x^{\frac{2}{3}}\cdot z^{\frac{4}{3}} = 2\sqrt[3]{x^2}\cdot\sqrt[3]{z^4} = 2\sqrt[3]{x^2}\cdot\sqrt[3]{z^{3+1}}$$

$$= 2\sqrt[3]{x^2}\cdot\sqrt[3]{z^3\cdot z^1} = 2\cdot z\cdot\sqrt[3]{x^2}\cdot\sqrt[3]{z} = \mathbf{2z\sqrt[3]{x^2z}}$$

Example 5.1-27

$$a\cdot\left(b^5\right)^{\frac{1}{2}} = a\cdot b^{5\times\frac{1}{2}} = a\cdot b^{\frac{5}{2}} = a\cdot\sqrt[2]{b^5} = a\cdot\sqrt{b^5} = a\cdot\sqrt{b^{2+2+1}} = a\cdot\sqrt{\left(b^2\cdot b^2\right)\cdot b^1} = a\cdot(b\cdot b)\cdot\sqrt{b^1}$$

$$= a\cdot b^{1+1}\cdot\sqrt{b} = \mathbf{ab^2\sqrt{b}}$$

Example 5.1-28

$$\left(x^3\right)^{\frac{2}{5}}\cdot\left(w^5\right)^{\frac{1}{3}} = x^{3\times\frac{2}{5}}\cdot w^{5\times\frac{1}{3}} = x^{\frac{3}{1}\times\frac{2}{5}}\cdot w^{\frac{5}{1}\times\frac{1}{3}} = x^{\frac{6}{5}}\cdot w^{\frac{5}{3}} = \sqrt[5]{x^6}\sqrt[3]{w^5} = \sqrt[5]{x^{5+1}}\cdot\sqrt[3]{w^{3+2}}$$

$$= \sqrt[5]{x^5\cdot x^1}\cdot\sqrt[3]{w^3\cdot w^2} = x\cdot w\cdot\sqrt[5]{x^1}\cdot\sqrt[3]{w^2} = \mathbf{xw\sqrt[5]{x}\sqrt[3]{w^2}}$$

Example 5.1-29

$$25^{\frac{1}{6}}\cdot\left(x^{\frac{2}{3}}\right)^{\frac{1}{4}}\cdot\left(y^{\frac{1}{5}}\right)^{\frac{5}{3}} = 25^{\frac{1}{6}}\cdot x^{\frac{2}{3}\times\frac{1}{4}}\cdot y^{\frac{1}{5}\times\frac{5}{3}} = 25^{\frac{1}{6}}\cdot x^{\frac{2\times 1}{3\times 4}}\cdot y^{\frac{1\times 5}{5\times 3}} = 25^{\frac{1}{6}}\cdot x^{\frac{2}{12}=\frac{1}{6}}\cdot y^{\frac{5}{15}=\frac{1}{3}} = 25^{\frac{1}{6}}\cdot x^{\frac{1}{6}}\cdot y^{\frac{1}{3}}$$

$$= (25x)^{\frac{1}{6}}\cdot y^{\frac{1}{3}} = \mathbf{\sqrt[6]{25x}\sqrt[3]{y}}$$

Example 5.1-30

$$\left(z^{\frac{2}{5}}\right)^6 = z^{\frac{2}{5}\times 6} = z^{\frac{2}{5}\times\frac{6}{1}} = z^{\frac{2\times 6}{5\times 1}} = z^{\frac{12}{5}} = \sqrt[5]{z^{12}} = \sqrt[5]{z^{5+5+2}} = \sqrt[5]{\left(z^5\cdot z^5\right)\cdot z^2} = (z\cdot z)\cdot\sqrt[5]{z^2}$$

$$= z^{1+1}\cdot\sqrt[5]{z^2} = \mathbf{z^2\sqrt[5]{z^2}}$$

Practice Problems - Variables Raised to Positive Fractional Exponents

Section 5.1 Case II Practice Problems - Solve the following exponential expressions with variables raised to positive fractional exponents:

1. $x^{\frac{5}{3}} =$
2. $-\left(w^2 \cdot z\right)^{\frac{2}{3}} =$
3. $(xy)^{\frac{2}{3}} \cdot \left(z^2\right)^{\frac{3}{5}} =$
4. $\left(b^2\right)^{\frac{1}{3}} =$
5. $\left(x^4\right)^{\frac{2}{5}} =$
6. $\left(a^{\frac{1}{2}}\right)^{\frac{2}{3}} =$
7. $x \cdot \left(y^{\frac{1}{2}}\right)^3 =$
8. $\left(c^4\right)^{\frac{4}{7}} =$
9. $\left(x^3\right)^{\frac{1}{4}} \cdot \left(y^{\frac{5}{2}}\right)^4 =$
10. $\left(x^3\right)^{\frac{2}{5}} =$

5.2 Negative Fractional Exponents

Negative fractional exponents are defined as $a^{-\frac{n}{m}}$ where a is referred to as the **base**, and $\frac{n}{m}$ is the **integer fractional exponent**. Again, note that the base a can be a real number or a variable. The fractional exponent $\frac{n}{m}$ can be a positive or a negative integer fraction. In this section, real numbers raised to negative fractional exponents (Case I) and variables raised to negative fractional exponents (Case II) are addressed.

Case I Real Numbers Raised to Negative Fractional Exponents

In the exponential expression $a^{-\frac{n}{m}}$ the base a can be a real number such as 2, 5, 7, 10, 30, or 45. Real numbers raised to a negative fractional exponent are solved using the following steps:

Step 1 Change the real number a raised to a negative fractional exponent, i.e., $a^{-\frac{n}{m}}$ to a positive integer fraction of the form $\frac{1}{a^{\frac{n}{m}}}$. For example, change $2^{-\frac{3}{5}}$ to $\frac{1}{2^{\frac{3}{5}}}$.

Step 2 Change the positive integer fraction $\frac{1}{a^{\frac{n}{m}}}$ to a fractional radical expression of the form $\frac{1}{\sqrt[m]{a^n}}$. For example, rewrite $\frac{1}{2^{\frac{3}{5}}}$ as $\frac{1}{\sqrt[5]{2^3}}$.

Step 3 Simplify the radical expression in the denominator (see Section 4.1, Case III). Note that to simplify radical expressions with real numbers as radicand we need to refer to Table 4-1 in Section 4.1, Case I.

Examples with Steps

The following examples show the steps as to how real numbers raised to negative fractional exponents are solved:

Example 5.2-1

$$\boxed{81^{-\frac{1}{4}}} =$$

Solution:

Step 1

$$\boxed{81^{-\frac{1}{4}}} = \boxed{\frac{1}{81^{\frac{1}{4}}}}$$

Step 2

$$\boxed{\frac{1}{81^{\frac{1}{4}}}} = \boxed{\frac{1}{\sqrt[4]{81^1}}} = \boxed{\frac{1}{\sqrt[4]{81}}}$$

Step 3

$$\boxed{\frac{1}{\sqrt[4]{81}}} = \boxed{\frac{1}{\sqrt[4]{3^4}}} = \boxed{\mathbf{\frac{1}{3}}}$$

Example 5.2-2

$$\boxed{8^{-\frac{1}{3}}} =$$

Solution:

Step 1 $\boxed{8^{-\frac{1}{3}}} = \boxed{\dfrac{1}{8^{\frac{1}{3}}}}$

Step 2 $\boxed{\dfrac{1}{8^{\frac{1}{3}}}} = \boxed{\dfrac{1}{\sqrt[3]{8^1}}} = \boxed{\dfrac{1}{\sqrt[3]{8}}}$

Step 3 $\boxed{\dfrac{1}{\sqrt[3]{8}}} = \boxed{\dfrac{1}{\sqrt[3]{2^3}}} = \boxed{\dfrac{1}{2}}$

Example 5.2-3

$$\boxed{16^{-\frac{2}{3}}} =$$

Solution:

Step 1 $\boxed{16^{-\frac{2}{3}}} = \boxed{\dfrac{1}{16^{\frac{2}{3}}}}$

Step 2 $\boxed{\dfrac{1}{16^{\frac{2}{3}}}} = \boxed{\dfrac{1}{\sqrt[3]{16^2}}} = \boxed{\dfrac{1}{\sqrt[3]{256}}}$

Step 3 $\boxed{\dfrac{1}{\sqrt[3]{256}}} = \boxed{\dfrac{1}{\sqrt[3]{4^4}}} = \boxed{\dfrac{1}{\sqrt[3]{4^{3+1}}}} = \boxed{\dfrac{1}{\sqrt[3]{4^3 \cdot 4^1}}} = \boxed{\dfrac{1}{4 \cdot \sqrt[3]{4^1}}} = \boxed{\dfrac{1}{4\sqrt[3]{4}}}$

Example 5.2-4

$$\boxed{32^{-\frac{1}{5}}} =$$

Solution:

Step 1 $\boxed{32^{-\frac{1}{5}}} = \boxed{\dfrac{1}{32^{\frac{1}{5}}}}$

Step 2 $\boxed{\dfrac{1}{32^{\frac{1}{5}}}} = \boxed{\dfrac{1}{\sqrt[5]{32^1}}} = \boxed{\dfrac{1}{\sqrt[5]{32}}}$

Step 3 $\boxed{\dfrac{1}{\sqrt[5]{32}}} = \boxed{\dfrac{1}{\sqrt[5]{2^5}}} = \boxed{\dfrac{1}{2}}$

Example 5.2-5

$$(-2)^{-\frac{4}{3}} =$$

Solution:

Step 1

$$(-2)^{-\frac{4}{3}} = \frac{1}{(-2)^{\frac{4}{3}}}$$

Step 2

$$\frac{1}{(-2)^{\frac{4}{3}}} = \frac{1}{\sqrt[3]{(-2)^4}} = \frac{1}{\sqrt[3]{(-2)\cdot(-2)\cdot(-2)\cdot(-2)}} = \frac{1}{\sqrt[3]{+2^4}} = \frac{1}{\sqrt[3]{2^4}}$$

Step 3

$$\frac{1}{\sqrt[3]{2^4}} = \frac{1}{\sqrt[3]{2^{3+1}}} = \frac{1}{\sqrt[3]{2^3\cdot 2^1}} = \frac{1}{2\cdot\sqrt[3]{2^1}} = \mathbf{\frac{1}{2\sqrt[3]{2}}}$$

Additional Examples - Real Numbers Raised to Negative Fractional Exponents

The following examples further illustrate how to solve real numbers raised to negative fractional exponents:

Example 5.2-6

$$64^{-\frac{2}{3}} = \frac{1}{64^{\frac{2}{3}}} = \frac{1}{\sqrt[3]{64^2}} = \frac{1}{\sqrt[3]{\left(8^2\right)^2}} = \frac{1}{\sqrt[3]{8^4}} = \frac{1}{\sqrt[3]{8^{3+1}}} = \frac{1}{\sqrt[3]{8^3\cdot 8^1}} = \frac{1}{8\cdot\sqrt[3]{8}} = \frac{1}{8\cdot\sqrt[3]{2^3}} = \frac{1}{8\cdot 2} = \mathbf{\frac{1}{16}}$$

Example 5.2-7

$$256^{-\frac{1}{4}} = \frac{1}{256^{\frac{1}{4}}} = \frac{1}{\sqrt[4]{256^1}} = \frac{1}{\sqrt[4]{256}} = \frac{1}{\sqrt[4]{4^4}} = \mathbf{\frac{1}{4}}$$

Example 5.2-8

$$(-9)^{-\frac{2}{3}} = \frac{1}{(-9)^{\frac{2}{3}}} = \frac{1}{\sqrt[3]{(-9)^2}} = \frac{1}{\sqrt[3]{-9\cdot -9}} = \frac{1}{\sqrt[3]{81}} = \frac{1}{\sqrt[3]{3^4}} = \frac{1}{\sqrt[3]{3^{3+1}}} = \frac{1}{\sqrt[3]{3^3\cdot 3^1}} = \frac{1}{3\cdot\sqrt[3]{3}} = \mathbf{\frac{1}{3\sqrt[3]{3}}}$$

Example 5.2-9

$$25^{-\frac{3}{2}} = \frac{1}{25^{\frac{3}{2}}} = \frac{1}{\sqrt[2]{25^3}} = \frac{1}{\sqrt{25^{2+1}}} = \frac{1}{\sqrt{25^2\cdot 25^1}} = \frac{1}{25\cdot\sqrt{25}} = \frac{1}{25\cdot\sqrt{5^2}} = \frac{1}{25\cdot 5} = \mathbf{\frac{1}{125}}$$

Example 5.2-10

$$10000^{-\frac{1}{4}} = \frac{1}{10000^{\frac{1}{4}}} = \frac{1}{\sqrt[4]{10000^1}} = \frac{1}{\sqrt[4]{10000}} = \frac{1}{\sqrt[4]{10^4}} = \mathbf{\frac{1}{10}}$$

Example 5.2-11

$$\boxed{2500^{-\frac{1}{2}}} = \boxed{\frac{1}{2500^{\frac{1}{2}}}} = \boxed{\frac{1}{\sqrt[2]{2500^1}}} = \boxed{\frac{1}{\sqrt{2500}}} = \boxed{\frac{1}{\sqrt{50 \cdot 50}}} = \boxed{\frac{1}{\sqrt{50^2}}} = \boxed{\mathbf{\frac{1}{50}}}$$

Example 5.2-12

$$\boxed{1215^{-\frac{1}{5}}} = \boxed{\frac{1}{1215^{\frac{1}{5}}}} = \boxed{\frac{1}{\sqrt[5]{1215^1}}} = \boxed{\frac{1}{\sqrt[5]{1215}}} = \boxed{\frac{1}{\sqrt[5]{243 \cdot 5}}} = \boxed{\frac{1}{\sqrt[5]{3^5 \cdot 5}}} = \boxed{\frac{1}{3 \cdot \sqrt[5]{5}}} = \boxed{\mathbf{\frac{1}{3\sqrt[5]{5}}}}$$

Example 5.2-13

$$\boxed{32^{-\frac{2}{5}}} = \boxed{\frac{1}{32^{\frac{2}{5}}}} = \boxed{\frac{1}{\sqrt[5]{32^2}}} = \boxed{\frac{1}{\sqrt[5]{\left(2^5\right)^2}}} = \boxed{\frac{1}{\sqrt[5]{2^{10}}}} = \boxed{\frac{1}{\sqrt[5]{2^{5+5}}}} = \boxed{\frac{1}{\sqrt[5]{2^5 \cdot 2^5}}} = \boxed{\frac{1}{2 \cdot 2}} = \boxed{\mathbf{\frac{1}{4}}}$$

Example 5.2-14

$$\boxed{81^{-\frac{3}{4}}} = \boxed{\frac{1}{81^{\frac{3}{4}}}} = \boxed{\frac{1}{\sqrt[4]{81^3}}} = \boxed{\frac{1}{\sqrt[4]{\left(3^4\right)^3}}} = \boxed{\frac{1}{\sqrt[4]{3^{12}}}} = \boxed{\frac{1}{\sqrt[4]{3^{4+4+4}}}} = \boxed{\frac{1}{\sqrt[4]{3^4 \cdot 3^4 \cdot 3^4}}} = \boxed{\frac{1}{3 \cdot 3 \cdot 3}} = \boxed{\mathbf{\frac{1}{27}}}$$

Example 5.2-15

$$\boxed{175^{-\frac{1}{2}}} = \boxed{\frac{1}{175^{\frac{1}{2}}}} = \boxed{\frac{1}{\sqrt[2]{175^1}}} = \boxed{\frac{1}{\sqrt{175}}} = \boxed{\frac{1}{\sqrt{25 \cdot 7}}} = \boxed{\frac{1}{\sqrt{5^2 \cdot 7}}} = \boxed{\mathbf{\frac{1}{5\sqrt{7}}}}$$

Practice Problems - Real Numbers Raised to Negative Fractional Exponents

Section 5.2 Case I Practice Problems - Solve the following exponential expressions with real numbers raised to negative fractional exponents:

1. $125^{-\frac{1}{2}} =$
2. $-(343)^{-\frac{1}{3}} =$
3. $4 \cdot (16)^{-\frac{1}{2}} =$
4. $49^{-\frac{1}{2}} =$
5. $-(8)^{-\frac{2}{3}} =$
6. $32^{-\frac{2}{5}} =$
7. $10^{-\frac{4}{3}} =$
8. $625^{-\frac{1}{4}} =$
9. $2^{-\frac{5}{4}} =$
10. $(9)^{-\frac{3}{2}} =$

Case II Variables Raised to Negative Fractional Exponents

In the exponential expression $a^{-\frac{n}{m}}$ the base a can be a variable such as x, y, z, k, or m. Variables raised to a negative fractional exponent are solved using the following steps:

Step 1 Change the negative fractional exponent $x^{-\frac{n}{m}}$, where x is a variable, to a positive integer fraction of the form $\frac{1}{x^{\frac{n}{m}}}$. For example, change $k^{-\frac{2}{7}}$ to $\frac{1}{k^{\frac{2}{7}}}$.

Step 2 Change the positive integer fraction $\frac{1}{x^{\frac{n}{m}}}$ to a fractional radical expression of the form $\frac{1}{\sqrt[m]{x^n}}$. For example, rewrite $\frac{1}{k^{\frac{2}{7}}}$ as $\frac{1}{\sqrt[7]{k^2}}$.

Step 3 Simplify the radical expression in the denominator (see Section 4.1, Case IV).

Examples with Steps

The following examples show the step as to how variables are raised to negative fractional exponents:

Example 5.2-16

$$y^{-\frac{9}{4}} =$$

Solution:

Step 1

$$y^{-\frac{9}{4}} = \frac{1}{y^{\frac{9}{4}}}$$

Step 2

$$\frac{1}{y^{\frac{9}{4}}} = \frac{1}{\sqrt[4]{y^9}}$$

Step 3

$$\frac{1}{\sqrt[4]{y^9}} = \frac{1}{\sqrt[4]{y^{4+4+1}}} = \frac{1}{\sqrt[4]{\left(y^4 \cdot y^4\right) \cdot y^1}} = \frac{1}{(y \cdot y) \cdot \sqrt[4]{y^1}} = \frac{1}{y^{1+1} \cdot \sqrt[4]{y}} = \frac{\mathbf{1}}{\mathbf{y^2 \sqrt[4]{y}}}$$

Example 5.2-17

$$\left(a^{\frac{3}{5}}\right)^{-\frac{1}{3}} =$$

Solution:

Step 1

$$\left(a^{\frac{3}{5}}\right)^{-\frac{1}{3}} = \frac{1}{\left(a^{\frac{3}{5}}\right)^{\frac{1}{3}}} = \frac{1}{a^{\frac{3}{5} \times \frac{1}{3}}} = \frac{1}{a^{\frac{3}{15}}}$$

Step 2 $\dfrac{1}{a^{\frac{3}{15}}} = \dfrac{1}{a^{\frac{\cancel{3}}{\cancel{15}}=\frac{1}{5}}} = \dfrac{1}{a^{\frac{1}{5}}} = \boxed{\dfrac{1}{\sqrt[5]{a}}}$

Step 3 *Not Applicable*

Example 5.2-18

$w^{-\frac{11}{4}} =$

Solution:

Step 1 $w^{-\frac{11}{4}} = \dfrac{1}{w^{\frac{11}{4}}}$

Step 2 $\dfrac{1}{w^{\frac{11}{4}}} = \dfrac{1}{\sqrt[4]{w^{11}}}$

Step 3 $\dfrac{1}{\sqrt[4]{w^{11}}} = \dfrac{1}{\sqrt[4]{w^{4+4+3}}} = \dfrac{1}{\sqrt[4]{\left(w^4 \cdot w^4\right) \cdot w^3}} = \dfrac{1}{w \cdot w \cdot \sqrt[4]{w^3}} = \dfrac{1}{w^{1+1} \cdot \sqrt[4]{w^3}} = \boxed{\dfrac{1}{w^2 \sqrt[4]{w^3}}}$

Example 5.2-19

$-\left(x^3\right)^{-\frac{2}{3}} =$

Solution:

Step 1 $-\left(x^3\right)^{-\frac{2}{3}} = -\dfrac{1}{\left(x^3\right)^{\frac{2}{3}}} = -\dfrac{1}{x^{3 \times \frac{2}{3}}} = -\dfrac{1}{x^{\frac{6}{3}}}$

Step 2 $-\dfrac{1}{x^{\frac{6}{3}}} = -\dfrac{1}{x^{\frac{\cancel{6}}{\cancel{3}}=\frac{2}{1}}} = -\dfrac{1}{x^{\frac{2}{1}}} = \boxed{-\dfrac{1}{x^2}}$

Step 3 *Not Applicable*

Example 5.2-20

$a^2 \cdot z^{-\frac{13}{4}} =$

Solution:

Step 1 $a^2 \cdot z^{-\frac{13}{4}} = a^2 \cdot \dfrac{1}{z^{\frac{13}{4}}} = \dfrac{a^2}{1} \cdot \dfrac{1}{z^{\frac{13}{4}}} = \dfrac{a^2 \cdot 1}{1 \cdot z^{\frac{13}{4}}} = \dfrac{a^2}{z^{\frac{13}{4}}}$

Step 2 $\boxed{\dfrac{a^2}{z^{\frac{13}{4}}}} = \boxed{\dfrac{a^2}{\sqrt[4]{z^{13}}}}$

Step 3 $\boxed{\dfrac{a^2}{\sqrt[4]{z^{13}}}} = \boxed{\dfrac{a^2}{\sqrt[4]{z^{4+4+4+1}}}} = \boxed{\dfrac{a^2}{\sqrt[4]{\left(z^4 \cdot z^4 \cdot z^4\right) \cdot z^1}}} = \boxed{\dfrac{a^2}{z \cdot z \cdot z \cdot \sqrt[4]{z^1}}} = \boxed{\dfrac{a^2}{z^{1+1+1} \cdot \sqrt[4]{z}}} = \boxed{\mathbf{\dfrac{a^2}{z^3 \sqrt[4]{z}}}}$

Additional Examples - Variables Raised to Negative Fractional Exponents

The following examples further illustrate how to solve variables raised to negative fractional exponents:

Example 5.2-21

$$\boxed{\left(a^5\right)^{-\frac{1}{3}}} = \boxed{\frac{1}{\left(a^5\right)^{\frac{1}{3}}}} = \boxed{\frac{1}{a^{5 \times \frac{1}{3}}}} = \boxed{\frac{1}{a^{\frac{5}{3}}}} = \boxed{\frac{1}{\sqrt[3]{a^5}}} = \boxed{\frac{1}{\sqrt[3]{a^{3+2}}}} = \boxed{\frac{1}{\sqrt[3]{a^3 \cdot a^2}}} = \boxed{\frac{1}{a \cdot \sqrt[3]{a^2}}} = \boxed{\mathbf{\frac{1}{a\sqrt[3]{a^2}}}}$$

Example 5.2-22

$$\boxed{\left(w^2\right)^{-\frac{3}{5}}} = \boxed{\frac{1}{\left(w^2\right)^{\frac{3}{5}}}} = \boxed{\frac{1}{w^{2 \times \frac{3}{5}}}} = \boxed{\frac{1}{w^{\frac{2}{1} \times \frac{3}{5}}}} = \boxed{\frac{1}{w^{\frac{6}{5}}}} = \boxed{\frac{1}{\sqrt[5]{w^6}}} = \boxed{\frac{1}{\sqrt[5]{w^{5+1}}}} = \boxed{\frac{1}{\sqrt[5]{w^5 \cdot w^1}}} = \boxed{\frac{1}{w \cdot \sqrt[5]{w}}} = \boxed{\mathbf{\frac{1}{w\sqrt[5]{w}}}}$$

Example 5.2-23

$$\boxed{\left(b^{-\frac{1}{4}}\right)^2} = \boxed{b^{-\frac{1}{4} \times 2}} = \boxed{b^{-\frac{1}{4} \times \frac{2}{1}}} = \boxed{b^{-\frac{2}{4}}} = \boxed{\frac{1}{b^{\frac{2}{4} = \frac{1}{2}}}} = \boxed{\frac{1}{b^{\frac{1}{2}}}} = \boxed{\frac{1}{\sqrt[2]{b^1}}} = \boxed{\mathbf{\frac{1}{\sqrt{b}}}}$$

Example 5.2-24

$$\boxed{b^{-\frac{5}{2}} \cdot c^3} = \boxed{\frac{1}{b^{\frac{5}{2}}} \cdot c^3} = \boxed{\frac{1}{b^{\frac{5}{2}}} \cdot \frac{c^3}{1}} = \boxed{\frac{1 \cdot c^3}{b^{\frac{5}{2}} \cdot 1}} = \boxed{\frac{c^3}{b^{\frac{5}{2}}}} = \boxed{\frac{c^3}{\sqrt[2]{b^5}}} = \boxed{\frac{c^3}{\sqrt{b^5}}} = \boxed{\frac{c^3}{\sqrt{b^{2+2+1}}}} = \boxed{\frac{c^3}{\sqrt{\left(b^2 \cdot b^2\right) \cdot b^1}}}$$

$$= \boxed{\frac{c^3}{(b \cdot b) \cdot \sqrt{b^1}}} = \boxed{\frac{c^3}{b^{1+1} \cdot \sqrt{b}}} = \boxed{\mathbf{\frac{c^3}{b^2 \sqrt{b}}}}$$

Example 5.2-25

$$\boxed{\left(y^{-\frac{2}{3}}\right)^4} = \boxed{y^{-\frac{2}{3} \times 4}} = \boxed{y^{-\frac{2}{3} \times \frac{4}{1}}} = \boxed{y^{-\frac{8}{3}}} = \boxed{\frac{1}{y^{\frac{8}{3}}}} = \boxed{\frac{1}{\sqrt[3]{y^8}}} = \boxed{\frac{1}{\sqrt[3]{y^{3+3+2}}}} = \boxed{\frac{1}{\sqrt[3]{\left(y^3 \cdot y^3\right) \cdot y^2}}} = \boxed{\frac{1}{y \cdot y \cdot \sqrt[3]{y^2}}}$$

$$= \boxed{\frac{1}{y^{1+1} \cdot \sqrt[3]{y^2}}} = \boxed{\mathbf{\frac{1}{y^2 \sqrt[3]{y^2}}}}$$

Example 5.2-26

$$\left[(x\cdot y)^2\cdot\left(z^{-\frac14}\right)^{\frac83}\right] = \left[\left(x^{1\times2}\cdot y^{1\times2}\right)\cdot z^{-\frac14\times\frac83}\right] = \left[\left(x^2\cdot y^2\right)\cdot z^{-\frac{8}{12}=-\frac23}\right] = \left[x^2\cdot y^2\cdot z^{-\frac23}\right] = \left[x^2\cdot y^2\cdot\frac{1}{z^{\frac23}}\right]$$

$$= \left[\frac{x^2\cdot y^2}{1}\cdot\frac{1}{z^{\frac23}}\right] = \left[\frac{x^2\cdot y^2}{z^{\frac23}}\right] = \boldsymbol{\left[\frac{x^2y^2}{\sqrt[3]{z^2}}\right]}$$

Example 5.2-27

$$\left[\left(a^2\cdot b^3\right)^{-\frac12}\cdot\left(a^3\right)^{-\frac23}\right] = \left[\left(a^{2\times-\frac12}\cdot b^{3\times-\frac12}\right)\cdot a^{3\times-\frac23}\right] = \left[\left(a^{\frac21\times-\frac12}\cdot b^{\frac31\times-\frac12}\right)\cdot a^{\frac31\times-\frac23}\right] = \left[a^{-\frac22=-\frac11}\cdot b^{-\frac32}\cdot a^{-\frac63=-\frac21}\right]$$

$$= \left[a^{-\frac11}\cdot b^{-\frac32}\cdot a^{-\frac21}\right] = \left[\left(a^{-1}\cdot a^{-2}\right)\cdot b^{-\frac32}\right] = \left[a^{-1-2}\cdot b^{-\frac32}\right] = \left[a^{-3}\cdot b^{-\frac32}\right] = \left[\frac{1}{a^3}\cdot\frac{1}{b^{\frac32}}\right] = \left[\frac{1\cdot1}{a^3\cdot b^{\frac32}}\right] = \left[\frac{1}{a^3\cdot\sqrt[2]{b^3}}\right]$$

$$= \left[\frac{1}{a^3\cdot\sqrt[2]{b^{2+1}}}\right] = \left[\frac{1}{a^3\cdot\sqrt{b^2\cdot b^1}}\right] = \left[\frac{1}{a^3\cdot b\cdot\sqrt{b^1}}\right] = \boldsymbol{\left[\frac{1}{a^3b\sqrt{b}}\right]}$$

Example 5.2-28

$$\left[w\cdot y^{\frac23}\cdot\left(w^{-\frac12}\cdot y^2\right)^{-\frac25}\right] = \left[w\cdot y^{\frac23}\cdot\left(w^{-\frac12\times-\frac25}\cdot y^{2\times-\frac25}\right)\right] = \left[w\cdot y^{\frac23}\cdot\left(w^{\frac{-1\times-2}{2\times5}}\cdot y^{\frac21\times-\frac25}\right)\right] = \left[w\cdot y^{\frac23}\cdot w^{\frac{2}{10}=\frac15}\cdot y^{\frac{2\times-2}{1\times5}}\right]$$

$$= \left[w\cdot y^{\frac23}\cdot w^{\frac15}\cdot y^{\frac{-4}{5}}\right] = \left[w\cdot y^{\frac23}\cdot w^{\frac15}\cdot y^{\frac{-4}{5}}\right] = \left[w^1\cdot w^{\frac15}\cdot y^{\frac23}\cdot y^{\frac{-4}{5}}\right] = \left[w^{\frac11+\frac15}\cdot y^{\frac23-\frac45}\right] = \left[w^{\frac{(1\cdot5)+(1\cdot1)}{1\cdot5}}\cdot y^{\frac{(2\cdot5)-(4\cdot3)}{3\cdot5}}\right]$$

$$= \left[w^{\frac{5+1}{5}}\cdot y^{\frac{10-12}{15}}\right] = \left[w^{\frac65}\cdot y^{\frac{-2}{15}}\right] = \left[w^{\frac65}\cdot\frac{1}{y^{\frac{2}{15}}}\right] = \left[\frac{w^{\frac65}}{y^{\frac{2}{15}}}\right] = \left[\frac{\sqrt[5]{w^6}}{\sqrt[15]{y^2}}\right] = \left[\frac{\sqrt[5]{w^{5+1}}}{\sqrt[15]{y^2}}\right] = \left[\frac{\sqrt[5]{w^5\cdot w^1}}{\sqrt[15]{y^2}}\right] = \left[\frac{w\cdot\sqrt[5]{w^1}}{\sqrt[15]{y^2}}\right] = \boldsymbol{\left[\frac{w\sqrt[5]{w}}{\sqrt[15]{y^2}}\right]}$$

Example 5.2-29

$$\left[\left(b^{\frac23}\cdot a^{\frac12}\right)^{-\frac14}\cdot a\right] = \left[\left(b^{\frac23\times-\frac14}\cdot a^{\frac12\times-\frac14}\right)\cdot a\right] = \left[\left(b^{\frac{2\times-1}{3\times4}}\cdot a^{\frac{1\times-1}{2\times4}}\right)\cdot a\right] = \left[b^{-\frac{2}{12}=-\frac16}\cdot a^{-\frac18}\cdot a\right] = \left[b^{-\frac16}\cdot a^{-\frac18}\cdot a^1\right]$$

$$= \left[b^{-\frac16}\cdot a^{-\frac18+1}\right] = \left[b^{-\frac16}\cdot a^{-\frac18+\frac11}\right] = \left[b^{-\frac16}\cdot a^{\frac{(-1\cdot1)+(1\cdot8)}{8\cdot1}}\right] = \left[b^{-\frac16}\cdot a^{\frac{-1+8}{8}}\right] = \left[b^{-\frac16}\cdot a^{\frac78}\right] = \left[\frac{1}{b^{\frac16}}\cdot a^{\frac78}\right] = \left[\frac{a^{\frac78}}{b^{\frac16}}\right] = \boldsymbol{\left[\frac{\sqrt[8]{a^7}}{\sqrt[6]{b}}\right]}$$

Example 5.2-30

$$\left(x\cdot y^{-\frac{1}{3}}\cdot z\right)\cdot\left(x\cdot y^{2}\right)^{-\frac{1}{4}}=\left(x\cdot y^{-\frac{1}{3}}\cdot z\right)\cdot\left(x^{1\times-\frac{1}{4}}\cdot y^{2\times-\frac{1}{4}}\right)=\left(x\cdot y^{-\frac{1}{3}}\cdot z\right)\cdot\left(x^{\frac{1}{1}\times-\frac{1}{4}}\cdot y^{\frac{2}{1}\times-\frac{1}{4}}\right)$$

$$=x\cdot y^{-\frac{1}{3}}\cdot z\cdot x^{-\frac{1}{4}}\cdot y^{-\frac{2}{4}=-\frac{1}{2}}=x\cdot y^{-\frac{1}{3}}\cdot z\cdot x^{-\frac{1}{4}}\cdot y^{-\frac{1}{2}}=x\cdot x^{-\frac{1}{4}}\cdot y^{-\frac{1}{3}}\cdot y^{-\frac{1}{2}}\cdot z=x^{1-\frac{1}{4}}\cdot y^{-\frac{1}{3}-\frac{1}{2}}\cdot z$$

$$=x^{\frac{1}{1}-\frac{1}{4}}\cdot y^{\frac{(-1\cdot 2)-(1\cdot 3)}{3\cdot 2}}\cdot z=x^{\frac{(1\cdot 4)-(1\cdot 1)}{1\cdot 4}}\cdot y^{\frac{-2-3}{6}}\cdot z=x^{\frac{4-1}{4}}\cdot y^{\frac{-5}{6}}\cdot z=x^{\frac{3}{4}}\cdot y^{-\frac{5}{6}}\cdot z=x^{\frac{3}{4}}\cdot\frac{1}{y^{\frac{5}{6}}}\cdot z$$

$$=\frac{x^{\frac{3}{4}}}{1}\cdot\frac{1}{y^{\frac{5}{6}}}\cdot\frac{z}{1}=\frac{x^{\frac{3}{4}}\cdot 1\cdot z}{1\cdot y^{\frac{5}{6}}\cdot 1}=\frac{x^{\frac{3}{4}}\cdot z}{y^{\frac{5}{6}}}=\frac{\sqrt[4]{x^{3}}\,z}{\sqrt[6]{y^{5}}}$$

Note: In cases where the solution has a radical expression in the denominator, the solution can be further simplified by rationalizing the denominator (see Section 4.3, Cases I and II). Since the objective of this chapter is not to repeat rationalization of radical expressions, therefore this process is not shown. The primary intent is to teach students how fractional expressions can be represented in radical form.

Practice Problems - Variables Raised to Negative Fractional Exponents

Section 5.2 Case II Practice Problems - Solve the following exponential expressions with variables raised to negative fractional exponents:

1. $z^{-\frac{7}{3}}=$
2. $-(a\cdot b)^{-\frac{2}{3}}=$
3. $\left(k\cdot l^{2}\right)^{-\frac{4}{5}}=$
4. $\left(y^{2}\right)^{-\frac{1}{3}}=$
5. $\left(x^{-3}\right)^{\frac{2}{5}}=$
6. $\left(a^{-\frac{1}{2}}\right)^{\frac{4}{3}}=$
7. $x^{2}\cdot\left(y^{-\frac{1}{2}}\right)^{3}=$
8. $\left(c^{4}\right)^{-\frac{2}{3}}=$
9. $x^{3}\cdot y^{-\frac{7}{3}}=$
10. $y^{2}\cdot\left(x^{3}\right)^{-\frac{2}{5}}=$

5.3 Operations with Positive Fractional Exponents

The following laws of exponents, which were introduced in Chapter 3 for solving integer exponential expressions, are needed to proceed with simplification of fractional exponents. These laws are used to simplify the work in solving fractional exponential expressions and should be memorized

Table 5.3-1: Fractional Exponent Laws 1 through 6 (positive Fractional Exponents)

I. **Multiplication**	$x^{\frac{a}{b}} \cdot x^{\frac{c}{d}} = x^{\frac{a}{b}+\frac{c}{d}}$	When multiplying fractional exponential terms, if bases x are the same, add the exponents $\frac{a}{b}$ and $\frac{c}{d}$.
II. **Power of a Power**	$\left(x^{\frac{a}{b}}\right)^{\frac{c}{d}} = x^{\frac{a}{b}\times\frac{c}{d}}$	When raising a fractional exponential term to a fractional power, multiply the exponents $\frac{a}{b}$ and $\frac{c}{d}$.
III. **Power of a Product**	$(x \cdot y)^{\frac{a}{b}} = x^{\frac{a}{b}} \cdot y^{\frac{a}{b}}$	When raising a product to a fractional power, raise each factor x and y to the fractional exponent $\frac{a}{b}$.
IV. **Power of a Fraction**	$\left(\frac{x}{y}\right)^{\frac{a}{b}} = \frac{x^{\frac{a}{b}}}{y^{\frac{a}{b}}}$	When raising a fraction to a fractional power, raise the numerator and the denominator to the fractional exponent $\frac{a}{b}$.
V. **Division**	$\frac{x^{\frac{a}{b}}}{x^{\frac{c}{d}}} = x^{\frac{a}{b}} \cdot x^{-\frac{c}{d}} = x^{\frac{a}{b}-\frac{c}{d}}$	When dividing fractional exponential terms, if the bases x are the same, subtract the fractional exponents $\frac{a}{b}$ and $\frac{c}{d}$.
VI. **Negative Power**	$x^{-\frac{a}{b}} = \frac{1}{x^{\frac{a}{b}}}$	A non-zero based x raised to the $-\frac{a}{b}$ power equals 1 divided by the base x raised to the $\frac{a}{b}$ power.

In this section students learn how to multiply (Case I), divide (Case II), and add or subtract (Case III) positive fractional exponents by one another.

Case I Multiplying Positive Fractional Exponents

Positive fractional exponents are multiplied by each other using the exponent laws I through III shown in Table 5.3-2.

Table 5.3-2: Exponent Laws 1 through 3 (positive Fractional Exponents)

I. **Multiplication**	$x^{\frac{a}{b}} \cdot x^{\frac{c}{d}} = x^{\frac{a}{b}+\frac{c}{d}}$	When multiplying fractional exponential terms, if bases x are the same, add the exponents $\frac{a}{b}$ and $\frac{c}{d}$.
II. **Power of a Power**	$\left(x^{\frac{a}{b}}\right)^{\frac{c}{d}} = x^{\frac{a}{b}\times\frac{c}{d}}$	When raising a fractional exponential term to a fractional power, multiply the exponents $\frac{a}{b}$ and $\frac{c}{d}$.
III. **Power of a Product**	$(x \cdot y)^{\frac{a}{b}} = x^{\frac{a}{b}} \cdot y^{\frac{a}{b}}$	When raising a product to a fractional power, raise each factor x and y to the fractional exponent $\frac{a}{b}$.

Positive fractional exponents are multiplied by each other using the following steps:

Step 1 Apply the Power of a Power and/or the Power of a Product Law (Laws II and III) from Table 5.3-2.

Step 2 Apply the Multiplication Law (Law I) from Table 5.3-2 and simplify the fractional exponential expressions by adding the exponents with similar bases. (Review sections 2.2, 2.3, and 2.4 for addition, subtraction, and multiplication of integer fractions.)

Step 3 Change the fractional exponential expressions to radical expressions (see Section 5.1).

Examples with Steps

The following examples show the steps as to how positive fractional exponents are multiplied by one another:

Example 5.3-1

$$\left(y^{\frac{2}{3}}\right)^3 \cdot \left(z^3\right)^{\frac{2}{5}} \cdot y =$$

Solution:

Step 1

$$\left(y^{\frac{2}{3}}\right)^3 \cdot \left(z^3\right)^{\frac{2}{5}} \cdot y = y^{\frac{2}{3}\times 3} \cdot z^{3\times\frac{2}{5}} \cdot y = y^{\frac{2}{3}\times\frac{3}{1}} \cdot z^{\frac{3}{1}\times\frac{2}{5}} \cdot y = y^{\frac{2\times 3}{3\times 1}} \cdot z^{\frac{3\times 2}{1\times 5}} \cdot y$$

$$= y^{\frac{6}{3}} \cdot z^{\frac{6}{5}} \cdot y$$

Step 2

$$y^{\frac{6}{3}} \cdot z^{\frac{6}{5}} \cdot y = y^{\frac{6}{3}=\frac{2}{1}} \cdot y \cdot z^{\frac{6}{5}} = y^{\frac{2}{1}} \cdot y \cdot z^{\frac{6}{5}} = y^2 \cdot y^1 \cdot z^{\frac{6}{5}} = y^{2+1} \cdot z^{\frac{6}{5}} = y^3 \cdot z^{\frac{6}{5}}$$

Step 3

$$y^3 \cdot z^{\frac{6}{5}} = y^3 \cdot \sqrt[5]{z^6} = y^3 \cdot \sqrt[5]{z^{5+1}} = y^3 \cdot \sqrt[5]{z^5 \cdot z^1} = y^3 \cdot z \cdot \sqrt[5]{z} = y^3 z \sqrt[5]{z}$$

Example 5.3-2

$$2^{\frac{1}{3}}\cdot\left(2\cdot x^{\frac{1}{5}}\right)^{\frac{1}{3}}=$$

Solution:

Step 1 $$2^{\frac{1}{3}}\cdot\left(2\cdot x^{\frac{1}{5}}\right)^{\frac{1}{3}}=2^{\frac{1}{3}}\cdot\left(2^{1\times\frac{1}{3}}\cdot x^{\frac{1}{5}\times\frac{1}{3}}\right)=2^{\frac{1}{3}}\cdot\left(2^{\frac{1}{1}\times\frac{1}{3}}\cdot x^{\frac{1\times1}{5\times3}}\right)=2^{\frac{1}{3}}\cdot 2^{\frac{1}{3}}\cdot x^{\frac{1}{15}}$$

Step 2 $$2^{\frac{1}{3}}\cdot 2^{\frac{1}{3}}\cdot x^{\frac{1}{15}}=2^{\frac{1}{3}+\frac{1}{3}}\cdot x^{\frac{1}{15}}=2^{\frac{1+1}{3}}\cdot x^{\frac{1}{15}}=2^{\frac{2}{3}}\cdot x^{\frac{1}{15}}$$

Step 3 $$2^{\frac{2}{3}}\cdot x^{\frac{1}{15}}=\sqrt[3]{2^2}\cdot\sqrt[15]{x^1}=\boldsymbol{\sqrt[3]{4}\,\sqrt[15]{x}}$$

Example 5.3-3

$$\left(2^{\frac{a}{2}}\right)^4\cdot\left(3^a\cdot 3^{3a}\right)^{\frac{2}{5}}=$$

Solution:

Step 1 $$\left(2^{\frac{a}{2}}\right)^4\cdot\left(3^a\cdot 3^{3a}\right)^{\frac{2}{5}}=2^{\frac{a}{2}\times4}\cdot\left(3^{a\times\frac{2}{5}}\cdot 3^{3a\times\frac{2}{5}}\right)=2^{\frac{a}{2}\times\frac{4}{1}}\cdot\left(3^{\frac{a}{1}\times\frac{2}{5}}\cdot 3^{\frac{3a}{1}\times\frac{2}{5}}\right)$$

$$=2^{\frac{4a}{2}}\cdot 3^{\frac{2a}{5}}\cdot 3^{\frac{6a}{5}}$$

Step 2 $$2^{\frac{4a}{2}}\cdot 3^{\frac{2a}{5}}\cdot 3^{\frac{6a}{5}}=2^{\frac{4a}{2}=\frac{2a}{1}}\cdot 3^{\frac{2a}{5}+\frac{6a}{5}}=2^{\frac{2a}{1}}\cdot 3^{\frac{2a+6a}{5}}=2^{2a}\cdot 3^{\frac{8a}{5}}$$

Step 3 $$2^{2a}\cdot 3^{\frac{8a}{5}}=\boldsymbol{2^{2a}\sqrt[5]{3^{8a}}}$$

Example 5.3-4

$$\left(b^{\frac{1}{2}}\cdot c^{\frac{3}{2}}\right)^4\cdot\left(b^{\frac{2}{3}}\right)^2=$$

Solution:

Step 1 $$\left(b^{\frac{1}{2}}\cdot c^{\frac{3}{2}}\right)^4\cdot\left(b^{\frac{2}{3}}\right)^2=\left(b^{\frac{1}{2}\times4}\cdot c^{\frac{3}{2}\times4}\right)\cdot b^{\frac{2}{3}\times2}=b^{\frac{4}{2}}\cdot c^{\frac{12}{2}}\cdot b^{\frac{4}{3}}$$

Step 2
$$\boxed{b^{\frac{4}{2}}\cdot c^{\frac{12}{2}}\cdot b^{\frac{4}{3}}}=\boxed{b^{\frac{4}{2}=\frac{2}{1}}\cdot c^{\frac{12}{2}=\frac{6}{1}}\cdot b^{\frac{4}{3}}}=\boxed{b^{\frac{2}{1}}\cdot c^{\frac{6}{1}}\cdot b^{\frac{4}{3}}}=\boxed{b^{\frac{2}{1}}\cdot b^{\frac{4}{3}}\cdot c^{6}}=\boxed{b^{\frac{2}{1}+\frac{4}{3}}\cdot c^{6}}$$

$$=\boxed{b^{\frac{(2\cdot3)+(4\cdot1)}{1\cdot3}}\cdot c^{6}}=\boxed{b^{\frac{6+4}{3}}\cdot c^{6}}=\boxed{b^{\frac{10}{3}}\cdot c^{6}}$$

Step 3
$$\boxed{b^{\frac{10}{3}}\cdot c^{6}}=\boxed{\sqrt[3]{b^{10}}\cdot c^{6}}=\boxed{\sqrt[3]{b^{3+3+3+1}}\cdot c^{6}}=\boxed{\sqrt[3]{\left(b^3\cdot b^3\cdot b^3\right)\cdot b^1}\cdot c^{6}}=\boxed{(b\cdot b\cdot b)\cdot\sqrt[3]{b^1}\cdot c^{6}}$$

$$=\boxed{b^{1+1+1}\cdot\sqrt[3]{b}\cdot c^{6}}=\boxed{\mathbf{b^3c^6\sqrt[3]{b}}}$$

Example 5.3-5

$$\boxed{\left[\left(2\cdot x^2\cdot y^{\frac{1}{4}}\right)^2\cdot x^3\right]^{\frac{2}{3}}}=$$

Solution:

Step 1
$$\boxed{\left[\left(2\cdot x^2\cdot y^{\frac{1}{4}}\right)^2\cdot x^3\right]^{\frac{2}{3}}}=\boxed{\left[\left(2^{1\times2}\cdot x^{2\times2}\cdot y^{\frac{1}{4}\times2}\right)\cdot x^3\right]^{\frac{2}{3}}}=\boxed{\left[2^2\cdot x^4\cdot y^{\frac{2}{4}}\cdot x^3\right]^{\frac{2}{3}}}$$

$$=\boxed{2^{2\times\frac{2}{3}}\cdot x^{4\times\frac{2}{3}}\cdot y^{\frac{2}{4}\times\frac{2}{3}}\cdot x^{3\times\frac{2}{3}}}=\boxed{2^{\frac{4}{3}}\cdot x^{\frac{8}{3}}\cdot y^{\frac{4}{12}}\cdot x^{\frac{6}{3}}}$$

Step 2
$$\boxed{2^{\frac{4}{3}}\cdot x^{\frac{8}{3}}\cdot y^{\frac{4}{12}}\cdot x^{\frac{6}{3}}}=\boxed{2^{\frac{4}{3}}\cdot x^{\frac{8}{3}}\cdot y^{\frac{4}{12}=\frac{1}{3}}\cdot x^{\frac{6}{3}=\frac{2}{1}}}=\boxed{2^{\frac{4}{3}}\cdot x^{\frac{8}{3}}\cdot y^{\frac{1}{3}}\cdot x^{\frac{2}{1}}}=\boxed{2^{\frac{4}{3}}\cdot x^{\frac{8}{3}}\cdot x^{\frac{2}{1}}\cdot y^{\frac{1}{3}}}$$

$$=\boxed{2^{\frac{4}{3}}\cdot x^{\frac{8}{3}+\frac{2}{1}}\cdot y^{\frac{1}{3}}}=\boxed{2^{\frac{4}{3}}\cdot x^{\frac{(8\cdot1)+(2\cdot3)}{3\cdot1}}\cdot y^{\frac{1}{3}}}=\boxed{2^{\frac{4}{3}}\cdot x^{\frac{8+6}{3}}\cdot y^{\frac{1}{3}}}=\boxed{2^{\frac{4}{3}}\cdot x^{\frac{14}{3}}\cdot y^{\frac{1}{3}}}$$

Step 3
$$\boxed{2^{\frac{4}{3}}\cdot x^{\frac{14}{3}}\cdot y^{\frac{1}{3}}}=\boxed{\sqrt[3]{2^4}\cdot\sqrt[3]{x^{14}}\cdot\sqrt[3]{y}}=\boxed{\sqrt[3]{2^{3+1}}\cdot\sqrt[3]{x^{3+3+3+3+2}}\cdot\sqrt[3]{y}}$$

$$=\boxed{\sqrt[3]{2^3\cdot2^1}\cdot\sqrt[3]{\left(x^3\cdot x^3\cdot x^3\cdot x^3\right)\cdot x^2}\cdot\sqrt[3]{y}}=\boxed{2\cdot\sqrt[3]{2}\cdot(x\cdot x\cdot x\cdot x)\cdot\sqrt[3]{x^2}\cdot\sqrt[3]{y}}$$

$$=\boxed{2\cdot\sqrt[3]{2}\cdot\left(x^{1+1+1+1}\cdot\sqrt[3]{x^2}\cdot\sqrt[3]{y}\right)}=\boxed{2\sqrt[3]{2}\left(x^4\sqrt[3]{x^2}\sqrt[3]{y}\right)}=\boxed{\mathbf{2x^4\sqrt[3]{2x^2y}}}$$

Additional Examples - Multiplying Positive Fractional Exponents

The following examples further illustrate how to multiply positive fractional exponents by one another:

Example 5.3-6

$$\boxed{y^{\frac{2}{3}}\cdot y^{\frac{1}{3}}} = \boxed{y^{\frac{2}{3}+\frac{1}{3}}} = \boxed{y^{\frac{2+1}{3}}} = \boxed{y^{\frac{3}{3}=\frac{1}{1}}} = \boxed{y^{\frac{1}{1}}} = \boxed{y^1} = \boxed{\mathbf{y}}$$

Example 5.3-7

$$\boxed{\left(2^3\cdot 2^2\right)^{\frac{2}{3}}\cdot\left(a^{\frac{3}{5}}\right)^2} = \boxed{\left(2^{3+2}\right)^{\frac{2}{3}}\cdot a^{\frac{3}{5}\times 2}} = \boxed{\left(2^5\right)^{\frac{2}{3}}\cdot a^{\frac{6}{5}}} = \boxed{2^{5\times\frac{2}{3}}\cdot a^{\frac{6}{5}}} = \boxed{2^{\frac{10}{3}}\cdot a^{\frac{6}{5}}} = \boxed{\sqrt[3]{2^{10}}\cdot\sqrt[5]{a^6}}$$

$$\boxed{\sqrt[3]{2^{3+3+3+1}}\cdot\sqrt[5]{a^{5+1}}} = \boxed{\sqrt[3]{\left(2^3\cdot 2^3\cdot 2^3\right)\cdot 2^1}\cdot\sqrt[5]{a^5\cdot a^1}} = \boxed{(2\cdot 2\cdot 2)\cdot\sqrt[3]{2}\cdot\left(a\cdot\sqrt[5]{a}\right)} = \boxed{8\sqrt[3]{2}\cdot a\sqrt[5]{a}} = \boxed{\mathbf{8\sqrt[3]{2}\left(a\sqrt[5]{a}\right)}}$$

Example 5.3-8

$$\boxed{\left(5^{2a}\right)^{\frac{3}{4}}\cdot\left(2\cdot 5^{2a}\right)^{\frac{1}{4}}} = \boxed{5^{2a\times\frac{3}{4}}\cdot 2^{1\times\frac{1}{4}}\cdot 5^{2a\times\frac{1}{4}}} = \boxed{5^{\frac{6a}{4}}\cdot 2^{\frac{1}{4}}\cdot 5^{\frac{2a}{4}}} = \boxed{2^{\frac{1}{4}}\cdot 5^{\frac{6a}{4}}\cdot 5^{\frac{2a}{4}}} = \boxed{2^{\frac{1}{4}}\cdot 5^{\frac{6a}{4}+\frac{2a}{4}}} = \boxed{2^{\frac{1}{4}}\cdot 5^{\frac{6a+2a}{4}}}$$

$$= \boxed{2^{\frac{1}{4}}\cdot 5^{\frac{8a}{4}=\frac{2a}{1}}} = \boxed{2^{\frac{1}{4}}\cdot 5^{\frac{2a}{1}}} = \boxed{\sqrt[4]{2^1}\cdot 5^{2a}} = \boxed{\mathbf{\sqrt[4]{2}\,5^{2a}}}$$

Example 5.3-9

$$\boxed{w^{\frac{3}{5}}\cdot\left(2\cdot w^2\right)^{\frac{1}{3}}} = \boxed{w^{\frac{3}{5}}\cdot 2^{1\times\frac{1}{3}}\cdot w^{2\times\frac{1}{3}}} = \boxed{w^{\frac{3}{5}}\cdot 2^{\frac{1}{3}}\cdot w^{\frac{2}{3}}} = \boxed{2^{\frac{1}{3}}\cdot w^{\frac{3}{5}}\cdot w^{\frac{2}{3}}} = \boxed{2^{\frac{1}{3}}\cdot w^{\frac{3}{5}+\frac{2}{3}}} = \boxed{2^{\frac{1}{3}}\cdot w^{\frac{(3\cdot 3)+(2\cdot 5)}{5\cdot 3}}}$$

$$= \boxed{2^{\frac{1}{3}}\cdot w^{\frac{9+10}{15}}} = \boxed{2^{\frac{1}{3}}\,w^{\frac{19}{15}}} = \boxed{\sqrt[3]{2^1}\cdot\sqrt[15]{w^{19}}} = \boxed{\sqrt[3]{2}\cdot\sqrt[15]{w^{15+4}}} = \boxed{\sqrt[3]{2}\cdot\sqrt[15]{w^{15}\cdot w^4}} = \boxed{\sqrt[3]{2}\cdot w\cdot\sqrt[15]{w^4}} = \boxed{\mathbf{\sqrt[3]{2}\,w\sqrt[15]{w^4}}}$$

Example 5.3-10

$$\boxed{y^2\cdot\left(a\cdot y^3\right)^{\frac{1}{2}}} = \boxed{y^2\cdot a^{1\times\frac{1}{2}}\cdot y^{3\times\frac{1}{2}}} = \boxed{y^2\cdot a^{\frac{1}{2}}\cdot y^{\frac{3}{2}}} = \boxed{a^{\frac{1}{2}}\cdot y^2\cdot y^{\frac{3}{2}}} = \boxed{a^{\frac{1}{2}}\cdot y^{2+\frac{3}{2}}} = \boxed{a^{\frac{1}{2}}\cdot y^{\frac{2}{1}+\frac{3}{2}}}$$

$$= \boxed{a^{\frac{1}{2}}\cdot y^{\frac{(2\cdot 2)+(3\cdot 1)}{1\cdot 2}}} = \boxed{a^{\frac{1}{2}}\cdot y^{\frac{4+3}{2}}} = \boxed{a^{\frac{1}{2}}\cdot y^{\frac{7}{2}}} = \boxed{\sqrt{a}\cdot\sqrt{y^7}} = \boxed{\sqrt{a}\cdot\sqrt{y^{2+2+2+1}}} = \boxed{\sqrt{a}\cdot\sqrt{\left(y^2\cdot y^2\cdot y^2\right)\cdot y^1}}$$

$$= \boxed{\sqrt{a}\cdot(y\cdot y\cdot y)\cdot\sqrt{y}} = \boxed{y^{1+1+1}\cdot\sqrt{a\cdot y}} = \boxed{\mathbf{y^3\sqrt{ay}}}$$

Example 5.3-11

$$\boxed{a^{\frac{2}{3}}\cdot\left(a^2\cdot b^{\frac{1}{2}}\right)^{\frac{2}{3}}\cdot\left(a^3\cdot b\right)^{\frac{1}{2}}} = \boxed{a^{\frac{2}{3}}\cdot\left(a^{2\times\frac{2}{3}}\cdot b^{\frac{1}{2}\times\frac{2}{3}}\right)\cdot\left(a^{3\times\frac{1}{2}}\cdot b^{1\times\frac{1}{2}}\right)} = \boxed{a^{\frac{2}{3}}\cdot\left(a^{\frac{4}{3}}\cdot b^{\frac{2}{6}}\right)\cdot\left(a^{\frac{3}{2}}\cdot b^{\frac{1}{2}}\right)}$$

$$= \boxed{\left(a^{\frac{2}{3}} \cdot a^{\frac{4}{3}} \cdot b^{\frac{2}{6}}\right) \cdot \left(a^{\frac{3}{2}} \cdot b^{\frac{1}{2}}\right)} = \boxed{\left(a^{\frac{2}{3}} \cdot a^{\frac{4}{3}} \cdot a^{\frac{3}{2}}\right) \cdot \left(b^{\frac{2}{6}} \cdot b^{\frac{1}{2}}\right)} = \boxed{\left(a^{\frac{2}{3}+\frac{4}{3}} \cdot a^{\frac{3}{2}}\right) \cdot \left(b^{\frac{1}{3}+\frac{1}{2}}\right)} = \boxed{\left(a^{\frac{2+4}{3}} \cdot a^{\frac{3}{2}}\right) \cdot b^{\frac{(1\cdot2)+(1\cdot3)}{3\cdot2}}}$$

$$= \boxed{a^{\frac{6}{3}=\frac{2}{1}} \cdot a^{\frac{3}{2}} \cdot b^{\frac{2+3}{6}}} = \boxed{a^{\frac{2}{1}} \cdot a^{\frac{3}{2}} \cdot b^{\frac{5}{6}}} = \boxed{a^{\frac{2}{1}+\frac{3}{2}} \cdot b^{\frac{5}{6}}} = \boxed{a^{\frac{(2\cdot2)+(3\cdot1)}{1\cdot2}} \cdot b^{\frac{5}{6}}} = \boxed{a^{\frac{4+3}{2}} \cdot b^{\frac{5}{6}}} = \boxed{a^{\frac{7}{2}} \cdot b^{\frac{5}{6}}} = \boxed{\sqrt[2]{a^7} \cdot \sqrt[6]{b^5}}$$

$$= \boxed{\sqrt[2]{a^{2+2+2+1}} \cdot \sqrt[6]{b^5}} = \boxed{\sqrt[2]{a^2 \cdot a^2 \cdot a^2 \cdot a^1} \cdot \sqrt[6]{b^5}} = \boxed{(a \cdot a \cdot a)\sqrt{a} \cdot \sqrt[6]{b^5}} = \boxed{\mathbf{a^3 \sqrt{a}\, \sqrt[6]{b^5}}}$$

Example 5.3-12

$$\boxed{\left(x^3 \cdot y\right)^{\frac{2}{3}} \cdot \left(x^2 \cdot y^{\frac{1}{5}} \cdot z\right)^{\frac{1}{2}}} = \boxed{\left(x^{3\times\frac{2}{3}} \cdot y^{1\times\frac{2}{3}}\right) \cdot \left(x^{2\times\frac{1}{2}} \cdot y^{\frac{1}{5}\times\frac{1}{2}} \cdot z^{1\times\frac{1}{2}}\right)} = \boxed{\left(x^{\frac{6}{3}=\frac{2}{1}} \cdot y^{\frac{2}{3}}\right) \cdot \left(x^{\frac{2}{2}=\frac{1}{1}} \cdot y^{\frac{1}{10}} \cdot z^{\frac{1}{2}}\right)}$$

$$= \boxed{x^{\frac{2}{1}} \cdot y^{\frac{2}{3}} \cdot x^{\frac{1}{1}} \cdot y^{\frac{1}{10}} \cdot z^{\frac{1}{2}}} = \boxed{x^{\frac{2}{1}} \cdot x^{\frac{1}{1}} \cdot y^{\frac{2}{3}} \cdot y^{\frac{1}{10}} \cdot z^{\frac{1}{2}}} = \boxed{x^{\frac{2+1}{1}} \cdot y^{\frac{2}{3}+\frac{1}{10}} \cdot z^{\frac{1}{2}}} = \boxed{x^{\frac{3}{1}} \cdot y^{\frac{(2\cdot10)+(1\cdot3)}{3\cdot10}} \cdot z^{\frac{1}{2}}}$$

$$= \boxed{x^3 \cdot y^{\frac{20+3}{30}} \cdot z^{\frac{1}{2}}} = \boxed{x^3 \cdot y^{\frac{23}{30}} \cdot z^{\frac{1}{2}}} = \boxed{x^3 \cdot \sqrt[30]{y^{23}} \cdot \sqrt{z}} = \boxed{\mathbf{x^3\, \sqrt[30]{y^{23}}\, \sqrt{z}}}$$

Example 5.3-13

$$\boxed{\left(x^{\frac{1}{2}} \cdot y\right)^2 \cdot \left(y \cdot x^3\right)^{\frac{2}{3}} \cdot y} = \boxed{\left(x^{\frac{1}{2}\times2} \cdot y^{1\times2}\right) \cdot \left(y^{1\times\frac{2}{3}} \cdot x^{3\times\frac{2}{3}}\right) \cdot y} = \boxed{\left(x^{\frac{2}{2}=\frac{1}{1}} \cdot y^2\right) \cdot \left(y^{\frac{2}{3}} \cdot x^{\frac{6}{3}=\frac{2}{1}}\right) \cdot y}$$

$$= \boxed{x^{\frac{1}{1}} \cdot y^2 \cdot y^{\frac{2}{3}} \cdot x^{\frac{2}{1}} \cdot y^1} = \boxed{x^{\frac{1}{1}} \cdot x^{\frac{2}{1}} \cdot y^2 \cdot y^1 \cdot y^{\frac{2}{3}}} = \boxed{x^{\frac{1}{1}+\frac{2}{1}} \cdot y^{2+1} \cdot y^{\frac{2}{3}}} = \boxed{x^{\frac{3}{1}} \cdot y^3 \cdot y^{\frac{2}{3}}} = \boxed{x^3 \cdot y^{3+\frac{2}{3}}} = \boxed{x^3 \cdot y^{\frac{3}{1}+\frac{2}{3}}}$$

$$= \boxed{x^3 \cdot y^{\frac{(3\cdot3)+(2\cdot1)}{1\cdot3}}} = \boxed{x^3 \cdot y^{\frac{9+2}{3}}} = \boxed{x^3 \cdot y^{\frac{11}{3}}} = \boxed{x^3 \cdot \sqrt[3]{y^{11}}} = \boxed{x^3 \cdot \sqrt[3]{y^{3+3+3+2}}} = \boxed{x^3 \cdot \sqrt[3]{\left(y^3 \cdot y^3 \cdot y^3\right) \cdot y^2}}$$

$$= \boxed{x^3 \cdot (y \cdot y \cdot y) \cdot \sqrt[3]{y^2}} = \boxed{x^3 \cdot y^3 \cdot \sqrt[3]{y^2}} = \boxed{\mathbf{x^3 y^3 \sqrt[3]{y^2}}}$$

Example 5.3-14

$$\boxed{\left(2^3 \cdot a^2\right)^{\frac{1}{3}} \cdot \left(3 \cdot a \cdot x^7\right)^{\frac{2}{7}}} = \boxed{\left(2^{3\times\frac{1}{3}} \cdot a^{2\times\frac{1}{3}}\right) \cdot \left(3^{1\times\frac{2}{7}} \cdot a^{1\times\frac{2}{7}} \cdot x^{7\times\frac{2}{7}}\right)} = \boxed{\left(2^{\frac{3}{3}=\frac{1}{1}} \cdot a^{\frac{2}{3}}\right) \cdot \left(3^{\frac{2}{7}} \cdot a^{\frac{2}{7}} \cdot x^{\frac{14}{7}=\frac{2}{1}}\right)}$$

$$= \boxed{2^{\frac{1}{1}} \cdot a^{\frac{2}{3}} \cdot 3^{\frac{2}{7}} \cdot a^{\frac{2}{7}} \cdot x^{\frac{2}{1}}} = \boxed{2 \cdot 3^{\frac{2}{7}} \cdot a^{\frac{2}{3}} \cdot a^{\frac{2}{7}} \cdot x^2} = \boxed{2 \cdot 3^{\frac{2}{7}} \cdot a^{\frac{2}{3}+\frac{2}{7}} \cdot x^2} = \boxed{2 \cdot 3^{\frac{2}{7}} \cdot a^{\frac{(2\times7)+(2\times3)}{3\times7}} \cdot x^2}$$

$$= \boxed{2 \cdot 3^{\frac{2}{7}} \cdot a^{\frac{14+6}{21}} \cdot x^2} = \boxed{2 \cdot 3^{\frac{2}{7}} \cdot a^{\frac{20}{21}} \cdot x^2} = \boxed{2 \cdot \sqrt[7]{3^2} \cdot \sqrt[21]{a^{20}} \cdot x^2} = \boxed{\mathbf{2\sqrt[7]{9}\left(\sqrt[21]{a^{20}}\, x^2\right)}}$$

Example 5.3-15

$$\boxed{\left(5^2\cdot y^4\cdot z^2\right)^{\frac{1}{4}}\cdot\left(y^2\cdot z^3\right)^{\frac{1}{3}}}=\boxed{\left(5^{2\times\frac{1}{4}}\cdot y^{4\times\frac{1}{4}}\cdot z^{2\times\frac{1}{4}}\right)\cdot\left(y^{2\times\frac{1}{3}}\cdot z^{3\times\frac{1}{3}}\right)}=\boxed{\left(5^{\frac{2}{4}=\frac{1}{2}}\cdot y^{\frac{4}{4}=\frac{1}{1}}\cdot z^{\frac{2}{4}=\frac{1}{2}}\right)\cdot\left(y^{\frac{2}{3}}\cdot z^{\frac{3}{3}=\frac{1}{1}}\right)}$$

$$=\boxed{5^{\frac{1}{2}}\cdot y^{\frac{1}{1}}\cdot z^{\frac{1}{2}}\cdot y^{\frac{2}{3}}\cdot z^{\frac{1}{1}}}=\boxed{5^{\frac{1}{2}}\cdot y^{\frac{1}{1}}\cdot y^{\frac{2}{3}}\cdot z^{\frac{1}{2}}\cdot z^{\frac{1}{1}}}=\boxed{5^{\frac{1}{2}}\cdot y^{\frac{1}{1}+\frac{2}{3}}\cdot z^{\frac{1}{2}+\frac{1}{1}}}=\boxed{5^{\frac{1}{2}}\cdot y^{\frac{(1\cdot3)+(2\cdot1)}{1\cdot3}}\cdot z^{\frac{(1\cdot1)+(1\cdot2)}{2\cdot1}}}$$

$$=\boxed{5^{\frac{1}{2}}\cdot y^{\frac{5}{3}}\cdot z^{\frac{3}{2}}}=\boxed{\sqrt{5}\cdot\sqrt[3]{y^5}\cdot\sqrt[2]{z^3}}=\boxed{\sqrt{5}\cdot\sqrt[3]{y^{3+2}}\cdot\sqrt[2]{z^{2+1}}}=\boxed{\sqrt{5}\cdot\sqrt[3]{y^3\cdot y^2}\cdot\sqrt{z^2\cdot z}}=\boxed{\sqrt{5}\cdot y\cdot\sqrt[3]{y^2}\cdot z\cdot\sqrt{z}}$$

$$=\boxed{\sqrt{5}\,yz\sqrt[3]{y^2}\,\sqrt{z}}$$

Practice Problems - Multiplying Positive Fractional Exponents

Section 5.3 Case I Practice Problems - Multiply the following positive fractional exponents:

1. $\left(x^2\cdot x^3\right)^{\frac{1}{2}}=$
2. $2\cdot\left(a^{\frac{2}{3}}\cdot b^0\right)^3=$
3. $\left(a^2\cdot b^3\right)^{\frac{2}{3}}\cdot\left(a^{\frac{1}{3}}\cdot b^4\right)=$
4. $\left(a^3\cdot a^{\frac{2}{3}}\right)\cdot\left(x^2\cdot x^3\right)^{\frac{2}{3}}=$
5. $\left(x\cdot y^2\cdot z^3\right)^{\frac{0}{2}}\cdot\left(x^2\right)^{\frac{1}{2}}=$
6. $2^{\frac{1}{3}}\cdot 3^{\frac{2}{3}}\cdot 3^5\cdot 2^{\frac{2}{3}}=$
7. $x^2\cdot\left(x^{\frac{3}{4}}\right)^4\cdot\left(x^{\frac{1}{4}}\cdot y^{\frac{3}{5}}\right)^2=$
8. $\left(x^0\cdot y^2\cdot z\right)^{\frac{3}{2}}\cdot\left(x^{\frac{1}{3}}\cdot y^{\frac{2}{5}}\cdot z\right)=$
9. $5^0\cdot\left(x^2\cdot y\right)^{\frac{2}{3}}\cdot\left(3^4\cdot x\cdot y^3\right)^{\frac{1}{3}}=$
10. $\left(x^{\frac{2}{3}}\right)^3\cdot\left[\left(2\cdot x\cdot y^2\right)^3\cdot x\right]^{\frac{1}{5}}=$

Case II Dividing Positive Fractional Exponents

Positive fractional exponents are divided by one another using the exponent laws I through VI shown in Table 5.3-1. These laws are used in order to simplify division of positive fractional exponents by each other. Positive fractional exponents are divided by one another using the following steps:

Step 1 Apply the Power of a Power, Power of a Product, and/or the Power of a Fraction Law (Laws II, III, and IV) from Table 5.3-1.

Step 2 Simplify the fractional expression by applying the appropriate exponent laws (Laws I, V, or VI) from Table 5.3-1 and the fractional techniques learned in Chapter 2.

Step 3 Change the fractional exponential expressions to radical expressions (see Section 5.1).

Examples with Steps

The following examples show the steps as to how positive fractional exponents are divided by one another:

Example 5.3-16

$$\boxed{\left(\frac{x^3}{x^2}\right)^{\frac{2}{3}}} =$$

Solution:

Step 1

$$\boxed{\left(\frac{x^3}{x^2}\right)^{\frac{2}{3}}} = \boxed{\frac{x^{3\times\frac{2}{3}}}{x^{2\times\frac{2}{3}}}} = \boxed{\frac{x^{\frac{3}{1}\times\frac{2}{3}}}{x^{\frac{2}{1}\times\frac{2}{3}}}} = \boxed{\frac{x^{\frac{3\times 2}{1\times 3}}}{x^{\frac{2\times 2}{1\times 3}}}} = \boxed{\frac{x^{\frac{6}{3}}}{x^{\frac{4}{3}}}}$$

Step 2

$$\boxed{\frac{x^{\frac{6}{3}}}{x^{\frac{4}{3}}}} = \boxed{\frac{x^{\frac{6}{3}=\frac{2}{1}}}{x^{\frac{4}{3}}}} = \boxed{\frac{x^{\frac{2}{1}}}{x^{\frac{4}{3}}}} = \boxed{\frac{x^{\frac{2}{1}}\cdot x^{-\frac{4}{3}}}{1}} = \boxed{x^{\frac{2}{1}-\frac{4}{3}}} = \boxed{x^{\frac{(2\cdot3)-(4\cdot1)}{1\cdot3}}} = \boxed{x^{\frac{6-4}{3}}} = \boxed{x^{\frac{2}{3}}}$$

Step 3

$$\boxed{x^{\frac{2}{3}}} = \boxed{\sqrt[3]{x^2}}$$

Example 5.3-17

$$\boxed{\left(\frac{2^2\cdot a^{\frac{1}{3}}}{a^{\frac{4}{5}}}\right)^3} =$$

Solution:

Step 1

$$\boxed{\left(\frac{2^2\cdot a^{\frac{1}{3}}}{a^{\frac{4}{5}}}\right)^3} = \boxed{\frac{2^{2\times3}\cdot a^{\frac{1}{3}\times3}}{a^{\frac{4}{5}\times3}}} = \boxed{\frac{2^6\cdot a^{\frac{3}{3}=\frac{1}{1}}}{a^{\frac{12}{5}}}} = \boxed{\frac{2^6\cdot a^{\frac{1}{1}}}{a^{\frac{12}{5}}}}$$

Step 2

$$\boxed{\frac{2^6\cdot a^{\frac{1}{1}}}{a^{\frac{12}{5}}}}=\boxed{\frac{2^6}{a^{\frac{12}{5}}\cdot a^{-\frac{1}{1}}}}=\boxed{\frac{2^6}{a^{\frac{12}{5}-\frac{1}{1}}}}=\boxed{\frac{2^6}{a^{\frac{(12\cdot1)-(1\cdot5)}{5\cdot1}}}}=\boxed{\frac{2^6}{a^{\frac{12-5}{5}}}}=\boxed{\frac{2^6}{a^{\frac{7}{5}}}}$$

Step 3

$$\boxed{\frac{2^6}{a^{\frac{7}{5}}}}=\boxed{\frac{2^6}{\sqrt[5]{a^7}}}=\boxed{\frac{2^6}{\sqrt[5]{a^{5+2}}}}=\boxed{\frac{2^6}{\sqrt[5]{a^5\cdot a^2}}}=\boxed{\frac{2^6}{a\cdot\sqrt[5]{a^2}}}=\boxed{\mathbf{\frac{64}{a\sqrt[5]{a^2}}}}$$

Example 5.3-18

$$\boxed{\left(\frac{3\cdot c^2}{256\cdot b^3}\right)^{\frac{1}{8}}}=$$

Solution:

Step 1

$$\boxed{\left(\frac{3\cdot c^2}{256\cdot b^3}\right)^{\frac{1}{8}}}=\boxed{\frac{3^{\frac{1}{8}}\cdot c^{2\times\frac{1}{8}}}{256^{\frac{1}{8}}\cdot b^{3\times\frac{1}{8}}}}=\boxed{\frac{3^{\frac{1}{8}}\cdot c^{\frac{2}{1}\times\frac{1}{8}}}{256^{\frac{1}{8}}\cdot b^{3\times\frac{1}{8}}}}=\boxed{\frac{3^{\frac{1}{8}}\cdot c^{\frac{2}{8}}}{256^{\frac{1}{8}}\cdot b^{\frac{3}{8}}}}$$

Step 2

$$\boxed{\frac{3^{\frac{1}{8}}\cdot c^{\frac{2}{8}}}{256^{\frac{1}{8}}\cdot b^{\frac{3}{8}}}}=\boxed{\frac{3^{\frac{1}{8}}\cdot c^{\frac{2}{8}=\frac{1}{4}}}{256^{\frac{1}{8}}\cdot b^{\frac{3}{8}}}}=\boxed{\frac{3^{\frac{1}{8}}\cdot c^{\frac{1}{4}}}{256^{\frac{1}{8}}\cdot b^{\frac{3}{8}}}}$$

Step 3

$$\boxed{\frac{3^{\frac{1}{8}}\cdot c^{\frac{1}{4}}}{256^{\frac{1}{8}}\cdot b^{\frac{3}{8}}}}=\boxed{\frac{\sqrt[8]{3}\cdot\sqrt[4]{c^1}}{\sqrt[8]{256}\cdot\sqrt[8]{b^3}}}=\boxed{\frac{\sqrt[8]{3}\cdot\sqrt[4]{c}}{\sqrt[8]{2^8}\cdot\sqrt[8]{b^3}}}=\boxed{\mathbf{\frac{\sqrt[8]{3}\sqrt[4]{c}}{2\sqrt[8]{b^3}}}}$$

Example 5.3-19

$$\boxed{\frac{z^{\frac{1}{3}}\cdot(a\cdot b)^2}{z^2\cdot a^{\frac{2}{3}}}}=$$

Solution:

Step 1

$$\boxed{\frac{z^{\frac{1}{3}}\cdot(a\cdot b)^2}{z^2\cdot a^{\frac{2}{3}}}}=\boxed{\frac{z^{\frac{1}{3}}\cdot\left(a^1\cdot b^1\right)^2}{z^2\cdot a^{\frac{2}{3}}}}=\boxed{\frac{z^{\frac{1}{3}}\cdot a^{1\times2}\cdot b^{1\times2}}{z^2\cdot a^{\frac{2}{3}}}}=\boxed{\frac{z^{\frac{1}{3}}\cdot a^2\cdot b^2}{z^2\cdot a^{\frac{2}{3}}}}$$

Step 2

$$\boxed{\frac{z^{\frac{1}{3}}\cdot a^2\cdot b^2}{z^2\cdot a^{\frac{2}{3}}}}=\boxed{\frac{a^2\cdot a^{-\frac{2}{3}}\cdot b^2}{z^2\cdot z^{-\frac{1}{3}}}}=\boxed{\frac{a^{\frac{2}{1}}\cdot a^{-\frac{2}{3}}\cdot b^2}{z^{\frac{2}{1}}\cdot z^{-\frac{1}{3}}}}=\boxed{\frac{a^{\frac{2}{1}-\frac{2}{3}}\cdot b^2}{z^{\frac{2}{1}-\frac{1}{3}}}}=\boxed{\frac{a^{\frac{(2\cdot3)-(2\cdot1)}{1\cdot3}}\cdot b^2}{z^{\frac{(2\cdot3)-(1\cdot1)}{1\cdot3}}}}$$

$$= \frac{a^{\frac{6-2}{3}} \cdot b^2}{z^{\frac{6-1}{3}}} = \frac{a^{\frac{4}{3}} \cdot b^2}{z^{\frac{5}{3}}}$$

Step 3

$$\frac{a^{\frac{4}{3}} \cdot b^2}{z^{\frac{5}{3}}} = \frac{\sqrt[3]{a^4} \cdot b^2}{\sqrt[3]{z^5}} = \frac{\sqrt[3]{a^{3+1}} \cdot b^2}{\sqrt[3]{z^{3+2}}} = \frac{\sqrt[3]{a^3 \cdot a^1} \cdot b^2}{\sqrt[3]{z^3 \cdot z^2}} = \frac{a \cdot \sqrt[3]{a^1} \cdot b^2}{z \cdot \sqrt[3]{z^2}} = \frac{\boldsymbol{ab^2 \sqrt[3]{a}}}{\boldsymbol{z\sqrt[3]{z^2}}}$$

Example 5.3-20

$$\left(\frac{\left(x^2 \cdot y^3\right)^4}{\left(x^3 \cdot y \cdot z\right)^2}\right)^{\frac{1}{2}} =$$

Solution:

Step 1

$$\left(\frac{\left(x^2 \cdot y^3\right)^4}{\left(x^3 \cdot y \cdot z\right)^2}\right)^{\frac{1}{2}} = \left(\frac{x^{2\times4} \cdot y^{3\times4}}{x^{3\times2} \cdot y^{1\times2} \cdot z^{1\times2}}\right)^{\frac{1}{2}} = \left(\frac{x^8 \cdot y^{12}}{x^6 \cdot y^2 \cdot z^2}\right)^{\frac{1}{2}} = \frac{x^{8\times\frac{1}{2}} \cdot y^{12\times\frac{1}{2}}}{x^{6\times\frac{1}{2}} \cdot y^{2\times\frac{1}{2}} \cdot z^{2\times\frac{1}{2}}}$$

$$= \frac{x^{\frac{8}{1}\times\frac{1}{2}} \cdot y^{\frac{12}{1}\times\frac{1}{2}}}{x^{\frac{6}{1}\times\frac{1}{2}} \cdot y^{\frac{2}{1}\times\frac{1}{2}} \cdot z^{\frac{2}{1}\times\frac{1}{2}}} = \frac{x^{\frac{8}{2}} \cdot y^{\frac{12}{2}}}{x^{\frac{6}{2}} \cdot y^{\frac{2}{2}} \cdot z^{\frac{2}{2}}}$$

Step 2

$$\frac{x^{\frac{8}{2}} \cdot y^{\frac{12}{2}}}{x^{\frac{6}{2}} \cdot y^{\frac{2}{2}} \cdot z^{\frac{2}{2}}} = \frac{x^{\frac{8}{2}=\frac{4}{1}} \cdot y^{\frac{12}{2}=\frac{6}{1}}}{x^{\frac{6}{2}=\frac{3}{1}} \cdot y^{\frac{2}{2}=\frac{1}{1}} \cdot z^{\frac{2}{2}=\frac{1}{1}}} = \frac{x^{\frac{4}{1}} \cdot y^{\frac{6}{1}}}{x^{\frac{3}{1}} \cdot y^{\frac{1}{1}} \cdot z^{\frac{1}{1}}} = \frac{x^4 \cdot y^6}{x^3 \cdot y^1 \cdot z^1} = \frac{x^4 \cdot x^{-3} \cdot y^6 \cdot y^{-1}}{z}$$

$$= \frac{x^{4-3} \cdot y^{6-1}}{z} = \frac{x^1 \cdot y^5}{z} = \frac{\boldsymbol{x\,y^5}}{\boldsymbol{z}}$$

Step 3 *Not Applicable*

Additional Examples - Dividing Positive Fractional Exponents

The following examples further illustrate how to divide positive fractional exponents by one another:

Example 5.3-21

$$\left(\frac{1}{2}\right)^{\frac{1}{2}} \cdot \left(\frac{a^2}{a^{\frac{3}{4}}}\right) = \frac{1^{\frac{1}{2}}}{2^{\frac{1}{2}}} \cdot a^2 \cdot a^{-\frac{3}{4}} = \frac{\sqrt{1}}{\sqrt{2}} \cdot a^{2-\frac{3}{4}} = \sqrt{\frac{1}{2}} \cdot a^{\frac{2}{1}-\frac{3}{4}} = \frac{1}{\sqrt{2}} \cdot a^{\frac{(2\cdot4)-(3\cdot1)}{1\cdot4}} = \frac{1}{\sqrt{2}} \cdot a^{\frac{8-3}{4}} = \frac{1}{\sqrt{2}} \cdot a^{\frac{5}{4}}$$

$$= \boxed{\frac{1}{\sqrt{2}}\cdot\sqrt[4]{a^5}} = \boxed{\frac{1}{\sqrt{2}}\cdot\sqrt[4]{a^{4+1}}} = \boxed{\frac{1}{\sqrt{2}}\cdot a\cdot\sqrt[4]{a^1}} = \boxed{\frac{1}{\sqrt{2}}\left(a\sqrt[4]{a}\right)}$$

Example 5.3-22

$$\boxed{\left(\frac{a^0\cdot x^2}{a^{\frac{3}{2}}\cdot x^{\frac{5}{4}}}\right)^{\frac{2}{5}}} = \boxed{\frac{a^{0\times\frac{2}{5}}\cdot x^{2\times\frac{2}{5}}}{a^{\frac{3}{2}\times\frac{2}{5}}\cdot x^{\frac{5}{4}\times\frac{2}{5}}}} = \boxed{\frac{a^0\cdot x^{\frac{4}{5}}}{a^{\frac{6}{10}=\frac{3}{5}}\cdot x^{\frac{10}{20}=\frac{1}{2}}}} = \boxed{\frac{1\cdot x^{\frac{4}{5}}}{a^{\frac{3}{5}}\cdot x^{\frac{1}{2}}}} = \boxed{\frac{x^{\frac{4}{5}}\cdot x^{-\frac{1}{2}}}{a^{\frac{3}{5}}}} = \boxed{\frac{x^{\frac{4}{5}-\frac{1}{2}}}{a^{\frac{3}{5}}}} = \boxed{\frac{x^{\frac{(4\cdot2)-(1\cdot5)}{5\cdot2}}}{a^{\frac{3}{5}}}} = \boxed{\frac{x^{\frac{8-5}{10}}}{a^{\frac{3}{5}}}}$$

$$= \boxed{\frac{x^{\frac{3}{10}}}{a^{\frac{3}{5}}}} = \boxed{\frac{\sqrt[10]{x^3}}{\sqrt[5]{a^3}}}$$

Example 5.3-23

$$\boxed{\frac{a^{\frac{2}{3}}\cdot b^{\frac{5}{2}}\cdot c^3}{a\cdot b^{\frac{2}{5}}\cdot c^{\frac{3}{4}}}} = \boxed{\frac{b^{\frac{5}{2}}\cdot b^{-\frac{2}{5}}\cdot c^3\cdot c^{-\frac{3}{4}}}{a\cdot a^{-\frac{2}{3}}}} = \boxed{\frac{b^{\frac{5}{2}}\cdot b^{-\frac{2}{5}}\cdot c^{\frac{3}{1}}\cdot c^{-\frac{3}{4}}}{a^{\frac{1}{1}}\cdot a^{-\frac{2}{3}}}} = \boxed{\frac{b^{\frac{5}{2}-\frac{2}{5}}\cdot c^{\frac{3}{1}-\frac{3}{4}}}{a^{\frac{1}{1}-\frac{2}{3}}}} = \boxed{\frac{b^{\frac{(5\cdot5)-(2\cdot2)}{2\cdot5}}\cdot c^{\frac{(3\cdot4)-(3\cdot1)}{1\cdot4}}}{a^{\frac{(1\cdot3)-(1\cdot2)}{3\cdot1}}}}$$

$$= \boxed{\frac{b^{\frac{25-4}{10}}\cdot c^{\frac{12-3}{4}}}{a^{\frac{3-2}{3}}}} = \boxed{\frac{b^{\frac{21}{10}}\cdot c^{\frac{9}{4}}}{a^{\frac{1}{3}}}} = \boxed{\frac{\sqrt[10]{b^{21}}\cdot\sqrt[4]{c^9}}{\sqrt[3]{a}}} = \boxed{\frac{\sqrt[10]{b^{10+10+1}}\cdot\sqrt[4]{c^{4+4+1}}}{\sqrt[3]{a}}} = \boxed{\frac{\sqrt[10]{b^{10}\cdot b^{10}\cdot b^1}\cdot\sqrt[4]{c^4\cdot c^4\cdot c^1}}{\sqrt[3]{a}}}$$

$$= \boxed{\frac{(b\cdot b)\cdot\sqrt[10]{b}\cdot(c\cdot c)\cdot\sqrt[4]{c}}{\sqrt[3]{a}}} = \boxed{\frac{b^2c^2\sqrt[10]{b}\sqrt[4]{c}}{\sqrt[3]{a}}}$$

Example 5.3-24

$$\boxed{\left(\frac{3^5\cdot2^0}{3^2}\right)^{\frac{3}{5}}} = \boxed{\frac{3^{5\times\frac{3}{5}}\cdot2^{0\times\frac{3}{5}}}{3^{2\times\frac{3}{5}}}} = \boxed{\frac{3^{\frac{15}{5}}\cdot2^0}{3^{\frac{6}{5}}}} = \boxed{\frac{3^{\frac{15}{5}}\cdot1}{3^{\frac{6}{5}}}} = \boxed{\frac{3^{\frac{15}{5}}}{3^{\frac{6}{5}}}} = \boxed{\frac{3^{\frac{15}{5}}\cdot3^{-\frac{6}{5}}}{1}} = \boxed{3^{\frac{15}{5}-\frac{6}{5}}} = \boxed{3^{\frac{15-6}{5}}} = \boxed{3^{\frac{9}{5}}} = \boxed{\sqrt[5]{3^9}}$$

$$= \boxed{\sqrt[5]{3^{5+4}}} = \boxed{\sqrt[5]{3^5\cdot3^4}} = \boxed{3\cdot\sqrt[5]{3^4}} = \boxed{3\sqrt[5]{81}}$$

Example 5.3-25

$$\boxed{\frac{x^{\frac{2}{3}}\cdot y^{\frac{3}{5}}}{x^{\frac{1}{2}}\cdot y^{\frac{2}{5}}}} = \boxed{x^{\frac{2}{3}}\cdot x^{-\frac{1}{2}}\cdot y^{\frac{3}{5}}\cdot y^{-\frac{2}{5}}} = \boxed{x^{\frac{2}{3}-\frac{1}{2}}\cdot y^{\frac{3}{5}-\frac{2}{5}}} = \boxed{x^{\frac{(2\cdot2)-(1\cdot3)}{3\cdot2}}\cdot y^{\frac{3-2}{5}}} = \boxed{x^{\frac{4-3}{6}}\cdot y^{\frac{1}{5}}} = \boxed{x^{\frac{1}{6}}\cdot y^{\frac{1}{5}}} = \boxed{\sqrt[6]{x}\sqrt[5]{y}}$$

Example 5.3-26

$$\boxed{\left(\frac{5\cdot a^3\cdot b^{\frac{3}{2}}}{a^2\cdot b^4}\right)^{\frac{3}{4}}} = \boxed{\frac{5^{1\times\frac{3}{4}}\cdot a^{3\times\frac{3}{4}}\cdot b^{\frac{3}{2}\times\frac{3}{4}}}{a^{2\times\frac{3}{4}}\cdot b^{4\times\frac{3}{4}}}} = \boxed{\frac{5^{\frac{3}{4}}\cdot a^{\frac{9}{4}}\cdot b^{\frac{9}{8}}}{a^{\frac{6}{4}}\cdot b^{\frac{12}{4}}}} = \boxed{\frac{5^{\frac{3}{4}}\cdot a^{\frac{9}{4}}\cdot b^{\frac{9}{8}}}{a^{\frac{6}{4}=\frac{3}{2}}\cdot b^{\frac{12}{4}=\frac{3}{1}}}} = \boxed{\frac{5^{\frac{3}{4}}\cdot a^{\frac{9}{4}}\cdot b^{\frac{9}{8}}}{a^{\frac{3}{2}}\cdot b^{\frac{3}{1}}}} = \boxed{\frac{5^{\frac{3}{4}}\cdot a^{\frac{9}{4}}\cdot a^{-\frac{3}{2}}}{b^{\frac{3}{1}}\cdot b^{-\frac{9}{8}}}}$$

$$= \boxed{\frac{5^{\frac{3}{4}} \cdot a^{\frac{9}{4}-\frac{3}{2}}}{b^{\frac{3}{1}-\frac{9}{8}}}} = \boxed{\frac{5^{\frac{3}{4}} \cdot a^{\frac{(9\cdot 2)-(3\cdot 4)}{4\cdot 2}}}{b^{\frac{(3\cdot 8)-(9\cdot 1)}{8\cdot 1}}}} = \boxed{\frac{5^{\frac{3}{4}} \cdot a^{\frac{18-12}{8}}}{b^{\frac{24-9}{8}}}} = \boxed{\frac{5^{\frac{3}{4}} \cdot a^{\frac{6}{8}}}{b^{\frac{15}{8}}}} = \boxed{\frac{5^{\frac{3}{4}} \cdot a^{\frac{3}{4}}}{b^{\frac{15}{8}}}} = \boxed{\frac{\sqrt[4]{5^3} \cdot \sqrt[4]{a^3}}{\sqrt[8]{b^{15}}}} = \boxed{\frac{\sqrt[4]{5^3} \cdot \sqrt[4]{a^3}}{\sqrt[8]{b^{8+7}}}}$$

$$= \boxed{\frac{\sqrt[4]{125} \cdot \sqrt[4]{a^3}}{\sqrt[8]{b^8 \cdot b^7}}} = \boxed{\mathbf{\frac{\sqrt[4]{125a^3}}{b\sqrt[8]{b^7}}}}$$

Example 5.3-27

$$\boxed{\left(\frac{2^5 \cdot w^2}{2^4 \cdot w \cdot w^3}\right)^{\frac{1}{3}} \cdot \left(2^2 \cdot w^3\right)^{\frac{1}{2}}} = \boxed{\left(\frac{2^5 \cdot 2^{-4}}{w^{-2} \cdot w^1 \cdot w^3}\right)^{\frac{1}{3}} \cdot \left(2^{2\times\frac{1}{2}} \cdot w^{3\times\frac{1}{2}}\right)} = \boxed{\left(\frac{2^{5-4}}{w^{-2+1+3}}\right)^{\frac{1}{3}} \cdot \left(2^{\frac{2}{2}=\frac{1}{1}} \cdot w^{\frac{3}{2}}\right)}$$

$$= \boxed{\left(\frac{2^1}{w^2}\right)^{\frac{1}{3}} \cdot \left(2^{\frac{1}{1}} \cdot w^{\frac{3}{2}}\right)} = \boxed{\frac{2^{1\times\frac{1}{3}}}{w^{2\times\frac{1}{3}}} \cdot \left(2^{\frac{1}{1}} \cdot w^{\frac{3}{2}}\right)} = \boxed{\frac{2^{\frac{1}{3}}}{w^{\frac{2}{3}}} \cdot \frac{2^{\frac{1}{1}} \cdot w^{\frac{3}{2}}}{1}} = \boxed{\frac{2^{\frac{1}{3}} \cdot 2^{\frac{1}{1}} \cdot w^{\frac{3}{2}}}{w^{\frac{2}{3}} \cdot 1}} = \boxed{\frac{2^{\frac{1}{3}} \cdot 2^{\frac{1}{1}} \cdot w^{\frac{3}{2}}}{w^{\frac{2}{3}}}}$$

$$= \boxed{2^{\frac{1}{3}} \cdot 2^{\frac{1}{1}} \cdot w^{\frac{3}{2}} \cdot w^{-\frac{2}{3}}} = \boxed{2^{\frac{1}{3}+\frac{1}{1}} \cdot w^{\frac{3}{2}-\frac{2}{3}}} = \boxed{2^{\frac{(1\cdot 1)+(1\cdot 3)}{3\cdot 1}} \cdot w^{\frac{(3\cdot 3)-(2\cdot 2)}{2\cdot 3}}} = \boxed{2^{\frac{1+3}{3}} \cdot w^{\frac{9-4}{6}}} = \boxed{2^{\frac{4}{3}} \cdot w^{\frac{5}{6}}} = \boxed{\sqrt[3]{2^4} \cdot \sqrt[6]{w^5}}$$

$$= \boxed{\sqrt[3]{2^{3+1}} \cdot \sqrt[6]{w^5}} = \boxed{\sqrt[3]{2^3 \cdot 2^1} \cdot \sqrt[6]{w^5}} = \boxed{2 \cdot \sqrt[3]{2^1} \cdot \sqrt[6]{w^5}} = \boxed{\mathbf{2\sqrt[3]{2}\,\sqrt[6]{w^5}}}$$

Example 5.3-28

$$\boxed{\frac{\left(a \cdot b^{\frac{2}{3}}\right)^4 \cdot x^{\frac{1}{4}}}{\left(a^3 \cdot x^2\right)^{\frac{1}{3}}}} = \boxed{\frac{\left(a^{1\times 4} \cdot b^{\frac{2}{3}\times 4}\right) \cdot x^{\frac{1}{4}}}{a^{3\times\frac{1}{3}} \cdot x^{2\times\frac{1}{3}}}} = \boxed{\frac{a^4 \cdot b^{\frac{8}{3}} \cdot x^{\frac{1}{4}}}{a^{\frac{3}{3}=\frac{1}{1}} \cdot x^{\frac{2}{3}}}} = \boxed{\frac{a^4 \cdot b^{\frac{8}{3}} \cdot x^{\frac{1}{4}}}{a^{\frac{1}{1}} \cdot x^{\frac{2}{3}}}} = \boxed{\frac{a^4 \cdot b^{\frac{8}{3}} \cdot x^{\frac{1}{4}}}{a^1 \cdot x^{\frac{2}{3}}}} = \boxed{\frac{a^4 \cdot a^{-1} \cdot b^{\frac{8}{3}}}{x^{\frac{2}{3}} \cdot x^{-\frac{1}{4}}}}$$

$$= \boxed{\frac{a^{4-1} \cdot b^{\frac{8}{3}}}{x^{\frac{2}{3}-\frac{1}{4}}}} = \boxed{\frac{a^3 \cdot b^{\frac{8}{3}}}{x^{\frac{(2\cdot 4)-(1\cdot 3)}{3\cdot 4}}}} = \boxed{\frac{a^3 \cdot b^{\frac{8}{3}}}{x^{\frac{8-3}{12}}}} = \boxed{\frac{a^3 \cdot b^{\frac{8}{3}}}{x^{\frac{5}{12}}}} = \boxed{\frac{a^3 \cdot \sqrt[3]{b^8}}{\sqrt[12]{x^5}}} = \boxed{\frac{a^3 \cdot \sqrt[3]{b^{3+3+2}}}{\sqrt[12]{x^5}}} = \boxed{\frac{a^3 \cdot \sqrt[3]{b^3 \cdot b^3 \cdot b^2}}{\sqrt[12]{x^5}}}$$

$$= \boxed{\frac{a^3 \cdot b \cdot b \cdot \sqrt[3]{b^2}}{\sqrt[12]{x^5}}} = \boxed{\mathbf{\frac{a^3 b^2 \sqrt[3]{b^2}}{\sqrt[12]{x^5}}}}$$

Example 5.3-29

$$\boxed{\frac{\left(a^5 \cdot x\right)^{\frac{1}{2}}}{a^{\frac{3}{4}} \cdot x}} = \boxed{\frac{a^{5\times\frac{1}{2}} \cdot x^{1\times\frac{1}{2}}}{a^{\frac{3}{4}} \cdot x}} = \boxed{\frac{a^{\frac{5}{2}} \cdot x^{\frac{1}{2}}}{a^{\frac{3}{4}} \cdot x}} = \boxed{\frac{a^{\frac{5}{2}} \cdot a^{-\frac{3}{4}}}{x \cdot x^{-\frac{1}{2}}}} = \boxed{\frac{a^{\frac{5}{2}-\frac{3}{4}}}{x^{\frac{1}{1}} \cdot x^{-\frac{1}{2}}}} = \boxed{\frac{a^{\frac{(5\cdot 4)-(3\cdot 2)}{2\cdot 4}}}{x^{\frac{(1\cdot 2)-(1\cdot 1)}{1\cdot 2}}}} = \boxed{\frac{a^{\frac{20-6}{8}}}{x^{\frac{2-1}{2}}}} = \boxed{\frac{a^{\frac{14}{8}=\frac{7}{4}}}{x^{\frac{1}{2}}}}$$

$$= \frac{a^{\frac{7}{4}}}{x^{\frac{1}{2}}} = \frac{\sqrt[4]{a^7}}{\sqrt{x}} = \frac{\sqrt[4]{a^{4+3}}}{\sqrt{x}} = \frac{\sqrt[4]{a^4 \cdot a^3}}{\sqrt{x}} = \boxed{\frac{a\sqrt[4]{a^3}}{\sqrt{x}}}$$

Example 5.3-30

$$\frac{\left(-5 \cdot x^{\frac{2}{3}}\right)^3 \cdot x^{\frac{1}{2}}}{3^2 \cdot \left(x^3 \cdot y\right)^{\frac{2}{3}}} = \frac{\left(-5^{1\times 3} \cdot x^{\frac{2}{3}\times 3}\right) \cdot x^{\frac{1}{2}}}{9 \cdot x^{3\times\frac{2}{3}} \cdot y^{1\times\frac{2}{3}}} = \frac{-5^3 \cdot x^{\frac{2}{3}\times 3} \cdot x^{\frac{1}{2}}}{9 \cdot x^{3\times\frac{2}{3}} \cdot y^{1\times\frac{2}{3}}} = \frac{(-5 \cdot -5 \cdot -5) \cdot x^{\frac{6}{3}=\frac{2}{1}} \cdot x^{\frac{1}{2}}}{9 \cdot x^{\frac{6}{3}=\frac{2}{1}} \cdot y^{\frac{2}{3}}} = \frac{-125 \cdot x^{\frac{2}{1}} \cdot x^{\frac{1}{2}}}{9 \cdot x^{\frac{2}{1}} \cdot y^{\frac{2}{3}}}$$

$$= -\frac{125 \cdot x^2 \cdot x^{\frac{1}{2}}}{9 \cdot x^2 \cdot y^{\frac{2}{3}}} = -\frac{125 \cdot x^{\frac{1}{2}}}{9 \cdot y^{\frac{2}{3}}} = -\frac{125}{9}\left(\frac{\sqrt{x}}{\sqrt[3]{y^2}}\right) = \boxed{-13\frac{8}{9}\left(\frac{\sqrt{x}}{\sqrt[3]{y^2}}\right)}$$

Practice Problems - Dividing Positive Fractional Exponents

Section 5.3 Case II Practice Problems - Divide the following positive fractional exponents:

1. $\dfrac{x^{\frac{2}{3}}}{x^2} =$

2. $\dfrac{a^2 \cdot b^{\frac{1}{3}}}{a^{\frac{1}{4}} \cdot b} =$

3. $\left(\dfrac{a^2 \cdot b^5 \cdot c^2}{a^{\frac{2}{3}}}\right)^{\frac{0}{4}} =$

4. $\dfrac{y^{\frac{3}{4}} \cdot (z \cdot w)^{\frac{5}{2}}}{y^2 \cdot w^{\frac{1}{2}}} =$

5. $\dfrac{(a \cdot b)^{\frac{1}{2}} \cdot (x \cdot y)^{\frac{2}{3}}}{a^3 \cdot y^2 \cdot x^{\frac{1}{2}}} =$

6. $\dfrac{(x \cdot y)^{\frac{2}{3}} \cdot x^2}{(x \cdot y)^{\frac{3}{2}}} =$

7. $\dfrac{2 \cdot a^{\frac{3}{2}} \cdot b^2 \cdot c^{\frac{1}{5}} \cdot a}{b \cdot c^{\frac{2}{5}}} =$

8. $\left(\dfrac{z^{\frac{1}{2}} \cdot w^3 \cdot (a \cdot b)^{\frac{3}{2}}}{a^2 \cdot w^0 \cdot z}\right)^{\frac{2}{3}} =$

9. $\dfrac{2^3 \cdot y^4 \cdot b^{\frac{3}{5}}}{2^2 \cdot b^2 \cdot y^{\frac{1}{2}}} =$

10. $\left(\dfrac{a^3 \cdot b^{\frac{3}{2}} \cdot (c \cdot d)^6}{a \cdot b^4 \cdot c^3 \cdot d}\right)^{\frac{2}{3}} =$

Case III Adding and Subtracting Positive Fractional Exponents

Addition and subtraction of positive fractional exponential expressions use the fraction techniques, outlined in chapter 2, and the exponent laws. Positive fractional exponents are added and subtracted using the following steps: (Again, note that the objective is to write the final answer in its simplified form and without a negative exponent.)

Step 1 Change the fractional exponent $x^{\frac{a}{b}}$, where x is a real number or a variable, to $\frac{x^{\frac{a}{b}}}{1}$.

For example, change $z^{\frac{2}{3}}$ and $5^{\frac{1}{4}}$ to $\frac{z^{\frac{2}{3}}}{1}$ and $\frac{5^{\frac{1}{4}}}{1}$, respectively.

Step 2 Simplify the exponential expressions by:

a. Using the fractional techniques learned in Chapter 2, and

b. Using appropriate exponent laws such as the Multiplication Law (Law I) from Table 5.3-1.

Step 3 a. Change the fractional exponents of the form $x^{\frac{a}{b}}$ to radical expressions of the form $\sqrt[b]{x^a}$. For example, change $w^{\frac{4}{5}}$ to $\sqrt[5]{w^4}$.

b. Simplify the radical expressions (see Section 4.1, Cases III and IV).

Examples with Steps

The following examples show the steps as to how fractional exponential expressions are added and subtracted:

Example 5.3-31

$$\boxed{2^{\frac{3}{5}} + \frac{1}{2^{\frac{1}{2}}}} =$$

Solution:

Step 1

$$\boxed{2^{\frac{3}{5}} + \frac{1}{2^{\frac{1}{2}}}} = \boxed{\frac{2^{\frac{3}{5}}}{1} + \frac{1}{2^{\frac{1}{2}}}}$$

Step 2

$$\boxed{\frac{2^{\frac{3}{5}}}{1} + \frac{1}{2^{\frac{1}{2}}}} = \boxed{\frac{2^{\frac{3}{5}} \cdot 2^{\frac{1}{2}} + (1 \cdot 1)}{1 \cdot 2^{\frac{1}{2}}}} = \boxed{\frac{2^{\frac{3}{5}+\frac{1}{2}} + 1}{2^{\frac{1}{2}}}} = \boxed{\frac{2^{\frac{(3\cdot 2)+(1\cdot 5)}{5\cdot 2}} + 1}{2^{\frac{1}{2}}}} = \boxed{\frac{2^{\frac{6+5}{10}} + 1}{2^{\frac{1}{2}}}} = \boxed{\frac{2^{\frac{11}{10}} + 1}{2^{\frac{1}{2}}}}$$

Step 3

$$\frac{2^{\frac{11}{10}}+1}{2^{\frac{1}{2}}}=\frac{\sqrt[10]{2^{11}}+1}{\sqrt{2}}=\frac{\sqrt[10]{2^{10+1}}+1}{\sqrt{2}}=\frac{\sqrt[10]{2^{10}\cdot 2^{1}}+1}{\sqrt{2}}=\frac{2\cdot\sqrt[10]{2^{1}}+1}{\sqrt{2}}=\frac{\mathbf{2\sqrt[10]{2}+1}}{\mathbf{\sqrt{2}}}$$

Example 5.3-32

$$a^{\frac{2}{3}}-\frac{5}{3-a^{\frac{1}{2}}}=$$

Solution:

Step 1

$$a^{\frac{2}{3}}-\frac{5}{3-a^{\frac{1}{2}}}=\frac{a^{\frac{2}{3}}}{1}-\frac{5}{3-a^{\frac{1}{2}}}$$

Step 2

$$\frac{a^{\frac{2}{3}}}{1}-\frac{5}{3-a^{\frac{1}{2}}}=\frac{\left[a^{\frac{2}{3}}\cdot\left(3-a^{\frac{1}{2}}\right)\right]-(5\cdot 1)}{1\cdot\left(3-a^{\frac{1}{2}}\right)}=\frac{\left(3\cdot a^{\frac{2}{3}}-a^{\frac{2}{3}}\cdot a^{\frac{1}{2}}\right)-5}{3-a^{\frac{1}{2}}}=\frac{3a^{\frac{2}{3}}-a^{\frac{2}{3}+\frac{1}{2}}-5}{3-a^{\frac{1}{2}}}$$

$$=\frac{3a^{\frac{2}{3}}-a^{\frac{2\cdot 2+1\cdot 3}{3\cdot 2}}-5}{3-a^{\frac{1}{2}}}=\frac{3a^{\frac{2}{3}}-a^{\frac{4+3}{6}}-5}{3-a^{\frac{1}{2}}}=\frac{3a^{\frac{2}{3}}-a^{\frac{7}{6}}-5}{3-a^{\frac{1}{2}}}$$

Step 3

$$\frac{3a^{\frac{2}{3}}-a^{\frac{7}{6}}-5}{3-a^{\frac{1}{2}}}=\frac{3\sqrt[3]{a^{2}}-\sqrt[6]{a^{6+1}}-5}{3-\sqrt{a}}=\frac{3\sqrt[3]{a^{2}}-\sqrt[6]{a^{6}\cdot a^{1}}-5}{3-\sqrt{a}}=\frac{\mathbf{3\sqrt[3]{a^{2}}-a\sqrt[6]{a}-5}}{\mathbf{3-\sqrt{a}}}$$

Example 5.3-33

$$\frac{1}{x^{\frac{2}{3}}}+x^{\frac{4}{3}}=$$

Solution:

Step 1

$$\frac{1}{x^{\frac{2}{3}}}+x^{\frac{4}{3}}=\frac{1}{x^{\frac{2}{3}}}+\frac{x^{\frac{4}{3}}}{1}$$

Step 2

$$\frac{1}{x^{\frac{2}{3}}}+\frac{x^{\frac{4}{3}}}{1}=\frac{(1\cdot 1)+\left(x^{\frac{2}{3}}\cdot x^{\frac{4}{3}}\right)}{x^{\frac{2}{3}}\cdot 1}=\frac{1+x^{\frac{2+4}{3}}}{x^{\frac{2}{3}}}=\frac{1+x^{\frac{6}{3}=\frac{2}{1}}}{x^{\frac{2}{3}}}=\frac{1+x^{\frac{2}{1}}}{x^{\frac{2}{3}}}$$

Step 3

$$\frac{1+x^{\frac{2}{1}}}{x^{\frac{2}{3}}} = \boxed{\frac{1+x^2}{\sqrt[3]{x^2}}}$$

Example 5.3-34

$$\frac{2}{x^{\frac{2}{3}}+2}+x^{\frac{2}{3}} =$$

Solution:

Step 1

$$\frac{2}{x^{\frac{2}{3}}+2}+x^{\frac{2}{3}} = \frac{2}{x^{\frac{2}{3}}+2}+\frac{x^{\frac{2}{3}}}{1}$$

Step 2

$$\frac{2}{x^{\frac{2}{3}}+2}+\frac{x^{\frac{2}{3}}}{1} = \frac{(2\cdot 1)+x^{\frac{2}{3}}\left(x^{\frac{2}{3}}+2\right)}{x^{\frac{2}{3}}+2} = \frac{2+\left(x^{\frac{2}{3}}\cdot x^{\frac{2}{3}}+2\cdot x^{\frac{2}{3}}\right)}{x^{\frac{2}{3}}+2} = \frac{2+\left(x^{\frac{2}{3}+\frac{2}{3}}+2x^{\frac{2}{3}}\right)}{x^{\frac{2}{3}}+2}$$

$$= \frac{2+x^{\frac{2+2}{3}}+2x^{\frac{2}{3}}}{x^{\frac{2}{3}}+2} = \frac{2+x^{\frac{4}{3}}+2x^{\frac{2}{3}}}{x^{\frac{2}{3}}+2}$$

Step 3

$$\frac{2+x^{\frac{4}{3}}+2x^{\frac{2}{3}}}{x^{\frac{2}{3}}+2} = \frac{2+\sqrt[3]{x^4}+2\sqrt[3]{x^2}}{\sqrt[3]{x^2}+2} = \frac{2+\sqrt[3]{x^{3+1}}+2\sqrt[3]{x^2}}{\sqrt[3]{x^2}+2} = \frac{2+\sqrt[3]{x^3\cdot x^1}+2\sqrt[3]{x^2}}{\sqrt[3]{x^2}+2}$$

$$= \boxed{\frac{2+x\sqrt[3]{x}+2\sqrt[3]{x^2}}{\sqrt[3]{x^2}+2}}$$

Example 5.3-35

$$\frac{2^{\frac{2}{3}}+5^{\frac{2}{3}}}{2}+\frac{2^{\frac{3}{5}}}{3} =$$

Solution:

Step 1

Not Applicable

Step 2

$$\frac{2^{\frac{2}{3}}+5^{\frac{2}{3}}}{2}+\frac{2^{\frac{3}{5}}}{3} = \frac{\left[3\cdot\left(2^{\frac{2}{3}}+5^{\frac{2}{3}}\right)\right]+\left(2^{\frac{3}{5}}\cdot 2\right)}{2\cdot 3} = \frac{3\cdot\left(2^{\frac{2}{3}}+5^{\frac{2}{3}}\right)+\left(2^{\frac{3}{5}}\cdot 2^{\frac{1}{1}}\right)}{6}$$

$$= \frac{3\cdot\left(2^{\frac{2}{3}}+5^{\frac{2}{3}}\right)+2^{\frac{3}{5}+\frac{1}{1}}}{6} = \frac{3\cdot\left(2^{\frac{2}{3}}+5^{\frac{2}{3}}\right)+2^{\frac{(3\cdot1)+(1\cdot5)}{5\cdot1}}}{6} = \frac{3\cdot\left(2^{\frac{2}{3}}+5^{\frac{2}{3}}\right)+2^{\frac{3+5}{5}}}{6}$$

$$= \frac{3\cdot\left(2^{\frac{2}{3}}+5^{\frac{2}{3}}\right)+2^{\frac{8}{5}}}{6}$$

Step 3

$$\frac{3\cdot\left(2^{\frac{2}{3}}+5^{\frac{5}{3}}\right)+2^{\frac{8}{5}}}{6} = \frac{3\left(\sqrt[3]{2^2}+\sqrt[3]{5^2}\right)+\sqrt[5]{2^8}}{6} = \frac{3\sqrt[3]{4}+3\sqrt[3]{25}+\sqrt[5]{2^{5+3}}}{6}$$

$$= \frac{3\sqrt[3]{4}+3\sqrt[3]{25}+\sqrt[5]{2^5\cdot2^3}}{6} = \frac{3\sqrt[3]{4}+3\sqrt[3]{25}+2\sqrt[5]{2^3}}{6} = \mathbf{\frac{3\sqrt[3]{4}+3\sqrt[3]{25}+2\sqrt[5]{8}}{6}}$$

Additional Examples - Adding and Subtracting Positive Fractional Exponents

The following examples further illustrate addition and subtraction of positive fractional exponential expressions:

Example 5.3-36

$$\frac{1}{z^{\frac{2}{3}}}+z^{\frac{2}{5}} = \frac{1}{z^{\frac{2}{3}}}+\frac{z^{\frac{2}{5}}}{1} = \frac{(1\cdot1)+z^{\frac{2}{5}}\cdot z^{\frac{2}{3}}}{z^{\frac{2}{3}}\cdot1} = \frac{1+z^{\frac{2}{5}+\frac{2}{3}}}{z^{\frac{2}{3}}} = \frac{1+z^{\frac{(2\cdot3)+(2\cdot5)}{5\cdot3}}}{z^{\frac{2}{3}}} = \frac{1+z^{\frac{6+10}{15}}}{z^{\frac{2}{3}}} = \frac{1+z^{\frac{16}{15}}}{z^{\frac{2}{3}}}$$

$$= \frac{1+\sqrt[15]{z^{16}}}{\sqrt[3]{z^2}} = \frac{1+\sqrt[15]{z^{15+1}}}{\sqrt[3]{z^2}} = \frac{1+\sqrt[15]{z^{15}\cdot z^1}}{\sqrt[3]{z^2}} = \mathbf{\frac{1+z\sqrt[15]{z}}{\sqrt[3]{z^2}}}$$

Example 5.3-37

$$\frac{a^{\frac{3}{5}}-b^{\frac{1}{4}}}{a}+\frac{1}{a^{\frac{2}{3}}} = \frac{a^{\frac{2}{3}}\cdot\left(a^{\frac{3}{5}}-b^{\frac{1}{4}}\right)+(1\cdot a)}{a\cdot a^{\frac{2}{3}}} = \frac{a^{\frac{2}{3}}\cdot a^{\frac{3}{5}}-a^{\frac{2}{3}}\cdot b^{\frac{1}{4}}+a}{a^{\frac{1}{1}}\cdot a^{\frac{2}{3}}} = \frac{a^{\frac{2}{3}+\frac{3}{5}}-a^{\frac{2}{3}}\cdot b^{\frac{1}{4}}+a}{a^{\frac{1}{1}+\frac{2}{3}}}$$

$$= \frac{a^{\frac{(2\cdot5)+(3\cdot3)}{3\cdot5}}-a^{\frac{2}{3}}\cdot b^{\frac{1}{4}}+a}{a^{\frac{(1\cdot3)+(2\cdot1)}{1\cdot3}}} = \frac{a^{\frac{10+9}{15}}-a^{\frac{2}{3}}\cdot b^{\frac{1}{4}}+a}{a^{\frac{3+2}{3}}} = \frac{a^{\frac{19}{15}}-a^{\frac{2}{3}}\cdot b^{\frac{1}{4}}+a}{a^{\frac{5}{3}}} = \frac{\sqrt[15]{a^{19}}-\sqrt[3]{a^2}\sqrt[4]{b}+a}{\sqrt[3]{a^5}}$$

$$= \frac{\sqrt[15]{a^{15+4}} - \sqrt[3]{a^2}\sqrt[4]{b} + a}{\sqrt[3]{a^{3+2}}} = \frac{\sqrt[15]{a^{15} \cdot a^4} - \sqrt[3]{a^2}\sqrt[4]{b} + a}{\sqrt[3]{a^3 \cdot a^2}} = \boxed{\frac{a\sqrt[15]{a^4} - \sqrt[3]{a^2}\sqrt[4]{b} + a}{a\sqrt[3]{a^2}}}$$

Example 5.3-38

$$\frac{x + 3^{\frac{2}{5}}}{x^{\frac{2}{3}}} + \frac{1}{x^{\frac{1}{3}}} = \frac{x^{\frac{1}{3}} \cdot \left(x + 3^{\frac{2}{5}}\right) + \left(1 \cdot x^{\frac{2}{3}}\right)}{x^{\frac{2}{3}} \cdot x^{\frac{1}{3}}} = \frac{x^{\frac{1}{3}} \cdot x + 3^{\frac{2}{5}} \cdot x^{\frac{1}{3}} + x^{\frac{2}{3}}}{x^{\frac{2}{3}+\frac{1}{3}}} = \frac{x^{\frac{1}{3}} \cdot x^{1} + 3^{\frac{2}{5}} \cdot x^{\frac{1}{3}} + x^{\frac{2}{3}}}{x^{\frac{2+1}{3}}}$$

$$= \frac{x^{\frac{1}{3}+\frac{1}{1}} + 3^{\frac{2}{5}} \cdot x^{\frac{1}{3}} + x^{\frac{2}{3}}}{x^{\frac{3}{3}=\frac{1}{1}}} = \frac{x^{\frac{(1\cdot1)+(1\cdot3)}{3\cdot1}} + 3^{\frac{2}{5}} \cdot x^{\frac{1}{3}} + x^{\frac{2}{3}}}{x^{\frac{1}{1}}} = \frac{x^{\frac{1+3}{3}} + 3^{\frac{2}{5}} \cdot x^{\frac{1}{3}} + x^{\frac{2}{3}}}{x} = \frac{x^{\frac{4}{3}} + 3^{\frac{2}{5}} \cdot x^{\frac{1}{3}} + x^{\frac{2}{3}}}{x}$$

$$= \frac{\sqrt[3]{x^4} + \sqrt[5]{3^2}\sqrt[3]{x} + \sqrt[3]{x^2}}{x} = \frac{\sqrt[3]{x^{3+1}} + \sqrt[5]{9}\sqrt[3]{x} + \sqrt[3]{x^2}}{x} = \frac{\sqrt[3]{x^3 \cdot x^1} + \sqrt[5]{9}\sqrt[3]{x} + \sqrt[3]{x^2}}{x} = \boxed{\frac{x\sqrt[3]{x} + \sqrt[5]{9}\sqrt[3]{x} + \sqrt[3]{x^2}}{x}}$$

Example 5.3-39

$$\frac{x^{\frac{3}{4}}}{x^{\frac{2}{3}} - y} - y^{\frac{3}{5}} = \frac{x^{\frac{3}{4}}}{x^{\frac{2}{3}} - y} - \frac{y^{\frac{3}{5}}}{1} = \frac{\left(x^{\frac{3}{4}} \cdot 1\right) - y^{\frac{3}{5}} \cdot \left(x^{\frac{2}{3}} - y\right)}{\left(x^{\frac{2}{3}} - y\right) \cdot 1} = \frac{x^{\frac{3}{4}} - x^{\frac{2}{3}} \cdot y^{\frac{3}{5}} + y^{\frac{3}{5}} \cdot y}{x^{\frac{2}{3}} - y}$$

$$= \frac{x^{\frac{3}{4}} - x^{\frac{2}{3}} \cdot y^{\frac{3}{5}} + y^{\frac{3}{5}} \cdot y^{\frac{1}{1}}}{x^{\frac{2}{3}} - y} = \frac{x^{\frac{3}{4}} - x^{\frac{2}{3}}\, y^{\frac{3}{5}} + y^{\frac{3}{5}+\frac{1}{1}}}{x^{\frac{2}{3}} - y} = \frac{x^{\frac{3}{4}} - x^{\frac{2}{3}}\, y^{\frac{3}{5}} + y^{\frac{(3\cdot1)+(1\cdot5)}{5\cdot1}}}{x^{\frac{2}{3}} - y} = \frac{x^{\frac{3}{4}} - x^{\frac{2}{3}}\, y^{\frac{3}{5}} + y^{\frac{3+5}{5}}}{x^{\frac{2}{3}} - y}$$

$$= \frac{x^{\frac{3}{4}} - x^{\frac{2}{3}}\, y^{\frac{3}{5}} + y^{\frac{8}{5}}}{x^{\frac{2}{3}} - y} = \frac{\sqrt[4]{x^3} - \sqrt[3]{x^2}\sqrt[5]{y^3} + \sqrt[5]{y^8}}{\sqrt[3]{x^2} - y} = \frac{\sqrt[4]{x^3} - \sqrt[3]{x^2}\sqrt[5]{y^3} + \sqrt[5]{y^{5+3}}}{\sqrt[3]{x^2} - y}$$

$$= \frac{\sqrt[4]{x^3} - \sqrt[3]{x^2}\sqrt[5]{y^3} + \sqrt[5]{y^5 \cdot y^3}}{\sqrt[3]{x^2} - y} = \boxed{\frac{\sqrt[4]{x^3} - \sqrt[3]{x^2}\sqrt[5]{y^3} + y\sqrt[5]{y^3}}{\sqrt[3]{x^2} - y}}$$

Example 5.3-40

$$\frac{x^{\frac{1}{3}} + x^{\frac{2}{3}}}{x^{\frac{1}{2}}} + \frac{1}{x^{\frac{2}{5}}} = \frac{x^{\frac{2}{5}} \cdot \left(x^{\frac{1}{3}} + x^{\frac{2}{3}}\right) + 1 \cdot x^{\frac{1}{2}}}{x^{\frac{1}{2}} \cdot x^{\frac{2}{5}}} = \frac{x^{\frac{2}{5}} \cdot x^{\frac{1}{3}} + x^{\frac{2}{5}} \cdot x^{\frac{2}{3}} + x^{\frac{1}{2}}}{x^{\frac{1}{2}+\frac{2}{5}}} = \frac{x^{\frac{2}{5}+\frac{1}{3}} + x^{\frac{2}{5}+\frac{2}{3}} + x^{\frac{1}{2}}}{x^{\frac{(1\cdot5)+(2\cdot2)}{2\cdot5}}}$$

$$= \frac{x^{\frac{(2\cdot3)+(1\cdot5)}{5\cdot3}} + x^{\frac{(2\cdot3)+(2\cdot5)}{5\cdot3}} + x^{\frac{1}{2}}}{x^{\frac{5+4}{10}}} = \frac{x^{\frac{6+5}{15}} + x^{\frac{6+10}{15}} + x^{\frac{1}{2}}}{x^{\frac{9}{10}}} = \frac{x^{\frac{11}{15}} + x^{\frac{16}{15}} + x^{\frac{1}{2}}}{x^{\frac{9}{10}}} = \frac{\sqrt[15]{x^{11}} + \sqrt[15]{x^{16}} + \sqrt{x}}{\sqrt[10]{x^9}}$$

$$= \frac{\sqrt[15]{x^{11}} + \sqrt[15]{x^{15+1}} + \sqrt{x}}{\sqrt[10]{x^9}} = \frac{\sqrt[15]{x^{11}} + \sqrt[15]{x^{15}\cdot x^1} + \sqrt{x}}{\sqrt[10]{x^9}} = \boxed{\frac{\sqrt[15]{x^{11}} + x\sqrt[15]{x} + \sqrt{x}}{\sqrt[10]{x^9}}}$$

Example 5.3-41

$$\frac{1-y^{\frac{1}{4}}}{3^{\frac{2}{3}}} + \frac{x^{\frac{2}{3}}}{3^{\frac{2}{3}}} - \frac{2}{5} = \left(\frac{1-y^{\frac{1}{4}}}{3^{\frac{2}{3}}} + \frac{x^{\frac{2}{3}}}{3^{\frac{2}{3}}}\right) - \frac{2}{5} = \left(\frac{1-y^{\frac{1}{4}}+x^{\frac{2}{3}}}{3^{\frac{2}{3}}}\right) - \frac{2}{5} = \frac{5\cdot\left(1-y^{\frac{1}{4}}+x^{\frac{2}{3}}\right) - \left(2\cdot3^{\frac{2}{3}}\right)}{5\cdot3^{\frac{2}{3}}}$$

$$= \frac{5-5y^{\frac{1}{4}}+5x^{\frac{2}{3}}-2\cdot3^{\frac{2}{3}}}{5\cdot3^{\frac{2}{3}}} = \frac{5-5\sqrt[4]{y}+5\sqrt[3]{x^2}-2\sqrt[3]{3^2}}{5\sqrt[3]{3^2}} = \boxed{\frac{5\sqrt[3]{x^2}-5\sqrt[4]{y}-2\sqrt[3]{9}+5}{5\sqrt[3]{9}}}$$

Example 5.3-42

$$x^{\frac{2}{5}} + \frac{xy^{\frac{2}{3}}}{(x+y)^{\frac{1}{2}}} = \frac{x^{\frac{2}{5}}}{1} + \frac{xy^{\frac{2}{3}}}{(x+y)^{\frac{1}{2}}} = \frac{x^{\frac{2}{5}}\cdot(x+y)^{\frac{1}{2}} + xy^{\frac{2}{3}}\cdot1}{1\cdot(x+y)^{\frac{1}{2}}} = \frac{x^{\frac{2}{5}}(x+y)^{\frac{1}{2}} + xy^{\frac{2}{3}}}{(x+y)^{\frac{1}{2}}} = \boxed{\frac{\sqrt[5]{x^2}\left(\sqrt{x+y}\right)+x\sqrt[3]{y^2}}{\sqrt{x+y}}}$$

Example 5.3-43

$$\frac{x^{\frac{2}{3}}}{x^{\frac{1}{3}}-y} + \frac{x}{y^{\frac{1}{2}}} = \frac{\left(x^{\frac{2}{3}}\cdot y^{\frac{1}{2}}\right)+x\cdot\left(x^{\frac{1}{3}}-y\right)}{\left(x^{\frac{1}{3}}-y\right)\cdot y^{\frac{1}{2}}} = \frac{x^{\frac{2}{3}}\cdot y^{\frac{1}{2}}+x\cdot x^{\frac{1}{3}}-x\cdot y}{x^{\frac{1}{3}}\cdot y^{\frac{1}{2}}-y\cdot y^{\frac{1}{2}}} = \frac{x^{\frac{2}{3}}\cdot y^{\frac{1}{2}}+x^{\frac{1}{1}}\cdot x^{\frac{1}{3}}-x\cdot y}{x^{\frac{1}{3}}\cdot y^{\frac{1}{2}}-y^{\frac{1}{1}}\cdot y^{\frac{1}{2}}}$$

$$= \frac{x^{\frac{2}{3}}y^{\frac{1}{2}}+x^{\frac{1}{1}+\frac{1}{3}}-xy}{x^{\frac{1}{3}}y^{\frac{1}{2}}-y^{\frac{1}{1}+\frac{1}{2}}} = \frac{x^{\frac{2}{3}}y^{\frac{1}{2}}+x^{\frac{(1\cdot3)+(1\cdot1)}{1\cdot3}}-xy}{x^{\frac{1}{3}}y^{\frac{1}{2}}-y^{\frac{(1\cdot2)+(1\cdot1)}{1\cdot2}}} = \frac{x^{\frac{2}{3}}y^{\frac{1}{2}}+x^{\frac{3+1}{3}}-xy}{x^{\frac{1}{3}}y^{\frac{1}{2}}-y^{\frac{2+1}{2}}} = \frac{x^{\frac{2}{3}}y^{\frac{1}{2}}+x^{\frac{4}{3}}-xy}{x^{\frac{1}{3}}y^{\frac{1}{2}}-y^{\frac{3}{2}}}$$

$$= \frac{\sqrt[3]{x^2}\sqrt{y}+\sqrt[3]{x^4}-xy}{\sqrt[3]{x}\sqrt{y}-\sqrt[2]{y^3}} = \frac{\sqrt[3]{x^2}\sqrt{y}+\sqrt[3]{x^{3+1}}-xy}{\sqrt[3]{x}\sqrt{y}-\sqrt[2]{y^{2+1}}} = \frac{\sqrt[3]{x^2}\sqrt{y}+\sqrt[3]{x^3\cdot x^1}-xy}{\sqrt[3]{x}\sqrt{y}-\sqrt{y^2\cdot y^1}} = \boxed{\frac{\sqrt[3]{x^2}\sqrt{y}+x\sqrt[3]{x}-xy}{\sqrt[3]{x}\sqrt{y}-y\sqrt{y}}}$$

Example 5.3-44

$$\boxed{\frac{m+n}{m^{\frac{1}{2}}-n^{\frac{2}{3}}}+\frac{m^{\frac{2}{3}}}{n^{\frac{2}{3}}}}=\boxed{\frac{\left[n^{\frac{2}{3}}\cdot(m+n)\right]+m^{\frac{2}{3}}\cdot\left(m^{\frac{1}{2}}-n^{\frac{2}{3}}\right)}{\left(m^{\frac{1}{2}}-n^{\frac{2}{3}}\right)\cdot n^{\frac{2}{3}}}}=\boxed{\frac{m\cdot n^{\frac{2}{3}}+n\cdot n^{\frac{2}{3}}+m^{\frac{2}{3}}\cdot m^{\frac{1}{2}}-m^{\frac{2}{3}}\cdot n^{\frac{2}{3}}}{m^{\frac{1}{2}}\cdot n^{\frac{2}{3}}-n^{\frac{2}{3}}\cdot n^{\frac{2}{3}}}}$$

$$=\boxed{\frac{m\cdot n^{\frac{2}{3}}+n^{1}\cdot n^{\frac{2}{3}}+m^{\frac{2}{3}+\frac{1}{2}}-m^{\frac{2}{3}}\cdot n^{\frac{2}{3}}}{m^{\frac{1}{2}}\cdot n^{\frac{2}{3}}-n^{\frac{2}{3}+\frac{2}{3}}}}=\boxed{\frac{mn^{\frac{2}{3}}+n^{\frac{1}{1}+\frac{2}{3}}+m^{\frac{(2\cdot2)+(1\cdot3)}{3\cdot2}}-m^{\frac{2}{3}}n^{\frac{2}{3}}}{m^{\frac{1}{2}}n^{\frac{2}{3}}-n^{\frac{2+2}{3}}}}$$

$$=\boxed{\frac{mn^{\frac{2}{3}}+n^{\frac{(1\cdot3)+(1\cdot2)}{1\cdot3}}+m^{\frac{4+3}{6}}-m^{\frac{2}{3}}n^{\frac{2}{3}}}{m^{\frac{1}{2}}n^{\frac{2}{3}}-n^{\frac{4}{3}}}}=\boxed{\frac{mn^{\frac{2}{3}}+n^{\frac{3+2}{1\cdot3}}+m^{\frac{7}{6}}-m^{\frac{2}{3}}n^{\frac{2}{3}}}{m^{\frac{1}{2}}n^{\frac{2}{3}}-n^{\frac{4}{3}}}}=\boxed{\frac{mn^{\frac{2}{3}}+n^{\frac{5}{3}}+m^{\frac{7}{6}}-m^{\frac{2}{3}}n^{\frac{2}{3}}}{m^{\frac{1}{2}}n^{\frac{2}{3}}-n^{\frac{4}{3}}}}$$

$$=\boxed{\frac{m\sqrt[3]{n^2}+\sqrt[3]{n^5}+\sqrt[6]{m^7}-\sqrt[3]{m^2}\sqrt[3]{n^2}}{\sqrt{m}\sqrt[3]{n^2}-\sqrt[3]{n^4}}}=\boxed{\frac{m\sqrt[3]{n^2}+\sqrt[3]{n^{3+2}}+\sqrt[6]{m^{6+1}}-\sqrt[3]{m^2}\sqrt[3]{n^2}}{\sqrt{m}\sqrt[3]{n^2}-\sqrt[3]{n^{3+1}}}}$$

$$=\boxed{\frac{m\sqrt[3]{n^2}+\sqrt[3]{n^3\cdot n^2}+\sqrt[6]{m^6\cdot m^1}-\sqrt[3]{m^2}\sqrt[3]{n^2}}{\sqrt{m}\sqrt[3]{n^2}-\sqrt[3]{n^3\cdot n^1}}}=\boxed{\frac{m\sqrt[3]{n^2}+n\sqrt[3]{n^2}+m\sqrt[6]{m}-\sqrt[3]{m^2}\sqrt[3]{n^2}}{\sqrt{m}\sqrt[3]{n^2}-n\sqrt[3]{n}}}$$

$$=\boxed{\boldsymbol{\frac{m\sqrt[3]{n^2}+n\sqrt[3]{n^2}+m\sqrt[6]{m}-\sqrt[3]{m^2n^2}}{\sqrt{m}\sqrt[3]{n^2}-n\sqrt[3]{n}}}}$$

Example 5.3-45

$$\boxed{\frac{w^{\frac{2}{5}}}{3+w^{\frac{2}{3}}}-\frac{w^2}{w^{\frac{2}{3}}+1}}=\boxed{\frac{w^{\frac{2}{5}}\cdot\left(w^{\frac{2}{3}}+1\right)-w^2\cdot\left(3+w^{\frac{2}{3}}\right)}{\left(3+w^{\frac{2}{3}}\right)\cdot\left(w^{\frac{2}{3}}+1\right)}}=\boxed{\frac{w^{\frac{2}{5}}\cdot w^{\frac{2}{3}}+w^{\frac{2}{5}}\cdot 1-3\cdot w^2-w^2\cdot w^{\frac{2}{3}}}{3\cdot w^{\frac{2}{3}}+3\cdot 1+w^{\frac{2}{3}}\cdot w^{\frac{2}{3}}+w^{\frac{2}{3}}\cdot 1}}$$

$$=\boxed{\frac{w^{\frac{2}{5}+\frac{2}{3}}+w^{\frac{2}{5}}-3w^2-w^{2+\frac{2}{3}}}{3w^{\frac{2}{3}}+3+w^{\frac{2}{3}+\frac{2}{3}}+w^{\frac{2}{3}}}}=\boxed{\frac{w^{\frac{(2\cdot3)+(2\cdot5)}{5\cdot3}}+w^{\frac{2}{5}}-3w^2-w^{\frac{2}{1}+\frac{2}{3}}}{3+3w^{\frac{2}{3}}+w^{\frac{2}{3}}+w^{\frac{2+2}{3}}}}=\boxed{\frac{w^{\frac{6+10}{15}}+w^{\frac{2}{5}}-3w^2-w^{\frac{(2\cdot3)+(2\cdot1)}{1\cdot3}}}{3+(3+1)w^{\frac{2}{3}}+w^{\frac{4}{3}}}}$$

$$=\boxed{\frac{w^{\frac{16}{15}}+w^{\frac{2}{5}}-3w^2-w^{\frac{6+2}{3}}}{3+4w^{\frac{2}{3}}+w^{\frac{4}{3}}}}=\boxed{\frac{w^{\frac{16}{15}}+w^{\frac{2}{5}}-w^{\frac{8}{3}}-3w^2}{3+4w^{\frac{2}{3}}+w^{\frac{4}{3}}}}=\boxed{\frac{\sqrt[15]{w^{16}}+\sqrt[5]{w^2}-\sqrt[3]{w^8}-3w^2}{3+4\sqrt[3]{w^2}+\sqrt[3]{w^4}}}$$

$$= \boxed{\frac{\sqrt[15]{w^{15+1}} + \sqrt[5]{w^2} - \sqrt[3]{w^{3+3+2}} - 3w^2}{3 + 4\sqrt[3]{w^2} + \sqrt[3]{w^{3+1}}}} = \boxed{\frac{\sqrt[15]{w^{15} \cdot w^1} + \sqrt[5]{w^2} - \sqrt[3]{\left(w^3 \cdot w^3\right) \cdot w^2} - 3w^2}{3 + 4\sqrt[3]{w^2} + \sqrt[3]{w^3 \cdot w^1}}}$$

$$= \boxed{\frac{w^1\sqrt[15]{w} + \sqrt[5]{w^2} - (w \cdot w)\sqrt[3]{w^2} - 3w^2}{3 + 4\sqrt[3]{w^2} + w\sqrt[3]{w}}} = \boxed{\frac{\mathbf{w\sqrt[15]{w} + \sqrt[5]{w^2} - w^2\sqrt[3]{w^2} - 3w^2}}{\mathbf{3 + 4\sqrt[3]{w^2} + w\sqrt[3]{w}}}}$$

Practice Problems - Adding and Subtracting Positive Fractional Exponents

Section 5.3 Case III Practice Problems - Simplify the following positive fractional exponential expressions:

1. $\dfrac{4^{\frac{1}{2}} - 4^{\frac{3}{5}}}{5} + 2^{\frac{2}{3}} =$
2. $\dfrac{1}{x^{\frac{3}{5}} + x^{\frac{2}{3}}} + x^{\frac{2}{3}} =$
3. $\dfrac{2}{x^{\frac{2}{3}}} + x^{\frac{1}{3}} =$
4. $\dfrac{n^{\frac{2}{3}}}{n^{\frac{1}{3}} + n^{\frac{2}{3}}} + n^2 =$
5. $\dfrac{y^{\frac{2}{3}}}{y - y^{\frac{3}{4}}} + \dfrac{y^{\frac{1}{2}}}{2} =$
6. $\dfrac{w^{\frac{2}{5}}}{w^{\frac{1}{2}} + w^{\frac{2}{3}}} + \dfrac{1}{w} =$
7. $\dfrac{x + y^{\frac{2}{3}}}{x - y^{\frac{2}{5}}} + \dfrac{1}{x^{\frac{2}{3}}} =$
8. $\dfrac{1}{a^{\frac{1}{2}} - b^{\frac{1}{2}}} + \dfrac{b^{\frac{3}{4}}}{3} =$
9. $\dfrac{x^{\frac{2}{5}}}{x^{\frac{2}{3}} - 3} + \dfrac{1}{y^{\frac{1}{3}}} =$
10. $\dfrac{a^{\frac{2}{3}} + b^{\frac{1}{2}}}{a - b} - a^{\frac{2}{5}} =$

5.4 Operations with Negative Fractional Exponents

The Negative Power Law is needed, in addition to the other exponent laws (shown in Table 5.4 - 1), to proceed with simplification of negative fractional exponents. The Negative Power Law states that a base raised to a negative fractional exponent is equal to one divided by the same base raised to the positive fractional exponent, or vice versa (see Section 5.2, Cases I and II for examples).

$$x^{-\frac{a}{b}} = \frac{1}{x^{\frac{a}{b}}}$$

and

$$x^{\frac{a}{b}} = \frac{1}{x^{-\frac{a}{b}}} \quad \text{since} \quad x^{\frac{a}{b}} = \frac{1}{x^{-\frac{a}{b}}} = \frac{1}{\frac{1}{x^{\frac{a}{b}}}} = \frac{\frac{1}{1}}{\frac{1}{x^{\frac{a}{b}}}} = \frac{1 \times x^{\frac{a}{b}}}{1 \times 1} = \frac{x^{\frac{a}{b}}}{1} = x^{\frac{a}{b}}.$$

Note that the objective is to write the final answer without a negative fractional exponent. To achieve this, the exponent laws are used when simplifying negative fractional exponents.

Table 5.4-1: Fractional Exponent Laws 1 through 6 (Negative Fractional Exponents)

I. **Multiplication**	$x^{-\frac{a}{b}} \cdot x^{-\frac{c}{d}} = x^{-\frac{a}{b}-\frac{c}{d}}$	When multiplying negative fractional exponential terms, if bases x are the same, add the negative exponents $-\frac{a}{b}$ and $-\frac{c}{d}$.
II. **Power of a Power**	$\left(x^{-\frac{a}{b}}\right)^{-\frac{c}{d}} = x^{-\frac{a}{b} \times -\frac{c}{d}}$	When raising a negative fractional exponential term to a negative fractional power, multiply the negative exponents $-\frac{a}{b}$ and $-\frac{c}{d}$.
III. **Power of a Product**	$(x \cdot y)^{-\frac{a}{b}} = x^{-\frac{a}{b}} \cdot y^{-\frac{a}{b}}$	When raising a product to a negative fractional power, raise each factor x and y to the negative exponent $-\frac{a}{b}$.
IV. **Power of a Fraction**	$\left(\frac{x}{y}\right)^{-\frac{a}{b}} = \frac{x^{-\frac{a}{b}}}{y^{-\frac{a}{b}}}$	When raising a fraction to a negative fractional power, raise the numerator and the denominator to the negative exponent $-\frac{a}{b}$.

Table 5.4-1: Fractional Exponent Laws 1 through 6 (Negative Fractional Exponents) - Continued

V. **Division**	$\dfrac{x^{-\frac{a}{b}}}{x^{-\frac{c}{d}}} = x^{-\frac{a}{b}} \cdot x^{+\frac{c}{d}}$ $= x^{-\frac{a}{b}+\frac{c}{d}}$	When dividing negative fractional exponential terms, if the bases x are the same, add the exponents $-\frac{a}{b}$ and $\frac{c}{d}$.
VI. **Negative Power**	$x^{-\frac{a}{b}} = \dfrac{1}{x^{\frac{a}{b}}}$	A non-zero based x raised to the $-\frac{a}{b}$ power equals 1 divided by the base x raised to the $\frac{a}{b}$ power.

In this section students learn how to multiply (Case I), divide (Case II), and add or subtract (Case III) negative fractional exponents by one another.

Case I Multiplying Negative Fractional Exponents

Negative fractional exponents are multiplied by one another using the following steps:

Step 1 Apply the Power of a Power and/or the Power of a Product Law (Laws II and III) from Table 5.4-1.

Step 2 Apply the Multiplication Law (Law I) from Table 5.4-1 and simplify the fractional exponential expressions by adding the exponents with similar bases. (Review sections 2.2, 2.3, and 2.4 for addition, subtraction, and multiplication of integer fractions.)

Step 3 Change the negative fractional exponents to positive fractional exponents.

Step 4 Change the fractional exponential expressions to radical expressions (see Section 5.1).

Examples with Steps

The following examples show the steps as to how negative fractional exponents are multiplied by one another:

Example 5.4-1

$$\left(a^3\right)^{-\frac{2}{3}} \cdot \left(a^2 \cdot b^{-3}\right)^{\frac{1}{6}} =$$

Solution:

Step 1

$$\left(a^3\right)^{-\frac{2}{3}} \cdot \left(a^2 \cdot b^{-3}\right)^{\frac{1}{6}} = \left(a^{3 \times -\frac{2}{3}}\right) \cdot \left(a^{2 \times \frac{1}{6}} \cdot b^{-3 \times \frac{1}{6}}\right) = a^{-\frac{6}{3}} \cdot \left(a^{\frac{2}{6}} \cdot b^{-\frac{3}{6}}\right)$$

Step 2

$$a^{-\frac{6}{3}} \cdot \left(a^{\frac{2}{6}} \cdot b^{-\frac{3}{6}}\right) = a^{-\frac{6}{3}=-\frac{2}{1}} \cdot a^{\frac{2}{6}=\frac{1}{3}} \cdot b^{-\frac{3}{6}=-\frac{1}{2}} = a^{-\frac{2}{1}} \cdot a^{\frac{1}{3}} \cdot b^{-\frac{1}{2}} = a^{-\frac{2}{1}+\frac{1}{3}} \cdot b^{-\frac{1}{2}}$$

$$= \boxed{a^{\frac{(-2\cdot3)+(1\cdot1)}{1\cdot3}} \cdot b^{-\frac{1}{2}}} = \boxed{a^{\frac{-6+1}{3}} \cdot b^{-\frac{1}{2}}} = \boxed{a^{-\frac{5}{3}} \cdot b^{-\frac{1}{2}}}$$

Step 3

$$\boxed{a^{-\frac{5}{3}} \cdot b^{-\frac{1}{2}}} = \boxed{\frac{1}{a^{\frac{5}{3}}} \cdot \frac{1}{b^{\frac{1}{2}}}}$$

Step 4

$$\boxed{\frac{1}{a^{\frac{5}{3}}} \cdot \frac{1}{b^{\frac{1}{2}}}} = \boxed{\frac{1}{\sqrt[3]{a^5}} \cdot \frac{1}{\sqrt{b}}} = \boxed{\frac{1}{\sqrt[3]{a^{3+2}}} \cdot \frac{1}{\sqrt[2]{b^1}}} = \boxed{\frac{1}{\sqrt[3]{a^3 \cdot a^2}} \cdot \frac{1}{\sqrt{b}}} = \boxed{\mathbf{\frac{1}{a\sqrt[3]{a^2}\sqrt{b}}}}$$

Example 5.4-2

$$\boxed{(64)^{-\frac{1}{3}} \cdot \left(16 \cdot b^{-2}\right)^{-\frac{1}{4}}} =$$

Solution:

Step 1

$$\boxed{(64)^{-\frac{1}{3}} \cdot \left(16 \cdot b^{-2}\right)^{-\frac{1}{4}}} = \boxed{(64)^{-\frac{1}{3}} \cdot \left(16^{1\times-\frac{1}{4}} \cdot b^{-2\times-\frac{1}{4}}\right)} = \boxed{(64)^{-\frac{1}{3}} \cdot \left(16^{-\frac{1}{4}} \cdot b^{\frac{2}{4}}\right)}$$

Step 2

$$\boxed{(64)^{-\frac{1}{3}} \cdot \left(16^{-\frac{1}{4}} \cdot b^{\frac{2}{4}}\right)} = \boxed{\left(4^3\right)^{-\frac{1}{3}} \cdot \left(2^4\right)^{-\frac{1}{4}} \cdot b^{\frac{2}{4}=\frac{1}{2}}} = \boxed{\left(4^3\right)^{-\frac{1}{3}} \cdot \left(2^4\right)^{-\frac{1}{4}} \cdot b^{\frac{1}{2}}}$$

$$= \boxed{4^{3\times-\frac{1}{3}} \cdot 2^{4\times-\frac{1}{4}} \cdot b^{\frac{1}{2}}} = \boxed{4^{-\frac{3}{3}=-\frac{1}{1}} \cdot 2^{-\frac{4}{4}=-\frac{1}{1}} \cdot b^{\frac{1}{2}}} = \boxed{4^{-\frac{1}{1}} \cdot 2^{-\frac{1}{1}} \cdot b^{\frac{1}{2}}} = \boxed{4^{-1} \cdot 2^{-1} \cdot b^{\frac{1}{2}}}$$

Step 3

$$\boxed{4^{-1} \cdot 2^{-1} \cdot b^{\frac{1}{2}}} = \boxed{\frac{1}{4} \cdot \frac{1}{2} \cdot b^{\frac{1}{2}}} = \boxed{\frac{1}{8} \cdot b^{\frac{1}{2}}} = \boxed{\frac{b^{\frac{1}{2}}}{8}}$$

Step 4

$$\boxed{\frac{b^{\frac{1}{2}}}{8}} = \boxed{\frac{\sqrt[2]{b^1}}{8}} = \boxed{\mathbf{\frac{\sqrt{b}}{8}}}$$

Example 5.4-3

$$\boxed{\left(x^a\right)^{-\frac{3}{2}} \cdot \left(x^{-b} \cdot x^{3a}\right)^{-\frac{1}{3}}} =$$

Solution:

Step 1

$$\boxed{\left(x^a\right)^{-\frac{3}{2}} \cdot \left(x^{-b} \cdot x^{3a}\right)^{-\frac{1}{3}}} = \boxed{\left(x^{a\times-\frac{3}{2}}\right) \cdot \left(x^{-b\times-\frac{1}{3}} \cdot x^{3a\times-\frac{1}{3}}\right)} = \boxed{x^{-\frac{3a}{2}} \cdot x^{\frac{b}{3}} \cdot x^{-\frac{3a}{3}}}$$

Step 2

$$\boxed{x^{-\frac{3a}{2}} \cdot x^{\frac{b}{3}} \cdot x^{-\frac{3a}{3}}} = \boxed{x^{-\frac{3a}{2}} \cdot x^{-\frac{3a}{3}} \cdot x^{\frac{b}{3}}} = \boxed{x^{-\frac{3a}{2}-\frac{3a}{3}} \cdot x^{\frac{b}{3}}} = \boxed{x^{\frac{(-3a\cdot3)-(3a\cdot2)}{2\cdot3}} \cdot x^{\frac{b}{3}}}$$

$$= \boxed{x^{\frac{-9a-6a}{6}} \cdot x^{\frac{b}{3}}} = \boxed{x^{-\frac{15a}{6}=-\frac{5a}{2}} \cdot x^{\frac{b}{3}}} = \boxed{x^{-\frac{5a}{2}} \cdot x^{\frac{b}{3}}}$$

Step 3

$$\boxed{x^{-\frac{5a}{2}} \cdot x^{\frac{b}{3}}} = \boxed{\frac{x^{\frac{b}{3}}}{x^{\frac{5a}{2}}}}$$

Step 4

$$\boxed{\frac{x^{\frac{b}{3}}}{x^{\frac{5a}{2}}}} = \boxed{\frac{\sqrt[3]{x^b}}{\sqrt[2]{x^{5a}}}} = \boxed{\frac{\sqrt[3]{x^b}}{\sqrt{x^{5a}}}}$$

Example 5.4-4

$$\boxed{\left(81 \cdot a^{-1} \cdot c^{-3}\right)^{-\frac{1}{4}} \cdot \left(a^{-2} \cdot c^{-3}\right)^{-\frac{1}{2}}} =$$

Solution:

Step 1

$$\boxed{\left(81 \cdot a^{-1} \cdot c^{-3}\right)^{-\frac{1}{4}} \cdot \left(a^{-2} \cdot c^{-3}\right)^{-\frac{1}{2}}} = \boxed{\left(81^{-\frac{1}{4}} \cdot a^{-1\times-\frac{1}{4}} \cdot c^{-3\times-\frac{1}{4}}\right) \cdot \left(a^{-2\times-\frac{1}{2}} \cdot c^{-3\times-\frac{1}{2}}\right)}$$

$$= \boxed{\left(81^{-\frac{1}{4}} \cdot a^{\frac{1}{4}} \cdot c^{\frac{3}{4}}\right) \cdot \left(a^{\frac{2}{2}} \cdot c^{\frac{3}{2}}\right)}$$

Step 2

$$\boxed{\left(81^{-\frac{1}{4}} \cdot a^{\frac{1}{4}} \cdot c^{\frac{3}{4}}\right) \cdot \left(a^{\frac{2}{2}} \cdot c^{\frac{3}{2}}\right)} = \boxed{\left(3^4\right)^{-\frac{1}{4}} \cdot \left(a^{\frac{1}{4}} \cdot a^{\frac{2}{2}}\right) \cdot \left(c^{\frac{3}{4}} \cdot c^{\frac{3}{2}}\right)} = \boxed{3^{4\times-\frac{1}{4}} \cdot a^{\frac{1}{4}+\frac{2}{2}} \cdot c^{\frac{3}{4}+\frac{3}{2}}}$$

$$= \boxed{3^{-\frac{4}{4}} \cdot a^{\frac{(1\cdot2)+(2\cdot4)}{4\cdot2}} \cdot c^{\frac{(3\cdot2)+(3\cdot4)}{4\cdot2}}} = \boxed{3^{-\frac{4}{4}} \cdot a^{\frac{2+8}{8}} \cdot c^{\frac{6+12}{8}}} = \boxed{3^{-\frac{4}{4}=-\frac{1}{1}} \cdot a^{\frac{10}{8}=\frac{5}{4}} \cdot c^{\frac{18}{8}=\frac{9}{4}}}$$

$$= \boxed{3^{-\frac{1}{1}} \cdot a^{\frac{5}{4}} \cdot c^{\frac{9}{4}}} = \boxed{3^{-1} \cdot a^{\frac{5}{4}} \cdot c^{\frac{9}{4}}}$$

Step 3

$$\boxed{3^{-1} \cdot a^{\frac{5}{4}} \cdot c^{\frac{9}{4}}} = \boxed{\frac{1}{3^1} \cdot a^{\frac{5}{4}} \cdot c^{\frac{9}{4}}} = \boxed{\frac{a^{\frac{5}{4}} c^{\frac{9}{4}}}{3}}$$

Step 4

$$\boxed{\frac{a^{\frac{5}{4}} c^{\frac{9}{4}}}{3}} = \boxed{\frac{\sqrt[4]{a^5}\,\sqrt[4]{c^9}}{3}} = \boxed{\frac{\sqrt[4]{a^{4+1}}\,\sqrt[4]{c^{4+4+1}}}{3}} = \boxed{\frac{\sqrt[4]{a^4 \cdot a^1}\left[\sqrt[4]{\left(c^4 \cdot c^4\right) \cdot c^1}\right]}{3}}$$

$$= \boxed{\frac{a\sqrt[4]{a^1}\left[(c \cdot c) \cdot \sqrt[4]{c^1}\right]}{3}} = \boxed{\frac{ac^2\left(\sqrt[4]{a}\,\sqrt[4]{c}\right)}{3}} = \boxed{\frac{ac^2\sqrt[4]{ac}}{3}}$$

Example 5.4-5

$$\left[\left(27\cdot x^{2}\cdot w^{3}\right)^{\frac{1}{3}}\cdot w^{-1}\right]^{-\frac{1}{2}} =$$

Solution:

Step 1

$$\left[\left(27\cdot x^{2}\cdot w^{3}\right)^{\frac{1}{3}}\cdot w^{-1}\right]^{-\frac{1}{2}} = \left[\left(27^{\frac{1}{3}}\cdot x^{2\times\frac{1}{3}}\cdot w^{3\times\frac{1}{3}}\right)\cdot w^{-1}\right]^{-\frac{1}{2}} = \left[27^{\frac{1}{3}}\cdot x^{\frac{2}{3}}\cdot w^{\frac{3}{3}}\cdot w^{-1}\right]^{-\frac{1}{2}}$$

$$= 27^{\frac{1}{3}\times-\frac{1}{2}}\cdot x^{\frac{2}{3}\times-\frac{1}{2}}\cdot w^{\frac{3}{3}\times-\frac{1}{2}}\cdot w^{-1\times-\frac{1}{2}} = 27^{-\frac{1}{6}}\cdot x^{-\frac{2}{6}}\cdot w^{-\frac{3}{6}}\cdot w^{\frac{1}{2}}$$

Step 2

$$27^{-\frac{1}{6}}\cdot x^{-\frac{2}{6}}\cdot w^{-\frac{3}{6}}\cdot w^{\frac{1}{2}} = 27^{-\frac{1}{6}}\cdot x^{-\frac{2}{6}=-\frac{1}{3}}\cdot w^{-\frac{3}{6}=-\frac{1}{2}}\cdot w^{\frac{1}{2}} = 27^{-\frac{1}{6}}\cdot x^{-\frac{1}{3}}\cdot w^{-\frac{1}{2}}\cdot w^{\frac{1}{2}}$$

$$= 27^{-\frac{1}{6}}\cdot x^{-\frac{1}{3}}\cdot w^{-\frac{1}{2}+\frac{1}{2}} = 27^{-\frac{1}{6}}\cdot x^{-\frac{1}{3}}\cdot w^{\frac{-1+1}{2}} = 27^{-\frac{1}{6}}\cdot x^{-\frac{1}{3}}\cdot w^{0} = 27^{-\frac{1}{6}}\cdot x^{-\frac{1}{3}}\cdot 1$$

$$= 27^{-\frac{1}{6}}\cdot x^{-\frac{1}{3}} = \left(3^{3}\right)^{-\frac{1}{6}}\cdot x^{-\frac{1}{3}} = 3^{3\times-\frac{1}{6}}\cdot x^{-\frac{1}{3}} = 3^{-\frac{3}{6}=-\frac{1}{2}}\cdot x^{-\frac{1}{3}} = 3^{-\frac{1}{2}}\cdot x^{-\frac{1}{3}}$$

Step 3

$$3^{-\frac{1}{2}}\cdot x^{-\frac{1}{3}} = \frac{1}{3^{\frac{1}{2}}\cdot x^{\frac{1}{3}}}$$

Step 4

$$\frac{1}{3^{\frac{1}{2}}\cdot x^{\frac{1}{3}}} = \frac{1}{\sqrt[2]{3^{1}}\cdot\sqrt[3]{x^{1}}} = \boxed{\frac{1}{\sqrt{3}\,\sqrt[3]{x}}}$$

Additional Examples - Multiplying Negative Fractional Exponents

The following examples further illustrate how to multiply negative exponential expressions by one another:

Example 5.4-6

$$x^{-\frac{2}{3}}\cdot x^{-\frac{3}{4}}\cdot x^{-\frac{1}{2}} = x^{-\frac{2}{3}-\frac{3}{4}}\cdot x^{-\frac{1}{2}} = x^{\frac{(-2\cdot4)-(3\cdot3)}{3\cdot4}}\cdot x^{-\frac{1}{2}} = x^{\frac{-8-9}{12}}\cdot x^{-\frac{1}{2}} = x^{\frac{-17}{12}}\cdot x^{-\frac{1}{2}} = x^{\frac{-17}{12}-\frac{1}{2}}$$

$$= x^{\frac{(-17\cdot2)-(1\cdot12)}{12\cdot2}} = x^{\frac{-34-12}{24}} = x^{-\frac{46}{24}=-\frac{23}{12}} = x^{-\frac{23}{12}} = \frac{1}{x^{\frac{23}{12}}} = \frac{1}{\sqrt[12]{x^{23}}} = \frac{1}{\sqrt[12]{x^{12+11}}} = \frac{1}{\sqrt[12]{x^{12}\cdot x^{11}}}$$

$$= \boxed{\frac{1}{x\,\sqrt[12]{x^{11}}}}$$

Example 5.4-7

$$\left(4^{-3}\cdot 2^{2}\right)^{-\frac{2}{3}}\cdot\left(2^{-1}\right)^{\frac{2}{3}} = \left(4^{-3\times-\frac{2}{3}}\cdot 2^{2\times-\frac{2}{3}}\right)\cdot\left(2^{-1\times\frac{2}{3}}\right) = 4^{\frac{6}{3}=\frac{2}{1}}\cdot 2^{-\frac{4}{3}}\cdot 2^{-\frac{2}{3}} = 4^{\frac{2}{1}}\cdot 2^{-\frac{4}{3}-\frac{2}{3}} = 4^{2}\cdot 2^{\frac{-4-2}{3}}$$

$$= 16\cdot 2^{-\frac{6}{3}=-\frac{2}{1}} = 16\cdot 2^{-\frac{2}{1}} = 16\cdot 2^{-2} = 16\cdot\frac{1}{2^{2}} = 16\cdot\frac{1}{4} = \frac{16}{1}\cdot\frac{1}{4} = \frac{16\cdot 1}{1\cdot 4} = \frac{\overset{4}{\cancel{16}}}{\underset{1}{\cancel{4}}} = \frac{4}{1} = \mathbf{4}$$

Example 5.4-8

$$\left(3^{2a}\right)^{-\frac{1}{4}}\cdot\left(2^{-1}\cdot 3^{-3a}\right)^{\frac{2}{3}} = \left(3^{2a\times-\frac{1}{4}}\right)\cdot\left(2^{-1\times\frac{2}{3}}\cdot 3^{-3a\times\frac{2}{3}}\right) = 3^{-\frac{2a}{4}}\cdot 2^{-\frac{2}{3}}\cdot 3^{-\frac{6a}{3}} = 3^{-\frac{2a}{4}}\cdot 3^{-\frac{2a}{1}}\cdot 2^{-\frac{2}{3}}$$

$$= 3^{-\frac{2a}{4}-\frac{2a}{1}}\cdot 2^{-\frac{2}{3}} = 3^{\frac{(-2a\cdot 1)-(2a\cdot 4)}{4\cdot 1}}\cdot 2^{-\frac{2}{3}} = \frac{3^{\frac{-2a-8a}{4}}}{2^{\frac{2}{3}}} = \frac{3^{-\frac{10a}{4}}}{2^{\frac{2}{3}}} = \frac{3^{-\frac{10a}{4}=-\frac{5a}{2}}}{2^{\frac{2}{3}}} = \frac{3^{-\frac{5a}{2}}}{2^{\frac{2}{3}}} = \frac{1}{3^{\frac{5a}{2}}\cdot 2^{\frac{2}{3}}}$$

$$= \frac{1}{\sqrt{3^{5a}}\cdot\sqrt[3]{2^{2}}} = \mathbf{\frac{1}{\sqrt{3^{5a}}\,\sqrt[3]{4}}}$$

Example 5.4-9

$$w^{3}\cdot\left(216\cdot w^{2}\right)^{-\frac{1}{3}} = w^{3}\cdot\left(216^{1\times-\frac{1}{3}}\cdot w^{2\times-\frac{1}{3}}\right) = w^{3}\cdot 216^{-\frac{1}{3}}\cdot w^{-\frac{2}{3}} = \frac{w^{3}\cdot w^{-\frac{2}{3}}}{216^{\frac{1}{3}}} = \frac{w^{3-\frac{2}{3}}}{216^{\frac{1}{3}}} = \frac{w^{\frac{3}{1}-\frac{2}{3}}}{\sqrt[3]{216}}$$

$$= \frac{w^{\frac{(3\cdot 3)-(2\cdot 1)}{1\cdot 3}}}{\sqrt[3]{216}} = \frac{w^{\frac{9-2}{3}}}{\sqrt[3]{6^{3}}} = \frac{w^{\frac{7}{3}}}{\sqrt[3]{6^{3}}} = \frac{\sqrt[3]{w^{7}}}{6} = \frac{\sqrt[3]{w^{3+3+1}}}{6} = \frac{\sqrt[3]{\left(w^{3}\cdot w^{3}\right)\cdot w^{1}}}{6} = \frac{(w\cdot w)\cdot\sqrt[3]{w^{1}}}{6} = \mathbf{\frac{w^{2}\sqrt[3]{w}}{6}}$$

Example 5.4-10

$$16^{-\frac{1}{4}}\cdot\left(2\cdot x^{\frac{2}{3}}\right)^{0}\cdot x^{-\frac{3}{5}}\cdot x = 16^{-\frac{1}{4}}\cdot 1\cdot x^{-\frac{3}{5}}\cdot x = 16^{-\frac{1}{4}}\cdot x^{-\frac{3}{5}}\cdot x = \frac{x^{-\frac{3}{5}}\cdot x^{1}}{16^{\frac{1}{4}}} = \frac{x^{-\frac{3}{5}}\cdot x^{1}}{\sqrt[4]{16}} = \frac{x^{-\frac{3}{5}+\frac{1}{1}}}{\sqrt[4]{2^{4}}}$$

$$= \frac{x^{\frac{(-3\cdot 1)+(1\cdot 5)}{5\cdot 1}}}{2} = \frac{x^{\frac{-3+5}{5}}}{2} = \frac{x^{\frac{2}{5}}}{2} = \mathbf{\frac{\sqrt[5]{x^{2}}}{2}}$$

Example 5.4-11

$$\left(a^{-2}\cdot b^{2}\right)^{-\frac{2}{5}}\cdot\left(a^{-4}\cdot b\right) = \left(a^{-2\times-\frac{2}{5}}\cdot b^{2\times-\frac{2}{5}}\right)\cdot\left(a^{-4}\cdot b\right) = \left(a^{\frac{4}{5}}\cdot b^{-\frac{4}{5}}\right)\cdot\left(a^{-\frac{4}{1}}\cdot b^{1}\right) = \left(a^{\frac{4}{5}}\cdot a^{-\frac{4}{1}}\right)\cdot\left(b^{-\frac{4}{5}}\cdot b^{1}\right)$$

$$= \boxed{a^{\frac{4}{5}-\frac{4}{1}} \cdot b^{-\frac{4}{5}+\frac{1}{1}}} = \boxed{a^{\frac{(4\cdot1)-(4\cdot5)}{5\cdot1}} \cdot b^{\frac{(-4\cdot1)+(1\cdot5)}{5\cdot1}}} = \boxed{a^{\frac{4-20}{5}} \cdot b^{\frac{-4+5}{5}}} = \boxed{a^{-\frac{16}{5}} \cdot b^{\frac{1}{5}}} = \boxed{\frac{b^{\frac{1}{5}}}{a^{\frac{16}{5}}}} = \boxed{\frac{\sqrt[5]{b}}{\sqrt[5]{a^{16}}}} = \boxed{\frac{\sqrt[5]{b}}{\sqrt[5]{a^{5+5+5+1}}}}$$

$$= \boxed{\frac{\sqrt[5]{b}}{\sqrt[5]{\left(a^5 \cdot a^5 \cdot a^5\right) \cdot a^1}}} = \boxed{\frac{\sqrt[5]{b}}{(a \cdot a \cdot a) \cdot \sqrt[5]{a^1}}} = \boxed{\frac{\sqrt[5]{b}}{a^3 \sqrt[5]{a}}} = \boxed{\frac{1}{a^3}\left(\sqrt[5]{\frac{b}{a}}\right)}$$

Example 5.4-12

$$\boxed{100^{-\frac{1}{2}} \cdot \left(x^2 \cdot y^{-1} \cdot z^{-2}\right)^{\frac{4}{3}} \cdot \left(x \cdot z^3\right)^{-\frac{1}{3}}} = \boxed{100^{-\frac{1}{2}} \cdot \left(x^{2\times\frac{4}{3}} \cdot y^{-1\times\frac{4}{3}} \cdot z^{-2\times\frac{4}{3}}\right) \cdot \left(x^{1\times-\frac{1}{3}} \cdot z^{3\times-\frac{1}{3}}\right)}$$

$$= \boxed{100^{-\frac{1}{2}} \cdot x^{\frac{8}{3}} \cdot y^{-\frac{4}{3}} \cdot z^{-\frac{8}{3}} \cdot x^{-\frac{1}{3}} \cdot z^{-\frac{3}{3}}} = \boxed{100^{-\frac{1}{2}} \cdot x^{\frac{8}{3}} \cdot x^{-\frac{1}{3}} \cdot y^{-\frac{4}{3}} \cdot z^{-\frac{8}{3}} \cdot z^{-\frac{3}{3}}} = \boxed{100^{-\frac{1}{2}} \cdot x^{\frac{8-1}{3}} \cdot y^{-\frac{4}{3}} \cdot z^{\frac{-8-3}{3}}}$$

$$= \boxed{100^{-\frac{1}{2}} \cdot x^{\frac{7}{3}} \cdot y^{-\frac{4}{3}} \cdot z^{-\frac{11}{3}}} = \boxed{\frac{x^{\frac{7}{3}}}{100^{\frac{1}{2}} \cdot y^{\frac{4}{3}} \cdot z^{\frac{11}{3}}}} = \boxed{\frac{\sqrt[3]{x^7}}{\sqrt{100} \cdot \sqrt[3]{y^4} \cdot \sqrt[3]{z^{11}}}} = \boxed{\frac{\sqrt[3]{x^{3+3+1}}}{\sqrt{10^2} \cdot \sqrt[3]{y^{3+1}} \cdot \sqrt[3]{z^{3+3+3+2}}}}$$

$$= \boxed{\frac{\sqrt[3]{\left(x^3 \cdot x^3\right) \cdot x^1}}{10 \cdot \sqrt[3]{y^3 \cdot y^1} \cdot \left[\sqrt[3]{\left(z^3 \cdot z^3 \cdot z^3\right) \cdot z^2}\right]}} = \boxed{\frac{(x \cdot x) \cdot \sqrt[3]{x^1}}{10y\sqrt[3]{y^1} \cdot \left[(z \cdot z \cdot z) \cdot \sqrt[3]{z^2}\right]}} = \boxed{\frac{x^2 \cdot \sqrt[3]{x}}{10y\sqrt[3]{y} \cdot z^3 \cdot \sqrt[3]{z^2}}} = \boxed{\frac{x^2 \sqrt[3]{x}}{10\,y z^3 \sqrt[3]{y}\, \sqrt[3]{z^2}}}$$

$$= \boxed{\frac{x^2 \sqrt[3]{x}}{10\,y z^3 \sqrt[3]{y z^2}}} = \boxed{\frac{\mathbf{x^2}}{\mathbf{10\,y z^3}}\left(\sqrt[3]{\frac{\mathbf{x}}{\mathbf{y z^2}}}\right)}$$

Example 5.4-13

$$\boxed{y^{-\frac{1}{3}} \cdot \left(x^{-3} \cdot y\right)^{-\frac{0}{2}} \cdot \left(y \cdot x^{-3}\right)^{-\frac{2}{3}} \cdot y^{-1}} = \boxed{y^{-\frac{1}{3}} \cdot 1 \cdot \left(y^{1\times-\frac{2}{3}} \cdot x^{-3\times-\frac{2}{3}}\right) \cdot y^{-1}} = \boxed{y^{-\frac{1}{3}} \cdot y^{-\frac{2}{3}} \cdot x^{\frac{6}{3}=\frac{2}{1}} \cdot y^{-1}}$$

$$= \boxed{y^{-\frac{1}{3}} \cdot y^{-\frac{2}{3}} \cdot y^{-1} \cdot x^{\frac{2}{1}}} = \boxed{y^{-\frac{1}{3}-\frac{2}{3}} \cdot y^{-1} \cdot x^2} = \boxed{y^{\frac{-1-2}{3}} \cdot y^{-1} \cdot x^2} = \boxed{y^{-\frac{3}{3}=-\frac{1}{1}} \cdot y^{-1} \cdot x^2} = \boxed{y^{-\frac{1}{1}} \cdot y^{-1} \cdot x^2}$$

$$= \boxed{y^{-1} \cdot y^{-1} \cdot x^2} = \boxed{y^{-1-1} \cdot x^2} = \boxed{y^{-2} \cdot x^2} = \boxed{\frac{1}{y^2} \cdot x^2} = \boxed{\frac{1}{y^2} \cdot \frac{x^2}{1}} = \boxed{\frac{1 \cdot x^2}{y^2 \cdot 1}} = \boxed{\frac{\mathbf{x^2}}{\mathbf{y^2}}}$$

Example 5.4-14

$$\boxed{\left(5 \cdot a^{\frac{2}{3}} \cdot w\right)^{-\frac{1}{2}} \cdot \left(32 \cdot a \cdot w^2\right)^{-\frac{1}{5}}} = \boxed{\left(5^{-\frac{1}{2}} \cdot a^{\frac{2}{3}\times-\frac{1}{2}} \cdot w^{1\times-\frac{1}{2}}\right) \cdot \left(32^{-\frac{1}{5}} \cdot a^{-\frac{1}{5}} \cdot w^{2\times-\frac{1}{5}}\right)}$$

$$= \boxed{5^{-\frac{1}{2}} \cdot a^{-\frac{2}{6}=-\frac{1}{3}} \cdot w^{-\frac{1}{2}} \cdot 32^{-\frac{1}{5}} \cdot a^{-\frac{1}{5}} \cdot w^{-\frac{2}{5}}} = \boxed{\left(5^{-\frac{1}{2}} \cdot 32^{-\frac{1}{5}}\right) \cdot \left(a^{-\frac{1}{3}} \cdot a^{-\frac{1}{5}}\right) \cdot \left(w^{-\frac{1}{2}} \cdot w^{-\frac{2}{5}}\right)} = \boxed{\frac{a^{-\frac{1}{3}-\frac{1}{5}} \cdot w^{-\frac{1}{2}-\frac{2}{5}}}{5^{\frac{1}{2}} \cdot 32^{\frac{1}{5}}}}$$

$$= \boxed{\frac{a^{\frac{(-1\cdot5)-(1\cdot3)}{3\cdot5}} \cdot w^{\frac{(-1\cdot5)-(2\cdot2)}{2\cdot5}}}{5^{\frac{1}{2}} \cdot 32^{\frac{1}{5}}}} = \boxed{\frac{a^{\frac{-5-3}{15}} \cdot w^{\frac{-5-4}{10}}}{\sqrt{5} \cdot \sqrt[5]{32}}} = \boxed{\frac{a^{-\frac{8}{15}} \cdot w^{-\frac{9}{10}}}{\sqrt{5} \cdot \sqrt[5]{2^5}}} = \boxed{\frac{a^{-\frac{8}{15}} \cdot w^{-\frac{9}{10}}}{2\sqrt{5}}} = \boxed{\frac{1}{2\sqrt{5} \cdot a^{\frac{8}{15}} \cdot w^{\frac{9}{10}}}}$$

$$= \boxed{\frac{1}{2\sqrt{5} \cdot \sqrt[15]{a^8} \cdot \sqrt[10]{w^9}}} = \boxed{\frac{1}{2\sqrt{5}\ \sqrt[15]{a^8}\ \sqrt[10]{w^9}}}$$

Example 5.4-15

$$\boxed{\left(9^3 \cdot y^{-2} \cdot z^4\right)^{-\frac{1}{3}} \cdot \left[\left(y^{-2} \cdot z^3\right)^{-\frac{2}{3}}\right]^{-1}} = \boxed{\left(9^{3\times-\frac{1}{3}} \cdot y^{-2\times-\frac{1}{3}} \cdot z^{4\times-\frac{1}{3}}\right) \cdot \left(y^{-2\times-\frac{2}{3}} \cdot z^{3\times-\frac{2}{3}}\right)^{-1}}$$

$$= \boxed{9^{-\frac{3}{3}=-\frac{1}{1}} \cdot y^{\frac{2}{3}} \cdot z^{-\frac{4}{3}} \cdot \left(y^{\frac{4}{3}} \cdot z^{-\frac{6}{3}}\right)^{-1}} = \boxed{9^{-\frac{1}{1}} \cdot y^{\frac{2}{3}} \cdot z^{-\frac{4}{3}} \cdot \left(y^{\frac{4}{3}\times-1} \cdot z^{-\frac{6}{3}\times-1}\right)} = \boxed{9^{-1} \cdot y^{\frac{2}{3}} \cdot z^{-\frac{4}{3}} \cdot y^{-\frac{4}{3}} \cdot z^{\frac{6}{3}}}$$

$$= \boxed{9^{-1} \cdot \left(y^{\frac{2}{3}} \cdot y^{-\frac{4}{3}}\right) \cdot \left(z^{-\frac{4}{3}} \cdot z^{\frac{6}{3}}\right)} = \boxed{\frac{y^{\frac{2}{3}-\frac{4}{3}} \cdot z^{-\frac{4}{3}+\frac{6}{3}}}{9}} = \boxed{\frac{y^{\frac{2-4}{3}} \cdot z^{\frac{-4+6}{3}}}{9}} = \boxed{\frac{y^{-\frac{2}{3}} \cdot z^{\frac{2}{3}}}{9}} = \boxed{\frac{z^{\frac{2}{3}}}{9 \cdot y^{\frac{2}{3}}}} = \boxed{\frac{\sqrt[3]{z^2}}{9\sqrt[3]{y^2}}} = \boxed{\frac{1}{9}\sqrt[3]{\frac{z^2}{y^2}}}$$

Practice Problems - Multiplying Negative Fractional Exponents

Section 5.4 Case I Practice Problems - Multiply the following negative fractional exponents:

1. $\left(a^2 \cdot a^3\right)^{-\frac{1}{3}} =$
2. $\left(a^2 \cdot y^{-1}\right)^{-\frac{1}{4}} \cdot y^{-3} =$
3. $\left(a^{-2} \cdot b^{-3}\right)^{-\frac{1}{2}} \cdot \left(a \cdot b^2\right) =$
4. $512^{-\frac{1}{3}} \cdot \left(x^{2a} \cdot x^{3a}\right)^{-\frac{1}{5}} =$
5. $\left(x \cdot z^2\right)^{-\frac{3}{2}} \cdot \left(x^2\right)^{-\frac{2}{3}} =$
6. $27^{-\frac{2}{3}} \cdot 4^2 \cdot 4^{-\frac{1}{2}} \cdot 4^0 =$
7. $y^2 \cdot x^{-\frac{2}{3}} \cdot \left(x \cdot y^{-1}\right)^{-\frac{2}{3}} =$
8. $\left(a^2 \cdot b^2\right)^{-\frac{3}{4}} \cdot \left(a^5 \cdot b^2\right)^{-\frac{1}{4}} =$
9. $x^{-\frac{2}{3}} \cdot \left(64^{-1} \cdot x \cdot y^3\right)^{-\frac{2}{3}} =$
10. $\left(4 \cdot m^2\right)^{-\frac{1}{2}} \cdot \left[\left(m \cdot p^2\right)^{-1} \cdot m\right]^{\frac{1}{2}} =$

Case II Dividing Negative Fractional Exponents

Negative fractional exponents are divided by one another using the following steps:

Step 1 Apply exponent laws such as the Power of a Fraction Law (Law IV), the Power of a Power, and the Power of a Product (Laws II and III) from Table 5.4-1 in both the numerator and the denominator.

Step 2 a. Apply the Division and/or the Negative Power Law (Laws V, and VI) from Table 5.4-1.

b. Group the exponential terms with similar bases.

c. Apply the Multiplication Law (Law I) from Table 5.4-1 and simplify the exponential expressions by adding the exponents with similar bases.

Step 3 Change the negative fractional exponents to positive fractional exponents.

Step 4 Change the fractional exponential expressions to radical expressions (see Section 5.1).

Examples with Steps

The following examples show the steps as to how negative fractional exponents are divided by each other:

Example 5.4-16

$$\left(\frac{a^3}{a^2}\right)^{-\frac{2}{3}} =$$

Solution:

Step 1

$$\left(\frac{a^3}{a^2}\right)^{-\frac{2}{3}} = \frac{a^{3\times-\frac{2}{3}}}{a^{2\times-\frac{2}{3}}} = \frac{a^{\frac{3}{1}\times-\frac{2}{3}}}{a^{\frac{2}{1}\times-\frac{2}{3}}} = \frac{a^{-\frac{6}{3}=-\frac{2}{1}}}{a^{-\frac{4}{3}}} = \frac{a^{-\frac{2}{1}}}{a^{-\frac{4}{3}}}$$

Step 2

$$\frac{a^{-\frac{2}{1}}}{a^{-\frac{4}{3}}} = \frac{1}{a^{-\frac{4}{3}}\cdot a^{\frac{2}{1}}} = \frac{1}{a^{-\frac{4}{3}+\frac{2}{1}}} = \frac{1}{a^{\frac{(-4\cdot1)+(2\cdot3)}{3\cdot1}}} = \frac{1}{a^{\frac{-4+6}{3}}} = \frac{1}{a^{\frac{2}{3}}}$$

Step 3 *Not Applicable*

Step 4

$$\frac{1}{a^{\frac{2}{3}}} = \frac{1}{\sqrt[3]{a^2}}$$

Example 5.4-17

$$\left(\frac{216\cdot b^3}{b^{-4}}\right)^{-\frac{1}{3}} =$$

Solution:

Step 1

$$\boxed{\left(\frac{216\cdot b^{3}}{b^{-4}}\right)^{-\frac{1}{3}}} = \boxed{\frac{216^{1\times-\frac{1}{3}}\cdot b^{3\times-\frac{1}{3}}}{b^{-4\times-\frac{1}{3}}}} = \boxed{\frac{216^{-\frac{1}{3}}\cdot b^{-\frac{3}{3}=-\frac{1}{1}}}{b^{+\frac{4}{3}}}} = \boxed{\frac{216^{-\frac{1}{3}}\cdot b^{-\frac{1}{1}}}{b^{\frac{4}{3}}}}$$

Step 2

$$\boxed{\frac{216^{-\frac{1}{3}}\cdot b^{-\frac{1}{1}}}{b^{\frac{4}{3}}}} = \boxed{\frac{216^{-\frac{1}{3}}}{b^{\frac{4}{3}}\cdot b^{\frac{1}{1}}}} = \boxed{\frac{216^{-\frac{1}{3}}}{b^{\frac{4}{3}+\frac{1}{1}}}} = \boxed{\frac{216^{-\frac{1}{3}}}{b^{\frac{(4\cdot1)+(1\cdot3)}{3\cdot1}}}} = \boxed{\frac{216^{-\frac{1}{3}}}{b^{\frac{4+3}{3}}}} = \boxed{\frac{216^{-\frac{1}{3}}}{b^{\frac{7}{3}}}}$$

Step 3

$$\boxed{\frac{216^{-\frac{1}{3}}}{b^{\frac{7}{3}}}} = \boxed{\frac{1}{216^{\frac{1}{3}}\cdot b^{\frac{7}{3}}}}$$

Step 4

$$\boxed{\frac{1}{216^{\frac{1}{3}}\cdot b^{\frac{7}{3}}}} = \boxed{\frac{1}{\sqrt[3]{216}\cdot\sqrt[3]{b^{7}}}} = \boxed{\frac{1}{\sqrt[3]{6^{3}}\cdot\sqrt[3]{b^{3+3+1}}}} = \boxed{\frac{1}{6\cdot\sqrt[3]{\left(b^{3}\cdot b^{3}\right)\cdot b^{1}}}} = \boxed{\frac{1}{6\cdot(b\cdot b)\cdot\sqrt[3]{b}}}$$

$$= \boxed{\frac{1}{6\cdot b^{2}\cdot\sqrt[3]{b}}} = \boxed{\mathbf{\frac{1}{6b^{2}\sqrt[3]{b}}}}$$

Example 5.4-18A

$$\boxed{\frac{x^{-\frac{3}{4}}\cdot y^{-2}}{x^{-\frac{1}{4}}}} =$$

Solution:

Step 1 Not Applicable

Step 2

$$\boxed{\frac{x^{-\frac{3}{4}}\cdot y^{-2}}{x^{-\frac{1}{4}}}} = \boxed{\frac{y^{-2}}{x^{\frac{3}{4}}\cdot x^{-\frac{1}{4}}}} = \boxed{\frac{y^{-2}}{x^{\frac{3}{4}-\frac{1}{4}}}} = \boxed{\frac{y^{-2}}{x^{\frac{3-1}{4}}}} = \boxed{\frac{y^{-2}}{x^{\frac{2}{4}=\frac{1}{2}}}} = \boxed{\frac{y^{-2}}{x^{\frac{1}{2}}}}$$

Step 3

$$\boxed{\frac{y^{-2}}{x^{\frac{1}{2}}}} = \boxed{\frac{1}{x^{\frac{1}{2}}\cdot y^{2}}}$$

Step 4

$$\boxed{\frac{1}{x^{\frac{1}{2}}\cdot y^{2}}} = \boxed{\frac{1}{\sqrt{x}\cdot y^{2}}} = \boxed{\mathbf{\frac{1}{\sqrt{x}\,y^{2}}}}$$

Example 5.4-18B

$$\boxed{\left(\frac{243\cdot a}{c^{-3}}\right)^{-\frac{4}{3}}\cdot\frac{c^{2}}{a}} =$$

Solution:

Step 1

$$\left(\frac{243\cdot a}{c^{-3}}\right)^{-\frac{4}{3}}\cdot\frac{c^2}{a}=\frac{243^{1\times-\frac{4}{3}}\cdot a^{1\times-\frac{4}{3}}}{c^{-3\times-\frac{4}{3}}}\cdot\frac{c^2}{a}=\frac{243^{-\frac{4}{3}}\cdot a^{-\frac{4}{3}}}{c^{\frac{12}{3}}}\cdot\frac{c^2}{a}$$

Step 2

$$\frac{243^{-\frac{4}{3}}\cdot a^{-\frac{4}{3}}}{c^{\frac{12}{3}}}\cdot\frac{c^2}{a}=\frac{243^{-\frac{4}{3}}\cdot a^{-\frac{4}{3}}}{c^{\frac{12}{3}=\frac{4}{1}}}\cdot\frac{c^2}{a^1}=\frac{243^{-\frac{4}{3}}\cdot a^{-\frac{4}{3}}\cdot a^{-1}}{c^{\frac{4}{1}}\cdot c^{-2}}=\frac{243^{-\frac{4}{3}}\cdot a^{-\frac{4}{3}}\cdot a^{-\frac{1}{1}}}{c^4\cdot c^{-2}}$$

$$=\frac{243^{-\frac{4}{3}}\cdot a^{-\frac{4}{3}-\frac{1}{1}}}{c^{4-2}}=\frac{243^{-\frac{4}{3}}\cdot a^{\frac{(-4\cdot1)-(1\cdot3)}{3\cdot1}}}{c^2}=\frac{243^{-\frac{4}{3}}\cdot a^{\frac{-4-3}{3}}}{c^2}=\frac{243^{-\frac{4}{3}}\cdot a^{-\frac{7}{3}}}{c^2}$$

Step 3

$$\frac{243^{-\frac{4}{3}}\cdot a^{-\frac{7}{3}}}{c^2}=\frac{1}{243^{\frac{4}{3}}\cdot c^2\cdot a^{\frac{7}{3}}}$$

Step 4

$$\frac{1}{243^{\frac{4}{3}}\cdot c^2\cdot a^{\frac{7}{3}}}=\frac{1}{\sqrt[3]{243^4}\cdot c^2\cdot\sqrt[3]{a^7}}=\frac{1}{\sqrt[3]{243^{3+1}}\cdot c^2\cdot\sqrt[3]{a^{3+3+1}}}$$

$$=\frac{1}{\sqrt[3]{243^3\cdot 243^1}\cdot c^2\cdot\sqrt[3]{\left(a^3\cdot a^3\right)\cdot a^1}}=\frac{1}{243\cdot\sqrt[3]{243^1}\cdot c^2\cdot(a\cdot a)\cdot\sqrt[3]{a^1}}$$

$$=\frac{1}{243\sqrt[3]{27\cdot 9}\cdot c^2\cdot a^2\cdot\sqrt[3]{a}}=\frac{1}{243\sqrt[3]{3^3\cdot 9}\cdot c^2\cdot a^2\cdot\sqrt[3]{a}}=\frac{1}{(243\cdot 3)\sqrt[3]{9}\cdot c^2\cdot a^2\cdot\sqrt[3]{a}}$$

$$=\frac{1}{729a^2c^2\sqrt[3]{9}\sqrt[3]{a}}=\mathbf{\frac{1}{729a^2c^2\sqrt[3]{9a}}}$$

Example 5.4-19

$$\frac{z^2\cdot(a\cdot b)^{-\frac{2}{3}}}{z^{-\frac{1}{4}}\cdot a^{-2}}=$$

Solution:

Step 1

$$\frac{z^2\cdot(a\cdot b)^{-\frac{2}{3}}}{z^{-\frac{1}{4}}\cdot a^{-2}}=\frac{z^2\cdot a^{1\times-\frac{2}{3}}\cdot b^{1\times-\frac{2}{3}}}{z^{-\frac{1}{4}}\cdot a^{-2}}=\frac{z^2\cdot a^{-\frac{2}{3}}\cdot b^{-\frac{2}{3}}}{z^{-\frac{1}{4}}\cdot a^{-2}}$$

Step 2

$$\frac{z^2\cdot a^{-\frac{2}{3}}\cdot b^{-\frac{2}{3}}}{z^{-\frac{1}{4}}\cdot a^{-2}}=\frac{z^2\cdot z^{\frac{1}{4}}\cdot b^{-\frac{2}{3}}}{a^{\frac{2}{3}}\cdot a^{-2}}=\frac{z^{\frac{2}{1}}\cdot z^{\frac{1}{4}}\cdot b^{-\frac{2}{3}}}{a^{\frac{2}{3}}\cdot a^{-\frac{2}{1}}}=\frac{z^{\frac{2}{1}+\frac{1}{4}}\cdot b^{-\frac{2}{3}}}{a^{\frac{2}{3}-\frac{2}{1}}}=\frac{z^{\frac{(2\cdot4)+(1\cdot1)}{1\cdot4}}\cdot b^{-\frac{2}{3}}}{a^{\frac{(2\cdot1)-(2\cdot3)}{3\cdot1}}}$$

$$= \boxed{\frac{z^{\frac{8+1}{4}} \cdot b^{-\frac{2}{3}}}{a^{\frac{2-6}{3}}}} = \boxed{\frac{z^{\frac{9}{4}} \cdot b^{-\frac{2}{3}}}{a^{-\frac{4}{3}}}}$$

Step 3

$$\boxed{\frac{z^{\frac{9}{4}} \cdot b^{-\frac{2}{3}}}{a^{-\frac{4}{3}}}} = \boxed{\frac{z^{\frac{9}{4}} \cdot a^{\frac{4}{3}}}{b^{\frac{2}{3}}}}$$

Step 4

$$\boxed{\frac{z^{\frac{9}{4}} \cdot a^{\frac{4}{3}}}{b^{\frac{2}{3}}}} = \boxed{\frac{\sqrt[4]{z^9} \cdot \sqrt[3]{a^4}}{\sqrt[3]{b^2}}} = \boxed{\frac{\sqrt[4]{z^{4+4+1}} \cdot \sqrt[3]{a^{3+1}}}{\sqrt[3]{b^2}}} = \boxed{\frac{\sqrt[4]{\left(z^4 \cdot z^4\right) \cdot z^1} \cdot \sqrt[3]{a^3 \cdot a^1}}{\sqrt[3]{b^2}}}$$

$$= \boxed{\frac{(z \cdot z) \cdot \sqrt[4]{z^1} \cdot a \cdot \sqrt[3]{a^1}}{\sqrt[3]{b^2}}} = \boxed{\frac{a \cdot z^2 \cdot \sqrt[4]{z} \cdot \sqrt[3]{a}}{\sqrt[3]{b^2}}} = \boxed{\frac{\mathbf{a z^2 \sqrt[3]{a} \sqrt[4]{z}}}{\mathbf{\sqrt[3]{b^2}}}}$$

Example 5.4-20

$$\boxed{\frac{\left(a^2 \cdot b^3\right)^{-\frac{1}{2}} \cdot c}{\left(a^{-3} \cdot b \cdot c\right)^{-\frac{2}{3}}}} =$$

Solution:

Step 1

$$\boxed{\frac{\left(a^2 \cdot b^3\right)^{-\frac{1}{2}} \cdot c}{\left(a^{-3} \cdot b \cdot c\right)^{-\frac{2}{3}}}} = \boxed{\frac{a^{2\times-\frac{1}{2}} \cdot b^{3\times-\frac{1}{2}} \cdot c}{a^{-3\times-\frac{2}{3}} \cdot b^{1\times-\frac{2}{3}} \cdot c^{1\times-\frac{2}{3}}}} = \boxed{\frac{a^{-\frac{2}{2}=-\frac{1}{1}} \cdot b^{-\frac{3}{2}} \cdot c}{a^{\frac{6}{3}=\frac{2}{1}} \cdot b^{-\frac{2}{3}} \cdot c^{-\frac{2}{3}}}} = \boxed{\frac{a^{-\frac{1}{1}} \cdot b^{-\frac{3}{2}} \cdot c^{\frac{1}{1}}}{a^{\frac{2}{1}} \cdot b^{-\frac{2}{3}} \cdot c^{-\frac{2}{3}}}}$$

Step 2

$$\boxed{\frac{a^{-\frac{1}{1}} \cdot b^{-\frac{3}{2}} \cdot c^{\frac{1}{1}}}{a^{\frac{2}{1}} \cdot b^{-\frac{2}{3}} \cdot c^{-\frac{2}{3}}}} = \boxed{\frac{a^{-1} \cdot b^{-\frac{3}{2}} \cdot c^{\frac{1}{1}}}{a^2 \cdot b^{-\frac{2}{3}} \cdot c^{-\frac{2}{3}}}} = \boxed{\frac{c^{\frac{1}{1}} \cdot c^{\frac{2}{3}}}{a^2 \cdot a^1 \cdot b^{-\frac{2}{3}} \cdot b^{\frac{3}{2}}}} = \boxed{\frac{c^{\frac{1}{1}+\frac{2}{3}}}{a^{2+1} \cdot b^{-\frac{2}{3}+\frac{3}{2}}}}$$

$$= \boxed{\frac{c^{\frac{(1\cdot3)+(2\cdot1)}{1\cdot3}}}{a^3 \cdot b^{\frac{(-2\cdot2)+(3\cdot3)}{3\cdot2}}}} = \boxed{\frac{c^{\frac{3+2}{3}}}{a^3 \cdot b^{\frac{-4+9}{6}}}} = \boxed{\frac{c^{\frac{5}{3}}}{a^3 \cdot b^{\frac{5}{6}}}}$$

Step 3

Not Applicable

Step 4

$$\boxed{\frac{c^{\frac{5}{3}}}{a^3 \cdot b^{\frac{5}{6}}}} = \boxed{\frac{\sqrt[3]{c^5}}{a^3 \cdot \sqrt[6]{b^5}}} = \boxed{\frac{\sqrt[3]{c^{3+2}}}{a^3 \cdot \sqrt[6]{b^5}}} = \boxed{\frac{\sqrt[3]{c^3 \cdot c^2}}{a^3 \cdot \sqrt[6]{b^5}}} = \boxed{\frac{c \cdot \sqrt[3]{c^2}}{a^3 \cdot \sqrt[6]{b^5}}} = \boxed{\frac{\mathbf{c\sqrt[3]{c^2}}}{\mathbf{a^3 \sqrt[6]{b^5}}}}$$

Additional Examples - Dividing Negative Fractional Exponents

The following examples further illustrate how to divide negative exponential expressions by one another. Note that these problems do not follow the same steps as outlined above. The exponent laws do not necessarily have to be applied in a specific order. The following problems are solved by using exponent laws in different order to strengthen students knowledge in applying these laws. (It is recommended that students exercise solving problems 5.4-21 through 5.4-30 using the steps outlined above. The final answers should agree with the answers given below.)

Example 5.4-21

$$32^{-\frac{1}{5}} \cdot \left(\frac{a^2}{a^3}\right)^{-\frac{1}{3}} = \frac{1}{32^{\frac{1}{5}}} \cdot \frac{1}{\left(\frac{a^2}{a^3}\right)^{\frac{1}{3}}} = \frac{1}{\sqrt[5]{32^1}} \cdot \frac{1}{\frac{a^{2\times\frac{1}{3}}}{a^{3\times\frac{1}{3}}}} = \frac{1}{\sqrt[5]{32}} \cdot \frac{1}{\frac{a^{\frac{2}{3}}}{a^{\frac{3}{3}}}} = \frac{1}{\sqrt[5]{2^5}} \cdot \frac{1}{\frac{a^{\frac{2}{3}}}{a^{\frac{1}{1}}}} = \frac{1}{2} \cdot \frac{1}{\frac{a^{\frac{2}{3}}}{a}} = \frac{1}{2} \cdot \frac{1\cdot a}{1\cdot a^{\frac{2}{3}}}$$

$$= \frac{1}{2} \cdot \frac{a^1}{a^{\frac{2}{3}}} = \frac{1}{2} \cdot \frac{a^1 \cdot a^{-\frac{2}{3}}}{1} = \frac{1}{2} \cdot a^{\frac{1}{1}-\frac{2}{3}} = \frac{1}{2} \cdot a^{\frac{(1\cdot3)-(1\cdot2)}{1\cdot3}} = \frac{1}{2} \cdot a^{\frac{3-2}{3}} = \frac{1}{2} \cdot a^{\frac{1}{3}} = \frac{1}{2} \cdot \sqrt[3]{a^1} = \mathbf{\frac{1}{2}\sqrt[3]{a}}$$

Example 5.4-22

$$\left(\frac{3^{-2}\cdot a^0 \cdot x^2}{a^2 \cdot x^{-3}}\right)^{-\frac{1}{2}} = \left(\frac{3^{-2}\cdot 1 \cdot x^2 \cdot x^3}{a^2}\right)^{-\frac{1}{2}} = \left(\frac{3^{-2}\cdot x^{2+3}}{a^2}\right)^{-\frac{1}{2}} = \left(\frac{3^{-2}\cdot x^5}{a^2}\right)^{-\frac{1}{2}} = \frac{3^{-2\times-\frac{1}{2}} \cdot x^{5\times-\frac{1}{2}}}{a^{2\times-\frac{1}{2}}}$$

$$= \frac{3^{\frac{2}{2}=\frac{1}{1}} \cdot x^{-\frac{5}{2}}}{a^{-\frac{2}{2}=-\frac{1}{1}}} = \frac{3^1 \cdot x^{-\frac{5}{2}}}{a^{-\frac{1}{1}}} = \frac{3^1}{a^{-1}\cdot x^{\frac{5}{2}}} = \frac{3\cdot a^1}{x^{\frac{5}{2}}} = \frac{3\cdot a}{\sqrt{x^5}} = \frac{3\cdot a}{\sqrt{x^{2+2+1}}} = \frac{3\cdot a}{\sqrt{\left(x^2\cdot x^2\right)\cdot x^1}} = \frac{3\cdot a}{(x\cdot x)\sqrt{x}}$$

$$= \mathbf{\frac{3a}{x^2\sqrt{x}}}$$

Example 5.4-23

$$\frac{a^2\cdot b^{-4}\cdot c^3}{a^{-\frac{1}{4}}\cdot b^{-\frac{2}{3}}\cdot c^{-\frac{1}{5}}} = \frac{a^2\cdot a^{\frac{1}{4}}\cdot c^3\cdot c^{\frac{1}{5}}}{b^4\cdot b^{-\frac{2}{3}}} = \frac{a^{2+\frac{1}{4}}\cdot c^{3+\frac{1}{5}}}{b^{4-\frac{2}{3}}} = \frac{a^{\frac{2}{1}+\frac{1}{4}}\cdot c^{\frac{3}{1}+\frac{1}{5}}}{b^{\frac{4}{1}-\frac{2}{3}}} = \frac{a^{\frac{(2\cdot4)+(1\cdot1)}{1\cdot4}}\cdot c^{\frac{(3\cdot5)+(1\cdot1)}{1\cdot5}}}{b^{\frac{(4\cdot3)-(2\cdot1)}{1\cdot3}}}$$

$$= \frac{a^{\frac{8+1}{4}}\cdot c^{\frac{15+1}{5}}}{b^{\frac{12-2}{3}}} = \frac{a^{\frac{9}{4}}\cdot c^{\frac{16}{5}}}{b^{\frac{10}{3}}} = \frac{\sqrt[4]{a^9}\cdot\sqrt[5]{c^{16}}}{\sqrt[3]{b^{10}}} = \frac{\sqrt[4]{a^{4+4+1}}\cdot\sqrt[5]{c^{5+5+5+1}}}{\sqrt[3]{b^{3+3+3+1}}} = \frac{\left[\sqrt[4]{\left(a^4\cdot a^4\right)\cdot a^1}\right]\cdot\left[\sqrt[5]{\left(c^5\cdot c^5\cdot c^5\right)\cdot c^1}\right]}{\sqrt[3]{\left(b^3\cdot b^3\cdot b^3\right)\cdot b^1}}$$

$$= \frac{\left[(a\cdot a)\cdot\sqrt[4]{a^1}\right]\cdot\left[(c\cdot c\cdot c)\cdot\sqrt[5]{c^1}\right]}{(b\cdot b\cdot b)\cdot\sqrt[3]{b^1}} = \frac{\left(a^2\cdot\sqrt[4]{a}\right)\cdot\left(c^3\cdot\sqrt[5]{c}\right)}{b^3\cdot\sqrt[3]{b}} = \frac{a^2\cdot c^3\cdot\sqrt[4]{a}\cdot\sqrt[5]{c}}{b^3\cdot\sqrt[3]{b}} = \boldsymbol{\frac{a^2c^3\sqrt[4]{a}\sqrt[5]{c}}{b^3\sqrt[3]{b}}}$$

Example 5.4-24

$$\left(\frac{2\cdot c^2}{3\cdot b^3}\right)^{-\frac{2}{3}} = \frac{1}{\left(\frac{2\cdot c^2}{3\cdot b^3}\right)^{\frac{2}{3}}} = \frac{1}{\frac{2^{\frac{2}{3}}\cdot c^{2\times\frac{2}{3}}}{3^{\frac{2}{3}}\cdot b^{3\times\frac{2}{3}}}} = \frac{1}{\frac{2^{\frac{2}{3}}\cdot c^{\frac{4}{3}}}{3^{\frac{2}{3}}\cdot b^{\frac{6}{3}}}} = \frac{1}{\frac{\sqrt[3]{2^2}\cdot\sqrt[3]{c^4}}{\sqrt[3]{3^2}\cdot b^2}} = \frac{\frac{1}{1}}{\frac{\sqrt[3]{2^2}\cdot\sqrt[3]{c^4}}{\sqrt[3]{3^2}\cdot b^2}} = \frac{1\times\sqrt[3]{3^2}\cdot b^2}{1\times\sqrt[3]{2^2}\cdot\sqrt[3]{c^4}}$$

$$= \frac{\sqrt[3]{3^2}\cdot b^2}{\sqrt[3]{2^2}\cdot\sqrt[3]{c^{3+1}}} = \frac{\sqrt[3]{3^2}\cdot b^2}{\sqrt[3]{2^2}\cdot\sqrt[3]{c^3}\cdot c^1} = \frac{\sqrt[3]{3^2}\cdot b^2}{\sqrt[3]{2^2}\cdot c\cdot\sqrt[3]{c}} = \frac{\sqrt[3]{3^2}\, b^2}{\sqrt[3]{2^2}\, c\sqrt[3]{c}} = \boldsymbol{\sqrt[3]{\frac{9}{4}}\left(\frac{b^2}{c\sqrt[3]{c}}\right)}$$

Example 5.4-25

$$\left(\frac{m^{-1}\cdot n^{-4}}{m^{-2}\cdot n^2}\right)^{-\frac{1}{4}} = \left(\frac{m^{-1}\cdot m^2}{n^2\cdot n^4}\right)^{-\frac{1}{4}} = \left(\frac{m^{-1+2}}{n^{2+4}}\right)^{-\frac{1}{4}} = \left(\frac{m^1}{n^6}\right)^{-\frac{1}{4}} = \frac{1}{\left(\frac{m^1}{n^6}\right)^{\frac{1}{4}}} = \frac{1}{\frac{m^{1\times\frac{1}{4}}}{n^{6\times\frac{1}{4}}}} = \frac{1}{\frac{m^{\frac{1}{4}}}{n^{\frac{6}{4}=\frac{3}{2}}}} = \frac{1}{\frac{m^{\frac{1}{4}}}{n^{\frac{3}{2}}}}$$

$$= \frac{1}{\frac{\sqrt{m}}{\sqrt{n^3}}} = \frac{\frac{1}{1}}{\frac{\sqrt{m}}{\sqrt{n^3}}} = \frac{1\cdot\sqrt{n^3}}{1\cdot\sqrt{m}} = \frac{\sqrt{n^{2+1}}}{\sqrt{m}} = \frac{\sqrt{n^2\cdot n^1}}{\sqrt{m}} = \frac{n\cdot\sqrt{n}}{\sqrt{m}} = \boldsymbol{\frac{n\sqrt{n}}{\sqrt{m}}}$$

Example 5.4-26

$$\left(\frac{125\cdot a^3\cdot b^{-3}}{a^{-2}\cdot b^4}\right)^{-\frac{1}{3}} = \left(\frac{125\cdot a^3\cdot a^2}{b^4\cdot b^3}\right)^{-\frac{1}{3}} = \left(\frac{125\cdot a^{3+2}}{b^{4+3}}\right)^{-\frac{1}{3}} = \left(\frac{125\cdot a^5}{b^7}\right)^{-\frac{1}{3}} = \frac{1}{\left(\frac{125\cdot a^5}{b^7}\right)^{\frac{1}{3}}} = \frac{1}{\frac{125^{1\times\frac{1}{3}}\cdot a^{5\times\frac{1}{3}}}{b^{7\times\frac{1}{3}}}}$$

$$= \frac{1}{\frac{125^{\frac{1}{3}}\cdot a^{\frac{5}{3}}}{b^{\frac{7}{3}}}} = \frac{1}{\frac{\sqrt[3]{125}\cdot\sqrt[3]{a^5}}{\sqrt[3]{b^7}}} = \frac{1}{\frac{\sqrt[3]{5^3}\cdot\sqrt[3]{a^{3+2}}}{\sqrt[3]{b^{3+3+1}}}} = \frac{1}{\frac{5\cdot\sqrt[3]{a^3\cdot a^2}}{\sqrt[3]{\left(b^3\cdot b^3\right)\cdot b^1}}} = \frac{1}{\frac{5\cdot a\cdot\sqrt[3]{a^2}}{(b\cdot b)\cdot\sqrt[3]{b^1}}} = \frac{1}{\frac{5a\sqrt[3]{a^2}}{b^2\sqrt[3]{b}}} = \frac{\frac{1}{1}}{\frac{5a\sqrt[3]{a^2}}{b^2\sqrt[3]{b}}}$$

$$= \frac{1\times\left(b^2\sqrt[3]{b}\right)}{1\times\left(5a\sqrt[3]{a^2}\right)} = \frac{b^2\sqrt[3]{b}}{5a\sqrt[3]{a^2}} = \boldsymbol{\frac{b^2}{5a}\left(\sqrt[3]{\frac{b}{a^2}}\right)}$$

Example 5.4-27

$$\left(\frac{(bxy)^{-\frac{1}{4}}}{y^{-\frac{2}{3}}}\right)\cdot\frac{x^3}{y^2} = \left(\frac{b^{-\frac{1}{4}}\cdot x^{-\frac{1}{4}}\cdot y^{-\frac{1}{4}}}{y^{-\frac{2}{3}}}\right)\cdot\frac{x^3}{y^2} = \frac{x^{-\frac{1}{4}}\cdot x^3}{b^{\frac{1}{4}}\cdot y^{-\frac{2}{3}}\cdot y^{\frac{1}{4}}\cdot y^2} = \frac{x^{-\frac{1}{4}}\cdot x^{\frac{3}{1}}}{b^{\frac{1}{4}}\cdot y^{-\frac{2}{3}+\frac{1}{4}}\cdot y^2} = \frac{x^{\frac{(-1\cdot1)+(3\cdot4)}{4\cdot1}}}{b^{\frac{1}{4}}\cdot y^{\frac{(-2\cdot4)+(1\cdot3)}{3\cdot4}}\cdot y^2}$$

$$= \frac{x^{\frac{-1+12}{4}}}{b^{\frac{1}{4}}\cdot y^{\frac{-8+3}{12}}\cdot y^2} = \frac{x^{\frac{11}{4}}}{b^{\frac{1}{4}}\cdot y^{\frac{-5}{12}}\cdot y^2} = \frac{x^{\frac{11}{4}}}{b^{\frac{1}{4}}\cdot y^{\frac{-5}{12}+2}} = \frac{x^{\frac{11}{4}}}{b^{\frac{1}{4}}\cdot y^{\frac{-5}{12}+\frac{2}{1}}} = \frac{x^{\frac{11}{4}}}{b^{\frac{1}{4}}\cdot y^{\frac{(-5\cdot1)+(2\cdot12)}{12\cdot1}}} = \frac{x^{\frac{11}{4}}}{b^{\frac{1}{4}}\cdot y^{\frac{-5+24}{12}}}$$

$$= \frac{x^{\frac{11}{4}}}{b^{\frac{1}{4}}\cdot y^{\frac{19}{12}}} = \frac{\sqrt[4]{x^{11}}}{\sqrt[4]{b}\cdot\sqrt[12]{y^{19}}} = \frac{\sqrt[4]{x^{4+4+3}}}{\sqrt[4]{b}\cdot\sqrt[12]{y^{12+7}}} = \frac{\sqrt[4]{\left(x^4\cdot x^4\right)\cdot x^3}}{\sqrt[4]{b}\cdot\sqrt[12]{y^{12}\cdot y^7}} = \frac{(x\cdot x)\cdot\sqrt[4]{x^3}}{\sqrt[4]{b}\cdot y\cdot\sqrt[12]{y^7}} = \boxed{\frac{x^2\sqrt[4]{x^3}}{\sqrt[4]{b}\,y\sqrt[12]{y^7}}}$$

Example 5.4-28

$$\frac{625^{-\frac{1}{4}}\cdot x^{\frac{1}{3}}}{125^{\frac{1}{3}}\cdot\left(x^2\cdot a^3\right)^{-\frac{1}{3}}} = \frac{\left(x^2\cdot a^3\right)^{\frac{1}{3}}\cdot x^{\frac{1}{3}}}{125^{\frac{1}{3}}\cdot 625^{\frac{1}{4}}} = \frac{x^{2\times\frac{1}{3}}\cdot a^{3\times\frac{1}{3}}\cdot x^{\frac{1}{3}}}{\sqrt[3]{125}\cdot\sqrt[4]{625}} = \frac{x^{\frac{2}{3}}\cdot a^{\frac{3}{3}=\frac{1}{1}}\cdot x^{\frac{1}{3}}}{\sqrt[3]{125}\cdot\sqrt[4]{625}} = \frac{a^{\frac{1}{1}}\cdot x^{\frac{2}{3}}\cdot x^{\frac{1}{3}}}{\sqrt[3]{5^3}\cdot\sqrt[4]{5^4}} = \frac{a^{\frac{1}{1}}\cdot x^{\frac{2}{3}+\frac{1}{3}}}{5\cdot5}$$

$$= \frac{a^1\cdot x^{\frac{2+1}{3}}}{25} = \frac{a\cdot x^{\frac{3}{3}=\frac{1}{1}}}{25} = \frac{a\cdot x^{\frac{1}{1}}}{25} = \boxed{\frac{ax}{25}}$$

Example 5.4-29

$$\left(\frac{2^{-2}\cdot2^4\cdot a^{-3}}{a}\right)^{-\frac{1}{4}}\cdot\left(\frac{b^{-2}}{b^2}\right)^{-\frac{1}{2}} = \left(\frac{2^{-2}\cdot2^4}{a^3\cdot a}\right)^{-\frac{1}{4}}\cdot\left(\frac{1}{b^2\cdot b^2}\right)^{-\frac{1}{2}} = \left(\frac{2^{-2+4}}{a^{3+1}}\right)^{-\frac{1}{4}}\cdot\left(\frac{1}{b^{2+2}}\right)^{-\frac{1}{2}} = \left(\frac{2^2}{a^4}\right)^{-\frac{1}{4}}\cdot\left(\frac{1}{b^4}\right)^{-\frac{1}{2}}$$

$$= \left(\frac{4}{a^4}\right)^{-\frac{1}{4}}\cdot\left(\frac{1}{b^4}\right)^{-\frac{1}{2}} = \frac{1}{\left(\frac{4}{a^4}\right)^{\frac{1}{4}}}\cdot\frac{1}{\left(\frac{1}{b^4}\right)^{\frac{1}{2}}} = \frac{1}{\frac{4^{1\times\frac{1}{4}}}{a^{4\times\frac{1}{4}}}}\cdot\frac{1}{\frac{1^{1\times\frac{1}{2}}}{b^{4\times\frac{1}{2}}}} = \frac{1}{\frac{4^{\frac{1}{4}}}{a^{\frac{4}{4}=\frac{1}{1}}}}\cdot\frac{1}{\frac{1^{\frac{1}{2}}}{b^{\frac{4}{2}=\frac{2}{1}}}} = \frac{1}{\frac{\sqrt[4]{4}}{a^{\frac{1}{1}}}}\cdot\frac{1}{\frac{\sqrt{1}}{b^{\frac{2}{1}}}} = \frac{1}{\frac{\sqrt[4]{4}}{a^1}}\cdot\frac{1}{\frac{1}{b^2}}$$

$$= \frac{\frac{1}{1}}{\frac{\sqrt[4]{4}}{a}}\cdot\frac{\frac{1}{1}}{\frac{1}{b^2}} = \frac{1\cdot a}{1\cdot\sqrt[4]{4}}\cdot\frac{1\cdot b^2}{1\cdot1} = \frac{a}{\sqrt[4]{4}}\cdot\frac{b^2}{1} = \frac{ab^2}{\sqrt[4]{4}} = \frac{ab^2}{4^{\frac{1}{4}}} = \frac{ab^2}{\left(2^2\right)^{\frac{1}{4}}} = \frac{ab^2}{2^{2\times\frac{1}{4}}} = \frac{ab^2}{2^{\frac{2}{4}=\frac{1}{2}}} = \frac{ab^2}{2^{\frac{1}{2}}} = \boxed{\frac{ab^2}{\sqrt{2}}}$$

Example 5.4-30

$$\left(\frac{a^{-2}}{w^{-1}\cdot z^2}\right)^{-\frac{3}{2}}\cdot\frac{w}{a^{-\frac{1}{2}}} = \frac{a^{-2\times-\frac{3}{2}}}{w^{-1\times-\frac{3}{2}}\cdot z^{2\times-\frac{3}{2}}}\cdot\frac{w\cdot a^{\frac{1}{2}}}{1} = \frac{a^{\frac{6}{2}}}{w^{\frac{3}{2}}\cdot z^{-\frac{6}{2}}}\cdot\frac{w\cdot a^{\frac{1}{2}}}{1} = \frac{a^{\frac{3}{1}}}{w^{\frac{3}{2}}\cdot z^{-\frac{3}{1}}}\cdot\frac{w\cdot a^{\frac{1}{2}}}{1} = \frac{a^{\frac{3}{1}}\cdot a^{\frac{1}{2}}\cdot z^{\frac{3}{1}}}{w^{\frac{3}{2}}\cdot w^{-1}}$$

$$= \boxed{\frac{a^{\frac{3}{1}+\frac{1}{2}} \cdot z^3}{w^{\frac{3}{2}-1}}} = \boxed{\frac{a^{\frac{3}{1}+\frac{1}{2}} \cdot z^3}{w^{\frac{3}{2}-\frac{1}{1}}}} = \boxed{\frac{a^{\frac{(3\cdot2)+(1\cdot1)}{1\cdot2}} \cdot z^3}{w^{\frac{(3\cdot1)-(1\cdot2)}{2\cdot1}}}} = \boxed{\frac{a^{\frac{6+1}{2}} \cdot z^3}{w^{\frac{3-2}{2}}}} = \boxed{\frac{a^{\frac{7}{2}} \cdot z^3}{w^{\frac{1}{2}}}} = \boxed{\frac{\sqrt{a^7} \cdot z^3}{\sqrt{w}}} = \boxed{\frac{\sqrt{a^{2+2+2+1}} \cdot z^3}{\sqrt{w}}}$$

$$= \boxed{\frac{\sqrt{\left(a^2 \cdot a^2 \cdot a^2\right) \cdot a^1} \cdot z^3}{\sqrt{w}}} = \boxed{\frac{(a \cdot a \cdot a) \cdot \sqrt{a^1} \cdot z^3}{\sqrt{w}}} = \boxed{\frac{a^3 \cdot \sqrt{a} \cdot z^3}{\sqrt{w}}} = \boxed{\frac{a^3 z^3 \sqrt{a}}{\sqrt{w}}} = \boxed{\boldsymbol{a^3 z^3 \left(\sqrt{\frac{a}{w}}\right)}}$$

Practice Problems - Dividing Negative Fractional Exponents

Section 5.4 Case II Practice Problems - Divide the following negative fractional exponents:

1. $\dfrac{a^{-\frac{1}{2}}}{a^2} =$

2. $\dfrac{a^{-\frac{2}{3}} \cdot b^3}{a^{-\frac{1}{3}}} =$

3. $\left(\dfrac{a^3 \cdot b^2}{a^2 \cdot b^5}\right)^{-\frac{2}{3}} =$

4. $\dfrac{d^3 \cdot (a \cdot b \cdot c)^{-\frac{1}{4}}}{d^{\frac{2}{3}} \cdot b^{-3}} =$

5. $\dfrac{(a \cdot b)^{-\frac{2}{3}} \cdot (x \cdot y)^{-\frac{1}{4}}}{a^2 \cdot y^{-\frac{3}{4}}} =$

6. $\dfrac{(x \cdot y)^0 \cdot x^2}{\left(x \cdot y^{-1}\right)^{-\frac{3}{4}}} =$

7. $\dfrac{1000^{-\frac{1}{3}} \cdot b^{-2} \cdot c^{-\frac{1}{3}}}{(b \cdot c)^{-\frac{2}{3}}} =$

8. $\left(\dfrac{z^2 \cdot w^{-3} \cdot (a \cdot b)^{-2}}{a^2 \cdot w^{-3} \cdot z^0}\right)^{-\frac{1}{2}} =$

9. $\dfrac{2^{-\frac{3}{4}} \cdot 2^{\frac{1}{4}} \cdot (y \cdot z)^{-3}}{2^{-\frac{1}{4}} \cdot b^{-1} \cdot y^{-\frac{2}{3}}} =$

10. $\dfrac{a^{\frac{1}{2}} \cdot b^{\frac{4}{3}} \cdot \left(c^5 \cdot d\right)^{-\frac{4}{5}}}{c^3 \cdot d^{\frac{1}{5}}} =$

Case III Adding and Subtracting Negative Fractional Exponents

Negative fractional exponents are added and subtracted using the following steps:

Step 1 Change the negative exponential expression $x^{-\frac{a}{b}}$ to $\frac{1}{x^{\frac{a}{b}}}$.

Step 2 Simplify the exponential expressions by:

a. Using fraction techniques learned in Chapter 2.

b. Using appropriate exponent laws such as the Multiplication Law (Law I) from Table 5.4-1.

Step 3 Change the fractional exponential expressions to radical expressions (see Section 5.1).

Examples with Steps

The following examples show the steps as to how negative fractional exponents are added or subtracted:

Example 5.4-31

$$\boxed{5^{-\frac{3}{2}}+5^{-\frac{2}{3}}} =$$

Solution:

Step 1

$$\boxed{5^{-\frac{3}{2}}+5^{-\frac{2}{3}}} = \boxed{\frac{1}{5^{\frac{3}{2}}}+\frac{1}{5^{\frac{2}{3}}}}$$

Step 2

$$\boxed{\frac{1}{5^{\frac{3}{2}}}+\frac{1}{5^{\frac{2}{3}}}} = \boxed{\frac{1\cdot 5^{\frac{2}{3}}+1\cdot 5^{\frac{3}{2}}}{5^{\frac{3}{2}}\cdot 5^{\frac{2}{3}}}} = \boxed{\frac{5^{\frac{2}{3}}+5^{\frac{3}{2}}}{5^{\frac{3}{2}+\frac{2}{3}}}} = \boxed{\frac{5^{\frac{2}{3}}+5^{\frac{3}{2}}}{5^{\frac{(3\cdot 3)+(2\cdot 2)}{2\cdot 3}}}} = \boxed{\frac{5^{\frac{2}{3}}+5^{\frac{3}{2}}}{5^{\frac{9+4}{6}}}} = \boxed{\frac{5^{\frac{2}{3}}+5^{\frac{3}{2}}}{5^{\frac{13}{6}}}}$$

Step 3

$$\boxed{\frac{5^{\frac{2}{3}}+5^{\frac{3}{2}}}{5^{\frac{13}{6}}}} = \boxed{\frac{\sqrt[3]{5^2}+\sqrt{5^3}}{\sqrt[6]{5^{13}}}} = \boxed{\frac{\sqrt[3]{5^2}+\sqrt{5^{2+1}}}{\sqrt[6]{5^{6+6+1}}}} = \boxed{\frac{\sqrt[3]{5^2}+\sqrt{5^2\cdot 5^1}}{\sqrt[6]{\left(5^6\cdot 5^6\right)\cdot 5^1}}} = \boxed{\frac{\sqrt[3]{5^2}+5\cdot\sqrt{5^1}}{(5\cdot 5)\cdot\sqrt[6]{5^1}}}$$

$$= \boxed{\frac{\sqrt[3]{25}+5\sqrt{5}}{25\sqrt[6]{5}}}$$

Note: The problems in this section are solved with the assumption that students are thoroughly familiar with the subject of fractions. In case of difficulty, students are encouraged to review Chapters 4, 6, 8, and 9 of the *Mastering Fractions* book by the author.

Example 5.4-32

$$\boxed{\frac{x^{\frac{1}{2}}-x^{-\frac{1}{2}}}{x^{-\frac{2}{3}}}} =$$

Solution:

Step 1

$$\frac{x^{\frac{1}{2}}-x^{-\frac{1}{2}}}{x^{-\frac{2}{3}}}=\frac{x^{\frac{1}{2}}-\frac{1}{x^{\frac{1}{2}}}}{\frac{1}{x^{\frac{2}{3}}}}$$

Step 2

$$\frac{x^{\frac{1}{2}}-\frac{1}{x^{\frac{1}{2}}}}{\frac{1}{x^{\frac{2}{3}}}}=\frac{\frac{x^{\frac{1}{2}}}{1}-\frac{1}{x^{\frac{1}{2}}}}{\frac{1}{x^{\frac{2}{3}}}}=\frac{\frac{x^{\frac{1}{2}}\cdot x^{\frac{1}{2}}-(1\cdot 1)}{1\cdot x^{\frac{1}{2}}}}{\frac{1}{x^{\frac{2}{3}}}}=\frac{\frac{x^{\frac{1}{2}+\frac{1}{2}}-1}{x^{\frac{1}{2}}}}{\frac{1}{x^{\frac{2}{3}}}}=\frac{\frac{x^{\frac{(1\cdot2)+(1\cdot2)}{2\cdot2}}-1}{x^{\frac{1}{2}}}}{\frac{1}{x^{\frac{2}{3}}}}$$

$$=\frac{\frac{x^{\frac{2+2}{4}}-1}{x^{\frac{1}{2}}}}{\frac{1}{x^{\frac{2}{3}}}}=\frac{\frac{x^{\frac{4}{4}=\frac{1}{1}}-1}{x^{\frac{1}{2}}}}{\frac{1}{x^{\frac{2}{3}}}}=\frac{\frac{x^{\frac{1}{1}}-1}{x^{\frac{1}{2}}}}{\frac{1}{x^{\frac{2}{3}}}}=\frac{\frac{x-1}{x^{\frac{1}{2}}}}{\frac{1}{x^{\frac{2}{3}}}}=\frac{(x-1)\cdot x^{\frac{2}{3}}}{x^{\frac{1}{2}}\cdot 1}=\frac{x\cdot x^{\frac{2}{3}}-1\cdot x^{\frac{2}{3}}}{x^{\frac{1}{2}}}$$

$$=\frac{x^{1+\frac{2}{3}}-x^{\frac{2}{3}}}{x^{\frac{1}{2}}}=\frac{x^{\frac{1}{1}+\frac{2}{3}}-x^{\frac{2}{3}}}{x^{\frac{1}{2}}}=\frac{x^{\frac{(1\cdot3)+(2\cdot1)}{1\cdot3}}-x^{\frac{2}{3}}}{x^{\frac{1}{2}}}=\frac{x^{\frac{3+2}{3}}-x^{\frac{2}{3}}}{x^{\frac{1}{2}}}=\frac{x^{\frac{5}{3}}-x^{\frac{2}{3}}}{x^{\frac{1}{2}}}$$

$$=\left(x^{\frac{5}{3}}-x^{\frac{2}{3}}\right)\cdot x^{-\frac{1}{2}}=\left(x^{\frac{5}{3}}\cdot x^{-\frac{1}{2}}\right)-\left(x^{\frac{2}{3}}\cdot x^{-\frac{1}{2}}\right)=\left(x^{\frac{5}{3}-\frac{1}{2}}\right)-\left(x^{\frac{2}{3}-\frac{1}{2}}\right)$$

$$=x^{\frac{(5\cdot2)-(1\cdot3)}{3\cdot2}}-x^{\frac{(2\cdot2)-(1\cdot3)}{3\cdot2}}=x^{\frac{10-3}{6}}-x^{\frac{4-3}{6}}=x^{\frac{7}{6}}-x^{\frac{1}{6}}$$

Step 3

$$x^{\frac{7}{6}}-x^{\frac{1}{6}}=\sqrt[6]{x^7}-\sqrt[6]{x}=\sqrt[6]{x^{6+1}}-\sqrt[6]{x}=\sqrt[6]{x^6\cdot x^1}-\sqrt[6]{x}=x\cdot\sqrt[6]{x}-\sqrt[6]{x}=\boxed{\sqrt[6]{x}\,(x-1)}$$

Example 5.4-33

$$\frac{(xy)^2-(xy)^{\frac{2}{3}}}{(xy)^{-\frac{1}{2}}}=$$

Solution:

Step 1

$$\frac{(xy)^2-(xy)^{\frac{2}{3}}}{(xy)^{-\frac{1}{2}}}=\frac{(xy)^2-(xy)^{\frac{2}{3}}}{\frac{1}{(xy)^{\frac{1}{2}}}}$$

Step 2

$$\frac{(xy)^2-(xy)^{\frac{2}{3}}}{\frac{1}{(xy)^{\frac{1}{2}}}}=\frac{\frac{(xy)^2-(xy)^{\frac{2}{3}}}{1}}{\frac{1}{(xy)^{\frac{1}{2}}}}=\frac{\left[(xy)^2-(xy)^{\frac{2}{3}}\right]\cdot(xy)^{\frac{1}{2}}}{1\cdot 1}$$

$$=\frac{(xy)^2\cdot(xy)^{\frac{1}{2}}-(xy)^{\frac{2}{3}}\cdot(xy)^{\frac{1}{2}}}{1}=(xy)^{2+\frac{1}{2}}-(xy)^{\frac{2}{3}+\frac{1}{2}}=(xy)^{\frac{2}{1}+\frac{1}{2}}-(xy)^{\frac{2}{3}+\frac{1}{2}}$$

$$=(xy)^{\frac{(2\cdot2)+(1\cdot1)}{1\cdot2}}-(xy)^{\frac{(2\cdot2)+(1\cdot3)}{3\cdot2}}=(xy)^{\frac{4+1}{2}}-(xy)^{\frac{4+3}{6}}=(xy)^{\frac{5}{2}}-(xy)^{\frac{7}{6}}$$

Step 3

$$(xy)^{\frac{5}{2}}-(xy)^{\frac{7}{6}}=\sqrt{(xy)^5}-\sqrt[6]{(xy)^7}=\sqrt{(xy)^{2+2+1}}-\sqrt[6]{(xy)^{6+1}}$$

$$=\sqrt{(xy)^2\cdot(xy)^2\cdot(xy)^1}-\sqrt[6]{(xy)^6\cdot(xy)^1}=xy\cdot xy\cdot\sqrt{(xy)^1}-xy\cdot\sqrt[6]{(xy)^1}$$

$$=(xy)^2\cdot\sqrt{xy}-xy\cdot\sqrt[6]{xy}=\mathbf{xy\left(xy\sqrt{xy}-\sqrt[6]{xy}\right)}$$

Example 5.4-34

$$\left(\frac{1}{x+1}\right)^{-\frac{1}{2}}=$$

Solution:

Step 1

$$\left(\frac{1}{x+1}\right)^{-\frac{1}{2}}=\frac{1}{\left(\frac{1}{x+1}\right)^{\frac{1}{2}}}$$

Step 2

$$\frac{1}{\left(\frac{1}{x+1}\right)^{\frac{1}{2}}}=\frac{1}{\frac{1^{\frac{1}{2}}}{(x+1)^{\frac{1}{2}}}}=\frac{\frac{1}{1}}{\frac{1}{(x+1)^{\frac{1}{2}}}}=\frac{1\cdot(x+1)^{\frac{1}{2}}}{1\cdot1}=\frac{(x+1)^{\frac{1}{2}}}{1}=(x+1)^{\frac{1}{2}}$$

Step 3 $\boxed{(x+1)^{\frac{1}{2}}} = \boxed{\sqrt{x+1}}$

Example 5.4-35

$$\boxed{\frac{2^{-\frac{3}{2}}+4}{2}} =$$

Solution:

Step 1 $\boxed{\frac{2^{-\frac{3}{2}}+4}{2}} = \boxed{\frac{\frac{1}{2^{\frac{3}{2}}}+4}{2}}$

Step 2 $\boxed{\frac{\frac{1}{2^{\frac{3}{2}}}+4}{2}} = \boxed{\frac{\frac{1}{2^{\frac{3}{2}}}+\frac{4}{1}}{2}} = \boxed{\frac{\frac{(1\cdot 1)+4\cdot 2^{\frac{3}{2}}}{2^{\frac{3}{2}}\cdot 1}}{2}} = \boxed{\frac{\frac{1+4\cdot 2^{\frac{3}{2}}}{2^{\frac{3}{2}}}}{\frac{2}{1}}} = \boxed{\frac{\left(1+4\cdot 2^{\frac{3}{2}}\right)\cdot 1}{2^{\frac{3}{2}}\cdot 2}} = \boxed{\frac{1+4\cdot 2^{\frac{3}{2}}}{2^{\frac{3}{2}+1}}}$

$= \boxed{\frac{1+4\cdot 2^{\frac{3}{2}}}{2^{\frac{3}{2}+\frac{1}{1}}}} = \boxed{\frac{1+4\cdot 2^{\frac{3}{2}}}{2^{\frac{(3\cdot 1)+(1\cdot 2)}{2\cdot 1}}}} = \boxed{\frac{1+4\cdot 2^{\frac{3}{2}}}{2^{\frac{3+2}{2}}}} = \boxed{\frac{1+4\cdot 2^{\frac{3}{2}}}{2^{\frac{5}{2}}}}$

Step 3 $\boxed{\frac{1+4\cdot 2^{\frac{3}{2}}}{2^{\frac{5}{2}}}} = \boxed{\frac{1+4\sqrt[2]{2^3}}{\sqrt[2]{2^5}}} = \boxed{\frac{1+4\sqrt{2^{2+1}}}{\sqrt{2^{2+2+1}}}} = \boxed{\frac{1+4\sqrt{2^2\cdot 2^1}}{\sqrt{\left(2^2\cdot 2^2\right)\cdot 2^1}}} = \boxed{\frac{1+(4\cdot 2)\sqrt{2}}{(2\cdot 2)\cdot\sqrt{2}}} = \boxed{\frac{\mathbf{1+8\sqrt{2}}}{\mathbf{4\sqrt{2}}}}$

Additional Examples - Adding and Subtracting Negative Fractional Exponents

The following examples further illustrate addition and subtraction of negative fractional exponents:

Example 5.4-36

$\boxed{(a+b)^{-\frac{3}{4}}} = \boxed{\frac{1}{(a+b)^{\frac{3}{4}}}} = \boxed{\frac{\mathbf{1}}{\sqrt[4]{(a+b)^3}}}$

Note: $\boxed{(a\cdot b)^{-\frac{3}{4}}} = \boxed{\frac{1}{(a\cdot b)^{\frac{3}{4}}}} = \boxed{\frac{1}{a^{\frac{3}{4}}\, b^{\frac{3}{4}}}} = \boxed{\frac{1}{\sqrt[4]{a^3}\,\sqrt[4]{b^3}}} = \boxed{\frac{\mathbf{1}}{\sqrt[4]{a^3b^3}}}$

Example 5.4-37

$\boxed{a^{-\frac{1}{5}}-a^{-\frac{2}{3}}} = \boxed{\frac{1}{a^{\frac{1}{5}}}-\frac{1}{a^{\frac{2}{3}}}} = \boxed{\frac{1\cdot a^{\frac{2}{3}}-1\cdot a^{\frac{1}{5}}}{a^{\frac{1}{5}}\cdot a^{\frac{2}{3}}}} = \boxed{\frac{a^{\frac{2}{3}}-a^{\frac{1}{5}}}{a^{\frac{1}{5}+\frac{2}{3}}}} = \boxed{\frac{a^{\frac{2}{3}}-a^{\frac{1}{5}}}{a^{\frac{(1\cdot 3)+(2\cdot 5)}{5\cdot 3}}}} = \boxed{\frac{a^{\frac{2}{3}}-a^{\frac{1}{5}}}{a^{\frac{3+10}{15}}}} = \boxed{\frac{a^{\frac{2}{3}}-a^{\frac{1}{5}}}{a^{\frac{13}{15}}}} = \boxed{\frac{\sqrt[3]{a^2}-\sqrt[5]{a}}{\sqrt[15]{a^{13}}}}$

Example 5.4-38

$\boxed{x^{-\frac{1}{2}}+y^{-2}+z^{-\frac{5}{2}}} = \boxed{\left(x^{-\frac{1}{2}}+y^{-2}\right)+z^{-\frac{5}{2}}} = \boxed{\left(\frac{1}{x^{\frac{1}{2}}}+\frac{1}{y^2}\right)+\frac{1}{z^{\frac{5}{2}}}} = \boxed{\frac{\left(1\cdot y^2\right)+1\cdot x^{\frac{1}{2}}}{x^{\frac{1}{2}}\cdot y^2}+\frac{1}{z^{\frac{5}{2}}}} = \boxed{\frac{y^2+x^{\frac{1}{2}}}{x^{\frac{1}{2}}y^2}+\frac{1}{z^{\frac{5}{2}}}}$

$$= \boxed{\frac{z^{\frac{5}{2}} \cdot \left(y^2 + x^{\frac{1}{2}}\right) + \left(1 \cdot x^{\frac{1}{2}} y^2\right)}{x^{\frac{1}{2}} y^2 \cdot z^{\frac{5}{2}}}} = \boxed{\frac{\sqrt{z^5}\left(y^2 + \sqrt{x}\right) + \sqrt{x}\, y^2}{\sqrt{x}\, y^2 \sqrt{z^5}}} = \boxed{\frac{\sqrt{z^{2+2+1}}\left(y^2 + \sqrt{x}\right) + \sqrt{x}\, y^2}{\sqrt{x}\, y^2 \sqrt{z^{2+2+1}}}}$$

$$= \boxed{\frac{\sqrt{\left(z^2 \cdot z^2\right) \cdot z^1}\left(y^2 + \sqrt{x}\right) + \sqrt{x}\, y^2}{\sqrt{x}\, y^2 \sqrt{\left(z^2 \cdot z^2\right) \cdot z^1}}} = \boxed{\frac{\left[(z \cdot z) \cdot \sqrt{z}\right] \cdot \left(y^2 + \sqrt{x}\right) + \sqrt{x}\, y^2}{\sqrt{x}\, y^2 (z \cdot z)\sqrt{z}}} = \boxed{\frac{z^2 \sqrt{z}\left(y^2 + \sqrt{x}\right) + \sqrt{x}\, y^2}{\sqrt{x}\, y^2 z^2 \sqrt{z}}}$$

Example 5.4-39

$$\boxed{\frac{x^{\frac{1}{3}}}{x^{-\frac{2}{3}} - y}} = \boxed{\frac{x^{\frac{1}{3}}}{\frac{1}{x^{\frac{2}{3}}} - y}} = \boxed{\frac{x^{\frac{1}{3}}}{\frac{1}{x^{\frac{2}{3}}} - \frac{y}{1}}} = \boxed{\frac{x^{\frac{1}{3}}}{\frac{(1 \cdot 1) - x^{\frac{2}{3}} \cdot y}{x^{\frac{2}{3}} \cdot 1}}} = \boxed{\frac{x^{\frac{1}{3}}}{\frac{1 - x^{\frac{2}{3}} \cdot y}{x^{\frac{2}{3}}}}} = \boxed{\frac{\frac{x^{\frac{1}{3}}}{1}}{\frac{1 - x^{\frac{2}{3}} \cdot y}{x^{\frac{2}{3}}}}} = \boxed{\frac{x^{\frac{1}{3}} \cdot x^{\frac{2}{3}}}{1 \cdot \left(1 - x^{\frac{2}{3}} \cdot y\right)}} = \boxed{\frac{x^{\frac{1}{3}+\frac{2}{3}}}{1 - x^{\frac{2}{3}} \cdot y}}$$

$$= \boxed{\frac{x^{\frac{3}{3}=\frac{1}{1}}}{1 - \sqrt[3]{x^2} \cdot y}} = \boxed{\frac{x^{\frac{1}{1}}}{1 - \sqrt[3]{x^2} \cdot y}} = \boxed{\frac{x}{1 - \sqrt[3]{x^2}\, y}}$$

Example 5.4-40

$$\boxed{\left(a^{-\frac{2}{3}} \cdot a\right) - \left(a^{-\frac{1}{5}} \cdot a^3\right)} = \boxed{a^{-\frac{2}{3}+1} - a^{-\frac{1}{5}+3}} = \boxed{a^{-\frac{2}{3}+\frac{1}{1}} - a^{-\frac{1}{5}+\frac{3}{1}}} = \boxed{a^{\frac{(-2\cdot1)+(1\cdot3)}{3\cdot1}} - a^{\frac{(-1\cdot1)+(3\cdot5)}{5}}} = \boxed{a^{\frac{-2+3}{3}} - a^{\frac{-1+15}{5}}}$$

$$= \boxed{a^{\frac{1}{3}} - a^{\frac{14}{5}}} = \boxed{\sqrt[3]{a} - \sqrt[5]{a^{14}}} = \boxed{\sqrt[3]{a} - \sqrt[5]{a^{5+5+4}}} = \boxed{\sqrt[3]{a} - \sqrt[5]{\left(a^5 \cdot a^5\right) \cdot a^4}} = \boxed{\sqrt[3]{a} - (a \cdot a) \cdot \sqrt[5]{a^4}} = \boxed{\sqrt[3]{a} - a^2 \sqrt[5]{a^4}}$$

Example 5.4-41

$$\boxed{\frac{5^{\frac{1}{2}} - 6^{-\frac{2}{3}}}{5^{\frac{1}{5}}}} = \boxed{\frac{5^{\frac{1}{2}} - \frac{1}{6^{\frac{2}{3}}}}{5^{\frac{1}{5}}}} = \boxed{\frac{\frac{5^{\frac{1}{2}}}{1} - \frac{1}{6^{\frac{2}{3}}}}{5^{\frac{1}{5}}}} = \boxed{\frac{\frac{5^{\frac{1}{2}} \cdot 6^{\frac{2}{3}} - (1 \cdot 1)}{1 \cdot 6^{\frac{2}{3}}}}{5^{\frac{1}{5}}}} = \boxed{\frac{\frac{5^{\frac{1}{2}} \cdot 6^{\frac{2}{3}} - 1}{6^{\frac{2}{3}}}}{\frac{5^{\frac{1}{5}}}{1}}} = \boxed{\frac{\left(5^{\frac{1}{2}} \cdot 6^{\frac{2}{3}} - 1\right) \times 1}{6^{\frac{2}{3}} \cdot 5^{\frac{1}{5}}}} = \boxed{\frac{\left(5^{\frac{1}{2}} \cdot 6^{\frac{2}{3}}\right) - 1}{6^{\frac{2}{3}} \cdot 5^{\frac{1}{5}}}}$$

$$= \boxed{\frac{\left(\sqrt{5}\,\sqrt[3]{6^2}\right) - 1}{\sqrt[3]{6^2}\,\sqrt[5]{5}}} = \boxed{\frac{\left(\sqrt{5}\,\sqrt[3]{36}\right) - 1}{\sqrt[3]{36}\,\sqrt[5]{5}}}$$

Example 5.4-42

$$\frac{1}{a^{\frac{1}{3}}+b^{-\frac{2}{3}}} = \frac{1}{a^{\frac{1}{3}}+\frac{1}{b^{\frac{2}{3}}}} = \frac{1}{\frac{a^{\frac{1}{3}}}{1}+\frac{1}{b^{\frac{2}{3}}}} = \frac{1}{\frac{a^{\frac{1}{3}}\cdot b^{\frac{2}{3}}+(1\cdot1)}{1\cdot b^{\frac{2}{3}}}} = \frac{\frac{1}{1}}{\frac{a^{\frac{1}{3}}\cdot b^{\frac{2}{3}}+1}{b^{\frac{2}{3}}}} = \frac{1\cdot b^{\frac{2}{3}}}{1\cdot\left(a^{\frac{1}{3}}\,b^{\frac{2}{3}}+1\right)} = \frac{b^{\frac{2}{3}}}{a^{\frac{1}{3}}\,b^{\frac{2}{3}}+1}$$

$$= \frac{\sqrt[3]{b^2}}{\sqrt[3]{a}\,\sqrt[3]{b^2}+1} = \boxed{\frac{\sqrt[3]{b^2}}{\sqrt[3]{ab^2}+1}}$$

Example 5.4-43

$$\frac{x^{\frac{1}{2}}}{x^{-\frac{1}{2}}-y} = \frac{x^{\frac{1}{2}}}{\frac{1}{x^{\frac{1}{2}}}-y} = \frac{x^{\frac{1}{2}}}{\frac{1}{x^{\frac{1}{2}}}-\frac{y}{1}} = \frac{x^{\frac{1}{2}}}{\frac{(1\cdot1)-y\cdot x^{\frac{1}{2}}}{x^{\frac{1}{2}}\cdot1}} = \frac{x^{\frac{1}{2}}}{\frac{1-yx^{\frac{1}{2}}}{x^{\frac{1}{2}}}} = \frac{\frac{x^{\frac{1}{2}}}{1}}{\frac{1-yx^{\frac{1}{2}}}{x^{\frac{1}{2}}}} = \frac{x^{\frac{1}{2}}\cdot x^{\frac{1}{2}}}{1\cdot\left(1-yx^{\frac{1}{2}}\right)} = \frac{x^{\frac{1}{2}+\frac{1}{2}}}{1-x^{\frac{1}{2}}y}$$

$$= \frac{x^{\frac{1+1}{2}}}{1-x^{\frac{1}{2}}y} = \frac{x^{\frac{2}{2}}}{1-x^{\frac{1}{2}}y} = \frac{x^{\frac{1}{1}}}{1-x^{\frac{1}{2}}y} = \frac{x}{1-x^{\frac{1}{2}}y} = \boxed{\frac{x}{1-\sqrt{x}\,y}}$$

Example 5.4-44

$$\left(x^{-\frac{3}{6}}+x^{-1}\right)\cdot x^{\frac{2}{3}} = \left(x^{-\frac{3}{6}=-\frac{1}{2}}+x^{-1}\right)\cdot x^{\frac{2}{3}} = \left(x^{-\frac{1}{2}}+x^{-\frac{1}{1}}\right)\cdot x^{\frac{2}{3}} = x^{-\frac{1}{2}}\cdot x^{\frac{2}{3}}+x^{-\frac{1}{1}}\cdot x^{\frac{2}{3}} = x^{-\frac{1}{2}+\frac{2}{3}}+x^{-\frac{1}{1}+\frac{2}{3}}$$

$$= x^{\frac{(-1\cdot3)+(2\cdot2)}{2\cdot3}}+x^{\frac{(-1\cdot3)+(2\cdot1)}{1\cdot3}} = x^{\frac{-3+4}{6}}+x^{\frac{-3+2}{3}} = x^{\frac{1}{6}}+x^{-\frac{1}{3}} = x^{\frac{1}{6}}+\frac{1}{x^{\frac{1}{3}}} = \frac{x^{\frac{1}{6}}}{1}+\frac{1}{x^{\frac{1}{3}}} = \frac{x^{\frac{1}{6}}\cdot x^{\frac{1}{3}}+(1\cdot1)}{1\cdot x^{\frac{1}{3}}}$$

$$= \frac{x^{\frac{1}{6}+\frac{1}{3}}+1}{x^{\frac{1}{3}}} = \frac{x^{\frac{(1\cdot3)+(1\cdot6)}{6\cdot3}}+1}{x^{\frac{1}{3}}} = \frac{x^{\frac{3+6}{18}}+1}{x^{\frac{1}{3}}} = \frac{x^{\frac{9}{18}=\frac{1}{2}}+1}{x^{\frac{1}{3}}} = \frac{x^{\frac{1}{2}}+1}{x^{\frac{1}{3}}} = \boxed{\frac{\sqrt{x}+1}{\sqrt[3]{x}}}$$

Example 5.4-45

$$\frac{x-x^{-\frac{2}{3}}}{x+x^{\frac{1}{2}}} = \frac{x-\frac{1}{x^{\frac{2}{3}}}}{x+x^{\frac{1}{2}}} = \frac{\frac{x}{1}-\frac{1}{x^{\frac{2}{3}}}}{x+x^{\frac{1}{2}}} = \frac{\frac{x\cdot x^{\frac{2}{3}}-(1\cdot1)}{1\cdot x^{\frac{2}{3}}}}{x+x^{\frac{1}{2}}} = \frac{\frac{x^{1+\frac{2}{3}}-1}{x^{\frac{2}{3}}}}{x+x^{\frac{1}{2}}} = \frac{\frac{x^{\frac{1}{1}+\frac{2}{3}}-1}{x^{\frac{2}{3}}}}{x+x^{\frac{1}{2}}} = \frac{\frac{x^{\frac{(1\cdot3)+(2\cdot1)}{1\cdot3}}-1}{x^{\frac{2}{3}}}}{x+x^{\frac{1}{2}}} = \frac{\frac{x^{\frac{3+2}{3}}-1}{x^{\frac{2}{3}}}}{x+x^{\frac{1}{2}}}$$

$$= \boxed{\frac{\frac{x^{\frac{5}{3}}-1}{x^{\frac{2}{3}}}}{\frac{x+x^{\frac{1}{2}}}{1}}} = \boxed{\frac{\left(x^{\frac{5}{3}}-1\right)\cdot 1}{x^{\frac{2}{3}}\cdot\left(x+x^{\frac{1}{2}}\right)}} = \boxed{\frac{x^{\frac{5}{3}}-1}{x^{\frac{2}{3}}\cdot x+x^{\frac{2}{3}}\cdot x^{\frac{1}{2}}}} = \boxed{\frac{x^{\frac{5}{3}}-1}{x^{\frac{2}{3}+1}+x^{\frac{2}{3}+\frac{1}{2}}}} = \boxed{\frac{x^{\frac{5}{3}}-1}{x^{\frac{2}{3}+\frac{1}{1}}+x^{\frac{(2\cdot 2)+(1\cdot 3)}{3\cdot 2}}}} = \boxed{\frac{x^{\frac{5}{3}}-1}{x^{\frac{(2\cdot 1)+(1\cdot 3)}{3\cdot 1}}+x^{\frac{4+3}{6}}}}$$

$$= \boxed{\frac{x^{\frac{5}{3}}-1}{x^{\frac{2+3}{3}}+x^{\frac{4+3}{6}}}} = \boxed{\frac{x^{\frac{5}{3}}-1}{x^{\frac{5}{3}}+x^{\frac{7}{6}}}} = \boxed{\frac{\sqrt[3]{x^5}-1}{\sqrt[3]{x^5}+\sqrt[6]{x^7}}} = \boxed{\frac{\sqrt[3]{x^{3+2}}-1}{\sqrt[3]{x^{3+2}}+\sqrt[6]{x^{6+1}}}} = \boxed{\frac{\sqrt[3]{x^3\cdot x^2}-1}{\sqrt[3]{x^3\cdot x^2}+\sqrt[6]{x^6\cdot x^1}}} = \boxed{\frac{\mathbf{x\sqrt[3]{x^2}-1}}{\mathbf{x\sqrt[3]{x^2}+x\sqrt[6]{x}}}}$$

Practice Problems - Adding and Subtracting Negative Fractional Exponents

Section 5.4 Case III Practice Problems - Simplify the following negative fractional exponential expressions:

1. $2^{-\frac{1}{2}}-2^3 =$

2. $\left(x^3+x^2\right)^{-\frac{1}{3}}-3 =$

3. $\dfrac{2}{a^{-\frac{1}{2}}}+a^{-\frac{1}{2}} =$

4. $\dfrac{\frac{3}{5}}{x^2+x^{-\frac{2}{5}}} =$

5. $\dfrac{y^{-\frac{1}{2}}}{y-y^{-\frac{1}{2}}} =$

6. $\dfrac{b^{-\frac{2}{3}}}{b^{-1}+b^{-2}} =$

7. $\dfrac{x^{-1}+y^{\frac{2}{3}}}{x^{-\frac{2}{3}}} =$

8. $\dfrac{\frac{1}{2}}{a^{-\frac{1}{2}}-b^{-\frac{1}{2}}} =$

9. $\dfrac{x^{-1}}{(x-3)^{-\frac{1}{4}}}+y^{-\frac{1}{3}} =$

10. $\dfrac{\left(a^2\right)^{-\frac{2}{3}}+\left(b^{-2}\right)^{-\frac{1}{3}}}{(ab)^{\frac{1}{5}}} =$

Chapter 5 Appendix: Scientific Notation

Scientific notation is used as a method for writing very large and very small numbers. How numbers are changed to scientific notation form is shown in Case I below. The conversion of scientific notation numbers to expanded forms is described in Case II. Multiplication and division of scientific numbers are addressed in Cases III and IV, respectively.

Case I Changing Numbers to Scientific Notation Form

Numbers are changed to scientific notation form using the following steps:

Step 1
a. Place a decimal point to the right of the first non-zero digit of the number.
b. Count the number of digits after the decimal point.
c. Use the number as the exponent of base 10.

Step 2
a. Multiply the number by base 10 raised to "+" or " − " the number counted in Step 1b above.
b. Use " − " sign in the exponent if the decimal point is moved to the right.
c. Use "+" sign in the exponent if the decimal point is moved to the left.

Examples with Steps

The following examples show the steps as to how numbers are changed to scientific notation form:

Example 5A-1

$\boxed{250000} =$

Solution:

Step 1 $\boxed{250000} = \boxed{250000.0}$; *Move the decimal point 5 places to the left.*

Step 2 $\boxed{250000.0} = \boxed{2.50000 \times 10^{+5}} = \boxed{2.5 \times 10^{+5}} = \boxed{\mathbf{2.5 \times 10^{5}}}$

Example 5A-2

$\boxed{0.00064} =$

Solution:

Step 1 $\boxed{0.00064}$; *Move the decimal point 4 places to the right.*

Step 2 $\boxed{0.00064} = \boxed{0006.4 \times 10^{-4}} = \boxed{\mathbf{6.4 \times 10^{-4}}}$

Example 5A-3

$\boxed{473200000} =$

Solution:

Step 1 $\boxed{473200000} = \boxed{473200000.0}$; *Move the decimal point 8 places to the left.*

Step 2 $\boxed{473200000.0} = \boxed{4.73200000 \times 10^{+8}} = \boxed{4.732 \times 10^{+8}} = \boxed{\mathbf{4.732 \times 10^{8}}}$

Example 5A-4

$\boxed{125.0547} =$

Solution:

Step 1 $\boxed{125.0547}$; $\boxed{\textit{Move the decimal point 2 places to the left.}}$

Step 2 $\boxed{125.0547} = \boxed{1.250547 \times 10^{+2}} = \boxed{\mathbf{1.250547 \times 10^{2}}}$

Example 5A-5

$\boxed{0.000000000096} =$

Solution:

Step 1 $\boxed{0.000000000096}$; $\boxed{\textit{Move the decimal point 11 places to the right.}}$

Step 2 $\boxed{0.000000000096} = \boxed{00000000009.6 \times 10^{-11}} = \boxed{\mathbf{9.6 \times 10^{-11}}}$

Additional Examples - Changing Numbers to Scientific Notation Form

The following examples further illustrate how to change numbers to scientific notation form:

Example 5A-6

$\boxed{1234.56} = \boxed{1.23456 \times 10^{+3}} = \boxed{\mathbf{1.23456 \times 10^{3}}}$

Example 5A-7

$\boxed{0.0002456} = \boxed{0002.456 \times 10^{-4}} = \boxed{\mathbf{2.456 \times 10^{-4}}}$

Example 5A-8

$\boxed{2345896} = \boxed{2345896.0} = \boxed{2.345896 \times 10^{+6}} = \boxed{\mathbf{2.345896 \times 10^{6}}}$

Example 5A-9

$\boxed{13} = \boxed{13.0} = \boxed{1.30 \times 10^{+1}} = \boxed{\mathbf{1.30 \times 10^{1}}}$

Example 5A-10

$\boxed{0.00013} = \boxed{0001.3 \times 10^{-4}} = \boxed{\mathbf{1.3 \times 10^{-4}}}$

Example 5A-11

$\boxed{34567.45} = \boxed{3.456745 \times 10^{+4}} = \boxed{\mathbf{3.456745 \times 10^{4}}}$

Example 5A-12

$\boxed{145.5} = \boxed{1.455 \times 10^{+2}} = \boxed{\mathbf{1.455 \times 10^{2}}}$

Example 5A-13

$\boxed{0.1800000} = \boxed{1.800000 \times 10^{-1}} = \boxed{\mathbf{1.8 \times 10^{-1}}}$

Example 5A-14

$\boxed{0.000256} = \boxed{0002.56 \times 10^{-4}} = \boxed{\mathbf{2.56 \times 10^{-4}}}$

Example 5A-15

$\boxed{3600} = \boxed{3.600 \times 10^{+3}} = \boxed{\mathbf{3.6 \times 10^{3}}}$

Practice Problems - Changing Numbers to Scientific Notation Form

Chapter 5 Appendix Case I Practice Problems - Change the following numbers to scientific notation form:

1. 0.00047 =
2. 1245.78 =
3. 0.000000456 =
4. 45789.456 =
5. 23 =
6. 45.8 =
7. 3445.38 =
8. 51244 =
9. 0.0058 =
10. 456794324.0=

Case II Changing Scientific Notation Numbers to Expanded Form

Scientific notation numbers are changed to expanded form using the following steps:

Step 1 a. Move the decimal point to the right the same number of places as the exponent number, if the base 10 exponent is positive.

b. Move the decimal point to the left the same number of places as the exponent number, if the base 10 exponent is negative.

Step 2 Add zeros to the expanded number as needed.

Examples with Steps

The following examples show the steps as to how scientific notations are changed to expanded form:

Example 5A-16

$\boxed{2.45 \times 10^{+3}} =$

Solution:

Step 1a $\boxed{2.45 \times 10^{+3}} = \boxed{2.45 \times 10^{3}}$; *Move the decimal point 3 places to the right. Add one zero.*

Step 2 $\boxed{2.45 \times 10^{3}} = \boxed{\mathbf{2450.}}$

Example 5A-17

$\boxed{3.008 \times 10^{-4}} =$

Solution:

Step 1b $\boxed{3.008 \times 10^{-4}}$; *Move the decimal point 4 places to the left. Add three zeros.*

Step 2 $\boxed{3.008 \times 10^{-4}} = \boxed{\mathbf{0.0003008}}$

Example 5A-18

$\boxed{3.7896845 \times 10^{+6}} =$

Solution:

Step 1a $\boxed{3.7896845 \times 10^{+6}} = \boxed{3.7896845 \times 10^{6}}$; *Move the decimal point 6 places to the right.*

Step 2 $\boxed{3.7896845 \times 10^{6}} = \boxed{\mathbf{3789684.5}}$

Example 5A-19

$\boxed{8.6 \times 10^{-7}} =$

Solution:

Step 1b $\boxed{8.6 \times 10^{-7}}$; *Move the decimal point 7 places to the left. Add six zeros.*

Step 2 $\boxed{8.6 \times 10^{-7}} = \boxed{\mathbf{0.00000086}}$

Example 5A-20

$\boxed{2.3 \times 10^{-1}} =$

Solution:

Step 1b $\boxed{2.3 \times 10^{-1}}$; $\boxed{\textit{Move the decimal point 1 place to the left.}}$

Step 2 $\boxed{2.3 \times 10^{-1}} = \boxed{\mathbf{0.23}}$

Additional Examples - Changing Scientific Notation Numbers to Expanded Form

The following examples further illustrate how to change scientific notation numbers to expanded form:

Example 5A-21

$\boxed{5.436 \times 10^{+4}} = \boxed{54360.0} = \boxed{\mathbf{54360}}$

Example 5A-22

$\boxed{9.4 \times 10^{-3}} = \boxed{\mathbf{0.0094}}$

Example 5A-23

$\boxed{6.4578936 \times 10^{+6}} = \boxed{\mathbf{6457893.6}}$

Example 5A-24

$\boxed{6.459 \times 10^{+2}} = \boxed{\mathbf{645.9}}$

Example 5A-25

$\boxed{9.265 \times 10^{-8}} = \boxed{\mathbf{0.00000009265}}$

Example 5A-26

$\boxed{5.0 \times 10^{+5}} = \boxed{500000.0} = \boxed{\mathbf{500000}}$

Example 5A-27

$\boxed{7.438 \times 10^{-2}} = \boxed{\mathbf{0.07438}}$

Example 5A-28

$\boxed{8.2 \times 10^{-3}} = \boxed{\mathbf{0.0082}}$

Example 5A-29

$\boxed{7.89675 \times 10^{+9}} = \boxed{7896750000.0} = \boxed{\mathbf{7896750000}}$

Example 5A-30

$\boxed{3.2358 \times 10^{-1}} = \boxed{\mathbf{0.32358}}$

Practice Problems - Changing Scientific Notation Numbers to Expanded Form

Chapter 5 Appendix Case II Practice Problems - Change the following scientific notation numbers to expanded forms:

1. $3.8 \times 10^{+4} =$
2. $12.783 \times 10^{-3} =$
3. $2.36796 \times 10^{+7} =$
4. $1.0035 \times 10^{-5} =$
5. $2.5 \times 10^{+2} =$
6. $3.7865 \times 10^{+5} =$
7. $1.00004 \times 10^{-3} =$
8. $2.04506 \times 10^{-1} =$
9. $9.34587654 \times 10^{+9} =$
10. $3.0500 \times 10^{+2} =$

Case III Multiplying Scientific Notation Numbers

Scientific notation numbers are multiplied using the following exponent law:

$$a^m \cdot a^n = a^{m+n}$$

Note that the above exponent law can only be used when the bases are the same. Scientific notation numbers are multiplied by each other using the following steps:

Step 1 Multiply the numbers and apply the Multiplication Law of exponents by adding or subtracting the base 10 exponents.

Step 2 Change the product to scientific notation form.

Examples with Steps

The following examples show the steps as to how scientific notation numbers are multiplied by each other:

Example 5A-31

$$\boxed{\left(4\times 10^{+3}\right)\cdot\left(6\times 10^{+2}\right)} =$$

Solution:

Step 1 $\boxed{\left(4\times 10^{+3}\right)\cdot\left(6\times 10^{+2}\right)} = \boxed{(4\times 6)\cdot\left(10^{+3}\times 10^{+2}\right)} = \boxed{(24)\cdot\left(10^{+3+2}\right)} = \boxed{24\times 10^{+5}}$

Step 2 $\boxed{24\times 10^{+5}} = \boxed{\left(2.4\times 10^{+1}\right)\times 10^{+5}} = \boxed{2.4\times 10^{+1}\times 10^{+5}} = \boxed{2.4\times 10^{+1+5}} = \boxed{\mathbf{2.4\times 10^{6}}}$

Example 5A-32

$$\boxed{\left(5.3\times 10^{+4}\right)\cdot\left(6.8\times 10^{-3}\right)} =$$

Solution:

Step 1 $\boxed{\left(5.3\times 10^{+4}\right)\cdot\left(6.8\times 10^{-3}\right)} = \boxed{(5.3\times 6.8)\cdot\left(10^{+4}\times 10^{-3}\right)} = \boxed{(36.04)\cdot\left(10^{+4-3}\right)} = \boxed{36.04\times 10^{+1}}$

Step 2 $\boxed{36.04\times 10^{+1}} = \boxed{\left(3.604\times 10^{+1}\right)\times 10^{+1}} = \boxed{3.604\times 10^{+1+1}} = \boxed{\mathbf{3.604\times 10^{2}}}$

Example 5A-33

$$\boxed{\left(2\times 10^{-2}\right)\cdot\left(6.6\times 10^{-3}\right)} =$$

Solution:

Step 1 $\boxed{\left(2\times 10^{-2}\right)\cdot\left(6.6\times 10^{-3}\right)} = \boxed{(2\times 6.6)\cdot\left(10^{-2}\times 10^{-3}\right)} = \boxed{(13.2)\cdot\left(10^{-2-3}\right)} = \boxed{13.2\times 10^{-5}}$

Step 2 $\boxed{13.2\times 10^{-5}} = \boxed{\left(1.32\times 10^{+1}\right)\times 10^{-5}} = \boxed{1.32\times 10^{+1-5}} = \boxed{\mathbf{1.32\times 10^{-4}}}$

Example 5A-34

$$\boxed{\left(2.343\times 10^{+4}\right)\cdot\left(6.01\times 10^{-2}\right)} =$$

Solution:

Step 1 $\boxed{\left(2.343\times10^{+4}\right)\cdot\left(6.01\times10^{-2}\right)}=\boxed{(2.343\times6.01)\cdot\left(10^{+4}\times10^{-2}\right)}=\boxed{(14.08143)\cdot\left(10^{+4-2}\right)}$

$=\boxed{14.08143\times10^{+2}}$

Step 2 $\boxed{14.08143\times10^{+2}}=\boxed{\left(1.408143\times10^{+1}\right)\times10^{+2}}=\boxed{1.408143\times10^{+1+2}}=\boxed{\mathbf{1.408143\times10^{3}}}$

Example 5A-35

$\boxed{\left(4\times10^{-3}\right)\cdot\left(2\times10^{+2}\right)\cdot\left(2.6\times10^{-4}\right)}=$

Solution:

Step 1 $\boxed{\left(4\times10^{-3}\right)\cdot\left(2\times10^{+2}\right)\cdot\left(2.6\times10^{-4}\right)}=\boxed{(4\times2\times2.6)\cdot\left(10^{-3}\times10^{+2}\times10^{-4}\right)}$

$=\boxed{(20.8)\cdot\left(10^{-3+2-4}\right)}=\boxed{20.8\times10^{-5}}$

Step 2 $\boxed{20.8\times10^{-5}}=\boxed{\left(2.08\times10^{+1}\right)\times10^{-5}}=\boxed{2.08\times10^{+1-5}}=\boxed{\mathbf{2.08\times10^{-4}}}$

Note: Exponential notation numbers are expressed as the product of the factor and 10 raised to some power. The factor is either a whole number or a decimal number. For example, the exponential notation form of 0.0353, 0.048, 489, 3987 are 353×10^{-4}, 48×10^{-3}, 489×10^{0}, and 39.87×10^{2}, respectively. Scientific notation numbers are also expressed as a product of the factor and 10 raised to some power. However, the factor is always of the form where the decimal point is to the right of the first non-zero digit. For example, 3.48×10^{-1}, and 4.345×10^{2} are in scientific notation form where as 48×10^{-3} is in exponential notation form. The scientific notation form of 48×10^{-3} is 4.8×10^{-2}.

Additional Examples - Multiplying Scientific Notation Numbers

The following examples further illustrate how to multiply scientific notation and exponential notation numbers:

Example 5A-36

$\boxed{\left(2.34\times10^{-2}\right)\cdot\left(9.4\times10^{+3}\right)}=\boxed{(2.34\times9.4)\cdot\left(10^{-2}\times10^{+3}\right)}=\boxed{(21.996)\cdot\left(10^{-2+3}\right)}=\boxed{21.996\times10^{+1}}$

$=\boxed{\left(2.1996\times10^{+1}\right)\times10^{+1}}=\boxed{2.1996\times10^{+1+1}}=\boxed{\mathbf{2.1996\times10^{2}}}$

Example 5A-37

$\boxed{\left(24.6\times10^{-1}\right)\cdot\left(9\times10^{+2}\right)}=\boxed{(24.6\times9)\cdot\left(10^{-1}\times10^{+2}\right)}=\boxed{(221.4)\cdot\left(10^{-1+2}\right)}=\boxed{221.4\times10^{+1}}$

$= \boxed{\left(2.214 \times 10^{+2}\right) \times 10^{+1}} = \boxed{2.214 \times 10^{+2+1}} = \boxed{\mathbf{2.214 \times 10^{3}}}$

Example 5A-38

$\boxed{\left(-33.5 \times 10^{-2}\right) \cdot \left(-8 \times 10^{-2}\right)} = \boxed{(-33.5 \times -8) \cdot \left(10^{-2} \times 10^{-2}\right)} = \boxed{(268) \cdot \left(10^{-2-2}\right)} = \boxed{268 \times 10^{-4}}$

$= \boxed{\left(2.68 \times 10^{+2}\right) \times 10^{-4}} = \boxed{2.68 \times 10^{+2-4}} = \boxed{\mathbf{2.68 \times 10^{-2}}}$

Example 5A-39

$\boxed{\left(-2.02 \times 10^{-3}\right) \cdot \left(99 \times 10^{+3}\right)} = \boxed{(-2.02 \times 99) \cdot \left(10^{-3} \times 10^{+3}\right)} = \boxed{(-199.98) \cdot \left(10^{-3+3}\right)} = \boxed{-199.98 \times 10^{0}}$

$= \boxed{-199.98 \times 1} = \boxed{-199.98} = \boxed{\mathbf{-1.9998 \times 10^{2}}}$

Example 5A-40

$\boxed{\left(2.875 \times 10^{-2}\right) \cdot \left(1.2 \times 10^{-3}\right)} = \boxed{(2.875 \times 1.2) \cdot \left(10^{-2} \times 10^{-3}\right)} = \boxed{(3.45) \cdot \left(10^{-2-3}\right)} = \boxed{\mathbf{3.45 \times 10^{-5}}}$

Example 5A-41

$\boxed{\left(3.456 \times 10^{-2}\right) \cdot \left(2.544 \times 10^{+5}\right)} = \boxed{(3.456 \times 2.544) \cdot \left(10^{-2} \times 10^{+5}\right)} = \boxed{(8.792064) \cdot \left(10^{-2+5}\right)} = \boxed{\mathbf{8.792064 \times 10^{3}}}$

Example 5A-42

$\boxed{\left(2.44 \times 10^{0}\right) \cdot \left(7.4 \times 10^{+1}\right)} = \boxed{(2.44 \times 7.4) \cdot \left(10^{0} \times 10^{+1}\right)} = \boxed{(18.056) \cdot \left(1 \times 10^{+1}\right)} = \boxed{18.056 \times 10^{+1}}$

$= \boxed{\left(1.8056 \times 10^{+1}\right) \times 10^{+1}} = \boxed{1.8056 \times 10^{1+1}} = \boxed{\mathbf{1.8056 \times 10^{2}}}$

Example 5A-43

$\boxed{\left(23.5 \times 10^{-2}\right) \cdot \left(44.2 \times 10^{+4}\right)} = \boxed{(23.5 \times 44.2) \cdot \left(10^{-2} \times 10^{+4}\right)} = \boxed{(1038.7) \cdot \left(10^{-2+4}\right)} = \boxed{1038.7 \times 10^{+2}}$

$= \boxed{\left(1.0387 \times 10^{+3}\right) \times 10^{+2}} = \boxed{1.0387 \times 10^{+3+2}} = \boxed{\mathbf{1.0387 \times 10^{5}}}$

Example 5A-44

$\boxed{\left(4.04 \times 10^{-1}\right) \cdot \left(2.4 \times 10^{-1}\right)} = \boxed{(4.04 \times 2.4) \cdot \left(10^{-1} \times 10^{-1}\right)} = \boxed{(9.696) \cdot \left(10^{-1-1}\right)} = \boxed{\mathbf{9.696 \times 10^{-2}}}$

Example 5A-45

$\boxed{\left(12 \times 10^{-4}\right) \cdot \left(36 \times 10^{+6}\right)} = \boxed{(12 \times 36) \cdot \left(10^{-4} \times 10^{+6}\right)} = \boxed{(432) \cdot \left(10^{-4+6}\right)} = \boxed{432 \times 10^{+2}} = \boxed{\left(4.32 \times 10^{+2}\right) \times 10^{+2}}$

$= \boxed{4.32 \times 10^{+2+2}} = \boxed{\mathbf{4.32 \times 10^{4}}}$

Practice Problems - Multiplying Scientific Notation Numbers

Chapter 5 Appendix Case III Practice Problems - Multiply the following scientific notation and exponential notation numbers. Show the answers in scientific notation form:

1. $\left(5.4\times10^{-3}\right)\cdot\left(1.2\times10^{+3}\right)=$
2. $\left(12.564\times10^{+3}\right)\cdot\left(9\times10^{+2}\right)=$
3. $\left(2.002\times10^{-2}\right)\cdot\left(3\times10^{-2}\right)=$
4. $\left(5\times10^{-2}\right)\cdot\left(8\times10^{+6}\right)=$
5. $\left(22.34\times10^{-4}\right)\cdot\left(39.4\times10^{-3}\right)=$
6. $\left(4.334\times10^{-2}\right)\cdot\left(2.94\times10^{+4}\right)=$
7. $\left(2\times10^{-4}\right)\cdot\left(9\times10^{+5}\right)=$
8. $\left(8.01\times10^{-5}\right)\cdot\left(3.4\times10^{-1}\right)=$
9. $\left(4.4\times10^{+1}\right)\cdot\left(5.4\times10^{+1}\right)=$
10. $\left(2.889\times10^{-1}\right)\cdot\left(9\times10^{-2}\right)=$

Case IV Dividing Scientific Notation Numbers

Scientific notation numbers are divided using the following exponent law:

$$\frac{a^m}{a^n} = a^m \cdot a^{-n} = a^{m-n}$$

Note that the above exponent law can only be used when the bases are the same. Scientific notation numbers are divided by one another using the following steps:

Step 1 Divide the numerator by the denominator and apply the Division Law of exponents by subtracting the base 10 exponents.

Step 2 Change the quotient to scientific notation form.

Examples with Steps

The following examples show the steps as to how scientific notation numbers are divided by each other:

Example 5A-46

$$\boxed{\frac{2.4 \times 10^{+5}}{2 \times 10^{+2}}} =$$

Solution:

Step 1 $\boxed{\frac{2.4 \times 10^{+5}}{2 \times 10^{+2}}} = \boxed{\left(\frac{\overset{1.2}{\cancel{2.4}}}{\underset{1}{\cancel{2}}}\right) \cdot \left(\frac{10^{+5}}{10^{+2}}\right)} = \boxed{\left(\frac{1.2}{1}\right) \cdot \left(10^{+5} \times 10^{-2}\right)} = \boxed{(1.2) \cdot \left(10^{+5-2}\right)} = \boxed{\mathbf{1.2 \times 10^3}}$

Step 2 *Not Applicable*

Example 5A-47

$$\boxed{\frac{2.346 \times 10^{-4}}{4 \times 10^{-2}}} =$$

Solution:

Step 1 $\boxed{\frac{2.346 \times 10^{-4}}{4 \times 10^{-2}}} = \boxed{\left(\frac{\overset{0.5865}{\cancel{2.346}}}{\underset{1}{\cancel{4}}}\right) \cdot \left(\frac{10^{-4}}{10^{-2}}\right)} = \boxed{\left(\frac{0.5865}{1}\right) \cdot \left(10^{-4} \times 10^{+2}\right)} = \boxed{(0.5865) \cdot \left(10^{-4+2}\right)}$

$= \boxed{0.5865 \times 10^{-2}}$

Step 2 $\boxed{0.5865 \times 10^{-2}} = \boxed{\left(5.865 \times 10^{-1}\right) \times 10^{-2}} = \boxed{5.865 \times 10^{-1-2}} = \boxed{\mathbf{5.865 \times 10^{-3}}}$

Example 5A-48

$$\boxed{\frac{3.65 \times 10^{-2}}{5.5 \times 10^{+1}}} =$$

Solution:

Step 1

$$\frac{3.65 \times 10^{-2}}{5.5 \times 10^{+1}} = \left(\frac{\overset{0.6636}{\cancel{3.65}}}{\underset{1}{\cancel{5.5}}}\right) \cdot \left(\frac{10^{-2}}{10^{+1}}\right) = \left(\frac{0.6636}{1}\right) \cdot \left(10^{-2} \times 10^{-1}\right) = (0.6636) \cdot \left(10^{-2-1}\right)$$

$$= 0.6636 \times 10^{-3}$$

Step 2

$$0.6636 \times 10^{-3} = \left(6.636 \times 10^{-1}\right) \times 10^{-3} = 6.636 \times 10^{-1-3} = \mathbf{6.636 \times 10^{-4}}$$

Example 5A-49

$$\frac{3.8 \times 10^{+6}}{6 \times 10^{+2}} =$$

Solution:

Step 1

$$\frac{3.8 \times 10^{+6}}{6 \times 10^{+2}} = \left(\frac{\overset{0.6333}{\cancel{3.8}}}{\underset{1}{\cancel{6}}}\right) \cdot \left(\frac{10^{+6}}{10^{+2}}\right) = \left(\frac{0.6333}{1}\right) \cdot \left(10^{+6} \times 10^{-2}\right) = (0.6333) \cdot \left(10^{+6-2}\right)$$

$$= 0.6333 \times 10^{4}$$

Step 2

$$0.6333 \times 10^{4} = \left(6.333 \times 10^{-1}\right) \times 10^{4} = 6.333 \times 10^{-1+4} = \mathbf{6.333 \times 10^{3}}$$

Example 5A-50

$$\frac{1.24 \times 10^{+7}}{8.248 \times 10^{-3}} =$$

Solution:

Step 1

$$\frac{1.24 \times 10^{+7}}{8.248 \times 10^{-3}} = \left(\frac{\overset{0.15034}{\cancel{1.24}}}{\underset{1}{\cancel{8.248}}}\right) \cdot \left(\frac{10^{+7}}{10^{-3}}\right) = \left(\frac{0.15034}{1}\right) \cdot \left(10^{+7} \times 10^{+3}\right) = (0.15034) \cdot \left(10^{+7+3}\right)$$

$$= 0.15034 \times 10^{10}$$

Step 2

$$0.15034 \times 10^{10} = \left(1.5034 \times 10^{-1}\right) \times 10^{10} = 1.5034 \times 10^{-1+10} = \mathbf{1.5034 \times 10^{9}}$$

Additional Examples - Dividing Scientific Notation Numbers

The following examples further illustrate how to divide scientific notation and exponential notation numbers:

Example 5A-51

$$\frac{2.857 \times 10^{+2}}{8 \times 10^{0}} = \left(\frac{2.857}{8}\right) \cdot \left(\frac{10^{+2}}{10^{0}}\right) = (0.35712) \cdot \left(\frac{10^{+2}}{1}\right) = 0.35712 \times 10^{+2} = \left(3.5712 \times 10^{-1}\right) \times 10^{+2}$$

$$= \boxed{3.5712 \times 10^{-1+2}} = \boxed{\mathbf{3.5712 \times 10^{1}}}$$

Example 5A-52

$$\boxed{\frac{2352.4 \times 10^{+4}}{4.4 \times 10^{+3}}} = \boxed{\left(\frac{2352.4}{4.4}\right) \cdot \left(\frac{10^{+4}}{10^{+3}}\right)} = \boxed{(534.636) \cdot \left(10^{+4} \times 10^{-3}\right)} = \boxed{534.636 \times 10^{+4-3}} = \boxed{534.636 \times 10^{+1}}$$

$$= \boxed{\left(5.34636 \times 10^{+2}\right) \times 10^{+1}} = \boxed{5.34636 \times 10^{2+1}} = \boxed{\mathbf{5.34636 \times 10^{3}}}$$

Example 5A-53

$$\boxed{\frac{0.454 \times 10^{+5}}{0.04 \times 10^{-1}}} = \boxed{\left(\frac{0.454}{0.04}\right) \cdot \left(\frac{10^{+5}}{10^{-1}}\right)} = \boxed{(11.35) \cdot \left(10^{+5} \times 10^{+1}\right)} = \boxed{11.35 \times 10^{+5+1}} = \boxed{11.35 \times 10^{+6}}$$

$$= \boxed{\left(1.135 \times 10^{+1}\right) \times 10^{6}} = \boxed{1.135 \times 10^{1+6}} = \boxed{\mathbf{1.135 \times 10^{7}}}$$

Example 5A-54

$$\boxed{\frac{45 \times 10^{+6}}{5 \times 10^{+4}}} = \boxed{\left(\frac{45}{5}\right) \cdot \left(\frac{10^{+6}}{10^{+4}}\right)} = \boxed{(9) \cdot \left(10^{+6} \times 10^{-4}\right)} = \boxed{9 \times 10^{+6-4}} = \boxed{9 \times 10^{+2}} = \boxed{\mathbf{9.0 \times 10^{2}}}$$

Example 5A-55

$$\boxed{\frac{4555 \times 10^{-3}}{3.5 \times 10^{-2}}} = \boxed{\left(\frac{4555}{3.5}\right) \cdot \left(\frac{10^{-3}}{10^{-2}}\right)} = \boxed{(1301.429) \cdot \left(10^{-3} \times 10^{+2}\right)} = \boxed{1301.429 \times 10^{-3+2}} = \boxed{1301.429 \times 10^{-1}}$$

$$= \boxed{\left(1.301429 \times 10^{+3}\right) \times 10^{-1}} = \boxed{1.301429 \times 10^{+3-1}} = \boxed{\mathbf{1.301429 \times 10^{2}}}$$

Example 5A-56

$$\boxed{\frac{232.44 \times 10^{+1}}{5 \times 10^{+2}}} = \boxed{\left(\frac{232.44}{5}\right) \cdot \left(\frac{10^{+1}}{10^{+2}}\right)} = \boxed{(46.488) \cdot \left(10^{+1} \times 10^{-2}\right)} = \boxed{46.488 \times 10^{+1-2}} = \boxed{46.488 \times 10^{-1}}$$

$$= \boxed{\left(4.6488 \times 10^{+1}\right) \times 10^{-1}} = \boxed{4.6488 \times 10^{+1-1}} = \boxed{\mathbf{4.6488 \times 10^{0}}}$$

Example 5A-57

$$\boxed{\frac{52.44 \times 10^{0}}{2 \times 10^{0}}} = \boxed{\frac{52.44 \times 1}{2 \times 1}} = \boxed{\frac{52.44}{2}} = \boxed{26.22} = \boxed{2.622 \times 10^{+1}} = \boxed{\mathbf{2.622 \times 10^{1}}}$$

Example 5A-58

$$\boxed{\frac{452.24 \times 10^{+3}}{4.2 \times 10^{+1}}} = \boxed{\left(\frac{452.24}{4.2}\right) \cdot \left(\frac{10^{+3}}{10^{+1}}\right)} = \boxed{(107.676) \cdot \left(10^{+3} \times 10^{-1}\right)} = \boxed{107.676 \times 10^{+3-1}} = \boxed{107.676 \times 10^{+2}}$$

$$= \boxed{\left(1.07676 \times 10^{+2}\right) \times 10^{+2}} = \boxed{1.07676 \times 10^{+2+2}} = \boxed{\mathbf{1.07676 \times 10^{4}}}$$

Example 5A-59

$$\boxed{\frac{48\times10^{+2}}{4.8\times10^{+3}}}=\boxed{\left(\frac{48}{4.8}\right)\cdot\left(\frac{10^{+2}}{10^{+3}}\right)}=\boxed{(10)\cdot\left(10^{+2}\times10^{-3}\right)}=\boxed{10\times10^{+2-3}}=\boxed{10\times10^{+2-3}}=\boxed{10\times10^{-1}}$$

$$=\boxed{\left(1.0\times10^{+1}\right)\times10^{-1}}=\boxed{1.0\times10^{+1-1}}=\boxed{\mathbf{1.0\times10^{0}}}$$

Example 5A-60

$$\boxed{\frac{452.44\times10^{0}}{8.8\times10^{+2}}}=\boxed{\left(\frac{452.44}{8.8}\right)\cdot\left(\frac{10^{0}}{10^{+2}}\right)}=\boxed{(51.414)\cdot\left(10^{0}\times10^{-2}\right)}=\boxed{51.414\times10^{0-2}}=\boxed{51.414\times10^{-2}}$$

$$=\boxed{\left(5.1414\times10^{+1}\right)\times10^{-2}}=\boxed{5.1414\times10^{+1-2}}=\boxed{\mathbf{5.1414\times10^{-1}}}$$

Practice Problems - Dividing Scientific Notation Numbers

Chapter 5 Appendix Case IV Practice Problems - Divide the following scientific notation and exponential notation numbers. Show the answers in scientific notation form.

1. $\frac{48.4\times10^{+4}}{12\times10^{+3}}=$
2. $\frac{8.4\times10^{-3}}{2.2\times10^{+5}}=$
3. $\frac{235.5\times10^{+3}}{5\times10^{+3}}=$
4. $\frac{1.84\times10^{+2}}{0.2\times10^{-1}}=$
5. $\frac{14.484\times10^{-3}}{2\times10^{-2}}=$
6. $\frac{2444.4\times10^{+5}}{4.4\times10^{+4}}=$
7. $\frac{35.745\times10^{+5}}{0.35\times10^{+2}}=$
8. $\frac{8.45\times10^{-4}}{5.5\times10^{-3}}=$
9. $\frac{2.4\times10^{0}}{4\times10^{+1}}=$
10. $\frac{22.45\times10^{+6}}{2.2\times10^{0}}=$

Chapter 6
Polynomials

Quick Reference to Chapter 6 Case Problems

$\boxed{x^2y^2z} = ;\quad \boxed{x^3 - 2x^2} = ;\quad \boxed{w^5 - 2w^3 + 4w^2 + 7} =$

$\boxed{18x^3 + 2x^2 - 5x^3 - 2x - x^2} = ;\quad \boxed{-8y^3 + 4y^5 - 5y^3 - 2y^5 + 3} = ;\quad \boxed{2w^4 + 4w^3 - w^4 - 8 + 2w - w^3 + 4} =$

$\boxed{\left(5x^3y^2\right)\cdot\left(3x^3y^2z\right)} = ;\quad \boxed{\left(3a^2b^3c^5\right)\cdot\left(5b^2c^4\right)\cdot\left(4a^3b^0c^3\right)} = ;\quad \boxed{\left(3x^2\right)\cdot(5xy)\cdot\left(2x^2y^3\right)} =$

$\boxed{\left(2x^4 + 3x^2 + 5x - x^4 + x^2 - 3\right)\cdot\left(3x^2\right)} = ;\quad \boxed{\left(5m^3n^3 + 2m^2n^2 - 3mn + mn + 2\right)\cdot(5mn)} = ;$

$\boxed{\left(\sqrt{27}x^2 - 2 + \sqrt{8}x + \sqrt{36}\right)\cdot\sqrt{125}x} =$

$\boxed{\left(x^2 + 3x\right)(x + 8)} = ;\quad \boxed{\left(\sqrt{225}x + 2\right)\left(5x - \sqrt{81}\right)} = ;\quad \boxed{\left(\frac{2}{3}x^3 + \frac{1}{3}x\right)\left(\frac{1}{2}x^2 - \frac{3}{8}\right)} =$

$\boxed{\left(x^3 + 5x^2 - 3x + 3\right)\left(x^2 - x + 1\right)} = ;\quad \boxed{\left(-x^5 + 3x^3 - x + 5\right)\left(x^3 - 2x - 3\right)} = ;\quad \boxed{\left(a^4 + 3a^3 - 2a^2 + 5\right)(a - 1)} =$

$\boxed{\frac{\sqrt{8x^3y^2}}{\sqrt{243xy^3}}} = ;\quad \boxed{\frac{-\sqrt{12a^2b^2c}}{\sqrt{225abc^4}}} = ;\quad \boxed{\frac{\sqrt[3]{16u^2v^2}}{-\sqrt[3]{27uv^3}}} =$

$\boxed{\frac{8x^3 - 16x^2}{-8x}} = ;\quad \boxed{\frac{-15\sqrt{a^3} + 10\sqrt{a^2}}{-5a^2}} = ;\quad \boxed{\frac{\sqrt[3]{x^3y^5} - 4\sqrt[4]{x^8y^4}}{2\sqrt[3]{x^3y^6}}} =$

Case I c - Dividing Polynomials by Monomials, p. 393

$\boxed{\dfrac{-\sqrt{144}x^5+6x^3+\sqrt{16}x^2-24}{2x^2}}=$; $\boxed{\dfrac{x^2y-4xy^2+2x-4}{-2\sqrt[3]{x^3y^3}}}=$; $\boxed{\dfrac{\sqrt{32}c^3-\sqrt{8}c^2-8c+\sqrt{72}}{\sqrt{2}c^3}}=$

Case II - Dividing Polynomials by Polynomials, *p. 399*

$\boxed{Divide\ x^4+8x^3+16x^2+5x\ \ by\ \ x^2+3x+1}=$; $\boxed{Divide\ 6x^2+19x+18\ \ by\ \ 3x+5}=$;

$\boxed{Divide\ x^4+2x^3+2x^2+2x+6\ \ by\ \ x+1}=$

Case I - Adding and Subtracting Polynomials Horizontally, *p. 409*

$\boxed{\left(x^2+3x^3+5\right)+\left(x^3+8x+2x^2\right)}=$; $\boxed{\left(y+y^2+3y^3+3\right)-\left(3y^2+2y-y^3\right)}=$;

$\boxed{\left(k^2l+kl^2+k+2k^2l\right)-\left(3k-2kl^2-k^2l\right)}=$

Case II - Adding and Subtracting Polynomials Vertically, *p. 414*

$\boxed{\left(x^4+x+3x^3+4x\right)+\left(x^2+3x^4-x^3+x\right)}=$; $\boxed{\left(w^4+2w^3+w+2w^2\right)+\left(2w+4w^2+6w\right)}=$;

$\boxed{\left(a^2+3a+2a^2-a-2\right)-\left(4-4a^2-3a-6\right)}=$

Chapter 6 - Polynomials

The objective of this chapter is to improve the student's ability to solve and simplify mathematical expressions involving various classes of polynomials. In Section 6.1 the student is introduced to different classes of polynomials and learns how to identify a polynomial's degree and write a polynomial in standard form. Multiplication of polynomials is addressed in Section 6.2. How monomials, binomials, and polynomials are multiplied by one another is also addressed in this section. Division of polynomials is addressed in Section 6.3. In this section the student learns how to divide a monomial, binomial, or a polynomial by another monomial, binomial, or polynomial. The steps as to how polynomials are added and subtracted horizontally and vertically are addressed in Section 6.4. Each section is concluded by solving additional examples with practice problems given in each section to further enhance the student's ability on the subject.

6.1 Introduction to Polynomials

A polynomial is an algebraic expression that can be expressed in the following general form:

$$P(x) = a_n x^n + a_{n-1} x^{n-1} + a_{n-2} x^{n-2} + \cdots + a_0$$

where a_n, a_{n-1}, a_{n-2}, ..., and a_0 are real numbers, n is a positive integer number, and x is a variable. Note that in the above algebraic expression the $+$ or $-$ signs separate the polynomial to **terms**, i.e., $a_n x^n$, $a_{n-1} x^{n-1}$, $a_{n-2} x^{n-2}$, and a_0 are each referred to as a polynomial term. Classification of polynomials and how polynomials are simplified is discussed in the following two cases.

Case I Polynomials Classification

Polynomials are usually named by their number of terms and are stated by the degree of the highest power of the variable in the polynomial. A polynomial is defined in the following way:

1. Definition of a Polynomial

A polynomial is a variable expression consisting of one or more terms. Note that in a polynomial the variable in each term has positive integer exponent. For example,

$x^3 + 5x$, $x^2 + 2x + 5$, $\frac{4}{5}x^2 + \frac{2}{3}x + \frac{1}{6}$, $3x^2$,

$9u^5 + 8u^3 - 6u - 5$, $x^3 - 1$, and $y^3 + 2y^2$

are polynomials. However,

$2x^4 - 5x^3 + 2x^{-2} + 5$, $6x^4 + \frac{2}{x^3} + 2x^2 - 5x$, $6w^6 + \frac{2}{w^4} - \frac{5}{w} + 3$, $\frac{5}{x^2} - 3$,

$2m^{-8} - \frac{5}{6}m^{-4} + \frac{2}{3}m^{-1}$, and $y^{-5} + 3y^{-2} - 2y + 6$

are not polynomials since the variable in one or more terms of the polynomials contain negative integer exponents.

Note that polynomial terms can have one or more variables. For example,

$x^4y^3 + 2x^3y^3 + 3x^2y^2 + 2xy - 5$, $\sqrt{5}a^4b^3 + \sqrt{3}a^2b^2 - \sqrt{2}ab + 12$,

$8x^4y^2z^3 + 3x^2y^2z - 2xyz + 1$, and $r^5s^4t^3 + 3rs^2t + 2r^2st - 4rst + 3$

are polynomials with two and three variable terms. In these instances a polynomial can be written in standard form in different ways depending on the variable selected.

For example, the polynomial $x^4y^3 + 2xy^2 - x^5 - 3x^2y^4 + 5$ is written in standard form as:

- $-x^5 + \left(y^3\right)x^4 + \left(-3y^4\right)x^2 + \left(2y^2\right)x + 5$ for the variable x, and
- $\left(-3x^2\right)y^4 + \left(x^4\right)y^3 + (2x)y^2 - x^5 + 5$ for the variable y

(See additional examples 6.1-7, 6.1-9, 6.1-12, and 6.1-15 in Section 6.1, Case II).

2. *Classification of Polynomials*

Polynomials are named by their number of terms. For example, a polynomial with one term only is called a **monomial**. A polynomial with two terms is called a **binomial** and a polynomial with three terms is called a **trinomial**. A polynomial with more than three terms is simply called a **polynomial**. For example,

- $5x^2$, 50, $\sqrt{2}y^3$, x^3y^3, x^2y^2z, $\frac{2}{3}y^2z^3$, and $8w$ are referred to as monomial expressions.
- $x^3 - 2x^2$, $y^5 + 1$, $2a^3b^4 + \sqrt{5}$, $\frac{2}{3}a^4 + \frac{1}{4}$, $\frac{1}{2}u^2w^4 + 4uw$, and $x + 2$ are referred to as binomial expressions.
- $x^5 - 2x^2 + 6x$, $y^7 + 4y^3 + 2y$, $x^4y^3 - 2x^2y + \sqrt{3}$, $a^8 - 4a^3 + 6a$, and $-\frac{1}{3}m^4 - \frac{4}{5}m^3 + 6m$ are referred to as trinomial expressions.
- $x^6 - 4x^2 + 6x + 1$, $w^5 - 2w^3 + 4w^2 + 7$, $x^5y^6 - 2x^2y^3 + 6xy + 1$, and $-x^4 - 2x^3 + 6x^2 + 7x - 5$ are referred to as polynomial expressions.

3. *Degree of Polynomials*

The degree of a polynomial is determined by the highest power of the variable in the polynomial. For example,

- $25x^0 = 25$ is a zero degree polynomial.
- $2x^1 + 1 = 2x + 1$ is a first degree polynomial.
- $3z^2 + 6z - 4$ is a second degree polynomial.
- $-3 + 5n^3$ is a third degree polynomial.
- $-4a^4 + 2a^3 + 2a^2 - 6a + 2$ is a fourth degree polynomial.
- $2u - \sqrt{3}u^6 - 3u^2 + 2$ is a sixth degree polynomial.
- $m^4 + 2m^5 + 3m^8 - m + 2$ is an eighth degree polynomial.

In general, the degree of a polynomial is an indication of the number of roots that polynomial has. Solving for polynomial roots is a subject which is beyond the scope of this book and will be addressed in the future series of *Mastering Algebra* books.

4. *Polynomials in Standard Form*

A polynomial in **standard form** is defined as a polynomial in which the terms of the polynomial are written in order from the highest to the lowest power of the variable. For example,

$y^6 + 3y^5 - 2y^3 + 6$, $x^5 - 2x^2 + 6x + 1$, $x^4 + 2x - 1$, and $a^3 + a + 1$

are polynomials written in standard form. Note that the powers in a polynomial written in standard form decreases as we go from left to right.

In general, when a polynomial is written with the highest power of the variable first, followed by the second, third, fourth, fifth, etc. highest power of the variable, the polynomial is said to be in **descending order**.

Table 6-1 show examples of polynomials indicating their type, degree, and number of terms.

Table 6-1: Polynomials

Polynomial in Standard Form	Type	Degree	Number of Terms
$x^3 + 6x^2$	binomial	3	2
$x^3 + 2x + 5$	trinomial	3	3
$x + 1$	binomial	1	2
$5y^5$	monomial	5	1
$35 = 35x^0$	monomial	0	1
$5x^5 - 2x^4 - 8x^2 + 3$	polynomial	5	4
$x^4 + 2x^3 + 5$	trinomial	4	3
$x^{-4} + 2x^2 + 5x - 3$	not a polynomial		
$w^6 - 4w^5 + 2w^3 + 5w^2 - 1$	polynomial	6	5
$x^2 + 3x - 2$	trinomial	2	3
$x^4 + \frac{1}{x^3} - 5$	not a polynomial		

Practice Problems - Polynomials Classification

Section 6.1 Case I Practice Problems - Write the following polynomials in standard form and identify each polynomial type, its degree, and number of terms.

1. $3x + 2x^3 - 6$
2. $-6y^8 + 2$
3. $2w + 6w^2 + 8w^5$
4. $6y$
5. $\sqrt{72}$
6. $-16 + 2x^4$
7. $x^5 + 8x^4 + 2x - x^3 - 5$
8. $x^4 + 2x^2 - x^{-1} + 1$
9. $\frac{2}{3}y + \frac{5}{8}y^2 + \frac{1}{4}y^3 - \frac{1}{2}$
10. $x^4 - 2x^3 + \frac{5}{x} - 4$

Case II Simplifying Polynomials

Polynomials are simplified using the following steps:

Step 1 Group like terms.

Step 2 Combine like terms and write the polynomial in standard form.

Note that **like terms** are defined as polynomial terms having the same variables raised to the same power. For example, in the polynomial expression:

$8y^3 + 5y^2 - 2y^3 - y + 5y^3 - 20 + y^2 - 3y + 4$

$8y^3$, $-2y^3$, *and* $5y^3$; $5y^2$ *and* y^2; $-y$ *and* $-3y$; -20 *and* $+4$;

are like terms of one another.

Examples with Steps

The following examples show the steps as to how polynomials are simplified:

Example 6.1-1

$\boxed{18x^3 + 2x^2 - 5x^3 - 2x - x^2} =$

Solution:

Step 1 $\boxed{18x^3 + 2x^2 - 5x^3 - 2x - x^2} = \boxed{\left(18x^3 - 5x^3\right) + \left(2x^2 - x^2\right) - 2x}$

Step 2 $\boxed{\left(18x^3 - 5x^3\right) + \left(2x^2 - x^2\right) - 2x} = \boxed{(18-5)x^3 + (2-1)x^2 - 2x} = \boxed{\mathbf{13x^3 + x^2 - 2x}}$

Example 6.1-2

$\boxed{-8y^3 + 4y^5 - 5y^3 - 2y^5 + 3} =$

Solution:

Step 1 $\boxed{-8y^3 + 4y^5 - 5y^3 - 2y^5 + 3} = \boxed{\left(-8y^3 - 5y^3\right) + \left(4y^5 - 2y^5\right) + 3}$

Step 2 $\boxed{\left(-8y^3 - 5y^3\right) + \left(4y^5 - 2y^5\right) + 3} = \boxed{(-8-5)y^3 + (4-2)y^5 + 3} = \boxed{-13y^3 + 2y^5 + 3}$

$= \boxed{\mathbf{2y^5 - 13y^3 + 3}}$

Example 6.1-3

$\boxed{2w^4 + 4w^3 - w^4 - 8 + 2w - w^3 + 4} =$

Solution:

Step 1 $\boxed{2w^4 + 4w^3 - w^4 - 8 + 2w - w^3 + 4} = \boxed{\left(2w^4 - w^4\right) + \left(4w^3 - w^3\right) + (-8+4) + 2w}$

Step 2 $\boxed{\left(2w^4 - w^4\right) + \left(4w^3 - w^3\right) + (-8+4) + 2w} = \boxed{(2-1)w^4 + (4-1)w^3 - 4 + 2w}$

$= \boxed{w^4 + 3w^3 - 4 + 2w} = \boxed{\mathbf{w^4 + 3w^3 + 2w - 4}}$

Example 6.1-4

$\boxed{2a^8 + 4a^5 - 8 + 2a - 5a^8 + 4 - a - a^5} =$

Solution:

Step 1 $\boxed{2a^8 + 4a^5 - 8 + 2a - 5a^8 + 4 - a - a^5} = \boxed{\left(2a^8 - 5a^8\right) + \left(4a^5 - a^5\right) + (-8+4) + (2a - a)}$

Step 2 $\boxed{\left(2a^8 - 5a^8\right) + \left(4a^5 - a^5\right) + (-8+4) + (2a - a)} = \boxed{(2-5)a^8 + (4-1)a^5 - 4 + (2-1)a}$

$= \boxed{-3a^8 + 3a^5 - 4 + a} = \boxed{\mathbf{-3a^8 + 3a^5 + a - 4}}$

Example 6.1-5

$\boxed{2x^4y^4 + 4x^3y^3 - x^4y^4 + 5 - x^3y^3 + 3} =$

Solution:

Step 1 $\boxed{2x^4y^4 + 4x^3y^3 - x^4y^4 + 5 - x^3y^3 + 3} = \boxed{\left(2x^4y^4 - x^4y^4\right) + \left(4x^3y^3 - x^3y^3\right) + (5+3)}$

Step 2 $\boxed{\left(2x^4y^4 - x^4y^4\right) + \left(4x^3y^3 - x^3y^3\right) + (5+3)} = \boxed{(2-1)x^4y^4 + (4-1)x^3y^3 + 8}$

$= \boxed{\mathbf{x^4y^4 + 3x^3y^3 + 8}}$

Additional Examples - Simplifying Polynomials

The following examples further illustrate how to simplify and write polynomials in standard form:

Example 6.1-6

$\boxed{-4w^7 + 3w^3 - 5 + 2w^7 + 2w^3 - 5w^7 + w^3 - 3} = \boxed{\left(-4w^7 + 2w^7 - 5w^7\right) + \left(3w^3 + 2w^3 + w^3\right) + (-5-3)}$

$= \boxed{(-4+2-5)w^7 + (3+2+1)w^3 - 8} = \boxed{\mathbf{-7w^7 + 6w^3 - 8}}$

Example 6.1-7

$\boxed{-2x^4y^3 + x^2y + 5x^3y^5 - 8xy^2 + 3x^3y^5 - 6x^2y} = \boxed{-2x^4y^3 + \left(5x^3y^5 + 3x^3y^5\right) - 8xy^2 + \left(-6x^2y + x^2y\right)}$

$= \boxed{-2x^4y^3 + (5+3)x^3y^5 - 8xy^2 + (-6+1)x^2y} = \boxed{-2x^4y^3 + 8x^3y^5 - 8xy^2 - 5x^2y}$

$= \boxed{\mathbf{\left(-2y^3\right)x^4 + \left(8y^5\right)x^3 + (-5y)x^2 + \left(-8y^2\right)x} \quad \textit{in standard form for the variable } x}$

$$= \boxed{\left(8x^3\right)y^5 + \left(-2x^4\right)y^3 + (-8x)y^2 + \left(-5x^2\right)y \quad \textit{in standard form for the variable } y}$$

Example 6.1-8

$$\boxed{-5y^7 + y + 5y^5 + 12y^7 - 5y^4 + y^5 - 3y^4 - 5y} = \boxed{\left(-5y^7 + 12y^7\right) + (y - 5y) + \left(5y^5 + y^5\right) + \left(-5y^4 - 3y^4\right)}$$

$$= \boxed{(-5+12)y^7 + (1-5)y + (5+1)y^5 + (-5-3)y^4} = \boxed{7y^7 - 4y + 6y^5 - 8y^4} = \boxed{7y^7 + 6y^5 - 8y^4 - 4y}$$

Example 6.1-9

$$\boxed{-8 + 2u^5v^5 + 6u^3v^3 - 5 + 2u^4v - 8u^2v^2 + 2u^3v^3} = \boxed{(-8-5) + 2u^5v^5 + \left(6u^3v^3 + 2u^3v^3\right) + 2u^4v - 8u^2v^2}$$

$$= \boxed{-13 + 2u^5v^5 + (6+2)u^3v^3 + 2u^4v - 8u^2v^2} = \boxed{-13 + 2u^5v^5 + 8u^3v^3 + 2u^4v - 8u^2v^2}$$

$$= \boxed{\left(2v^5\right)u^5 + (2v)u^4 + \left(8v^3\right)u^3 + \left(-8v^2\right)u^2 - 13 \quad \textit{in standard form for the variable } u}$$

$$= \boxed{\left(2u^5\right)v^5 + \left(8u^3\right)v^3 + \left(-8u^2\right)v^2 + \left(2u^4\right)v - 13 \quad \textit{in standard form for the variable } v}$$

Example 6.1-10

$$\boxed{5m^5 + 5m^3 + 10m^5 - 6 - m^5 + 3m^3 + 4m + 9} = \boxed{\left(5m^5 + 10m^5 - m^5\right) + \left(5m^3 + 3m^3\right) + (-6+9) + 4m}$$

$$= \boxed{(5+10-1)m^5 + (5+3)m^3 + 3 + 4m} = \boxed{14m^5 + 8m^3 + 3 + 4m} = \boxed{14m^5 + 8m^3 + 4m + 3}$$

Example 6.1-11

$$\boxed{a^8 + 5a^4 + 4a^6 - 2a^8 - a^6 + 9a^4} = \boxed{\left(a^8 - 2a^8\right) + \left(5a^4 + 9a^4\right) + \left(4a^6 - a^6\right)} = \boxed{(1-2)a^8 + (5+9)a^4 + (4-1)a^6}$$

$$= \boxed{-a^8 + 14a^4 + 3a^6} = \boxed{-a^8 + 3a^6 + 14a^4}$$

Example 6.1-12

$$\boxed{5x^3y^4z^2 + 10xyz + 2x^2y^5z^3 + 2x^3y^4z - 2xyz} = \boxed{5x^3y^4z^2 + (10xyz - 2xyz) + 2x^2y^5z^3 + 2x^3y^4z}$$

$$= \boxed{5x^3y^4z^2 + (10-2)xyz + 2x^2y^5z^3 + 2x^3y^4z} = \boxed{5x^3y^4z^2 + 8xyz + 2x^2y^5z^3 + 2x^3y^4z}$$

The following is the polynomial in standard form with respect to the variables x, y, and z:

A. $\boxed{5x^3y^4z^2 + 8xyz + 2x^2y^5z^3 + 2x^3y^4z} = \boxed{\left(5x^3y^4z^2 + 2x^3y^4z\right) + 8xyz + 2x^2y^5z^3}$

$$= \boxed{\left(5y^4z^2 + 2y^4z\right)x^3 + 8xyz + 2x^2y^5z^3} = \boxed{\left(5y^4z^2 + 2y^4z\right)x^3 + \left(2y^5z^3\right)x^2 + (8yz)x} \quad \textbf{for the variable x}$$

B. $\boxed{5x^3y^4z^2 + 8xyz + 2x^2y^5z^3 + 2x^3y^4z} = \boxed{\left(5x^3y^4z^2 + 2x^3y^4z\right) + 8xyz + 2x^2y^5z^3}$

$= \boxed{\left(5x^3z^2 + 2x^3z\right)y^4 + 8xyz + 2x^2y^5z^3} = \boxed{\left(\mathbf{2x^2z^3}\right)\mathbf{y^5} + \left(\mathbf{5x^3z^2 + 2x^3z}\right)\mathbf{y^4} + \left(\mathbf{8xz}\right)\mathbf{y}}$ **for the variable y**

C. $\boxed{5x^3y^4z^2 + 8xyz + 2x^2y^5z^3 + 2x^3y^4z} = \boxed{5x^3y^4z^2 + \left(8xyz + 2x^3y^4z\right) + 2x^2y^5z^3}$

$= \boxed{5x^3y^4z^2 + \left(8xy + 2x^3y^4\right)z + 2x^2y^5z^3} = \boxed{\left(\mathbf{2x^2y^5}\right)\mathbf{z^3} + \left(\mathbf{5x^3y^4}\right)\mathbf{z^2} + \left(\mathbf{8xy + 2x^3y^4}\right)\mathbf{z}}$ **for the variable z**

Example 6.1-13

$\boxed{\sqrt{8r^3} - \sqrt{32r} - \sqrt{75} + \sqrt{2r^3} - 3\sqrt{2r} + \sqrt{108}} = \boxed{\sqrt{4\cdot 2r^3} - \sqrt{16\cdot 2r} - \sqrt{25\cdot 3} + \sqrt{2r^3} - 3\sqrt{2r} + \sqrt{36\cdot 3}}$

$= \boxed{\sqrt{2^2\cdot 2r^3} - \sqrt{4^2\cdot 2r} - \sqrt{5^2\cdot 3} + \sqrt{2r^3} - 3\sqrt{2r} + \sqrt{6^2\cdot 3}} = \boxed{2\sqrt{2r^3} - 4\sqrt{2r} - 5\sqrt{3} + \sqrt{2r^3} - 3\sqrt{2r} + 6\sqrt{3}}$

$= \boxed{\left(2\sqrt{2r^3} + \sqrt{2r^3}\right) + \left(-4\sqrt{2r} - 3\sqrt{2r}\right) + \left(-5\sqrt{3} + 6\sqrt{3}\right)} = \boxed{(2+1)\sqrt{2r^3} - (4+3)\sqrt{2r} + (-5+6)\sqrt{3}}$

$= \boxed{\mathbf{3\sqrt{2r^3} - 7\sqrt{2r} + \sqrt{3}}}$

Example 6.1-14

$\boxed{\frac{2}{3}x + \frac{1}{2} + \frac{1}{4}x^2 + \frac{1}{3}x + \frac{3}{4}x^2 - \frac{2}{3}} = \boxed{\left(\frac{2}{3}x + \frac{1}{3}x\right) + \left(\frac{1}{2} - \frac{2}{3}\right) + \left(\frac{1}{4}x^2 + \frac{3}{4}x^2\right)}$

$= \boxed{\left(\frac{2}{3} + \frac{1}{3}\right)x + \left(\frac{1}{2} - \frac{2}{3}\right) + \left(\frac{1}{4} + \frac{3}{4}\right)x^2} = \boxed{\left(\frac{2+1}{3}\right)x + \left(\frac{(1\cdot 3) - (2\cdot 2)}{2\cdot 3}\right) + \left(\frac{1+3}{4}\right)x^2} = \boxed{\frac{\overset{1}{\not 3}}{\underset{1}{\not 3}}x + \frac{3-4}{6} + \frac{\overset{1}{\not 4}}{\underset{1}{\not 4}}x^2}$

$= \boxed{\frac{1}{1}x - \frac{1}{6} + \frac{1}{1}x^2} = \boxed{x - \frac{1}{6} + x^2} = \boxed{\mathbf{x^2 + x - \frac{1}{6}}}$

Example 6.1-15

$\boxed{4abc + ab^2c^3 + a^3b^2c - a^2bc^3 + 3abc - 5ab^2c^3} = \boxed{(4abc + 3abc) + \left(ab^2c^3 - 5ab^2c^3\right) + a^3b^2c - a^2bc^3}$

$= \boxed{(4+3)abc + (1-5)ab^2c^3 + a^3b^2c - a^2bc^3} = \boxed{7abc - 4ab^2c^3 + a^3b^2c - a^2bc^3}$

The following is the polynomial in standard form with respect to the variables a, b, and c:

A. $\boxed{7abc - 4ab^2c^3 + a^3b^2c - a^2bc^3} = \boxed{\left(7abc - 4ab^2c^3\right) + a^3b^2c - a^2bc^3}$

$= \boxed{\left(7bc - 4b^2c^3\right)a + \left(b^2c\right)a^3 - \left(bc^3\right)a^2} = \boxed{\left(\mathbf{b^2c}\right)\mathbf{a^3} + \left(\mathbf{-bc^3}\right)\mathbf{a^2} + \left(\mathbf{7bc - 4b^2c^3}\right)\mathbf{a}}$ **for the variable a**

B. $\boxed{7abc - 4ab^2c^3 + a^3b^2c - a^2bc^3} = \boxed{\left(7abc - a^2bc^3\right) + \left(-4ab^2c^3 + a^3b^2c\right)}$

$= \boxed{\left(7ac - a^2c^3\right)b + \left(-4ac^3 + a^3c\right)b^2} = \boxed{\left(\mathbf{-4ac^3 + a^3c}\right)\mathbf{b^2} + \left(\mathbf{7ac - a^2c^3}\right)\mathbf{b}}$ **for the variable b**

C. $7abc - 4ab^2c^3 + a^3b^2c - a^2bc^3 = \left(7abc + a^3b^2c\right) + \left(-4ab^2c^3 - a^2bc^3\right)$

$= \left(7ab + a^3b^2\right)c + \left(-4ab^2 - a^2b\right)c^3 = +\left(-4ab^2 - a^2b\right)c^3 + \left(7ab + a^3b^2\right)c$ **for the variable c**

Practice Problems - Simplifying Polynomials

Section 6.1 Case II Practice Problems - Simplify the following polynomial expressions. Write the answer in standard form.

1. $-x^3 + 4x - 8x^2 + 3x - 5x^3 - 5x =$
2. $2y + 2y^3 - 5 + 4y - 5y^3 + 1 + y =$
3. $2a^5 + 2a^2 - 3 + 4a^5 + a^2 =$
4. $3x + 2x^4 + 2x^3 - 7x - 5x^4 =$
5. $2rs + 4r^3s^3 - 20 + 2rs - 5r^3s^3 - 3 =$
6. $2xyz + 2x^3y^3z^3 + 10 - 4xyz - 4 =$
7. $-8 + 2u^2v^2 + 6uv - 5 + 2uv - 8u^2v^2 =$
8. $2x + 7x - 8 + 2x^2 - 3 - 5x + 2 =$
9. $-2y^8 + 5y^5 - 8 + 12y^8 - 5y^8 + y^5 - 3 =$
10. $2m^3 + 4m^3 - 4 + 2m - 5m + 3 =$

6.2 Multiplying Polynomials

Polynomials are multiplied by each other using the general product rule (see Section 1.4). Note that polynomials with like terms are multiplied by one another using the multiplication law for exponents (see Section 3.3). In this section, students learn how to multiply monomials (Case I), binomials (Case II), and polynomials (Case III) by one another.

Case I Multiplying Monomials

Monomial expressions are multiplied by each other using the general exponent rule, i.e.,

$$\left(a_0x^m\right)\cdot\left(a_1x^n\right) = \left(a_0a_1\right)\cdot\left(x^mx^n\right) = \left(a_0a_1\right)\cdot\left(x^{m+n}\right)$$ When monomial terms have the same variable.

or,

$$\left(a_0x^m\right)\cdot\left(a_1y^n\right) = \left(a_0a_1\right)\cdot\left(x^my^n\right)$$ When monomial terms have different variables.

where a_0, and a_1 are real numbers, x and y are variables, and m and n are integer numbers. Multiplication of monomial expressions is divided to two cases. Case I a - multiplication of monomials by monomials, and Case I b - multiplication of polynomials by monomials.

Case I a Multiplying Monomials by Monomials

Monomials are multiplied by one another using the following steps:

Step 1 Group like terms with each other.

Step 2 a. Multiply the numerical coefficients (see Section 1.4).
b. Multiply the variables using the exponent rule $x^n \cdot x^m = x^{n+m}$ (see Section 3.3, Case I).

Examples with Steps

The following examples show the steps as to how monomials are multiplied by one another:

Example 6.2-1

$$\left(3x^3\right)\cdot\left(2x^2y\right) =$$

Solution:

Step 1 $\left(3x^3\right)\cdot\left(2x^2y\right) = (3\cdot 2)\cdot\left(x^3\cdot x^2\right)\cdot y$

Step 2 $(3\cdot 2)\cdot\left(x^3\cdot x^2\right)\cdot y = 6\cdot x^{3+2}\cdot y = \mathbf{6x^5y}$

Example 6.2-2

$$\left(5x^3y^2\right)\cdot\left(3x^3y^2z\right) =$$

Solution:

Step 1 $\left(5x^3y^2\right)\cdot\left(3x^3y^2z\right) = (5\cdot 3)\cdot\left(x^3\cdot x^3\right)\cdot\left(y^2\cdot y^2\right)\cdot z$

Step 2 $(5\cdot 3)\cdot\left(x^3\cdot x^3\right)\cdot\left(y^2\cdot y^2\right)\cdot z = 15\cdot x^{3+3}\cdot y^{2+2}\cdot z = \mathbf{15x^6y^4z}$

Example 6.2-3

$$\left(2a^3b^2c\right)\cdot\left(3a^2b\right) =$$

Solution:

Step 1 $\left(2a^3b^2c\right)\cdot\left(3a^2b\right) = (2\cdot 3)\cdot\left(a^3\cdot a^2\right)\cdot\left(b^2\cdot b\right)\cdot c$

Step 2 $(2\cdot 3)\cdot\left(a^3\cdot a^2\right)\cdot\left(b^2\cdot b\right)\cdot c = 6\cdot a^{3+2}\cdot b^{2+1}\cdot c = \mathbf{6a^5b^3c}$

Example 6.2-4

$$\left(20x^0y^2\right)\cdot\left(5xy^0\right)\cdot\left(x^3\right) =$$

Solution:

Step 1 $\left(20x^0y^2\right)\cdot\left(5xy^0\right)\cdot\left(x^3\right) = (20\cdot 5)\cdot\left(x^3\cdot x\cdot x^0\right)\cdot\left(y^2\cdot y^0\right)$

Step 2 $(20\cdot 5)\cdot\left(x^3\cdot x\cdot x^0\right)\cdot\left(y^2\cdot y^0\right) = 100\cdot x^{3+1+0}\cdot y^{2+0} = \mathbf{100x^4y^2}$

Example 6.2-5

$$\left(3a^2b^3c^5\right)\cdot\left(5b^2c^4\right)\cdot\left(4a^3b^0c^3\right) =$$

Solution:

Step 1 $\left(3a^2b^3c^5\right)\cdot\left(5b^2c^4\right)\cdot\left(4a^3b^0c^3\right) = (3\cdot 5\cdot 4)\cdot\left(a^3\cdot a^2\right)\cdot\left(b^3\cdot b^2\cdot b^0\right)\cdot\left(c^5\cdot c^4\cdot c^3\right)$

Step 2 $(3\cdot 5\cdot 4)\cdot\left(a^3\cdot a^2\right)\cdot\left(b^3\cdot b^2\cdot b^0\right)\cdot\left(c^5\cdot c^4\cdot c^3\right) = 60\cdot a^{3+2}\cdot b^{3+2+0}\cdot c^{5+4+3} = \mathbf{60a^5b^5c^{12}}$

Additional Examples - Multiplying Monomials by Monomials

The following examples further illustrate how to multiply monomials by monomials:

Example 6.2-6

$$\left(8x^3y^2\right)\cdot\left(2x^2y\right) = (8\cdot 2)\cdot\left(x^3\cdot x^2\right)\cdot\left(y^2\cdot y\right) = 16\cdot x^{3+2}\cdot y^{2+1} = \mathbf{16x^5y^3}$$

Example 6.2-7

$$(5ax)\cdot(2ay) = (5\cdot 2)\cdot(a\cdot a)\cdot x\cdot y = 10\cdot a^{1+1}\cdot x\cdot y = \mathbf{10a^2xy}$$

Example 6.2-8

$$\left(3x^2\right)\cdot(5xy)\cdot\left(2x^2y^3\right) = (3\cdot 5\cdot 2)\cdot\left(x^2\cdot x\cdot x^2\right)\cdot\left(y^3\cdot y\right) = 30\cdot x^{2+1+2}\cdot y^{3+1} = \mathbf{30x^5y^4}$$

Example 6.2-9

$\boxed{(4xy)\cdot(5x^2y)\cdot(3x^0)} = \boxed{(4\cdot5\cdot3)\cdot(x\cdot x^2\cdot x^0)\cdot(y\cdot y)} = \boxed{60\cdot x^{1+2+0}\cdot y^{1+1}} = \boxed{\mathbf{60x^3y^2}}$

Example 6.2-10

$\boxed{(2x^2y^2z^2)\cdot(3xyz)\cdot(4xz)} = \boxed{(2\cdot3\cdot4)\cdot(x^2\cdot x\cdot x)\cdot(y^2\cdot y)\cdot(z^2\cdot z\cdot z)} = \boxed{24\cdot x^{2+1+1}\cdot y^{2+1}\cdot z^{2+1+1}} = \boxed{\mathbf{24x^4y^3z^4}}$

Example 6.2-11

$\boxed{(5x^2y)^0\cdot(3xy)\cdot x^2y^3} = \boxed{1\cdot(3xy)\cdot x^2y^3} = \boxed{3\cdot(x\cdot x^2)\cdot(y\cdot y^3)} = \boxed{3\cdot x^{1+2}\cdot y^{1+3}} = \boxed{\mathbf{3x^3y^4}}$

Example 6.2-12

$\boxed{(8x^2y)\cdot(2x)\cdot(4z)} = \boxed{(8\cdot2\cdot4)\cdot(x^2\cdot x)\cdot yz} = \boxed{64\cdot x^{2+1}\cdot yz} = \boxed{\mathbf{64x^3yz}}$

Example 6.2-13

$\boxed{(15x^2)\cdot(3x^2y^2)^0\cdot(5x)} = \boxed{(15x^2)\cdot1\cdot(5x)} = \boxed{(15\cdot5)\cdot(x^2\cdot x)} = \boxed{75\cdot x^{2+1}} = \boxed{\mathbf{75x^3}}$

Example 6.2-14

$\boxed{(3a^2b^2)\cdot(2ab)\cdot(3a^0)} = \boxed{(3\cdot2\cdot3)\cdot(a^2\cdot a\cdot a^0)\cdot(b^2\cdot b)} = \boxed{18\cdot a^{2+1+0}\cdot b^{2+1}} = \boxed{\mathbf{18a^3b^3}}$

Example 6.2-15

$\boxed{(5abc)\cdot(3a^2b^2c)\cdot(2c^0)} = \boxed{(5\cdot3\cdot2)\cdot(a\cdot a^2)\cdot(b\cdot b^2)\cdot(c\cdot c\cdot c^0)} = \boxed{30\cdot a^{1+2}\cdot b^{1+2}\cdot c^{1+1+0}} = \boxed{\mathbf{30a^3b^3c^2}}$

Practice Problems - Multiplying Monomials by Monomials

Section 6.2 Case I a Practice Problems - Multiply the following monomials by each other:

1. $(2ax)\cdot(3a^2x^2) =$
2. $(5x^2y^2)\cdot(2x)\cdot(4y) =$
3. $(6x^2)^0\cdot(3x^2)\cdot(-2x) =$
4. $(x^2y)\cdot(3xy)\cdot(4x^3y^2) =$
5. $(3x^2y^2)\cdot(2xy^0)\cdot(5x^0y) =$
6. $(8a^2b^2)\cdot(2a)\cdot(3a^2b^3) =$
7. $(4m^2n^2)\cdot(3n^2) =$
8. $(3m^2n^3)\cdot(2mn^2)\cdot(4n) =$
9. $(6x^3y^6z^2)\cdot(3x^5y^0z)\cdot(z^3) =$
10. $(2w^2z^3)\cdot(2w^3z)^0\cdot(5wz^2) =$

Case I b Multiplying Polynomials by Monomials

Polynomials are multiplied by monomials using the following steps:

Step 1 Group like terms with each other.

Step 2 Multiply each term of the polynomial by the monomial by:

a. Multiplying the numerical coefficients (see Section 1.4).

b. Multiplying the variables using the exponent rule $x^n \cdot x^m = x^{n+m}$ (see Section 3.3, Case I).

Examples with Steps

The following examples show the steps as to how polynomials are multiplied by monomials:

Example 6.2-16

$$\left(2x^4 + 3x^2 + 5x - x^4 + x^2 - 3\right) \cdot \left(3x^2\right) =$$

Solution:

Step 1

$$\left(2x^4 + 3x^2 + 5x - x^4 + x^2 - 3\right) \cdot \left(3x^2\right) = \left(2x^4 - x^4 + 3x^2 + x^2 + 5x - 3\right) \cdot \left(3x^2\right)$$

$$= \left[\left(2x^4 - x^4\right) + \left(3x^2 + x^2\right) + 5x - 3\right] \cdot \left(3x^2\right) = \left[(2-1)x^4 + (3+1)x^2 + 5x - 3\right] \cdot \left(3x^2\right)$$

$$= \left[x^4 + 4x^2 + 5x - 3\right] \cdot \left(3x^2\right)$$

Step 2

$$\left[x^4 + 4x^2 + 5x - 3\right] \cdot \left(3x^2\right) = 3\left(x^4 \cdot x^2\right) + (4 \cdot 3)\left(x^2 \cdot x^2\right) + (5 \cdot 3)\left(x \cdot x^2\right) - (3 \cdot 3)x^2$$

$$= 3x^{4+2} + 12x^{2+2} + 15x^{1+2} - 9x^2 = \mathbf{3x^6 + 12x^4 + 15x^3 - 9x^2}$$

Example 6.2-17

$$\left(5m^3n^3 + 2m^2n^2 - 3mn + mn + 2\right) \cdot (5mn) =$$

Solution:

Step 1

$$\left(5m^3n^3 + 2m^2n^2 - 3mn + mn + 2\right) \cdot (5mn) = \left[5m^3n^3 + 2m^2n^2 + (-3mn + mn) + 2\right] \cdot (5mn)$$

$$= \left[5m^3n^3 + 2m^2n^2 + (-3+1)mn + 2\right] \cdot (5mn) = \left(5m^3n^3 + 2m^2n^2 - 2mn + 2\right) \cdot (5mn)$$

Step 2

$$\left(5m^3n^3 + 2m^2n^2 - 2mn + 2\right) \cdot (5mn)$$

$$= (5 \cdot 5)\left(m^3m\right)\left(n^3 \cdot n\right) + (2 \cdot 5)\left(m^2 \cdot m\right)\left(n^2 \cdot n\right) - (2 \cdot 5)(m \cdot m)(n \cdot n) + (2 \cdot 5)mn$$

$$= \boxed{25m^{3+1}n^{3+1} + 10m^{2+1}n^{2+1} - 10m^{1+1}n^{1+1} + 10mn} = \boxed{\mathbf{25m^4n^4 + 10m^3n^3 - 10m^2n^2 + 10mn}}$$

Example 6.2-18

$$\boxed{\left(\frac{8}{9}w^3 + \frac{5}{6}w^2 + \frac{1}{9}w^3 + \frac{4}{6}\right)\cdot\left(\frac{3}{4}w\right)} =$$

Solution:

Step 1

$$\boxed{\left(\frac{8}{9}w^3 + \frac{5}{6}w^2 + \frac{1}{9}w^3 + \frac{4}{6}\right)\cdot\left(\frac{3}{4}w\right)} = \boxed{\left[\left(\frac{8}{9}w^3 + \frac{1}{9}w^3\right) + \frac{5}{6}w^2 + \frac{4}{6}\right]\cdot\left(\frac{3}{4}w\right)}$$

$$= \boxed{\left[\left(\frac{8}{9} + \frac{1}{9}\right)w^3 + \frac{5}{6}w^2 + \frac{4}{6}\right]\cdot\left(\frac{3}{4}w\right)} = \boxed{\left[\left(\frac{8+1}{9}\right)w^3 + \frac{5}{6}w^2 + \frac{4}{6}\right]\cdot\left(\frac{3}{4}w\right)}$$

$$= \boxed{\left[\frac{\overset{1}{\cancel{9}}}{\underset{1}{\cancel{9}}}w^3 + \frac{5}{6}w^2 + \frac{4}{6}\right]\cdot\left(\frac{3}{4}w\right)} = \boxed{\left[w^3 + \frac{5}{6}w^2 + \frac{4}{6}\right]\cdot\left(\frac{3}{4}w\right)}$$

Step 2

$$\boxed{\left[w^3 + \frac{5}{6}w^2 + \frac{4}{6}\right]\cdot\left(\frac{3}{4}w\right)} = \boxed{\frac{3}{4}\left(w^3 w\right) + \left(\frac{5}{6}\cdot\frac{3}{4}\right)\left(w^2 w\right) + \left(\frac{4}{6}\cdot\frac{3}{4}\right)w}$$

$$= \boxed{\frac{3}{4}w^4 + \left(\frac{5\cdot 3}{6\cdot 4}\right)w^3 + \left(\frac{4\cdot 3}{6\cdot 4}\right)w} = \boxed{\frac{3}{4}w^4 + \frac{\overset{5}{\cancel{15}}}{\underset{8}{\cancel{24}}}w^3 + \frac{\overset{1}{\cancel{12}}}{\underset{2}{\cancel{24}}}w} = \boxed{\mathbf{\frac{3}{4}w^4 + \frac{5}{8}w^3 + \frac{1}{2}w}}$$

Example 6.2-19

$$\boxed{\left(\sqrt{27}x^2 - 2 + \sqrt{8}x + \sqrt{36}\right)\cdot\sqrt{125}x} =$$

Solution:

Step 1

$$\boxed{\left(\sqrt{27}x^2 - 2 + \sqrt{8}x + \sqrt{36}\right)\cdot\sqrt{125}x} = \boxed{\left(\sqrt{9\cdot 3}x^2 - 2 + \sqrt{4\cdot 2}x + \sqrt{6^2}\right)\cdot\sqrt{25\cdot 5}x}$$

$$= \boxed{\left[\sqrt{3^2\cdot 3}x^2 + \sqrt{2^2\cdot 2}x + (-2+6)\right]\cdot\sqrt{5^2\cdot 5}x} = \boxed{\left(3\sqrt{3}x^2 + 2\sqrt{2}x + 4\right)\cdot 5\sqrt{5}x}$$

Step 2

$$\boxed{\left(3\sqrt{3}x^2 + 2\sqrt{2}x + 4\right)\cdot 5\sqrt{5}x} = \boxed{(3\cdot 5)\left(\sqrt{3}\cdot\sqrt{5}\right)\left(x^2\cdot x\right) + (2\cdot 5)\left(\sqrt{2}\cdot\sqrt{5}\right)(x\cdot x) + (4\cdot 5)\sqrt{5}x}$$

$$= \boxed{15\left(\sqrt{3\cdot 5}\right)x^{2+1} + 10\left(\sqrt{2\cdot 5}\right)x^{1+1} + 20\sqrt{5}x} = \boxed{15\left(\sqrt{3\cdot 5}\right)x^{2+1} + 10\left(\sqrt{2\cdot 5}\right)x^{1+1} + 20\sqrt{5}x}$$

$$= \boxed{\mathbf{15\sqrt{15}x^3 + 10\sqrt{10}x^2 + 20\sqrt{5}x}}$$

Example 6.2-20

$$\boxed{\left(10x^5 + 3x^4 + 2x^5 - 8 + 4x + 3x^0\right)\cdot\left(8x^2\right)} =$$

Solution:

Step 1 $\left(10x^5+3x^4+2x^5-8+4x+3x^0\right)\cdot\left(8x^2\right)=\left(10x^5+3x^4+2x^5-8+4x+3\right)\cdot\left(8x^2\right)$

$=\left[\left(10x^5+2x^5\right)+3x^4+4x+(-8+3)\right]\cdot\left(8x^2\right)=\left[(10+2)x^5+3x^4+4x-5\right]\cdot\left(8x^2\right)$

$=\left[12x^5+3x^4+4x-5\right]\cdot\left(8x^2\right)$

Step 2 $\left[12x^5+3x^4+4x-5\right]\cdot\left(8x^2\right)=(12\cdot 8)\left(x^5\cdot x^2\right)+(3\cdot 8)\left(x^4\cdot x^2\right)+(4\cdot 8)\left(x\cdot x^2\right)-(5\cdot 8)x^2$

$=96x^{5+2}+24x^{4+2}+32x^{1+2}-40x^2=\mathbf{96x^7+24x^6+32x^3-40x^2}$

Additional Examples - Multiplying Polynomials by Monomials

The following examples further illustrate how to multiply polynomial expressions by monomials:

Example 6.2-21

$\left(5a^2b^2+3ab-2a^2b^2-ab+1\right)\cdot(3ab)=\left[\left(5a^2b^2-2a^2b^2\right)+(3ab-ab)+1\right]\cdot(3ab)$

$=\left[(5-2)a^2b^2+(3-1)ab+1\right]\cdot(3ab)=\left[3a^2b^2+2ab+1\right]\cdot(3ab)=(3\cdot 3)\left(a^2\cdot a\right)\left(b^2\cdot b\right)+(2\cdot 3)(a\cdot a)(b\cdot b)+3ab$

$=9a^{2+1}b^{2+1}+6a^{1+1}b^{1+1}+3ab=\mathbf{9a^3b^3+6a^2b^2+3ab}$

Example 6.2-22

$\left(2x+4y+3x^2y^2-2y+x^2y^2-5x+4\right)\cdot(4xy)=\left[\left(3x^2y^2+x^2y^2\right)+(2x-5x)+(4y-2y)+4\right]\cdot(4xy)$

$=\left[(3+1)x^2y^2+(2-5)x+(4-2)y+4\right]\cdot(4xy)=\left[4x^2y^2-3x+2y+4\right]\cdot(4xy)$

$=(4\cdot 4)\left(x^2\cdot x\right)\left(y^2\cdot y\right)-(3\cdot 4)(x\cdot x)y+(2\cdot 4)x(y\cdot y)+(4\cdot 4)xy=16x^{2+1}y^{2+1}-12x^{1+1}y+8xy^{1+1}+16xy$

$=\mathbf{16x^3y^3-12x^2y+8xy^2+16xy}$

Example 6.2-23

$\left(-3x^2+4x^3+x-5+2x^3\right)\cdot\left(-2x^2\right)=\left[\left(4x^3+2x^3\right)-3x^2+x-5\right]\cdot\left(-2x^2\right)=\left[(4+2)x^3-3x^2+x-5\right]\cdot\left(-2x^2\right)$

$=\left(6x^3-3x^2+x-5\right)\cdot\left(-2x^2\right)=-(6\cdot 2)\left(x^3\cdot x^2\right)+(3\cdot 2)\left(x^2\cdot x^2\right)-2\left(x\cdot x^2\right)+(5\cdot 2)x^2$

$$= \boxed{-12x^{3+2} + 6x^{2+2} - 2x^{1+2} + 10x^2} = \boxed{\mathbf{-12x^5 + 6x^4 - 2x^3 + 10x^2}}$$

Example 6.2-24

$$\boxed{\left(\sqrt{25}z^3 + \sqrt{36}z^2 - 2z^3 - \sqrt{27}z + 2z^2\right)\cdot\left(-z^3\right)} = \boxed{\left(\sqrt{5^2}z^3 - 2z^3 + \sqrt{6^2}z^2 + 2z^2 - \sqrt{3^2\cdot 3z}\right)\cdot\left(-z^3\right)}$$

$$= \boxed{\left(5z^3 - 2z^3 + 6z^2 + 2z^2 - 3\sqrt{3z}\right)\cdot\left(-z^3\right)} = \boxed{\left[(5-2)z^3 + (6+2)z^2 - 3\sqrt{3z}\right]\cdot\left(-z^3\right)}$$

$$= \boxed{\left[3z^3 + 8z^2 - 3\sqrt{3z}\right]\cdot\left(-z^3\right)} = \boxed{-3\left(z^3\cdot z^3\right) - 8\left(z^2\cdot z^3\right) + 3\sqrt{3}\left(z\cdot z^3\right)} = \boxed{-3z^{3+3} - 8z^{2+3} + 3\sqrt{3}z^{1+3}}$$

$$= \boxed{\mathbf{-3z^6 - 8z^5 + 3\sqrt{3}z^4}}$$

Example 6.2-25

$$\boxed{\left(4x^3 + 3x^2 - 3 + 3x^5 + (3x)^0\right)\cdot\left(2x^3\right)} = \boxed{\left(4x^3 + 3x^2 - 3 + 3x^5 + 1\right)\cdot\left(2x^3\right)} = \boxed{\left[3x^5 + 4x^3 + 3x^2(-3+1)\right]\cdot\left(2x^3\right)}$$

$$= \boxed{\left[3x^5 + 4x^3 + 3x^2 - 2\right]\cdot\left(2x^3\right)} = \boxed{(3\cdot 2)\left(x^5\cdot x^3\right) + (4\cdot 2)\left(x^3\cdot x^3\right) + (3\cdot 2)\left(x^2\cdot x^3\right) - (2\cdot 2)x^3}$$

$$= \boxed{6x^{5+3} + 8x^{3+3} + 6x^{2+3} - 4x^3} = \boxed{\mathbf{6x^8 + 8x^6 + 6x^5 - 4x^3}}$$

Example 6.2-26

$$\boxed{\left(\sqrt[3]{24}m^3 + 3\sqrt[3]{m^3} - \sqrt[3]{54}m^2 - \sqrt[3]{125}\right)\cdot\left(3m^2\right)} = \boxed{\left(\sqrt[3]{2^3\cdot 3}m^3 + 3m - \sqrt[3]{3^3\cdot 2}m^2 - \sqrt[3]{5^3}\right)\cdot\left(3m^2\right)}$$

$$= \boxed{\left(2\sqrt[3]{3}m^3 + 3m - 3\sqrt[3]{2}m^2 - 5\right)\cdot\left(3m^2\right)} = \boxed{\left(2\sqrt[3]{3}m^3 - 3\sqrt[3]{2}m^2 + 3m - 5\right)\cdot\left(3m^2\right)}$$

$$= \boxed{(2\cdot 3)\sqrt[3]{3}\left(m^3\cdot m^2\right) - (3\cdot 3)\sqrt[3]{2}\left(m^2\cdot m^2\right) + (3\cdot 3)\left(m\cdot m^2\right) - (5\cdot 3)m^2} = \boxed{6\sqrt[3]{3}m^{3+2} - 9\sqrt[3]{2}m^{2+2} + 9m^{1+2} - 15m^2}$$

$$= \boxed{\mathbf{6\sqrt[3]{3}m^5 - 9\sqrt[3]{2}m^4 + 9m^3 - 15m^2}}$$

Example 6.2-27

$$\boxed{\left(3l^2 + 3l^3 - 5l + 2l^4 + 2l - 2\right)\cdot l^2} = \boxed{\left[2l^4 + 3l^3 + 3l^2 + (2l - 5l) - 2\right]\cdot l^2} = \boxed{\left[2l^4 + 3l^3 + 3l^2 + (2-5)l - 2\right]\cdot l^2}$$

$$= \boxed{\left(2l^4 + 3l^3 + 3l^2 - 3l - 2\right)\cdot l^2} = \boxed{2\left(l^4\cdot l^2\right) + 3\left(l^3\cdot l^2\right) + 3\left(l^2\cdot l^2\right) - 3\left(l\cdot l^2\right) - 2l^2}$$

$$= \boxed{2l^{4+2} + 3l^{3+2} + 3l^{2+2} - 3l^{1+2} - 2l^2} = \boxed{\mathbf{2l^6 + 3l^5 + 3l^4 - 3l^3 - 2l^2}}$$

Example 6.2-28

$$\boxed{\left(2x^3 + 5x^2 - 2x + 4x^3 - 20x^2 + 8x\right)\cdot\sqrt{4x^2}} = \boxed{\left[\left(2x^3 + 4x^3\right) + \left(5x^2 - 20x^2\right) + (-2x + 8x)\right]\cdot 2x}$$

$$= \boxed{\left[(2+4)x^3+(5-20)x^2+(-2+8)x\right]\cdot 2x} = \boxed{\left[6x^3-15x^2+6x\right]\cdot 2x} = \boxed{(6\cdot 2)\left(x^3\cdot x\right)-(15\cdot 2)\left(x^2\cdot x\right)+(6\cdot 2)(x\cdot x)}$$

$$= \boxed{12x^{3+1}-30x^{2+1}+12x^{1+1}} = \boxed{\mathbf{12x^4-30x^3+12x^2}}$$

Example 6.2-29

$$\boxed{\left(w+3w^2-4w^3-w^4-4w+6w^4\right)\cdot\left(2w^2\right)} = \boxed{\left[\left(6w^4-w^4\right)-4w^3+3w^2+(w-4w)\right]\cdot 2w^2}$$

$$= \boxed{\left[(6-1)w^4-4w^3+3w^2+(1-4)w\right]\cdot 2w^2} = \boxed{\left[5w^4-4w^3+3w^2-3w\right]\cdot 2w^2}$$

$$= \boxed{(5\cdot 2)\left(w^4\cdot w^2\right)-(4\cdot 2)\left(w^3\cdot w^2\right)+(3\cdot 2)\left(w^2\cdot w^2\right)-(3\cdot 2)\left(w\cdot w^2\right)} = \boxed{10w^{4+2}-8w^{3+2}+6w^{2+2}-6w^{1+2}}$$

$$= \boxed{\mathbf{10w^6-8w^5+6w^4-6w^3}}$$

Example 6.2-30

$$\boxed{\left(5x^2+4x-2x-6\right)\cdot\left(4x^2\right)\cdot(-2x)} = \boxed{\left[5x^2+(4-2)x-6\right](-4\cdot 2)\left(x^2\cdot x\right)} = \boxed{\left[5x^2+2x-6\right]\left(-8x^{2+1}\right)}$$

$$= \boxed{\left[5x^2+2x-6\right]\left(-8x^3\right)} = \boxed{-(5\cdot 8)\left(x^3\cdot x^2\right)-(2\cdot 8)\left(x^3\cdot x\right)+(6\cdot 8)x^3} = \boxed{-40x^{3+2}-16x^{3+1}+48x^3}$$

$$= \boxed{\mathbf{-40x^5-16x^4+48x^3}}$$

Practice Problems - Multiplying Polynomials by Monomials

Section 6.2 Case I b Practice Problems - Multiply the following polynomial expressions by monomials:

1. $2\cdot\left(5x^2+6x-2x^2-x+5\right) =$
2. $\left(2x^2y-5y^2+3x^2y-2y^2+3\right)\cdot\left(3x^2y^2\right) =$
3. $\left(5x^3+2x^2-5+3x-2x^3\right)\cdot(-2x)^2 =$
4. $6w\cdot\left(4w+2w^2+2-3w+w^2\right) =$
5. $2x\cdot\left(2x^2\right)^2\cdot\left(5x^2+3x-2x^2+x-2\right) =$
6. $\left(\sqrt{162}+\sqrt{9}x-2x^2+\sqrt{16}x^3\right)\cdot\left(2x^3\right) =$
7. $\left(5y-3y^2+2y-4\right)\cdot\left(3y^2\right) =$
8. $9x\cdot\left(2x^2+5x-5x^2+6\right)\cdot\left(3x^3\right) =$
9. $\left(5x^2\right)\cdot\left(2x^3-4x+2+x^3-x\right)\cdot 2x =$
10. $\left(\sqrt[3]{8}x^2-4x-2x^2+8x-\sqrt[3]{125}+28\right)\cdot\left(2x^3\right) =$

Case II Multiplying Binomials by Binomials

Binomials are multiplied by one another using the multiplication method known as the FOIL method (see Section 4.2, Case II). In general, binomials are multiplied by each other in the following way:

$$\left(a_0x^n + a_1x^{n-m}\right)\left(b_0x^n + b_1x^{n-m}\right)$$

$$= \left(a_0 \cdot b_0\right)\cdot\left(x^n \cdot x^n\right) + \left(a_0 \cdot b_1\right)\cdot\left(x^n \cdot x^{n-m}\right) + \left(a_1 \cdot b_0\right)\cdot\left(x^{n-m} \cdot x^n\right) + \left(a_1 \cdot b_1\right)\cdot\left(x^{n-m} \cdot x^{n-m}\right)$$

$$= a_0b_0\left(x^{n+n}\right) + a_0b_1\left(x^{n+n-m}\right) + a_1b_0\left(x^{n-m+n}\right) + a_1b_1\left(x^{n-m+n-m}\right)$$

$$= a_0b_0\left(x^{2n}\right) + a_0b_1\left(x^{2n-m}\right) + a_1b_0\left(x^{2n-m}\right) + a_1b_1\left(x^{2n-2m}\right) = \mathbf{a_0b_0x^{2n} + \left(a_0b_1 + a_1b_0\right)x^{2n-m} + a_1b_1x^{2n-2m}}$$

where n and m are positive integer numbers and $n \geq m$.

Binomials are multiplied by one another using the following steps:

Step 1 a. Simplify each binomial term, if possible.

b. Multiply the terms of the first binomial by each term of the second binomial using the FOIL method.

Step 2 Group like terms with each other.

Examples with Steps

The following examples show the steps as to how binomials are multiplied by each other:

Example 6.2-31

$$\boxed{\left(x^2 + 3x\right)\left(x + 8\right)} =$$

Solution:

Step 1

$$\boxed{\left(x^2 + 3x\right)\left(x + 8\right)} = \boxed{\left(x^2 \cdot x\right) + \left(8 \cdot x^2\right) + 3(x \cdot x) + (3 \cdot 8)x} = \boxed{x^3 + 8x^2 + 3x^2 + 24x}$$

Step 2

$$\boxed{x^3 + 8x^2 + 3x^2 + 24x} = \boxed{x^3 + (8 + 3)x^2 + 24x} = \boxed{\mathbf{x^3 + 11x^2 + 24x}}$$

Example 6.2-32

$$\boxed{\left(\sqrt{225}x + 2\right)\left(5x - \sqrt{81}\right)} =$$

Solution:

Step 1

$$\boxed{\left(\sqrt{225}x + 2\right)\left(5x - \sqrt{81}\right)} = \boxed{\left(\sqrt{15^2}x + 2\right)\left(5x - \sqrt{9^2}\right)} = \boxed{(15x + 2)(5x - 9)}$$

$$= \boxed{(15 \cdot 5)(x \cdot x) - (15 \cdot 9)x + (2 \cdot 5)x - (2 \cdot 9)} = \boxed{75x^2 - 135x + 10x - 18}$$

Step 2 $\boxed{75x^2 - 135x + 10x - 18} = \boxed{75x^2 + (-135 + 10)x - 18} = \boxed{\mathbf{75x^2 - 125x - 18}}$

Example 6.2-33

$$\boxed{(3x^2 + 6x)(x^3 + 2x^2)} =$$

Solution:

Step 1 $\boxed{(3x^2 + 6x)(x^3 + 2x^2)} = \boxed{3(x^2x^3) + (3\cdot 2)(x^2\cdot x^2) + 6(x\cdot x^3) + (6\cdot 2)(x\cdot x^2)}$

$= \boxed{3x^{2+3} + 6x^{2+2} + 6x^{1+3} + 12x^{1+2}} = \boxed{3x^5 + 6x^4 + 6x^4 + 12x^3}$

Step 2 $\boxed{3x^5 + 6x^4 + 6x^4 + 12x^3} = \boxed{3x^5 + (6+6)x^4 + 12x^3} = \boxed{\mathbf{3x^5 + 12x^4 + 12x^3}}$

Example 6.2-34

$$\boxed{(a^2 - 5)(-a^2 + 3)} =$$

Solution:

Step 1 $\boxed{(a^2 - 5)(-a^2 + 3)} = \boxed{-(a^2\cdot a^2) + (3\cdot a^2) + (5\cdot a^2) - (5\cdot 3)} = \boxed{-a^4 + 3a^2 + 5a^2 - 15}$

Step 2 $\boxed{-a^4 + 3a^2 + 5a^2 - 15} = \boxed{-a^4 + (3+5)a^2 - 15} = \boxed{\mathbf{-a^4 + 8a^2 - 15}}$

Example 6.2-35

$$\boxed{\left(\frac{2}{3}x^3 + \frac{1}{3}x\right)\left(\frac{1}{2}x^2 - \frac{3}{8}\right)} =$$

Solution:

Step 1 $\boxed{\left(\frac{2}{3}x^3 + \frac{1}{3}x\right)\left(\frac{1}{2}x^2 - \frac{3}{8}\right)} = \boxed{\left(\frac{2}{3}\cdot\frac{1}{2}\right)(x^3\cdot x^2) - \left(\frac{2}{3}\cdot\frac{3}{8}\right)x^3 + \left(\frac{1}{3}\cdot\frac{1}{2}\right)(x\cdot x^2) - \left(\frac{1}{3}\cdot\frac{3}{8}\right)x}$

$= \boxed{\frac{\overset{1}{\cancel{2}}}{\underset{3}{\cancel{6}}}x^5 - \frac{\overset{1}{\cancel{6}}}{\underset{4}{\cancel{24}}}x^3 + \frac{1}{6}x^3 - \frac{\overset{1}{\cancel{3}}}{\underset{8}{\cancel{24}}}x} = \boxed{\frac{1}{3}x^5 - \frac{1}{4}x^3 + \frac{1}{6}x^3 - \frac{1}{8}x}$

Step 2 $\boxed{\frac{1}{3}x^5 - \frac{1}{4}x^3 + \frac{1}{6}x^3 - \frac{1}{8}x} = \boxed{\frac{1}{3}x^5 + \left(-\frac{1}{4} + \frac{1}{6}\right)x^3 - \frac{1}{8}x} = \boxed{\frac{1}{3}x^5 + \left(\frac{-(1\cdot 6) + (1\cdot 4)}{4\cdot 6}\right)x^3 - \frac{1}{8}x}$

$= \boxed{\frac{1}{3}x^5 + \left(\frac{-6+4}{24}\right)x^3 - \frac{1}{8}x} = \boxed{\frac{1}{3}x^5 - \frac{\overset{1}{\cancel{2}}}{\underset{12}{\cancel{24}}}x^3 - \frac{1}{8}x} = \boxed{\mathbf{\frac{1}{3}x^5 - \frac{1}{12}x^3 - \frac{1}{8}x}}$

Additional Examples - Multiplying Binomials by Binomials

The following examples further illustrate how to multiply binomials by one another:

Example 6.2-36

$$\left(x+\sqrt{98}\right)\left(x-2\sqrt{162}\right)=\left(x+\sqrt{49\cdot 2}\right)\left(x-2\sqrt{81\cdot 2}\right)=\left(x+\sqrt{7^2\cdot 2}\right)\left(x-2\sqrt{9^2\cdot 2}\right)=\left(x+7\sqrt{2}\right)\left(x-(2\cdot 9)\sqrt{2}\right)$$

$$=\left(x+7\sqrt{2}\right)\left(x-18\sqrt{2}\right)=(x\cdot x)-\left(18\sqrt{2}\cdot x\right)+\left(7\sqrt{2}\cdot x\right)-(7\cdot 18)\cdot\left(\sqrt{2}\cdot\sqrt{2}\right)=x^2-18\sqrt{2}x+7\sqrt{2}x-126\left(\sqrt{2\cdot 2}\right)$$

$$=x^2+(-18+7)\sqrt{2}x-126\sqrt{2^2}=x^2-11\sqrt{2}x-(126\cdot 2)=\mathbf{x^2-11\sqrt{2}x-252}$$

Example 6.2-37

$$\left(y^2-3y\right)\left(y^2+y\right)=\left(y^2\cdot y^2\right)+\left(y^2\cdot y\right)-3\left(y\cdot y^2\right)-3(y\cdot y)=y^4+y^3-3y^3-3y^2=y^4+(1-3)y^3-3y^2$$

$$=\mathbf{y^4-2y^3-3y^2}$$

Example 6.2-38

$$\left(a\sqrt{a^3}+a\sqrt{a}\right)\left(\sqrt{a^5}-a\right)=\left(a\sqrt{a^{2+1}}+a\sqrt{a}\right)\left(\sqrt{a^{2+2+1}}-a\right)=\left(a\sqrt{a^2\cdot a^1}+a\sqrt{a}\right)\left(\sqrt{a^2\cdot a^2\cdot a^1}-a\right)$$

$$=\left[(a\cdot a)\sqrt{a}+a\sqrt{a}\right]\left[(a\cdot a)\sqrt{a}-a\right]=\left(a^2\sqrt{a}+a\sqrt{a}\right)\left(a^2\sqrt{a}-a\right)$$

$$=\left(a^2\cdot a^2\right)\left(\sqrt{a}\cdot\sqrt{a}\right)-\left(a^2\cdot a\right)\sqrt{a}+\left(a\cdot a^2\right)\left(\sqrt{a}\cdot\sqrt{a}\right)-(a\cdot a)\sqrt{a}=a^4\left(\sqrt{a\cdot a}\right)-a^3\sqrt{a}+a^3\left(\sqrt{a\cdot a}\right)-a^2\sqrt{a}$$

$$=a^4\sqrt{a^2}-a^3\sqrt{a}+a^3\sqrt{a^2}-a^2\sqrt{a}=\left(a^4\cdot a\right)-a^3\sqrt{a}+\left(a^3\cdot a\right)-a^2\sqrt{a}=a^5-a^3\sqrt{a}+a^4-a^2\sqrt{a}$$

$$=a^5+a^4-a^3\sqrt{a}-a^2\sqrt{a}=a^4(a+1)-a^2\sqrt{a}(a+1)=(a+1)\left[a^4-a^2\sqrt{a}\right]=\mathbf{a^2(a+1)\left[a^2-\sqrt{a}\right]}$$

Example 6.2-39

$$\left(\sqrt[3]{l^5}-\sqrt[3]{l^3}\right)\left(l+3\sqrt[3]{l^6}\right)=\left(\sqrt[3]{l^{3+2}}-l\right)\left(l+3\sqrt[3]{l^{3+3}}\right)=\left(\sqrt[3]{l^3\cdot l^2}-l\right)\left(l+3\sqrt[3]{l^3\cdot l^3}\right)=\left(l\sqrt[3]{l^2}-l\right)(l+3(l\cdot l))$$

$$=\left(l\sqrt[3]{l^2}-l\right)\left(l+3l^2\right)=(l\cdot l)\sqrt[3]{l^2}+3\left(l^2\cdot l\right)\sqrt[3]{l^2}-(l\cdot l)-3\left(l\cdot l^2\right)=l^2\sqrt[3]{l^2}+3l^3\sqrt[3]{l^2}-l^2-3l^3$$

$$=\left(3l^3\sqrt[3]{l^2}-3l^3\right)+\left(l^2\sqrt[3]{l^2}-l^2\right)=3l^3\left(\sqrt[3]{l^2}-1\right)+l^2\left(\sqrt[3]{l^2}-1\right)=\mathbf{\left[l^2\sqrt[3]{l^2}-1\right](3l+1)}$$

Example 6.2-40

$$\left(b^3-\sqrt{8}\right)\left(b+\sqrt{50b^2}\right)=\left(b^3-\sqrt{4\cdot 2}\right)\left(b+\sqrt{25\cdot 2b^2}\right)=\left(b^3-\sqrt{2^2\cdot 2}\right)\left(b+\sqrt{5^2\cdot 2b^2}\right)=\left(b^3-2\sqrt{2}\right)\left(b+5\sqrt{2b^2}\right)$$

$$= \boxed{\left(b^3 \cdot b\right) + 5\sqrt{2}\left(b^3 \cdot b^2\right) - 2\sqrt{2} \cdot b - \left(2\sqrt{2} \cdot 5\sqrt{2}\right)b^2} = \boxed{b^4 + 5\sqrt{2}b^5 - 2\sqrt{2}b - (2 \cdot 5)\sqrt{2 \cdot 2}b^2}$$

$$= \boxed{b^4 + 5\sqrt{2}b^5 - 2\sqrt{2}b - 10\sqrt{2^2}b^2} = \boxed{b^4 + 5\sqrt{2}b^5 - 2\sqrt{2}b - (10 \cdot 2)b^2} = \boxed{\mathbf{5\sqrt{2}b^5 + b^4 - 20b^2 - 2\sqrt{2}b}}$$

Example 6.2-41

$$\boxed{\left(u^2 + 3u\right)\left(-u + u^2\right)} = \boxed{-\left(u^2 \cdot u\right) + \left(u^2 \cdot u^2\right) - 3(u \cdot u) + 3\left(u \cdot u^2\right)} = \boxed{-u^3 + u^4 - 3u^2 + 3u^3} = \boxed{u^4 + 3u^3 - u^3 - 3u^2}$$

$$= \boxed{u^4 + (3-1)u^3 - 3u^2} = \boxed{\mathbf{u^4 + 2u^3 - 3u^2}}$$

Example 6.2-42

$$\boxed{\left(\frac{a^2}{\sqrt{36}} - \frac{a}{4}\right)\left(\frac{3a}{\sqrt{4}} - \frac{6a^3}{\sqrt{81}}\right)} = \boxed{\left(\frac{a^2}{\sqrt{6^2}} - \frac{a}{4}\right)\left(\frac{3a}{\sqrt{2^2}} - \frac{6a^3}{\sqrt{9^2}}\right)} = \boxed{\left(\frac{a^2}{6} - \frac{a}{4}\right)\left(\frac{3a}{2} - \frac{6a^3}{9}\right)}$$

$$= \boxed{\left(\frac{a^2}{6} \cdot \frac{3a}{2}\right) - \left(\frac{a^2}{6} \cdot \frac{6a^3}{9}\right) - \left(\frac{a}{4} \cdot \frac{3a}{2}\right) + \left(\frac{a}{4} \cdot \frac{6a^3}{9}\right)} = \boxed{\left(\frac{3a^2 \cdot a}{6 \cdot 2}\right) - \left(\frac{6a^3 \cdot a^2}{6 \cdot 9}\right) - \left(\frac{3a \cdot a}{4 \cdot 2}\right) + \left(\frac{6a^3 \cdot a}{4 \cdot 9}\right)}$$

$$= \boxed{\frac{\overset{1}{\cancel{3}}a^3}{\underset{4}{\cancel{12}}} - \frac{\overset{1}{\cancel{6}}a^5}{\underset{9}{\cancel{54}}} - \frac{3a^2}{8} + \frac{\overset{1}{\cancel{6}}a^4}{\underset{6}{\cancel{36}}}} = \boxed{\frac{a^3}{4} - \frac{a^5}{9} - \frac{3a^2}{8} + \frac{a^4}{6}} = \boxed{\mathbf{-\frac{a^5}{9} + \frac{a^4}{6} + \frac{a^3}{4} - \frac{3a^2}{8}}}$$

Example 6.2-43

$$\boxed{\left(\sqrt[5]{q^{10}} + 3q\right)\left(4q^2 - 3\sqrt[5]{q^5}\right)} = \boxed{\left(\sqrt[5]{q^{5+5}} + 3q\right)\left(4q^2 - 3q\right)} = \boxed{\left(\sqrt[5]{q^5 \cdot q^5} + 3q\right)\left(4q^2 - 3q\right)}$$

$$= \boxed{[(q \cdot q) + 3q]\left(4q^2 - 3q\right)} = \boxed{\left(q^2 + 3q\right)\left(4q^2 - 3q\right)} = \boxed{4\left(q^2 \cdot q^2\right) - 3\left(q^2 \cdot q\right) + (3 \cdot 4)\left(q \cdot q^2\right) - (3 \cdot 3)(q \cdot q)}$$

$$= \boxed{4q^4 - 3q^3 + 12q^3 - 9q^2} = \boxed{4q^4 + (-3+12)q^3 - 9q^2} = \boxed{\mathbf{4q^4 + 9q^3 - 9q^2}}$$

Example 6.2-44

$$\boxed{\left(\sqrt[3]{32} - 4\right)\left(\sqrt[3]{54} + 5\right)} = \boxed{\left(\sqrt[3]{8 \cdot 4} - 4\right)\left(\sqrt[3]{27 \cdot 2} + 5\right)} = \boxed{\left(\sqrt[3]{2^3 \cdot 4} - 4\right)\left(\sqrt[3]{3^3 \cdot 2} + 5\right)} = \boxed{\left(2\sqrt[3]{4} - 4\right)\left(3\sqrt[3]{2} + 5\right)}$$

$$= \boxed{(2 \cdot 3)\left(\sqrt[3]{4} \cdot \sqrt[3]{2}\right) + (2 \cdot 5)\sqrt[3]{4} - (4 \cdot 3)\sqrt[3]{2} - (4 \cdot 5)} = \boxed{6\left(\sqrt[3]{4 \cdot 2}\right) + 10\sqrt[3]{4} - 12\sqrt[3]{2} - 20} = \boxed{6\sqrt[3]{8} + 10\sqrt[3]{4} - 12\sqrt[3]{2} - 20}$$

$$= \boxed{6\sqrt[3]{2^3} + 10\sqrt[3]{4} - 12\sqrt[3]{2} - 20} = \boxed{(6 \cdot 2) + 10\sqrt[3]{4} - 12\sqrt[3]{2} - 20} = \boxed{12 + 10\sqrt[3]{4} - 12\sqrt[3]{2} - 20}$$

$$= \boxed{10\sqrt[3]{4} - 12\sqrt[3]{2} + (-20 + 12)} = \boxed{\mathbf{10\sqrt[3]{4} - 12\sqrt[3]{2} - 8}}$$

Example 6.2-45

$$\boxed{\left(b^3-\sqrt{b}\right)\left(-\sqrt{b^3}+\sqrt{b^4}\right)} = \boxed{\left(b^3-\sqrt{b}\right)\left(-\sqrt{b^{2+1}}+\sqrt{b^{2+2}}\right)} = \boxed{\left(b^3-\sqrt{b}\right)\left(-\sqrt{b^2\cdot b^1}+\sqrt{b^2\cdot b^2}\right)}$$

$$= \boxed{\left(b^3-\sqrt{b}\right)\left[-b\sqrt{b}+(b\cdot b)\right]} = \boxed{\left(b^3-\sqrt{b}\right)\left(-b\sqrt{b}+b^2\right)} = \boxed{-\left(b^3\cdot b\right)\sqrt{b}+\left(b^3\cdot b^2\right)+b\left(\sqrt{b}\cdot\sqrt{b}\right)-\left(b^2\cdot\sqrt{b}\right)}$$

$$= \boxed{-b^4\sqrt{b}+b^5+b\left(\sqrt{b\cdot b}\right)-b^2\sqrt{b}} = \boxed{-b^4\sqrt{b}+b^5-b^2\sqrt{b}+b\sqrt{b^2}} = \boxed{-b^4\sqrt{b}+b^5-b^2\sqrt{b}+(b\cdot b)}$$

$$= \boxed{-b^4\sqrt{b}+b^5-b^2\sqrt{b}+b^2} = \boxed{\mathbf{b^5-b^4\sqrt{b}-b^2\sqrt{b}+b^2}}$$

Practice Problems - Multiplying Binomials by Binomials

Section 6.2 Case II Practice Problems - Multiply the following binomial expressions:

1. $(x+3)(x-2) =$
2. $(-y+8)(y-6) =$
3. $\left(x^2-2xy\right)\left(-y^2+2xy\right) =$
4. $\left(a^3-a^2\right)(a-6) =$
5. $\left(\sqrt{x^3}-2x\sqrt{x^5}\right)\left(\sqrt{x}-4\right) =$
6. $\left(\sqrt[3]{y^5}-\sqrt[3]{y^2}\right)\left(\sqrt[3]{y^7}-\sqrt[3]{y}\right) =$
7. $\left(\sqrt{81}-\sqrt{72}\right)\left(\sqrt{36}+\sqrt{18}\right) =$
8. $\left(5-3\sqrt{x^4}\right)\left(\sqrt{x^2}-2\right) =$
9. $\left(u^2-\sqrt[3]{27}\right)\left(u^3-\sqrt[3]{8}\right) =$
10. $\left(\sqrt{a^2c}+2\right)\left(-\sqrt{a^4c}+8\right) =$

Case III Multiplying Polynomials by Polynomials

In general, polynomials are multiplied by each other in the following way:

$$\left(a_n x^n + a_{n-1}x^{n-1} + \ldots + a_0\right)\left(b_n x^n + b_{n-1}x^{n-1} + \ldots + b_0\right)$$

$$= (a_n \cdot b_n)\cdot\left(x^n \cdot x^n\right) + (a_n \cdot b_{n-1})\cdot\left(x^n \cdot x^{n-1}\right) + \ldots + (a_n \cdot b_0)x^n$$

$$+ (a_{n-1} \cdot b_n)\cdot\left(x^{n-1} \cdot x^n\right) + (a_{n-1} \cdot b_{n-1})\cdot\left(x^{n-1} \cdot x^{n-1}\right) + \ldots + (a_{n-1} \cdot b_0)x^{n-1} + \ldots$$

$$+ (a_0 \cdot b_n)x^n + (a_0 \cdot b_{n-1})x^{n-1} + \ldots + (a_0 \cdot b_0)$$

$$= a_n b_n\left(x^{n+n}\right) + a_n b_{n-1}\left(x^{n+n-1}\right) + \ldots + a_n b_0\left(x^n\right) + a_{n-1}b_n\left(x^{n-1+n}\right) + a_{n-1}b_{n-1}\left(x^{n-1+n-1}\right) + \ldots + a_{n-1}b_0\left(x^{n-1}\right) + \ldots$$

$$+ a_0 b_n\left(x^n\right) + a_0 b_{n-1}\left(x^{n-1}\right) + \ldots + a_0 b_0$$

$$= a_n b_n\left(x^{2n}\right) + a_n b_{n-1}\left(x^{2n-1}\right) + \ldots + a_n b_0 x^n + a_{n-1}b_n\left(x^{2n-1}\right) + a_{n-1}b_{n-1}\left(x^{2n-2}\right) + \ldots + a_{n-1}b_0\left(x^{n-1}\right) + \ldots$$

$$+ a_0 b_n\left(x^n\right) + a_0 b_{n-1}\left(x^{n-1}\right) + \ldots + a_0 b_0$$

$$= \mathbf{a_n b_n x^{2n} + (a_n b_{n-1} + a_{n-1}b_n)x^{2n-1} + \ldots + (a_n b_0 + a_0 b_n)x^n}$$

$$\mathbf{+ a_{n-1}b_{n-1}x^{2n-2} + \ldots + (a_{n-1}b_0 + a_0 b_{n-1})x^{n-1} + \ldots + a_0 b_0}$$

Polynomials are multiplied by one another using the following steps:

Step 1 Write the polynomials vertically. Multiply each term of one polynomial by each term of the other polynomial.

Step 2 Write like terms under one another.

Step 3 Combine like terms by adding or subtracting like terms.

Examples with Steps

The following examples show the steps as to how polynomials are multiplied by each other:

Example 6.2-46

Multiply $\left(x^3 + 5x^2 - 3x + 3\right)$ by $\left(x^2 - x + 1\right)$.

Solution:

Step 1

$$\begin{array}{r} x^3 + 5x^2 - 3x + 3 \\ x^2 - x + 1 \\ \hline \end{array}$$

Step 2

$$\begin{array}{rrrrrr} x^5 & +5x^4 & -3x^3 & +3x^2 & & \\ & -x^4 & -5x^3 & +3x^2 & -3x & \\ & & x^3 & +5x^2 & -3x & +3 \\ \hline \end{array}$$

Step 3 $\mathbf{x^5 + 4x^4 - 7x^3 + 11x^2 - 6x + 3}$

Example 6.2-47

Multiply $\left(-x^5 + 3x^3 - x + 5\right)$ by $\left(x^3 - 2x - 3\right)$.

Solution:

Step 1

$$\begin{array}{l} -x^5 + 3x^3 - x + 5 \\ x^3 - 2x - 3 \\ \hline \end{array}$$

Step 2

$$\begin{array}{l} -x^8 + 3x^6 - x^4 + 5x^3 \\ +2x^6 - 6x^4 + 2x^2 - 10x \\ +3x^5 - 9x^3 + 3x - 15 \\ \hline \end{array}$$

Step 3 $\mathbf{-x^8 + 5x^6 + 3x^5 - 7x^4 - 4x^3 + 2x^2 - 7x - 15}$

Example 6.2-48

Multiply $(a^4 + 3a^3 - 2a^2 + 5)$ by $(a - 1)$.

Solution:

Step 1

$$\begin{array}{l} a^4 + 3a^3 - 2a^2 + 5 \\ a - 1 \\ \hline \end{array}$$

Step 2

$$\begin{array}{l} a^5 + 3a^4 - 2a^3 + 5a \\ - a^4 - 3a^3 + 2a^2 - 5 \\ \hline \end{array}$$

Step 3 $\mathbf{a^5 + 2a^4 - 5a^3 + 2a^2 + 5a - 5}$

Example 6.2-49

Multiply $(x^6 - 2x^5 - 3x^4 + 3x^3 - 2)$ by $(x^2 - 2x - 4)$.

Solution:

Step 1

$$\begin{array}{l} x^6 - 2x^5 - 3x^4 + 3x^3 - 2 \\ x^2 - 2x - 4 \\ \hline \end{array}$$

Step 2

$$\begin{array}{l} x^8 - 2x^7 - 3x^6 + 3x^5 - 2x^2 \\ -2x^7 + 4x^6 + 6x^5 - 6x^4 + 4x \\ -4x^6 + 8x^5 + 12x^4 - 12x^3 + 8 \\ \hline \end{array}$$

Step 3 $\mathbf{x^8 - 4x^7 - 3x^6 + 17x^5 + 6x^4 - 12x^3 - 2x^2 + 4x + 8}$

Example 6.2-50

Multiply $(2y^2 - 6y - 9)$ by $(y^3 - 2y^2 + 3)$.

Solution:

Step 1

$$\begin{array}{l} 2y^2 - 6y - 9 \\ y^3 - 2y^2 + 3 \\ \hline \end{array}$$

Step 2

$$\begin{array}{l} 2y^5 - 6y^4 - 9y^3 \\ -4y^4 + 12y^3 + 18y^2 \\ + 6y^2 - 18y - 27 \\ \hline \end{array}$$

Step 3 $\mathbf{2y^5 - 10y^4 + 3y^3 + 24y^2 - 18y - 27}$

Additional Examples - Multiplying Polynomials by Polynomials

The following examples further illustrate how to multiply monomials by monomials:

Example 6.2-51 Multiply $\left(2x^4 - 2x + 5\right)$ by $\left(x^2 - 2\right)$.

Solution:

$$\begin{array}{llllll} 2x^4 & & -2x & +5 & & \\ \underline{x^2 - 2} & & & & & \\ 2x^6 & & -2x^3 & +5x^2 & & \\ & -4x^4 & & & +4x & -10 \\ \hline \mathbf{2x^6} & \mathbf{-4x^4} & \mathbf{-2x^3} & \mathbf{+5x^2} & \mathbf{+4x} & \mathbf{-10} \end{array}$$

Example 6.2-52 Multiply $\left(-3x^4 - 2x^2 + 1\right)$ by $\left(x^3 + 2x - 3\right)$.

Solution:

$$\begin{array}{llllllll} -3x^4 - 2x^2 + 1 & & & & & & & \\ \underline{x^3 + 2x - 3} & & & & & & & \\ -3x^7 & -2x^5 & & +x^3 & & & & \\ & -6x^5 & & -4x^3 & & +2x & & \\ & & +9x^4 & & +6x^2 & & -3 & \\ \hline \mathbf{-3x^7} & \mathbf{-8x^5} & \mathbf{+9x^4} & \mathbf{-3x^3} & \mathbf{+6x^2} & \mathbf{+2x} & \mathbf{-3} & \end{array}$$

Example 6.2-53 Multiply $\left(-x^4 + 2x^3 - 2\right)$ by $(-3x + 2)$.

Solution:

$$\begin{array}{lllll} -x^4 + 2x^3 - 2 & & & & \\ \underline{-3x + 2} & & & & \\ +3x^5 - 6x^4 & & & +6x & \\ & -2x^4 & +4x^3 & & -4 \\ \hline \mathbf{+3x^5 - 8x^4} & & \mathbf{+4x^3} & \mathbf{+6x} & \mathbf{-4} \end{array}$$

Example 6.2-54 Multiply $\left(a^4 - 3a^3 + 10\right)$ by $\left(a^2 - 3a + 2\right)$.

Solution:

$$\begin{array}{lllllll} a^4 - 3a^3 + 10 & & & & & & \\ \underline{a^2 - 3a + 2} & & & & & & \\ a^6 & -3a^5 & & & +10a^2 & & \\ & -3a^5 & +9a^4 & & & -30a & \\ & & +2a^4 & -6a^3 & & & +20 \\ \hline \mathbf{a^6} & \mathbf{-6a^5} & \mathbf{+11a^4} & \mathbf{-6a^3} & \mathbf{+10a^2} & \mathbf{-30a} & \mathbf{+20} \end{array}$$

Example 6.2-55 Multiply $\left(x^5 - 2x^3 - x^2 + 2\right)$ by $(-x + 3)$.

Solution:

$$\begin{array}{l} x^5 - 2x^3 - x^2 + 2 \\ \underline{-x + 3} \\ -x^6 \quad\quad + 2x^4 + x^3 \quad\quad - 2x \\ \quad\quad + 3x^5 \quad\quad - 6x^3 - 3x^2 \quad\quad + 6 \\ \hline \mathbf{-x^6 + 3x^5 + 2x^4 - 5x^3 - 3x^2 - 2x + 6} \end{array}$$

Example 6.2-56 Multiply $\left(w^3 - w + 2\right)$ by $\left(w^2 - w - 1\right)$.

Solution:

$$\begin{array}{l} w^3 - w + 2 \\ \underline{w^2 - w - 1} \\ w^5 \quad\quad - w^3 + 2w^2 \\ \quad\quad - w^4 \quad\quad + w^2 - 2w \\ \quad\quad\quad\quad - w^3 \quad\quad + w - 2 \\ \hline \mathbf{w^5 - w^4 - 2w^3 + 3w^2 - w - 2} \end{array}$$

Example 6.2-57 Multiply $\left(x^2 - 3x + 6\right)$ by $(x + 5)$.

Solution:

$$\begin{array}{l} x^2 - 3x + 6 \\ \underline{x + 5} \\ x^3 - 3x^2 + 6x \\ \quad\quad + 5x^2 - 15x + 30 \\ \hline \mathbf{x^3 + 2x^2 - 9x + 30} \end{array}$$

Example 6.2-58 Multiply $\left(b^3 - b^2 - 1\right)$ by $(b + 1)$.

Solution:

$$\begin{array}{l} b^3 - b^2 - 1 \\ \underline{b + 1} \\ b^4 \; - b^3 \quad\quad - b \\ \quad\quad + b^3 \; - b^2 \quad - 1 \\ \hline \mathbf{b^4 + 0b^3 - b^2 - b - 1} \end{array}$$

Example 6.2-59 Multiply $\left(b^5 - 4b^3 - 2b + 3\right)$ by $\left(b^2 - b + 1\right)$.

Solution:

$$\begin{array}{l} b^5 - 4b^3 - 2b + 3 \\ \underline{b^2 - b + 1} \\ b^7 \quad\quad - 4b^5 \quad\quad - 2b^3 + 3b^2 \\ \quad\quad - b^6 \quad\quad + 4b^4 \quad\quad + 2b^2 - 3b \\ \quad\quad\quad\quad b^5 \quad\quad - 4b^3 \quad\quad - 2b + 3 \\ \hline \mathbf{b^7 - b^6 - 3b^5 + 4b^4 - 6b^3 + 5b^2 - 5b + 3} \end{array}$$

Example 6.2-60 Multiply $\left(a^2 - 2a - 3\right)$ by $(a+2)$.

Solution:

$$\begin{array}{l} a^2 - 2a - 3 \\ \underline{a + 2} \\ a^3 - 2a^2 - 3a \\ \underline{\quad + 2a^2 - 4a - 6} \\ \mathbf{a^3 + 0a^2 - 7a - 6} \end{array}$$

Practice Problems - Multiplying Polynomials by Polynomials

Section 6.2 Case III Practice Problems - Multiply the following polynomials:

1. $\left(6x^2 - 5x - 2\right)(3x+5) =$
2. $\left(b^3 - b^2 + 6\right)\left(b^2 - b + 2\right) =$
3. $\left(u^6 - 3u^4 - 2u^2 + 2\right)\left(u^3 - 2u^2 + 1\right) =$
4. $\left(x^5 - 3x^4 + x^3 - 2x\right)(x+3) =$
5. $\left(x^2 - 3x - 8\right)\left(x^3 - 2x\right) =$
6. $\left(-3x^3 + x^2 - 2x + 1\right)\left(x^2 - x + 1\right) =$
7. $\left(v^5 - 3v^3 + 2v - 1\right)\left(v^2 + v - 2\right) =$
8. $\left(x^3 + 5x^2 - 5x + 6\right)(x-5) =$
9. $\left(3y^3 + 2y^2 - 2\right)\left(-2y^2 + 1\right) =$
10. $\left(-3a^3 - 2a^2 + 1\right)\left(-a^2 + a - 2\right) =$

6.3 Dividing Polynomials

Polynomials are divided by one another using a similar method like the long division used in arithmetic operations. In this section students learn how to divide polynomials by monomials (Case I), and how to divide polynomials by polynomials (Case II).

Case I Dividing by Monomials

To divide a polynomial, a trinomial, or a binomial by a monomial we **divide each term** in the numerator which is separated by a + or a − sign by the denominator. In general, polynomials are divided by monomials in the following way:

$$\frac{a_n x^n + a_{n-1}x^{n-1} + a_{n-2}x^{n-2} + \ldots + a_0}{bx^m} = \frac{a_n x^n}{bx^m} + \frac{a_{n-1}x^{n-1}}{bx^m} + \frac{a_{n-2}x^{n-2}}{bx^m} + \ldots + \frac{a_0}{bx^m}$$

where a_n, a_{n-1}, a_{n-2}, ..., a_0, and b are real numbers, n and m are positive integer numbers, and x is a variable. For example,

$$\frac{16y^4 + 5y^3 + 4y^2 + 8y + 20}{4} = \frac{16y^4}{4} + \frac{5y^3}{4} + \frac{4y^2}{4} + \frac{8y}{4} + \frac{20}{4}$$

Note that we **can not divide out only one term** of the polynomial in the numerator by the denominator, i.e., we can not do the following:

$$\frac{4x^3 - 5x^2 + 6x + 12}{2} \neq \frac{\overset{2}{\cancel{4}}x^3 - 5x^2 + 6x + 12}{\underset{1}{\cancel{2}}}$$

instead,

$$\frac{4x^3 - 5x^2 + 6x + 12}{2} = \frac{\overset{2}{\cancel{4}}x^3}{\underset{1}{\cancel{2}}} - \frac{5x^2}{2} + \frac{\overset{3}{\cancel{6}}x}{\underset{1}{\cancel{2}}} + \frac{\overset{6}{\cancel{12}}}{\underset{1}{\cancel{2}}} = 2x^3 - \frac{5}{2}x^2 + 3x + 6$$

Division by monomial expressions is divided to three cases. Case I a - dividing monomial by monomials, Case I b - dividing binomials by monomials, and Case I c - dividing polynomials by monomials.

Case I a - Dividing Monomials by Monomials

Monomials are divided by one another using the following steps:

Step 1 Simplify the monomials in both the numerator and the denominator.

Step 2 Divide the numerator by the denominator using the exponent rule (see Section 3.3, Case II) to divide the variables, i.e.,

$$\frac{x^n}{x^m} = \frac{x^{n-m}}{1} = x^{n-m} \qquad \text{if } n > m$$

or,

$$\frac{x^n}{x^m} = \frac{1}{x^{m-n}} \qquad \text{if } n < m$$

where n and m are positive integer numbers and x is a variable.

Examples with Steps

The following examples show the steps as to how monomials are divided by each other:

Example 6.3-1

$$\frac{\sqrt{8x^3y^2}}{\sqrt{243xy^3}} =$$

Solution:

Step 1

$$\frac{\sqrt{8x^3y^2}}{\sqrt{243xy^3}} = \frac{\sqrt{4\cdot 2x^3y^2}}{\sqrt{81\cdot 3xy^3}} = \frac{\sqrt{2^2\cdot 2x^3y^2}}{\sqrt{9^2\cdot 3xy^3}} = \frac{2\sqrt{2x^3y^2}}{9\sqrt{3xy^3}}$$

Step 2

$$\frac{2\sqrt{2x^3y^2}}{9\sqrt{3xy^3}} = \frac{2\sqrt{2}}{9\sqrt{3}}\frac{x^3y^2}{x^1y^3} = \frac{2\sqrt{2}}{9\sqrt{3}}\frac{x^3x^{-1}}{y^3y^{-2}} = \frac{2\sqrt{2}}{9\sqrt{3}}\frac{x^{3-1}}{y^{3-2}} = \frac{2\sqrt{2}}{9\sqrt{3}}\frac{x^2}{y^1} = \boldsymbol{\frac{2\sqrt{2}x^2}{9\sqrt{3}y}}$$

Example 6.3-2

$$\frac{-\sqrt{12a^2b^2c}}{\sqrt{225abc^4}} =$$

Solution:

Step 1

$$\frac{-\sqrt{12a^2b^2c}}{\sqrt{225abc^4}} = -\frac{\sqrt{4\cdot 3a^2b^2c}}{\sqrt{15\cdot 15abc^4}} = -\frac{\sqrt{2^2\cdot 3a^2b^2c}}{\sqrt{15^2abc^4}} = -\frac{2\sqrt{3a^2b^2c}}{15abc^4}$$

Step 2

$$-\frac{2\sqrt{3a^2b^2c}}{15abc^4} = -\frac{2\sqrt{3}}{15}\frac{a^2b^2c^1}{a^1b^1c^4} = -\frac{2\sqrt{3}}{15}\frac{\left(a^2a^{-1}\right)\cdot\left(b^2b^{-1}\right)}{c^4c^{-1}} = -\frac{2\sqrt{3}}{15}\frac{\left(a^{2-1}\right)\cdot\left(b^{2-1}\right)}{c^{4-1}}$$

$$= -\frac{2\sqrt{3}}{15}\frac{a^1b^1}{c^3} = \boldsymbol{-\frac{2\sqrt{3}ab}{15c^3}}$$

Example 6.3-3

$$\frac{\sqrt[3]{16u^2v^2}}{-\sqrt[3]{27uv^3}} =$$

Solution:

Step 1

$$\frac{\sqrt[3]{16u^2v^2}}{-\sqrt[3]{27uv^3}} = -\frac{\sqrt[3]{8\cdot 2u^2v^2}}{\sqrt[3]{3^3uv^3}} = -\frac{\sqrt[3]{2^3\cdot 2u^2v^2}}{3uv^3} = -\frac{2\sqrt[3]{2u^2v^2}}{3uv^3}$$

Step 2

$$-\frac{2\sqrt[3]{2u^2v^2}}{3uv^3} = -\frac{2\sqrt[3]{2}}{3}\frac{u^2v^2}{u^1v^3} = -\frac{2\sqrt[3]{2}}{3}\frac{u^2u^{-1}}{v^3v^{-2}} = -\frac{2\sqrt[3]{2}}{3}\frac{u^{2-1}}{v^{3-2}} = -\frac{2\sqrt[3]{2}}{3}\frac{u^1}{v^1} = \boldsymbol{-\frac{2\sqrt[3]{2u}}{3v}}$$

Example 6.3-4

$$\frac{-\sqrt[5]{3^5}\cdot 5\sqrt{u^3w^5}}{-\sqrt[4]{2^4}u^2w} =$$

Solution:

Step 1

$$\boxed{\frac{-\sqrt[5]{3^5}\cdot 5\sqrt{u^3w^5}}{-\sqrt[4]{2^4}u^2w}}=\boxed{+\frac{3\sqrt[5]{5}\sqrt{u^{2+1}w^{2+2+1}}}{2u^2w}}=\boxed{\frac{3\sqrt[5]{5}\sqrt{\left(u^2\cdot u^1\right)\cdot\left(w^2\cdot w^2\cdot w^1\right)}}{2u^2w}}$$

$$=\boxed{\frac{3\sqrt[5]{5}}{2}\,\frac{u(w\cdot w)\sqrt{uw}}{u^2w}}=\boxed{\frac{3\sqrt[5]{5}}{2}\,\frac{uw^2\sqrt{uw}}{u^2w}}$$

Step 2

$$\boxed{\frac{3\sqrt[5]{5}}{2}\,\frac{uw^2\sqrt{uw}}{u^2w}}=\boxed{\frac{3\sqrt[5]{5}}{2}\,\frac{u^1w^2\sqrt{uw}}{u^2w^1}}=\boxed{\frac{3\sqrt[5]{5}}{2}\,\frac{w^2w^{-1}\sqrt{uw}}{u^2u^{-1}}}=\boxed{\frac{3\sqrt[5]{5}}{2}\,\frac{w^{2-1}\sqrt{uw}}{u^{2-1}}}$$

$$=\boxed{\frac{3\sqrt[5]{5}}{2}\,\frac{w^1\sqrt{uw}}{u^1}}=\boxed{\mathbf{\frac{3\sqrt[5]{5}w\sqrt{uw}}{2u}}}$$

Example 6.3-5

$$\boxed{\frac{\sqrt[4]{3^4x^5y^6}}{\sqrt[3]{64x^6y^3}}}=$$

Solution:

Step 1

$$\boxed{\frac{\sqrt[4]{3^4x^5y^6}}{\sqrt[3]{64x^6y^3}}}=\boxed{\frac{3\sqrt[4]{x^{4+1}y^{4+2}}}{y\sqrt[3]{4^3x^{3+3}}}}=\boxed{\frac{3\sqrt[4]{\left(x^4\cdot x^1\right)\cdot\left(y^4\cdot y^2\right)}}{4y\sqrt[3]{x^3\cdot x^3}}}=\boxed{\frac{3xy\sqrt[4]{xy^2}}{4y(x\cdot x)}}=\boxed{\frac{3xy\sqrt[4]{xy^2}}{4x^2y}}$$

Step 2

$$\boxed{\frac{3xy\sqrt[4]{xy^2}}{4x^2y}}=\boxed{\frac{3x^1y^1\sqrt[4]{xy^2}}{4x^2y^1}}=\boxed{\frac{3\left(y^1y^{-1}\right)\sqrt[4]{xy^2}}{4\left(x^2x^{-1}\right)}}=\boxed{\frac{3y^{1-1}\sqrt[4]{xy^2}}{4x^{2-1}}}=\boxed{\frac{3y^0\sqrt[4]{xy^2}}{4x^1}}=\boxed{\mathbf{\frac{3\sqrt[4]{xy^2}}{4x}}}$$

Additional Examples - Dividing Monomials by Monomials

The following examples further illustrate how to divide monomials by each other:

Example 6.3-6

$$\boxed{\frac{x^6y^5}{x^4y}}=\boxed{\frac{x^6y^5}{x^4y^1}}=\boxed{\frac{\left(x^6x^{-4}\right)\cdot\left(y^5y^{-1}\right)}{1}}=\boxed{\frac{x^{6-4}y^{5-1}}{1}}=\boxed{\frac{x^2y^4}{1}}=\boxed{\mathbf{x^2y^4}}$$

Example 6.3-7

$$\boxed{\frac{4xy^2z^3}{-16x^2yz^2}}=\boxed{-\frac{\overset{1}{\cancel{4}}}{\underset{4}{\cancel{16}}}\,\frac{x^1y^2z^3}{x^2y^1z^2}}=\boxed{-\frac{1}{4}\,\frac{\left(y^2y^{-1}\right)\cdot\left(z^3z^{-2}\right)}{x^2x^{-1}}}=\boxed{-\frac{1}{4}\,\frac{y^{2-1}z^{3-2}}{x^{2-1}}}=\boxed{-\frac{1}{4}\,\frac{y^1z^1}{x^1}}=\boxed{-\frac{1}{4}\,\frac{yz}{x}}=\boxed{-\mathbf{\frac{yz}{4x}}}$$

Example 6.3-8

$$\boxed{\frac{-5a^2b}{15a^2b^3}}=\boxed{-\frac{\overset{1}{\cancel{5}}}{\underset{3}{\cancel{15}}}\,\frac{a^2b^1}{a^2b^3}}=\boxed{-\frac{1}{3}\,\frac{a^2a^{-2}}{b^3b^{-1}}}=\boxed{-\frac{1}{3}\,\frac{a^{2-2}}{b^{3-1}}}=\boxed{-\frac{1}{3}\,\frac{a^0}{b^2}}=\boxed{-\frac{1}{3}\,\frac{1}{b^2}}=\boxed{-\mathbf{\frac{1}{3b^2}}}$$

Example 6.3-9

$$\boxed{\frac{-4xy^4}{-\sqrt{64x^3y^2}}} = \boxed{+\frac{4}{\sqrt{64}}\frac{xy^4}{x^3y^2}} = \boxed{\frac{4}{\sqrt{8^2}}\frac{xy^4}{x^3y^2}} = \boxed{\frac{\overset{1}{\cancel{4}}}{\underset{2}{\cancel{8}}}\frac{x^1y^4}{x^3y^2}} = \boxed{\frac{1}{2}\frac{y^4y^{-2}}{x^3x^{-1}}} = \boxed{\frac{1}{2}\frac{y^{4-2}}{x^{3-1}}} = \boxed{\frac{1}{2}\frac{y^2}{x^2}} = \boxed{\mathbf{\frac{y^2}{2x^2}}}$$

Example 6.3-10

$$\boxed{\frac{56}{-\sqrt{36}}} = \boxed{-\frac{56}{\sqrt{6^2}}} = \boxed{-\frac{\overset{28}{\cancel{56}}}{\underset{3}{\cancel{6}}}} = \boxed{-\frac{28}{3}} = \boxed{\mathbf{-\left(9\frac{1}{3}\right)}}$$

Example 6.3-11

$$\boxed{\frac{49\sqrt{x^2y^3z^4}}{-7\sqrt{x^2y^4z^6}}} = \boxed{-\frac{49}{7}\frac{x\sqrt{y^{2+1}z^{2+2}}}{x\sqrt{y^{2+2}z^{2+2+2}}}} = \boxed{-\frac{\overset{7}{\cancel{49}}}{\underset{1}{\cancel{7}}}\frac{x\sqrt{(y^2\cdot y^1)\cdot(z^2\cdot z^2)}}{x\sqrt{(y^2\cdot y^2)\cdot(z^2\cdot z^2\cdot z^2)}}} = \boxed{-\frac{7}{1}\frac{x\cdot y(z\cdot z)\sqrt{y}}{x\cdot(y\cdot y)\cdot(z\cdot z\cdot z)\sqrt{1}}}$$

$$= \boxed{-\frac{7}{1}\frac{xyz^2\sqrt{y}}{xy^2z^3}} = \boxed{-\frac{7}{1}\frac{x^1y^1z^2\sqrt{y}}{x^1y^2z^3}} = \boxed{-\frac{7}{1}\frac{(x^1x^{-1})\sqrt{y}}{(y^2y^{-1})\cdot(z^3z^{-2})}} = \boxed{-\frac{7}{1}\frac{x^{1-1}\sqrt{y}}{y^{2-1}z^{3-2}}} = \boxed{-\frac{7}{1}\frac{x^0\sqrt{y}}{y^1z^1}} = \boxed{\mathbf{-\frac{7\sqrt{y}}{yz}}}$$

Example 6.3-12

$$\boxed{\frac{36\sqrt[3]{x^5y^4z^6}}{-\sqrt[3]{81x^2yz^2}}} = \boxed{-\frac{36\sqrt[3]{x^{3+2}y^{3+1}z^{3+3}}}{\sqrt[3]{27\cdot 3x^2yz^2}}} = \boxed{-\frac{36\sqrt[3]{(x^3\cdot x^2)\cdot(y^3\cdot y^1)\cdot(z^3\cdot z^3)}}{\sqrt[3]{27\cdot 3x^2yz^2}}} = \boxed{-\frac{36xy(z\cdot z)\sqrt[3]{x^2y}}{\sqrt[3]{3^3\cdot 3x^2yz^2}}}$$

$$= \boxed{-\frac{\overset{12}{\cancel{36}}xyz^2\sqrt[3]{x^2y}}{\underset{1}{\cancel{3}}\sqrt[3]{3x^2yz^2}}} = \boxed{-\frac{12x^1y^1z^2\sqrt[3]{x^2y}}{\sqrt[3]{3x^2y^1z^2}}} = \boxed{-\frac{12}{\sqrt[3]{3}}\frac{(y^1y^{-1})\cdot(z^2z^{-2})\sqrt[3]{x^2y}}{x^2x^{-1}}} = \boxed{-\frac{12}{\sqrt[3]{3}}\frac{y^{1-1}z^{2-2}\sqrt[3]{x^2y}}{x^{2-1}}}$$

$$= \boxed{-\frac{12}{\sqrt[3]{3}}\frac{y^0z^0\sqrt[3]{x^2y}}{x^1}} = \boxed{-\frac{12}{\sqrt[3]{3}}\frac{(1\cdot 1)\sqrt[3]{x^2y}}{x}} = \boxed{-\frac{12}{\sqrt[3]{3}}\frac{\sqrt[3]{x^2y}}{x}} = \boxed{\mathbf{-\frac{12\sqrt[3]{x^2y}}{\sqrt[3]{3}x}}}$$

Example 6.3-13

$$\boxed{\frac{uvw}{-2u^2vw^4}} = \boxed{-\frac{1}{2}\frac{u^1v^1w^1}{u^2v^1w^4}} = \boxed{-\frac{1}{2}\frac{v^1v^{-1}}{(u^2u^{-1})\cdot(w^4w^{-1})}} = \boxed{-\frac{1}{2}\frac{v^{1-1}}{u^{2-1}w^{4-1}}} = \boxed{-\frac{1}{2}\frac{v^0}{u^1w^3}} = \boxed{-\frac{1}{2}\frac{1}{uw^3}} = \boxed{\mathbf{-\frac{1}{2uw^3}}}$$

Example 6.3-14

$$\boxed{\frac{\sqrt[3]{l^5m^3n^2}}{-lm^2n^3}} = \boxed{-\frac{m\sqrt[3]{l^{3+2}n^2}}{lm^2n^3}} = \boxed{-\frac{m\sqrt[3]{(l^3\cdot l^2)n^2}}{lm^2n^3}} = \boxed{-\frac{lm\sqrt[3]{l^2n^2}}{lm^2n^3}} = \boxed{-\frac{l^1m^1\sqrt[3]{l^2n^2}}{l^1m^2n^3}} = \boxed{-\frac{(l^1l^{-1})\sqrt[3]{l^2n^2}}{(m^2m^{-1})n^3}}$$

$$= \boxed{-\frac{l^{1-1}\sqrt[3]{l^2n^2}}{m^{2-1}n^3}} = \boxed{-\frac{l^0\sqrt[3]{l^2n^2}}{m^1n^3}} = \boxed{-\frac{1\cdot\sqrt[3]{l^2n^2}}{mn^3}} = \boxed{\mathbf{-\frac{\sqrt[3]{l^2n^2}}{mn^3}}}$$

Example 6.3-15

$$\boxed{\frac{-\sqrt{72xy}}{12x^2y^3}} = \boxed{-\frac{\sqrt{36\cdot 2xy}}{12x^2y^3}} = \boxed{-\frac{\sqrt{6^2\cdot 2x^1y^1}}{12x^2y^3}} = \boxed{-\frac{\overset{1}{\cancel{6}}\sqrt{2}}{\underset{2}{\cancel{12}}\left(x^2x^{-1}\right)\cdot\left(y^3y^{-1}\right)}} = \boxed{-\frac{\sqrt{2}}{2x^{2-1}y^{3-1}}} = \boxed{-\frac{\sqrt{2}}{2x^1y^2}} = \boxed{-\frac{\sqrt{2}}{2xy^2}}$$

Practice Problems - Dividing Monomials by Monomials

Section 6.3 Case I a Practice Problems - Divide the following monomial expressions:

1. $\frac{-4xyz}{-8xyz} =$
2. $\frac{u^2v^3w}{-uw^4} =$
3. $\frac{\sqrt{72x^2y^4}}{-12xy^2} =$
4. $\frac{-36x^3y^3z^4}{-\sqrt{25xyz^3}} =$
5. $\frac{-9a^2b^2c^3}{\sqrt[3]{27a^6b^3c^3}} =$
6. $\frac{-24lm^3n^2}{12l^2mn} =$
7. $\frac{xyz^2}{-xy^2z^3} =$
8. $\frac{\sqrt[4]{3^4x^5y^4z^6}}{\sqrt{18x^6y^2}} =$
9. $\frac{\sqrt[3]{27x^3y^6}}{-9x^2y} =$
10. $\frac{\sqrt{x^2y^3}}{-\sqrt[4]{2^4x^4y^8}} =$

Case I b Dividing Binomials by Monomials

Binomials are divided by monomial expressions using the following steps:

Step 1 Simplify each term in the numerator and the denominator.

Step 2 Divide each binomial term by the denominator using the exponent rule (see Section 3.3, Case II) to divide the variables, i.e.,

$$\frac{x^n}{x^m} = \frac{x^{n-m}}{1} = x^{n-m} \qquad \textit{if } n > m$$

or,

$$\frac{x^n}{x^m} = \frac{1}{x^{m-n}} \qquad \textit{if } n < m$$

where n and m are positive integer numbers and x is a variable.

Examples with Steps

The following examples show the steps as to how binomials are divided by monomials:

Example 6.3-16

$$\frac{8x^3 - 16x^2}{-8x} =$$

Solution:

Step 1 *Not Applicable*

Step 2

$$\frac{8x^3 - 16x^2}{-8x} = \frac{8x^3}{-8x} + \frac{-16x^2}{-8x} = -\frac{\overset{1}{\cancel{8}}x^3}{\underset{1}{\cancel{8}}x^1} + \frac{\overset{2}{\cancel{16}}x^2}{\underset{1}{\cancel{8}}x^1} = -\frac{x^3x^{-1}}{1} + \frac{2x^2x^{-1}}{1}$$

$$= -\frac{x^{3-1}}{1} + \frac{2x^{2-1}}{1} = -\frac{x^2}{1} + \frac{2x^1}{1} = \frac{-x^2 + 2x}{1} = \mathbf{-x^2 + 2x}$$

Example 6.3-17

$$\frac{-15\sqrt{a^3} + 10\sqrt{a^2}}{-5a^2} =$$

Solution:

Step 1

$$\frac{-15\sqrt{a^3} + 10\sqrt{a^2}}{-5a^2} = \frac{-15\sqrt{a^{2+1}} + 10a}{-5a^2} = \frac{-15\sqrt{a^2a^1} + 10a}{-5a^2} = \frac{-15a\sqrt{a} + 10a}{-5a^2}$$

Step 2

$$\frac{-15a\sqrt{a} + 10a}{-5a^2} = \frac{-15a\sqrt{a}}{-5a^2} + \frac{10a}{-5a^2} = \frac{\overset{3}{\cancel{15}}a^1\sqrt{a}}{\underset{1}{\cancel{5}}a^2} - \frac{\overset{2}{\cancel{10}}a^1}{\underset{1}{\cancel{5}}a^2} = \frac{3\sqrt{a}}{a^2a^{-1}} - \frac{2}{a^2a^{-1}}$$

$$= \boxed{\frac{3\sqrt{a}}{a^{2-1}} - \frac{2}{a^{2-1}}} = \boxed{\frac{3\sqrt{a}}{a^1} - \frac{2}{a^1}} = \boxed{\frac{3\sqrt{a}}{a} - \frac{2}{a}} = \boxed{\mathbf{\frac{3\sqrt{a}-2}{a}}}$$

Example 6.3-18

$$\boxed{\frac{x^5 - y^6}{x^3 y^3}} =$$

Solution:

Step 1 *Not Applicable*

Step 2

$$\boxed{\frac{x^5 - y^6}{x^3 y^3}} = \boxed{\frac{x^5}{x^3 y^3} + \frac{-y^6}{x^3 y^3}} = \boxed{\frac{x^5 x^{-3}}{y^3} - \frac{y^6 y^{-3}}{x^3}} = \boxed{\frac{x^{5-3}}{y^3} - \frac{y^{6-3}}{x^3}} = \boxed{\mathbf{\frac{x^2}{y^3} - \frac{y^3}{x^3}}}$$

Example 6.3-19

$$\boxed{\frac{\sqrt[3]{x^3 y^5} - 4\sqrt[4]{x^8 y^4}}{2\sqrt[3]{x^3 y^6}}} =$$

Solution:

Step 1

$$\boxed{\frac{\sqrt[3]{x^3 y^5} - 4\sqrt[4]{x^8 y^4}}{2\sqrt[3]{x^3 y^6}}} = \boxed{\frac{x\sqrt[3]{y^{3+2}} - 4y\sqrt[4]{x^{4+4}}}{2x\sqrt[3]{y^{3+3}}}} = \boxed{\frac{x\sqrt[3]{y^3 \cdot y^2} - 4y\sqrt[4]{x^4 \cdot x^4}}{2x\sqrt[3]{y^3 \cdot y^3}}}$$

$$= \boxed{\frac{xy\sqrt[3]{y^2} - 4(x \cdot x)y}{2x(y \cdot y)}} = \boxed{\frac{xy\sqrt[3]{y^2} - 4x^2 y}{2xy^2}}$$

Step 2

$$\boxed{\frac{xy\sqrt[3]{y^2} - 4x^2 y}{2xy^2}} = \boxed{\frac{xy\sqrt[3]{y^2}}{2xy^2} + \frac{-4x^2 y}{2xy^2}} = \boxed{\frac{x^1 y^1 \sqrt[3]{y^2}}{2x^1 y^2} - \frac{\overset{2}{\cancel{4}} x^2 y^1}{\underset{1}{\cancel{2}} x^1 y^2}} = \boxed{\frac{x^1 x^{-1} \sqrt[3]{y^2}}{2y^2 y^{-1}} - \frac{2x^2 x^{-1}}{y^2 y^{-1}}}$$

$$= \boxed{\frac{x^{1-1}\sqrt[3]{y^2}}{2y^{2-1}} - \frac{2x^{2-1}}{y^{2-1}}} = \boxed{\frac{x^0 \sqrt[3]{y^2}}{2y^1} - \frac{2x^1}{y^1}} = \boxed{\mathbf{\frac{\sqrt[3]{y^2}}{2y} - \frac{2x}{y}}}$$

Example 6.3-20

$$\boxed{\frac{5u\sqrt{v^2 w^4} - u^2 v^2 \sqrt{w^3}}{10u^2 v^2 w^2}} =$$

Solution:

Step 1

$$\boxed{\frac{5u\sqrt{v^2 w^4} - u^2 v^2 \sqrt{w^3}}{10u^2 v^2 w^2}} = \boxed{\frac{5uv\sqrt{w^{2+2}} - u^2 v^2 \sqrt{w^{2+1}}}{10u^2 v^2 w^2}} = \boxed{\frac{5uv\sqrt{w^2 \cdot w^2} - u^2 v^2 \sqrt{w^2 \cdot w^1}}{10u^2 v^2 w^2}}$$

$$= \boxed{\frac{5uv(w \cdot w) - u^2 v^2 w\sqrt{w}}{10u^2 v^2 w^2}} = \boxed{\frac{5uvw^2 - u^2 v^2 w\sqrt{w}}{10u^2 v^2 w^2}}$$

Step 2

$$\frac{5uvw^2 - u^2v^2w\sqrt{w}}{10u^2v^2w^2} = \frac{5uvw^2}{10u^2v^2w^2} + \frac{-u^2v^2w\sqrt{w}}{10u^2v^2w^2} = \frac{\overset{1}{\cancel{5}}u^1v^1w^2}{\underset{2}{\cancel{10}}u^2v^2w^2} - \frac{u^2v^2w^1\sqrt{w}}{10u^2v^2w^2}$$

$$= \frac{w^2w^{-2}}{2\left(u^2u^{-1}\right)\cdot\left(v^2v^{-1}\right)} - \frac{u^2u^{-2}\sqrt{w}}{10\left(v^2v^{-2}\right)\cdot\left(w^2w^{-1}\right)} = \frac{w^{2-2}}{2u^{2-1}v^{2-1}} - \frac{u^{2-2}\sqrt{w}}{10v^{2-2}w^{2-1}}$$

$$= \frac{w^0}{2u^1v^1} - \frac{u^0\sqrt{w}}{10v^0w^1} = \mathbf{\frac{1}{2uv} - \frac{\sqrt{w}}{10w}}$$

Additional Examples - Dividing Binomials by Monomials

The following examples further illustrate how binomials are divided by monomials:

Example 6.3-21

$$\frac{-7x^2 + 14x}{-7} = \frac{-7x^2}{-7} + \frac{14x}{-7} = +\frac{\overset{1}{\cancel{7}}x^2}{\underset{1}{\cancel{7}}} - \frac{\overset{2}{\cancel{14}}x}{\underset{1}{\cancel{7}}} = \frac{x^2}{1} - \frac{2x}{1} = \frac{x^2 - 2x}{1} = \mathbf{x^2 - 2x}$$

Example 6.3-22

$$\frac{10x^3 - 5x^2}{5x^2} = \frac{10x^3}{5x^2} + \frac{-5x^2}{5x^2} = \frac{10x^3x^{-2}}{5} - \frac{5x^2x^{-2}}{5} = \frac{\overset{2}{\cancel{10}}x^{3-2}}{\underset{1}{\cancel{5}}} - \frac{\overset{1}{\cancel{5}}x^{2-2}}{\underset{1}{\cancel{5}}} = \frac{2x^1}{1} - \frac{x^0}{1} = \frac{2x}{1} - \frac{1}{1} = \mathbf{2x - 1}$$

Example 6.3-23

$$\frac{\sqrt{x^2y^4} - 4\sqrt{x^3y^4}}{xy} = \frac{x\sqrt{y^{2+2}} - 4\sqrt{x^{2+1}y^{2+2}}}{xy} = \frac{x\sqrt{y^2\cdot y^2} - 4\sqrt{\left(x^2\cdot x^1\right)\left(y^2\cdot y^2\right)}}{xy} = \frac{x(y\cdot y) - 4x(y\cdot y)\sqrt{x}}{xy}$$

$$= \frac{xy^2 - 4xy^2\sqrt{x}}{xy} = \frac{xy^2}{xy} + \frac{-4xy^2\sqrt{x}}{xy} = \frac{y^2y^{-1}}{x^1x^{-1}} - \frac{4\left(y^2y^{-1}\right)\sqrt{x}}{x^1x^{-1}} = \frac{y^{2-1}}{x^{1-1}} - \frac{4y^{2-1}\sqrt{x}}{x^{1-1}} = \frac{y^1}{x^0} - \frac{4y^1\sqrt{x}}{x^0}$$

$$= \frac{y}{1} - \frac{4y\sqrt{x}}{1} = \frac{y - 4y\sqrt{x}}{1} = \mathbf{y - 4y\sqrt{x}}$$

Example 6.3-24

$$\frac{-\sqrt[3]{x^3y^6} + \sqrt{x^4y^5}}{-x^2y^2} = \frac{-x\sqrt[3]{y^{3+3}} + \sqrt{x^{2+2}y^{2+2+1}}}{-x^2y^2} = \frac{-x\sqrt[3]{y^3\cdot y^3} + \sqrt{\left(x^2\cdot x^2\right)\cdot\left(y^2\cdot y^2\cdot y^1\right)}}{-x^2y^2}$$

$$= \frac{-x(y\cdot y) + (x\cdot x)\cdot(y\cdot y)\sqrt{y}}{-x^2y^2} = \frac{-xy^2 + x^2y^2\sqrt{y}}{-x^2y^2} = \frac{-xy^2}{-x^2y^2} + \frac{x^2y^2\sqrt{y}}{-x^2y^2} = +\frac{x^1y^2}{x^2y^2} - \frac{x^2y^2\sqrt{y}}{x^2y^2}$$

$$=\boxed{\frac{y^2y^{-2}}{x^2x^{-1}}-\frac{y^2y^{-2}\sqrt{y}}{x^2x^{-2}}}=\boxed{\frac{y^{2-2}}{x^{2-1}}-\frac{y^{2-2}\sqrt{y}}{x^{2-2}}}=\boxed{\frac{y^0}{x^1}-\frac{y^0\sqrt{y}}{x^0}}=\boxed{\frac{1}{x}-\frac{\sqrt{y}}{1}}=\boxed{\mathbf{\frac{1}{x}-\sqrt{y}}}$$

Example 6.3-25

$$\boxed{\frac{36x^5-4x}{8x^2}}=\boxed{\frac{36x^5}{8x^2}+\frac{-4x}{8x^2}}=\boxed{\frac{\overset{9}{\cancel{36}}x^5}{\underset{2}{\cancel{8}}x^2}-\frac{\overset{1}{\cancel{4}}x}{\underset{2}{\cancel{8}}x^2}}=\boxed{\frac{9x^5x^{-2}}{2}-\frac{1}{2x^2x^{-1}}}=\boxed{\frac{9x^{5-2}}{2}-\frac{1}{2x^{2-1}}}=\boxed{\mathbf{\frac{9x^3}{2}-\frac{1}{2x}}}$$

Example 6.3-26

$$\boxed{\frac{m^3+n^4}{m^2n^2}}=\boxed{\frac{m^3}{m^2n^2}+\frac{n^4}{m^2n^2}}=\boxed{\frac{m^3m^{-2}}{n^2}+\frac{n^4n^{-2}}{m^2}}=\boxed{\frac{m^{3-2}}{n^2}+\frac{n^{4-2}}{m^2}}=\boxed{\frac{m^1}{n^2}+\frac{n^2}{m^2}}=\boxed{\mathbf{\frac{m}{n^2}+\frac{n^2}{m^2}}}$$

Example 6.3-27

$$\boxed{\frac{x^3y^2z+x^4yz^2}{x^2y^3z^5}}=\boxed{\frac{x^3y^2z}{x^2y^3z^5}+\frac{x^4yz^2}{x^2y^3z^5}}=\boxed{\frac{x^3y^2z^1}{x^2y^3z^5}+\frac{x^4y^1z^2}{x^2y^3z^5}}=\boxed{\frac{x^3x^{-2}}{\left(y^3y^{-2}\right)\cdot\left(z^5z^{-1}\right)}+\frac{x^4x^{-2}}{\left(y^3y^{-1}\right)\cdot\left(z^5z^{-2}\right)}}$$

$$=\boxed{\frac{x^{3-2}}{y^{3-2}z^{5-1}}+\frac{x^{4-2}}{y^{3-1}z^{5-2}}}=\boxed{\frac{x^1}{y^1z^4}+\frac{x^2}{y^2z^3}}=\boxed{\mathbf{\frac{x}{yz^4}+\frac{x^2}{y^2z^3}}}$$

Example 6.3-28

$$\boxed{\frac{\sqrt[3]{a^3b^4}+\sqrt[2]{a^2b^4}}{a^2b^2}}=\boxed{\frac{a\sqrt[3]{b^{3+1}}+a\sqrt{b^{2+2}}}{a^2b^2}}=\boxed{\frac{a\sqrt[3]{b^3\cdot b^1}+a\sqrt{b^2\cdot b^2}}{a^2b^2}}=\boxed{\frac{ab\sqrt[3]{b}+a(b\cdot b)}{a^2b^2}}=\boxed{\frac{ab\sqrt[3]{b}+ab^2}{a^2b^2}}$$

$$=\boxed{\frac{ab\sqrt[3]{b}}{a^2b^2}+\frac{ab^2}{a^2b^2}}=\boxed{\frac{a^1b^1\sqrt[3]{b}}{a^2b^2}+\frac{a^1b^2}{a^2b^2}}=\boxed{\frac{\sqrt[3]{b}}{a^2a^{-1}\cdot b^2b^{-1}}+\frac{b^2b^{-2}}{a^2a^{-1}}}=\boxed{\frac{\sqrt[3]{b}}{a^{2-1}b^{2-1}}+\frac{b^{2-2}}{a^{2-1}}}=\boxed{\frac{\sqrt[3]{b}}{a^1b^1}+\frac{b^0}{a^1}}=\boxed{\mathbf{\frac{\sqrt[3]{b}}{ab}+\frac{1}{a}}}$$

Example 6.3-29

$$\boxed{\frac{3\sqrt{125}-\sqrt{27}}{\sqrt{225}}}=\boxed{\frac{3\sqrt{25\cdot5}-\sqrt{9\cdot3}}{\sqrt{15\cdot15}}}=\boxed{\frac{3\sqrt{5^2\cdot5}-\sqrt{3^2\cdot3}}{\sqrt{15^2}}}=\boxed{\frac{(3\cdot5)\sqrt{5}-3\sqrt{3}}{15}}=\boxed{\frac{15\sqrt{5}-3\sqrt{3}}{15}}=\boxed{\frac{15\sqrt{5}}{15}+\frac{-3\sqrt{3}}{15}}$$

$$=\boxed{\frac{\overset{1}{\cancel{15}}\sqrt{5}}{\underset{1}{\cancel{15}}}-\frac{\overset{1}{\cancel{3}}\sqrt{3}}{\underset{5}{\cancel{15}}}}=\boxed{\frac{\sqrt{5}}{1}-\frac{\sqrt{3}}{5}}=\boxed{\mathbf{\sqrt{5}-\frac{\sqrt{3}}{5}}}$$

Example 6.3-30

$$\boxed{\frac{4m^2n^3l+16mn^4l^3}{-8m^3n^3l^2}}=\boxed{\frac{4m^2n^3l}{-8m^3n^3l^2}+\frac{16mn^4l^3}{-8m^3n^3l^2}}=\boxed{-\frac{\overset{1}{\cancel{4}}m^2n^3l}{\underset{2}{\cancel{8}}m^3n^3l^2}-\frac{\overset{2}{\cancel{16}}mn^4l^3}{\underset{1}{\cancel{8}}m^3n^3l^2}}=\boxed{-\frac{m^2n^3l^1}{2m^3n^3l^2}-\frac{2m^1n^4l^3}{m^3n^3l^2}}$$

$$=\boxed{-\frac{n^3n^{-3}}{2\left(m^3m^{-2}\right)\cdot\left(l^2l^{-1}\right)}-\frac{2\left(n^4n^{-3}\right)\cdot\left(l^3l^{-2}\right)}{m^3m^{-1}}}=\boxed{-\frac{n^{3-3}}{2m^{3-2}l^{2-1}}-\frac{2n^{4-3}l^{3-2}}{m^{3-1}}}=\boxed{-\frac{n^0}{2m^1l^1}-\frac{2n^1l^1}{m^2}}=\boxed{\mathbf{-\frac{1}{2ml}-\frac{2nl}{m^2}}}$$

Practice Problems - Dividing Binomials by Monomials

Section 6.3 Case I b Practice Problems - Divide the following binomial expressions by monomials:

1. $\dfrac{98-46}{-12} =$
2. $\dfrac{x^3y^3z+4x^2y^2}{-2xy^2z} =$
3. $\dfrac{-a^3b^3c+a^2bc^2}{-a^2b^2c^2} =$
4. $\dfrac{\sqrt[4]{a^5b^4c^3}-\sqrt[3]{a^3b^6c}}{\sqrt{a^2b^4c^6}} =$
5. $\dfrac{m^3n^2l+ml^2}{mnl} =$
6. $\dfrac{36y^2-18y^3}{-9y} =$
7. $\dfrac{-5\sqrt{150}+10\sqrt{125}}{\sqrt{25}} =$
8. $\dfrac{99x^2-18x^3}{\sqrt{81x^4}} =$
9. $\dfrac{\sqrt{8x^2y^3}-\sqrt{9xy^2}}{\sqrt{4x^4y^6}} =$
10. $\dfrac{\sqrt{100w^5}+\sqrt{75w^3}}{-\sqrt{25w^2}} =$

Case I c Dividing Polynomials by Monomials

Polynomials are divided by monomials using the following steps:

Step 1 Simplify each term in the numerator and the denominator.

Step 2 Divide each polynomial term by the denominator using the exponent rule (see Section 3.3, Case II) to divide the variables, i.e.,

$$\frac{x^n}{x^m} = \frac{x^{n-m}}{1} = x^{n-m} \qquad if\ n > m$$

or,

$$\frac{x^n}{x^m} = \frac{1}{x^{m-n}} \qquad if\ n < m$$

where n and m are positive integer numbers and x is a variable.

Examples with Steps

The following examples show the steps as to how polynomials are divided by monomials:

Example 6.3-31

$$\frac{-\sqrt{144}x^5 + 6x^3 + \sqrt{16}x^2 - 24}{2x^2} =$$

Solution:

Step 1

$$\frac{-\sqrt{144}x^5 + 6x^3 + \sqrt{16}x^2 - 24}{2x^2} = \frac{-\sqrt{12 \cdot 12}x^5 + 6x^3 + \sqrt{4 \cdot 4}x^2 - 24}{2x^2}$$

$$= \frac{-\sqrt{12^2}x^5 + 6x^3 + \sqrt{4^2}x^2 - 24}{2x^2} = \frac{-12x^5 + 6x^3 + 4x^2 - 24}{2x^2}$$

Step 2

$$\frac{-12x^5 + 6x^3 + 4x^2 - 24}{2x^2} = \frac{-12x^5}{2x^2} + \frac{6x^3}{2x^2} + \frac{4x^2}{2x^2} + \frac{-24}{2x^2} = -\frac{\overset{6}{\cancel{12}}x^5}{\underset{1}{\cancel{2}}x^2} + \frac{\overset{3}{\cancel{6}}x^3}{\underset{1}{\cancel{2}}x^2} + \frac{\overset{2}{\cancel{4}}x^2}{\underset{1}{\cancel{2}}x^2} - \frac{\overset{12}{\cancel{24}}}{\underset{1}{\cancel{2}}x^2}$$

$$= -\frac{6x^5}{x^2} + \frac{3x^3}{x^2} + \frac{2x^2}{x^2} - \frac{12}{x^2} = -\frac{6x^5x^{-2}}{1} + \frac{3x^3x^{-2}}{1} + \frac{2x^2x^{-2}}{1} - \frac{12}{x^2}$$

$$= -\frac{6x^{5-2}}{1} + \frac{3x^{3-2}}{1} + \frac{2x^{2-2}}{1} - \frac{12}{x^2} = -\frac{6x^3}{1} + \frac{3x^1}{1} + \frac{2x^0}{1} - \frac{12}{x^2} = \boldsymbol{-6x^3 + 3x + 2 - \frac{12}{x^2}}$$

Example 6.3-32

$$\frac{x^2y - 4xy^2 + 2x - 4}{-2\sqrt[3]{x^3y^3}} =$$

Solution:

Step 1

$$\frac{x^2y-4xy^2+2x-4}{-2\sqrt[3]{x^3y^3}}=\frac{x^2y-4xy^2+2x-4}{-2xy}$$

Step 2

$$\frac{x^2y-4xy^2+2x-4}{-2xy}=\frac{x^2y}{-2xy}+\frac{-4xy^2}{-2xy}+\frac{+2x}{-2xy}+\frac{-4}{-2xy}=-\frac{x^2y}{2xy}+\frac{\overset{2}{\cancel{4}}xy^2}{\underset{1}{\cancel{2}}xy}-\frac{\overset{1}{\cancel{2}}x}{\underset{1}{\cancel{2}}xy}+\frac{\overset{2}{\cancel{4}}}{\underset{1}{\cancel{2}}xy}$$

$$-\frac{x^2y}{2xy}+\frac{2\cancel{x}y^2}{\cancel{x}y}-\frac{\cancel{x}}{\cancel{x}y}+\frac{2}{xy}=-\frac{x^2}{2x}+\frac{2y^2}{y}-\frac{1}{y}+\frac{2}{xy}=-\frac{x^{2-1}}{2}+\frac{2y^{2-1}}{1}-\frac{1}{y}+\frac{2}{xy}$$

$$=-\frac{x^1}{2}+\frac{2y^1}{1}-\frac{1}{y}+\frac{2}{xy}=\boldsymbol{-\frac{x}{2}+2y-\frac{1}{y}+\frac{2}{xy}}$$

Example 6.3-33

$$\frac{\sqrt{32c^3}-\sqrt{8c^2}-8c+\sqrt{72}}{\sqrt{2c^3}}=$$

Solution:

Step 1

$$\frac{\sqrt{32c^3}-\sqrt{8c^2}-8c+\sqrt{72}}{\sqrt{2c^3}}=\frac{\sqrt{16\cdot 2c^3}-\sqrt{4\cdot 2c^2}-8c+\sqrt{36\cdot 2}}{\sqrt{2c^3}}$$

$$=\frac{\sqrt{4^2\cdot 2c^3}-\sqrt{2^2\cdot 2c^2}-8c+\sqrt{6^2\cdot 2}}{\sqrt{2c^3}}=\frac{4\sqrt{2c^3}-2\sqrt{2c^2}-8c+6\sqrt{2}}{\sqrt{2c^3}}$$

Step 2

$$\frac{4\sqrt{2c^3}-2\sqrt{2c^2}-8c+6\sqrt{2}}{\sqrt{2c^3}}=\frac{4\sqrt{2c^3}}{\sqrt{2c^3}}+\frac{-2\sqrt{2c^2}}{\sqrt{2c^3}}+\frac{-8c}{\sqrt{2c^3}}+\frac{6\sqrt{2}}{\sqrt{2c^3}}$$

$$=\frac{4\sqrt{\cancel{2}c^3}}{\sqrt{\cancel{2}c^3}}-\frac{2\sqrt{\cancel{2}c^2}}{\sqrt{\cancel{2}c^3}}-\frac{8c}{\sqrt{2c^3}}+\frac{6\sqrt{\cancel{2}}}{\sqrt{\cancel{2}c^3}}=\frac{4c^3}{c^3}-\frac{2c^2}{c^3}-\frac{8c}{\sqrt{2c^3}}+\frac{6}{c^3}$$

$$=\frac{4}{c^3c^{-3}}-\frac{2}{c^3c^{-2}}-\frac{8}{\sqrt{2c^3c^{-1}}}+\frac{6}{c^3}=\frac{4}{c^{3-3}}-\frac{2}{c^{3-2}}-\frac{8}{\sqrt{2c^{3-1}}}+\frac{6}{c^3}$$

$$=\frac{4}{c^0}-\frac{2}{c^1}-\frac{8}{\sqrt{2c^2}}+\frac{6}{c^3}=\boldsymbol{4-\frac{2}{c}-\frac{8}{\sqrt{2c^2}}+\frac{6}{c^3}}$$

Example 6.3-34

$$\frac{y^6-8y^5-4\sqrt{y^4}+120}{-\sqrt{144y^2}}=$$

Solution:

Step 1

$$\frac{y^6-8y^5-4\sqrt{y^4}+120}{-\sqrt{144y^2}} = \frac{y^6-8y^5-4\sqrt{y^{2+2}}+120}{-y\sqrt{12\cdot 12}} = \frac{y^6-8y^5-4\sqrt{y^2\cdot y^2}+120}{-y\sqrt{12^2}}$$

$$= \frac{y^6-8y^5-4(y\cdot y)+120}{-12y} = \frac{y^6-8y^5-4y^2+120}{-12y}$$

Step 2

$$\frac{y^6-8y^5-4y^2+120}{-12y} = \frac{y^6}{-12y}+\frac{-8y^5}{-12y}+\frac{-4y^2}{-12y}+\frac{120}{-12y} = -\frac{y^6}{12y}+\frac{\overset{2}{\not 8}y^5}{\underset{3}{\not 12}y}+\frac{\overset{1}{\not 4}y^2}{\underset{3}{\not 12}y}-\frac{\overset{10}{\not 120}}{\underset{1}{\not 12}y}$$

$$= -\frac{y^6}{12y^1}+\frac{2y^5}{3y^1}+\frac{y^2}{3y^1}-\frac{10}{y} = -\frac{y^6y^{-1}}{12}+\frac{2y^5y^{-1}}{3}+\frac{y^2y^{-1}}{3}-\frac{10}{y}$$

$$= -\frac{y^{6-1}}{12}+\frac{2y^{5-1}}{3}+\frac{y^{2-1}}{3}-\frac{10}{y} = \mathbf{-\frac{y^5}{12}+\frac{2y^4}{3}+\frac{y}{3}-\frac{10}{y}}$$

Example 6.3-35

$$\frac{28n^5-8n^3+2n^2-4}{2n^4} =$$

Solution:

Step 1

Not Applicable

Step 2

$$\frac{28n^5-8n^3+2n^2-4}{2n^4} = \frac{28n^5}{2n^4}+\frac{-8n^3}{2n^4}+\frac{2n^2}{2n^4}+\frac{-4}{2n^4} = \frac{\overset{14}{\not 28}n^5}{\underset{1}{\not 2}n^4}-\frac{\overset{4}{\not 8}n^3}{\underset{1}{\not 2}n^4}+\frac{\overset{1}{\not 2}n^2}{\underset{1}{\not 2}n^4}-\frac{\overset{2}{\not 4}}{\underset{1}{\not 2}n^4}$$

$$= \frac{14n^5}{n^4}-\frac{4n^3}{n^4}+\frac{n^2}{n^4}-\frac{2}{n^4} = \frac{14n^5n^{-4}}{1}-\frac{4}{n^4n^{-3}}+\frac{1}{n^4n^{-2}}-\frac{2}{n^4}$$

$$= \frac{14n^{5-4}}{1}-\frac{4}{n^{4-3}}+\frac{1}{n^{4-2}}-\frac{2}{n^4} = \frac{14n^1}{1}-\frac{4}{n^1}+\frac{1}{n^2}-\frac{2}{n^4} = \mathbf{14n-\frac{4}{n}+\frac{1}{n^2}-\frac{2}{n^4}}$$

Additional Examples - Dividing Polynomials by Monomials

The following examples further illustrate how to divide polynomials by monomials:

Example 6.3-36

$$\frac{6u^5-4u^4+12u^3-8}{-2} = \frac{6u^5}{-2}+\frac{-4u^4}{-2}+\frac{12u^3}{-2}+\frac{-8}{-2} = -\frac{\overset{3}{\not 6}u^5}{\underset{1}{\not 2}}+\frac{\overset{2}{\not 4}u^4}{\underset{1}{\not 2}}-\frac{\overset{6}{\not 12}u^3}{\underset{1}{\not 2}}+\frac{\overset{4}{\not 8}}{\underset{1}{\not 2}} = -\frac{3u^5}{1}+\frac{2u^4}{1}-\frac{6u^3}{1}+\frac{4}{1}$$

$$= \mathbf{-3u^5+2u^4-6u^3+4}$$

Example 6.3-37

$$\boxed{\frac{7a^3-14a^2+21a-49}{7a}} = \boxed{\frac{7a^3}{7a}+\frac{-14a^2}{7a}+\frac{21a}{7a}+\frac{-49}{7a}} = \boxed{\frac{\overset{1}{\cancel{7}}a^3}{\underset{1}{\cancel{7}}a}-\frac{\overset{2}{\cancel{14}}a^2}{\underset{1}{\cancel{7}}a}+\frac{\overset{3}{\cancel{21}}a}{\underset{1}{\cancel{7}}a}-\frac{\overset{7}{\cancel{49}}}{\underset{1}{\cancel{7}}a}} = \boxed{\frac{a^3}{a}-\frac{2a^2}{a}+\frac{3a}{a}-\frac{7}{a}}$$

$$= \boxed{\frac{a^3}{a^1}-\frac{2a^2}{a^1}+\frac{3a^1}{a^1}-\frac{7}{a}} = \boxed{\frac{a^3a^{-1}}{1}-\frac{2a^2a^{-1}}{1}+\frac{3a^1a^{-1}}{1}-\frac{7}{a}} = \boxed{\frac{a^{3-1}}{1}-\frac{2a^{2-1}}{1}+\frac{3a^{1-1}}{1}-\frac{7}{a}}$$

$$= \boxed{\frac{a^2}{1}-\frac{2a^1}{1}+\frac{3a^0}{1}-\frac{7}{a}} = \boxed{\boldsymbol{a^2-2a+3-\frac{7}{a}}}$$

Example 6.3-38

$$\boxed{\frac{36y^5-12y^4+8y+24}{-6y^3}} = \boxed{\frac{36y^5}{-6y^3}+\frac{-12y^4}{-6y^3}+\frac{8y}{-6y^3}+\frac{24}{-6y^3}} = \boxed{-\frac{\overset{6}{\cancel{36}}y^5}{\underset{1}{\cancel{6}}y^3}+\frac{\overset{2}{\cancel{12}}y^4}{\underset{1}{\cancel{6}}y^3}-\frac{\overset{4}{\cancel{8}}y}{\underset{3}{\cancel{6}}y^3}-\frac{\overset{4}{\cancel{24}}}{\underset{1}{\cancel{6}}y^3}}$$

$$= \boxed{-\frac{6y^5}{y^3}+\frac{2y^4}{y^3}-\frac{4y^1}{3y^3}-\frac{4}{y^3}} = \boxed{-\frac{6y^5y^{-3}}{1}+\frac{2y^4y^{-3}}{1}-\frac{4}{3y^3y^{-1}}-\frac{4}{y^3}} = \boxed{-\frac{6y^{5-3}}{1}+\frac{2y^{4-3}}{1}-\frac{4}{3y^{3-1}}-\frac{4}{y^3}}$$

$$= \boxed{-\frac{6y^2}{1}+\frac{2y^1}{1}-\frac{4}{3y^2}-\frac{4}{y^3}} = \boxed{\boldsymbol{-6y^2+2y-\frac{4}{3y^2}-\frac{4}{y^3}}}$$

Example 6.3-39

$$\boxed{\frac{48x^5+8x^3-4x^2+2}{-4x^4}} = \boxed{\frac{48x^5}{-4x^4}+\frac{8x^3}{-4x^4}+\frac{-4x^2}{-4x^4}+\frac{2}{-4x^4}} = \boxed{-\frac{\overset{12}{\cancel{48}}x^5}{\underset{1}{\cancel{4}}x^4}-\frac{\overset{2}{\cancel{8}}x^3}{\underset{1}{\cancel{4}}x^4}+\frac{\overset{1}{\cancel{4}}x^2}{\underset{1}{\cancel{4}}x^4}-\frac{\overset{1}{\cancel{2}}}{\underset{2}{\cancel{4}}x^4}}$$

$$= \boxed{-\frac{12x^5}{x^4}-\frac{2x^3}{x^4}+\frac{x^2}{x^4}-\frac{1}{2x^4}} = \boxed{-\frac{12x^5x^{-4}}{1}-\frac{2}{x^4x^{-3}}+\frac{1}{x^4x^{-2}}-\frac{1}{2x^4}} = \boxed{-\frac{12x^{5-4}}{1}-\frac{2}{x^{4-3}}+\frac{1}{x^{4-2}}-\frac{1}{2x^4}}$$

$$= \boxed{-\frac{12x^1}{1}-\frac{2}{x^1}+\frac{1}{x^2}-\frac{1}{2x^4}} = \boxed{\boldsymbol{-12x-\frac{2}{x}+\frac{1}{x^2}-\frac{1}{2x^4}}}$$

Example 6.3-40

$$\boxed{\frac{x^5+9x^4-3x^3+2x-6}{3x^2}} = \boxed{\frac{x^5}{3x^2}+\frac{9x^4}{3x^2}+\frac{-3x^3}{3x^2}+\frac{2x}{3x^2}+\frac{-6}{3x^2}} = \boxed{\frac{x^5}{3x^2}+\frac{\overset{3}{\cancel{9}}x^4}{\underset{1}{\cancel{3}}x^2}-\frac{\overset{1}{\cancel{3}}x^3}{\underset{1}{\cancel{3}}x^2}+\frac{2x}{3x^2}-\frac{\overset{2}{\cancel{6}}}{\underset{1}{\cancel{3}}x^2}}$$

$$= \boxed{\frac{x^5x^{-2}}{3}+\frac{3x^4x^{-2}}{1}-\frac{x^3x^{-2}}{1}+\frac{2}{3x^2x^{-1}}-\frac{2}{x^2}} = \boxed{\frac{x^{5-2}}{3}+\frac{3x^{4-2}}{1}-\frac{x^{3-2}}{1}+\frac{2}{3x^{2-1}}-\frac{2}{x^2}}$$

$$= \boxed{\frac{x^3}{3}+\frac{3x^2}{1}-\frac{x^1}{1}+\frac{2}{3x^1}-\frac{2}{x^2}} = \boxed{\boldsymbol{\frac{1}{3}x^3+3x^2-x+\frac{2}{3x}-\frac{2}{x^2}}}$$

Example 6.3-41

$$\boxed{\frac{xy^2-2x^2y^2+4x^2y+16}{8xy}}=\boxed{\frac{xy^2}{8xy}+\frac{-2x^2y^2}{8xy}+\frac{4x^2y}{8xy}+\frac{16}{8xy}}=\boxed{\frac{xy^2}{8xy}-\frac{\overset{1}{\cancel{2}}x^2y^2}{\underset{4}{\cancel{8}}xy}+\frac{\overset{1}{\cancel{4}}x^2y}{\underset{2}{\cancel{8}}xy}+\frac{\overset{2}{\cancel{16}}}{\underset{1}{\cancel{8}}xy}}$$

$$=\boxed{\frac{xy^2}{8xy}-\frac{x^2y^2}{4xy}+\frac{x^2y}{2xy}+\frac{2}{xy}}=\boxed{\frac{y^2y^{-1}}{8x^1x^{-1}}-\frac{x^2x^{-1}y^2y^{-1}}{4}+\frac{x^2x^{-1}}{2y^1y^{-1}}+\frac{2}{xy}}=\boxed{\frac{y^{2-1}}{8x^{1-1}}-\frac{x^{2-1}y^{2-1}}{4}+\frac{x^{2-1}}{2y^{1-1}}+\frac{2}{xy}}$$

$$=\boxed{\frac{y^1}{8x^0}-\frac{x^1y^1}{4}+\frac{x^1}{2y^0}+\frac{2}{xy}}=\boxed{\mathbf{\frac{y}{8}-\frac{xy}{4}+\frac{x}{2}+\frac{2}{xy}}}$$

Example 6.3-42

$$\boxed{\frac{\sqrt{w^3}-2\sqrt{u^3}+8u\sqrt{u^2w^3}-u^2w}{2uw^2}}=\boxed{\frac{\sqrt{w^{2+1}}-2\sqrt{u^{2+1}}+8(u\cdot u)\sqrt{w^{2+1}}-u^2w}{2uw^2}}=\boxed{\frac{\sqrt{w^2w}-2\sqrt{u^2u}+8u^2\sqrt{w^2w}-u^2w}{2uw^2}}$$

$$=\boxed{\frac{w\sqrt{w}-2u\sqrt{u}+8u^2w\sqrt{w}-u^2w}{2uw^2}}=\boxed{\frac{w\sqrt{w}}{2uw^2}+\frac{-2u\sqrt{u}}{2uw^2}+\frac{8u^2w\sqrt{w}}{2uw^2}+\frac{-u^2w}{2uw^2}}=\boxed{\frac{w\sqrt{w}}{2uw^2}-\frac{\overset{1}{\cancel{2}}u\sqrt{u}}{\underset{1}{\cancel{2}}uw^2}+\frac{\overset{4}{\cancel{8}}u^2w\sqrt{w}}{\underset{1}{\cancel{2}}uw^2}-\frac{u^2w}{2uw^2}}$$

$$=\boxed{\frac{w\sqrt{w}}{2uw^2}-\frac{u\sqrt{u}}{uw^2}+\frac{4u^2w\sqrt{w}}{uw^2}-\frac{u^2w}{2uw^2}}=\boxed{\frac{\sqrt{w}}{2uw^2w^{-1}}-\frac{\sqrt{u}}{u^1u^{-1}w^2}+\frac{4u^2u^{-1}\sqrt{w}}{w^2w^{-1}}-\frac{u^2u^{-1}}{2w^2w^{-1}}}$$

$$=\boxed{\frac{\sqrt{w}}{2uw^{2-1}}-\frac{\sqrt{u}}{u^{1-1}w^2}+\frac{4u^{2-1}\sqrt{w}}{w^{2-1}}-\frac{u^{2-1}}{2w^{2-1}}}=\boxed{\frac{\sqrt{w}}{2uw^1}-\frac{\sqrt{u}}{u^0w^2}+\frac{4u^1\sqrt{w}}{w^1}-\frac{u^1}{2w^1}}=\boxed{\mathbf{\frac{\sqrt{w}}{2uw}-\frac{\sqrt{u}}{w^2}+\frac{4u\sqrt{w}}{w}-\frac{u}{2w}}}$$

Example 6.3-43

$$\boxed{\frac{x^2y^2z+xy^2z-x^2yz-2z^3}{4xyz}}=\boxed{\frac{x^2y^2z}{4xyz}+\frac{xy^2z}{4xyz}+\frac{-x^2yz}{4xyz}+\frac{-2z^3}{4xyz}}=\boxed{\frac{x^2y^2z}{4xyz}+\frac{xy^2z}{4xyz}-\frac{x^2yz}{4xyz}-\frac{\overset{1}{\cancel{2}}z^3}{\underset{2}{\cancel{4}}xyz}}$$

$$=\boxed{\frac{x^2y^2z}{4xyz}+\frac{xy^2z}{4xyz}-\frac{x^2yz}{4xyz}-\frac{z^3}{2xyz}}=\boxed{\frac{x^2y^2\cancel{z}}{4xy\cancel{z}}+\frac{\cancel{x}y^2\cancel{z}}{4\cancel{x}y\cancel{z}}-\frac{x^2\cancel{y}\cancel{z}}{4x\cancel{y}\cancel{z}}-\frac{z^3}{2xyz}}=\boxed{\frac{x^2y^2}{4xy}+\frac{y^2}{4y}-\frac{x^2}{4x}-\frac{z^3}{2xyz}}$$

$$=\boxed{\frac{x^2x^{-1}y^2y^{-1}}{4}+\frac{y^2y^{-1}}{4}-\frac{x^2x^{-1}}{4}-\frac{z^3z^{-1}}{2xy}}=\boxed{\frac{x^{2-1}y^{2-1}}{4}+\frac{y^{2-1}}{4}-\frac{x^{2-1}}{4}-\frac{z^{3-1}}{2xy}}=\boxed{\frac{x^1y^1}{4}+\frac{y^1}{4}-\frac{x^1}{4}-\frac{z^2}{2xy}}$$

$$=\boxed{\mathbf{\frac{xy}{4}+\frac{y}{4}-\frac{x}{4}-\frac{z^2}{2xy}}}$$

Example 6.3-44

$$\boxed{\frac{\sqrt[5]{64}-\sqrt[4]{48}+\sqrt[5]{486}+\sqrt[4]{243}-16}{\sqrt[5]{32}}}=\boxed{\frac{\sqrt[5]{32\cdot 2}-\sqrt[4]{16\cdot 3}+\sqrt[5]{243\cdot 2}+\sqrt[4]{81\cdot 3}-16}{\sqrt[5]{32}}}$$

$$= \boxed{\frac{\sqrt[5]{2^5\cdot 2}-\sqrt[4]{2^4\cdot 3}+\sqrt[5]{3^5\cdot 2}+\sqrt[4]{3^4\cdot 3}-16}{\sqrt[5]{2^5}}} = \boxed{\frac{2\sqrt[5]{2}-2\sqrt[4]{3}+3\sqrt[5]{2}+3\sqrt[4]{3}-16}{2}} = \boxed{\frac{\left(2\sqrt[5]{2}+3\sqrt[5]{2}\right)+\left(3\sqrt[4]{3}-2\sqrt[4]{3}\right)-16}{2}}$$

$$= \boxed{\frac{(2+3)\sqrt[5]{2}+(3-2)\sqrt[4]{3}-16}{2}} = \boxed{\frac{5\sqrt[5]{2}+\sqrt[4]{3}-16}{2}} = \boxed{\frac{5\sqrt[5]{2}}{2}+\frac{\sqrt[4]{3}}{2}+\frac{-16}{2}} = \boxed{\frac{5\sqrt[5]{2}}{2}+\frac{\sqrt[4]{3}}{2}-\frac{\overset{8}{\cancel{16}}}{\underset{1}{\cancel{2}}}} = \boxed{\frac{5\sqrt[5]{2}}{2}+\frac{\sqrt[4]{3}}{2}-\frac{8}{1}}$$

$$= \boxed{\mathbf{\frac{5\sqrt[5]{2}}{2}+\frac{\sqrt[4]{3}}{2}-8}}$$

Example 6.3-45

$$\boxed{\frac{\sqrt{125y^3}-\sqrt{72y^2}+\sqrt{18y}-4}{\sqrt{64y}}} = \boxed{\frac{\sqrt{25\cdot 5y^3}-\sqrt{36\cdot 2y^2}+\sqrt{9\cdot 2y}-4}{\sqrt{8\cdot 8y}}} = \boxed{\frac{\sqrt{5^2\cdot 5y^3}-\sqrt{6^2\cdot 2y^2}+\sqrt{3^2\cdot 2y}-4}{\sqrt{8^2 y}}}$$

$$= \boxed{\frac{5\sqrt{5y^3}-6\sqrt{2y^2}+3\sqrt{2y}-4}{8y}} = \boxed{\frac{5\sqrt{5y^3}}{8y}+\frac{-6\sqrt{2y^2}}{8y}+\frac{3\sqrt{2y}}{8y}+\frac{-4}{8y}} = \boxed{\frac{5\sqrt{5y^3}}{8y}-\frac{\overset{3}{\cancel{6}}\sqrt{2y^2}}{\underset{4}{\cancel{8}}y}+\frac{3\sqrt{2y}}{8y}-\frac{\overset{1}{\cancel{4}}}{\underset{2}{\cancel{8}}y}}$$

$$= \boxed{\frac{5\sqrt{5y^3}}{8y}-\frac{3\sqrt{2y^2}}{4y}+\frac{3\sqrt{2y}}{8y}-\frac{1}{2y}} = \boxed{\frac{5\sqrt{5}y^3y^{-1}}{8}-\frac{3\sqrt{2}y^2y^{-1}}{4}+\frac{3\sqrt{2}}{8y^1y^{-1}}-\frac{1}{2y}}$$

$$= \boxed{\frac{5\sqrt{5}y^{3-1}}{8}-\frac{3\sqrt{2}y^{2-1}}{4}+\frac{3\sqrt{2}}{8y^{1-1}}-\frac{1}{2y}} = \boxed{\frac{5\sqrt{5}y^2}{8}-\frac{3\sqrt{2}y^1}{4}+\frac{3\sqrt{2}}{8y^0}-\frac{1}{2y}} = \boxed{\mathbf{\frac{5\sqrt{5}y^2}{8}-\frac{3\sqrt{2}y}{4}+\frac{3\sqrt{2}}{8}-\frac{1}{2y}}}$$

Practice Problems - Dividing Polynomials by Monomials

Section 6.3 Case I c Practice Problems - Divide the following polynomials by monomials:

1. $\frac{x^6-x^4+2x^2-6}{4} =$
2. $\frac{y^3-2y^2+4y-12}{-2} =$
3. $\frac{x^2y-2xy+4xy^3-8}{6xy} =$
4. $\frac{\sqrt[3]{375y^3}-\sqrt[3]{108y^2}-3y+5}{15y} =$
5. $\frac{-w^6+2w^3+4w^2-8}{-2w} =$
6. $\frac{3\sqrt{a^{10}}-6a^4-\sqrt{64a^2}-12}{4a^2} =$
7. $\frac{-m^6-3m^3+2m-8}{-m^2} =$
8. $\frac{m^3n^4-m^3n^2+mn-6n}{2m^2n^2} =$
9. $\frac{y^7-3y^5-5y^3+6y-12}{15y^2} =$
10. $\frac{\sqrt[3]{32}-\sqrt{72}+\sqrt[3]{108}+\sqrt{162}}{-\sqrt{81}} =$

Case II Dividing Polynomials by Polynomials

Whole numbers are divided by one another using the long division method which can be summarized as: selecting a quotient, multiplying the quotient by the divisor to obtain a product, subtracting the product from the dividend, and bringing down the next digit/dividend term. Polynomials are divided by one another in a similar way as the long division method used for whole numbers. The following are the steps for dividing two polynomials by each other:

Step 1 a. Select the first term for the quotient which divides the first term of the dividend by the first term of the divisor.

b. Multiply the selected first term of the quotient by the divisor.

c. Write the product under the dividend.

Step 1 a d. Change the signs of the product written under the dividend.

e. Subtract the product from the dividend.

f. Bring down the next term from the dividend to obtain a new dividend.

Step 2 a. Select the second term for the quotient which divides the first term of the new dividend by the first term of the divisor.

b. Multiply the selected second term of the quotient by the divisor.

c. Write the product under the new dividend.

Step 2 a d. Change the signs of the product written under the new dividend.

e. Subtract the product from the new dividend to obtain a remainder. If a remainder is not obtained, proceed with the next step.

f. Bring down the next term from the dividend to obtain another new dividend.

g. Repeat Steps $2a.$ through $2f.$ until a remainder is obtained.

To **check** the answer multiply the quotient by the divisor and add in the remainder. The result should match the dividend.

Examples with Steps

The following examples show the steps as to how polynomials are divided by one another:

Example 6.3-46: Divide $x^4+8x^3+16x^2+5x$ by x^2+3x+1.

Solution:

Step 1

$$\begin{array}{r l l}
 & x^2 & \textit{first term of the quotient} \\
x^2+3x+1 \,\big) & +x^4+8x^3+16x^2+5x & \textit{dividend} \\
\textit{divisor} & +x^4+3x^3+\;\;x^2 &
\end{array}$$

Step 1 a

$$\begin{array}{r l l}
 & x^2 & \\
x^2+3x+1 \,\big) & +x^4+8x^3+16x^2+5x & \\
 & \overset{-}{+}x^4\overset{-}{+}3x^3\overset{-}{+}\;\;x^2 & \\
\hline
 & \qquad +5x^3+15x^2+5x & \textit{new dividend}
\end{array}$$

Step 2

$$\begin{array}{r|l} & x^2 + 5x \\ \hline x^2 + 3x + 1 & +x^4 + 8x^3 + 16x^2 + 5x \\ & \overset{-}{+}x^4 \overset{-}{+}3x^3 \overset{-}{+} x^2 \\ \hline & +5x^3 + 15x^2 + 5x \\ & +5x^3 + 15x^2 + 5x \end{array}$$

first and final term of the quotient

Step 2 a

$$\begin{array}{r|l} & x^2 + 5x \quad \textit{quotient} \\ \hline x^2 + 3x + 1 & +x^4 + 8x^3 + 16x^2 + 5x \quad \textit{dividend} \\ \textit{divisor} & \overset{-}{+}x^4 \overset{-}{+}3x^3 \overset{-}{+} x^2 \\ \hline & +5x^3 + 15x^2 + 5x \\ & \overset{-}{+}5x^3 \overset{-}{+}15x^2 \overset{-}{+} 5x \\ \hline & 0 \quad \textit{remainder} \end{array}$$

The answer is $x^2 + 5x$ with remainder of zero.

Check:

$$\left(x^2 + 5x\right)\left(x^2 + 3x + 1\right) = \left(x^2 \cdot x^2\right) + 3\left(x^2 \cdot x\right) + \left(x^2 \cdot 1\right) + 5\left(x \cdot x^2\right) + (5 \cdot 3)(x \cdot x) + (5 \cdot 1)x$$

$$= x^4 + 3x^3 + x^2 + 5x^3 + 15x^2 + 5x = x^4 + \left(3x^3 + 5x^3\right) + \left(x^2 + 15x^2\right) + 5x = x^4 + (3+5)x^3 + (1+15)x^2 + 5x$$

$$= x^4 + 8x^3 + 16x^2 + 5x \quad \textit{which is the same as the dividend}$$

Example 6.3-47: Divide $6x^2 + 19x + 18$ by $3x + 5$.

Solution:

Step 1

$$\begin{array}{r|l} & 2x \quad \textit{first term of the quotient} \\ \hline 3x + 5 & +6x^2 + 19x + 18 \quad \textit{dividend} \\ \textit{divisor} & +6x^2 + 10x \end{array}$$

Step 1 a

$$\begin{array}{r|l} & 2x \\ \hline 3x + 5 & +6x^2 + 19x + 18 \\ & \overset{-}{+}6x^2 \overset{-}{+} 10x \\ \hline & +9x + 18 \quad \textit{new dividend} \end{array}$$

Step 2

$$\begin{array}{r|l} & 2x + 3 \quad \textit{first and final term of the quotient} \\ \hline 3x + 5 & +6x^2 + 19x + 18 \\ & \overset{-}{+}6x^2 \overset{-}{+} 10x \\ \hline & +9x + 18 \\ & +9x + 15 \end{array}$$

Step 2 a

$$\begin{array}{r|l} & 2x + 3 \quad \textit{quotient} \\ \hline 3x + 5 & +6x^2 + 19x + 18 \quad \textit{dividend} \\ \textit{divisor} & \overset{-}{+}6x^2 \overset{-}{+} 10x \\ \hline \end{array}$$

$$\begin{array}{r}
+9x+18 \\
\overset{-}{+}9x\overset{-}{+}15 \\
\hline
+\ 3
\end{array} \quad \textit{remainder}$$

The answer is $2x+3$ with remainder of $+3$, or $2x+3+\dfrac{3}{3x+5}$.

Check:

$(2x+3)(3x+5)+3 = (2\cdot3)(x\cdot x)+(2\cdot5)x+(3\cdot3)x+(3\cdot5)+3 = 6x^2+10x+9x+15+3$

$= 6x^2+(10x+9x)+(15+3) = 6x^2+19x+18$ *which is the same as the dividend*

Example 6.3-48: Divide $x^3+8x^2+25x+50$ by $x+5$.

Solution:

Step 1

$$\begin{array}{rl}
 & x^2 \\
x+5 & \overline{)\,+x^3+8x^2+25x+50} \\
\textit{divisor} & +x^3+5x^2
\end{array}$$

first term of the quotient

dividend

Step 1 a

$$\begin{array}{rl}
 & x^2 \\
x+5 & \overline{)\,+x^3+8x^2+25x+50} \\
 & \overset{-}{+}x^3\overset{-}{+}5x^2 \\
\hline
 & +3x^2+25x
\end{array}$$

new dividend

Step 2

$$\begin{array}{rl}
 & x^2+3x \\
x+5 & \overline{)\,+x^3+8x^2+25x+50} \\
 & \overset{-}{+}x^3\overset{-}{+}5x^2 \\
\hline
 & +3x^2+25x \\
 & +3x^2+15x
\end{array}$$

first and second term of the quotient

Step 2 a

$$\begin{array}{rl}
 & x^2+3x \\
x+5 & \overline{)\,+x^3+8x^2+25x+50} \\
 & \overset{-}{+}x^3\overset{-}{+}5x^2 \\
\hline
 & +3x^2+25x \\
 & \overset{-}{+}3x^2\overset{-}{+}15x \\
\hline
 & +10x+50
\end{array}$$

new dividend

Step 3

$$\begin{array}{rl}
 & x^2+3x+10 \\
x+5 & \overline{)\,+x^3+8x^2+25x+50} \\
 & \overset{-}{+}x^3\overset{-}{+}5x^2 \\
\hline
 & +3x^2+25x \\
 & \overset{-}{+}3x^2\overset{-}{+}15x \\
\hline
 & +10x+50 \\
 & +10x+50
\end{array}$$

first, second, and final term of the quotient

Step 3 a

$$
\begin{array}{rl}
 & x^2+3x+10 \quad \textit{quotient} \\
x+5 & \overline{)+x^3+8x^2+25x+50} \quad \textit{dividend} \\
\textit{divisor} & \underline{\overset{-}{+}x^3 \overset{-}{+}5x^2} \\
 & +3x^2+25x \\
 & \underline{\overset{-}{+}3x^2 \overset{-}{+}15x} \\
 & +10x+50 \\
 & \underline{\overset{-}{+}10x \overset{-}{+}50} \\
 & 0 \quad \textit{remainder}
\end{array}
$$

The answer is $x^2+3x+10$ with remainder of zero.

Check:

$$\left(x^2+3x+10\right)(x+5) = \left(x^2\cdot x\right)+\left(5\cdot x^2\right)+3(x\cdot x)+(3\cdot 5)x+(10\cdot x)+(10\cdot 5)$$

$$= x^3+5x^2+3x^2+15x+10x+50 = x^3+\left(5x^2+3x^2\right)+(15x+10x)+50 = x^3+(5+3)x^2+(15+10)x+50$$

$$= x^3+8x^2+25x+50 \quad \textit{which is the same as the dividend}$$

Example 6.3-49: Divide $-x^4+3x^3-2x^2-5x+10$ by $-x+1$.

Solution:

Step 1

$$
\begin{array}{rl}
 & x^3 \quad \textit{first term of the quotient} \\
-x+1 & \overline{)-x^4+3x^3-2x^2-5x+10} \quad \textit{dividend} \\
\textit{divisor} & -x^4+x^3
\end{array}
$$

Step 1 a

$$
\begin{array}{rl}
 & x^3 \\
-x+1 & \overline{)-x^4+3x^3-2x^2-5x+10} \\
 & \underline{\pm x^4 \overset{-}{+} x^3} \\
 & +2x^3-2x^2 \quad \textit{new dividend}
\end{array}
$$

Step 2

$$
\begin{array}{rl}
 & x^3-2x^2 \quad \textit{first and second term of the quotient} \\
-x+1 & \overline{)-x^4+3x^3-2x^2-5x+10} \\
 & \underline{\pm x^4 \overset{-}{+} x^3} \\
 & +2x^3-2x^2 \\
 & +2x^3-2x^2
\end{array}
$$

Step 2 a

$$
\begin{array}{rl}
 & x^3-2x^2 \\
-x+1 & \overline{)-x^4+3x^3-2x^2-5x+10} \\
 & \underline{\pm x^4 \overset{-}{+} x^3} \\
 & +2x^3-2x^2 \\
 & \underline{\overset{-}{+}2x^3 \pm 2x^2} \\
 & -5x+10 \quad \textit{new dividend}
\end{array}
$$

Step 3

$$
\begin{array}{r|l}
 & x^3 - 2x^2 + 5 \\
\hline
-x+1 & -x^4 + 3x^3 - 2x^2 - 5x + 10 \\
 & \pm x^4 \mp x^3 \\
\hline
 & +2x^3 - 2x^2 \\
 & \mp 2x^3 \pm 2x^2 \\
\hline
 & -5x + 10 \\
 & -5x + 5
\end{array}
$$

first, second, and final term of the quotient

Step 3 a

$$
\begin{array}{r|l}
 & x^3 - 2x^2 + 5 \\
\hline
-x+1 & -x^4 + 3x^3 - 2x^2 - 5x + 10 \\
 & \pm x^4 \mp x^3 \\
\hline
 & +2x^3 - 2x^2 \\
 & \mp 2x^3 \pm 2x^2 \\
\hline
 & -5x + 10 \\
 & \pm 5x \mp 5 \\
\hline
 & +5
\end{array}
$$

quotient; *dividend*; *divisor*; *remainder*

The answer is $x^3 - 2x^2 + 5$ with remainder of $+5$, or $x^3 - 2x^2 + 5 + \dfrac{5}{-x+1}$.

Check:

$$\left(x^3 - 2x^2 + 5\right)(-x+1) + 5 = -\left(x^3 \cdot x\right) + \left(x^3 \cdot 1\right) + 2\left(x^2 \cdot x\right) - \left(2x^2 \cdot 1\right) - (5 \cdot x) + (5 \cdot 1) + 5$$

$$= -x^4 + x^3 + 2x^3 - 2x^2 - 5x + 5 + 5 = -x^4 + \left(x^3 + 2x^3\right) - 2x^2 - 5x + (5+5) = -x^4 + (1+2)x^3 - 2x^2 - 5x + 10$$

$$= -x^4 + 3x^3 - 2x^2 - 5x + 10$$ *which is the same as the dividend*

Example 6.3-50: Divide $x^4 + 2x^3 + 2x^2 + 2x + 6$ by $x + 1$.

Solution:

Step 1

$$
\begin{array}{r|l}
 & x^3 \\
\hline
x+1 & +x^4 + 2x^3 + 2x^2 + 2x + 6 \\
 & +x^4 + x^3
\end{array}
$$

first term of the quotient; *dividend*; *divisor*

Step 1 a

$$
\begin{array}{r|l}
 & x^3 \\
\hline
x+1 & +x^4 + 2x^3 + 2x^2 + 2x + 6 \\
 & \mp x^4 \mp x^3 \\
\hline
 & +x^3 + 2x^2
\end{array}
$$

new dividend

Step 2

$$
\begin{array}{r|l}
 & x^3 + x^2 \\
\hline
x+1 & +x^4 + 2x^3 + 2x^2 + 2x + 6 \\
 & \mp x^4 \mp x^3 \\
\hline
 & +x^3 + 2x^2 \\
 & +x^3 + x^2
\end{array}
$$

first and second term of the quotient

Step 2 a

$$
\begin{array}{rl}
 & x^3 + x^2 \\
x+1 & \overline{)\,+x^4 + 2x^3 + 2x^2 + 2x + 6} \\
 & \underline{\overset{-}{+}x^4 \overset{-}{+} x^3} \\
 & \quad + x^3 + 2x^2 \\
 & \quad \underline{\overset{-}{+} x^3 \overset{-}{+} x^2} \\
 & \qquad + x^2 + 2x
\end{array}
$$

new dividend

Step 3

first, second, and third term of the quotient

$$
\begin{array}{rl}
 & x^3 + x^2 + x \\
x+1 & \overline{)\,+x^4 + 2x^3 + 2x^2 + 2x + 6} \\
 & \underline{\overset{-}{+}x^4 \overset{-}{+} x^3} \\
 & \quad + x^3 + 2x^2 \\
 & \quad \underline{\overset{-}{+} x^3 \overset{-}{+} x^2} \\
 & \qquad + x^2 + 2x \\
 & \qquad + x^2 + x
\end{array}
$$

Step 3 a

$$
\begin{array}{rl}
 & x^3 + x^2 + x \\
x+1 & \overline{)\,+x^4 + 2x^3 + 2x^2 + 2x + 6} \\
 & \underline{\overset{-}{+}x^4 \overset{-}{+} x^3} \\
 & \quad + x^3 + 2x^2 \\
 & \quad \underline{\overset{-}{+} x^3 \overset{-}{+} x^2} \\
 & \qquad + x^2 + 2x \\
 & \qquad \underline{\overset{-}{+} x^2 \overset{-}{+} x} \\
 & \qquad\qquad + x + 6
\end{array}
$$

new dividend

Step 4

first, second, third, and final term of the quotient

$$
\begin{array}{rl}
 & x^3 + x^2 + x + 1 \\
x+1 & \overline{)\,+x^4 + 2x^3 + 2x^2 + 2x + 6} \\
 & \underline{\overset{-}{+}x^4 \overset{-}{+} x^3} \\
 & \quad + x^3 + 2x^2 \\
 & \quad \underline{\overset{-}{+} x^3 \overset{-}{+} x^2} \\
 & \qquad + x^2 + 2x \\
 & \qquad \underline{\overset{-}{+} x^2 \overset{-}{+} x} \\
 & \qquad\qquad + x + 6 \\
 & \qquad\qquad + x + 1
\end{array}
$$

Step 4 a

$$
\begin{array}{rll}
 & x^3 + x^2 + x + 1 & \textit{quotient} \\
x+1 & \overline{)\,+x^4 + 2x^3 + 2x^2 + 2x + 6} & \textit{dividend} \\
\textit{divisor} & \underline{\overset{-}{+}x^4 \overset{-}{+} x^3} &
\end{array}
$$

$$\begin{array}{r}
+\,x^3+2x^2 \qquad\qquad \\
\overset{-}{+}\,x^3 \overset{-}{+}\,x^2 \qquad\qquad \\
\hline
+x^2+2x \qquad \\
\overset{-}{+}x^2\overset{-}{+}\,x \qquad \\
\hline
+\,x+6 \\
\overset{-}{+}\,x\overset{-}{+}1 \\
\hline
+5 \quad \textit{remainder}
\end{array}$$

The answer is x^3+x^2+x+1 with remainder of $+5$, or $x^3+x^2+x+1+\dfrac{5}{x+1}$.

Check:

$$\left(x^3+x^2+x+1\right)(x+1)+5 = \left(x^3\cdot x\right)+x^3+\left(x^2\cdot x\right)+\left(x^2\cdot 1\right)+(x\cdot x)+(x\cdot 1)+(1\cdot x)+(1\cdot 1)+5$$

$$= x^4+x^3+x^3+x^2+x^2+x+x+1+5 = x^4+\left(x^3+x^3\right)+\left(x^2+x^2\right)+(x+x)+(1+5)$$

$$= x^4+(1+1)x^3+(1+1)x^2+(1+1)x+6 = x^4+2x^3+2x^2+2x+6 \quad \textit{which is the same as the dividend}$$

Additional Examples - Dividing Polynomials by Polynomials

The following examples further illustrate how to divide two polynomials by each other: (As an exercise, the students are encouraged to check the result of the following additional examples.)

Example 6.3-51: Divide $x^3+6x^2+14x+20$ by $x+3$.

Solution:

$$\begin{array}{rl}
 & x^2+3x+5 \\
x+3 & \overline{\big)\,+x^3+6x^2+14x+20} \\
 & \overset{-}{+}x^3\overset{-}{+}3x^2 \\
\hline
 & \qquad +3x^2+14x \\
 & \qquad \overset{-}{+}3x^2\overset{-}{+}9\,x \\
\hline
 & \qquad\qquad +5x+\;20 \\
 & \qquad\qquad \overset{-}{+}5x\overset{-}{+}\;15 \\
\hline
 & \qquad\qquad\qquad +5
\end{array}$$

The answer is x^2+3x+5 with remainder of $+5$, or $x^2+3x+5+\dfrac{5}{x+3}$.

Example 6.3-52: Divide $2x^5+3x^4-9x^3+12x-18$ by $2x-3$.

Solution:

$$\begin{array}{rl}
 & x^4+3x^3+6 \\
2x-3 & \overline{\big)\,+2x^5+3x^4-9x^3+12x-18} \\
 & \overset{-}{+}2x^5\pm 3x^4 \\
\hline
 & \qquad +6x^4-9x^3 \\
 & \qquad \overset{-}{+}6x^4\overset{+}{-}9x^3 \\
\hline
 & \qquad\qquad +12x-18 \\
 & \qquad\qquad \overset{-}{+}12x\pm 18 \\
\hline
 & \qquad\qquad\qquad 0
\end{array}$$

The answer is $x^4 + 3x^3 + 6$ with remainder of zero.

Example 6.3-53: Divide $-x^4 + x^3 + 2x^2 - 2x + 4$ by $-x + 2$.

Solution:

$$
\begin{array}{r|l}
 & x^3 + x^2 + 2 \\
\hline
-x+2 & -x^4 + x^3 + 2x^2 - 2x + 4 \\
 & \pm x^4 \mp 2x^3 \\
\hline
 & \quad - x^3 + 2x^2 \\
 & \quad \pm x^3 \mp 2x^2 \\
\hline
 & \qquad\qquad -2x + 4 \\
 & \qquad\qquad \pm 2x \mp 4 \\
\hline
 & \qquad\qquad\qquad 0
\end{array}
$$

The answer is $x^3 + x^2 + 2$ with remainder of zero.

Example 6.3-54: Divide $x^5 + 2x^4 + 9x^3 + 13x^2 + 17x + 14$ by $x^2 + x + 2$.

Solution:

$$
\begin{array}{r|l}
 & x^3 + x^2 + 6x + 5 \\
\hline
x^2 + x + 2 & +x^5 + 2x^4 + 9x^3 + 13x^2 + 17x + 14 \\
 & \mp x^5 \mp x^4 \mp 2x^3 \\
\hline
 & \quad +x^4 + 7x^3 + 13x^2 \\
 & \quad \mp x^4 \mp x^3 \mp 2x^2 \\
\hline
 & \qquad +6x^3 + 11x^2 + 17x \\
 & \qquad \mp 6x^3 \mp 6x^2 \mp 12x \\
\hline
 & \qquad\quad +5x^2 + 5x + 14 \\
 & \qquad\quad \mp 5x^2 \mp 5x \mp 10 \\
\hline
 & \qquad\qquad\qquad 4
\end{array}
$$

The answer is $x^3 + x^2 + 6x + 5$ with remainder of 4, or $x^3 + x^2 + 6x + 5 + \frac{4}{x^2 + x + 2}$.

Example 6.3-55: Divide $3x^2 + 5x + 2$ by $x + 1$.

Solution:

$$
\begin{array}{r|l}
 & 3x + 2 \\
\hline
x+1 & +3x^2 + 5x + 2 \\
 & \mp 3x^2 \mp 3x \\
\hline
 & \quad +2x + 2 \\
 & \quad \mp 2x \mp 2 \\
\hline
 & \qquad\quad 0
\end{array}
$$

The answer is $3x + 2$ with remainder of zero.

Example 6.3-56: Divide $-2x^7 + 6x^5 - 4x^3 + 5x^2 - 15$ by $x^2 - 2$.

Solution:

$$
\begin{array}{r|l}
 & -2x^5 + 2x^3 + 5 \\
\hline
x^2 - 2 & -2x^7 + 6x^5 - 4x^3 + 5x^2 - 15 \\
 & \pm 2x^7 \mp 4x^5 \\
\hline
 & +2x^5 - 4x^3 \\
 & \mp 2x^5 \pm 4x^3 \\
\hline
 & +5x^2 - 15 \\
 & \mp 5x^2 \pm 10 \\
\hline
 & -5
\end{array}
$$

The answer is $-2x^5 + 2x^3 + 5$ **with remainder of** -5**, or** $-2x^5 + 2x^3 + 5 - \frac{5}{x^2 - 2}$.

Example 6.3-57: Divide $x^3 - 2x^2 + 5$ by $x - 1$.

Solution:

$$
\begin{array}{r|l}
 & x^2 - x - 1 \\
\hline
x - 1 & +x^3 - 2x^2 + 0x + 5 \\
 & \mp x^3 \pm x^2 \\
\hline
 & -x^2 + 0x \\
 & \pm x^2 \mp x \\
\hline
 & -x + 5 \\
 & \pm x \mp 1 \\
\hline
 & 4
\end{array}
$$

The answer is $x^2 - x - 1$ **with remainder of** 4**, or** $x^2 - x - 1 + \frac{4}{x - 1}$.

Example 6.3-58: Divide $x^5 - 3x^4 - 2x^2 + 8x - 6$ by $x - 3$.

Solution:

$$
\begin{array}{r|l}
 & x^4 - 2x + 2 \\
\hline
x - 3 & +x^5 - 3x^4 - 2x^2 + 8x - 6 \\
 & \mp x^5 \pm 3x^4 \\
\hline
 & -2x^2 + 8x \\
 & \pm 2x^2 \mp 6x \\
\hline
 & +2x - 6 \\
 & \mp 2x \pm 6 \\
\hline
 & 0
\end{array}
$$

The answer is $x^4 - 2x + 2$ **with remainder of zero.**

Example 6.3-59: Divide $12x^3 - 4x^2 - 3x + 1$ by $x - 1$.

Solution:

$$\begin{array}{r|l} & \quad 12x^2 + 8x + 5 \\ \hline x-1 & +12x^3 - 4x^2 - 3x + 1 \\ & \overset{-}{+}12x^3 \pm 12x^2 \\ \hline & \qquad +8x^2 - 3x \\ & \qquad \overset{-}{+}8x^2 \pm 8x \\ \hline & \qquad\qquad +5x + 1 \\ & \qquad\qquad \overset{-}{+}5x \pm 5 \\ \hline & \qquad\qquad\qquad +6 \end{array}$$

The answer is $12x^2 + 8x + 5$ with remainder of $+6$, or $12x^2 + 8x + 5 + \dfrac{6}{x-1}$.

Example 6.3-60: Divide $18x^3 - 12x^2 - 11x + 10$ by $x - 2$.

Solution:

$$\begin{array}{r|l} & \quad 18x^2 + 24x + 37 \\ \hline x-2 & +18x^3 - 12x^2 - 11x + 10 \\ & \overset{-}{+}18x^3 \pm 36x^2 \\ \hline & \qquad +24x^2 - 11x \\ & \qquad \overset{-}{+}24x^2 \pm 48x \\ \hline & \qquad\qquad +37x + 10 \\ & \qquad\qquad \overset{-}{+}37x \pm 74 \\ \hline & \qquad\qquad\qquad +84 \end{array}$$

The answer is $18x^2 + 24x + 37$ with remainder of $+84$, or $18x^2 + 24x + 37 + \dfrac{84}{x-2}$.

Practice Problems - Dividing Polynomials by Polynomials

Section 6.3 Case II Practice Problems - Divide the following polynomial expressions:

1. $3x^2 + 10x + 3$ by $x + 3$
2. $x^4 + 7x^3 + 13x^2 + 17x + 10$ by $x + 5$
3. $x^6 - x^5 - 2x^4 - x^3 + 2x^2 + 5x - 10$ by $x - 2$
4. $-2x^4 + 5x^3 - 4x^2 + 16x - 15$ by $-2x + 5$
5. $2x^4 - 13x^3 + 13x^2 + 15x - 25$ by $x - 5$
6. $-2x^4 + 7x^3 - 6x^2 - 2x + 3$ by $-2x + 3$
7. $3x^7 + 3x^6 - 2x^3 - 2x^2 + 5x + 5$ by $x + 1$
8. $2y^3 + 5y^2 - 4y - 12$ by $y + 2$
9. $x^3 + 2x^2 - 18x + 14$ by $x - 3$
10. $x^4 + 2x^3 + 2x^2 + 5x + 2$ by $x + 2$

6.4 Adding and Subtracting Polynomials

Polynomials are added and subtracted by combining their numerical coefficients while keeping the like terms. Polynomials can be added horizontally or vertically as described in the following two cases:

Case I Adding and Subtracting Polynomials Horizontally

Polynomials are horizontally added and subtracted using the following steps:

Step 1 Write the polynomial in descending order.

Step 2 Group the like terms. (Note: In the case of subtraction, change the sign in each term of the polynomial being subtracted before grouping the like terms.)

Step 3 Add or subtract the like terms.

Examples with Steps

The following examples show the steps as to how polynomials are added and subtracted horizontally:

Example 6.4-1

$$\left(x^2+3x^3+5\right)+\left(x^3+8x+2x^2\right)=$$

Solution:

Step 1 $\left(x^2+3x^3+5\right)+\left(x^3+8x+2x^2\right)=\left(3x^3+x^2+5\right)+\left(x^3+2x^2+8x\right)$

Step 2 $\left(3x^3+x^2+5\right)+\left(x^3+2x^2+8x\right)=\left(3x^3+x^3\right)+\left(x^2+2x^2\right)+8x+5$

Step 3 $\left(3x^3+x^3\right)+\left(x^2+2x^2\right)+8x+5=(3+1)x^3+(1+2)x^2+8x+5=\mathbf{4x^3+3x^2+8x+5}$

Example 6.4-2

$$\left(y+y^2+3y^3+3\right)-\left(3y^2+2y-y^3\right)=$$

Solution:

Step 1 $\left(y+y^2+3y^3+3\right)-\left(3y^2+2y-y^3\right)=\left(3y^3+y^2+y+3\right)-\left(-y^3+3y^2+2y\right)$

Step 2 $\left(3y^3+y^2+y+3\right)-\left(-y^3+3y^2+2y\right)=\left(3y^3+y^2+y+3\right)+\left(+y^3-3y^2-2y\right)$

$=\left(3y^3+y^3\right)+\left(y^2-3y^2\right)+(y-2y)+3$

Step 3 $\left(3y^3+y^3\right)+\left(y^2-3y^2\right)+(y-2y)+3=(3+1)y^3+(1-3)y^2+(1-2)y+3$

$= \boxed{\mathbf{4y^3 - 2y^2 - y + 3}}$

Example 6.4-3

$$\boxed{\left(-2x + x^4 + 5x^2 + 1\right) + \left(3x + 8x^2 + 3x^4 - 3\right)} =$$

Solution:

Step 1 $\boxed{\left(-2x + x^4 + 5x^2 + 1\right) + \left(3x + 8x^2 + 3x^4 - 3\right)} = \boxed{\left(x^4 + 5x^2 - 2x + 1\right) + \left(3x^4 + 8x^2 + 3x - 3\right)}$

Step 2 $\boxed{\left(x^4 + 5x^2 - 2x + 1\right) + \left(3x^4 + 8x^2 + 3x - 3\right)} = \boxed{\left(x^4 + 3x^4\right) + \left(5x^2 + 8x^2\right) + (3x - 2x) + (1 - 3)}$

Step 3 $\boxed{\left(x^4 + 3x^4\right) + \left(5x^2 + 8x^2\right) + (3x - 2x) + (1 - 3)} = \boxed{(1+3)x^4 + (5+8)x^2 + (3-2)x + (-2)}$

$= \boxed{\mathbf{4x^4 + 13x^2 + x - 2}}$

Example 6.4-4

$$\boxed{\left(k^2l + kl^2 + k + 2k^2l\right) - \left(3k - 2kl^2 - k^2l\right)} =$$

Solution:

Step 1 $\boxed{\textit{Not Applicable}}$

Step 2 $\boxed{\left(k^2l + kl^2 + k + 2k^2l\right) - \left(3k - 2kl^2 - k^2l\right)} = \boxed{\left(k^2l + kl^2 + k + 2k^2l\right) + \left(-3k + 2kl^2 + k^2l\right)}$

$= \boxed{\left(k^2l + 2k^2l + k^2l\right) + \left(kl^2 + 2kl^2\right) + (k - 3k)}$

Step 3 $\boxed{\left(k^2l + 2k^2l + k^2l\right) + \left(kl^2 + 2kl^2\right) + (k - 3k)} = \boxed{(1+2+1)k^2l + (1+2)kl^2 + (1-3)k}$

$= \boxed{4k^2l + 3kl^2 + (-2)k} = \boxed{\mathbf{4k^2l + 3kl^2 - 2k}}$

Example 6.4-5

$$\boxed{\left(m^3 + m + 4m^2\right) + \left(m + m^2\right) + \left(3m + 4m^3\right)} =$$

Solution:

Step 1 $\boxed{\left(m^3 + m + 4m^2\right) + \left(m + m^2\right) + \left(3m + 4m^3\right)} = \boxed{\left(m^3 + 4m^2 + m\right) + \left(m^2 + m\right) + \left(4m^3 + 3m\right)}$

Step 2 $\boxed{\left(m^3 + 4m^2 + m\right) + \left(m^2 + m\right) + \left(4m^3 + 3m\right)} = \boxed{\left(m^3 + 4m^3\right) + \left(4m^2 + m^2\right) + (m + m + 3m)}$

Step 3 $\boxed{(m^3+4m^3)+(4m^2+m^2)+(m+m+3m)} = \boxed{(1+4)m^3+(4+1)m^2+(1+1+3)m}$

$= \boxed{\mathbf{5m^3+5m^2+5m}}$

Additional Examples - Adding and Subtracting Polynomials Horizontally

The following examples further illustrate how to add and subtract polynomials horizontally:

Example 6.4-6

$\boxed{(5x^2+9+6x)+(-2x-3x^2)} = \boxed{(5x^2+6x+9)+(-3x^2-2x)} = \boxed{(5x^2-3x^2)+(6x-2x)+9}$

$= \boxed{(5-3)x^2+(6-2)x+9} = \boxed{\mathbf{2x^2+4x+9}}$

Example 6.4-7

$\boxed{(-3x^3-5x-6x+7x^0)-(12x^3+5x^2-3x)} = \boxed{(-3x^3-5x-6x+(7\cdot 1))+(-12x^3-5x^2+3x)}$

$= \boxed{(-3x^3-12x^3)-5x^2+(-5x-6x+3x)+7} = \boxed{(-3-12)x^3-5x^2+(-5-6+3)x+7} = \boxed{\mathbf{-15x^3-5x^2-8x+7}}$

Example 6.4-8

$\boxed{(3x^2y-3xy^2+2x)+(-2x^2y-5+2xy^2)} = \boxed{(3x^2y-2x^2y)+(-3xy^2+2xy^2)+2x-5}$

$= \boxed{(3-2)x^2y+(-3+2)xy^2+2x-5} = \boxed{\mathbf{x^2y-xy^2+2x-5}}$

Example 6.4-9

$\boxed{-(x^3+3x^2+6x^4-5)+(5x-3x^3-2x^2+2x^0)} = \boxed{-(6x^4+x^3+3x^2-5)+(-3x^3-2x^2+5x+2x^0)}$

$= \boxed{(-6x^4-x^3-3x^2+5)+(-3x^3-2x^2+5x+(2\cdot 1))} = \boxed{-6x^4+(-x^3-3x^3)+(-3x^2-2x^2)+5x+(5+2)}$

$= \boxed{-6x^4+(-1-3)x^3+(-3-2)x^2+5x+7} = \boxed{\mathbf{-6x^4-4x^3-5x^2+5x+7}}$

Example 6.4-10

$\boxed{(4x^3-3x^4+2x+10)+(-3-x^2+5x+2x^4)} = \boxed{(-3x^4+4x^3+2x+10)+(2x^4-x^2+5x-3)}$

$= \boxed{(-3x^4+2x^4)+4x^3-x^2+(2x+5x)+(10-3)} = \boxed{(-3+2)x^4+4x^3-x^2+(2+5)x+7}$

$= \boxed{\mathbf{-x^4+4x^3-x^2+7x+7}}$

Example 6.4-11

$$\boxed{\left(6x^3-4x+5+3x^2\right)+\left(-2x-5x^2+3x-x^2+2^3x\right)}=\boxed{\left(6x^3+3x^2-4x+5\right)+\left(-5x^2-x^2-2x+3x+8x\right)}$$

$$=\boxed{6x^3+\left(3x^2-5x^2-x^2\right)+\left(-4x-2x+3x+8x\right)+5}=\boxed{6x^3+\left(3-5-1\right)x^2+\left(-4-2+3+8\right)x+5}$$

$$=\boxed{\mathbf{6x^3-3x^2+5x+5}}$$

Example 6.4-12

$$\boxed{\left(3x^5+4x^0+2x^4+2x^2\right)+\left(5x-3x^2-3x^4+3\right)}=\boxed{\left(3x^5+2x^4+2x^2+4x^0\right)+\left(-3x^4-3x^2+5x+3\right)}$$

$$=\boxed{\left(3x^5+2x^4+2x^2+4\right)+\left(-3x^4-3x^2+5x+3\right)}=\boxed{3x^5+\left(2x^4-3x^4\right)+\left(2x^2-3x^2\right)+5x+\left(4+3\right)}$$

$$=\boxed{3x^5+\left(2-3\right)x^4+\left(2-3\right)x^2+5x+7}=\boxed{\mathbf{3x^5-x^4-x^2+5x+7}}$$

Example 6.4-13

$$\boxed{\left(7x+3x^3-2x^2+5\right)+\left(2x-3x^2+x^0\right)-\left(x^2+4x^3\right)}=\boxed{\left(3x^3-2x^2+7x+5\right)+\left(-3x^2+2x+x^0\right)-\left(4x^3+x^2\right)}$$

$$=\boxed{\left(3x^3-2x^2+7x+5\right)+\left(-3x^2+2x+1\right)+\left(-4x^3-x^2\right)}=\boxed{\left(3x^3-4x^3\right)+\left(-2x^2-3x^2-x^2\right)+\left(7x+2x\right)+\left(5+1\right)}$$

$$=\boxed{\left(3-4\right)x^3+\left(-2-3-1\right)x^2+\left(7+2\right)x+6}=\boxed{\mathbf{-x^3-6x^2+9x+6}}$$

Example 6.4-14

$$\boxed{\left(3a-5ab+3a^2+6a\right)+\left(-3ab+2a^2+2a+2ab-5a^2\right)}=\boxed{\left(3a^2+6a+3a-5ab\right)+\left(2a^2-5a^2+2a-3ab+2ab\right)}$$

$$=\boxed{\left(3a^2+2a^2-5a^2\right)+\left(-5ab-3ab+2ab\right)+\left(6a+3a+2a\right)}=\boxed{\left(3+2-5\right)a^2+\left(-5-3+2\right)ab+\left(6+3+2\right)a}$$

$$=\boxed{0a^2-6ab+11a}=\boxed{\mathbf{-6ab+11a}}$$

Example 6.4-15

$$\boxed{\left(u^2v-2u^2v+5uv^2\right)-\left(uv^2+8u^2v\right)+\left(2u^2v-2uv^2\right)}=\boxed{\left(u^2v-2u^2v+5uv^2\right)+\left(-uv^2-8u^2v\right)+\left(2u^2v-2uv^2\right)}$$

$$=\boxed{\left(u^2v-2u^2v-8u^2v+2u^2v\right)+\left(5uv^2-uv^2-2uv^2\right)}=\boxed{\left(1-2-8+2\right)u^2v+\left(5-1-2\right)uv^2}=\boxed{\mathbf{-7u^2v+2uv^2}}$$

Practice Problems - Adding and Subtracting Polynomials Horizontally

Section 6.4 Case I Practice Problems - Add or subtract the following polynomials horizontally:

1. $\left(x^3+2x^5-3x+2\right)+\left(3x^3+x-x^5\right)=$

2. $\left(y-y^2+2y^4+3y^2-3\right)+\left(2y^4+y^3+5-y^2\right)=$

3. $\left(3x-3x^2+5x-3\right)-\left(-2x+5-x^2+2\right)=$

4. $\left(xyz+2x^2yz+4xyz\right)+\left(4x^2yz-x^2yz+2xyz\right)=$

5. $\left(-2ab-3+2a^2b^2\right)+\left(-3ab+a^2b^2+2(ab)^0\right)=$

6. $\left(5x^6-x^5-4x^4+3x+x^2\right)-\left(x-3x^2+x^4-3x^6\right)=$

7. $-\left(w^2+w+2w^4+8\right)+\left(w-3w^4-2w^2-8\right)=$

8. $\left(u^2v+2uv+5u^2v-6+uv\right)+\left(uv+2-3u^2v\right)=$

9. $\left(x^3+x\right)-\left(3x^2+x^3+5x\right)+\left(-x^2-6x-4x^3+x\right)=$

10. $\left(x^5+x^4+2\right)-\left(x^4+3-2x^5+x^3\right)+\left(-4x^4-8\right)=$

$(3a-5ab+3a^2+6a)+(-3ab+2a^2+2a+2ab-5a^2)=$

$=3a^2+6a+3a-5ab-5a^2+2a^2-3ab+2a+2ab=$

$=(3a^2-5a^2+2a^2)+(6a+3a+2a)+(-5ab-3ab+2ab)=$

$=11a-6ab$

Case II Adding and Subtracting Polynomials Vertically

Polynomials are vertically added and subtracted using the following steps:

Step 1 Write the polynomials in descending order.

Step 2 Group the like terms in each polynomial separately. (Note: In the case of subtraction, change the sign in each term of the polynomial being subtracted before grouping the like terms.)

Step 3 Write the like terms under one another.

Step 4 Add or subtract the like terms.

Examples with Steps

The following examples show the steps as to how polynomials are added and subtracted vertically:

Example 6.4-16

$$\left(x^4+x+3x^3+4x\right)+\left(x^2+3x^4-x^3+x\right)=$$

Solution:

Step 1

$$\left(x^4+x+3x^3+4x\right)+\left(x^2+3x^4-x^3+x\right)=\left(x^4+3x^3+4x+x\right)+\left(3x^4-x^3+x^2+x\right)$$

Step 2

$$\left(x^4+3x^3+4x+x\right)+\left(3x^4-x^3+x^2+x\right)=\left[x^4+3x^3+(4+1)x\right]+\left(3x^4-x^3+x^2+x\right)$$

$$=\left(x^4+3x^3+5x\right)+\left(3x^4-x^3+x^2+x\right)$$

Step 3

$$\left(x^4+3x^3+5x\right)+\left(3x^4-x^3+x^2+x\right)=\begin{array}{rrrr} x^4 & +3x^3 & & +5x \\ 3x^4 & -\;x^3 & +x^2 & +\;x \end{array}$$

Step 4

$$\begin{array}{rrrr} x^4 & +3x^3 & & +5x \\ 3x^4 & -\;x^3 & +x^2 & +\;x \end{array} = \begin{array}{rrrr} x^4 & +3x^3 & & +5x \\ 3x^4 & -\;x^3 & +x^2 & +\;x \\ \hline \mathbf{4x^4} & \mathbf{+2x^3} & \mathbf{+x^2} & \mathbf{+6x} \end{array}$$

Example 6.4-17

$$\left(x^5+x^4+2x+5\right)-\left(5x-3x^4-x+3x+6\right)=$$

Solution:

Step 1

$$\left(x^5+x^4+2x+5\right)-\left(5x-3x^4-x+3x+6\right)=\left(x^5+x^4+2x+5\right)-\left(-3x^4+5x+3x-x+6\right)$$

Step 2

$$\left(x^5+x^4+2x+5\right)-\left(-3x^4+5x+3x-x+6\right)=\left(x^5+x^4+2x+5\right)+\left(3x^4-5x-3x+x-6\right)$$

$$=\left(x^5+x^4+2x+5\right)+\left[3x^4+(-5-3+1)x-6\right]=\left(x^5+x^4+2x+5\right)+\left(3x^4-7x-6\right)$$

Step 3 $\boxed{\left(x^5+x^4+2x+5\right)+\left(3x^4-7x-6\right)} = \boxed{\begin{array}{r} x^5+x^4+2x+5 \\ +3x^4-7x-6 \end{array}}$

Step 4 $\boxed{\begin{array}{r} x^5+x^4+2x+5 \\ +3x^4-7x-6 \end{array}} = \begin{array}{r} x^5+x^4+2x+5 \\ +3x^4-7x-6 \\ \hline \mathbf{x^5+4x^4-5x-1} \end{array}$

Example 6.4-18

$\boxed{\left(x^2+9x-2x+5\right)+\left(3+2x^2+6x+5x^2\right)} =$

Solution:

Step 1 $\boxed{\left(x^2+9x-2x+5\right)+\left(3+2x^2+6x+5x^2\right)} = \boxed{\left(x^2+9x-2x+5\right)+\left(2x^2+5x^2+6x+3\right)}$

Step 2 $\boxed{\left(x^2+9x-2x+5\right)+\left(2x^2+5x^2+6x+3\right)} = \boxed{\left[x^2+(9-2)x+5\right]+\left[(2+5)x^2+6x+3\right]}$

$= \boxed{\left(x^2+7x+5\right)+\left(7x^2+6x+3\right)}$

Step 3 $\boxed{\left(x^2+7x+5\right)+\left(7x^2+6x+3\right)} = \boxed{\begin{array}{r} x^2+7x+5 \\ 7x^2+6x+3 \end{array}}$

Step 4 $\boxed{\begin{array}{r} x^2+7x+5 \\ 7x^2+6x+3 \end{array}} = \begin{array}{r} x^2+7x+5 \\ 7x^2+6x+3 \\ \hline \mathbf{8x^2+13x+8} \end{array}$

Example 6.4-19

$\boxed{\left(w^4+2w^3+w+2w^2\right)+\left(2w+4w^2+6w\right)} =$

Solution:

Step 1 $\boxed{\left(w^4+2w^3+w+2w^2\right)+\left(2w+4w^2+6w\right)} = \boxed{\left(w^4+2w^3+2w^2+w\right)+\left(4w^2+2w+6w\right)}$

Step 2 $\boxed{\left(w^4+2w^3+2w^2+w\right)+\left(4w^2+2w+6w\right)} = \boxed{\left(w^4+2w^3+2w^2+w\right)+\left[4w^2+(2+6)w\right]}$

$= \boxed{\left(w^4+2w^3+2w^2+w\right)+\left(4w^2+8w\right)}$

Step 3 $\boxed{\left(w^4+2w^3+2w^2+w\right)+\left(4w^2+8w\right)} = \boxed{\begin{array}{r} w^4+2w^3+2w^2+w \\ +4w^2+8w \end{array}}$

Step 4 $\boxed{\begin{array}{r} w^4+2w^3+2w^2+w \\ +4w^2+8w \end{array}} = \begin{array}{r} w^4+2w^3+2w^2+w \\ +4w^2+8w \\ \hline \mathbf{w^4+2w^3+6w^2+9w} \end{array}$

Example 6.4-20

$$\boxed{\left(a^2+3a+2a^2-a-2\right)-\left(4-4a^2-3a-6\right)}=$$

Solution:

Step 1

$$\boxed{\left(a^2+3a+2a^2-a-2\right)-\left(4-4a^2-3a-6\right)}=\boxed{\left(a^2+2a^2+3a-a-2\right)-\left(-4a^2-3a-6+4\right)}$$

Step 2

$$\boxed{\left(a^2+2a^2+3a-a-2\right)-\left(-4a^2-3a-6+4\right)}=\boxed{\left(a^2+2a^2+3a-a-2\right)+\left(4a^2+3a+6-4\right)}$$

$$\boxed{\left[(1+2)a^2+(3-1)a-2\right]+\left[4a^2+3a+(6-4)\right]}=\boxed{\left(3a^2+2a-2\right)+\left(4a^2+3a+2\right)}$$

Step 3

$$\boxed{\left(3a^2+2a-2\right)+\left(4a^2+3a+2\right)}=\boxed{\begin{array}{l}3a^2+2a-2\\4a^2+3a+2\end{array}}$$

Step 4

$$\boxed{\begin{array}{l}3a^2+2a-2\\4a^2+3a+2\end{array}}=\begin{array}{r}3a^2+2a-2\\\underline{4a^2+3a+2}\\\mathbf{7a^2+5a+0}\end{array}$$

Additional Examples - Adding and Subtracting Polynomials Vertically

The following examples further illustrate how to add and subtract polynomials vertically:

Example 6.4-21

$$\boxed{\left(x^2+3+x\right)+\left(3x+3x^2+5\right)}=\boxed{\left(x^2+x+3\right)+\left(3x^2+3x+5\right)}=\begin{array}{r}x^2+\ x+3\\\underline{3x^2+3x+5}\\\mathbf{4x^2+4x+8}\end{array}$$

Example 6.4-22

$$\boxed{\left(x^3+2x+x^2+x^0\right)-\left(3x+2x^3-4x-2\right)}=\boxed{\left(x^3+x^2+2x+1\right)-\left(2x^3+3x-4x-2\right)}$$

$$=\boxed{\left(x^3+x^2+2x+1\right)+\left(-2x^3-3x+4x+2\right)}=\boxed{\left(x^3+x^2+2x+1\right)+\left(-2x^3+(-3+4)x+2\right)}$$

$$=\boxed{\left(x^3+x^2+2x+1\right)+\left(-2x^3+x+2\right)}=\begin{array}{r}+\ x^3+x^2+2x+1\\\underline{-2x^3\qquad\quad+x+2}\\-\ \mathbf{x^3+x^2+3x+3}\end{array}$$

Example 6.4-23

$$\boxed{\left(y^0+2y+4y^3+3y^2\right)-\left(y^3-2y^2-4y+3y+4\right)}=\boxed{\left(4y^3+3y^2+2y+y^0\right)-\left(y^3-2y^2-4y+3y+4\right)}$$

$$=\boxed{\left(4y^3+3y^2+2y+y^0\right)+\left(-y^3+2y^2+4y-3y-4\right)}=\boxed{\left(4y^3+3y^2+2y+1\right)+\left(-y^3+2y^2+(4-3)y-4\right)}$$

$$= \boxed{\left(4y^3+3y^2+2y+1\right)+\left(-y^3+2y^2+y-4\right)} = \begin{array}{r} 4y^3+3y^2+2y+1 \\ -y^3+2y^2+y-4 \\ \hline \mathbf{3y^3+5y^2+3y-3} \end{array}$$

Example 6.4-24

$$\boxed{\left(n^2+6n^3+n+2\right)+\left(2n^2-n-5\right)} = \boxed{\left(6n^3+n^2+n+2\right)+\left(2n^2-n-5\right)} = \begin{array}{r} 6n^3+n^2+n+2 \\ +2n^2-n-5 \\ \hline \mathbf{6n^3+3n^2+0-3} \end{array}$$

Example 6.4-25

$$\boxed{\left(3a^4+2a+6a^3+4a^2\right)+\left(a^2+a-a^4+a^3-3a\right)} = \boxed{\left(3a^4+6a^3+4a^2+2a\right)+\left(-a^4+a^3+a^2+a-3a\right)}$$

$$= \boxed{\left(3a^4+6a^3+4a^2+2a\right)+\left(-a^4+a^3+a^2+(a-3a)\right)} = \boxed{\left(3a^4+6a^3+4a^2+2a\right)+\left(-a^4+a^3+a^2-2a\right)}$$

$$= \begin{array}{r} 3a^4+6a^3+4a^2+2a \\ -a^4+a^3+a^2-2a \\ \hline \mathbf{2a^4+7a^3+5a^2+0} \end{array}$$

Example 6.4-26

$$\boxed{\left(3x+5x^3+x^2\right)-\left(3x+x^2-2x^3+5\right)+\left(6x-3+3x^2\right)} = \boxed{\left(5x^3+x^2+3x\right)-\left(-2x^3+x^2+3x+5\right)+\left(3x^2+6x-3\right)}$$

$$= \boxed{\left(5x^3+x^2+3x\right)+\left(+2x^3-x^2-3x-5\right)+\left(3x^2+6x-3\right)} = \begin{array}{r} 5x^3+x^2+3x \\ 2x^3-x^2-3x-5 \\ +3x^2+6x-3 \\ \hline \mathbf{7x^3+3x^2+6x-8} \end{array}$$

Example 6.4-27

$$\boxed{\left(y^2z^2+5y^2z^2-15+2yz\right)-\left(30+4y^2z^2-2yz+3\right)} = \boxed{\left(y^2z^2+5y^2z^2+2yz-15\right)-\left(4y^2z^2-2yz+30+3\right)}$$

$$= \boxed{\left(y^2z^2+5y^2z^2+2yz-15\right)+\left(-4y^2z^2+2yz-30-3\right)} = \boxed{\left((1+5)y^2z^2+2yz-15\right)+\left(-4y^2z^2+2yz-(30+3)\right)}$$

$$= \boxed{\left(6y^2z^2+2yz-15\right)+\left(-4y^2z^2+2yz-33\right)} = \begin{array}{r} 6y^2z^2+2yz-15 \\ -4y^2z^2+2yz-33 \\ \hline \mathbf{2y^2z^2+4yz-48} \end{array}$$

Example 6.4-28

$$\boxed{\left(w^5+3w^3+2w+5w^2\right)+\left(2w+5w^5-2w^3-5\right)} = \boxed{\left(w^5+3w^3+5w^2+2w\right)+\left(5w^5-2w^3+2w-5\right)}$$

$$= \begin{array}{r} w^5 + 3w^3 + 5w^2 + 2w \\ 5w^5 - 2w^3 \qquad + 2w - 5 \\ \hline \mathbf{6w^5 + w^3 + 5w^2 + 4w - 5} \end{array}$$

Example 6.4-29

$$\boxed{-(u^3 + 2u + 5 + 3u - 3) + (2u^2 + 2u^3 - 8 - 4u)} = \boxed{-(u^3 + 2u + 3u + 5 - 3) + (2u^3 + 2u^2 - 4u - 8)}$$

$$= \boxed{(-u^3 - 2u - 3u - 5 + 3) + (2u^3 + 2u^2 - 4u - 8)} = \boxed{(-u^3 - (2+3)u + (-5+3)) + (2u^3 + 2u^2 - 4u - 8)}$$

$$= \boxed{(-u^3 - 5u - 2) + (2u^3 + 2u^2 - 4u - 8)} = \begin{array}{r} -u^3 \qquad - 5u - 2 \\ +2u^3 + 2u^2 - 4u - 8 \\ \hline \mathbf{u^3 + 2u^2 - 9u - 10} \end{array}$$

Example 6.4-30

$$\boxed{(x^2y^2 + 2x^2y^2 + 2x + 5) + (3x^2y^2 - 3 - 6x + 5)} = \boxed{(x^2y^2 + 2x^2y^2 + 2x + 5) + (3x^2y^2 - 6x - 3 + 5)}$$

$$= \boxed{((1+2)x^2y^2 + 2x + 5) + (3x^2y^2 - 6x + (-3+5))} = \boxed{(3x^2y^2 + 2x + 5) + (3x^2y^2 - 6x + 2)} = \begin{array}{r} 3x^2y^2 + 2x + 5 \\ 3x^2y^2 - 6x + 2 \\ \hline \mathbf{6x^2y^2 - 4x + 7} \end{array}$$

Practice Problems - Adding and Subtracting Polynomials Vertically

Section 6.4 Case II Practice Problems - Add or subtract the following polynomials vertically:

1. $(x^2 + 2x + x^3) + (3x - 2x^3) =$
2. $(y + y^2 + 3y^3 + 4) + (-2 + y^2 + 3y^2 + 2y) =$
3. $(x^3 + x^2 - 3 + 3x^2) - (-2x^3 - 5x + 5) =$
4. $(z^5 + 3z^2 + z - 2z^2 - 4z + 2) + (z^2 + 4z^5 + z^0) =$
5. $-(a^3 - 2a + a + 2 - 3a^3) + (-2a^3 - 4a - 3) =$
6. $(u^2 + 2u + u + 5) + (-2u^2 - 3 - 5u - 8) =$
7. $(x - x^2 + 3x^4 - 5) - (-2x^4 + x - 3 + 4x^2) =$
8. $(a + 3a^2 - 60 + 5a) + (-4a - 5a + 5^0 + 2a^2) =$
9. $(x^2y^2 - 2xy - 8 + 2x^2y^2) + (3x^2y^2 + 4xy + 3 + 2xy) =$
10. $(3a^3 + 2a + a^6 + 1) + (8 + 3a^3 + a) - (2a^6 - 4 + 4a + 2a^3) =$

Appendix - Exercise Solutions

Chapter 1 Solutions:

Section 1.1 Solutions

1. $\frac{-95}{-5} = \frac{95}{5} = \mathbf{19}$

2. $(-20)\times(-8) = +160 = \mathbf{160}$

3. $(-33)+(-14) = -33-14 = \mathbf{-47}$

4. $(-18)-(-5) = (-18)+(5) = -18+5 = \mathbf{-13}$

5. $(-20)+8 = -20+8 = \mathbf{-12}$

6. $\frac{48}{-4} = -\frac{48}{4} = \mathbf{-12}$

7. $-15-32 = \mathbf{-47}$

8. $30+(-9) = 30-9 = \mathbf{21}$

9. $55-(-6) = 55+(6) = 55+6 = \mathbf{61}$

10. $8\times(-35) = -8\times 35 = \mathbf{-280}$

Section 1.2 Solutions

1. $2+3+5+6 = \mathbf{16}$

2. $(2+5)+(6+3)+9 = (7)+(9)+9 = 7+9+9 = \mathbf{25}$

3. $(6+3+8)+(2+3)+4 = (17)+(5)+4 = 17+5+4 = \mathbf{26}$

4. $8+[(1+3+4)+(1+2)] = 8+[(8)+(3)] = 8+[8+3] = 8+[11] = 8+11 = \mathbf{19}$

5. $[(18+4)+9]+[1+(2+3)] = [(22)+9]+[1+(5)] = [22+9]+[1+5] = [31]+[6] = 31+6 = \mathbf{37}$

6. $8+[(2+3)+(6+3)+15] = 8+[(5)+(9)+15] = 8+[5+9+15] = 8+[29] = 8+29 = \mathbf{37}$

7. $(7+3+8)+[(7+2+3)+5] = (18)+[(12)+5] = 18+[12+5] = 18+[17] = 18+17 = \mathbf{35}$

8. $[(3+9+4)+1+(1+8)]+(8+2) = [(16)+1+(9)]+(10) = [16+1+9]+10 = [26]+10 = 26+10 = \mathbf{36}$

9. $[(2+3+6)+(1+8)]+[(1+3)+4] = [(11)+(9)]+[(4)+4] = [11+9]+[4+4] = [20]+[8] = 20+8 = \mathbf{28}$

10. $[[(3+5)+(4+3)+5]+(2+3+5)]+6 = [[(8)+(7)+5]+(10)]+6 = [[8+7+5]+10]+6 = [[20]+10]+6 = [20+10]+6$
$= [30]+6 = 30+6 = \mathbf{36}$

Section 1.3 Solutions

1. $(55-5)-3-8 = (50)-11 = 50-11 = \mathbf{39}$

2. $59-38-12-(20-5) = 21-12-(15) = 9-15 = \mathbf{-6}$

3. $(20-5)-(11-2) = (15)-(9) = 15-9 = \mathbf{6}$

4. $[-25-(4-13)]-5 = [-25-(-9)]-5 = [-25+(9)]-5 = [-25+9]-5 = [-16]-5 = -16-5 = \mathbf{-21}$

5. $350-(25-38)-30 = 350-(-13)-30 = 350+(13)-30 = 350+13-30 = 363-30 = \mathbf{333}$

6. $[(-30-3)-8]-(16-9) = [(-33)-8]-(7) = [-33-8]-7 = [-41]-7 = -41-7 = \mathbf{-48}$

7. $[(40-4)-(8-10)]-9 = [(36)-(-2)]-9 = [36+(2)]-9 = [36+2]-9 = [38]-9 = 38-9 = \mathbf{29}$

8. $(35-56)-[(20-15)-8] = (-21)-[(5)-8] = -21-[5-8] = -21-[-3] = -21+[3] = -21+3 = \mathbf{-18}$

9. $[(-175-55)-245]-(5-6) = [(-230)-245]-(-1) = [-230-245]+(1) = [-475]+1 = -475+1 = \mathbf{-474}$

10. $(48-80)-[(12-2)-(15-37)] = (-32)-[(10)-(-22)] = -32-[10+(22)] = -32-[10+22] = -32-[32] = -32-32$

$= \mathbf{-64}$

Section 1.4 Solutions

1. $5\times 2\times 7\times 4 = \mathbf{280}$

2. $(3\times 5)\times(4\times 2)\times 7 = (15)\times(8)\times 7 = 15\times 8\times 7 = \mathbf{840}$

3. $(20\times 3\times 4)\times(1\times 2\times 6) = (240)\times(12) = 240\times 12 = \mathbf{2880}$

4. $8\times[(1\times 5\times 6)\times(7\times 2)] = 8\times[(30)\times(14)] = 8\times[30\times 14] = 8\times[420] = 8\times 420 = \mathbf{3360}$

5. $[(2\times 7)\times 4]\times[6\times(5\times 3)] = [(14)\times 4]\times[6\times(15)] = [14\times 4]\times[6\times 15] = [56]\times[90] = 56\times 90 = \mathbf{5040}$

6. $(6\times 8)\times[(2\times 3)\times 5]\times 10 = (48)\times[(6)\times 5]\times 10 = 48\times[6\times 5]\times 10 = 48\times[30]\times 10 = 48\times 30\times 10 = \mathbf{14400}$

7. $(2\times 3\times 9)\times[(4\times 5)\times 0]\times 7 = \mathbf{0}$

8. $[(1\times 6\times 3)\times[(7\times 3)\times 5]]\times 3 = [(18)\times[(21)\times 5]]\times 3 = [18\times[21\times 5]]\times 3 = [18\times[105]]\times 3 = [18\times 105]\times 3 = [1890]\times 3$

$= 1890\times 3 = \mathbf{5670}$

9. $[(2\times 3)\times(6\times 5\times 2)]\times[4\times(2\times 4)] = [(6)\times(60)]\times[4\times(8)] = [6\times 60]\times[4\times 8] = [360]\times[32] = 360\times 32 = \mathbf{11520}$

10. $[(2\times 3)\times(6\times 7)\times 2]\times[(4\times 2)\times 5] = [(6)\times(42)\times 2]\times[(8)\times 5] = [6\times 42\times 2]\times[8\times 5] = [504]\times[40] = 504\times 40$

$= \mathbf{20160}$

Section 1.5 Solutions

1. $(16 \div 2) \div 4 = (8) \div 4 = 8 \div 4 = \mathbf{2}$

2. $(125 \div 5) \div (15 \div 5) = (25) \div (3) = 25 \div 3 = \mathbf{8.33}$

3. $[25 \div (8 \div 2)] \div 3 = [25 \div (4)] \div 3 = [25 \div 4] \div 3 = [6.25] \div 3 = 6.25 \div 3 = \mathbf{2.08}$

4. $[(140 \div 10) \div 2] \div 6 = [(14) \div 2] \div 6 = [14 \div 2] \div 6 = [7] \div 6 = 7 \div 6 = \mathbf{1.17}$

5. $[155 \div (15 \div 3)] \div 9 = [155 \div (5)] \div 9 = [155 \div 5] \div 9 = [31] \div 9 = 31 \div 9 = \mathbf{3.44}$

6. $250 \div [(48 \div 2) \div 4] = 250 \div [(24) \div 4] = 250 \div [24 \div 4] = 250 \div [6] = 250 \div 6 = \mathbf{41.67}$

7. $[(28 \div 4) \div (16 \div 3)] \div 8 = [(7) \div (5.33)] \div 8 = [7 \div 5.33] \div 8 = [1.31] \div 8 = 1.31 \div 8 = \mathbf{0.164}$

8. $66 \div [48 \div (14 \div 2)] = 66 \div [48 \div (7)] = 66 \div [48 \div 7] = 66 \div [6.86] = 66 \div 6.86 = \mathbf{9.62}$

9. $(180 \div 2) \div [(88 \div 2) \div 4] = (90) \div [(44) \div 4] = 90 \div [44 \div 4] = 90 \div [11] = 90 \div 11 = \mathbf{8.18}$

10. $[(48 \div 4) \div 2] \div (18 \div 3) = [(12) \div 2] \div (6) = [12 \div 2] \div 6 = [6] \div 6 = 6 \div 6 = \mathbf{1}$

Section 1.6 Solutions

1. $(28 \div 4) \times 3 = (7) \times 3 = 7 \times 3 = \mathbf{21}$

2. $250 + (15 \div 3) = 250 + (5) = 250 + 5 = \mathbf{255}$

3. $28 \div [(23 + 5) \times 8] = 28 \div [(28) \times 8] = 28 \div [28 \times 8] = 28 \div [224] = 28 \div 224 = \mathbf{0.125}$

4. $[(255 - 15) \div 20] + 8 = [(240) \div 20] + 8 = [240 \div 20] + 8 = [12] + 8 = 12 + 8 = \mathbf{20}$

5. $[230 \div (15 \times 2)] + 12 = [230 \div (30)] + 12 = [230 \div 30] + 12 = [7.67] + 12 = 7.67 + 12 = \mathbf{19.67}$

6. $55 \times [(28 + 2) \div 3] = 55 \times [(30) \div 3] = 55 \times [30 \div 3] = 55 \times [10] = 55 \times 10 = \mathbf{550}$

7. $[(55 \div 5) + (18 - 4)] \times 4 = [(11) + (14)] \times 4 = [11 + 14] \times 4 = [25] \times 4 = 25 \times 4 = \mathbf{100}$

8. $35 - [400 \div (16 + 4)] = 35 - [400 \div (20)] = 35 - [400 \div 20] = 35 - [20] = 35 - 20 = \mathbf{15}$

9. $(230 + 5) \div [2 \times (18 + 2)] = (235) \div [2 \times (20)] = 235 \div [2 \times 20] = 235 \div [40] = 235 \div 40 = \mathbf{5.875}$

10. $[(38 \div 4) + 2] \times (15 - 3) = [(9.5) + 2] \times (12) = [9.5 + 2] \times 12 = [11.5] \times 12 = 11.5 \times 12 = \mathbf{138}$

Chapter 2 Solutions:

Section 2.1 Solutions

1. $\frac{60}{150} = \frac{60 \div 30}{150 \div 30} = \mathbf{\frac{2}{5}}$

2. $\frac{8}{18} = \frac{8 \div 2}{18 \div 2} = \mathbf{\frac{4}{9}}$

3. $\frac{355}{15} = \frac{355 \div 5}{15 \div 5} = \frac{71}{3} = \mathbf{23\frac{2}{3}}$

4. $\mathbf{\frac{3}{8}}$ is in its lowest term.

5. $\frac{27}{6} = \frac{27 \div 3}{6 \div 3} = \frac{9}{2} = \mathbf{4\frac{1}{2}}$

6. $\frac{33}{6} = \frac{33 \div 3}{6 \div 3} = \frac{11}{2} = \mathbf{5\frac{1}{2}}$

7. $\frac{250}{1000} = \frac{250 \div 250}{1000 \div 250} = \mathbf{\frac{1}{4}}$

8. $\frac{4}{32} = \frac{4 \div 4}{32 \div 4} = \mathbf{\frac{1}{8}}$

9. $\frac{284}{568} = \frac{284 \div 4}{568 \div 4} = \frac{71}{142} = \frac{71 \div 71}{142 \div 71} = \mathbf{\frac{1}{2}}$

10. $\frac{45}{75} = \frac{45 \div 15}{75 \div 15} = \mathbf{\frac{3}{5}}$

Section 2.1 Appendix Solutions

1. $\frac{83}{4} = \mathbf{20\frac{3}{4}}$

2. $\frac{13}{3} = \mathbf{4\frac{1}{3}}$

3. $-\frac{26}{5} = -\left(\mathbf{5\frac{1}{5}}\right)$

4. $\frac{67}{10} = \mathbf{6\frac{7}{10}}$

5. $\frac{9}{2} = \mathbf{4\frac{1}{2}}$

6. $-\frac{332}{113} = -\left(\mathbf{2\frac{106}{113}}\right)$

7. $\frac{205}{9} = \mathbf{22\frac{7}{9}}$

8. $-\frac{235}{14} = -\left(\mathbf{16\frac{11}{14}}\right)$

9. $\frac{207}{11} = \mathbf{18\frac{9}{11}}$

10. $-\frac{523}{101} = -\left(\mathbf{5\frac{18}{101}}\right)$

Section 2.2 Solutions

1. $\frac{4}{9} + \frac{2}{9} = \frac{4+2}{9} = \frac{\overset{2}{\cancel{6}}}{\underset{3}{\cancel{9}}} = \mathbf{\frac{2}{3}}$

2. $\frac{3}{8} + \frac{2}{5} = \frac{3}{8} + \frac{2}{5} = \frac{(3 \times 5) + (2 \times 8)}{8 \times 5} = \frac{15 + 16}{40} = \mathbf{\frac{31}{40}}$

3. $\frac{3}{8} + \frac{2}{4} + \frac{5}{6} = \left(\frac{3}{8} + \frac{2}{4}\right) + \frac{5}{6} = \left(\frac{(3 \times 4) + (2 \times 8)}{8 \times 4}\right) + \frac{5}{6} = \left(\frac{12+16}{32}\right) + \frac{5}{6} = \left(\frac{28}{32}\right) + \frac{5}{6} = \frac{28}{32} + \frac{5}{6} = \frac{(28 \times 6) + (5 \times 32)}{32 \times 6}$

$= \frac{168 + 160}{192} = \frac{\overset{41}{\cancel{328}}}{\underset{24}{\cancel{192}}} = \frac{41}{24} = \mathbf{1\frac{17}{24}}$

4. $\frac{4}{5} + \frac{2}{5} + \frac{3}{5} = \frac{4+2+3}{5} = \frac{9}{5} = \mathbf{1\frac{4}{5}}$

5. $5+\frac{0}{10}+\frac{6}{1}+\frac{4}{8} = \left(\frac{5}{1}+\frac{0}{10}\right)+\left(\frac{6}{1}+\frac{4}{8}\right) = \left(\frac{5}{1}+0\right)+\left(\frac{(6\times 8)+(4\times 1)}{1\times 8}\right) = \left(\frac{5}{1}\right)+\left(\frac{48+4}{8}\right) = \frac{5}{1}+\left(\frac{52}{8}\right) = \frac{5}{1}+\frac{52}{8}$

$= \frac{(5\times 8)+(52\times 1)}{1\times 8} = \frac{40+52}{8} = \frac{\overset{23}{\cancel{92}}}{\underset{2}{\cancel{8}}} = \frac{23}{2} = \mathbf{11\frac{1}{2}}$

6. $\left(\frac{3}{16}+\frac{1}{8}\right)+\frac{1}{6} = \left(\frac{(3\times 8)+(1\times 16)}{16\times 8}\right)+\frac{1}{6} = \left(\frac{24+16}{128}\right)+\frac{1}{6} = \left(\frac{40}{128}\right)+\frac{1}{6} = \frac{\overset{5}{\cancel{40}}}{\underset{16}{\cancel{128}}}+\frac{1}{6} = \frac{5}{16}+\frac{1}{6} = \frac{(5\times 6)+(1\times 16)}{16\times 6}$

$= \frac{30+16}{96} = \frac{\overset{23}{\cancel{46}}}{\underset{48}{\cancel{96}}} = \mathbf{\frac{23}{48}}$

7. $\left(\frac{4}{5}+\frac{2}{8}\right)+\left(\frac{2}{4}+\frac{1}{4}+\frac{3}{4}\right) = \left(\frac{(4\times 8)+(2\times 5)}{5\times 8}\right)+\left(\frac{2+1+3}{4}\right) = \left(\frac{32+10}{40}\right)+\left(\frac{6}{4}\right) = \left(\frac{42}{40}\right)+\frac{6}{4} = \frac{\overset{21}{\cancel{42}}}{\underset{20}{\cancel{40}}}+\frac{\overset{3}{\cancel{6}}}{\underset{2}{\cancel{4}}} = \frac{21}{20}+\frac{3}{2}$

$= \frac{(21\times 2)+(3\times 20)}{20\times 2} = \frac{42+60}{40} = \frac{\overset{51}{\cancel{102}}}{\underset{20}{\cancel{40}}} = \frac{51}{20} = \mathbf{2\frac{11}{20}}$

8. $\frac{2}{5}+\left(\frac{4}{9}+\frac{2}{9}+\frac{1}{9}\right) = \frac{2}{5}+\left(\frac{4+2+1}{9}\right) = \frac{2}{5}+\left(\frac{7}{9}\right) = \frac{2}{5}+\frac{7}{9} = \frac{(2\times 9)+(7\times 5)}{5\times 9} = \frac{18+35}{45} = \frac{53}{45} = \mathbf{1\frac{8}{45}}$

9. $\frac{2}{5}+\frac{1}{2}+\frac{4}{5}+\frac{2}{3}+12 = \left(\frac{2}{5}+\frac{1}{2}\right)+\left(\frac{4}{5}+\frac{2}{3}\right)+\frac{12}{1} = \left(\frac{(2\times 2)+(1\times 5)}{5\times 2}\right)+\left(\frac{(4\times 3)+(2\times 5)}{5\times 3}\right)+\frac{12}{1}$

$= \left(\frac{4+5}{10}\right)+\left(\frac{12+10}{15}\right)+\frac{12}{1} = \left(\frac{9}{10}\right)+\left(\frac{22}{15}\right)+\frac{12}{1} = \frac{9}{10}+\frac{22}{15}+\frac{12}{1} = \left(\frac{9}{10}+\frac{22}{15}\right)+\frac{12}{1} = \left(\frac{(9\times 15)+(22\times 10)}{10\times 15}\right)+\frac{12}{1}$

$= \left(\frac{135+220}{150}\right)+\frac{12}{1} = \left(\frac{355}{150}\right)+\frac{12}{1} = \frac{\overset{71}{\cancel{355}}}{\underset{30}{\cancel{150}}}+\frac{12}{1} = \frac{71}{30}+\frac{12}{1} = \frac{(71\times 1)+(12\times 30)}{30\times 1} = \frac{71+360}{30} = \frac{431}{30} = \mathbf{14\frac{11}{30}}$

10. $\left[\frac{5}{8}+\left(\frac{3}{5}+\frac{1}{8}\right)\right]+\left(\frac{1}{8}+\frac{3}{8}\right) = \left[\frac{5}{8}+\left(\frac{(3\times 8)+(1\times 5)}{5\times 8}\right)\right]+\left(\frac{1+3}{8}\right) = \left[\frac{5}{8}+\left(\frac{24+5}{40}\right)\right]+\left(\frac{4}{8}\right) = \left[\frac{5}{8}+\left(\frac{29}{40}\right)\right]+\frac{4}{8}$

$= \left[\frac{5}{8}+\frac{29}{40}\right]+\frac{\overset{1}{\cancel{4}}}{\underset{2}{\cancel{8}}} = \left[\frac{(5\times 40)+(29\times 8)}{8\times 40}\right]+\frac{1}{2} = \left[\frac{200+232}{320}\right]+\frac{1}{2} = \left[\frac{432}{320}\right]+\frac{1}{2} = \frac{\overset{27}{\cancel{432}}}{\underset{20}{\cancel{320}}}+\frac{1}{2} = \frac{27}{20}+\frac{1}{2}$

$= \frac{(27\times 2)+(1\times 20)}{20\times 2} = \frac{54+20}{40} = \frac{\overset{37}{\cancel{74}}}{\underset{20}{\cancel{40}}} = \frac{37}{20} = \mathbf{1\frac{17}{20}}$

Section 2.3 Solutions

1. $\frac{3}{5}-\frac{2}{5} = \frac{3-2}{5} = \mathbf{\frac{1}{5}}$

2. $\frac{2}{5}-\frac{3}{4} = \frac{(2\times 4)-(3\times 5)}{5\times 4} = \frac{8-15}{20} = \mathbf{-\frac{7}{20}}$

3. $\frac{12}{15}-\frac{3}{15}-\frac{6}{15}=\frac{12-3-6}{15}=\frac{\overset{1}{\cancel{3}}}{\underset{5}{\cancel{15}}}=\mathbf{\frac{1}{5}}$

4. $\frac{5}{8}-\frac{3}{4}-\frac{1}{3}=\left(\frac{5}{8}-\frac{3}{4}\right)-\frac{1}{3}=\left(\frac{(5\times4)-(3\times8)}{8\times4}\right)-\frac{1}{3}=\left(\frac{20-24}{32}\right)-\frac{1}{3}=\left(\frac{-4}{32}\right)-\frac{1}{3}=\frac{\overset{-1}{\cancel{-4}}}{\underset{8}{\cancel{32}}}-\frac{1}{3}=\frac{-1}{8}-\frac{1}{3}$

$=\frac{(-1\times3)-(1\times8)}{8\times3}=\frac{-3-8}{24}=-\mathbf{\frac{11}{24}}$

5. $\left(\frac{2}{8}-\frac{1}{6}\right)-\frac{2}{5}=\left(\frac{(2\times6)-(1\times8)}{8\times6}\right)-\frac{2}{5}=\left(\frac{12-8}{48}\right)-\frac{2}{5}=\left(\frac{4}{48}\right)-\frac{2}{5}=\frac{\overset{1}{\cancel{4}}}{\underset{12}{\cancel{48}}}-\frac{2}{5}=\frac{1}{12}-\frac{2}{5}=\frac{(1\times5)-(2\times12)}{12\times5}=\frac{5-24}{60}$

$=-\mathbf{\frac{19}{60}}$

6. $28-\left(\frac{1}{8}-\frac{2}{3}\right)=\frac{28}{1}-\left(\frac{(1\times3)-(2\times8)}{8\times3}\right)=\frac{28}{1}-\left(\frac{3-16}{24}\right)=\frac{28}{1}-\left(\frac{-13}{24}\right)=\frac{28}{1}-\frac{-13}{24}=\frac{28}{1}+\frac{13}{24}=\frac{(28\times24)+(13\times1)}{1\times24}$

$=\frac{672+13}{24}=\frac{685}{24}=\mathbf{28\frac{13}{24}}$

7. $\left(\frac{4}{6}-\frac{1}{8}\right)-\left(\frac{4}{5}-\frac{1}{2}\right)=\left(\frac{(4\times8)-(1\times6)}{6\times8}\right)-\left(\frac{(4\times2)-(1\times5)}{5\times2}\right)=\left(\frac{32-6}{48}\right)-\left(\frac{8-5}{10}\right)=\left(\frac{26}{48}\right)-\left(\frac{3}{10}\right)=\frac{\overset{13}{\cancel{26}}}{\underset{24}{\cancel{48}}}-\frac{3}{10}$

$=\frac{13}{24}-\frac{3}{10}=\frac{(13\times10)-(3\times24)}{24\times10}=\frac{130-72}{240}=\frac{\overset{29}{\cancel{58}}}{\underset{120}{\cancel{240}}}=\mathbf{\frac{29}{120}}$

8. $\left(20-\frac{1}{6}\right)-\left(\frac{3}{4}-\frac{1}{2}\right)=\left(\frac{20}{1}-\frac{1}{6}\right)-\left(\frac{(3\times2)-(1\times4)}{4\times2}\right)=\left(\frac{(20\times6)-(1\times1)}{1\times6}\right)-\left(\frac{6-4}{8}\right)=\left(\frac{120-1}{6}\right)-\left(\frac{2}{8}\right)=\left(\frac{119}{6}\right)-\frac{2}{8}$

$=\frac{119}{6}-\frac{\overset{1}{\cancel{2}}}{\underset{4}{\cancel{8}}}=\frac{119}{6}-\frac{1}{4}=\frac{(119\times4)-(1\times6)}{6\times4}=\frac{476-6}{24}=\frac{\overset{235}{\cancel{470}}}{\underset{12}{\cancel{24}}}=\frac{235}{12}=\mathbf{19\frac{7}{12}}$

9. $\left[\frac{18}{5}-\left(\frac{4}{3}-\frac{2}{3}\right)\right]-2=\left[\frac{18}{5}-\left(\frac{4-2}{3}\right)\right]-\frac{2}{1}=\left[\frac{18}{5}-\left(\frac{2}{3}\right)\right]-\frac{2}{1}=\left[\frac{18}{5}-\frac{2}{3}\right]-\frac{2}{1}=\left[\frac{(18\times3)-(2\times5)}{5\times3}\right]-\frac{2}{1}=\left[\frac{54-10}{15}\right]-\frac{2}{1}$

$=\left[\frac{44}{15}\right]-\frac{2}{1}=\frac{44}{15}-\frac{2}{1}=\frac{(44\times1)-(2\times15)}{15\times1}=\frac{44-30}{15}=\mathbf{\frac{14}{15}}$

10. $\left[\left(18-\frac{1}{2}\right)-\left(\frac{16}{2}-2\right)\right]-\frac{1}{5}=\left[\left(\frac{18}{1}-\frac{1}{2}\right)-\left(\frac{16}{2}-\frac{2}{1}\right)\right]-\frac{1}{5}=\left[\left(\frac{(18\times2)-(1\times1)}{1\times2}\right)-\left(\frac{(16\times1)-(2\times2)}{2\times1}\right)\right]-\frac{1}{5}$

$=\left[\left(\frac{36-1}{2}\right)-\left(\frac{16-4}{2}\right)\right]-\frac{1}{5}=\left[\left(\frac{35}{2}\right)-\left(\frac{12}{2}\right)\right]-\frac{1}{5}=\left[\frac{35}{2}-\frac{12}{2}\right]-\frac{1}{5}=\left[\frac{35-12}{2}\right]-\frac{1}{5}=\left[\frac{23}{2}\right]-\frac{1}{5}=\frac{23}{2}-\frac{1}{5}$

$=\frac{(23\times5)-(1\times2)}{2\times5}=\frac{115-2}{10}=\frac{113}{10}=\mathbf{11\frac{3}{10}}$

Section 2.4 Solutions

1. $\frac{4}{8}\times\frac{3}{5} = \frac{\overset{1}{\cancel{4}}\times 3}{\underset{2}{\cancel{8}}\times 5} = \frac{1\times 3}{2\times 5} = \mathbf{\frac{3}{10}}$

2. $\frac{4}{8}\times\frac{5}{6}\times 100 = \frac{4}{8}\times\frac{5}{6}\times\frac{100}{1} = \frac{4\times5\times100}{8\times6\times1} = \frac{\overset{125}{\cancel{2000}}}{\underset{3}{\cancel{48}}} = \frac{125}{3} = \mathbf{41\frac{2}{3}}$

3. $\frac{7}{3}\times\frac{9}{4}\times\frac{6}{3} = \frac{7\times\overset{3}{\cancel{9}}\times\overset{2}{\cancel{6}}}{\underset{1}{\cancel{3}}\times4\times\underset{1}{\cancel{3}}} = \frac{7\times3\times2}{1\times4\times1} = \frac{7\times3\times\overset{1}{\cancel{2}}}{1\times\underset{2}{\cancel{4}}\times1} = \frac{7\times3\times1}{1\times2\times1} = \frac{21}{2} = \mathbf{10\frac{1}{2}}$

4. $34\times\frac15\times\frac{3}{17}\times\frac18\times20 = \frac{34}{1}\times\frac15\times\frac{3}{17}\times\frac18\times\frac{20}{1} = \frac{\overset{2}{\cancel{34}}\times1\times3\times1\times\overset{4}{\cancel{20}}}{1\times\underset{1}{\cancel{5}}\times\underset{1}{\cancel{17}}\times8\times1} = \frac{2\times1\times3\times1\times4}{1\times1\times1\times8\times1} = \frac{\overset{3}{\cancel{24}}}{\underset{1}{\cancel{8}}} = \frac31 = \mathbf{3}$

5. $\left(\frac{2}{55}\times3\right)\times\left(\frac45\times\frac{25}{8}\right) = \left(\frac{2}{55}\times\frac31\right)\times\left(\frac{\overset{1}{\cancel{4}}\times\overset{5}{\cancel{25}}}{\underset{1}{\cancel{5}}\times\underset{2}{\cancel{8}}}\right) = \left(\frac{2\times3}{55\times1}\right)\times\left(\frac{1\times5}{1\times2}\right) = \left(\frac{6}{55}\right)\times\left(\frac52\right) = \frac{6}{55}\times\frac52 = \frac{\overset{3}{\cancel{6}}\times\overset{1}{\cancel{5}}}{\underset{11}{\cancel{55}}\times\underset{1}{\cancel{2}}} = \frac{3\times1}{11\times1} = \mathbf{\frac{3}{11}}$

6. $\left(1000\times\frac15\right)\times\left(\frac{25}{5}\times\frac18\right)\times\frac{0}{100} = \mathbf{0}$

7. $\frac26\times\frac{36}{1}\times\frac{1}{100}\times10\times\frac16 = \frac26\times\frac{36}{1}\times\frac{1}{100}\times\frac{10}{1}\times\frac16 = \frac{\overset{1}{\cancel{2}}\times\overset{6}{\cancel{36}}\times1\times\overset{1}{\cancel{10}}\times1}{\underset{1}{\cancel{6}}\times1\times\underset{10}{\cancel{100}}\times1\times\underset{3}{\cancel{6}}} = \frac{1\times6\times1\times1\times1}{1\times1\times10\times1\times3} = \frac{\overset{1}{\cancel{6}}}{\underset{5}{\cancel{30}}} = \mathbf{\frac15}$

8. $\left(\frac78\times\frac94\right)\times\left(\frac{4}{18}\times\frac{1}{14}\times\frac19\right) = \left(\frac{7\times9}{8\times4}\right)\times\left(\frac{\overset{2}{\cancel{4}}\times1\times1}{\underset{9}{\cancel{18}}\times14\times9}\right) = \left(\frac{63}{32}\right)\times\left(\frac{2\times1\times1}{9\times14\times9}\right) = \frac{63}{32}\times\left(\frac{\overset{1}{\cancel{2}}\times1\times1}{9\times\underset{7}{\cancel{14}}\times9}\right) = \frac{63}{32}\times\left(\frac{1\times1\times1}{9\times7\times9}\right)$

$= \frac{63}{32}\times\left(\frac{1}{567}\right) = \frac{63}{32}\times\frac{1}{567} = \frac{\overset{1}{\cancel{63}}\times1}{32\times\underset{9}{\cancel{567}}} = \frac{1\times1}{32\times9} = \mathbf{\frac{1}{288}}$

9. $\left[\left(18\times\frac28\right)\times\left(\frac15\times\frac{25}{3}\right)\right]\times\frac29 = \left[\left(\frac{18}{1}\times\frac28\right)\times\left(\frac{1\times\overset{5}{\cancel{25}}}{\underset{1}{\cancel{5}}\times3}\right)\right]\times\frac29 = \left[\left(\frac{18\times\overset{1}{\cancel{2}}}{1\times\underset{4}{\cancel{8}}}\right)\times\left(\frac{1\times5}{1\times3}\right)\right]\times\frac29 = \left[\left(\frac{18\times1}{1\times4}\right)\times\left(\frac53\right)\right]\times\frac29$

$= \left[\left(\frac{18}{4}\right)\times\frac53\right]\times\frac29 = \left[\frac{\overset{9}{\cancel{18}}}{\underset{2}{\cancel{4}}}\times\frac53\right]\times\frac29 = \left[\frac92\times\frac53\right]\times\frac29 = \left[\frac{9\times5}{2\times3}\right]\times\frac29 = \left[\frac{45}{6}\right]\times\frac29 = \frac{45}{6}\times\frac29 = \frac{\overset{5}{\cancel{45}}\times\overset{1}{\cancel{2}}}{\underset{3}{\cancel{6}}\times\underset{1}{\cancel{9}}} = \frac{5\times1}{3\times1} = \frac53$

$= \mathbf{1\frac23}$

10. $\left(\frac38\times\frac{4}{49}\times\frac65\right)\times\left(\frac73\times\frac48\right)\times\frac72 = \left(\frac{3\times4\times\overset{3}{\cancel{6}}}{\underset{4}{\cancel{8}}\times49\times5}\right)\times\left(\frac{7\times\overset{1}{\cancel{4}}}{3\times\underset{2}{\cancel{8}}}\right)\times\frac72 = \left(\frac{3\times4\times3}{4\times49\times5}\right)\times\left(\frac{7\times1}{3\times2}\right)\times\frac72 = \left(\frac{36}{980}\right)\times\left(\frac76\right)\times\frac72$

$= \frac{36}{980}\times\frac76\times\frac72 = \frac{\overset{6}{\cancel{36}}\times7\times7}{980\times\underset{1}{\cancel{6}}\times2} = \frac{6\times7\times7}{980\times1\times2} = \frac{\overset{147}{\cancel{294}}}{\underset{980}{\cancel{1960}}} = \mathbf{\frac{147}{980}}$

Section 2.5 Solutions

1. $\frac{8}{10}\div\frac{4}{30} = \frac{8}{10}\times\frac{30}{4} = \frac{\overset{2}{\cancel{8}}\times\overset{3}{\cancel{30}}}{\underset{1}{\cancel{10}}\times\underset{1}{\cancel{4}}} = \frac{2\times 3}{1\times 1} = \frac{6}{1} = \mathbf{6}$

2. $\left(\frac{3}{8}\div\frac{12}{16}\right)\div\frac{4}{8} = \left(\frac{3}{8}\times\frac{16}{12}\right)\div\frac{4}{8} = \left(\frac{\overset{1}{\cancel{3}}\times\overset{2}{\cancel{16}}}{\underset{1}{\cancel{8}}\times\underset{4}{\cancel{12}}}\right)\div\frac{\overset{1}{\cancel{4}}}{\underset{2}{\cancel{8}}} = \left(\frac{1\times 2}{1\times 4}\right)\div\frac{1}{2} = \left(\frac{2}{4}\right)\div\frac{1}{2} = \frac{2}{4}\div\frac{1}{2} = \frac{2}{4}\times\frac{2}{1} = \frac{2\times 2}{4\times 1} = \frac{\overset{1}{\cancel{4}}}{\underset{1}{\cancel{4}}} = \frac{1}{1} = \mathbf{1}$

3. $\left(\frac{4}{16}\div\frac{1}{32}\right)\div 8 = \left(\frac{4}{16}\times\frac{32}{1}\right)\div\frac{8}{1} = \left(\frac{4\times\overset{2}{\cancel{32}}}{\underset{1}{\cancel{16}}\times 1}\right)\div\frac{8}{1} = \left(\frac{4\times 2}{1\times 1}\right)\div\frac{8}{1} = \left(\frac{8}{1}\right)\div\frac{8}{1} = \frac{8}{1}\div\frac{8}{1} = \frac{8}{1}\times\frac{1}{8} = \frac{\overset{1}{\cancel{8}}\times 1}{1\times\underset{1}{\cancel{8}}} = \frac{1\times 1}{1\times 1}$

$= \frac{1}{1} = \mathbf{1}$

4. $12\div\left(\frac{9}{8}\div\frac{27}{16}\right) = \frac{12}{1}\div\left(\frac{9}{8}\times\frac{16}{27}\right) = \frac{12}{1}\div\left(\frac{\overset{1}{\cancel{9}}\times\overset{2}{\cancel{16}}}{\underset{1}{\cancel{8}}\times\underset{3}{\cancel{27}}}\right) = \frac{12}{1}\div\left(\frac{1\times 2}{1\times 3}\right) = \frac{12}{1}\div\left(\frac{2}{3}\right) = \frac{12}{1}\div\frac{2}{3} = \frac{12}{1}\times\frac{3}{2} = \frac{\overset{6}{\cancel{12}}\times 3}{1\times\underset{1}{\cancel{2}}} = \frac{6\times 3}{1\times 1}$

$= \frac{18}{1} = \mathbf{18}$

5. $\left(\frac{2}{20}\div\frac{4}{5}\right)\div 2 = \left(\frac{2}{20}\times\frac{5}{4}\right)\div\frac{2}{1} = \left(\frac{\overset{1}{\cancel{2}}\times\overset{1}{\cancel{5}}}{\underset{4}{\cancel{20}}\times\underset{2}{\cancel{4}}}\right)\div\frac{2}{1} = \left(\frac{1\times 1}{4\times 2}\right)\div\frac{2}{1} = \left(\frac{1}{8}\right)\div\frac{2}{1} = \frac{1}{8}\div\frac{2}{1} = \frac{1}{8}\times\frac{1}{2} = \frac{1\times 1}{8\times 2} = \mathbf{\frac{1}{16}}$

6. $\left(\frac{4}{15}\div\frac{8}{30}\right)\div\left(\frac{1}{5}\div\frac{4}{35}\right) = \left(\frac{4}{15}\times\frac{30}{8}\right)\div\left(\frac{1}{5}\times\frac{35}{4}\right) = \left(\frac{\overset{1}{\cancel{4}}\times\overset{2}{\cancel{30}}}{\underset{1}{\cancel{15}}\times\underset{2}{\cancel{8}}}\right)\div\left(\frac{1\times\overset{7}{\cancel{35}}}{\underset{1}{\cancel{5}}\times 4}\right) = \left(\frac{1\times 2}{1\times 2}\right)\div\left(\frac{1\times 7}{1\times 4}\right) = \left(\frac{2}{2}\right)\div\left(\frac{7}{4}\right) = \frac{2}{2}\div\frac{7}{4}$

$= \frac{2}{2}\times\frac{4}{7} = \frac{2\times\overset{2}{\cancel{4}}}{\underset{1}{\cancel{2}}\times 7} = \frac{2\times 2}{1\times 7} = \mathbf{\frac{4}{7}}$

7. $\left(\frac{2}{5}\div\frac{4}{10}\right)\div\left(\frac{9}{1}\div\frac{18}{4}\right) = \left(\frac{2}{5}\times\frac{10}{4}\right)\div\left(\frac{9}{1}\times\frac{4}{18}\right) = \left(\frac{\overset{1}{\cancel{2}}\times\overset{2}{\cancel{10}}}{\underset{1}{\cancel{5}}\times\underset{2}{\cancel{4}}}\right)\div\left(\frac{\overset{1}{\cancel{9}}\times 4}{1\times\underset{2}{\cancel{18}}}\right) = \left(\frac{1\times 2}{1\times 2}\right)\div\left(\frac{1\times 4}{1\times 2}\right) = \left(\frac{2}{2}\right)\div\left(\frac{4}{2}\right) = \frac{\overset{1}{\cancel{2}}}{\underset{1}{\cancel{2}}}\div\frac{\overset{2}{\cancel{4}}}{\underset{1}{\cancel{2}}}$

$= \frac{1}{1}\div\frac{2}{1} = \frac{1}{1}\times\frac{1}{2} = \frac{1\times 1}{1\times 2} = \mathbf{\frac{1}{2}}$

8. $\left(\frac{4}{5}\div\frac{2}{5}\right)\div\left(\frac{8}{5}\div 4\right) = \left(\frac{4}{5}\times\frac{5}{2}\right)\div\left(\frac{8}{5}\div\frac{4}{1}\right) = \left(\frac{\overset{2}{\cancel{4}}\times\overset{1}{\cancel{5}}}{\underset{1}{\cancel{5}}\times\underset{1}{\cancel{2}}}\right)\div\left(\frac{8}{5}\times\frac{1}{4}\right) = \left(\frac{2\times 1}{1\times 1}\right)\div\left(\frac{\overset{2}{\cancel{8}}\times 1}{5\times\underset{1}{\cancel{4}}}\right) = \left(\frac{2}{1}\right)\div\left(\frac{2\times 1}{5\times 1}\right) = \frac{2}{1}\div\left(\frac{2}{5}\right)$

$= \frac{2}{1}\div\frac{2}{5} = \frac{2}{1}\times\frac{5}{2} = \frac{\overset{1}{\cancel{2}}\times 5}{1\times\underset{1}{\cancel{2}}} = \frac{1\times 5}{1\times 1} = \frac{5}{1} = \mathbf{5}$

9. $\left(\frac{6}{10}\div 1\right)\div\left(\frac{4}{6}\div\frac{1}{3}\right) = \left(\frac{6}{10}\div\frac{1}{1}\right)\div\left(\frac{4}{6}\times\frac{3}{1}\right) = \left(\frac{6}{10}\times\frac{1}{1}\right)\div\left(\frac{4\times\overset{1}{\cancel{3}}}{\underset{2}{\cancel{6}}\times 1}\right) = \left(\frac{\overset{3}{\cancel{6}}\times 1}{\underset{5}{\cancel{10}}\times 1}\right)\div\left(\frac{\overset{2}{\cancel{4}}\times 1}{\underset{1}{\cancel{2}}\times 1}\right) = \left(\frac{3\times 1}{5\times 1}\right)\div\left(\frac{2\times 1}{1\times 1}\right) = \left(\frac{3}{5}\right)\div\left(\frac{2}{1}\right)$

$= \frac{3}{5}\div\frac{2}{1} = \frac{3}{5}\times\frac{1}{2} = \frac{3\times 1}{5\times 2} = \mathbf{\frac{3}{10}}$

10. $\left[\left(\frac{9}{8}\div\frac{18}{16}\right)\div\frac{4}{2}\right]\div\frac{1}{8}=\left[\left(\frac{9}{8}\times\frac{16}{18}\right)\div\frac{4}{2}\right]\div\frac{1}{8}=\left[\left(\frac{\overset{1}{\cancel{9}}\times\overset{2}{\cancel{16}}}{\underset{1}{\cancel{8}}\times\underset{2}{\cancel{18}}}\right)\div\frac{4}{2}\right]\div\frac{1}{8}=\left[\left(\frac{1\times\overset{1}{\cancel{2}}}{1\times\underset{1}{\cancel{2}}}\right)\div\frac{4}{2}\right]\div\frac{1}{8}=\left[\left(\frac{1\times 1}{1\times 1}\right)\div\frac{4}{2}\right]\div\frac{1}{8}$

$=\left[\left(\frac{1}{1}\right)\div\frac{4}{2}\right]\div\frac{1}{8}=\left[\frac{1}{1}\div\frac{4}{2}\right]\div\frac{1}{8}=\left[\frac{1}{1}\times\frac{2}{4}\right]\div\frac{1}{8}=\left[\frac{1\times\overset{1}{\cancel{2}}}{1\times\underset{2}{\cancel{4}}}\right]\div\frac{1}{8}=\left[\frac{1\times 1}{1\times 2}\right]\div\frac{1}{8}=\left[\frac{1}{2}\right]\div\frac{1}{8}=\frac{1}{2}\div\frac{1}{8}=\frac{1}{2}\times\frac{8}{1}=\frac{1\times\overset{4}{\cancel{8}}}{\underset{1}{\cancel{2}}\times 1}$

$=\frac{1\times 4}{1\times 1}=\frac{4}{1}=\mathbf{4}$

Chapter 3 Solutions:

Section 3.1 Case I Practice Problems

1. $4^3 = 4 \cdot 4 \cdot 4 = \mathbf{64}$

2. $(-10)^4 = -10 \cdot -10 \cdot -10 \cdot -10 = \mathbf{+10000}$

3. $0.25^3 = 0.25 \cdot 0.25 \cdot 0.25 = \mathbf{0.0156}$

4. $12^5 = 12 \cdot 12 \cdot 12 \cdot 12 \cdot 12 = \mathbf{248832}$

5. $-(3)^5 = -(3 \cdot 3 \cdot 3 \cdot 3 \cdot 3) = \mathbf{-243}$

6. $489^0 = \mathbf{1}$

7. $100^3 = 100 \cdot 100 \cdot 100 = \mathbf{1000000}$

8. $3.6^3 = 3.6 \cdot 3.6 \cdot 3.6 = \mathbf{46.656}$

9. $6^4 = 6 \cdot 6 \cdot 6 \cdot 6 = \mathbf{1296}$

10. $(-2.4)^4 = (-2.4) \cdot (-2.4) \cdot (-2.4) \cdot (-2.4) = \mathbf{+33.1776}$

Section 3.1 Case II Practice Problems

1. $c^5 = \boldsymbol{c \cdot c \cdot c \cdot c \cdot c}$

2. $w^4 z^2 = \boldsymbol{(w \cdot w \cdot w \cdot w) \cdot (z \cdot z)}$

3. $a^3 \cdot b^6 \cdot c^2 = \boldsymbol{(a \cdot a \cdot a) \cdot (b \cdot b \cdot b \cdot b \cdot b \cdot b) \cdot (c \cdot c)}$

4. $y^3 \cdot (zw)^2 = \boldsymbol{(y \cdot y \cdot y) \cdot (zw) \cdot (zw)}$

5. $(ab)^4 \cdot (xy)^2 = \boldsymbol{(ab) \cdot (ab) \cdot (ab) \cdot (ab) \cdot (xy) \cdot (xy)}$

6. $(xyz)^5 = \boldsymbol{(xyz) \cdot (xyz) \cdot (xyz) \cdot (xyz) \cdot (xyz)}$

7. $a^3 b^2 = \boldsymbol{(a \cdot a \cdot a) \cdot (b \cdot b)}$

8. $z^4 \cdot w^3 \cdot (ab)^2 = \boldsymbol{(z \cdot z \cdot z \cdot z) \cdot (w \cdot w \cdot w) \cdot (ab) \cdot (ab)}$

9. $(xyzw)^4 \cdot b^3 = \boldsymbol{(xyzw) \cdot (xyzw) \cdot (xyzw) \cdot (xyzw) \cdot (b \cdot b \cdot b)}$

10. $a^3 \cdot b^2 \cdot (cd)^4 = \boldsymbol{(a \cdot a \cdot a) \cdot (b \cdot b) \cdot (cd) \cdot (cd) \cdot (cd) \cdot (cd)}$

Section 3.2 Case I Practice Problems

1. $4^{-3} = \frac{1}{4^3} = \frac{1}{4 \cdot 4 \cdot 4} = \mathbf{\frac{1}{64}}$

2. $(-5)^{-4} = \frac{1}{(-5)^4} = \frac{1}{(-5) \cdot (-5) \cdot (-5) \cdot (-5)} = \mathbf{\frac{1}{625}}$

3. $0.25^{-3} = \frac{1}{0.25^3} = \frac{1}{(0.25) \cdot (0.25) \cdot (0.25)} = \mathbf{\frac{1}{0.0156}}$

4. $12^{-5} = \frac{1}{12^5} = \frac{1}{12 \cdot 12 \cdot 12 \cdot 12 \cdot 12} = \mathbf{\frac{1}{248832}}$

5. $-(3)^{-4} = -\frac{1}{3^4} = -\frac{1}{3 \cdot 3 \cdot 3 \cdot 3} = \mathbf{-\frac{1}{81}}$

6. $48^{-2} = \frac{1}{48^2} = \frac{1}{48 \cdot 48} = \mathbf{\frac{1}{2304}}$

7. $(-10)^{-3} = \frac{1}{(-10)^3} = \frac{1}{-10 \cdot -10 \cdot -10} = \frac{1}{-1000} = \mathbf{-\frac{1}{1000}}$

8. $3.2^{-1} = \frac{1}{3.2^1} = \mathbf{\frac{1}{3.2}}$

9. $6^{-3} = \frac{1}{6^3} = \frac{1}{6 \cdot 6 \cdot 6} = \mathbf{\frac{1}{216}}$

10. $(-4.5)^{-2} = \frac{1}{(-4.5)^2} = \frac{1}{(-4.5) \cdot (-4.5)} = \mathbf{\frac{1}{20.25}}$

Section 3.2 Case II Practice Problems

1. $c^{-6} = \frac{1}{c^6} = \frac{\mathbf{1}}{\mathbf{c \cdot c \cdot c \cdot c \cdot c \cdot c}}$

2. $a^{-1}w^{-3} = \frac{1}{a^1 w^3} = \frac{\mathbf{1}}{\mathbf{a \cdot (w \cdot w \cdot w)}}$

3. $a^{-3} \cdot b^{-4} \cdot c^0 = a^{-3} \cdot b^{-4} \cdot 1 = \frac{1}{a^3 \cdot b^4} = \frac{\mathbf{1}}{\mathbf{(a \cdot a \cdot a) \cdot (b \cdot b \cdot b \cdot b)}}$

4. $y^{-3} \cdot (zw)^{-4} = \frac{1}{y^3 \cdot (zw)^4} = \frac{\mathbf{1}}{\mathbf{(y \cdot y \cdot y) \cdot (zw \cdot zw \cdot zw \cdot zw)}}$

5. $(ab)^{-3} \cdot (xy)^{-1} \cdot z^{-2} = \frac{1}{(ab)^3 \cdot (xy)^1 \cdot z^2} = \frac{\mathbf{1}}{\mathbf{(ab \cdot ab \cdot ab) \cdot (xy) \cdot (z \cdot z)}}$

6. $c^{-2} \cdot (xyz)^{-4} = \frac{1}{c^2 \cdot (xyz)^4} = \frac{\mathbf{1}}{\mathbf{(c \cdot c) \cdot (xyz \cdot xyz \cdot xyz \cdot xyz)}}$

7. $a^{-2}b^{-1} = \frac{1}{a^2 b^1} = \frac{\mathbf{1}}{\mathbf{(a \cdot a) \cdot b}}$

8. $z^{-4} \cdot w^{-2} \cdot (abc)^2 = \frac{(abc)^2}{z^4 \cdot w^2} = \frac{\mathbf{(abc) \cdot (abc)}}{\mathbf{(z \cdot z \cdot z \cdot z) \cdot (w \cdot w)}}$

9. $(xyzw)^{-1} \cdot b^{-2} \cdot (ab)^0 = (xyzw)^{-1} \cdot b^{-2} \cdot 1 = (xyzw)^{-1} \cdot b^{-2} = \frac{1}{(xyzw)^1 \cdot b^2} = \frac{\mathbf{1}}{\mathbf{xyzw \cdot (b \cdot b)}}$

10. $(ad)^3 \cdot b^{-2} \cdot (xy)^{-4} = \frac{(ad)^3}{b^2 \cdot (xy)^4} = \frac{\mathbf{ad \cdot ad \cdot ad}}{\mathbf{(b \cdot b) \cdot (xy \cdot xy \cdot xy \cdot xy)}}$

Section 3.3 Case I a Practice Problems

1. $x^2 \cdot x^3 \cdot x = x^2 \cdot x^3 \cdot x^1 = x^{2+3+1} = \mathbf{x^6}$

2. $2 \cdot a^2 \cdot b^0 \cdot a^3 \cdot b^2 = 2 \cdot (a^2 \cdot a^3) \cdot (b^0 \cdot b^2) = 2 \cdot (a^{2+3}) \cdot (b^{0+2}) = \mathbf{2a^5b^2}$

3. $\frac{4}{-6}a^2b^3ab^4b^5 = -\frac{\overset{2}{\cancel{4}}}{\underset{3}{\cancel{6}}}a^2b^3a^1b^4b^5 = -\frac{2}{3}(a^2a^1) \cdot (b^3b^4b^5) = -\frac{2}{3}(a^{2+1}) \cdot (b^{3+4+5}) = -\frac{\mathbf{2}}{\mathbf{3}}\mathbf{a^3b^{12}}$

4. $2^3 \cdot 2^2 \cdot x^{2a} \cdot x^{3a} \cdot x^a = (2^3 \cdot 2^2) \cdot (x^{2a} \cdot x^{3a} \cdot x^a) = (2^{3+2}) \cdot (x^{2a+3a+a}) = 2^5 \cdot x^{6a} = \mathbf{32x^{6a}}$

5. $(x \cdot y^2 \cdot z^3)^0 \cdot w^2z^3zw^4z^2 = 1 \cdot w^2z^3z^1w^4z^2 = (w^2w^4) \cdot (z^3z^1z^2) = (w^{2+4}) \cdot (z^{3+1+2}) = \mathbf{w^6z^6}$

6. $2^0 \cdot 4^2 \cdot 4^2 \cdot 2^2 \cdot 4^1 = (2^0 \cdot 2^2) \cdot (4^2 \cdot 4^2 \cdot 4^1) = (2^{0+2}) \cdot (4^{2+2+1}) = 2^2 \cdot 4^5 = 4 \cdot 1024 = \mathbf{4096}$

7. $(x^2x^3) \cdot \left(\frac{2}{3}xy^2\right) \cdot (-2x^2y) = (x^2x^3) \cdot \left(\frac{2}{3}x^1y^2\right) \cdot \left(-\frac{2}{1}x^2y^1\right) = -\left(\frac{2}{1} \cdot \frac{2}{3}\right) \cdot (x^2x^3x^1x^2) \cdot (y^2y^1)$

$= -\left(\frac{2 \cdot 2}{1 \cdot 3}\right) \cdot (x^{2+3+1+2}) \cdot (y^{2+1}) = -\frac{4}{3}x^8y^3 = -\left(\mathbf{1\frac{1}{3}}\right)\mathbf{x^8y^3}$

8. $\left(p^3\cdot q^2\cdot r\right)\cdot\left(p\cdot q^2\cdot r^3\right) = \left(p^3\cdot q^2\cdot r^1\right)\cdot\left(p^1\cdot q^2\cdot r^3\right) = \left(p^3\cdot p^1\right)\cdot\left(q^2\cdot q^2\right)\cdot\left(r^1\cdot r^3\right) = \left(p^{3+1}\right)\cdot\left(q^{2+2}\right)\cdot\left(r^{1+3}\right) = \mathbf{p^4q^4r^4}$

9. $\frac{-2}{-8}\cdot r^2\cdot s\cdot 2^4\cdot r\cdot s^3 = +\frac{2}{8}\cdot r^2\cdot s^1\cdot 16\cdot r^1\cdot s^3 = \left(\frac{2}{8}\cdot\frac{16}{1}\right)\cdot\left(r^2\cdot r^1\right)\cdot\left(s^1\cdot s^3\right) = \left(\frac{2\cdot \overset{2}{\cancel{16}}}{\underset{1}{\cancel{8}}\cdot 1}\right)\cdot\left(r^{2+1}\right)\cdot\left(s^{1+3}\right) = \frac{4}{1}r^3s^4 = \mathbf{4r^3s^4}$

10. $-2\cdot k^2\cdot l\cdot\frac{3}{-4}\cdot k\cdot l^2\cdot k^3 = -\frac{2}{1}\cdot\frac{3}{-4}\cdot k^2\cdot l^1\cdot k^1\cdot l^2\cdot k^3 = +\left(\frac{2}{1}\cdot\frac{3}{4}\right)\cdot\left(k^2\cdot k^1\cdot k^3\right)\cdot\left(l^1\cdot l^2\right) = \left(\frac{\overset{1}{\cancel{2}}\cdot 3}{1\cdot\underset{2}{\cancel{4}}}\right)\cdot\left(k^{2+1+3}\right)\cdot\left(l^{1+2}\right)$

$= \frac{3}{2}k^6l^3 = \mathbf{1\frac{1}{2}k^6l^3}$

Section 3.3 Case I b Practice Problems

1. $\left(x^2\cdot x^3\right)^2\cdot x = \left(x^{2\times 2}\cdot x^{3\times 2}\right)\cdot x = x^4\cdot x^6\cdot x^1 = x^{4+6+1} = \mathbf{x^{11}}$

2. $2\cdot\left(p^2\cdot q^0\right)^3\cdot p^2q = 2\cdot\left(p^{2\times 3}\cdot q^{0\times 3}\right)\cdot p^2q = 2\cdot p^6\cdot q^0\cdot p^2\cdot q^1 = 2\cdot\left(p^6\cdot p^2\right)\cdot\left(q^0\cdot q^1\right) = 2\cdot\left(p^{6+2}\right)\cdot\left(q^{0+1}\right) = \mathbf{2p^8q}$

3. $\left(a^2\cdot b^3\right)^2\cdot\left(a\cdot b^4\right)^2 = \left(a^{2\times 2}\cdot b^{3\times 2}\right)\cdot\left(a^{1\times 2}\cdot b^{4\times 2}\right) = a^4\cdot b^6\cdot a^2\cdot b^8 = \left(a^4\cdot a^2\right)\cdot\left(b^6\cdot b^8\right) = \left(a^{4+2}\right)\cdot\left(b^{6+8}\right) = \mathbf{a^6b^{14}}$

4. $2^3\cdot 2^2\cdot\left(x^{2a}\cdot x^{3a}\right)^2 = \left(2^3\cdot 2^2\right)\cdot\left(x^{2a\times 2}\cdot x^{3a\times 2}\right) = 2^{3+2}\cdot\left(x^{4a}\cdot x^{6a}\right) = 2^5\cdot\left(x^{4a+6a}\right) = 2^5\cdot x^{10a} = \mathbf{32x^{10a}}$

5. $\left(h\cdot k^2\right)^0\cdot\left(h^2\right)^{3a}\cdot h^a\cdot k = 1\cdot\left(h^{2\times 3a}\right)\cdot h^a\cdot k = \left(h^{6a}\cdot h^a\right)\cdot k = \left(h^{6a+a}\right)\cdot k = \mathbf{h^{7a}k}$

6. $2^0\cdot 3^2\cdot 3^3\cdot 2^2\cdot 2 = \left(2^0\cdot 2^2\cdot 2^1\right)\cdot\left(3^2\cdot 3^3\right) = \left(2^{0+2+1}\right)\cdot\left(3^{2+3}\right) = 2^3\cdot 3^5 = 8\cdot 243 = \mathbf{1944}$

7. $u^2\cdot\left(u^3\cdot v\right)^4\cdot\left(u\cdot v^2\right)^2 = u^2\cdot\left(u^{3\times 4}\cdot v^{1\times 4}\right)\cdot\left(u^{1\times 2}\cdot v^{2\times 2}\right) = u^2\cdot\left(u^{12}\cdot v^4\right)\cdot\left(u^2\cdot v^4\right) = u^2\cdot u^{12}\cdot v^4\cdot u^2\cdot v^4$

$= \left(u^2\cdot u^{12}\cdot u^2\right)\cdot\left(v^4\cdot v^4\right) = \left(u^{2+12+2}\right)\cdot\left(v^{4+4}\right) = \mathbf{u^{16}v^8}$

8. $\left(x^3\cdot y^2\cdot z\right)^2\cdot\left(x\cdot y^2\cdot z^3\right) = \left(x^{3\times 2}\cdot y^{2\times 2}\cdot z^{1\times 2}\right)\cdot\left(x^1\cdot y^2\cdot z^3\right) = \left(x^6\cdot y^4\cdot z^2\right)\cdot\left(x^1\cdot y^2\cdot z^3\right) = x^6\cdot y^4\cdot z^2\cdot x^1\cdot y^2\cdot z^3$

$= \left(x^6\cdot x^1\right)\cdot\left(y^4\cdot y^2\right)\cdot\left(z^2\cdot z^3\right) = \left(x^{6+1}\right)\cdot\left(y^{4+2}\right)\cdot\left(z^{2+3}\right) = \mathbf{x^7y^6z^5}$

9. $5^0\cdot\left(r^2\cdot s\right)^2\cdot\left(3^2\cdot r\cdot s^3\right)^3 = 1\cdot\left(r^{2\times 2}\cdot s^{1\times 2}\right)\cdot\left(3^{2\times 3}\cdot r^{1\times 3}\cdot s^{3\times 3}\right) = \left(r^4\cdot s^2\right)\cdot\left(3^6\cdot r^3\cdot s^9\right) = 3^6\cdot r^4\cdot s^2\cdot r^3\cdot s^9$

$= 3^6\cdot\left(r^4\cdot r^3\right)\cdot\left(s^2\cdot s^9\right) = 3^6\cdot\left(r^{4+3}\right)\cdot\left(s^{2+9}\right) = \mathbf{729r^7s^{11}}$

10. $\left(-3\cdot x^2\right)^3\cdot\left[\left(2\cdot x\cdot y^2\right)^3\cdot x\right]^2 = \left(-3^{1\times 3}\cdot x^{2\times 3}\right)\cdot\left[\left(2^{1\times 3}\cdot x^{1\times 3}\cdot y^{2\times 3}\right)\cdot x\right]^2 = \left(-3^3\cdot x^6\right)\cdot\left[\left(2^3\cdot x^3\cdot y^6\right)\cdot x\right]^2$

$= \left(-27\cdot x^6\right)\cdot\left[2^3\cdot x^3\cdot y^6\cdot x^1\right]^2 = \left(-27\cdot x^6\right)\cdot\left[2^3\cdot\left(x^3\cdot x^1\right)\cdot y^6\right]^2 = \left(-27\cdot x^6\right)\cdot\left[2^3\cdot x^{3+1}\cdot y^6\right]^2 = \left(-27\cdot x^6\right)\cdot\left[2^3\cdot x^4\cdot y^6\right]^2$

$= \left(-27\cdot x^6\right)\cdot\left[2^{3\times 2}\cdot x^{4\times 2}\cdot y^{6\times 2}\right] = \left(-27\cdot x^6\right)\cdot\left(2^6\cdot x^8\cdot y^{12}\right) = -27\cdot x^6\cdot 2^6\cdot x^8\cdot y^{12} = -\left(27\cdot 2^6\right)\cdot\left(x^6\cdot x^8\right)\cdot y^{12}$

$= -\left(27\cdot 64\right)\cdot\left(x^{6+8}\right)\cdot y^{12} = -1728\cdot x^{14}\cdot y^{12} = \mathbf{-1728x^{14}y^{12}}$

Section 3.3 Case II a Practice Problems

1. $\frac{x^5}{x^3} = \frac{x^5x^{-3}}{1} = \frac{x^{5-3}}{1} = \frac{x^2}{1} = \mathbf{x^2}$

2. $\frac{a^2b^3}{a} = \frac{a^2b^3}{a^1} = \frac{\left(a^2a^{-1}\right)b^3}{1} = \frac{a^{2-1}b^3}{1} = \frac{a^1b^3}{1} = \frac{ab^3}{1} = \mathbf{ab^3}$

3. $\frac{a^3b^3c^2}{a^2b^6c} = \frac{a^3b^3c^2}{a^2b^6c^1} = \frac{\left(a^3a^{-2}\right)\cdot\left(c^2c^{-1}\right)}{b^6b^{-3}} = \frac{\left(a^{3-2}\right)\cdot\left(c^{2-1}\right)}{b^{6-3}} = \frac{a^1\cdot c^1}{b^3} = \mathbf{\frac{ac}{b^3}}$

4. $\frac{3^2\cdot\left(rs^2\right)}{(2rs)\cdot r^3} = \frac{9\cdot\left(rs^2\right)}{\left(2r^1s\right)\cdot r^3} = \frac{9rs^2}{2\left(r^3r^1\right)\cdot s} = \frac{9}{2}\frac{rs^2}{r^{3+1}s} = \frac{9}{2}\frac{r^1s^2}{r^4s^1} = \frac{9}{2}\frac{s^2s^{-1}}{r^4r^{-1}} = \frac{9}{2}\frac{s^{2-1}}{r^{4-1}} = \frac{9}{2}\frac{s^1}{r^3} = \mathbf{4\frac{1}{2}\left(\frac{s}{r^3}\right)}$

5. $\frac{2p^2q^3pr^4}{-6p^4q^2r} = -\frac{\overset{1}{\cancel{2}}\;p^2q^3p^1r^4}{\underset{3}{\cancel{6}}\;p^4q^2r^1} = -\frac{1}{3}\frac{\left(q^3q^{-2}\right)\cdot\left(r^4r^{-1}\right)}{p^4p^{-2}p^{-1}} = -\frac{1}{3}\frac{\left(q^{3-2}\right)\cdot\left(r^{4-1}\right)}{p^{4-2-1}} = -\frac{1}{3}\frac{q^1r^3}{p^1} = \mathbf{-\frac{1}{3}\left(\frac{qr^3}{p}\right)}$

6. $\frac{\left(k^2l^3\right)\cdot\left(kl^2m^0\right)}{k^4l^3m^5} = \frac{\left(k^2l^3\right)\cdot\left(kl^2\cdot 1\right)}{k^4l^3m^5} = \frac{\left(k^2l^3\right)\cdot\left(kl^2\right)}{k^4l^3m^5} = \frac{k^2l^3k^1l^2}{k^4l^3m^5} = \frac{l^3l^{-3}l^2}{\left(k^4k^{-2}k^{-1}\right)m^5} = \frac{l^{3-3+2}}{\left(k^{4-2-1}\right)m^5} = \frac{l^2}{k^1m^5} = \mathbf{\frac{l^2}{km^5}}$

7. $\frac{2\cdot a^5\cdot b^2\cdot c}{-a\cdot b\cdot c^3} = -\frac{2\cdot a^5\cdot b^2\cdot c^1}{a^1\cdot b\cdot c^3} = -\frac{2\cdot\left(a^5a^{-1}\right)\cdot\left(b^2b^{-1}\right)}{c^3c^{-1}} = -\frac{2\cdot\left(a^{5-1}\right)\cdot\left(b^{2-1}\right)}{c^3c^{-1}} = -\frac{2\cdot a^4\cdot b^1}{c^2} = \mathbf{-\frac{2a^4b}{c^2}}$

8. $\frac{-c^3d^6e^5}{8c^6d^2e^3e^2}\cdot\frac{-2c}{3d} = \frac{+2c^3d^6e^5}{(8\cdot 3)c^6d^2e^3e^2}\cdot\frac{c^1}{d^1} = \frac{\overset{1}{\cancel{2}}\left(c^3\cdot c^1\right)d^6e^5}{\underset{12}{\cancel{24}}\,c^6\left(d^2\cdot d^1\right)\cdot\left(e^3\cdot e^2\right)} = \frac{\left(c^{3+1}\right)d^6e^5}{12\,c^6\left(d^{2+1}\right)\cdot\left(e^{3+2}\right)} = \frac{c^4d^6e^5}{12\,c^6d^3e^5}$

$= \frac{\left(d^6d^{-3}\right)\cdot\left(e^5e^{-5}\right)}{12\left(c^6c^{-4}\right)} = \frac{\left(d^{6-3}\right)\cdot\left(e^{5-5}\right)}{12\left(c^{6-4}\right)} = \frac{d^3\cdot e^0}{12\,c^2} = \frac{d^3\cdot 1}{12\,c^2} = \frac{d^3}{12\,c^2} = \mathbf{\frac{1}{12}\left(\frac{d^3}{c^2}\right)}$

9. $\frac{-2\left(m^3n^3l^5\right)\cdot 3m^2}{\left(10n^2n\right)\cdot\left(l^2l^4\right)} = -\frac{(2\cdot 3)\left(m^3n^3l^5\right)\cdot m^2}{10\left(n^2n^1\right)\cdot\left(l^2l^4\right)} = -\frac{\overset{3}{\cancel{6}}\left(m^3\cdot m^2\right)n^3l^5}{\underset{5}{\cancel{10}}\left(n^{2+1}\right)\cdot\left(l^{2+4}\right)} = -\frac{3\left(m^{3+2}\right)n^3l^5}{5\,n^3l^6} = -\frac{3\,m^5n^3l^5}{5\,n^3l^6}$

$= -\frac{3\,m^5}{5\left(n^3n^{-3}\right)\cdot\left(l^6l^{-5}\right)} = -\frac{3\,m^5}{5\left(n^{3-3}\right)\cdot\left(l^{6-5}\right)} = -\frac{3}{5}\frac{m^5}{n^0\cdot l^1} = -\frac{3}{5}\frac{m^5}{1\cdot l} = \mathbf{-\frac{3}{5}\left(\frac{m^5}{l}\right)}$

10. $\frac{-5\left(x^2y^5z^4\right)\cdot(xyz)}{-2x^4y^2z^3} = +\frac{5}{2}\frac{\left(x^2y^5z^4\right)\cdot\left(x^1y^1z^1\right)}{x^4y^2z^3} = \frac{5}{2}\frac{\left(x^2\cdot x^1\right)\cdot\left(y^5\cdot y^1\right)\cdot\left(z^4\cdot z^1\right)}{x^4y^2z^3} = \frac{5}{2}\frac{\left(x^{2+1}\right)\cdot\left(y^{5+1}\right)\cdot\left(z^{4+1}\right)}{x^4y^2z^3}$

$= \frac{5}{2}\frac{x^3y^6z^5}{x^4y^2z^3} = \frac{5}{2}\frac{\left(y^6y^{-2}\right)\cdot\left(z^5z^{-3}\right)}{x^4x^{-3}} = \frac{5}{2}\frac{\left(y^{6-2}\right)\cdot\left(z^{5-3}\right)}{x^{4-3}} = \frac{5}{2}\frac{y^4\cdot z^2}{x^1} = \frac{5}{2}\left(\frac{y^4z^2}{x}\right) = \mathbf{2\frac{1}{2}\left(\frac{y^4z^2}{x}\right)}$

Section 3.3 Case II b Practice Problems

1. $\left(\frac{x^5}{x^3}\right)^3 = \frac{x^{5\times3}}{x^{3\times3}} = \frac{x^{15}}{x^9} = \frac{x^{15}x^{-9}}{1} = \frac{x^{15-9}}{1} = \frac{x^6}{1} = \boldsymbol{x^6}$

2. $\frac{\left(a^2\cdot b^3\right)^2}{a^3} = \frac{a^{2\times2}\cdot b^{3\times2}}{a^3} = \frac{a^4\cdot b^6}{a^3} = \frac{\left(a^4\cdot a^{-3}\right)b^6}{1} = \frac{\left(a^{4-3}\right)b^6}{1} = \frac{ab^6}{1} = \boldsymbol{ab^6}$

3. $\left(\frac{a^3\cdot b^6\cdot c^2}{a^2\cdot b^3}\right)^0 = \boldsymbol{1}$

4. $\frac{y^3\cdot(z\cdot w)^2}{-y^2\cdot w^3\cdot z^4} = -\frac{y^3\cdot\left(z^{1\times2}\cdot w^{1\times2}\right)}{y^2\cdot w^3\cdot z^4} = -\frac{y^3\cdot z^2\cdot w^2}{y^2\cdot w^3\cdot z^4} = -\frac{y^3y^{-2}}{\left(w^3w^2\right)\cdot\left(z^4z^{-2}\right)} = -\frac{y^{3-2}}{\left(w^{3-2}\right)\cdot\left(z^{4-2}\right)} = -\frac{y^1}{w^1\cdot z^2} = -\boldsymbol{\frac{y}{wz^2}}$

5. $\frac{(a\cdot b)^4\cdot(x\cdot y)^2}{a^3\cdot y^2\cdot x} = \frac{\left(a^{1\times4}\cdot b^{1\times4}\right)\cdot\left(x^{1\times2}\cdot y^{1\times2}\right)}{a^3\cdot y^2\cdot x} = \frac{a^4\cdot b^4\cdot x^2\cdot y^2}{a^3\cdot y^2\cdot x^1} = \frac{\left(a^4a^{-3}\right)\cdot b^4\cdot\left(x^2\cdot x^{-1}\right)}{y^2y^{-2}} = \frac{\left(a^{4-3}\right)\cdot b^4\cdot\left(x^{2-1}\right)}{y^{2-2}}$

$= \frac{a^1\cdot b^4\cdot x^1}{y^0} = \frac{ab^4x}{1} = \boldsymbol{ab^4x}$

6. $\frac{-3(x\cdot y\cdot z)^5\cdot x^2}{6(x\cdot y)^3\cdot z^7} = -\frac{\frac{1}{3}}{\frac{6}{2}}\frac{\left(x^{1\times5}\cdot y^{1\times5}\cdot z^{1\times5}\right)\cdot x^2}{\left(x^{1\times3}\cdot y^{1\times3}\right)\cdot z^7} = -\frac{1}{2}\frac{x^5\cdot y^5\cdot z^5\cdot x^2}{x^3\cdot y^3\cdot z^7} = -\frac{1}{2}\frac{\left(x^5\cdot x^2\cdot x^{-3}\right)\cdot\left(y^5\cdot y^{-3}\right)}{z^7\cdot z^{-5}}$

$= -\frac{1}{2}\frac{\left(x^{5+2-3}\right)\cdot\left(y^{5-3}\right)}{z^{7-5}} = -\frac{1}{2}\frac{x^4\cdot y^2}{z^2} = \boldsymbol{-\frac{1}{2}\left(\frac{x^4y^2}{z^2}\right)}$

7. $\frac{2^2\cdot a^3\cdot b^2\cdot c}{-2^3\cdot a\cdot\left(b\cdot c^2\right)^2} = -\frac{2^2\cdot a^3\cdot b^2\cdot c}{2^3\cdot a\cdot\left(b^{1\times2}\cdot c^{2\times2}\right)} = -\frac{2^2\cdot a^3\cdot b^2\cdot c^1}{2^3\cdot a^1\cdot b^2\cdot c^4} = -\frac{\left(a^3\cdot a^{-1}\right)\cdot\left(b^2\cdot b^{-2}\right)}{\left(2^3\cdot 2^{-2}\right)\cdot\left(c^4\cdot c^{-1}\right)} = -\frac{\left(a^{3-1}\right)\cdot\left(b^{2-2}\right)}{\left(2^{3-2}\right)\cdot\left(c^{4-1}\right)} = -\frac{a^2\cdot b^0}{2^1\cdot c^3}$

$= -\frac{a^2\cdot 1}{2\cdot c^3} = \boldsymbol{-\frac{a^2}{2c^3}}$

8. $\left(\frac{z^4\cdot w^3\cdot(a\cdot b)^2}{a^2\cdot w^0\cdot z}\right)^2 = \left(\frac{z^4\cdot w^3\cdot\left(a^{1\times2}\cdot b^{1\times2}\right)}{a^2\cdot 1\cdot z}\right)^2 = \left(\frac{z^4\cdot w^3\cdot a^2\cdot b^2}{a^2\cdot z^1}\right)^2 = \frac{z^{4\times2}\cdot w^{3\times2}\cdot a^{2\times2}\cdot b^{2\times2}}{a^{2\times2}\cdot z^{1\times2}} = \frac{z^8\cdot w^6\cdot a^4\cdot b^4}{a^4\cdot z^2}$

$= \frac{\left(z^8z^{-2}\right)\cdot w^6\cdot\left(a^4a^{-4}\right)\cdot b^4}{1} = \frac{\left(z^{8-2}\right)\cdot w^6\cdot\left(a^{4-4}\right)\cdot b^4}{1} = \frac{z^6\cdot w^6\cdot a^0\cdot b^4}{1} = \frac{z^6\cdot w^6\cdot 1\cdot b^4}{1} = \frac{z^6\cdot w^6\cdot b^4}{1} = \boldsymbol{b^4z^6w^6}$

9. $\frac{3^3\cdot 3^2\cdot(y\cdot z)^4\cdot b^3}{3^6\cdot b^2\cdot y^3} = \frac{\left(3^3\cdot 3^2\right)\cdot\left(y^{1\times4}\cdot z^{1\times4}\right)\cdot b^3}{3^6\cdot b^2\cdot y^3} = \frac{3^{3+2}\cdot\left(y^4\cdot z^4\right)\cdot b^3}{3^6\cdot b^2\cdot y^3} = \frac{3^5\cdot y^4\cdot z^4\cdot b^3}{3^6\cdot b^2\cdot y^3} = \frac{\left(y^4\cdot y^{-3}\right)\cdot z^4\cdot\left(b^3\cdot b^{-2}\right)}{3^6\cdot 3^{-5}}$

$= \frac{\left(y^{4-3}\right)\cdot z^4\cdot\left(b^{3-2}\right)}{3^{6-5}} = \frac{y^1\cdot z^4\cdot b^1}{3^1} = \boldsymbol{\frac{byz^4}{3}}$

10. $\left(\dfrac{a\cdot b^2\cdot(c\cdot d)^4}{c^5\cdot d^6}\right)^3\cdot a^2b^2 = \left(\dfrac{a\cdot b^2\cdot\left(c^{1\times4}\cdot d^{1\times4}\right)}{c^5\cdot d^6}\right)^3\cdot a^2b^2 = \left(\dfrac{a\cdot b^2\cdot c^4\cdot d^4}{c^5\cdot d^6}\right)^3\cdot a^2b^2 = \left(\dfrac{a\cdot b^2}{\left(c^5\cdot c^{-4}\right)\cdot\left(d^6\cdot d^{-4}\right)}\right)^3\cdot a^2b^2$

$= \left(\dfrac{a\cdot b^2}{c^{5-4}\cdot d^{6-4}}\right)^3\cdot a^2b^2 = \left(\dfrac{a\cdot b^2}{c^1\cdot d^2}\right)^3\cdot a^2b^2 = \left(\dfrac{a^{1\times3}\cdot b^{2\times3}}{c^{1\times3}\cdot d^{2\times3}}\right)\cdot a^2b^2 = \left(\dfrac{a^3\cdot b^6}{c^3\cdot d^6}\right)\cdot\dfrac{a^2b^2}{1} = \dfrac{\left(a^3\cdot a^2\right)\cdot\left(b^6\cdot b^2\right)}{c^3\cdot d^6\cdot 1}$

$= \dfrac{\left(a^{3+2}\right)\cdot\left(b^{6+2}\right)}{c^3\cdot d^6} = \dfrac{a^5\cdot b^8}{c^3\cdot d^6} = \mathbf{\dfrac{a^5b^8}{c^3d^6}}$

Section 3.3 Case III a Practice Problems

1. $x^2+4xy-2x^2-2xy+z^3 = \left(x^2-2x^2\right)+(4xy-2xy)+z^3 = (1-2)x^2+(4-2)xy+z^3 = \mathbf{-x^2+2xy+z^3}$

2. $\left(a^3+2a^2+4^3\right)-\left(4a^3+20\right) = \left(a^3+2a^2+4^3\right)+\left(-4a^3-20\right) = a^3+2a^2+64-4a^3-20 = \left(a^3-4a^3\right)+2a^2+(64-20)$

$= (1-4)a^3+2a^2+44 = \mathbf{-3a^3+2a^2+44}$

3. $3x^4+2x^2+2x^4-\left(x^4-2x^2+3\right) = 3x^4+2x^2+2x^4+\left(-x^4+2x^2-3\right) = 3x^4+2x^2+2x^4-x^4+2x^2-3$

$= \left(3x^4+2x^4-x^4\right)+\left(2x^2+2x^2\right)-3 = (3+2-1)x^4+(2+2)x^2-3 = \mathbf{4x^4+4x^2-3}$

4. $-\left(-2l^3a^3+2l^2a^2-5^3\right)-\left(4l^3a^3-20\right) = \left(+2l^3a^3-2l^2a^2+5^3\right)+\left(-4l^3a^3+20\right) = 2l^3a^3-2l^2a^2+125-4l^3a^3+20$

$= \left(2l^3a^3-4l^3a^3\right)-2l^2a^2+(125+20) = (2-4)l^3a^3-2l^2a^2+145 = \mathbf{-2l^3a^3-2l^2a^2+145}$

5. $\left(m^{3n}-4m^{2n}\right)-\left(2m^{3n}+3m^{2n}\right)+5m = \left(m^{3n}-4m^{2n}\right)+\left(-2m^{3n}-3m^{2n}\right)+5m = m^{3n}-4m^{2n}-2m^{3n}-3m^{2n}+5m$

$= \left(m^{3n}-2m^{3n}\right)+\left(-4m^{2n}-3m^{2n}\right)+5m = (1-2)m^{3n}+(-4-3)m^{2n}+5m = \mathbf{-m^{3n}-7m^{2n}+5m}$

6. $\left(-7z^3+3z-5\right)-\left(-3z^3+z-4\right)+5z+20 = \left(-7z^3+3z-5\right)+\left(3z^3-z+4\right)+5z+20 = -7z^3+3z-5+3z^3-z+4+5z+20$

$= \left(-7z^3+3z^3\right)+(3z-z+5z)+(-5+4+20) = (-7+3)z^3+(3-1+5)z+19 = \mathbf{-4z^3+7z+19}$

7. $\left(a^3\right)^2+\left(a^2\cdot b^2\right)^2-5a^6+3a^4b^4+2a^6 = \left(a^{3\times2}\right)+\left(a^{2\times2}\cdot b^{2\times2}\right)-5a^6+3a^4b^4+2a^6 = a^6+a^4b^4-5a^6+3a^4b^4+2a^6$

$= \left(a^6+2a^6-5a^6\right)+\left(a^4b^4+3a^4b^4\right) = (1+2-5)a^6+(1+3)a^4b^4 = \mathbf{-2a^6+4a^4b^4}$

8. $\left(k^5+10k^2+5\right)+\left(-2k^5-5k^2+5k\right)-4k^3-k = k^5+10k^2+5-2k^5-5k^2+5k-4k^3-k$

$= \left(k^5-2k^5\right)-4k^3+\left(10k^2-5k^2\right)+(5k-k)+5 = (1-2)k^5-4k^3+(10-5)k^2+(5-1)k+5 = \mathbf{-k^5-4k^3+5k^2+4k+5}$

9. $\left(3x^2+xy-x^2+3x^3\right)-\left(2x^3-y^3-4y^3+x^3\right) = \left(3x^2+xy-x^2+3x^3\right)+\left(-2x^3+y^3+4y^3-x^3\right)$

$= 3x^2+xy-x^2+3x^3-2x^3+y^3+4y^3-x^3 = \left(3x^3-2x^3-x^3\right)+\left(3x^2-x^2\right)+\left(y^3+4y^3\right)+xy$

$= (3-2-1)x^3+(3-1)x^2+(1+4)y^3+xy = 0x^3+2x^2+5y^3+xy = \mathbf{5y^3+2x^2+xy}$

10. $\left(xy^2+20x^2+5x\right)-\left(3xy^2+20x\right)+2^4 = \left(xy^2+20x^2+5x\right)+\left(-3xy^2-20x\right)+16 = xy^2+20x^2+5x-3xy^2-20x+16$
$= \left(xy^2-3xy^2\right)+20x^2+(5x-20x)+16 = (1-3)xy^2+20x^2+(5-20)x+16 = \mathbf{-2xy^2+20x^2-15x+16}$

Section 3.3 Case III b Practice Problems

1. $\frac{2-2^3}{7}+\frac{4^2}{3} = \frac{2-8}{7}+\frac{16}{3} = \frac{-6}{7}+\frac{16}{3} = \frac{-(6\cdot3)+(16\cdot7)}{7\cdot3} = \frac{-18+112}{21} = \frac{94}{21} = \mathbf{4\frac{10}{21}}$

2. $\frac{b^2+3b-4b^2}{c}-\frac{4b+b^2}{c} = \frac{\left(b^2+3b-4b^2\right)-\left(4b+b^2\right)}{c} = \frac{\left(b^2+3b-4b^2\right)+\left(-4b-b^2\right)}{c} = \frac{b^2+3b-4b^2-4b-b^2}{c}$
$= \frac{\left(b^2-4b^2-b^2\right)+(3b-4b)}{c} = \frac{(1-4-1)b^2+(3-4)b}{c} = \frac{-4b^2-b}{c} = \mathbf{\frac{-b(4b+1)}{c}}$

3. $\frac{2a^3-3b^3}{a^3+b^3}-2^2 = \frac{2a^3-3b^3}{a^3+b^3}-4 = \frac{2a^3-3b^3}{a^3+b^3}-\frac{4}{1} = \frac{\left[\left(2a^3-3b^3\right)\cdot1\right]-\left[4\cdot\left(a^3+b^3\right)\right]}{\left(a^3+b^3\right)\cdot1} = \frac{2a^3-3b^3-4a^3-4b^3}{a^3+b^3}$
$= \frac{\left(2a^3-4a^3\right)+\left(-3b^3-4b^3\right)}{a^3+b^3} = \frac{(2-4)a^3+(-3-4)b^3}{a^3+b^3} = \mathbf{\frac{-2a^3-7b^3}{a^3+b^3}}$

4. $\frac{3x^2+3x}{5}-\frac{x^2+x}{5} = \frac{\left(3x^2+3x\right)-\left(x^2+x\right)}{5} = \frac{\left(3x^2+3x\right)+\left(-x^2-x\right)}{5} = \frac{3x^2+3x-x^2-x}{5} = \frac{\left(3x^2-x^2\right)+(3x-x)}{5}$
$= \frac{(3-1)x^2+(3-1)x}{5} = \frac{2x^2+2x}{5} = \mathbf{\frac{2x(x+1)}{5}}$

5. $\frac{y^2}{y-y^3}+\frac{y}{2} = \frac{\left(2\cdot y^2\right)+\left[y\cdot\left(y-y^3\right)\right]}{2\cdot\left(y-y^3\right)} = \frac{2y^2+y^2-y^4}{2\left(y-y^3\right)} = \frac{\left(2y^2+y^2\right)-y^4}{2y\left(1-y^2\right)} = \frac{(2+1)y^2-y^4}{2y\left(1-y^2\right)} = \frac{3y^2-y^4}{2y\left(1-y^2\right)}$
$= \frac{y^2\left(3-y^2\right)}{2y\left(1-y^2\right)} = \frac{y^2\left(3-y^2\right)}{2y^1\left(1-y^2\right)} = \frac{\left(y^2y^{-1}\right)\cdot\left(3-y^2\right)}{2\left(1-y^2\right)} = \frac{y^{2-1}\cdot\left(3-y^2\right)}{2\left(1-y^2\right)} = \frac{y\left(3-y^2\right)}{2\left(1-y^2\right)} = \mathbf{\frac{y(3-y^2)}{2(1-y)(1+y)}}$

6. $\frac{b+2}{1+b}+\frac{1}{b} = \frac{b\cdot(b+2)+[1\cdot(1+b)]}{b\cdot(1+b)} = \frac{b^2+2b+1+b}{b(1+b)} = \frac{b^2+(2b+b)+1}{b(1+b)} = \frac{b^2+(2+1)b+1}{b(1+b)} = \mathbf{\frac{b^2+3b+1}{b(1+b)}}$

7. $\frac{x+y^2}{x-y^2}+\frac{2^2}{3} = \frac{x+y^2}{x-y^2}+\frac{4}{3} = \frac{3\cdot\left(x+y^2\right)+4\cdot\left(x-y^2\right)}{3\cdot\left(x-y^2\right)} = \frac{3x+3y^2+4x-4y^2}{3\left(x-y^2\right)} = \frac{(3x+4x)+\left(-4y^2+3y^2\right)}{3\left(x-y^2\right)}$
$= \frac{(3+4)x+(-4+3)y^2}{3\left(x-y^2\right)} = \mathbf{\frac{7x-y^2}{3(x-y^2)}}$

8. $\frac{4l^2+5}{lm}-\frac{5l}{m} = \frac{\left[\left(4l^2+5\right)\cdot m\right]-(5l\cdot lm)}{lm\cdot m} = \frac{4l^2m+5m-5l^2m}{lm^2} = \frac{\left(4l^2m-5l^2m\right)+5m}{lm^2} = \frac{(4-5)l^2m+5m}{lm^2}$
$= \frac{-l^2m+5m}{lm^2} = \frac{m\left(-l^2+5\right)}{lm^2} = \frac{-l^2+5}{l\left(m^2m^{-1}\right)} = \frac{-l^2+5}{lm^{2-1}} = \mathbf{\frac{-l^2+5}{lm}}$

9. $\frac{4xy^2}{2x-3}+y^2 = \frac{4xy^2}{2x-3}+\frac{y^2}{1} = \frac{\left[\left(4xy^2\right)\cdot 1\right]+\left[y^2\cdot(2x-3)\right]}{(2x-3)\cdot 1} = \frac{4xy^2+2xy^2-3y^2}{2x-3} = \frac{\left(4xy^2+2xy^2\right)-3y^2}{2x-3}$

$= \frac{(4+2)xy^2-3y^2}{2x-3} = \frac{6xy^2-3y^2}{2x-3} = \mathbf{\frac{3y^2(2x-1)}{2x-3}}$

10. $\frac{2a^3}{a+1}-\frac{a^2}{3} = \frac{\left(2a^3\cdot 3\right)-\left[a^2\cdot(a+1)\right]}{3\cdot(a+1)} = \frac{6a^3-\left(a^3+a^2\right)}{3(a+1)} = \frac{6a^3-a^3-a^2}{3(a+1)} = \frac{\left(6a^3-a^3\right)-a^2}{3(a+1)} = \frac{(6-1)a^3-a^2}{3(a+1)}$

$= \frac{5a^3-a^2}{3(a+1)} = \mathbf{\frac{a^2(5a-1)}{3(a+1)}}$

Section 3.4 Case I a Practice Problems

1. $\left(3^{-3}\cdot 2^{-1}\right)\cdot\left(2^{-3}\cdot 3^{-2}\cdot 2\right) = 3^{-3}\cdot 2^{-1}\cdot 2^{-3}\cdot 3^{-2}\cdot 2^{1} = \left(3^{-3}\cdot 3^{-2}\right)\cdot\left(2^{-1}\cdot 2^{-3}\cdot 2^{1}\right) = \left(3^{-3-2}\right)\cdot\left(2^{-1-3+1}\right) = 3^{-5}\cdot 2^{-3}$

$= \frac{1}{3^5\cdot 2^3} = \frac{1}{243\cdot 8} = \mathbf{\frac{1}{1944}}$

2. $a^{-6}\cdot b^{-4}\cdot a^{-1}\cdot b^{-2}\cdot a^{0} = \left(a^{-6}a^{-1}a^{0}\right)\cdot\left(b^{-2}b^{-4}\right) = \left(a^{-6-1+0}\right)\cdot\left(b^{-2-4}\right) = a^{-7}b^{-6} = \mathbf{\frac{1}{a^7b^6}}$

3. $\left(a^{-2}\cdot b^{-3}\right)^2\cdot\left(a\cdot b^{-2}\right) = \left(a^{-2\times 2}\cdot b^{-3\times 2}\right)\cdot\left(a\cdot b^{-2}\right) = \left(a^{-4}\cdot b^{-6}\right)\cdot\left(a\cdot b^{-2}\right) = a^{-4}\cdot b^{-6}\cdot a\cdot b^{-2} = \left(a^{-4}\cdot a^{1}\right)\cdot\left(b^{-6}\cdot b^{-2}\right)$

$= \left(a^{-4+1}\right)\cdot\left(b^{-6-2}\right) = a^{-3}\cdot b^{-8} = \mathbf{\frac{1}{a^3b^8}}$

4. $(-2)^{-4}\left(r^{-2}s^2t\right)\cdot\left(r^3st^{-2}s^{-1}\right) = \frac{1}{(-2)^4}r^{-2}s^2t^1r^3s^1t^{-2}s^{-1} = \frac{1}{(-2\cdot -2\cdot -2\cdot -2)}\left(r^{-2}r^3\right)\cdot\left(s^2s^1s^{-1}\right)\cdot\left(t^1t^{-2}\right)$

$= \frac{1}{+16}\left(r^{-2+3}\right)\cdot\left(s^{2+1-1}\right)\cdot\left(t^{1-2}\right) = \frac{1}{16}r^1\cdot s^2\cdot t^{-1} = \frac{1}{16}\cdot\frac{rs^2}{t^1} = \mathbf{\frac{1}{16}\left(\frac{rs^2}{t}\right)}$

5. $\left(\frac{4}{5}\right)^{-4}2^2v^{-5}2^{-4}v^3v^{-2} = \left(\frac{4^{1\times -4}}{5^{1\times -4}}\right)\cdot\left(2^22^{-4}\right)\cdot\left(v^{-5}v^3v^{-2}\right) = \left(\frac{4^{-4}}{5^{-4}}\right)\cdot\left(2^{2-4}\right)\cdot\left(v^{-5+3-2}\right) = \left(\frac{5^4}{4^4}\right)\cdot 2^{-2}\cdot v^{-4}$

$= \left(\frac{625}{256}\right)\cdot\frac{1}{2^2v^4} = \left(\frac{625}{256}\right)\cdot\frac{1}{4\cdot v^4} = \left(\frac{625}{256\cdot 4}\right)\cdot\frac{1}{v^4} = \left(\frac{625}{1024}\right)\cdot\frac{1}{v^4} = \mathbf{\frac{625}{1024\,v^4}}$

6. $2^{-1}\cdot 3^2\cdot 3^{-5}\cdot 2^2\cdot 2^0 = \left(2^{-1}\cdot 2^2\cdot 2^0\right)\cdot\left(3^2\cdot 3^{-5}\right) = \left(2^{-1+2+0}\right)\cdot\left(3^{2-5}\right) = 2^1\cdot 3^{-3} = 2\cdot 3^{-3} = \frac{2}{3^3} = \mathbf{\frac{2}{27}}$

7. $(-3)^{-3}\left(k^{-2}l^{-4}\right)\cdot\left(k^{-3}l^2\right) = (-3)^{-3}\left(k^{-2}l^{-4}k^{-3}l^2\right) = (-3)^{-3}\left(k^{-2}k^{-3}\right)\cdot\left(l^{-4}l^2\right) = (-3)^{-3}\left(k^{-2-3}\right)\cdot\left(l^{-4+2}\right) = (-3)^{-3}k^{-5}\cdot l^{-2}$

$= \frac{1}{(-3)^3k^5l^2} = \frac{1}{(-3\cdot -3\cdot -3)k^5l^2} = \frac{1}{-27k^5l^2} = \mathbf{-\frac{1}{27k^5l^2}}$

8. $-\left(2^{-2}\right)\cdot\left(h^{-3}m^{-3}n^{-4}hm^{-2}n\right) = -\left(\frac{1}{2^2}\right)\cdot\left(h^{-3}m^{-3}n^{-4}h^1m^{-2}n^1\right) = -\left(\frac{1}{4}\right)\cdot\left(h^{-3}h^1\right)\cdot\left(m^{-3}m^{-2}\right)\cdot\left(n^{-4}n^1\right)$

$= -\frac{1}{4}\cdot\left(h^{-3+1}\right)\cdot\left(m^{-3-2}\right)\cdot\left(n^{-4+1}\right) = -\frac{1}{4}\cdot h^{-2}m^{-5}n^{-3} = -\frac{1}{4}\cdot\frac{1}{h^2m^5n^3} = \mathbf{-\frac{1}{4}\left(\frac{1}{h^2m^5n^3}\right)}$

9. $\left(-\frac{1}{5}a^{-3}b^{-5}\right)\cdot\left(-ba^{-2}b^{3}\right) = \left(+\frac{1}{5}a^{-3}b^{-5}\right)\cdot\left(b^{1}a^{-2}b^{3}\right) = \frac{1}{5}a^{-3}b^{-5}b^{1}a^{-2}b^{3} = \frac{1}{5}\left(a^{-3}a^{-2}\right)\cdot\left(b^{3}b^{-5}b^{1}\right)$

$= \frac{1}{5}\left(a^{-3-2}\right)\cdot\left(b^{3-5+1}\right) = \frac{1}{5}a^{-5}b^{-1} = \frac{1}{5}\cdot\frac{1}{a^{5}b^{1}} = \mathbf{\frac{1}{5}\left(\frac{1}{a^{5}b}\right)}$

10. $\frac{-2}{-5}\cdot m^{3}\cdot m^{-5}\cdot r^{-2}\cdot m\cdot r^{3} = +\frac{2}{5}\cdot\left(m^{3}\cdot m^{-5}\cdot m^{1}\right)\cdot\left(r^{3}\cdot r^{-2}\right) = \frac{2}{5}\cdot\left(m^{3-5+1}\right)\cdot\left(r^{3-2}\right) = \frac{2}{5}m^{-1}r^{1} = \frac{2}{5}\cdot\frac{r}{m^{1}} = \mathbf{\frac{2}{5}\left(\frac{r}{m}\right)}$

Section 3.4 Case I b Practice Problems

1. $\left(a^{2}\cdot a^{3}\right)^{-2} = a^{2\times-2}\cdot a^{3\times-2} = a^{-4}\cdot a^{-6} = a^{-4-6} = a^{-10} = \mathbf{\frac{1}{a^{10}}}$

2. $2\cdot\left(a^{2}\cdot y^{-1}\right)^{-2}\cdot y^{-3} = 2\cdot\left(a^{2\times-2}\cdot y^{-1\times-2}\right)\cdot y^{-3} = 2\left(a^{-4}y^{2}\right)y^{-3} = 2a^{-4}\left(y^{2}y^{-3}\right) = 2a^{-4}\left(y^{2-3}\right) = 2a^{-4}y^{-1} = \mathbf{\frac{2}{a^{4}y}}$

3. $\left(a^{-2}\cdot b^{-3}\right)^{2}\cdot\left(a\cdot b^{-2}\right) = \left(a^{-2\times2}b^{-3\times2}\right)ab^{-2} = \left(a^{-4}b^{-6}\right)ab^{-2} = \left(a^{-4}a^{1}\right)\cdot\left(b^{-6}b^{-2}\right) = \left(a^{-4+1}\right)\cdot\left(b^{-6-2}\right) = a^{-3}b^{-8} = \mathbf{\frac{1}{a^{3}b^{8}}}$

4. $2^{-3}\cdot\left(x^{-2a}\cdot x^{3a}\right)^{-2} = 2^{-3}\cdot\left(x^{-2a\times-2}\cdot x^{3a\times-2}\right) = 2^{-3}\cdot\left(x^{4a}\cdot x^{-6a}\right) = 2^{-3}\cdot\left(x^{4a-6a}\right) = 2^{-3}\cdot x^{-2a} = \frac{1}{2^{3}x^{2a}} = \mathbf{\frac{1}{8x^{2a}}}$

5. $\left(x\cdot y^{-1}\cdot z^{3}\right)^{-3a}\cdot\left(x^{-2}\right)^{-2a} = \left(x^{1\times-3a}y^{-1\times-3a}z^{3\times-3a}\right)\cdot\left(x^{-2\times-2a}\right) = \left(x^{-3a}y^{3a}z^{-9a}\right)\cdot\left(x^{4a}\right) = \left(x^{-3a}x^{4a}\right)\cdot y^{3a}z^{-9a}$

$= \left(x^{-3a+4a}\right)\cdot y^{3a}z^{-9a} = x^{a}y^{3a}z^{-9a} = \mathbf{\frac{x^{a}y^{3a}}{z^{9a}}}$

6. $2^{-1}\cdot5^{2}\cdot5^{-5}\cdot2^{0} = 2^{-1}\cdot5^{2}\cdot5^{-5}\cdot1 = 2^{-1}\cdot5^{2-5} = 2^{-1}\cdot5^{-3} = \frac{1}{2\cdot5^{3}} = \frac{1}{2\cdot(5\cdot5\cdot5)} = \frac{1}{2\cdot125} = \mathbf{\frac{1}{250}}$

7. $y^{2}\cdot\left(x^{-2}\right)^{4}\cdot\left(-x\cdot y^{-4}\right)^{2} = y^{2}\cdot\left(x^{-2\times4}\right)\cdot\left[(-x)^{2}\cdot y^{-4\times2}\right] = y^{2}\cdot x^{-8}\cdot\left[(-x)\cdot(-x)\cdot y^{-8}\right] = y^{2}\cdot x^{-8}\cdot\left[\left(+x^{2}\right)\cdot y^{-8}\right]$

$= \left(y^{2}\cdot y^{-8}\right)\cdot\left(x^{-8}\cdot x^{2}\right) = \left(y^{2-8}\right)\cdot\left(x^{-8+2}\right) = y^{-6}x^{-6} = \frac{1}{y^{6}x^{6}} = \mathbf{\frac{1}{x^{6}y^{6}}}$

8. $\left(x^{3}\cdot y^{2}\right)^{-3}\cdot\left(x^{5}\cdot y^{2}\cdot z^{-3}\right) = \left(x^{3\times-3}\cdot y^{2\times-3}\right)\cdot\left(x^{5}\cdot y^{2}\cdot z^{-3}\right) = \left(x^{-9}\cdot y^{-6}\right)\cdot\left(x^{5}\cdot y^{2}\cdot z^{-3}\right) = \left(x^{-9}\cdot x^{5}\right)\cdot\left(y^{-6}\cdot y^{2}\right)\cdot z^{-3}$

$= \left(x^{-9+5}\right)\cdot\left(y^{-6+2}\right)\cdot z^{-3} = x^{-4}\cdot y^{-4}\cdot z^{-3} = x^{-4}y^{-4}z^{-3} = \mathbf{\frac{1}{x^{4}y^{4}z^{3}}}$

9. $\left(x^{2}\cdot y\right)^{-1}\cdot\left(3^{-1}\cdot x\cdot y^{3}\right)^{-4} = \left(x^{2\times-1}\cdot y^{-1}\right)\cdot\left(3^{-1\times-4}\cdot x^{-4}\cdot y^{3\times-4}\right) = \left(x^{-2}\cdot y^{-1}\right)\cdot\left(3^{4}\cdot x^{-4}\cdot y^{-12}\right) = 3^{4}\left(x^{-2}x^{-4}\right)\cdot\left(y^{-1}y^{-12}\right)$

$= 3^{4}\left(x^{-2-4}\right)\cdot\left(y^{-1-12}\right) = 3^{4}x^{-6}y^{-13} = \frac{3^{4}}{x^{6}y^{13}} = \mathbf{\frac{81}{x^{6}y^{13}}}$

10. $\left(5\cdot x^{2}\right)^{-1}\cdot\left[\left(x\cdot y^{2}\right)^{-2}\cdot x\right]^{-2} = \left(5^{-1}\cdot x^{2\times-1}\right)\cdot\left[\left(x^{-2}\cdot y^{2\times-2}\right)\cdot x\right]^{-2} = \left(5^{-1}\cdot x^{-2}\right)\cdot\left[\left(x^{-2}\cdot y^{-4}\right)\cdot x\right]^{-2}$

$= \left(5^{-1}\cdot x^{-2}\right)\cdot\left[x^{-2}\cdot y^{-4}\cdot x\right]^{-2} = \left(5^{-1}\cdot x^{-2}\right)\cdot\left[x^{-2\times-2}\cdot y^{-4\times-2}\cdot x^{-2}\right] = \left(5^{-1}\cdot x^{-2}\right)\cdot\left[x^{4}\cdot y^{8}\cdot x^{-2}\right] = 5^{-1}\cdot x^{-2}\cdot x^{4}\cdot y^{8}\cdot x^{-2}$

$= 5^{-1}\cdot\left(x^{-2}x^{4}x^{-2}\right)\cdot y^{8} = 5^{-1}\cdot\left(x^{-2+4-2}\right)\cdot y^{8} = 5^{-1}\cdot x^{0}\cdot y^{8} = 5^{-1}\cdot1\cdot y^{8} = 5^{-1}\cdot y^{8} = \frac{y^{8}}{5^{1}} = \mathbf{\frac{y^{8}}{5}}$

Section 3.4 Case II a Practice Problems

1. $\dfrac{x^{-2}x}{x^3x^0} = \dfrac{x^{-2}x^1}{x^3x^0} = \dfrac{x^{-2+1}}{x^{3+0}} = \dfrac{x^{-1}}{x^3} = \dfrac{1}{x^3x^1} = \dfrac{1}{x^{3+1}} = \mathbf{\dfrac{1}{x^4}}$

2. $\dfrac{-2a^{-2}b^3}{-6a^{-1}b^{-2}} = +\dfrac{\overset{1}{\cancel{2}}a^{-2}b^3}{\underset{3}{\cancel{6}}a^{-1}b^{-2}} = \dfrac{b^3b^2}{3a^2a^{-1}} = \dfrac{b^{3+2}}{3a^{2-1}} = \dfrac{b^5}{3a^1} = \mathbf{\dfrac{b^5}{3a}}$

3. $\dfrac{-(-3)^{-4}}{3\cdot(-3)^{-3}} = -\dfrac{(-3)^3}{3\cdot(-3)^4} = -\dfrac{-3\cdot-3\cdot-3}{3\cdot(-3\cdot-3\cdot-3\cdot-3)} = -\dfrac{-27}{3\cdot(+81)} = +\dfrac{27}{3\cdot 81} = \dfrac{\overset{1}{\cancel{27}}}{\underset{9}{\cancel{243}}} = \mathbf{\dfrac{1}{9}}$

4. $\dfrac{-3^3y^{-3}yw}{(-3)^{-2}y^2w^{-3}} = -\dfrac{3^3y^{-3}y^1w^1}{(-3)^{-2}y^2w^{-3}} = -\dfrac{27\cdot(-3)^2w^1w^3}{y^2y^3y^{-1}} = -\dfrac{27\cdot(-3\cdot-3)w^{1+3}}{y^{2+3-1}} = -\dfrac{(27\cdot 9)w^4}{y^4} = -\mathbf{\dfrac{243w^4}{y^4}}$

5. $\dfrac{a^{-2}b^2a^{-5}y^{-2}}{a^{-3}y} = \dfrac{a^{-2}b^2a^{-5}y^{-2}}{a^{-3}y^1} = \dfrac{b^2}{(a^{-3}a^2a^5)\cdot(y^2y^1)} = \dfrac{b^2}{a^{-3+2+5}\cdot y^{2+1}} = \dfrac{b^2}{a^4\cdot y^3} = \mathbf{\dfrac{b^2}{a^4y^3}}$

6. $\dfrac{(x\cdot y\cdot z)^0\cdot yx^{-2}}{x^{-4}y^{-1}} = \dfrac{1\cdot yx^{-2}}{x^{-4}y^{-1}} = \dfrac{y^1x^{-2}}{x^{-4}y^{-1}} = \dfrac{(x^4x^{-2})\cdot(y^1y^1)}{1} = \dfrac{(x^{4-2})\cdot(y^{1+1})}{1} = \dfrac{x^2\cdot y^2}{1} = \mathbf{x^2y^2}$

7. $\dfrac{2^{-1}a^3b^{-2}c}{8ab^{-1}c^{-2}} = \dfrac{a^3b^{-2}c^1}{(2\cdot 8)a^1b^{-1}c^{-2}} = \dfrac{(a^3a^{-1})\cdot(c^1c^2)}{16b^2b^{-1}} = \dfrac{a^{3-1}\cdot c^{1+2}}{16b^{2-1}} = \dfrac{a^2\cdot c^3}{16b^1} = \mathbf{\dfrac{a^2c^3}{16b}}$

8. $\dfrac{-4^{-2}z^4w^2a}{a^2w^{-2}z^0} = -\dfrac{4^{-2}z^4w^2a^1}{a^2w^{-2}\cdot 1} = -\dfrac{z^4w^2a^1}{4^2a^2w^{-2}} = -\dfrac{z^4(w^2w^2)}{16(a^2a^{-1})} = -\dfrac{z^4w^{2+2}}{16a^{2-1}} = -\dfrac{z^4w^4}{16a^1} = -\mathbf{\dfrac{w^4z^4}{16a}}$

9. $\dfrac{2^{-3}2^2y^{-3}yb^3}{2^2b^{-1}y^3} = \dfrac{2^{-3+2}y^{-3}y^1b^3}{2^2b^{-1}y^3} = \dfrac{2^{-1}y^{-3}y^1b^3}{2^2b^{-1}y^3} = \dfrac{b^3b^1}{(2^22^1)\cdot(y^3y^3y^{-1})} = \dfrac{b^{3+1}}{2^{2+1}\cdot y^{3+3-1}} = \dfrac{b^4}{2^3\cdot y^5} = \mathbf{\dfrac{b^4}{8y^5}}$

10. $\dfrac{2^{-3}a^{-3}b^2c^{-1}d^{-3}}{(-2)^{-3}b^{-1}c^3d} = \dfrac{(-2)^3a^{-3}b^2c^{-1}d^{-3}}{2^3b^{-1}c^3d^1} = \dfrac{(-2\cdot-2\cdot-2)a^{-3}b^2c^{-1}d^{-3}}{8b^{-1}c^3d^1} = \dfrac{-8(b^2b^1)}{8a^3(c^3c^1)\cdot(d^1d^3)} = -\dfrac{\overset{1}{\cancel{8}}b^{2+1}}{\underset{1}{\cancel{8}}a^3(c^{3+1})\cdot(d^{1+3})}$

$= -\mathbf{\dfrac{b^3}{a^3c^4d^4}}$

Section 3.4 Case II b Practice Problems

1. $\dfrac{(x^{-2}\cdot x)^{-3}}{x^{-3}} = \dfrac{x^{-2\times-3}\cdot x^{1\times-3}}{x^{-3}} = \dfrac{x^6\cdot x^{-3}}{x^{-3}} = \dfrac{x^{6-3}}{x^{-3}} = \dfrac{x^3}{x^{-3}} = \dfrac{x^3x^3}{1} = \dfrac{x^{3+3}}{1} = \dfrac{x^6}{1} = \mathbf{x^6}$

2. $\dfrac{-(2)^{-2}a^{-2}b^3}{(a^{-1}b)^{-1}} = -\dfrac{2^{-2}a^{-2}b^3}{a^{-1\times-1}b^{1\times-1}} = -\dfrac{2^{-2}a^{-2}b^3}{a^1b^{-1}} = -\dfrac{b^3b^1}{2^2a^2a^1} = -\dfrac{b^{3+1}}{4a^{2+1}} = -\mathbf{\dfrac{b^4}{4a^3}}$

3. $\left(\frac{a^3\cdot b^6\cdot c^2}{a^2\cdot b^3}\right)^{-2} = \frac{a^{3\times-2}\cdot b^{6\times-2}\cdot c^{2\times-2}}{a^{2\times-2}\cdot b^{3\times-2}} = \frac{a^{-6}\cdot b^{-12}\cdot c^{-4}}{a^{-4}\cdot b^{-6}} = \frac{1}{\left(a^6\cdot a^{-4}\right)\cdot\left(b^{12}\cdot b^{-6}\right)\cdot c^4} = \frac{1}{\left(a^{6-4}\right)\cdot\left(b^{12-6}\right)\cdot c^4}$

$= \frac{1}{a^2\cdot b^6\cdot c^4} = \mathbf{\frac{1}{a^2b^6c^4}}$

4. $\frac{y^3\cdot(yzw)^{-1}}{z\cdot y^{-2}\cdot w^{-3}} = \frac{y^3\cdot\left(y^{1\times-1}z^{1\times-1}w^{1\times-1}\right)}{z^1\cdot y^{-2}\cdot w^{-3}} = \frac{y^3y^{-1}z^{-1}w^{-1}}{z^1y^{-2}w^{-3}} = \frac{\left(y^3y^{-1}y^2\right)\cdot\left(w^3w^{-1}\right)}{z^1z^1} = \frac{\left(y^{3-1+2}\right)\cdot\left(w^{3-1}\right)}{z^{1+1}} = \mathbf{\frac{y^4w^2}{z^2}}$

5. $\frac{(a\cdot b)^{-2}\cdot(xy)^{-1}}{a^{-3}\cdot y} = \frac{\left(a^{1\times-2}\cdot b^{1\times-2}\right)\cdot\left(x^{1\times-1}y^{1\times-1}\right)}{a^{-3}\cdot y} = \frac{\left(a^{-2}\cdot b^{-2}\right)\cdot\left(x^{-1}y^{-1}\right)}{a^{-3}\cdot y} = \frac{a^{-2}b^{-2}x^{-1}y^{-1}}{a^{-3}y^1} = \frac{a^3a^{-2}}{b^2x^1y^1y^1} = \frac{a^{3-2}}{b^2x\,y^{1+1}}$

$= \frac{a^1}{b^2x\,y^2} = \mathbf{\frac{a}{b^2x\,y^2}}$

6. $\frac{(xyz)^0\cdot x^{-2}}{\left(x\cdot y^{-1}\right)^{-3}} = \frac{1\cdot x^{-2}}{\left(x\cdot y^{-1}\right)^{-3}} = \frac{x^{-2}}{x^{1\times-3}\cdot y^{-1\times-3}} = \frac{x^{-2}}{x^{-3}\cdot y^3} = \frac{x^3x^{-2}}{y^3} = \frac{x^{3-2}}{y^3} = \frac{x^1}{y^3} = \mathbf{\frac{x}{y^3}}$

7. $\frac{2^{-1}\cdot a^3\cdot b^{-2}\cdot c}{\left(b\cdot c^{-2}\right)^2} = \frac{a^3\cdot b^{-2}\cdot c}{2^1\cdot\left(b^{1\times2}\cdot c^{-2\times2}\right)} = \frac{a^3\cdot b^{-2}\cdot c^1}{2\cdot b^2\cdot c^{-4}} = \frac{a^3\cdot\left(c^1\cdot c^4\right)}{2\cdot\left(b^2\cdot b^2\right)} = \frac{a^3\cdot c^{1+4}}{2\cdot b^{2+2}} = \mathbf{\frac{a^3c^5}{2b^4}}$

8. $\left(\frac{z^4\cdot w^{-2}\cdot(a\cdot b)^{-1}}{a^2\cdot w^{-2}\cdot z^0}\right)^2 = \left(\frac{z^4\cdot w^{-2}\cdot\left(a^{1\times-1}\cdot b^{1\times-1}\right)}{a^2\cdot w^{-2}\cdot 1}\right)^2 = \left(\frac{z^4\cdot w^{-2}\cdot a^{-1}\cdot b^{-1}}{a^2\cdot w^{-2}}\right)^2 = \frac{z^{4\times2}\cdot w^{-2\times2}\cdot a^{-1\times2}\cdot b^{-1\times2}}{a^{2\times2}\cdot w^{-2\times2}}$

$= \frac{z^8\cdot w^{-4}\cdot a^{-2}\cdot b^{-2}}{a^4\cdot w^{-4}} = \frac{z^8\cdot\left(w^{-4}\cdot w^4\right)}{\left(a^4\cdot a^2\right)\cdot b^2} = \frac{z^8\cdot w^{-4+4}}{a^{4+2}\cdot b^2} = \frac{z^8\cdot w^0}{a^6\cdot b^2} = \frac{z^8\cdot 1}{a^6b^2} = \mathbf{\frac{z^8}{a^6b^2}}$

9. $\frac{2^{-3}\cdot 2^2\cdot(y\cdot z)^{-3}\cdot b^3}{b^{-1}\cdot y^3} = \frac{\left(2^{-3}\cdot 2^2\right)\cdot\left(y^{1\times-3}\cdot z^{1\times-3}\right)\cdot b^3}{b^{-1}\cdot y^3} = \frac{\left(2^{-3+2}\right)\cdot\left(y^{-3}\cdot z^{-3}\right)\cdot b^3}{b^{-1}\cdot y^3} = \frac{2^{-1}\cdot y^{-3}\cdot z^{-3}\cdot b^3}{b^{-1}\cdot y^3}$

$= \frac{b^3\cdot b^1}{2^1\cdot\left(y^3\cdot y^3\right)\cdot z^3} = \frac{b^{3+1}}{2\cdot y^{3+3}\cdot z^3} = \frac{b^4}{2\cdot y^6\cdot z^3} = \mathbf{\frac{b^4}{2y^6z^3}}$

10. $\left(\frac{a^{-3}\cdot b^2\cdot\left(c^{-1}\cdot d\right)^4}{c^3\cdot d}\right)^{-2} = \left(\frac{a^{-3}\cdot b^2\cdot\left(c^{-1\times4}\cdot d^{1\times4}\right)}{c^3\cdot d}\right)^{-2} = \left(\frac{a^{-3}\cdot b^2\cdot c^{-4}\cdot d^4}{c^3\cdot d}\right)^{-2} = \frac{a^{-3\times-2}\cdot b^{2\times-2}\cdot c^{-4\times-2}\cdot d^{4\times-2}}{c^{3\times-2}\cdot d^{1\times-2}}$

$= \frac{a^6\cdot b^{-4}\cdot c^8\cdot d^{-8}}{c^{-6}\cdot d^{-2}} = \frac{a^6\cdot\left(c^8\cdot c^6\right)}{b^4\cdot\left(d^{-2}\cdot d^8\right)} = \frac{a^6\cdot c^{8+6}}{b^4\cdot d^{-2+8}} = \frac{a^6\cdot c^{14}}{b^4\cdot d^6} = \mathbf{\frac{a^6c^{14}}{b^4d^6}}$

Section 3.4 Case III a Practice Problems

1. $x^{-1}+2x^{-2}+3x^{-1}-6x^{-2} = \left(2x^{-2}-6x^{-2}\right)+\left(x^{-1}+3x^{-1}\right) = (2-6)x^{-2}+(1+3)x^{-1} = -4x^{-2}+4x^{-1} = \frac{-4}{x^2}+\frac{4}{x^1}$

$= -\frac{4}{x^2}+\frac{4}{x} = \frac{(-4\cdot x)+\left(4\cdot x^2\right)}{x^2\cdot x} = \frac{4x^2-4x}{x^3} = \frac{4x(x-1)}{x^3} = \frac{4(x-1)}{x^3\cdot x^{-1}} = \frac{4(x-1)}{x^{3-1}} = \mathbf{\frac{4(x-1)}{x^2}}$

2. $\left(3a^{-4}-b^{-2}\right)+\left(-2a^{-4}+3b^{-2}\right) = 3a^{-4}-b^{-2}-2a^{-4}+3b^{-2} = \left(3a^{-4}-2a^{-4}\right)+\left(-b^{-2}+3b^{-2}\right) = (3-2)a^{-4}+(-1+3)b^{-2}$

$= a^{-4}+2b^{-2} = \frac{1}{a^4}+\frac{2}{b^2} = \frac{\left(b^2\cdot 1\right)+\left(2\cdot a^4\right)}{a^4\cdot b^2} = \mathbf{\frac{b^2+2a^4}{a^4b^2}}$

3. $(xy)^{-1}+y^{-2}+4(xy)^{-1}-3y^{-2}+2^{-3} = \left[(xy)^{-1}+4(xy)^{-1}\right]+\left(y^{-2}-3y^{-2}\right)+2^{-3} = [1+4](xy)^{-1}+(1-3)y^{-2}+2^{-3}$

$= 5(xy)^{-1}-2y^{-2}+2^{-3} = \frac{5}{xy}-\frac{2}{y^2}+\frac{1}{2^3} = \left(\frac{5}{xy}-\frac{2}{y^2}\right)+\frac{1}{8} = \left(\frac{\left(5\cdot y^2\right)-(2\cdot xy)}{xy\cdot y^2}\right)+\frac{1}{8} = \left(\frac{5y^2-2xy}{xy^3}\right)+\frac{1}{8}$

$= \frac{\left[8\cdot\left(5y^2-2xy\right)\right]+\left(1\cdot xy^3\right)}{8\cdot xy^3} = \frac{40y^2-16xy+xy^3}{8xy^3} = \frac{xy^3+40y^2-16xy}{8xy^3} = \frac{y\left(xy^2+40y-16x\right)}{8xy^3} = \frac{xy^2+40y-16x}{8xy^3y^{-1}}$

$= \frac{xy^2+40y-16x}{8xy^{3-1}} = \mathbf{\frac{xy^2+40y-16x}{8\,xy^2}}$

4. $4x^{-1}+y^{-3}+5y^{-3} = 4x^{-1}+\left(y^{-3}+5y^{-3}\right) = 4x^{-1}+(1+5)y^{-3} = 4x^{-1}+6y^{-3} = \frac{4}{x}+\frac{6}{y^3} = \frac{\left(4\cdot y^3\right)+(6\cdot x)}{x\cdot y^3} = \mathbf{\frac{4y^3+6x}{xy^3}}$

5. $m^{-5}-\left(m^{-2}-3m^{-5}+m^0\right)+3m^{-2} = m^{-5}-\left(m^{-2}-3m^{-5}+1\right)+3m^{-2} = m^{-5}+\left(-m^{-2}+3m^{-5}-1\right)+3m^{-2}$

$= m^{-5}-m^{-2}+3m^{-5}-1+3m^{-2} = \left(m^{-5}+3m^{-5}\right)+\left(-m^{-2}+3m^{-2}\right)-1 = (1+3)m^{-5}+(-1+3)m^{-2}-1 = 4m^{-5}+2m^{-2}-1$

$= \frac{4}{m^5}+\frac{2}{m^2}-1 = \left(\frac{4}{m^5}+\frac{2}{m^2}\right)-1 = \left(\frac{\left(4\cdot m^2\right)+\left(2\cdot m^5\right)}{m^5\cdot m^2}\right)-1 = \left(\frac{4m^2+2m^5}{m^{5+2}}\right)-1 = \frac{4m^2+2m^5}{m^7}-\frac{1}{1}$

$= \frac{\left[1\cdot\left(4m^2+2m^5\right)\right]-\left(1\cdot m^7\right)}{m^7\cdot 1} = \frac{4m^2+2m^5-m^7}{m^7} = \frac{m^2\left(4+2m^3-m^5\right)}{m^7} = \frac{4+2m^3-m^5}{m^7m^{-2}} = \frac{4+2m^3-m^5}{m^{7-2}} = \mathbf{\frac{-m^5+2m^3+4}{m^5}}$

6. $\left(a^3\right)^{-2}+\left(a^{-2}b\right)^2-6a^{-6}+3a^{-4}b^2 = \left(a^{3\times -2}\right)+\left(a^{-2\times 2}b^{1\times 2}\right)-6a^{-6}+3a^{-4}b^2 = a^{-6}+a^{-4}b^2-6a^{-6}+3a^{-4}b^2$

$= \left(a^{-6}-6a^{-6}\right)+\left(a^{-4}b^2+3a^{-4}b^2\right) = (1-6)a^{-6}+(1+3)a^{-4}b^2 = -5a^{-6}+4a^{-4}b^2 = -\frac{5}{a^6}+\frac{4b^2}{a^4} = \frac{-\left(5\cdot a^4\right)+\left(a^6\cdot 4b^2\right)}{a^6\cdot a^4}$

$= \frac{-5a^4+4a^6b^2}{a^{6+4}} = \frac{a^4\left(-5+4a^2b^2\right)}{a^{10}} = \frac{-5+4a^2b^2}{a^{10}a^{-4}} = \frac{-5+4a^2b^2}{a^{10-4}} = \mathbf{\frac{4a^2b^2-5}{a^6}}$

7. $3x^{-4}+3x^{-2}+2x^{-4}-\left(x^{-4}+3x^{-2}\right) = 3x^{-4}+3x^{-2}+2x^{-4}-x^{-4}-3x^{-2} = \left(3x^{-4}+2x^{-4}-x^{-4}\right)+\left(3x^{-2}-3x^{-2}\right)$

$= (3+2-1)x^{-4}+(3-3)x^{-2} = 4x^{-4}+0x^{-2} = 4x^{-4} = \mathbf{\frac{4}{x^4}}$

8. $k^{-2n}+k^{-3n}-3k^{-2n}+2^{-2} = k^{-3n}+\left(k^{-2n}-3k^{-2n}\right)+2^{-2} = k^{-3n}+(1-3)k^{-2n}+2^{-2} = k^{-3n}-2k^{-2n}+2^{-2}$

$= \frac{1}{k^{3n}}-\frac{2}{k^{2n}}+\frac{1}{2^2} = \left(\frac{1}{k^{3n}}-\frac{2}{k^{2n}}\right)+\frac{1}{4} = \frac{\left(1\cdot k^{2n}\right)-\left(2\cdot k^{3n}\right)}{k^{3n}\cdot k^{2n}}+\frac{1}{4} = \frac{k^{2n}-2k^{3n}}{k^{3n+2n}}+\frac{1}{4} = \frac{k^{2n}-2k^{3n}}{k^{5n}}+\frac{1}{4}$

$= \frac{\left[4\cdot\left(k^{2n}-2k^{3n}\right)\right]+\left(1\cdot k^{5n}\right)}{4\cdot k^{5n}} = \frac{4k^{2n}-8k^{3n}+k^{5n}}{4k^{5n}} = \frac{k^{2n}\left(4-8k^{n}+k^{3n}\right)}{4k^{5n}} = \frac{4-8k^{n}+k^{3n}}{4\left(k^{5n}\cdot k^{-2n}\right)} = \frac{4-8k^{n}+k^{3n}}{4k^{5n-2n}}$

$= \mathbf{\frac{k^{3n}-8k^{n}+4}{4k^{3n}}}$

9. $w^{-2}+3w^{-4}+2w^{-2}-\left(w^{0}-4w^{-2}\right) = w^{-2}+3w^{-4}+2w^{-2}+\left(-1+4w^{-2}\right) = w^{-2}+3w^{-4}+2w^{-2}-1+4w^{-2}$

$= 3w^{-4}+\left(2w^{-2}+w^{-2}+4w^{-2}\right)-1 = 3w^{-4}+(2+1+4)w^{-2}-1 = \frac{3}{w^4}+\frac{7}{w^2}-1 = \left(\frac{3}{w^4}+\frac{7}{w^2}\right)-\frac{1}{1}$

$= \frac{\left(3\cdot w^2\right)+\left(7\cdot w^4\right)}{w^4\cdot w^2}-\frac{1}{1} = \frac{3w^2+7w^4}{w^{4+2}}-\frac{1}{1} = \frac{3w^2+7w^4}{w^6}-\frac{1}{1} = \frac{\left[\left(3w^2+7w^4\right)\cdot 1\right]-\left(1\cdot w^6\right)}{w^6\cdot 1} = \frac{3w^2+7w^4-w^6}{w^6}$

$= \frac{w^2\left(3+7w^2-w^4\right)}{w^6} = \frac{3+7w^2-w^4}{w^6w^{-2}} = \frac{3+7w^2-w^4}{w^{6-2}} = \mathbf{\frac{-w^4+7w^2+3}{w^4}}$

10. $\left(-7z^{-3}+3z-5\right)-\left(-3z^{-3}+z-2\right)+4z = \left(-7z^{-3}+3z-5\right)+\left(+3z^{-3}-z+2\right)+4z = -7z^{-3}+3z-5+3z^{-3}-z+2+4z$

$= \left(-7z^{-3}+3z^{-3}\right)+\left(3z-z+4z\right)+(2-5) = (-7+3)z^{-3}+(3-1+4)z-3 = -4z^{-3}+6z-3 = -\frac{4}{z^3}+\frac{6z-3}{1}$

$= \frac{-(4\cdot 1)+\left[z^3\cdot(6z-3)\right]}{z^3\cdot 1} = \frac{-4+6z^4-3z^3}{z^3} = \mathbf{\frac{6z^4-3z^3-4}{z^3}}$

Section 3.4 Case III b Practice Problems

1. $\frac{2^{-1}-2^{-3}}{2} = \frac{\frac{1}{2^1}-\frac{1}{2^3}}{2} = \frac{\frac{1}{2}-\frac{1}{8}}{2} = \frac{\frac{(1\cdot 8)-(1\cdot 2)}{2\cdot 8}}{2} = \frac{\frac{8-2}{16}}{2} = \frac{\frac{6}{16}}{2} = \frac{\frac{6}{16}}{\frac{2}{1}} = \frac{6\cdot 1}{16\cdot 2} = \frac{\overset{3}{\cancel{6}}}{\underset{16}{\cancel{32}}} = \mathbf{\frac{3}{16}}$

2. $\left(a^{-3}+\frac{a^{-2}}{2}\right)^2 = \left(a^{-3\times 2}+\frac{a^{-2\times 2}}{2^2}\right) = a^{-6}+\frac{a^{-4}}{4} = \frac{1}{a^6}+\frac{1}{4a^4} = \frac{\left(1\cdot 4a^4\right)+\left(1\cdot a^6\right)}{4\cdot a^6\cdot a^4} = \frac{4a^4+a^6}{4a^{6+4}} = \frac{a^4\left(a^2+4\right)}{4a^{10}}$

$= \frac{a^2+4}{4a^{10}a^{-4}} = \frac{a^2+4}{4a^{10-4}} = \mathbf{\frac{a^2+4}{4a^6}}$

3. $\frac{2}{x^{-1}}+x^{-1} = \frac{2}{\frac{1}{x}}+\frac{1}{x^1} = \frac{\frac{2}{1}}{\frac{1}{x}}+\frac{1}{x} = \frac{2\cdot x}{1\cdot 1}+\frac{1}{x} = \frac{2x}{1}+\frac{1}{x} = \frac{(2x\cdot x)+(1\cdot 1)}{1\cdot x} = \mathbf{\frac{2x^2+1}{x}}$

4. $\frac{5}{x+x^{-1}} = \frac{5}{x+\frac{1}{x}} = \frac{5}{\frac{x}{1}+\frac{1}{x}} = \frac{5}{\frac{(x\cdot x)+(1\cdot 1)}{1\cdot x}} = \frac{5}{\frac{x^2+1}{x}} = \frac{\frac{5}{1}}{\frac{x^2+1}{x}} = \frac{5\cdot x}{1\cdot\left(x^2+1\right)} = \mathbf{\frac{5x}{x^2+1}}$

5. $\dfrac{y^{-1}}{y-y^{-1}} = \dfrac{\frac{1}{y}}{y-\frac{1}{y}} = \dfrac{\frac{1}{y}}{\frac{y}{1}-\frac{1}{y}} = \dfrac{\frac{1}{y}}{\frac{(y\cdot y)-(1\cdot 1)}{1\cdot y}} = \dfrac{\frac{1}{y}}{\frac{y^2-1}{y}} = \dfrac{1\cdot y}{y\cdot\left(y^2-1\right)} = \dfrac{\not{y}}{\not{y}\left(y^2-1\right)} = \dfrac{1}{y^2-1} = \dfrac{\mathbf{1}}{\mathbf{(y-1)(y+1)}}$

6. $\dfrac{b^{-2}}{b^{-1}+b^{-2}} = \dfrac{\frac{1}{b^2}}{\frac{1}{b}+\frac{1}{b^2}} = \dfrac{\frac{1}{b^2}}{\frac{\left(1\cdot b^2\right)+(1\cdot b)}{b\cdot b^2}} = \dfrac{\frac{1}{b^2}}{\frac{b^2+b}{b^3}} = \dfrac{1\cdot b^3}{b^2\cdot\left(b^2+b\right)} = \dfrac{b^3}{b^2\left(b^2+b\right)} = \dfrac{b^3b^{-2}}{b^2+b} = \dfrac{b^{3-2}}{b^2+b} = \dfrac{b}{b^2+b}$

$= \dfrac{b}{b(b+1)} = \dfrac{\not{b}}{\not{b}(b+1)} = \dfrac{\mathbf{1}}{\mathbf{b+1}}$

7. $\dfrac{x^{-1}+y^{-2}}{x^{-1}-y^{-2}} = \dfrac{\frac{1}{x}+\frac{1}{y^2}}{\frac{1}{x}-\frac{1}{y^2}} = \dfrac{\frac{\left(1\cdot y^2\right)+(1\cdot x)}{x\cdot y^2}}{\frac{\left(1\cdot y^2\right)-(1\cdot x)}{x\cdot y^2}} = \dfrac{\frac{y^2+x}{x\,y^2}}{\frac{y^2-x}{x\,y^2}} = \dfrac{\not{x}\,y^2\left(y^2+x\right)}{\not{x}\,y^2\left(y^2-x\right)} = \dfrac{\mathbf{y^2+x}}{\mathbf{y^2-x}}$

8. $\dfrac{3a}{a^{-1}-b^{-1}} = \dfrac{3a}{\frac{1}{a}-\frac{1}{b}} = \dfrac{3a}{\frac{(1\cdot b)-(1\cdot a)}{a\cdot b}} = \dfrac{3a}{\frac{b-a}{ab}} = \dfrac{\frac{3a}{1}}{\frac{b-a}{ab}} = \dfrac{3a\cdot ab}{1\cdot(b-a)} = \dfrac{\mathbf{3a^2b}}{\mathbf{b-a}}$

9. $\dfrac{x^{-1}}{(x-3)^{-1}}+y^{-1} = \dfrac{\frac{1}{x}}{\frac{1}{x-3}}+\dfrac{1}{y} = \dfrac{1\cdot(x-3)}{1\cdot x}+\dfrac{1}{y} = \dfrac{x-3}{x}+\dfrac{1}{y} = \dfrac{[y\cdot(x-3)]+(1\cdot x)}{x\cdot y} = \dfrac{\mathbf{xy-3y+x}}{\mathbf{xy}}$

10. $\dfrac{2a^{-1}\cdot b^{-1}}{a^{-1}-b} = \dfrac{\frac{2}{a\cdot b}}{\frac{1}{a}-b} = \dfrac{\frac{2}{ab}}{\frac{1}{a}-\frac{b}{1}} = \dfrac{\frac{2}{ab}}{\frac{(1\cdot 1)-(a\cdot b)}{1\cdot a}} = \dfrac{\frac{2}{ab}}{\frac{1-ab}{a}} = \dfrac{2\cdot a}{ab\cdot(1-ab)} = \dfrac{2\not{a}}{\not{a}b(1-ab)} = \dfrac{\mathbf{2}}{\mathbf{b(1-ab)}}$

Chapter 4 Solutions:

Section 4.1 Case I Practice Problems

1. $\sqrt[2]{98} = \sqrt{98} = \sqrt{49 \cdot 2} = \sqrt{7^2 \cdot 2} = \mathbf{7\sqrt{2}}$

2. $3\sqrt{75} = 3\sqrt{25 \cdot 3} = 3\sqrt{5^2 \cdot 3} = (3 \cdot 5)\sqrt{3} = \mathbf{15\sqrt{3}}$

3. $\sqrt[3]{125} = \sqrt[3]{5^3} = \mathbf{5}$

4. $\sqrt[5]{3125} = \sqrt[5]{5^5} = \mathbf{5}$

5. $\sqrt[4]{162} = \sqrt[4]{81 \cdot 2} = \sqrt[4]{3^4 \cdot 2} = \mathbf{3\sqrt[4]{2}}$

6. $\sqrt[2]{192} = \sqrt{192} = \sqrt{64 \cdot 3} = \sqrt{8^2 \cdot 3} = \mathbf{8\sqrt{3}}$

7. $-\sqrt[3]{64} = -\sqrt[3]{4^3} = \mathbf{-4}$

8. $\sqrt{250} = \sqrt{25 \cdot 10} = \sqrt{5^2 \cdot 10} = \mathbf{5\sqrt{10}}$

9. $\sqrt[3]{54} = \sqrt[3]{27 \cdot 2} = \sqrt[3]{3^3 \cdot 2} = \mathbf{3\sqrt[3]{2}}$

10. $\sqrt[5]{486} = \sqrt[5]{243 \cdot 2} = \sqrt[5]{3^5 \cdot 2} = \mathbf{3\sqrt[5]{2}}$

Section 4.1 Case II Practice Problems

1. $\frac{5}{8}$; is a rational and real number

2. $\sqrt{45} = \sqrt{9 \cdot 5} = 3\sqrt{5}$; is an irrational and real number

3. 450 ; is a rational and real number

4. $-\frac{2}{\sqrt{10}}$; is an irrational and real number

5. $-\sqrt{-5}$; is not a real number

6. $\frac{\sqrt{5}}{-2}$; is an irrational and real number

7. 0.1111111... ; is a rational and real number

8. −0.2367432... ; is an irrational and real number

9. $\sqrt[5]{7776} = \sqrt[5]{6^5} = 6$; is a rational and real number

10. −0.35 ; is a rational and real number

Section 4.1 Case III Practice Problems

1. $-\sqrt{49} = -\sqrt{7 \cdot 7} = -\sqrt{7 \cdot 7} = -\sqrt{7^1 \cdot 7^1} = -\sqrt{7^{1+1}} = -\sqrt{7^2} = \mathbf{-7}$

2. $\sqrt{54} = \sqrt{9 \cdot 5} = \sqrt{(3 \cdot 3) \cdot 5} = \sqrt{\left(3^1 \cdot 3^1\right) \cdot 5} = \sqrt{3^{1+1} \cdot 5} = \sqrt{3^2 \cdot 5} = \mathbf{3\sqrt{5}}$

3. $-\sqrt{500} = -\sqrt{100 \cdot 5} = -\sqrt{(10 \cdot 10) \cdot 5} = -\sqrt{\left(10^1 \cdot 10^1\right) \cdot 5} = -\sqrt{\left(10^{1+1}\right) \cdot 5} = -\sqrt{10^2 \cdot 5} = \mathbf{-10\sqrt{5}}$

4. $\sqrt[5]{3^5 \cdot 5} = \mathbf{3\sqrt[5]{5}}$

5. $\sqrt[2]{216} = \sqrt{216} = \sqrt{36 \cdot 6} = \sqrt{(6 \cdot 6) \cdot 6} = \sqrt{\left(6^1 \cdot 6^1\right) \cdot 6} = \sqrt{6^{1+1} \cdot 6} = \sqrt{6^2 \cdot 6} = \mathbf{6\sqrt{6}}$

6. $-\frac{1}{4}\sqrt[4]{4^5\cdot 2} = -\frac{1}{4}\sqrt[4]{4^{4+1}\cdot 2} = -\frac{1}{4}\sqrt[4]{\left(4^4\cdot 4^1\right)\cdot 2} = -\frac{1}{4}\cdot 4\sqrt[4]{(4\cdot 2)} = -\frac{\overset{1}{\cancel{4}}}{\underset{1}{\cancel{4}}}\sqrt[4]{8} = -\sqrt[4]{\mathbf{8}}$

7. $\sqrt[4]{162} = \sqrt[4]{81\cdot 2} = \sqrt[4]{3^4\cdot 2} = \mathbf{3\sqrt[4]{2}}$

8. $\frac{-2}{-9}\sqrt[4]{3^9} = +\frac{2}{9}\sqrt[4]{3^{4+4+1}} = \frac{2}{9}\sqrt[4]{3^4\cdot 3^4\cdot 3^1} = \frac{2}{9}\cdot\left(3^1\cdot 3^1\right)\sqrt[4]{3} = \frac{2}{9}\cdot\left(3^{1+1}\right)\sqrt[4]{3} = \frac{2}{9}\cdot 3^2\sqrt[4]{3} = \frac{2}{9}\cdot 9\sqrt[4]{3} = \frac{2}{9}\cdot\frac{9}{1}\sqrt[4]{3} = \frac{2\cdot\overset{1}{\cancel{9}}}{\underset{1}{\cancel{9}}\cdot 1}\sqrt[4]{3}$

$= \frac{2\cdot 1}{1\cdot 1}\sqrt[4]{3} = \frac{2}{1}\sqrt[4]{3} = \mathbf{2\sqrt[4]{3}}$

9. $-4\sqrt[2]{1800} = -4\sqrt{1800} = -4\sqrt{9\cdot 200} = -4\sqrt{9\cdot 100\cdot 2} = -4\sqrt{(3\cdot 3)\cdot(10\cdot 10)\cdot 2} = -4\sqrt{\left(3^1\cdot 3^1\right)\cdot\left(10^1\cdot 10^1\right)\cdot 2}$

$= -4\sqrt{\left(3^{1+1}\right)\cdot\left(10^{1+1}\right)\cdot 2} = -4\sqrt{3^2\cdot 10^2\cdot 2} = -(4\cdot 3\cdot 10)\sqrt{2} = \mathbf{-120\sqrt{2}}$

10. $\frac{1}{-6}\sqrt{100000} = -\frac{1}{6}\sqrt{10^5} = -\frac{1}{6}\sqrt{10^{2+2+1}} = -\frac{1}{6}\sqrt{10^2\cdot 10^2\cdot 10^1} = -\frac{1}{6}(10\cdot 10)\sqrt{10} = -\frac{(10\cdot 10)}{6}\sqrt{10} = -\frac{\overset{50}{\cancel{100}}}{\underset{3}{\cancel{6}}}\sqrt{10}$

$= -\frac{50}{3}\sqrt{10} = -\left(\mathbf{16\frac{2}{3}}\right)\mathbf{\sqrt{10}}$

Section 4.1 Case IV Practice Problems

1. $\sqrt{y^3} = \sqrt{y^{2+1}} = \sqrt{y^2\cdot y^1} = \boldsymbol{y\sqrt{y}}$

2. $x\sqrt[2]{x^4} = x\sqrt{x^4} = x\sqrt{x^{2+2}} = x\sqrt{x^2\cdot x^2} = x\cdot x\cdot x = x^1\cdot x^1\cdot x^1 = x^{1+1+1} = \boldsymbol{x^3}$

3. $x^3y^2\sqrt[2]{x^3y^5} = x^3y^2\sqrt{x^{2+1}y^{2+2+1}} = x^3y^2\sqrt{\left(x^2\cdot x^1\right)\cdot\left(y^2\cdot y^2\cdot y^1\right)} = \left(x^3\cdot x\right)\cdot\left(y^2\cdot y\cdot y\right)\sqrt{xy} = \boldsymbol{x^4y^4\sqrt{xy}}$

4. $-2x\sqrt{8xy^3} = -2x\sqrt{(4\cdot 2)xy^{2+1}} = -2x\sqrt{\left(2^2\cdot 2\right)\cdot x\cdot\left(y^2\cdot y^1\right)} = -(2\cdot 2)xy\sqrt{2xy} = \boldsymbol{-4xy\sqrt{2xy}}$

5. $\frac{1}{-12}\sqrt[3]{216a^5b^6c^7} = -\frac{1}{12}\sqrt[3]{6^3a^{3+2}b^{3+3}c^{3+3+1}} = -\frac{1}{12}\cdot 6\sqrt[3]{\left(a^3\cdot a^2\right)\cdot\left(b^3\cdot b^3\right)\cdot\left(c^3\cdot c^3\cdot c^1\right)} = -\frac{\overset{1}{\cancel{6}}}{\underset{2}{\cancel{12}}}a(b\cdot b)\cdot(c\cdot c)\sqrt[3]{a^2c}$

$= -\boldsymbol{\frac{1}{2}ab^2c^2\sqrt[3]{a^2c}}$

6. $uv^2\sqrt[5]{u^6v^8w^4} = uv^2\sqrt[5]{u^{5+1}v^{5+3}w^4} = uv^2\sqrt[5]{\left(u^5\cdot u^1\right)\cdot\left(v^5\cdot v^3\right)\cdot w^4} = (u\cdot u)\cdot\left(v^2\cdot v\right)\sqrt[5]{uv^3w^4} = \boldsymbol{u^2v^3\sqrt[5]{uv^3w^4}}$

7. $\frac{-3}{-10}\sqrt[4]{1250l^8m^7n^6} = +\frac{3}{10}\sqrt[4]{(625\cdot 2)l^{4+4}m^{4+3}n^{4+2}} = \frac{3}{10}\sqrt[4]{\left(5^4\cdot 2\right)\cdot\left(l^4\cdot l^4\right)\cdot\left(m^4\cdot m^3\right)\cdot\left(n^4\cdot n^2\right)}$

$= \frac{3}{10}\cdot 5(l\cdot l)\cdot mn\sqrt[4]{2m^3n^2} = \frac{3\cdot\overset{1}{\cancel{5}}}{\underset{2}{\cancel{10}}}l^2mn\sqrt[4]{2m^3n^2} = \frac{3\cdot 1}{2}l^2mn\sqrt[4]{2m^3n^2} = \frac{3}{2}\left(l^2mn\sqrt[4]{2m^3n^2}\right) = \boldsymbol{1\frac{1}{2}\left(l^2mn\sqrt[4]{2m^3n^2}\right)}$

8. $-x^2\sqrt[3]{729x^3y^5z^6} = -\left(x^2\cdot x\right)\sqrt[3]{9^3y^{3+2}z^{3+3}} = -9x^3\sqrt[3]{\left(y^3\cdot y^2\right)\cdot\left(z^3\cdot z^3\right)} = -9x^3y\cdot(z\cdot z)\sqrt[3]{y^2} = \mathbf{-9x^3yz^2\sqrt[3]{y^2}}$

9. $\frac{-5}{-6}\sqrt{72r^3s^5t^7} = +\frac{5}{6}\sqrt{(36\cdot 2)r^{2+1}s^{2+2+1}t^{2+2+2+1}} = \frac{5}{6}\sqrt{\left(6^2\cdot 2\right)\cdot\left(r^2\cdot r^1\right)\cdot\left(s^2\cdot s^2\cdot s^1\right)\cdot\left(t^2\cdot t^2\cdot t^2\cdot t^1\right)}$

$= \frac{5}{6}\cdot 6r(s\cdot s)\cdot(t\cdot t\cdot t)\sqrt{2rst} = \left(\frac{5}{6}\cdot\frac{6}{1}\right)rs^2t^3\sqrt{2rst} = \left(\frac{5\cdot \cancel{6}^{\,1}}{\cancel{6}_{\,1}\cdot 1}\right)rs^2t^3\sqrt{2rst} = \left(\frac{5\cdot 1}{1\cdot 1}\right)rs^2t^3\sqrt{2rst} = \frac{5}{1}rs^2t^3\sqrt{2rst} = \mathbf{5rs^2t^3\sqrt{2rst}}$

10. $\frac{-3x}{4}\sqrt[2]{100x^5y^6z^3} = -\frac{3}{4}x\sqrt{10^2x^{2+2+1}y^{2+2+2}z^{2+1}} = -\frac{3}{4}\cdot 10x\sqrt{\left(x^2\cdot x^2\cdot x^1\right)\cdot\left(y^2\cdot y^2\cdot y^2\right)\cdot\left(z^2\cdot z^1\right)}$

$= -\frac{3\cdot \cancel{10}^{\,5}}{\cancel{4}_{\,2}}(x\cdot x\cdot x)\cdot(y\cdot y\cdot y)\cdot z\sqrt{xz} = -\frac{3\cdot 5}{2}x^3y^3z\sqrt{xz} = -\frac{15}{2}x^3y^3z\sqrt{xz} = \mathbf{-\left(7\frac{1}{2}\right)x^3y^3z\sqrt{xz}}$

Section 4.2 Case I a Practice Problems

1. $\sqrt{72}\cdot\sqrt{75} = \sqrt{36\cdot 2}\cdot\sqrt{25\cdot 3} = \sqrt{6^2\cdot 2}\cdot\sqrt{5^2\cdot 3} = 6\sqrt{2}\cdot 5\sqrt{3} = (6\cdot 5)\sqrt{2\cdot 3} = \mathbf{30\sqrt{6}}$

2. $-3\sqrt{20}\cdot 2\sqrt{32} = -3\sqrt{4\cdot 5}\cdot 2\sqrt{16\cdot 2} = -3\sqrt{2^2\cdot 5}\cdot 2\sqrt{4^2\cdot 2} = -(3\cdot 2)\sqrt{5}\cdot(2\cdot 4)\sqrt{2} = -6\sqrt{5}\cdot 8\sqrt{2} = -(6\cdot 8)\sqrt{5\cdot 2} = \mathbf{-48\sqrt{10}}$

3. $\sqrt[2]{16}\cdot\sqrt[2]{27} = \sqrt{16}\cdot\sqrt{27} = \sqrt{4^2}\cdot\sqrt{9\cdot 3} = 4\cdot\sqrt{3^2\cdot 3} = (4\cdot 3)\sqrt{3} = \mathbf{12\sqrt{3}}$

4. $\sqrt{64}\cdot\sqrt{100}\cdot\sqrt{54} = \sqrt{8^2}\cdot\sqrt{10^2}\cdot\sqrt{9\cdot 6} = (8\cdot 10)\cdot\sqrt{3^2\cdot 6} = (80\cdot 3)\sqrt{6} = \mathbf{240\sqrt{6}}$

5. $-\sqrt{125}\cdot -2\sqrt{98} = +2\sqrt{25\cdot 5}\cdot\sqrt{49\cdot 2} = 2\sqrt{5^2\cdot 5}\cdot\sqrt{7^2\cdot 2} = (2\cdot 5)\sqrt{5}\cdot 7\sqrt{2} = (10\cdot 7)\sqrt{5\cdot 2} = \mathbf{70\sqrt{10}}$

6. $\sqrt[4]{625}\cdot\sqrt[4]{324}\cdot\sqrt[4]{48} = \sqrt[4]{5^4}\cdot\sqrt[4]{81\cdot 4}\cdot\sqrt[4]{16\cdot 3} = 5\cdot\sqrt[4]{3^4\cdot 4}\cdot\sqrt[4]{2^4\cdot 3} = 5\cdot 3\sqrt[4]{4}\cdot 2\sqrt[4]{3} = (5\cdot 3\cdot 2)\sqrt[4]{4\cdot 3} = \mathbf{30\sqrt[4]{12}}$

7. $\sqrt[2]{192}\cdot\sqrt[2]{48}\cdot\sqrt[2]{300} = \sqrt{192}\cdot\sqrt{48}\cdot\sqrt{300} = \sqrt{64\cdot 3}\cdot\sqrt{16\cdot 3}\cdot\sqrt{100\cdot 3} = \sqrt{8^2\cdot 3}\cdot\sqrt{4^2\cdot 3}\cdot\sqrt{10^2\cdot 3} = 8\sqrt{3}\cdot 4\sqrt{3}\cdot 10\sqrt{3}$

$= (8\cdot 4\cdot 10)\sqrt{3\cdot 3\cdot 3} = 320\sqrt{3^2\cdot 3} = (320\cdot 3)\sqrt{3} = \mathbf{960\sqrt{3}}$

8. $\sqrt{75}\cdot\sqrt{150} = \sqrt{25\cdot 3}\cdot\sqrt{25\cdot 6} = \sqrt{5^2\cdot 3}\cdot\sqrt{5^2\cdot 6} = 5\sqrt{3}\cdot 5\sqrt{6} = (5\cdot 5)\sqrt{3\cdot 6} = 25\sqrt{18} = 25\sqrt{9\cdot 2} = 25\sqrt{3^2\cdot 2}$

$= (25\cdot 3)\sqrt{2} = \mathbf{75\sqrt{2}}$

9. $\sqrt[3]{343}\cdot\sqrt[3]{128}\cdot\sqrt[3]{108} = \sqrt[3]{7^3}\cdot\sqrt[3]{64\cdot 2}\cdot\sqrt[3]{27\cdot 4} = 7\cdot\sqrt[3]{4^3\cdot 2}\cdot\sqrt[3]{3^3\cdot 4} = 7\cdot 4\sqrt[3]{2}\cdot 3\sqrt[3]{4} = (7\cdot 4\cdot 3)\sqrt[3]{2\cdot 4} = 84\sqrt[3]{8} = 84\sqrt[3]{2^3}$

$= (84\cdot 2) = \mathbf{168}$

10. $\sqrt{225}\cdot\sqrt{800}\cdot\sqrt{18} = \sqrt{15\cdot 15}\cdot\sqrt{400\cdot 2}\cdot\sqrt{9\cdot 2} = \sqrt{15^2}\cdot\sqrt{20^2\cdot 2}\cdot\sqrt{3^2\cdot 2} = 15\cdot 20\sqrt{2}\cdot 3\sqrt{2} = (15\cdot 20\cdot 3)\sqrt{2\cdot 2}$

$= 900\sqrt{2^2} = (900\cdot 2) = \mathbf{1800}$

Section 4.2 Case I b Practice Problems

1. $\sqrt{x^5y^6}\cdot\sqrt{x^2y^3} = \sqrt{x^{2+2+1}y^{2+2+2}}\cdot x\sqrt{y^{2+1}} = \sqrt{\left(x^2\cdot x^2\cdot x^1\right)\cdot\left(y^2\cdot y^2\cdot y^2\right)}\cdot x\sqrt{y^2\cdot y^1} = (x\cdot x)\cdot(y\cdot y\cdot y)\sqrt{x}\cdot xy\sqrt{y}$

$= x^2y^3\sqrt{x}\cdot xy\sqrt{y} = \left(x^2\cdot x\right)\cdot\left(y^3\cdot y\right)\sqrt{x\cdot y} = \mathbf{x^3y^4\sqrt{xy}}$

2. $\sqrt{a^5b^5}\cdot\sqrt{a^2b^3}\cdot b^2\sqrt{a^4} = \sqrt{a^{2+2+1}b^{2+2+1}}\cdot a\sqrt{b^{2+1}}\cdot b^2\sqrt{a^{2+2}} = \sqrt{\left(a^2\cdot a^2\cdot a^1\right)\cdot\left(b^2\cdot b^2\cdot b^1\right)}\cdot a\sqrt{b^2\cdot b^1}\cdot b^2\sqrt{a^2\cdot a^2}$

$= (a\cdot a)\cdot(b\cdot b)\sqrt{ab}\cdot ab\sqrt{b}\cdot(a\cdot a)b^2 = a^2b^2\sqrt{ab}\cdot ab\sqrt{b}\cdot a^2b^2 = \left(a^2\cdot a^2\cdot a\right)\cdot\left(b^2\cdot b^2\cdot b\right)\sqrt{a(b\cdot b)} = a^5b^5\sqrt{ab^2}$

$= a^5\left(b^5\cdot b\right)\sqrt{a} = \mathbf{a^5b^6\sqrt{a}}$

3. $\sqrt[5]{x^8y^4z^7}\cdot z^2\sqrt[5]{x^3y^2} = \sqrt[5]{x^{5+3}y^4z^{5+2}}\cdot z^2\sqrt[5]{x^3y^2} = \sqrt[5]{\left(x^5\cdot x^3\right)\cdot y^4\cdot\left(z^5\cdot z^2\right)}\cdot z^2\sqrt[5]{x^3y^2} = xz\sqrt[5]{x^3y^4z^2}\cdot z^2\sqrt[5]{x^3y^2}$

$= x\left(z\cdot z^2\right)\sqrt[5]{\left(x^3\cdot x^3\right)\cdot\left(y^4\cdot y^2\right)\cdot z^2} = xz^3\sqrt[5]{x^6y^6z^2} = xz^3\sqrt[5]{x^{5+1}y^{5+1}z^2} = xz^3\sqrt[5]{\left(x^5\cdot x^1\right)\cdot\left(y^5\cdot y^1\right)z^2} = (x\cdot x)yz^3\sqrt[5]{xyz^2}$

$= \mathbf{x^2yz^3\sqrt[5]{xyz^2}}$

4. $\sqrt{x^5y^3z^2}\cdot\sqrt{x^3y^2z} = z\sqrt{x^{2+2+1}y^{2+1}}\cdot y\sqrt{x^{2+1}z} = z\sqrt{\left(x^2\cdot x^2\cdot x^1\right)\cdot\left(y^2\cdot y^1\right)}\cdot y\sqrt{\left(x^2\cdot x^1\right)z} = (x\cdot x)yz\sqrt{x\cdot y}\cdot xy\sqrt{xz}$

$= \left(x^2\cdot x\right)\cdot(y\cdot y)\cdot z\sqrt{(x\cdot x)yz} = x^3y^2z\sqrt{x^2yz} = \left(x^3\cdot x\right)y^2z\sqrt{yz} = \mathbf{x^4y^2z\sqrt{yz}}$

5. $\sqrt[3]{x^5y^6z}\cdot\sqrt[3]{xyz}\cdot\sqrt[3]{x^4yz^4} = \sqrt[3]{x^{3+2}y^{3+3}z}\cdot\sqrt[3]{xyz}\cdot\sqrt[3]{x^{3+1}yz^{3+1}} = \sqrt[3]{\left(x^3\cdot x^2\right)\cdot\left(y^3\cdot y^3\right)\cdot z}\cdot\sqrt[3]{xyz}\cdot\sqrt[3]{\left(x^3\cdot x^1\right)\cdot y\cdot\left(z^3\cdot z^1\right)}$

$= x(y\cdot y)\sqrt[3]{x^2z}\cdot\sqrt[3]{xyz}\cdot xz\sqrt[3]{xyz} = (x\cdot x)\cdot y^2z\sqrt[3]{\left(x^2\cdot x\cdot x\right)\cdot(y\cdot y)\cdot(z\cdot z\cdot z)} = x^2y^2z\sqrt[3]{x^4y^2z^3} = x^2y^2(z\cdot z)\sqrt[3]{x^{3+1}y^2}$

$= x^2y^2z^2\sqrt[3]{\left(x^3\cdot x^1\right)y^2} = \left(x^2\cdot x\right)y^2z^2\sqrt[3]{xy^2} = \mathbf{x^3y^2z^2\sqrt[3]{xy^2}}$

6. $\sqrt[4]{u^5v^6}\cdot\sqrt[4]{uv^8}\cdot\sqrt[4]{u^2v^2} = \sqrt[4]{u^{4+1}v^{4+2}}\cdot\sqrt[4]{uv^{4+4}}\cdot\sqrt[4]{u^2v^2} = \sqrt[4]{\left(u^4\cdot u^1\right)\cdot\left(v^4\cdot v^2\right)}\cdot\sqrt[4]{u\left(v^4\cdot v^4\right)}\cdot\sqrt[4]{u^2v^2}$

$= uv\sqrt[4]{uv^2}\cdot(v\cdot v)\sqrt[4]{u}\cdot\sqrt[4]{u^2v^2} = u\left(v\cdot v^2\right)\sqrt[4]{uv^2}\cdot\sqrt[4]{u}\cdot\sqrt[4]{u^2v^2} = uv^3\sqrt[4]{\left(u\cdot u\cdot u^2\right)\cdot\left(v^2\cdot v^2\right)} = uv^3\sqrt[4]{u^4v^4}$

$= (u\cdot u)\cdot\left(v^3\cdot v\right) = \mathbf{u^2v^4}$

7. $\sqrt{40r^3s}\cdot\sqrt{36r^2s}\cdot\sqrt{4r^5s^5} = \sqrt{(4\cdot 10)r^{2+1}s}\cdot r\sqrt{6^2s}\cdot\sqrt{2^2r^{2+2+1}s^{2+2+1}} = 2\sqrt{10\left(r^2\cdot r^1\right)s}\cdot 6r\sqrt{s}\cdot 2\sqrt{\left(r^2\cdot r^2\cdot r^1\right)\cdot\left(s^2\cdot s^2\cdot s^1\right)}$

$= 2r\sqrt{10rs}\cdot 6r\sqrt{s}\cdot 2(r\cdot r)(s\cdot s)\sqrt{rs} = 2r\sqrt{10rs}\cdot 6r\sqrt{s}\cdot 2r^2s^2\sqrt{rs} = (2\cdot 6\cdot 2)\cdot\left(r\cdot r\cdot r^2\right)\cdot s^2\sqrt{10\cdot(r\cdot r)\cdot(s\cdot s\cdot s)}$

$= 24r^4s^2\sqrt{10r^2s^3} = 24\left(r^4\cdot r\right)\cdot s^2\sqrt{10s^{2+1}} = 24r^5\cdot s^2\sqrt{10\left(s^2\cdot s^1\right)} = 24r^5\cdot\left(s^2\cdot s\right)\sqrt{10s} = \mathbf{24r^5s^3\sqrt{10s}}$

8. $\sqrt[3]{125u^6v^8w}\cdot\sqrt[3]{54uv^2w^5} = \sqrt[3]{5^3u^{3+3}v^{3+3+2}w}\cdot\sqrt[3]{(27\cdot 2)uv^2w^{3+2}} = 5\sqrt[3]{\left(u^3\cdot u^3\right)\cdot\left(v^3\cdot v^3\cdot v^2\right)\cdot w}\cdot\sqrt[3]{\left(3^3\cdot 2\right)u\cdot v^2\cdot\left(w^3\cdot w^2\right)}$

$= 5(u\cdot u)(v\cdot v)\sqrt[3]{v^2\cdot w}\cdot 3w\sqrt[3]{2uv^2w^2} = 5u^2v^2\sqrt[3]{v^2\cdot w}\cdot 3w\sqrt[3]{2uv^2w^2} = (5\cdot 3)u^2v^2w\sqrt[3]{2u\cdot\left(v^2\cdot v^2\right)\cdot\left(w\cdot w^2\right)}$

$= 15u^2v^2w\sqrt[3]{2uv^4w^3} = 15u^2v^2(w\cdot w)\sqrt[3]{2uv^{3+1}} = 15u^2v^2w^2\sqrt[3]{2uv^3\cdot v^1} = 15u^2\left(v^2\cdot v\right)w^2\sqrt[3]{2uv} = \mathbf{15u^2v^3w^2\sqrt[3]{2uv}}$

9. $\sqrt[3]{m^4n^3l}\cdot\sqrt{m^2l}\cdot\sqrt[3]{m^4n^5l^4}\cdot\sqrt{mn^5l^3} = n\sqrt[3]{m^{3+1}l}\cdot m\sqrt{l}\cdot\sqrt[3]{m^{3+1}n^{3+2}l^{3+1}}\cdot\sqrt{mn^{2+2+1}l^{2+1}}$

$= n\sqrt[3]{\left(m^3\cdot m^1\right)\cdot l}\cdot m\sqrt{l}\cdot\sqrt[3]{\left(m^3\cdot m^1\right)\cdot\left(n^3\cdot n^2\right)\cdot\left(l^3\cdot l^1\right)}\cdot\sqrt{m\cdot\left(n^2\cdot n^2\cdot n^1\right)\cdot\left(l^2\cdot l^1\right)} = mn\sqrt[3]{ml}\cdot m\sqrt{l}\cdot mnl\sqrt[3]{mn^2l}\cdot(n\cdot n)l\sqrt{mnl}$

$= mn\sqrt[3]{ml}\cdot m\sqrt{l}\cdot mnl\sqrt[3]{mn^2l}\cdot n^2l\sqrt{mnl} = (m\cdot m\cdot m)\cdot\left(n\cdot n\cdot n^2\right)\cdot(l\cdot l)\left[\sqrt[3]{ml}\cdot\sqrt[3]{mn^2l}\cdot\sqrt{l}\cdot\sqrt{mnl}\right]$

$= m^3n^4l^2\left[\sqrt[3]{(m\cdot m)n^2(l\cdot l)}\cdot\sqrt{mn(l\cdot l)}\right] = m^3n^4l^2\left[\sqrt[3]{m^2n^2l^2}\cdot\sqrt{mnl^2}\right] = m^3n^4l^2\left[\sqrt[3]{m^2n^2l^2}\cdot l\sqrt{mn}\right]$

$= m^3n^4\left(l^2\cdot l\right)\left[\sqrt[3]{m^2n^2l^2}\cdot\sqrt{mn}\right] = \mathbf{m^3n^4l^3\left(\sqrt[3]{m^2n^2l^2}\cdot\sqrt{mn}\right)}$

10. $\sqrt[5]{a^{10}b^4c^2}\cdot\sqrt[5]{abc^3}\cdot\sqrt[5]{a^2b^{10}c^5} = \sqrt[5]{a^{5+5}b^4c^2}\cdot\sqrt[5]{abc^3}\cdot c\sqrt[5]{a^2b^{5+5}} = \sqrt[5]{\left(a^5\cdot a^5\right)b^4c^2}\cdot\sqrt[5]{abc^3}\cdot c\sqrt[5]{a^2\left(b^5\cdot b^5\right)}$

$= (a\cdot a)\sqrt[5]{b^4c^2}\cdot\sqrt[5]{abc^3}\cdot(b\cdot b)c\sqrt[5]{a^2} = a^2b^2c\sqrt[5]{\left(a\cdot a^2\right)\cdot\left(b^4\cdot b\right)\cdot\left(c^2\cdot c^3\right)} = a^2b^2c\sqrt[5]{a^3b^5c^5} = a^2\left(b^2\cdot b\right)\cdot(c\cdot c)\sqrt[5]{a^3}$

$= \mathbf{a^2b^3c^2\sqrt[5]{a^3}}$

Section 4.2 Case II a Practice Problems

1. $\left(2\sqrt{3}+1\right)\cdot\left(2+\sqrt{2}\right) = (2\cdot 2)\sqrt{3}+\left(2\sqrt{3}\cdot\sqrt{2}\right)+(1\cdot 2)+\left(1\cdot\sqrt{2}\right) = 4\sqrt{3}+2\sqrt{3\cdot 2}+2+\sqrt{2} = \mathbf{4\sqrt{3}+2\sqrt{6}+\sqrt{2}+2}$

2. $\left(1+\sqrt{5}\right)\cdot\left(\sqrt{8}+\sqrt{5}\right) = \left(1+\sqrt{5}\right)\cdot\left(\sqrt{4\cdot 2}+\sqrt{5}\right) = \left(1+\sqrt{5}\right)\cdot\left(\sqrt{2^2\cdot 2}+\sqrt{5}\right) = \left(1+\sqrt{5}\right)\cdot\left(2\sqrt{2}+\sqrt{5}\right)$

$= \left(1\cdot 2\sqrt{2}\right)+\left(1\cdot\sqrt{5}\right)+\left(2\sqrt{2}\cdot\sqrt{5}\right)+\left(\sqrt{5}\cdot\sqrt{5}\right) = 2\sqrt{2}+\sqrt{5}+2\sqrt{2\cdot 5}+\sqrt{5\cdot 5} = 2\sqrt{2}+\sqrt{5}+2\sqrt{10}+\sqrt{5^2}$

$= \mathbf{2\sqrt{2}+\sqrt{5}+2\sqrt{10}+5}$

3. $\left(2-\sqrt{2}\right)\cdot\left(3+\sqrt{2}\right) = (2\cdot 3)+\left(2\cdot\sqrt{2}\right)-\left(3\cdot\sqrt{2}\right)-\left(\sqrt{2}\cdot\sqrt{2}\right) = 6+2\sqrt{2}-3\sqrt{2}-\sqrt{2\cdot 2} = 6+(2-3)\sqrt{2}-\sqrt{2^2}$

$= 6-\sqrt{2}-2 = (6-2)-\sqrt{2} = \mathbf{4-\sqrt{2}}$

4. $\left(5+\sqrt{5}\right)\cdot\left(5-\sqrt{5^3}\right) = \left(5+\sqrt{5}\right)\cdot\left(5-\sqrt{5^{2+1}}\right) = \left(5+\sqrt{5}\right)\cdot\left(5-\sqrt{5^2\cdot5^1}\right) = \left(5+\sqrt{5}\right)\cdot\left(5-5\sqrt{5}\right)$

$= (5\cdot5)-(5\cdot5)\sqrt{5}+\left(5\cdot\sqrt{5}\right)-\left(5\sqrt{5}\cdot\sqrt{5}\right) = 25-25\sqrt{5}+5\sqrt{5}-5\sqrt{5\cdot5} = 25+(-25+5)\sqrt{5}-5\sqrt{5^2}$

$= 25-20\sqrt{5}-5\cdot5 = 25-20\sqrt{5}-25 = (25-25)-20\sqrt{5} = \mathbf{-20\sqrt{5}}$

5. $\left(2+\sqrt{6}\right)\cdot\left(\sqrt[4]{16}-\sqrt{18}\right) = \left(2+\sqrt{6}\right)\cdot\left(\sqrt[4]{2^4}-\sqrt{9\cdot2}\right) = \left(2+\sqrt{6}\right)\cdot\left(2-\sqrt{3^2\cdot2}\right) = \left(2+\sqrt{6}\right)\cdot\left(2-3\sqrt{2}\right)$

$= (2\cdot2)-(2\cdot3)\sqrt{2}+\left(2\cdot\sqrt{6}\right)-\left(3\sqrt{2}\cdot\sqrt{6}\right) = 4-6\sqrt{2}+2\sqrt{6}-3\sqrt{2\cdot6} = 4-6\sqrt{2}+2\sqrt{6}-3\sqrt{12}$

$= 4-6\sqrt{2}+2\sqrt{6}-3\sqrt{4\cdot3} = 4-6\sqrt{2}+2\sqrt{6}-3\sqrt{2^2\cdot3} = 4-6\sqrt{2}+2\sqrt{6}-(3\cdot2)\sqrt{3} = \mathbf{4-6\sqrt{2}+2\sqrt{6}-6\sqrt{3}}$

6. $\left(2-\sqrt{5}\right)\cdot\left(\sqrt{45}+\sqrt[4]{81}\right) = \left(2-\sqrt{5}\right)\cdot\left(\sqrt{9\cdot5}+\sqrt[4]{3^4}\right) = \left(2-\sqrt{5}\right)\cdot\left(\sqrt{3^2\cdot5}+3\right) = \left(2-\sqrt{5}\right)\cdot\left(3\sqrt{5}+3\right)$

$= (2\cdot3)\sqrt{5}+(2\cdot3)-\left(3\sqrt{5}\cdot\sqrt{5}\right)-\left(3\cdot\sqrt{5}\right) = 6\sqrt{5}+6-3\sqrt{5\cdot5}-3\sqrt{5} = 6\sqrt{5}+6-3\sqrt{5^2}-3\sqrt{5} = 6\sqrt{5}+6-(3\cdot5)-3\sqrt{5}$

$= 6\sqrt{5}-3\sqrt{5}+6-15 = (6-3)\sqrt{5}-9 = 3\sqrt{5}-9 = \mathbf{3\left(\sqrt{5}-3\right)}$

7. $\left(2-\sqrt{m}\right)\cdot\left(\sqrt{8}+\sqrt{m^3}\right) = \left(2-\sqrt{m}\right)\cdot\left(\sqrt{4\cdot2}+\sqrt{m^{2+1}}\right) = \left(2-\sqrt{m}\right)\cdot\left(\sqrt{2^2\cdot2}+\sqrt{m^2\cdot m^1}\right) = \left(2-\sqrt{m}\right)\cdot\left(2\sqrt{2}+m\sqrt{m}\right)$

$= (2\cdot2)\sqrt{2}+(2\cdot m)\sqrt{m}-2\left(\sqrt{2}\cdot\sqrt{m}\right)-m\left(\sqrt{m}\cdot\sqrt{m}\right) = 4\sqrt{2}+2m\sqrt{m}-2\sqrt{2\cdot m}-m\sqrt{m\cdot m}$

$= 4\sqrt{2}+2m\sqrt{m}-2\sqrt{2m}-m\sqrt{m^2} = 4\sqrt{2}+2m\sqrt{m}-2\sqrt{2m}-m\cdot m = \mathbf{4\sqrt{2}+2m\sqrt{m}-2\sqrt{2m}-m^2}$

8. $\left(\sqrt{32}-\sqrt{3}\right)\cdot\left(2+\sqrt{3}\right) = \left(\sqrt{16\cdot2}-\sqrt{3}\right)\cdot\left(2+\sqrt{3}\right) = \left(\sqrt{4^2\cdot2}-\sqrt{3}\right)\cdot\left(2+\sqrt{3}\right) = \left(4\sqrt{2}-\sqrt{3}\right)\cdot\left(2+\sqrt{3}\right)$

$= (4\cdot2)\sqrt{2}+4\left(\sqrt{2}\cdot\sqrt{3}\right)-\left(2\cdot\sqrt{3}\right)-\left(\sqrt{3}\cdot\sqrt{3}\right) = 8\sqrt{2}+4\sqrt{2\cdot3}-2\sqrt{3}-\sqrt{3\cdot3} = 8\sqrt{2}+4\sqrt{6}-2\sqrt{3}-\sqrt{3^2}$

$= \mathbf{8\sqrt{2}+4\sqrt{6}-2\sqrt{3}-3}$

9. $\left(a\sqrt{x}+\sqrt{x^2}\right)\cdot\left(\sqrt{a}-\sqrt{x}\right) = \left(a\sqrt{x}+x\right)\cdot\left(\sqrt{a}-\sqrt{x}\right) = \left(a\sqrt{x}\cdot\sqrt{a}\right)-\left(a\sqrt{x}\cdot\sqrt{x}\right)+\left(x\sqrt{a}\right)-\left(x\sqrt{x}\right)$

$= a\sqrt{a\cdot x}-a\sqrt{x\cdot x}+x\sqrt{a}-x\sqrt{x} = a\sqrt{ax}-ax+x\sqrt{a}-x\sqrt{x} = \mathbf{a\left(\sqrt{ax}-x\right)+x\left(\sqrt{a}-\sqrt{x}\right)}$

10. $\left(\sqrt{2}+\sqrt{3}\right)\cdot\left(\sqrt{32}-\sqrt{75}\right) = \left(\sqrt{2}+\sqrt{3}\right)\cdot\left(\sqrt{16\cdot2}-\sqrt{25\cdot3}\right) = \left(\sqrt{2}+\sqrt{3}\right)\cdot\left(\sqrt{4^2\cdot2}-\sqrt{5^2\cdot3}\right) = \left(\sqrt{2}+\sqrt{3}\right)\cdot\left(4\sqrt{2}-5\sqrt{3}\right)$

$= 4\left(\sqrt{2}\cdot\sqrt{2}\right)-5\left(\sqrt{2}\cdot\sqrt{3}\right)+4\left(\sqrt{2}\cdot\sqrt{3}\right)-5\left(\sqrt{3}\cdot\sqrt{3}\right) = 4\sqrt{2\cdot2}-5\sqrt{2\cdot3}+4\sqrt{2\cdot3}-5\sqrt{3\cdot3} = 4\sqrt{2^2}-5\sqrt{6}+4\sqrt{6}-5\sqrt{3^2}$

$= (4\cdot2)+(-5+4)\sqrt{6}-(5\cdot3) = 8-\sqrt{6}-15 = (8-15)-\sqrt{6} = -7-\sqrt{6} = \mathbf{-\left(7+\sqrt{6}\right)}$

Section 4.2 Case II b Practice Problems

1. $\left(a+\sqrt{b}\right)\cdot\left(a-\sqrt{b^3}\right) = \left(a+\sqrt{b}\right)\cdot\left(a-\sqrt{b^{2+1}}\right) = \left(a+\sqrt{b}\right)\cdot\left(a-\sqrt{b^2\cdot b^1}\right) = \left(a+\sqrt{b}\right)\cdot\left(a-b\sqrt{b}\right)$

$= (a\cdot a)-\left(a\cdot b\sqrt{b}\right)+\left(a\cdot\sqrt{b}\right)-\left(b\sqrt{b}\cdot\sqrt{b}\right) = a^2-ab\sqrt{b}+a\sqrt{b}-b\sqrt{b\cdot b} = a^2-ab\sqrt{b}+a\sqrt{b}-b\sqrt{b^2}$

$= a^2-ab\sqrt{b}+a\sqrt{b}-b\cdot b = a^2-ab\sqrt{b}+a\sqrt{b}-b^2 = \mathbf{a^2-b^2+a\sqrt{b}(1-b)}$

2. $\left(a+x\sqrt{x}\right)\cdot\left(a-\sqrt{x}\right) = (a\cdot a)-\left(a\cdot\sqrt{x}\right)+\left(a\cdot x\sqrt{x}\right)-\left(x\sqrt{x}\cdot\sqrt{x}\right) = a^2-a\sqrt{x}+ax\sqrt{x}-x\sqrt{x\cdot x}$

$= a^2-a\sqrt{x}+ax\sqrt{x}-x\sqrt{x^2} = a^2-a\sqrt{x}+ax\sqrt{x}-x\cdot x = a^2-a\sqrt{x}+ax\sqrt{x}-x^2 = \mathbf{a^2-x^2+a\sqrt{x}(x-1)}$

3. $\left(5a+\sqrt{x^5}\right)\cdot\left(2a-\sqrt{x}\right) = \left(5a+\sqrt{x^{2+2+1}}\right)\cdot\left(2a-\sqrt{x}\right) = \left(5a+\sqrt{x^2\cdot x^2\cdot x^1}\right)\cdot\left(2a-\sqrt{x}\right) = \left(5a+(x\cdot x)\sqrt{x}\right)\cdot\left(2a-\sqrt{x}\right)$

$= \left(5a+x^2\sqrt{x}\right)\cdot\left(2a-\sqrt{x}\right) = (5a\cdot 2a)-\left(5a\cdot\sqrt{x}\right)+\left(2a\cdot x^2\sqrt{x}\right)-\left(x^2\sqrt{x}\cdot\sqrt{x}\right) = 10a^2-5a\sqrt{x}+2ax^2\sqrt{x}-x^2\sqrt{x\cdot x}$

$= 10a^2-5a\sqrt{x}+2ax^2\sqrt{x}-x^2\cdot x = 10a^2-5a\sqrt{x}+2ax^2\sqrt{x}-x^3 = \mathbf{10a^2-x^3+a\sqrt{x}\left(2x^2-5\right)}$

4. $\left(4+\sqrt{r}\right)\cdot\left(7-\sqrt{r}\right) = (4\cdot 7)-\left(4\cdot\sqrt{r}\right)+\left(7\cdot\sqrt{r}\right)-\left(\sqrt{r}\cdot\sqrt{r}\right) = 28-4\sqrt{r}+7\sqrt{r}-\sqrt{r\cdot r} = 28+(-4+7)\sqrt{r}-\sqrt{r^2}$

$= \mathbf{28+3\sqrt{r}-r}$

5. $\left(2+\sqrt{x^3y^3}\right)\cdot\left(2-y\sqrt{x^3}\right) = \left(2+xy\sqrt{xy}\right)\cdot\left(2-xy\sqrt{x}\right) = (2\cdot 2)-\left(2\cdot xy\sqrt{x}\right)+\left(2\cdot xy\sqrt{xy}\right)-x^2y^2\left(\sqrt{xy}\cdot\sqrt{x}\right)$

$= 4-2xy\sqrt{x}+2xy\sqrt{xy}-x^2y^2\sqrt{(x\cdot x)y} = 4-2xy\sqrt{x}+2xy\sqrt{xy}-x^2y^2\sqrt{x^2y} = 4-2xy\sqrt{x}+2xy\sqrt{xy}-\left(x^2\cdot x\right)y^2\sqrt{y}$

$= 4-2xy\sqrt{x}+2xy\sqrt{xy}-x^3y^2\sqrt{y} = 2xy\left(\sqrt{xy}-\sqrt{x}\right)-x^3y^2\sqrt{y}+4 = \mathbf{2xy\sqrt{x}\left(\sqrt{y}-1\right)-x^3y^2\sqrt{y}+4}$

6. $\left(m+\sqrt[3]{m^4}\right)\cdot\left(m-\sqrt[3]{m^5}\right) = \left(m+\sqrt[3]{m^{3+1}}\right)\cdot\left(m-\sqrt[3]{m^{3+2}}\right) = \left(m+\sqrt[3]{m^3\cdot m^1}\right)\cdot\left(m-\sqrt[3]{m^3\cdot m^2}\right) = \left(m+m\sqrt[3]{m}\right)\cdot\left(m-m\sqrt[3]{m^2}\right)$

$= (m\cdot m)-\left(m\cdot m\sqrt[3]{m^2}\right)+\left(m\cdot m\sqrt[3]{m}\right)-(m\cdot m)\sqrt[3]{m}\cdot\sqrt[3]{m^2} = m^2-m^2\sqrt[3]{m^2}+m^2\sqrt[3]{m}-m^2\sqrt[3]{m\cdot m^2}$

$= m^2-m^2\sqrt[3]{m^2}+m^2\sqrt[3]{m}-m^2\sqrt[3]{m^3} = m^2-m^2\sqrt[3]{m^2}+m^2\sqrt[3]{m}-m^2\cdot m = m^2-m^3+m^2\sqrt[3]{m}-m^2\sqrt[3]{m^2}$

$= \mathbf{m^2\left(1-m+\sqrt[3]{m}-\sqrt[3]{m^2}\right)}$

7. $\left(\sqrt{r}+2\right)\cdot\left(4\sqrt{r^3}-2\right) = \left(\sqrt{r}+2\right)\cdot\left(4\sqrt{r^{2+1}}-2\right) = \left(\sqrt{r}+2\right)\cdot\left(4\sqrt{r^2\cdot r^1}-2\right) = \left(\sqrt{r}+2\right)\cdot\left(4r\sqrt{r}-2\right)$

$= \left(4r\sqrt{r}\cdot\sqrt{r}\right)-\left(2\cdot\sqrt{r}\right)+(2\cdot 4)r\sqrt{r}-(2\cdot 2) = 4r\sqrt{r\cdot r}-2\sqrt{r}+8r\sqrt{r}-4 = 4r\sqrt{r^2}+2\sqrt{r}(4r-1)-4$

$= 4r \cdot r + 2\sqrt{r}(4r-1) - 4 = 4r^2 - 4 + 2\sqrt{r}(4r-1) = 4(r^2-1) + 2\sqrt{r}(4r-1) = \mathbf{4(r-1)\cdot(r+1) + 2\sqrt{r}(4r-1)}$

8. $(4+\sqrt{a})\cdot(4-\sqrt{a}) = (4\cdot 4) - (4\cdot\sqrt{a}) + (4\cdot\sqrt{a}) - (\sqrt{a}\cdot\sqrt{a}) = 16 - 4\sqrt{a} + 4\sqrt{a} - \sqrt{a\cdot a} = 16 - \sqrt{a^2} = \mathbf{16 - a}$

9. $(3+\sqrt[4]{x^5})\cdot(3-\sqrt[4]{x^7}) = (3+\sqrt[4]{x^{4+1}})\cdot(3-\sqrt[4]{x^{4+3}}) = (3+\sqrt[4]{x^4\cdot x^1})\cdot(3-\sqrt[4]{x^4\cdot x^3}) = (3+x\sqrt[4]{x})\cdot(3-x\sqrt[4]{x^3})$

$= (3\cdot 3) - (3\cdot x\sqrt[4]{x^3}) + (3\cdot x\sqrt[4]{x}) - (x\cdot x)(\sqrt[4]{x}\cdot\sqrt[4]{x^3}) = 9 - 3x\sqrt[4]{x^3} + 3x\sqrt[4]{x} - x^2\sqrt[4]{x\cdot x^3} = 9 - 3x\sqrt[4]{x^3} + 3x\sqrt[4]{x} - x^2\sqrt[4]{x^4}$

$= 9 - 3x\sqrt[4]{x^3} + 3x\sqrt[4]{x} - x^2\cdot x = 9 - 3x\sqrt[4]{x^3} + 3x\sqrt[4]{x} - x^3 = \mathbf{9 - x^3 + 3x(\sqrt[4]{x} - \sqrt[4]{x^3})}$

10. $(1+\sqrt{xy})\cdot(1-\sqrt{x^3y^3}) = (1+\sqrt{xy})\cdot(1-\sqrt{x^{2+1}y^{2+1}}) = (1+\sqrt{xy})\cdot\left(1-\sqrt{(x^2\cdot x^1)\cdot(y^2\cdot y^1)}\right) = (1+\sqrt{xy})\cdot(1-xy\sqrt{xy})$

$= (1\cdot 1) - (1\cdot xy\sqrt{xy}) + (1\cdot\sqrt{xy}) - (xy\sqrt{xy}\cdot\sqrt{xy}) = 1 - xy\sqrt{xy} + \sqrt{xy} - xy\sqrt{xy\cdot xy} = 1 - xy\sqrt{xy} + \sqrt{xy} - xy\sqrt{x^2y^2}$

$= 1 - xy\sqrt{xy} + \sqrt{xy} - xy\cdot xy = 1 - xy\sqrt{xy} + \sqrt{xy} - x^2y^2 = \mathbf{1 - x^2y^2 + \sqrt{xy}(1-xy)}$

Section 4.2 Case III a Practice Problems

1. $2\sqrt{3}\cdot(2+\sqrt{2}) = (2\cdot 2)\sqrt{3} + (2\sqrt{3}\cdot\sqrt{2}) = 4\sqrt{3} + 2\sqrt{3\cdot 2} = 4\sqrt{3} + 2\sqrt{6} = \mathbf{2(2\sqrt{3} + \sqrt{6})}$

2. $\sqrt{5}\cdot(\sqrt{8}+\sqrt{5}) = (\sqrt{5}\cdot\sqrt{8}) + (\sqrt{5}\cdot\sqrt{5}) = (\sqrt{5\cdot 8}) + (\sqrt{5\cdot 5}) = \sqrt{40} + \sqrt{5^2} = \sqrt{4\cdot 10} + 5 = \sqrt{2^2\cdot 10} + 5 = \mathbf{5 + 2\sqrt{10}}$

3. $-\sqrt{8}\cdot(3-\sqrt{3}) = -\sqrt{4\cdot 2}\cdot(3-\sqrt{3}) = -\sqrt{2^2\cdot 2}\cdot(3-\sqrt{3}) = -2\sqrt{2}\cdot(3-\sqrt{3}) = (-(2\cdot 3)\cdot\sqrt{2}) + (2\sqrt{2}\cdot\sqrt{3})$

$= -6\sqrt{2} + 2\sqrt{2\cdot 3} = -6\sqrt{2} + 2\sqrt{6} = \mathbf{2(\sqrt{6} - 3\sqrt{2})}$

4. $4\sqrt{98}\cdot(3-\sqrt{2^3}) = 4\sqrt{49\cdot 2}\cdot(3-\sqrt{2^{2+1}}) = 4\sqrt{7^2\cdot 2}\cdot(3-\sqrt{2^2\cdot 2^1}) = (4\cdot 7)\sqrt{2}\cdot(3-2\sqrt{2}) = 28\sqrt{2}\cdot(3-2\sqrt{2})$

$= (28\cdot 3)\sqrt{2} - (28\cdot 2)\cdot(\sqrt{2}\cdot\sqrt{2}) = 84\sqrt{2} - 56(\sqrt{2\cdot 2}) = 84\sqrt{2} - 56\sqrt{2^2} = 84\sqrt{2} - (56\cdot 2) = 84\sqrt{2} - 112 = \mathbf{4(21\sqrt{2} - 28)}$

5. $\sqrt[4]{48}\cdot(\sqrt[4]{324} + \sqrt[4]{32}) = \sqrt[4]{16\cdot 3}\cdot(\sqrt[4]{81\cdot 4} + \sqrt[4]{16\cdot 2}) = \sqrt[4]{2^4\cdot 3}\cdot(\sqrt[4]{3^4\cdot 4} + \sqrt[4]{2^4\cdot 2}) = 2\sqrt[4]{3}\cdot(3\sqrt[4]{4} + 2\sqrt[4]{2})$

$= (2\cdot 3)\cdot(\sqrt[4]{3}\cdot\sqrt[4]{4}) + (2\cdot 2)\cdot(\sqrt[4]{3}\cdot\sqrt[4]{2}) = 6\cdot(\sqrt[4]{3\cdot 4}) + 4\cdot(\sqrt[4]{3\cdot 2}) = 6\sqrt[4]{12} + 4\sqrt[4]{6} = \mathbf{2(3\sqrt[4]{12} + 2\sqrt[4]{6})}$

6. $2\sqrt{5}\cdot(\sqrt{45} + \sqrt[4]{81}) = 2\sqrt{5}\cdot(\sqrt{9\cdot 5} + \sqrt[4]{3^4}) = 2\sqrt{5}\cdot(\sqrt{3^2\cdot 5} + 3) = 2\sqrt{5}\cdot(3\sqrt{5} + 3) = (2\cdot 3)(\sqrt{5}\cdot\sqrt{5}) + (2\cdot 3)\sqrt{5}$

$= 6(\sqrt{5\cdot 5}) + 6\sqrt{5} = 6\sqrt{5^2} + 6\sqrt{5} = (6\cdot 5) + 6\sqrt{5} = 30 + 6\sqrt{5} = \mathbf{6(5 + \sqrt{5})}$

7. $-\sqrt[5]{64}\cdot\left(-\sqrt[5]{486}+4\right) = -\sqrt[5]{32\cdot 2}\cdot\left(-\sqrt[5]{243\cdot 2}+4\right) = -\sqrt[5]{2^5\cdot 2}\cdot\left(-\sqrt[5]{3^5\cdot 2}+4\right) = -2\sqrt[5]{2}\cdot\left(-3\sqrt[5]{2}+4\right)$

$= +(2\cdot 3)\cdot\left(\sqrt[5]{2}\cdot\sqrt[5]{2}\right)-(2\cdot 4)\sqrt[5]{2} = 6\left(\sqrt[5]{2\cdot 2}\right)-8\sqrt[5]{2} = 6\sqrt[5]{2^2}-8\sqrt[5]{2} = 6\sqrt[5]{4}-8\sqrt[5]{2} = \mathbf{2\left(3\sqrt[5]{4}-4\sqrt[5]{2}\right)}$

8. $\sqrt{32}\cdot\left(2+\sqrt{3}\right) = \sqrt{16\cdot 2}\cdot\left(2+\sqrt{3}\right) = \sqrt{4^2\cdot 2}\cdot\left(2+\sqrt{3}\right) = 4\sqrt{2}\cdot\left(2+\sqrt{3}\right) = (4\cdot 2)\sqrt{2}+4\sqrt{2}\cdot\sqrt{3} = 8\sqrt{2}+4\sqrt{2\cdot 3}$

$= 8\sqrt{2}+4\sqrt{6} = \mathbf{4\left(2\sqrt{2}+\sqrt{6}\right)}$

9. $\left(3\sqrt{44}+\sqrt{27}\right)\cdot\sqrt{8} = \left(3\sqrt{4\cdot 11}+\sqrt{9\cdot 3}\right)\cdot\sqrt{4\cdot 2} = \left(3\sqrt{2^2\cdot 11}+\sqrt{3^2\cdot 3}\right)\cdot\sqrt{2^2\cdot 2} = \left[(3\cdot 2)\sqrt{11}+3\sqrt{3}\right]\cdot 2\sqrt{2}$

$= \left(6\sqrt{11}+3\sqrt{3}\right)2\sqrt{2} = (6\cdot 2)\cdot\left(\sqrt{11}\cdot\sqrt{2}\right)+(3\cdot 2)\cdot\left(\sqrt{3}\cdot\sqrt{2}\right) = 12\left(\sqrt{11\cdot 2}\right)+6\left(\sqrt{3\cdot 2}\right) = 12\sqrt{22}+6\sqrt{6} = \mathbf{6\left(2\sqrt{22}+\sqrt{6}\right)}$

10. $-\sqrt{2}\cdot\left(\sqrt{32}-2\sqrt{75}\right) = -\sqrt{2}\cdot\left(\sqrt{16\cdot 2}-2\sqrt{25\cdot 3}\right) = -\sqrt{2}\cdot\left(\sqrt{4^2\cdot 2}-2\sqrt{5^2\cdot 3}\right) = -\sqrt{2}\left(4\sqrt{2}-(2\cdot 5)\sqrt{3}\right) = -\sqrt{2}\left(4\sqrt{2}-10\sqrt{3}\right)$

$= -4\left(\sqrt{2}\cdot\sqrt{2}\right)+10\left(\sqrt{2}\cdot\sqrt{3}\right) = -4\left(\sqrt{2\cdot 2}\right)+10\left(\sqrt{2\cdot 3}\right) = -4\sqrt{2^2}+10\sqrt{6} = -(4\cdot 2)+10\sqrt{6} = -8+10\sqrt{6} = \mathbf{2\left(5\sqrt{6}-4\right)}$

Section 4.2 Case III b Practice Problems

1. $x\sqrt{x^3}\cdot\left(x^2+\sqrt{x}\right) = x\sqrt{x^{2+1}}\cdot\left(x^2+\sqrt{x}\right) = x\sqrt{x^2\cdot x^1}\cdot\left(x^2+\sqrt{x}\right) = (x\cdot x)\sqrt{x}\cdot\left(x^2+\sqrt{x}\right) = x^2\sqrt{x}\cdot\left(x^2+\sqrt{x}\right)$

$= \left(x^2\cdot x^2\right)\sqrt{x}+x^2\left(\sqrt{x}\cdot\sqrt{x}\right) = x^4\sqrt{x}+x^2\sqrt{x\cdot x} = x^4\sqrt{x}+x^2\sqrt{x^2} = x^4\sqrt{x}+x^2\cdot x = x^4\sqrt{x}+x^3 = \mathbf{x^3\left(1+x\sqrt{x}\right)}$

2. $\sqrt{a^5b^4}\cdot\left(\sqrt{a^2b}+\sqrt{ab^3}\right) = \sqrt{a^{2+2+1}b^{2+2}}\cdot\left(a\sqrt{b}+\sqrt{ab^{2+1}}\right) = \left[\sqrt{\left(a^2\cdot a^2\cdot a^1\right)\cdot\left(b^2\cdot b^2\right)}\right]\cdot\left[a\sqrt{b}+\sqrt{a\left(b^2\cdot b^1\right)}\right]$

$= \left[(a\cdot a)\cdot(b\cdot b)\sqrt{a}\right]\cdot\left[a\sqrt{b}+b\sqrt{ab}\right] = a^2b^2\sqrt{a}\cdot\left(a\sqrt{b}+b\sqrt{ab}\right) = \left[\left(a^2\cdot a\right)b^2\sqrt{a}\cdot\sqrt{b}\right]+\left[a^2\left(b^2\cdot b\right)\sqrt{a}\cdot\sqrt{ab}\right]$

$= a^3b^2\sqrt{a\cdot b}+a^2b^3\sqrt{(a\cdot a)b} = a^3b^2\sqrt{ab}+a^2b^3\sqrt{a^2b} = a^3b^2\sqrt{ab}+\left(a^2\cdot a\right)b^3\sqrt{b} = a^3b^2\sqrt{ab}+a^3b^3\sqrt{b}$

$= a^3b^2\left(\sqrt{ab}+b\sqrt{b}\right) = \mathbf{a^3b^2\sqrt{b}\left(\sqrt{a}+b\right)}$

3. $\left(w^3-\sqrt{u^3w^2}\right)\cdot\sqrt{u^4w^5} = \left(w^3-w\sqrt{u^{2+1}}\right)\cdot\sqrt{u^{2+2}w^{2+2+1}} = \left(w^3-w\sqrt{u^2\cdot u^1}\right)\cdot\sqrt{\left(u^2\cdot u^2\right)\cdot\left(w^2\cdot w^2\cdot w^1\right)}$

$= \left(w^3-uw\sqrt{u}\right)\cdot(u\cdot u)\cdot(w\cdot w)\sqrt{w} = \left(w^3-uw\sqrt{u}\right)\cdot u^2w^2\sqrt{w} = u^2\left(w^2\cdot w^3\right)\sqrt{w}-\left(u^2\cdot u\right)\cdot\left(w^2\cdot w\right)\left(\sqrt{u}\cdot\sqrt{w}\right)$

$= u^2w^5\sqrt{w}-u^3w^3\left(\sqrt{u\cdot w}\right) = u^2w^5\sqrt{w}-u^3w^3\sqrt{uw} = u^2w^3\left(w^2\sqrt{w}-u\sqrt{uw}\right) = \mathbf{u^2w^3\sqrt{w}\left(w^2-u\sqrt{u}\right)}$

4. $-\sqrt{m}\cdot\left(-\sqrt{m}+\sqrt{m^3}\right) = -\sqrt{m}\cdot\left(-\sqrt{m}+\sqrt{m^{2+1}}\right) = -\sqrt{m}\cdot\left(-\sqrt{m}+\sqrt{m^2\cdot m^1}\right) = -\sqrt{m}\cdot\left(-\sqrt{m}+m\sqrt{m}\right)$

$= \left(\sqrt{m}\cdot\sqrt{m}\right)-\left(m\sqrt{m}\cdot\sqrt{m}\right) = \sqrt{m\cdot m} - m\sqrt{m\cdot m} = \sqrt{m^2} - m\sqrt{m^2} = m-(m\cdot m) = m-m^2 = \mathbf{m(1-m)}$

5. $\left(\sqrt[5]{x^4y^6}+\sqrt[5]{x}\right)\cdot\sqrt[5]{x^{10}y^7} = \left(\sqrt[5]{x^4y^{5+1}}+\sqrt[5]{x}\right)\cdot\sqrt[5]{x^{5+5}y^{5+2}} = \left(\sqrt[5]{x^4\left(y^5\cdot y^1\right)}+\sqrt[5]{x}\right)\cdot\sqrt[5]{\left(x^5\cdot x^5\right)\cdot\left(y^5\cdot y^2\right)}$

$= \left(y\sqrt[5]{x^4y}+\sqrt[5]{x}\right)\cdot(x\cdot x)y\sqrt[5]{y^2} = \left(y\sqrt[5]{x^4y}+\sqrt[5]{x}\right)\cdot x^2y\sqrt[5]{y^2} = x^2(y\cdot y)\cdot\left(\sqrt[5]{y^2}\cdot\sqrt[5]{x^4y}\right)+x^2y\left(\sqrt[5]{x}\cdot\sqrt[5]{y^2}\right)$

$= x^2y^2\left(\sqrt[5]{x^4\left(y^2\cdot y\right)}\right)+x^2y\left(\sqrt[5]{x\cdot y^2}\right) = x^2y^2\left(\sqrt[5]{x^4y^3}\right)+x^2y\left(\sqrt[5]{xy^2}\right) = \mathbf{x^2y\left[y\sqrt[5]{x^4y^3}+\sqrt[5]{xy^2}\right]}$

6. $\sqrt[3]{p^5q^2}\cdot\left(p^2+\sqrt[3]{p^4q}\right) = \sqrt[3]{p^{3+2}q^2}\cdot\left(p^2+\sqrt[3]{p^{3+1}q}\right) = \sqrt[3]{\left(p^3\cdot p^2\right)q^2}\cdot\left(p^2+\sqrt[3]{\left(p^3\cdot p^1\right)q}\right) = p\sqrt[3]{p^2q^2}\cdot\left(p^2+p\sqrt[3]{pq}\right)$

$= \left(p\cdot p^2\right)\sqrt[3]{p^2q^2}+(p\cdot p)\cdot\left(\sqrt[3]{p^2q^2}\cdot\sqrt[3]{pq}\right) = p^3\sqrt[3]{p^2q^2}+p^2\sqrt[3]{\left(p^2\cdot p\right)\cdot\left(q^2\cdot q\right)} = p^3\sqrt[3]{p^2q^2}+p^2\sqrt[3]{p^3q^3}$

$= p^3\sqrt[3]{p^2q^2}+\left(p^2\cdot p\right)q = p^3\sqrt[3]{p^2q^2}+p^3q = \mathbf{p^3\left(\sqrt[3]{p^2q^2}+q\right)}$

7. $\sqrt[3]{xy^4}\cdot\left(\sqrt[3]{x^5y^3}+\sqrt[3]{x^2y^2}\right) = \sqrt[3]{xy^{3+1}}\cdot\left(y\sqrt[3]{x^{3+2}}+\sqrt[3]{x^2y^2}\right) = \sqrt[3]{x\left(y^3\cdot y^1\right)}\cdot\left(y\sqrt[3]{\left(x^3\cdot x^2\right)}+\sqrt[3]{x^2y^2}\right)$

$= y\sqrt[3]{xy}\cdot\left(xy\sqrt[3]{x^2}+\sqrt[3]{x^2y^2}\right) = x(y\cdot y)\cdot\left(\sqrt[3]{xy}\cdot\sqrt[3]{x^2}\right)+y\left(\sqrt[3]{xy}\cdot\sqrt[3]{x^2y^2}\right) = xy^2\sqrt[3]{\left(x\cdot x^2\right)y}+y\sqrt[3]{\left(x\cdot x^2\right)\cdot\left(y\cdot y^2\right)}$

$= xy^2\sqrt[3]{x^3y}+y\sqrt[3]{x^3y^3} = (x\cdot x)y^2\sqrt[3]{y}+x(y\cdot y) = x^2y^2\sqrt[3]{y}+xy^2 = \mathbf{xy^2\left(1+x\sqrt[3]{y}\right)}$

8. $2\sqrt[4]{r}\cdot\left(\sqrt[4]{r^5s^6}-3\sqrt[4]{r^2s^7}\right) = 2\sqrt[4]{r}\cdot\left(\sqrt[4]{r^{4+1}s^{4+2}}-3\sqrt[4]{r^2s^{4+3}}\right) = 2\sqrt[4]{r}\cdot\left[\sqrt[4]{\left(r^4\cdot r^1\right)\cdot\left(s^4\cdot s^2\right)}-3\sqrt[4]{r^2\left(s^4\cdot s^3\right)}\right]$

$= 2\sqrt[4]{r}\cdot\left[rs\sqrt[4]{rs^2}-3s\sqrt[4]{r^2s^3}\right] = 2rs\left(\sqrt[4]{rs^2}\cdot\sqrt[4]{r}\right)-(2\cdot 3)s\sqrt[4]{r^2s^3}\cdot\sqrt[4]{r} = 2rs\sqrt[4]{(r\cdot r)s^2}-6s\sqrt[4]{\left(r^2\cdot r\right)s^3} = 2rs\sqrt[4]{r^2s^2}-6s\sqrt[4]{r^3s^3}$

$= 2s\left(r\sqrt[4]{r^2s^2}-3\sqrt[4]{r^3s^3}\right) = \mathbf{2s\sqrt[4]{r^2s^2}\left(r-3\sqrt[4]{rs}\right)}$

9. $3a\sqrt{b}\cdot\left(a-\sqrt{b^3}\right) = 3a\sqrt{b}\cdot\left(a-\sqrt{b^{2+1}}\right) = 3a\sqrt{b}\cdot\left(a-\sqrt{b^2\cdot b^1}\right) = 3a\sqrt{b}\cdot\left(a-b\sqrt{b}\right) = 3(a\cdot a)\sqrt{b}-3ab\left(\sqrt{b}\cdot\sqrt{b}\right)$

$= 3a^2\sqrt{b}-3ab\sqrt{b\cdot b} = 3a^2\sqrt{b}-3ab\sqrt{b^2} = 3a^2\sqrt{b}-3a(b\cdot b) = 3a^2\sqrt{b}-3ab^2 = \mathbf{3a\left(a\sqrt{b}-b^2\right)}$

10. $-3\sqrt{m^5}\cdot\left(m-4\sqrt{m^3}\right) = -3\sqrt{m^{2+2+1}}\cdot\left(m-4\sqrt{m^{2+1}}\right) = -3\sqrt{m^2\cdot m^2\cdot m^1}\cdot\left(m-4\sqrt{m^2\cdot m^1}\right)$

$= -3(m\cdot m)\sqrt{m}\cdot\left(m-4m\sqrt{m}\right) = -3m^2\sqrt{m}\cdot\left(m-4m\sqrt{m}\right) = -3\left(m^2\cdot m\right)\sqrt{m}+(3\cdot 4)\cdot\left(m^2\cdot m\right)\cdot\left(\sqrt{m}\cdot\sqrt{m}\right)$

$= -3m^3\sqrt{m}+12m^3\left(\sqrt{m\cdot m}\right) = -3m^3\sqrt{m}+12m^3\sqrt{m^2} = -3m^3\sqrt{m}+12\left(m^3\cdot m\right) = -3m^3\sqrt{m}+12m^4 = \mathbf{3m^3\left(4m-\sqrt{m}\right)}$

Section 4.3 Case I a Practice Problems

1. $\sqrt{\frac{1}{8}} = \sqrt{\frac{1}{4\cdot 2}} = \sqrt{\frac{1}{2^2\cdot 2}} = \frac{1}{2}\cdot\sqrt{\frac{1}{2}} = \frac{1}{2}\cdot\frac{\sqrt{1}}{\sqrt{2}} = \frac{1}{2}\cdot\frac{1}{\sqrt{2}} = \frac{1}{2}\left(\frac{1}{\sqrt{2}}\times\frac{\sqrt{2}}{\sqrt{2}}\right) = \frac{1}{2}\left(\frac{1\times\sqrt{2}}{\sqrt{2}\times\sqrt{2}}\right) = \frac{1}{2}\left(\frac{\sqrt{2}}{\sqrt{2\cdot 2}}\right) = \frac{1}{2}\cdot\frac{\sqrt{2}}{\sqrt{2^2}}$

$= \frac{1}{2}\cdot\frac{\sqrt{2}}{2} = \frac{1\times\sqrt{2}}{2\cdot 2} = \mathbf{\frac{\sqrt{2}}{4}}$

2. $\sqrt[2]{\frac{50}{7}} = \sqrt{\frac{50}{7}} = \sqrt{\frac{25\cdot 2}{7}} = \sqrt{\frac{5^2\cdot 2}{7}} = 5\sqrt{\frac{2}{7}} = 5\frac{\sqrt{2}}{\sqrt{7}} = 5\frac{\sqrt{2}}{\sqrt{7}}\times\frac{\sqrt{7}}{\sqrt{7}} = 5\frac{\sqrt{2}\times\sqrt{7}}{\sqrt{7}\times\sqrt{7}} = 5\frac{\sqrt{2\cdot 7}}{\sqrt{7\cdot 7}} = 5\frac{\sqrt{14}}{\sqrt{7^2}} = \mathbf{5\frac{\sqrt{14}}{7}}$

3. $\frac{\sqrt{75}}{-5} = -\frac{\sqrt{25\cdot 3}}{5} = -\frac{\sqrt{5^2\cdot 3}}{5} = -\frac{\overset{1}{\cancel{5}}\sqrt{3}}{\underset{1}{\cancel{5}}} = -\frac{\sqrt{3}}{1} = \mathbf{-\sqrt{3}}$

4. $\sqrt[3]{\frac{25}{16}} = \sqrt[3]{\frac{25}{8\cdot 2}} = \sqrt[3]{\frac{25}{2^3\cdot 2}} = \frac{1}{2}\sqrt[3]{\frac{25}{2}} = \frac{1}{2}\frac{\sqrt[3]{25}}{\sqrt[3]{2^1}} = \frac{1}{2}\left(\frac{\sqrt[3]{25}}{\sqrt[3]{2^1}}\times\frac{\sqrt[3]{2^2}}{\sqrt[3]{2^2}}\right) = \frac{1}{2}\left(\frac{\sqrt[3]{25}\times\sqrt[3]{4}}{\sqrt[3]{2^1}\times\sqrt[3]{2^2}}\right) = \frac{1}{2}\left(\frac{\sqrt[3]{25\cdot 4}}{\sqrt[3]{2^1\cdot 2^2}}\right) = \frac{1}{2}\cdot\frac{\sqrt[3]{100}}{\sqrt[3]{2^{1+2}}}$

$= \frac{1}{2}\cdot\frac{\sqrt[3]{100}}{\sqrt[3]{2^3}} = \frac{1}{2}\cdot\frac{\sqrt[3]{100}}{2} = \frac{1\cdot\sqrt[3]{100}}{2\cdot 2} = \mathbf{\frac{\sqrt[3]{100}}{4}}$

5. $\sqrt[5]{\frac{32}{8}} = \sqrt[5]{\frac{2^5}{2^3}} = 2\sqrt[5]{\frac{1}{2^3}} = 2\frac{1}{\sqrt[5]{2^3}} = 2\frac{1}{\sqrt[5]{2^3}}\times\frac{\sqrt[5]{2^2}}{\sqrt[5]{2^2}} = 2\frac{1\times\sqrt[5]{2^2}}{\sqrt[5]{2^3}\times\sqrt[5]{2^2}} = 2\frac{\sqrt[5]{2^2}}{\sqrt[5]{2^3\cdot 2^2}} = 2\frac{\sqrt[5]{4}}{\sqrt[5]{2^{3+2}}} = 2\frac{\sqrt[5]{4}}{\sqrt[5]{2^5}} = \frac{\overset{1}{\cancel{2}}}{1}\cdot\frac{\sqrt[5]{4}}{\underset{1}{\cancel{2}}}$

$= \frac{1}{1}\cdot\frac{\sqrt[5]{4}}{1} = \frac{1\cdot\sqrt[5]{4}}{1\cdot 1} = \frac{\sqrt[5]{4}}{1} = \mathbf{\sqrt[5]{4}}$

The following are two other ways to solve this problem:

5. $\sqrt[5]{\frac{32}{8}} = \sqrt[5]{\frac{2^5}{2^3}} = \sqrt[5]{2^5\cdot 2^{-3}} = \sqrt[5]{2^{5-3}} = \sqrt[5]{2^2} = \mathbf{\sqrt[5]{4}}$ or, $\sqrt[5]{\frac{32}{8}} = \sqrt[5]{\frac{\overset{4}{\cancel{32}}}{\underset{1}{\cancel{8}}}} = \sqrt[5]{\frac{4}{1}} = \mathbf{\sqrt[5]{4}}$

6. $\frac{-3\sqrt{100}}{-5\sqrt{3000}} = +\frac{3\sqrt{10^2}}{5\sqrt{100\cdot 30}} = \frac{3\cdot 10}{5\sqrt{10^2\cdot 30}} = \frac{30}{(5\cdot 10)\sqrt{30}} = \frac{\overset{3}{\cancel{30}}}{\underset{5}{\cancel{50}}\sqrt{30}} = \frac{3}{5\sqrt{30}} = \frac{3}{5}\cdot\frac{1}{\sqrt{30}} = \frac{3}{5}\left(\frac{1}{\sqrt{30}}\times\frac{\sqrt{30}}{\sqrt{30}}\right)$

$= \frac{3}{5}\left(\frac{1\times\sqrt{30}}{\sqrt{30}\times\sqrt{30}}\right) = \frac{3}{5}\left(\frac{\sqrt{30}}{\sqrt{30\cdot 30}}\right) = \frac{3}{5}\cdot\frac{\sqrt{30}}{\sqrt{30^2}} = \frac{3}{5}\cdot\frac{\sqrt{30}}{30} = \frac{\overset{1}{\cancel{3}}\cdot\sqrt{30}}{5\cdot\underset{10}{\cancel{30}}} = \frac{1\cdot\sqrt{30}}{5\cdot 10} = \mathbf{\frac{\sqrt{30}}{50}}$

7. $\frac{-20}{\sqrt[2]{45}} = -\frac{20}{\sqrt{45}} = -\frac{20}{\sqrt{9\cdot 5}} = -\frac{20}{\sqrt{3^2\cdot 5}} = -\frac{20}{3\sqrt{5}} = -\frac{20}{3}\cdot\frac{1}{\sqrt{5}} = -\frac{20}{3}\left(\frac{1}{\sqrt{5}}\times\frac{\sqrt{5}}{\sqrt{5}}\right) = -\frac{20}{3}\left(\frac{1\times\sqrt{5}}{\sqrt{5}\times\sqrt{5}}\right) = -\frac{20}{3}\cdot\frac{\sqrt{5}}{\sqrt{5\cdot 5}}$

$= -\frac{20}{3}\cdot\frac{\sqrt{5}}{\sqrt{5^2}} = -\frac{\overset{4}{\cancel{20}}}{3}\cdot\frac{\sqrt{5}}{\underset{1}{\cancel{5}}} = -\frac{4}{3}\cdot\frac{\sqrt{5}}{1} = -\frac{4\cdot\sqrt{5}}{3\cdot 1} = -\frac{4\sqrt{5}}{3} = -\frac{4}{3}\sqrt{5} = \mathbf{-\left(1\frac{1}{3}\right)\sqrt{5}}$

8. $\frac{1}{\sqrt[4]{324}} = \frac{1}{\sqrt[4]{81\cdot4}} = \frac{1}{\sqrt[4]{3^4\cdot4}} = \frac{1}{3\sqrt[4]{4^1}} = \frac{1}{3}\frac{1}{\sqrt[4]{4^1}} = \frac{1}{3}\left(\frac{1}{\sqrt[4]{4^1}}\times\frac{\sqrt[4]{4^3}}{\sqrt[4]{4^3}}\right) = \frac{1}{3}\left(\frac{1\times\sqrt[4]{4^3}}{\sqrt[4]{4^1}\times\sqrt[4]{4^3}}\right) = \frac{1}{3}\left(\frac{\sqrt[4]{4^3}}{\sqrt[4]{4^1\cdot4^3}}\right) = \frac{1}{3}\left(\frac{\sqrt[4]{64}}{\sqrt[4]{4^{1+3}}}\right)$

$= \frac{1}{3}\cdot\frac{\sqrt[4]{64}}{\sqrt[4]{4^4}} = \frac{1}{3}\cdot\frac{\sqrt[4]{64}}{4} = \frac{1\cdot\sqrt[4]{64}}{3\cdot4} = \mathbf{\frac{\sqrt[4]{64}}{12}}$

9. $\sqrt[2]{\frac{3}{48}} = \sqrt{\frac{3}{48}} = \sqrt{\frac{\overset{1}{\cancel{3}}}{\underset{16}{\cancel{24}}}} = \sqrt{\frac{1}{16}} = \sqrt{\frac{1}{4\cdot4}} = \sqrt{\frac{1}{4^2}} = \frac{\sqrt{1}}{\sqrt{4^2}} = \mathbf{\frac{1}{4}}$

10. $\sqrt[5]{\frac{243}{256}} = \sqrt[5]{\frac{3^5}{32\cdot8}} = \sqrt[5]{\frac{3^5}{2^5\cdot8}} = \frac{\sqrt[5]{3^5}}{\sqrt[5]{2^5\cdot8}} = \frac{3}{2\sqrt[5]{8}} = \frac{1}{2}\frac{3}{\sqrt[5]{8^1}} = \frac{1}{2}\left(\frac{3}{\sqrt[5]{8^1}}\times\frac{\sqrt[5]{8^4}}{\sqrt[5]{8^4}}\right) = \frac{1}{2}\left(\frac{3\times\sqrt[5]{8^4}}{\sqrt[5]{8^1}\times\sqrt[5]{8^4}}\right) = \frac{1}{2}\left(\frac{3\sqrt[5]{8^4}}{\sqrt[5]{8^1\cdot8^4}}\right)$

$= \frac{1}{2}\cdot\frac{3\sqrt[5]{8^4}}{\sqrt[5]{8^{1+4}}} = \frac{1}{2}\cdot\frac{3\sqrt[5]{4096}}{\sqrt[5]{8^5}} = \frac{1}{2}\cdot\frac{3\sqrt[5]{4096}}{8} = \frac{1\cdot3\sqrt[5]{4096}}{2\cdot8} = \mathbf{\frac{3\sqrt[5]{4096}}{16}}$

Section 4.3 Case I b Practice Problems

1. $\frac{\sqrt{80x}}{\sqrt{4x^4}} = \sqrt{\frac{80x}{4x^4}} = \sqrt{\frac{\overset{20}{\cancel{80}}}{\underset{1}{\cancel{4}}x^4x^{-1}}} = \sqrt{\frac{20}{x^{4-1}}} = \sqrt{\frac{20}{x^3}} = \frac{\sqrt{20}}{\sqrt{x^3}} = \frac{\sqrt{4\cdot5}}{\sqrt{x^{2+1}}} = \frac{\sqrt{2^2\cdot5}}{\sqrt{x^2\cdot x^1}} = \frac{2\sqrt{5}}{x\sqrt{x}} = \frac{2}{x}\cdot\frac{\sqrt{5}}{\sqrt{x}} = \frac{2}{x}\left(\frac{\sqrt{5}}{\sqrt{x}}\times\frac{\sqrt{x}}{\sqrt{x}}\right)$

$= \frac{2}{x}\left(\frac{\sqrt{5}\times\sqrt{x}}{\sqrt{x}\times\sqrt{x}}\right) = \frac{2}{x}\left(\frac{\sqrt{5\cdot x}}{\sqrt{x\cdot x}}\right) = \frac{2}{x}\cdot\frac{\sqrt{5x}}{\sqrt{x^2}} = \frac{2\cdot\sqrt{5x}}{x\cdot x} = \mathbf{\frac{2\sqrt{5x}}{x^2}}$

An alternative way of solving this problem would be:

1. $\frac{\sqrt{80x}}{\sqrt{4x^4}} = \frac{\sqrt{80x}}{\sqrt{2^2x^{2+2}}} = \frac{\sqrt{(16\cdot5)x}}{2\sqrt{x^2\cdot x^2}} = \frac{\sqrt{(4^2\cdot5)x}}{2(x\cdot x)} = \frac{\overset{2}{\cancel{4}}\sqrt{5x}}{\underset{1}{\cancel{2}}x^2} = \mathbf{\frac{2\sqrt{5x}}{x^2}}$

2. $\frac{\sqrt{48u^3v^2}}{\sqrt{16uv^3}} = \sqrt{\frac{48u^3v^2}{16uv^3}} = \sqrt{\frac{\overset{3}{\cancel{48}}u^3u^{-1}}{\underset{1}{\cancel{16}}v^3v^{-2}}} = \sqrt{\frac{3u^{3-1}}{v^{3-2}}} = \sqrt{\frac{3u^2}{v^1}} = \frac{\sqrt{3u^2}}{\sqrt{v}} = \frac{u\sqrt{3}}{\sqrt{v}} = \frac{u\sqrt{3}}{\sqrt{v}}\times\frac{\sqrt{v}}{\sqrt{v}} = \frac{u\sqrt{3}\times\sqrt{v}}{\sqrt{v}\times\sqrt{v}} = \frac{u\sqrt{3\cdot v}}{\sqrt{v\cdot v}}$

$= \frac{u\sqrt{3v}}{\sqrt{v^2}} = \mathbf{\frac{u\sqrt{3v}}{v}}$

3. $\sqrt{\frac{100x^3y^5}{25x^5y}} = \sqrt{\frac{\overset{4}{\cancel{100}}y^5y^{-1}}{\underset{1}{\cancel{25}}x^5x^{-3}}} = \sqrt{\frac{4y^{5-1}}{x^{5-3}}} = \sqrt{\frac{2^2y^4}{x^2}} = 2\frac{\sqrt{y^4}}{\sqrt{x^2}} = 2\frac{\sqrt{y^{2+2}}}{x} = 2\frac{\sqrt{y^2\cdot y^2}}{x} = 2\frac{y\cdot y}{x} = \mathbf{2\frac{y^2}{x}}$

4. $\frac{\sqrt[4]{81x^3y^2z^2}}{\sqrt[4]{2x^6y^5z^3}} = \sqrt[4]{\frac{81x^3y^2z^2}{2x^6y^5z^3}} = \sqrt[4]{\frac{3^4}{2(x^6x^{-3})\cdot(y^5y^{-2})\cdot(z^3z^{-2})}} = \sqrt[4]{\frac{3^4}{2x^{6-3}y^{5-2}z^{3-2}}} = 3\sqrt[4]{\frac{1}{2x^3y^3z^1}} = 3\frac{\sqrt[4]{1}}{\sqrt[4]{2x^3y^3z^1}}$

$= \frac{3\cdot1}{\sqrt[4]{2x^3y^3z^1}} = \frac{3}{\sqrt[4]{2^1x^3y^3z^1}}\times\frac{\sqrt[4]{2^3x^1y^1z^3}}{\sqrt[4]{2^3x^1y^1z^3}} = \frac{3\times\sqrt[4]{2^3xyz^3}}{\sqrt[4]{2^1x^3y^3z^1}\times\sqrt[4]{2^3x^1y^1z^3}} = \frac{3\times\sqrt[4]{2^3xyz^3}}{\sqrt[4]{2^1\cdot2^3\cdot x^3\cdot x^1\cdot y^3\cdot y^1\cdot z^1\cdot z^3}}$

$$= \frac{3\sqrt[4]{8xyz^3}}{\sqrt[4]{2^{1+3}x^{3+1}y^{3+1}z^{1+3}}} = \frac{3\sqrt[4]{8xyz^3}}{\sqrt[4]{2^4x^4y^4z^4}} = \frac{3\sqrt[4]{8xyz^3}}{2xyz} = \frac{3}{2}\left(\frac{\sqrt[4]{8xyz^3}}{xyz}\right) = \mathbf{1\frac{1}{2}\left(\frac{\sqrt[4]{8xyz^3}}{xyz}\right)}$$

5. $$\sqrt[5]{\frac{32a^2b^3}{256a^6b}} = \sqrt[5]{\frac{2^5b^3b^{-1}}{4^4a^6a^{-2}}} = 2\sqrt[5]{\frac{b^{3-1}}{4^4a^{6-2}}} = 2\sqrt[5]{\frac{b^2}{4^4a^4}} = 2\frac{\sqrt[5]{b^2}}{\sqrt[5]{4^4a^4}} = 2\frac{\sqrt[5]{b^2}}{\sqrt[5]{4^4a^4}}\times\frac{\sqrt[5]{4^1a^1}}{\sqrt[5]{4^1a^1}} = 2\frac{\sqrt[5]{b^2}\times\sqrt[5]{4^1a^1}}{\sqrt[5]{4^4a^4}\times\sqrt[5]{4^1a^1}}$$

$$= 2\frac{\sqrt[5]{4a\cdot b^2}}{\sqrt[5]{\left(4^4\cdot4^1\right)\cdot\left(a^4\cdot a^1\right)}} = \frac{2\sqrt[5]{4ab^2}}{\sqrt[5]{4^{4+1}a^{4+1}}} = \frac{2\sqrt[5]{4ab^2}}{\sqrt[5]{4^5a^5}} = \frac{\overset{1}{\cancel{2}}\sqrt[5]{4ab^2}}{\underset{2}{\cancel{4}}a} = \mathbf{\frac{\sqrt[5]{4ab^2}}{2a}}$$

6. $$\frac{\sqrt{18u^3v^2}}{\sqrt{4uv^3}} = \sqrt{\frac{\overset{9}{\cancel{18}}u^3u^{-1}}{\underset{2}{\cancel{4}}v^3v^{-2}}} = \sqrt{\frac{9u^{3-1}}{2v^{3-2}}} = \sqrt{\frac{3^2u^2}{2v^1}} = 3u\sqrt{\frac{1}{2v}} = 3u\frac{\sqrt{1}}{\sqrt{2v}} = 3u\frac{1}{\sqrt{2v}} = \frac{3u}{\sqrt{2v}}\times\frac{\sqrt{2v}}{\sqrt{2v}} = \frac{3u\times\sqrt{2v}}{\sqrt{2v}\times\sqrt{2v}}$$

$$= \frac{3u\sqrt{2v}}{\sqrt{(2\cdot2)(v\cdot v)}} = \frac{3u\sqrt{2v}}{\sqrt{2^2v^2}} = \frac{3u\sqrt{2v}}{2v} = \frac{3}{2}\left(\frac{u\sqrt{2v}}{v}\right) = \mathbf{1\frac{1}{2}\left(\frac{u\sqrt{2v}}{v}\right)}$$

7. $$\frac{\sqrt{5k^5l^2}}{\sqrt{40k^3l}} = \sqrt{\frac{\overset{1}{\cancel{5}}k^5l^2}{\underset{8}{\cancel{40}}k^3l}} = \sqrt{\frac{k^5k^{-3}l^2l^{-1}}{8}} = \sqrt{\frac{k^{5-3}l^{2-1}}{8}} = \sqrt{\frac{k^2l^1}{8}} = \frac{\sqrt{k^2l}}{\sqrt{4\cdot2}} = \frac{k\sqrt{l}}{\sqrt{2^2\cdot2}} = \frac{k\sqrt{l}}{2\sqrt{2}} = \frac{k}{2}\cdot\frac{\sqrt{l}}{\sqrt{2}}$$

$$= \frac{k}{2}\left(\frac{\sqrt{l}}{\sqrt{2}}\times\frac{\sqrt{2}}{\sqrt{2}}\right) = \frac{k}{2}\left(\frac{\sqrt{l}\times\sqrt{2}}{\sqrt{2}\times\sqrt{2}}\right) = \frac{k}{2}\cdot\frac{\sqrt{l\cdot2}}{\sqrt{2\cdot2}} = \frac{k}{2}\cdot\frac{\sqrt{2l}}{\sqrt{2^2}} = \frac{k}{2}\cdot\frac{\sqrt{2l}}{2} = \frac{k\cdot\sqrt{2l}}{2\cdot2} = \mathbf{\frac{k\sqrt{2l}}{4}}$$

8. $$\frac{\sqrt[4]{625x^4y^2}}{-3\sqrt[4]{81x^6y^6}} = -\frac{x\sqrt[4]{5^4y^2}}{3\sqrt[4]{3^4x^{4+2}y^{4+2}}} = -\frac{5x\sqrt[4]{y^2}}{(3\cdot3)\sqrt[4]{\left(x^4\cdot x^2\right)\cdot\left(y^4\cdot y^2\right)}} = -\frac{5x\sqrt[4]{y^2}}{9xy\sqrt[4]{x^2y^2}} = -\frac{5\overset{1}{\cancel{x}}}{9\underset{1}{\cancel{x}}y}\cdot\frac{\sqrt[4]{y^2}}{\sqrt[4]{x^2y^2}}$$

$$= -\frac{5}{9y}\cdot\sqrt[4]{\frac{y^2}{x^2y^2}} = -\frac{5}{9y}\cdot\sqrt[4]{\frac{y^2y^{-2}}{x^2}} = -\frac{5}{9y}\cdot\sqrt[4]{\frac{y^{2-2}}{x^2}} = -\frac{5}{9y}\cdot\sqrt[4]{\frac{y^0}{x^2}} = -\frac{5}{9y}\cdot\sqrt[4]{\frac{1}{x^2}} = -\frac{5}{9y}\cdot\frac{\sqrt[4]{1}}{\sqrt[4]{x^2}} = -\frac{5}{9y}\cdot\frac{1}{\sqrt[4]{x^2}}$$

$$= -\frac{5}{9y}\left(\frac{1}{\sqrt[4]{x^2}}\times\frac{\sqrt[4]{x^2}}{\sqrt[4]{x^2}}\right) = -\frac{5}{9y}\left(\frac{1\times\sqrt[4]{x^2}}{\sqrt[4]{x^2}\times\sqrt[4]{x^2}}\right) = -\frac{5}{9y}\cdot\frac{\sqrt[4]{x^2}}{\sqrt[4]{x^2\cdot x^2}} = -\frac{5}{9y}\cdot\frac{\sqrt[4]{x^2}}{\sqrt[4]{x^{2+2}}} = -\frac{5}{9y}\cdot\frac{\sqrt[4]{x^2}}{\sqrt[4]{x^4}} = -\frac{5}{9y}\cdot\frac{\sqrt[4]{x^2}}{x}$$

$$= -\mathbf{\frac{5\sqrt[4]{x^2}}{9xy}}$$

9. $$\sqrt{\frac{x^3y^5z}{81xy^6z^3}} = \sqrt{\frac{x^3x^{-1}}{9^2\left(y^6y^{-5}\right)\cdot\left(z^3z^{-1}\right)}} = \frac{1}{9}\sqrt{\frac{x^{3-1}}{y^{6-5}z^{3-1}}} = \frac{1}{9}\sqrt{\frac{x^2}{y^1z^2}} = \frac{1}{9}\frac{\sqrt{x^2}}{\sqrt{yz^2}} = \frac{1}{9}\cdot\frac{x}{z\sqrt{y}} = \frac{1}{9z}\cdot\frac{x}{\sqrt{y}}$$

$$= \frac{1}{9z}\left(\frac{x}{\sqrt{y}}\times\frac{\sqrt{y}}{\sqrt{y}}\right) = \frac{1}{9z}\left(\frac{x\times\sqrt{y}}{\sqrt{y}\times\sqrt{y}}\right) = \frac{1}{9z}\cdot\frac{x\sqrt{y}}{\sqrt{y\cdot y}} = \frac{1}{9z}\cdot\frac{x\sqrt{y}}{\sqrt{y^2}} = \frac{1}{9z}\cdot\frac{x\sqrt{y}}{y} = \frac{1\cdot x\sqrt{y}}{9z\cdot y} = \mathbf{\frac{x\sqrt{y}}{9yz}}$$

10. $$\frac{\sqrt[3]{m^3n^2}}{\sqrt[3]{5^3m^7n^4}} = \frac{m\sqrt[3]{n^2}}{5\sqrt[3]{m^7n^4}} = \frac{m}{5}\sqrt[3]{\frac{n^2}{m^7n^4}} = \frac{m}{5}\sqrt[3]{\frac{1}{m^7\left(n^4n^{-2}\right)}} = \frac{m}{5}\sqrt[3]{\frac{1}{m^7n^{4-2}}} = \frac{m}{5}\sqrt[3]{\frac{1}{m^7n^2}} = \frac{m}{5}\cdot\frac{\sqrt[3]{1}}{\sqrt[3]{m^7n^2}}$$

$$= \frac{m}{5}\cdot\frac{1}{\sqrt[3]{m^{3+3+1}n^2}} = \frac{m}{5}\cdot\frac{1}{\sqrt[3]{m^3\cdot m^3\cdot m^1\cdot n^2}} = \frac{m}{5}\cdot\frac{1}{(m\cdot m)\sqrt[3]{m^1\cdot n^2}} = \frac{m}{5}\cdot\frac{1}{m^2\sqrt[3]{m^1\cdot n^2}} = \frac{m}{5m^2}\left(\frac{1}{\sqrt[3]{m^1\cdot n^2}}\times\frac{\sqrt[3]{m^2\cdot n^1}}{\sqrt[3]{m^2\cdot n^1}}\right)$$

$$= \frac{\frac{1}{\cancel{m}}}{\frac{5\cancel{m}^2}{m}}\left(\frac{1\times\sqrt[3]{m^2\cdot n^1}}{\sqrt[3]{m^1\cdot n^2}\times\sqrt[3]{m^2\cdot n^1}}\right) = \frac{1}{5m}\left(\frac{\sqrt[3]{m^2 n}}{\sqrt[3]{(m^1\cdot m^2)\cdot(n^2\cdot n^1)}}\right) = \frac{1}{5m}\cdot\frac{\sqrt[3]{m^2 n}}{\sqrt[3]{m^{1+2}n^{2+1}}} = \frac{1}{5m}\cdot\frac{\sqrt[3]{m^2 n}}{\sqrt[3]{m^3 n^3}} = \frac{1}{5m}\cdot\frac{\sqrt[3]{m^2 n}}{mn}$$

$$= \frac{1\cdot\sqrt[3]{m^2 n}}{5(m\cdot m)n} = \boldsymbol{\frac{\sqrt[3]{m^2 n}}{5m^2 n}}$$

Section 4.3 Case II a Practice Problems

1. $$\frac{7}{1+\sqrt{7}} = \frac{7}{1+\sqrt{7}}\times\frac{1-\sqrt{7}}{1-\sqrt{7}} = \frac{7\times(1-\sqrt{7})}{(1+\sqrt{7})\times(1-\sqrt{7})} = \frac{7(1-\sqrt{7})}{(1\cdot 1)+(1\cdot\sqrt{7})-(1\cdot\sqrt{7})-(\sqrt{7}\cdot\sqrt{7})} = \frac{7(1-\sqrt{7})}{1+\sqrt{7}-\sqrt{7}-\sqrt{7\cdot 7}}$$

$$= \frac{7(1-\sqrt{7})}{1-\sqrt{7^2}} = \frac{7(1-\sqrt{7})}{1-7} = \frac{7(1-\sqrt{7})}{-6} = \boldsymbol{-\frac{7(1-\sqrt{7})}{6}}$$

2. $$\frac{1-\sqrt{18}}{2+\sqrt{18}} = \frac{1-\sqrt{9\cdot 2}}{2+\sqrt{9\cdot 2}} = \frac{1-\sqrt{3^2\cdot 2}}{2+\sqrt{3^2\cdot 2}} = \frac{1-3\sqrt{2}}{2+3\sqrt{2}} = \frac{1-3\sqrt{2}}{2+3\sqrt{2}}\times\frac{2-3\sqrt{2}}{2-3\sqrt{2}} = \frac{(1-3\sqrt{2})\times(2-3\sqrt{2})}{(2+3\sqrt{2})\times(2-3\sqrt{2})}$$

$$= \frac{(1\cdot 2)-(1\cdot 3)\sqrt{2}-(2\cdot 3)\sqrt{2}+(3\cdot 3)\cdot(\sqrt{2}\cdot\sqrt{2})}{(2\cdot 2)-(2\cdot 3)\sqrt{2}+(2\cdot 3)\sqrt{2}-(3\cdot 3)\cdot(\sqrt{2}\cdot\sqrt{2})} = \frac{2-3\sqrt{2}-6\sqrt{2}+9\sqrt{2\cdot 2}}{4-6\sqrt{2}+6\sqrt{2}-9\sqrt{2\cdot 2}} = \frac{2-(3+6)\sqrt{2}+9\sqrt{2^2}}{4-9\sqrt{2^2}} = \frac{2-9\sqrt{2}+(9\cdot 2)}{4-(9\cdot 2)}$$

$$= \frac{2-9\sqrt{2}+18}{4-18} = \frac{(2+18)-9\sqrt{2}}{-14} = \boldsymbol{-\frac{20-9\sqrt{2}}{14}}$$

3. $$\frac{\sqrt{5}}{\sqrt{5}+\sqrt{2}} = \frac{\sqrt{5}}{\sqrt{5}+\sqrt{2}}\times\frac{\sqrt{5}-\sqrt{2}}{\sqrt{5}-\sqrt{2}} = \frac{\sqrt{5}\times(\sqrt{5}-\sqrt{2})}{(\sqrt{5}+\sqrt{2})\times(\sqrt{5}-\sqrt{2})} = \frac{(\sqrt{5}\cdot\sqrt{5})-(\sqrt{5}\cdot\sqrt{2})}{(\sqrt{5}\cdot\sqrt{5})-(\sqrt{5}\cdot\sqrt{2})+(\sqrt{2}\cdot\sqrt{5})-(\sqrt{2}\cdot\sqrt{2})}$$

$$= \frac{\sqrt{5\cdot 5}-\sqrt{5\cdot 2}}{\sqrt{5\cdot 5}-\sqrt{5\cdot 2}+\sqrt{2\cdot 5}-\sqrt{2\cdot 2}} = \frac{\sqrt{5^2}-\sqrt{10}}{\sqrt{5^2}-\sqrt{10}+\sqrt{10}-\sqrt{2^2}} = \frac{5-\sqrt{10}}{5-2} = \boldsymbol{\frac{5-\sqrt{10}}{3}}$$

4. $$\frac{3-\sqrt{5}}{\sqrt{7}-\sqrt{4}} = \frac{3-\sqrt{5}}{\sqrt{7}-\sqrt{2^2}} = \frac{3-\sqrt{5}}{\sqrt{7}-2} = \frac{3-\sqrt{5}}{\sqrt{7}-2}\times\frac{\sqrt{7}+2}{\sqrt{7}+2} = \frac{(3-\sqrt{5})\times(\sqrt{7}+2)}{(\sqrt{7}-2)\times(\sqrt{7}+2)} = \frac{(3\cdot\sqrt{7})+(3\cdot 2)-(\sqrt{5}\cdot\sqrt{7})-(2\cdot\sqrt{5})}{(\sqrt{7}\cdot\sqrt{7})+(2\cdot\sqrt{7})-(2\cdot\sqrt{7})-(2\cdot 2)}$$

$$= \frac{3\sqrt{7}+6-\sqrt{5\cdot 7}-2\sqrt{5}}{\sqrt{7\cdot 7}+2\sqrt{7}-2\sqrt{7}-4} = \frac{3\sqrt{7}+6-\sqrt{35}-2\sqrt{5}}{\sqrt{7^2}-4} = \frac{3\sqrt{7}+6-\sqrt{35}-2\sqrt{5}}{7-4} = \boldsymbol{\frac{3\sqrt{7}-\sqrt{35}-2\sqrt{5}+6}{3}}$$

5. $$\frac{-3+\sqrt{3}}{4+\sqrt{5}} = \frac{-3+\sqrt{3}}{4+\sqrt{5}}\times\frac{4-\sqrt{5}}{4-\sqrt{5}} = \frac{(-3+\sqrt{3})\times(4-\sqrt{5})}{(4+\sqrt{5})\times(4-\sqrt{5})} = \frac{-(3\cdot 4)+(3\cdot\sqrt{5})+(4\cdot\sqrt{3})-(\sqrt{3}\cdot\sqrt{5})}{(4\cdot 4)-(4\cdot\sqrt{5})+(4\cdot\sqrt{5})-(\sqrt{5}\cdot\sqrt{5})}$$

$$= \frac{-12+3\sqrt{5}+4\sqrt{3}-\sqrt{3\cdot 5}}{16-4\sqrt{5}+4\sqrt{5}-\sqrt{5\cdot 5}} = \frac{-12+3\sqrt{5}+4\sqrt{3}-\sqrt{15}}{16-\sqrt{5^2}} = \frac{3\sqrt{5}+4\sqrt{3}-\sqrt{15}-12}{16-5} = \boldsymbol{\frac{3\sqrt{5}+4\sqrt{3}-\sqrt{15}-12}{11}}$$

6. $\frac{3-\sqrt{3}}{3+\sqrt{3}} = \frac{3-\sqrt{3}}{3+\sqrt{3}} \times \frac{3-\sqrt{3}}{3-\sqrt{3}} = \frac{\left(3-\sqrt{3}\right)\times\left(3-\sqrt{3}\right)}{\left(3+\sqrt{3}\right)\times\left(3-\sqrt{3}\right)} = \frac{(3\cdot 3)-\left(3\cdot\sqrt{3}\right)-\left(3\cdot\sqrt{3}\right)+\left(\sqrt{3}\cdot\sqrt{3}\right)}{(3\cdot 3)-\left(3\cdot\sqrt{3}\right)+\left(3\cdot\sqrt{3}\right)-\left(\sqrt{3}\cdot\sqrt{3}\right)} = \frac{9-3\sqrt{3}-3\sqrt{3}+\sqrt{3\cdot 3}}{9-3\sqrt{3}+3\sqrt{3}-\sqrt{3\cdot 3}}$

$= \frac{9-(3+3)\sqrt{3}+\sqrt{3^2}}{9-\sqrt{3^2}} = \frac{9-6\sqrt{3}+3}{9-3} = \frac{(9+3)-6\sqrt{3}}{6} = \frac{12-6\sqrt{3}}{6} = \frac{\overset{1}{\cancel{6}}\left(2-\sqrt{3}\right)}{\underset{1}{\cancel{6}}} = \frac{2-\sqrt{3}}{1} = \mathbf{2-\sqrt{3}}$

7. $\frac{\sqrt{72}+\sqrt{20}}{\sqrt{50}+\sqrt{27}} = \frac{\sqrt{36\cdot 2}+\sqrt{4\cdot 5}}{\sqrt{25\cdot 2}+\sqrt{9\cdot 3}} = \frac{\sqrt{6^2\cdot 2}+\sqrt{2^2\cdot 5}}{\sqrt{5^2\cdot 2}+\sqrt{3^2\cdot 3}} = \frac{6\sqrt{2}+2\sqrt{5}}{5\sqrt{2}+3\sqrt{3}} = \frac{6\sqrt{2}+2\sqrt{5}}{5\sqrt{2}+3\sqrt{3}} \times \frac{5\sqrt{2}-3\sqrt{3}}{5\sqrt{2}-3\sqrt{3}}$

$= \frac{\left(6\sqrt{2}+2\sqrt{5}\right)\times\left(5\sqrt{2}-3\sqrt{3}\right)}{\left(5\sqrt{2}+3\sqrt{3}\right)\times\left(5\sqrt{2}-3\sqrt{3}\right)} = \frac{(6\cdot 5)\cdot\left(\sqrt{2}\cdot\sqrt{2}\right)-(6\cdot 3)\cdot\left(\sqrt{2}\cdot\sqrt{3}\right)+(2\cdot 5)\cdot\left(\sqrt{5}\cdot\sqrt{2}\right)-(2\cdot 3)\cdot\left(\sqrt{5}\cdot\sqrt{3}\right)}{(5\cdot 5)\cdot\left(\sqrt{2}\cdot\sqrt{2}\right)-(5\cdot 3)\cdot\left(\sqrt{2}\cdot\sqrt{3}\right)+(3\cdot 5)\cdot\left(\sqrt{3}\cdot\sqrt{2}\right)-(3\cdot 3)\cdot\left(\sqrt{3}\cdot\sqrt{3}\right)}$

$= \frac{30\sqrt{2\cdot 2}-18\sqrt{2\cdot 3}+10\sqrt{5\cdot 2}-6\sqrt{5\cdot 3}}{25\sqrt{2\cdot 2}-15\sqrt{2\cdot 3}+15\sqrt{3\cdot 2}-9\sqrt{3\cdot 3}} = \frac{30\sqrt{2^2}-18\sqrt{6}+10\sqrt{10}-6\sqrt{15}}{25\sqrt{2^2}-15\sqrt{6}+15\sqrt{6}-9\sqrt{3^2}} = \frac{(30\cdot 2)-18\sqrt{6}+10\sqrt{10}-6\sqrt{15}}{(25\cdot 2)-(9\cdot 3)}$

$= \frac{60-18\sqrt{6}+10\sqrt{10}-6\sqrt{15}}{50-27} = \mathbf{\frac{2\left(30-9\sqrt{6}+5\sqrt{10}-3\sqrt{15}\right)}{23}}$

8. $\frac{1-\sqrt{8}}{2+\sqrt{8}} = \frac{1-\sqrt{4\cdot 2}}{2+\sqrt{4\cdot 2}} = \frac{1-\sqrt{2^2\cdot 2}}{2+\sqrt{2^2\cdot 2}} = \frac{1-2\sqrt{2}}{2+2\sqrt{2}} = \frac{1-2\sqrt{2}}{2+2\sqrt{2}} \times \frac{2-2\sqrt{2}}{2-2\sqrt{2}} = \frac{\left(1-2\sqrt{2}\right)\times\left(2-2\sqrt{2}\right)}{\left(2+2\sqrt{2}\right)\times\left(2-2\sqrt{2}\right)}$

$= \frac{(1\cdot 2)-(1\cdot 2)\sqrt{2}-(2\cdot 2)\sqrt{2}+(2\cdot 2)\left(\sqrt{2}\cdot\sqrt{2}\right)}{(2\cdot 2)-(2\cdot 2)\sqrt{2}+(2\cdot 2)\sqrt{2}-(2\cdot 2)\left(\sqrt{2}\cdot\sqrt{2}\right)} = \frac{2-2\sqrt{2}-4\sqrt{2}+4\sqrt{2\cdot 2}}{4-4\sqrt{2}+4\sqrt{2}-4\sqrt{2\cdot 2}} = \frac{2-(2+4)\sqrt{2}+4\sqrt{2^2}}{4-4\sqrt{2^2}} = \frac{2-6\sqrt{2}+(4\cdot 2)}{4-(4\cdot 2)}$

$= \frac{2-6\sqrt{2}+8}{4-8} = \frac{(2+8)-6\sqrt{2}}{-4} = -\frac{10-6\sqrt{2}}{4} = -\frac{\overset{1}{\cancel{2}}\left(5-3\sqrt{2}\right)}{\underset{2}{\cancel{4}}} = \mathbf{-\frac{5-3\sqrt{2}}{2}}$

9. $\frac{5+5\sqrt{25}}{5-\sqrt{125}} = \frac{5+5\sqrt{5^2}}{5-\sqrt{25\cdot 5}} = \frac{5+(5\cdot 5)}{5-\sqrt{5^2\cdot 5}} = \frac{5+25}{5-5\sqrt{5}} = \frac{30}{5-5\sqrt{5}} = \frac{30}{5-5\sqrt{5}} \times \frac{5+5\sqrt{5}}{5+5\sqrt{5}} = \frac{30\times\left(5+5\sqrt{5}\right)}{\left(5-5\sqrt{5}\right)\times\left(5+5\sqrt{5}\right)}$

$= \frac{(30\cdot 5)+(30\cdot 5)\sqrt{5}}{(5\cdot 5)+(5\cdot 5)\sqrt{5}-(5\cdot 5)\sqrt{5}-(5\cdot 5)\left(\sqrt{5}\cdot\sqrt{5}\right)} = \frac{150+150\sqrt{5}}{25+25\sqrt{5}-25\sqrt{5}-25\sqrt{5\cdot 5}} = \frac{150\left(1+\sqrt{5}\right)}{25-25\sqrt{5^2}} = \frac{150\left(1+\sqrt{5}\right)}{25-(25\cdot 5)}$

$= \frac{150\left(1+\sqrt{5}\right)}{25-125} = \frac{150\left(1+\sqrt{5}\right)}{-100} = -\frac{\overset{3}{\cancel{150}}\left(1+\sqrt{5}\right)}{\underset{2}{\cancel{100}}} = -\frac{3\left(1+\sqrt{5}\right)}{2} = -\frac{3}{2}\left(1+\sqrt{5}\right) = \mathbf{-1\frac{1}{2}\left(1+\sqrt{5}\right)}$

10. $\frac{\sqrt{5}+\sqrt{3}}{\sqrt{5}-\sqrt{3}} = \frac{\sqrt{5}+\sqrt{3}}{\sqrt{5}-\sqrt{3}} \times \frac{\sqrt{5}+\sqrt{3}}{\sqrt{5}+\sqrt{3}} = \frac{\left(\sqrt{5}+\sqrt{3}\right)\times\left(\sqrt{5}+\sqrt{3}\right)}{\left(\sqrt{5}-\sqrt{3}\right)\times\left(\sqrt{5}+\sqrt{3}\right)} = \frac{\left(\sqrt{5}\cdot\sqrt{5}\right)+\left(\sqrt{5}\cdot\sqrt{3}\right)+\left(\sqrt{3}\cdot\sqrt{5}\right)+\left(\sqrt{3}\cdot\sqrt{3}\right)}{\left(\sqrt{5}\cdot\sqrt{5}\right)+\left(\sqrt{5}\cdot\sqrt{3}\right)-\left(\sqrt{3}\cdot\sqrt{5}\right)-\left(\sqrt{3}\cdot\sqrt{3}\right)}$

$= \frac{\sqrt{5\cdot 5}+\sqrt{5\cdot 3}+\sqrt{3\cdot 5}+\sqrt{3\cdot 3}}{\sqrt{5\cdot 5}+\sqrt{5\cdot 3}-\sqrt{3\cdot 5}-\sqrt{3\cdot 3}} = \frac{\sqrt{5^2}+\sqrt{15}+\sqrt{15}+\sqrt{3^2}}{\sqrt{5^2}+\sqrt{15}-\sqrt{15}-\sqrt{3^2}} = \frac{5+(1+1)\sqrt{15}+3}{5-3} = \frac{(5+3)+2\sqrt{15}}{2} = \frac{8+2\sqrt{15}}{2}$

$= \frac{\overset{1}{\cancel{2}}\left(4+\sqrt{15}\right)}{\underset{1}{\cancel{2}}} = \frac{4+\sqrt{15}}{1} = \mathbf{4+\sqrt{15}}$

Section 4.3 Case II b Practice Problems

1. $\frac{5x}{1+\sqrt{x}} = \frac{5x}{1+\sqrt{x}} \times \frac{1-\sqrt{x}}{1-\sqrt{x}} = \frac{5x \times \left(1-\sqrt{x}\right)}{\left(1+\sqrt{x}\right) \times \left(1-\sqrt{x}\right)} = \frac{5x\left(1-\sqrt{x}\right)}{(1 \cdot 1) - \left(1 \cdot \sqrt{x}\right) + \left(1 \cdot \sqrt{x}\right) - \left(\sqrt{x} \cdot \sqrt{x}\right)} = \frac{5x\left(1-\sqrt{x}\right)}{1-\sqrt{x}+\sqrt{x}-\sqrt{x \cdot x}}$

$= \frac{5x\left(1-\sqrt{x}\right)}{1-\sqrt{x^2}} = \mathbf{\frac{5x\left(1-\sqrt{x}\right)}{1-x}}$

2. $\frac{\sqrt{x}}{2-\sqrt{x}} = \frac{\sqrt{x}}{2-\sqrt{x}} \times \frac{2+\sqrt{x}}{2+\sqrt{x}} = \frac{\sqrt{x} \times \left(2+\sqrt{x}\right)}{\left(2-\sqrt{x}\right) \times \left(2+\sqrt{x}\right)} = \frac{\left(2 \cdot \sqrt{x}\right) + \left(\sqrt{x} \cdot \sqrt{x}\right)}{(2 \cdot 2) + \left(2 \cdot \sqrt{x}\right) - \left(2 \cdot \sqrt{x}\right) - \left(\sqrt{x} \cdot \sqrt{x}\right)} = \frac{2\sqrt{x}+\sqrt{x \cdot x}}{4+2\sqrt{x}-2\sqrt{x}-\sqrt{x \cdot x}}$

$= \frac{2\sqrt{x}+\sqrt{x^2}}{4-\sqrt{x^2}} = \mathbf{\frac{2\sqrt{x}+x}{4-x}}$

3. $\frac{1+3x}{1-2\sqrt{x}} = \frac{1+3x}{1-2\sqrt{x}} \times \frac{1+2\sqrt{x}}{1+2\sqrt{x}} = \frac{(1+3x) \times \left(1+2\sqrt{x}\right)}{\left(1-2\sqrt{x}\right) \times \left(1+2\sqrt{x}\right)} = \frac{(1 \cdot 1) + (1 \cdot 2)\sqrt{x} + (1 \cdot 3)x + (2 \cdot 3) \cdot \left(x \cdot \sqrt{x}\right)}{(1 \cdot 1) + (1 \cdot 2)\sqrt{x} - (1 \cdot 2)\sqrt{x} - (2 \cdot 2) \cdot \left(\sqrt{x} \cdot \sqrt{x}\right)}$

$= \frac{1+2\sqrt{x}+3x+6x\sqrt{x}}{1+2\sqrt{x}-2\sqrt{x}-4\sqrt{x \cdot x}} = \frac{1+3x+2\sqrt{x}+6x\sqrt{x}}{1-4\sqrt{x^2}} = \mathbf{\frac{1+3x+2\sqrt{x}+6x\sqrt{x}}{1-4x}}$

4. $\frac{a-b}{\sqrt{a}-\sqrt{b}} = \frac{a-b}{\sqrt{a}-\sqrt{b}} \times \frac{\sqrt{a}+\sqrt{b}}{\sqrt{a}+\sqrt{b}} = \frac{(a-b) \times \left(\sqrt{a}+\sqrt{b}\right)}{\left(\sqrt{a}-\sqrt{b}\right) \times \left(\sqrt{a}+\sqrt{b}\right)} = \frac{(a-b)\left(\sqrt{a}+\sqrt{b}\right)}{\left(\sqrt{a} \cdot \sqrt{a}\right) + \left(\sqrt{a} \cdot \sqrt{b}\right) - \left(\sqrt{a} \cdot \sqrt{b}\right) - \left(\sqrt{b} \cdot \sqrt{b}\right)}$

$= \frac{(a-b)\left(\sqrt{a}+\sqrt{b}\right)}{\sqrt{a \cdot a}+\sqrt{a \cdot b}-\sqrt{a \cdot b}-\sqrt{b \cdot b}} = \frac{(a-b)\left(\sqrt{a}+\sqrt{b}\right)}{\sqrt{a^2}+\sqrt{ab}-\sqrt{ab}-\sqrt{b^2}} = \frac{(a-b)\left(\sqrt{a}+\sqrt{b}\right)}{\sqrt{a^2}-\sqrt{b^2}} = \frac{(a-b)\left(\sqrt{a}+\sqrt{b}\right)}{(a-b)}$

$= \frac{\sqrt{a}+\sqrt{b}}{1} = \mathbf{\sqrt{a}+\sqrt{b}}$

5. $\frac{-a}{a-\sqrt{a}} = \frac{-a}{a-\sqrt{a}} \times \frac{a+\sqrt{a}}{a+\sqrt{a}} = \frac{-a \times \left(a+\sqrt{a}\right)}{\left(a-\sqrt{a}\right) \times \left(a+\sqrt{a}\right)} = \frac{-a\left(a+\sqrt{a}\right)}{(a \cdot a) + \left(a \cdot \sqrt{a}\right) - \left(a \cdot \sqrt{a}\right) - \left(\sqrt{a} \cdot \sqrt{a}\right)} = \frac{-a\left(a+\sqrt{a}\right)}{a^2+a\sqrt{a}-a\sqrt{a}-\sqrt{aa}}$

$= -\frac{a\left(a+\sqrt{a}\right)}{a^2-\sqrt{a^2}} = -\frac{a\left(a+\sqrt{a}\right)}{a^2-a} = -\frac{\overset{1}{\cancel{a}}\left(a+\sqrt{a}\right)}{\underset{1}{\cancel{a}}(a-1)} = \mathbf{-\frac{a+\sqrt{a}}{a-1}}$

6. $\frac{x+y}{x+\sqrt{y}} = \frac{x+y}{x+\sqrt{y}} \times \frac{x-\sqrt{y}}{x-\sqrt{y}} = \frac{(x+y) \times \left(x-\sqrt{y}\right)}{\left(x+\sqrt{y}\right) \times \left(x-\sqrt{y}\right)} = \frac{(x \cdot x) - \left(x \cdot \sqrt{y}\right) + (x \cdot y) - \left(y \cdot \sqrt{y}\right)}{(x \cdot x) - \left(x \cdot \sqrt{y}\right) + \left(x \cdot \sqrt{y}\right) - \left(\sqrt{y} \cdot \sqrt{y}\right)}$

$= \frac{x^2-x\sqrt{y}+xy-y\sqrt{y}}{x^2-x\sqrt{y}+x\sqrt{y}-\sqrt{y \cdot y}} = \frac{x^2+xy-x\sqrt{y}-y\sqrt{y}}{x^2-\sqrt{y^2}} = \frac{x(x+y)-\sqrt{y}(x+y)}{x^2-y} = \mathbf{\frac{(x+y)\left[x-\sqrt{y}\right]}{x^2-y}}$

7. $\frac{5+x}{2-\sqrt{x}} = \frac{5+x}{2-\sqrt{x}} \times \frac{2+\sqrt{x}}{2+\sqrt{x}} = \frac{(5+x) \times \left(2+\sqrt{x}\right)}{\left(2-\sqrt{x}\right) \times \left(2+\sqrt{x}\right)} = \frac{(5 \cdot 2) + \left(5 \cdot \sqrt{x}\right) + (2 \cdot x) + \left(x \cdot \sqrt{x}\right)}{(2 \cdot 2) + \left(2 \cdot \sqrt{x}\right) - \left(2 \cdot \sqrt{x}\right) - \left(\sqrt{x} \cdot \sqrt{x}\right)}$

$$= \frac{10+5\sqrt{x}+2x+x\sqrt{x}}{4+2\sqrt{x}-2\sqrt{x}-\sqrt{x\cdot x}} = \frac{(10+2x)+5\sqrt{x}+x\sqrt{x}}{4-\sqrt{x^2}} = \frac{2(5+x)+\sqrt{x}(5+x)}{4-x} = \frac{\mathbf{(5+x)\left[2+\sqrt{x}\right]}}{\mathbf{4-x}}$$

8. $$\frac{-w+\sqrt{w}}{w+\sqrt{w}} = \frac{-w+\sqrt{w}}{w+\sqrt{w}} \times \frac{w-\sqrt{w}}{w-\sqrt{w}} = \frac{\left(-w+\sqrt{w}\right)\times\left(w-\sqrt{w}\right)}{\left(w+\sqrt{w}\right)\times\left(w-\sqrt{w}\right)} = \frac{-(w\cdot w)+\left(w\cdot\sqrt{w}\right)+\left(w\cdot\sqrt{w}\right)-\left(\sqrt{w}\cdot\sqrt{w}\right)}{(w\cdot w)-\left(w\cdot\sqrt{w}\right)+\left(w\cdot\sqrt{w}\right)-\left(\sqrt{w}\cdot\sqrt{w}\right)}$$

$$= \frac{-w^2+w\sqrt{w}+w\sqrt{w}-\sqrt{w\cdot w}}{w^2-w\sqrt{w}+w\sqrt{w}-\sqrt{w\cdot w}} = \frac{-w^2+(1+1)w\sqrt{w}-w}{w^2-w} = \frac{-w^2+2w\sqrt{w}-w}{w^2-w} = \frac{\overset{1}{\cancel{w}}\left(-w+2\sqrt{w}-1\right)}{\underset{1}{\cancel{w}}(w-1)} = \frac{\mathbf{-w+2\sqrt{w}-1}}{\mathbf{w-1}}$$

9. $$\frac{\sqrt{k}-3}{1+\sqrt{k}} = \frac{\sqrt{k}-3}{1+\sqrt{k}} \times \frac{1-\sqrt{k}}{1-\sqrt{k}} = \frac{\left(\sqrt{k}-3\right)\times\left(1-\sqrt{k}\right)}{\left(1+\sqrt{k}\right)\times\left(1-\sqrt{k}\right)} = \frac{\left(1\cdot\sqrt{k}\right)-\left(\sqrt{k}\cdot\sqrt{k}\right)-(1\cdot 3)+\left(3\cdot\sqrt{k}\right)}{(1\cdot 1)-\left(1\cdot\sqrt{k}\right)+\left(1\cdot\sqrt{k}\right)-\left(\sqrt{k}\cdot\sqrt{k}\right)} = \frac{\sqrt{k}-\sqrt{k\cdot k}-3+3\sqrt{k}}{1-\sqrt{k}+\sqrt{k}-\sqrt{k\cdot k}}$$

$$= \frac{\sqrt{k}-\sqrt{k^2}-3+3\sqrt{k}}{1-\sqrt{k^2}} = \frac{\sqrt{k}-k-3+3\sqrt{k}}{1-k} = \frac{\sqrt{k}+3\sqrt{k}-k-3}{1-k} = \frac{(1+3)\sqrt{k}-k-3}{1-k} = \frac{\mathbf{4\sqrt{k}-k-3}}{\mathbf{1-k}}$$

10. $$\frac{m\sqrt{m}+\sqrt{n}}{\sqrt{m}-n\sqrt{n}} = \frac{m\sqrt{m}+\sqrt{n}}{\sqrt{m}-n\sqrt{n}} \times \frac{\sqrt{m}+n\sqrt{n}}{\sqrt{m}+n\sqrt{n}} = \frac{\left(m\sqrt{m}+\sqrt{n}\right)\times\left(\sqrt{m}+n\sqrt{n}\right)}{\left(\sqrt{m}-n\sqrt{n}\right)\times\left(\sqrt{m}+n\sqrt{n}\right)}$$

$$= \frac{m\left(\sqrt{m}\cdot\sqrt{m}\right)+(m\cdot n)\cdot\left(\sqrt{m}\cdot\sqrt{n}\right)+\left(\sqrt{n}\cdot\sqrt{m}\right)+n\left(\sqrt{n}\cdot\sqrt{n}\right)}{\left(\sqrt{m}\cdot\sqrt{m}\right)+n\left(\sqrt{m}\cdot\sqrt{n}\right)-n\left(\sqrt{n}\cdot\sqrt{m}\right)-(n\cdot n)\cdot\left(\sqrt{n}\cdot\sqrt{n}\right)} = \frac{m\sqrt{m\cdot m}+mn\sqrt{m\cdot n}+\sqrt{n\cdot m}+n\sqrt{n\cdot n}}{\sqrt{m\cdot m}+n\sqrt{m\cdot n}-n\sqrt{m\cdot n}-n^2\sqrt{n\cdot n}}$$

$$= \frac{m\sqrt{m^2}+mn\sqrt{mn}+\sqrt{mn}+n\sqrt{n^2}}{\sqrt{m^2}+n\sqrt{mn}-n\sqrt{mn}-n^2\sqrt{n^2}} = \frac{(m\cdot m)+mn\sqrt{mn}+\sqrt{mn}+(n\cdot n)}{m-\left(n^2\cdot n\right)} = \frac{\mathbf{m^2+n^2+\sqrt{mn}(mn+1)}}{\mathbf{m-n^3}}$$

Section 4.4 Case I Practice Problems

1. $5\sqrt{3}+8\sqrt{3} = (5+8)\sqrt{3} = \mathbf{13\sqrt{3}}$

2. $2\sqrt[3]{3}-4\sqrt[3]{3} = (2-4)\sqrt[3]{3} = \mathbf{-2\sqrt[3]{3}}$

3. $12\sqrt[4]{5}+8\sqrt[4]{5}+2\sqrt[4]{3} = (12+8+2)\sqrt[4]{5} = \mathbf{22\sqrt[4]{5}}$

4. $a\sqrt{ab}-b\sqrt{ab}+c\sqrt{ab} = \mathbf{(a-b+c)\sqrt{ab}}$

5. $3x\sqrt[3]{x}-2x\sqrt[3]{x}+4x\sqrt[3]{x^2} = (3x-2x)\sqrt[3]{x}+4x\sqrt[3]{x^2} = \mathbf{x\sqrt[3]{x}+4x\sqrt[3]{x^2}}$

6. $5\sqrt[3]{2}+8\sqrt[3]{5}$; ***can not be simplified***

7. $2\sqrt[5]{5}+8\sqrt[3]{5}-5\sqrt[5]{5}+2\sqrt{5} = \left(2\sqrt[5]{5}-5\sqrt[5]{5}\right)+8\sqrt[3]{5}+2\sqrt{5} = (2-5)\sqrt[5]{5}+8\sqrt[3]{5}+2\sqrt{5} = \mathbf{-3\sqrt[5]{5}+8\sqrt[3]{5}+2\sqrt{5}}$

8. $3\sqrt{a}+3a\sqrt{a}-4a\sqrt{a} = 3\sqrt{a}+(3a-4a)\sqrt{a} = 3\sqrt{a}-a\sqrt{a} = \mathbf{(3-a)\sqrt{a}}$

9. $2\sqrt[3]{x^2}+4\sqrt[4]{x^2}+3\sqrt[5]{x^2}$; ***can not be simplified***

10. $3\sqrt{ac}+4\sqrt{ac}-2\sqrt[3]{ac}+3\sqrt[3]{ac} = (3+4)\sqrt{ac}+(-2+3)\sqrt[3]{ac} = \mathbf{7\sqrt{ac}+\sqrt[3]{ac}}$

Section 4.4 Case II Practice Problems

1. $5\sqrt{2a}+\sqrt{32a} = 5\sqrt{2a}+\sqrt{(16\cdot 2)a} = 5\sqrt{2a}+\sqrt{\left(4^2\cdot 2\right)a} = 5\sqrt{2a}+4\sqrt{2a} = (5+4)\sqrt{2a} = \mathbf{9\sqrt{2a}}$

2. $\sqrt[3]{27x}-\sqrt[3]{375x^4}-x\sqrt[3]{24x^7} = \sqrt[3]{3^3 x}-\sqrt[3]{(125\cdot 3)x^{3+1}}-x\sqrt[3]{(8\cdot 3)x^{3+3+1}} = 3\sqrt[3]{x}-\sqrt[3]{\left(5^3\cdot 3\right)\cdot\left(x^3\cdot x^1\right)}-x\sqrt[3]{\left(2^3\cdot 3\right)\cdot\left(x^3\cdot x^3\cdot x^1\right)}$

$= 3\sqrt[3]{x}-5x\sqrt[3]{3x}-2(x\cdot x\cdot x)\sqrt[3]{3x} = 3\sqrt[3]{x}-5x\sqrt[3]{3x}-2x^3\sqrt[3]{3x} = \mathbf{3\sqrt[3]{x}-\sqrt[3]{3x}\left(2x^3+5x\right)}$

3. $2a^2\sqrt[4]{x^5}+4\sqrt[4]{81\cdot x^9}+\sqrt[4]{256x} = 2a^2\sqrt[4]{x^{4+1}}+4\sqrt[4]{3^4\cdot x^{4+4+1}}+\sqrt[4]{4^4 x} = 2a^2\sqrt[4]{x^4\cdot x^1}+(4\cdot 3)\sqrt[4]{x^4\cdot x^4\cdot x^1}+4\sqrt[4]{x}$

$= 2a^2x\sqrt[4]{x}+12(x\cdot x)\sqrt[4]{x}+4\sqrt[4]{x} = 2a^2x\sqrt[4]{x}+12x^2\sqrt[4]{x}+4\sqrt[4]{x} = \mathbf{2\sqrt[4]{x}\left(6x^2+a^2x+2\right)}$

4. $\sqrt[5]{w^{11}}+5\sqrt[5]{32w^6}-2a\sqrt[5]{w^{16}} = \sqrt[5]{w^{5+5+1}}+5\sqrt[5]{32w^{5+1}}-2a\sqrt[5]{w^{5+5+5+1}} = \sqrt[5]{w^5\cdot w^5\cdot w^1}+5\sqrt[5]{2^5 w^5\cdot w^1}-2a\sqrt[5]{w^5\cdot w^5\cdot w^5\cdot w^1}$

$= (w\cdot w)\sqrt[5]{w}+(5\cdot 2)w\sqrt[5]{w}-2a(w\cdot w\cdot w)\sqrt[5]{w} = w^2\sqrt[5]{w}+10w\sqrt[5]{w}-2aw^3\sqrt[5]{w} = \mathbf{w\sqrt[5]{w}\left(-2aw^2+w+10\right)}$

5. $\sqrt{4xy}-4\sqrt{(xy)^5}+2\sqrt{49\cdot(xy)^3} = \sqrt{2^2 xy}-4\sqrt{(xy)^{2+2+1}}+2\sqrt{7^2\cdot(xy)^{2+1}}$

$= 2\sqrt{xy}-4\sqrt{(xy)^2\cdot(xy)^2\cdot(xy)^1}+(2\cdot 7)\sqrt{(xy)^2\cdot(xy)^1} = 2\sqrt{xy}-4(xy)^2\sqrt{xy}+14(xy)\sqrt{xy} = \mathbf{2\sqrt{xy}\left[1-2(xy)^2+7(xy)\right]}$

6. $\sqrt[3]{x^5y}+5\sqrt[3]{x^2y^7}+3\sqrt[3]{x^8y^4} = \sqrt[3]{x^{3+2}y}+5\sqrt[3]{x^2y^{3+3+1}}+3\sqrt[3]{x^{3+3+2}y^{3+1}}$

$= \sqrt[3]{\left(x^3\cdot x^2\right)y}+5\sqrt[3]{x^2\left(y^3\cdot y^3\cdot y^1\right)}+3\sqrt[3]{\left(x^3\cdot x^3\cdot x^2\right)\cdot\left(y^3\cdot y^1\right)} = x\sqrt[3]{x^2y}+5(y\cdot y)\sqrt[3]{x^2y}+3(x\cdot x)y\sqrt[3]{x^2y}$

$= x\sqrt[3]{x^2y}+5y^2\sqrt[3]{x^2y}+3x^2y\sqrt[3]{x^2y} = \mathbf{\sqrt[3]{x^2y}\left(5y^2+3x^2y+x\right)}$

7. $\sqrt[5]{mn+3}+\sqrt[5]{(mn+3)^6}+2a\sqrt[5]{(mn+3)^{11}} = \sqrt[5]{(mn+3)}+\sqrt[5]{(mn+3)^{5+1}}+2a\sqrt[5]{(mn+3)^{5+5+1}}$

$= \sqrt[5]{(mn+3)}+\sqrt[5]{(mn+3)^5\cdot(mn+3)^1}+2a\sqrt[5]{(mn+3)^5\cdot(mn+3)^5\cdot(mn+3)^1}$

$= \sqrt[5]{(mn+3)}+(mn+3)\sqrt[5]{(mn+3)}+2a(mn+3)^2\sqrt[5]{(mn+3)} = \mathbf{\sqrt[5]{(mn+3)}\left[1+(mn+3)+2a(mn+3)^2\right]}$

8. $\sqrt{x^3}-\sqrt{125x^5}+3\sqrt{x^3} = (1+3)\sqrt{x^3}-\sqrt{125x^5} = 4\sqrt{x^{2+1}}-\sqrt{(25\cdot 5)x^{2+2+1}} = 4\sqrt{x^2\cdot x^1}-\sqrt{\left(5^2\cdot 5\right)x^2\cdot x^2\cdot x^1}$

$= 4x\sqrt{x}-5(x\cdot x)\sqrt{5x} = \mathbf{4x\sqrt{x}-5x^2\sqrt{5x}}$

9. $\sqrt[3]{x^7y^8z^3}+2\sqrt[3]{x^4y^5z^6}+8\sqrt[3]{64\cdot xy^2} = z\sqrt[3]{x^{3+3+1}y^{3+3+2}}+2\sqrt[3]{x^{3+1}y^{3+2}z^{3+3}}+8\sqrt[3]{4^3\cdot xy^2}$

$= z\sqrt[3]{\left(x^3 \cdot x^3 \cdot x^1\right) \cdot \left(y^3 \cdot y^3 \cdot y^2\right)} + 2\sqrt[3]{\left(x^3 \cdot x^1\right) \cdot \left(y^3 \cdot y^2\right) \cdot \left(z^3 \cdot z^3\right)} + (8 \cdot 4)\sqrt[3]{xy^2}$

$= (x \cdot x) \cdot (y \cdot y) \cdot z\sqrt[3]{xy^2} + 2xy(z \cdot z)\sqrt[3]{xy^2} + 32\sqrt[3]{xy^2} = x^2y^2z\sqrt[3]{xy^2} + 2xyz^2\sqrt[3]{xy^2} + 32\sqrt[3]{xy^2}$

$= \mathbf{\sqrt[3]{xy^2}\left(x^2y^2z + 2xyz^2 + 32\right)}$

10. $\sqrt[4]{512 \cdot x^5y^{10}} + \sqrt[4]{48 \cdot x\,y^6} - 2\sqrt[4]{81 \cdot x^9y^2} = \sqrt[4]{(256 \cdot 2) \cdot x^{4+1}y^{4+4+2}} + \sqrt[4]{(16 \cdot 3) \cdot x\,y^{4+2}} - 2\sqrt[4]{3^4 \cdot x^{4+4+1}y^2}$

$= \sqrt[4]{\left(4^4 \cdot 2\right) \cdot \left(x^4 \cdot x^1\right) \cdot \left(y^4 \cdot y^4 \cdot y^2\right)} + \sqrt[4]{\left(2^4 \cdot 3\right) \cdot x \cdot \left(y^4 \cdot y^2\right)} - (2 \cdot 3)\sqrt[4]{\left(x^4 \cdot x^4 \cdot x^1\right)y^2}$

$= 4x(y \cdot y)\sqrt[4]{2xy^2} + 2y\sqrt[4]{3xy^2} - 6(x \cdot x)\sqrt[4]{xy^2} = \mathbf{4xy^2\sqrt[4]{2xy^2} + 2y\sqrt[4]{3xy^2} - 6x^2\sqrt[4]{xy^2}}$

Chapter 5 Solutions:

Section 5.1 Case I Practice Problems

1. $75^{\frac{1}{2}} = \sqrt[2]{75^1} = \sqrt[2]{75} = \sqrt[2]{25 \cdot 3} = \sqrt[2]{5^2 \cdot 3} = 5\sqrt[2]{3} = \mathbf{5\sqrt{3}}$

2. $-8^{\frac{1}{3}} = -\sqrt[3]{8^1} = -\sqrt[3]{8} = -\sqrt[3]{2^3} = \mathbf{-2}$

3. $36^{\frac{5}{2}} = \sqrt[2]{36^5} = \sqrt[2]{36^{2+2+1}} = \sqrt[2]{36^2 \cdot 36^2 \cdot 36^1} = (36 \cdot 36)\sqrt{36} = 1296\sqrt{6^2} = 1296 \cdot 6 = \mathbf{7776}$

4. $72^{\frac{1}{2}} = \sqrt[2]{72^1} = \sqrt[2]{72} = \sqrt{72} = \sqrt{36 \cdot 2} = \sqrt{6^2 \cdot 2} = \mathbf{6\sqrt{2}}$

5. $5^{\frac{4}{3}} = \sqrt[3]{5^4} = \sqrt[3]{5^{3+1}} = \sqrt[3]{5^3 \cdot 5^1} = \mathbf{5\sqrt[3]{5}}$

6. $32^{\frac{1}{5}} = \sqrt[5]{32^1} = \sqrt[5]{32} = \sqrt[5]{2^5} = \mathbf{2}$

7. $64^{\frac{2}{3}} = \sqrt[3]{64^2} = \sqrt[3]{\left(4^3\right)^2} = \sqrt[3]{4^{3\times2}} = \sqrt[3]{4^6} = \sqrt[3]{4^{3+3}} = \sqrt[3]{4^3 \cdot 4^3} = 4 \cdot 4 = \mathbf{16}$

8. $125^{\frac{1}{3}} = \sqrt[3]{125^1} = \sqrt[3]{125} = \sqrt[3]{5^3} = \mathbf{5}$

9. $2^{\frac{5}{4}} = \sqrt[4]{2^5} = \sqrt[4]{2^{4+1}} = \sqrt[4]{2^4 \cdot 2^1} = \mathbf{2\sqrt[4]{2}}$

10. $343^{\frac{1}{3}} = \sqrt[3]{343^1} = \sqrt[3]{343} = \sqrt[3]{7^3} = \mathbf{7}$

Section 5.1 Case II Practice Problems

1. $x^{\frac{5}{3}} = \sqrt[3]{x^5} = \sqrt[3]{x^{3+2}} = \sqrt[3]{x^3 \cdot x^2} = \mathbf{x\sqrt[3]{x^2}}$

2. $-\left(w^2 \cdot z\right)^{\frac{2}{3}} = -w^{2\times\frac{2}{3}} \cdot z^{1\times\frac{2}{3}} = -w^{\frac{4}{3}} \cdot z^{\frac{2}{3}} = -\sqrt[3]{w^4} \cdot \sqrt[3]{z^2} = -\sqrt[3]{w^{3+1}} \cdot \sqrt[3]{z^2} = -\sqrt[3]{w^3 \cdot w^1} \cdot \sqrt[3]{z^2} = -w\sqrt[3]{w^1} \cdot \sqrt[3]{z^2}$

$= -w\sqrt[3]{w}\,\sqrt[3]{z^2} = \mathbf{-w\sqrt[3]{wz^2}}$

3. $(x\,y)^{\frac{2}{3}} \cdot \left(z^2\right)^{\frac{3}{5}} = (x\,y)^{\frac{2}{3}} \cdot z^{2\times\frac{3}{5}} = (x\,y)^{\frac{2}{3}} \cdot z^{\frac{6}{5}} = \sqrt[3]{(x\,y)^2} \cdot \sqrt[5]{z^6} = \sqrt[3]{(x\,y)^2} \cdot \sqrt[5]{z^{5+1}} = \sqrt[3]{(x\,y)^2} \cdot \sqrt[5]{z^5 \cdot z^1}$

$= \sqrt[3]{(x\,y)^2} \cdot z\sqrt[5]{z^1} = \mathbf{z\sqrt[3]{(xy)^2}\,\sqrt[5]{z}}$

4. $\left(b^2\right)^{\frac{1}{3}} = b^{2\times\frac{1}{3}} = b^{\frac{2}{3}} = \mathbf{\sqrt[3]{b^2}}$

5. $\left(x^4\right)^{\frac{2}{5}} = x^{4\times\frac{2}{5}} = x^{\frac{8}{5}} = \sqrt[5]{x^8} = \sqrt[5]{x^{5+3}} = \sqrt[5]{x^5 \cdot x^3} = \mathbf{x\sqrt[5]{x^3}}$

6. $\left(a^{\frac{1}{2}}\right)^{\frac{2}{3}} = a^{\frac{1}{2}\times\frac{2}{3}} = a^{\frac{\frac{1}{2}}{\frac{6}{3}}} = a^{\frac{1}{3}} = \sqrt[3]{a^1} = \mathbf{\sqrt[3]{a}}$

7. $x\cdot\left(y^{\frac{1}{2}}\right)^3 = x\cdot\left(y^{\frac{1}{2}\times 3}\right) = x\cdot y^{\frac{3}{2}} = x\cdot\sqrt[2]{y^3} = x\sqrt[2]{y^{2+1}} = x\sqrt{y^2\cdot y^1} = x\,y\sqrt{y^1} = \mathbf{xy\sqrt{y}}$

8. $\left(c^4\right)^{\frac{4}{7}} = c^{4\times\frac{4}{7}} = c^{\frac{16}{7}} = \sqrt[7]{c^{16}} = \sqrt[7]{c^{7+7+2}} = \sqrt[7]{\left(c^7\cdot c^7\right)\cdot c^2} = (c\cdot c)\sqrt[7]{c^2} = \mathbf{c^2\sqrt[7]{c^2}}$

9. $\left(x^3\right)^{\frac{1}{4}}\cdot\left(y^{\frac{5}{2}}\right)^4 = \left(x^{3\times\frac{1}{4}}\right)\cdot\left(y^{\frac{5}{2}\times 4}\right) = x^{\frac{3}{4}}\cdot y^{\frac{\frac{20}{10}}{\frac{2}{1}}} = x^{\frac{3}{4}}\cdot y^{\frac{10}{1}} = \sqrt[4]{x^3}\cdot y^{10} = \mathbf{\sqrt[4]{x^3}\,y^{10}}$

10. $\left(x^3\right)^{\frac{2}{5}} = x^{3\times\frac{2}{5}} = x^{\frac{6}{5}} = \sqrt[5]{x^6} = \sqrt[5]{x^{5+1}} = \sqrt[5]{x^5\cdot x^1} = x\sqrt[5]{x^1} = \mathbf{x\sqrt[5]{x}}$

Section 5.2 Case I Practice Problems

1. $125^{-\frac{1}{2}} = \frac{1}{125^{\frac{1}{2}}} = \frac{1}{\sqrt[2]{125^1}} = \frac{1}{\sqrt{125}} = \frac{1}{\sqrt{25\cdot 5}} = \frac{1}{\sqrt{5^2\cdot 5}} = \mathbf{\frac{1}{5\sqrt{5}}}$

2. $-(343)^{-\frac{1}{3}} = -\frac{1}{(343)^{\frac{1}{3}}} = -\frac{1}{\sqrt[3]{343^1}} = -\frac{1}{\sqrt[3]{7^3}} = \mathbf{-\frac{1}{7}}$

3. $4\cdot(16)^{-\frac{1}{2}} = 4\cdot\frac{1}{(16)^{\frac{1}{2}}} = \frac{4\cdot 1}{\sqrt[2]{16^1}} = \frac{4}{\sqrt{16}} = \frac{4}{\sqrt{4^2}} = \frac{\overset{1}{\cancel{4}}}{\underset{1}{\cancel{4}}} = \frac{1}{1} = \mathbf{1}$

4. $49^{-\frac{1}{2}} = \frac{1}{49^{\frac{1}{2}}} = \frac{1}{\sqrt[2]{49^1}} = \frac{1}{\sqrt{49}} = \frac{1}{\sqrt{7^2}} = \mathbf{\frac{1}{7}}$

5. $-(8)^{-\frac{2}{3}} = -\frac{1}{(8)^{\frac{2}{3}}} = -\frac{1}{\sqrt[3]{8^2}} = -\frac{1}{\sqrt[3]{64}} = -\frac{1}{\sqrt[3]{4^3}} = \mathbf{-\frac{1}{4}}$

6. $32^{-\frac{2}{5}} = \frac{1}{32^{\frac{2}{5}}} = \frac{1}{\sqrt[5]{32^2}} = \frac{1}{\sqrt[5]{\left(2^5\right)^2}} = \frac{1}{\sqrt[5]{2^{5\times 2}}} = \frac{1}{\sqrt[5]{2^{10}}} = \frac{1}{\sqrt[5]{2^{5+5}}} = \frac{1}{\sqrt[5]{2^5\cdot 2^5}} = \frac{1}{2\cdot 2} = \mathbf{\frac{1}{4}}$

7. $10^{-\frac{4}{3}} = \frac{1}{10^{\frac{4}{3}}} = \frac{1}{\sqrt[3]{10^4}} = \frac{1}{\sqrt[3]{10^{3+1}}} = \frac{1}{\sqrt[3]{10^3\cdot 10^1}} = \mathbf{\frac{1}{10\sqrt[3]{10}}}$

8. $625^{-\frac{1}{4}} = \frac{1}{625^{\frac{1}{4}}} = \frac{1}{\sqrt[4]{625^1}} = \frac{1}{\sqrt[4]{5^4}} = \mathbf{\frac{1}{5}}$

9. $2^{-\frac{5}{4}} = \frac{1}{2^{\frac{5}{4}}} = \frac{1}{\sqrt[4]{2^5}} = \frac{1}{\sqrt[4]{2^{4+1}}} = \frac{1}{\sqrt[4]{2^4\cdot 2^1}} = \mathbf{\frac{1}{2\sqrt[4]{2}}}$

10. $(9)^{-\frac{3}{2}} = \frac{1}{(9)^{\frac{3}{2}}} = \frac{1}{\sqrt[2]{9^3}} = \frac{1}{\sqrt[2]{9^{2+1}}} = \frac{1}{\sqrt{9^2 \cdot 9^1}} = \frac{1}{9\sqrt{9}} = \frac{1}{9\sqrt{3^2}} = \frac{1}{9 \cdot 3} = \mathbf{\frac{1}{27}}$

Section 5.2 Case II Practice Problems

1. $z^{-\frac{7}{3}} = \frac{1}{z^{\frac{7}{3}}} = \frac{1}{\sqrt[3]{z^7}} = \frac{1}{\sqrt[3]{z^{3+3+1}}} = \frac{1}{\sqrt[3]{z^3 \cdot z^3 \cdot z^1}} = \frac{1}{(z \cdot z)\sqrt[3]{z}} = \mathbf{\frac{1}{z^2\sqrt[3]{z}}}$

2. $-(a \cdot b)^{-\frac{2}{3}} = -\frac{1}{(a \cdot b)^{\frac{2}{3}}} = -\frac{1}{\sqrt[3]{(a \cdot b)^2}} = -\frac{1}{\sqrt[3]{a^{1\times 2} \cdot b^{1\times 2}}} = -\mathbf{\frac{1}{\sqrt[3]{a^2 b^2}}}$

3. $(k \cdot l^2)^{-\frac{4}{5}} = \frac{1}{(k \cdot l^2)^{\frac{4}{5}}} = \frac{1}{k^{1\times\frac{4}{5}} \cdot l^{2\times\frac{4}{5}}} = \frac{1}{k^{\frac{4}{5}} \cdot l^{\frac{8}{5}}} = \frac{1}{\sqrt[5]{k^4} \cdot \sqrt[5]{l^8}} = \frac{1}{\sqrt[5]{k^4} \cdot \sqrt[5]{l^{5+3}}} = \frac{1}{\sqrt[5]{k^4} \cdot \sqrt[5]{l^5 \cdot l^3}} = \frac{1}{\sqrt[5]{k^4} \cdot l\sqrt[5]{l^3}}$

$= \frac{1}{l\sqrt[5]{k^4 \cdot l^3}} = \mathbf{\frac{1}{l\sqrt[5]{k^4 l^3}}}$

4. $(y^2)^{-\frac{1}{3}} = \frac{1}{(y^2)^{\frac{1}{3}}} = \frac{1}{y^{2\times\frac{1}{3}}} = \frac{1}{y^{\frac{2}{3}}} = \mathbf{\frac{1}{\sqrt[3]{y^2}}}$

5. $(x^{-3})^{\frac{2}{5}} = x^{-3\times\frac{2}{5}} = x^{-\frac{6}{5}} = \frac{1}{x^{\frac{6}{5}}} = \frac{1}{\sqrt[5]{x^6}} = \frac{1}{\sqrt[5]{x^{5+1}}} = \frac{1}{\sqrt[5]{x^5 \cdot x^1}} = \mathbf{\frac{1}{x\sqrt[5]{x}}}$

6. $\left(a^{-\frac{1}{2}}\right)^{\frac{4}{3}} = a^{-\frac{1}{2}\times\frac{4}{3}} = a^{-\frac{4}{6}=-\frac{2}{3}} = a^{-\frac{2}{3}} = \frac{1}{a^{\frac{2}{3}}} = \mathbf{\frac{1}{\sqrt[3]{a^2}}}$

7. $x^2 \cdot \left(y^{-\frac{1}{2}}\right)^3 = x^2 \cdot y^{-\frac{1}{2}\times 3} = x^2 \cdot y^{-\frac{3}{2}} = x^2 \cdot \frac{1}{y^{\frac{3}{2}}} = \frac{x^2}{y^{\frac{3}{2}}} = \frac{x^2}{\sqrt[2]{y^3}} = \frac{x^2}{\sqrt[2]{y^{2+1}}} = \frac{x^2}{\sqrt[2]{y^2 \cdot y^1}} = \frac{x^2}{y\sqrt[2]{y^1}} = \mathbf{\frac{x^2}{y\sqrt{y}}}$

8. $(c^4)^{-\frac{2}{3}} = c^{4\times-\frac{2}{3}} = c^{-\frac{8}{3}} = \frac{1}{c^{\frac{8}{3}}} = \frac{1}{\sqrt[3]{c^8}} = \frac{1}{\sqrt[3]{c^{3+3+2}}} = \frac{1}{\sqrt[3]{c^3 \cdot c^3 \cdot c^2}} = \frac{1}{(c \cdot c)\sqrt[3]{c^2}} = \mathbf{\frac{1}{c^2\sqrt[3]{c^2}}}$

9. $x^3 \cdot y^{-\frac{7}{3}} = x^3 \cdot \frac{1}{y^{\frac{7}{3}}} = \frac{x^3}{y^{\frac{7}{3}}} = \frac{x^3}{\sqrt[3]{y^7}} = \frac{x^3}{\sqrt[3]{y^{3+3+1}}} = \frac{x^3}{\sqrt[3]{y^3 \cdot y^3 \cdot y^1}} = \frac{x^3}{(y \cdot y)\sqrt[3]{y^1}} = \mathbf{\frac{x^3}{y^2\sqrt[3]{y}}}$

10. $y^2 \cdot (x^3)^{-\frac{2}{5}} = y^2 \cdot x^{3\times-\frac{2}{5}} = y^2 \cdot x^{-\frac{6}{5}} = y^2 \cdot \frac{1}{x^{\frac{6}{5}}} = \frac{y^2}{x^{\frac{6}{5}}} = \frac{y^2}{\sqrt[5]{x^6}} = \frac{y^2}{\sqrt[5]{x^{5+1}}} = \frac{y^2}{\sqrt[5]{x^5 \cdot x^1}} = \mathbf{\frac{y^2}{x\sqrt[5]{x}}}$

Section 5.3 Case I Practice Problems

1. $\left(x^2\cdot x^3\right)^{\frac{1}{2}} = x^{2\times\frac{1}{2}}\cdot x^{3\times\frac{1}{2}} = x^{\frac{2}{2}=\frac{1}{1}}\cdot x^{\frac{3}{2}} = x^{\frac{1}{1}}\cdot x^{\frac{3}{2}} = x^{\frac{1}{1}+\frac{3}{2}} = x^{\frac{(1\cdot2)+(1\cdot3)}{1\cdot2}} = x^{\frac{5}{2}} = \sqrt[2]{x^5} = \sqrt[2]{x^{2+2+1}} = \sqrt{x^2\cdot x^2\cdot x^1}$

$= (x\cdot x)\sqrt{x} = \mathbf{x^2\sqrt{x}}$

2. $2\cdot\left(a^{\frac{2}{3}}\cdot b^0\right)^3 = 2\cdot\left(a^{\frac{2}{3}}\cdot 1\right)^3 = 2\cdot\left(a^{\frac{2}{3}}\right)^3 = 2a^{\frac{2}{3}\times 3} = 2a^{\frac{\cancel{6}}{\cancel{3}}=\frac{2}{1}} = 2a^{\frac{2}{1}} = \mathbf{2a^2}$

3. $\left(a^2\cdot b^3\right)^{\frac{2}{3}}\cdot\left(a^{\frac{1}{3}}\cdot b^4\right) = \left(a^{2\times\frac{2}{3}}\cdot b^{3\times\frac{2}{3}}\right)\cdot\left(a^{\frac{1}{3}}\cdot b^4\right) = \left(a^{\frac{4}{3}}\cdot b^{\frac{\cancel{6}}{\cancel{3}}=\frac{2}{1}}\right)\cdot\left(a^{\frac{1}{3}}\cdot b^4\right) = \left(a^{\frac{4}{3}}\cdot b^{\frac{2}{1}}\right)\cdot\left(a^{\frac{1}{3}}\cdot b^{\frac{4}{1}}\right)$

$= \left(a^{\frac{4}{3}}\cdot a^{\frac{1}{3}}\right)\cdot\left(b^{\frac{2}{1}}\cdot b^{\frac{4}{1}}\right) = \left(a^{\frac{4}{3}+\frac{1}{3}}\right)\cdot\left(b^{\frac{2}{1}+\frac{4}{1}}\right) = a^{\frac{4+1}{3}}\cdot b^{\frac{2+4}{1}} = a^{\frac{5}{3}}\cdot b^{\frac{6}{1}} = b^6\cdot a^{\frac{5}{3}} = b^6\cdot\sqrt[3]{a^5} = b^6\cdot\sqrt[3]{a^{3+2}}$

$= b^6\cdot\sqrt[3]{a^3\cdot a^2} = \mathbf{ab^6\sqrt[3]{a^2}}$

4. $\left(a^3\cdot a^{\frac{2}{3}}\right)\cdot\left(x^2\cdot x^3\right)^{\frac{2}{3}} = \left(a^{\frac{3}{1}}\cdot a^{\frac{2}{3}}\right)\cdot\left(x^{2\times\frac{2}{3}}\cdot x^{3\times\frac{2}{3}}\right) = \left(a^{\frac{3}{1}+\frac{2}{3}}\right)\cdot\left(x^{\frac{4}{3}}\cdot x^{\frac{\cancel{6}}{\cancel{3}}}\right) = \left(a^{\frac{(3\cdot3)+(1\cdot2)}{1\cdot3}}\right)\cdot\left(x^{\frac{4}{3}}\cdot x^{\frac{2}{1}}\right) = a^{\frac{9+2}{3}}\cdot\left(x^{\frac{4}{3}+\frac{2}{1}}\right)$

$= a^{\frac{11}{3}}\cdot\left(x^{\frac{(1\cdot4)+(3\cdot2)}{3}}\right) = a^{\frac{11}{3}}\cdot\left(x^{\frac{4+6}{3}}\right) = a^{\frac{11}{3}}\cdot x^{\frac{10}{3}} = \sqrt[3]{a^{11}}\cdot\sqrt[3]{x^{10}} = \sqrt[3]{a^{3+3+3+2}}\cdot\sqrt[3]{x^{3+3+3+1}}$

$= \sqrt[3]{\left(a^3\cdot a^3\cdot a^3\right)\cdot a^2}\cdot\sqrt[3]{\left(x^3\cdot x^3\cdot x^3\right)\cdot x^1} = \left[(a\cdot a\cdot a)\sqrt[3]{a^2}\right]\cdot\left[(x\cdot x\cdot x)\sqrt[3]{x^1}\right] = a^3x^3\left(\sqrt[3]{a^2}\sqrt[3]{x}\right) = \mathbf{a^3x^3\left(\sqrt[3]{a^2x}\right)}$

5. $\left(x\cdot y^2\cdot z^3\right)^{\frac{0}{2}}\cdot\left(x^2\right)^{\frac{1}{2}} = \left(x\cdot y^2\cdot z^3\right)^0\cdot x^{2\times\frac{1}{2}} = 1\cdot x^{\frac{2}{2}} = x^{\frac{1}{1}} = x^1 = \mathbf{x}$

6. $2^{\frac{1}{3}}\cdot 3^{\frac{2}{3}}\cdot 3^5\cdot 2^{\frac{2}{3}} = \left(2^{\frac{1}{3}}\cdot 2^{\frac{2}{3}}\right)\cdot\left(3^{\frac{2}{3}}\cdot 3^{\frac{5}{1}}\right) = \left(2^{\frac{1}{3}+\frac{2}{3}}\right)\cdot\left(3^{\frac{2}{3}+\frac{5}{1}}\right) = 2^{\frac{1+2}{3}}\cdot\left(3^{\frac{(2\cdot1)+(5\cdot3)}{3}}\right) = 2^{\frac{3}{3}=\frac{1}{1}}\cdot\left(3^{\frac{2+15}{3}}\right) = 2^{\frac{1}{1}}\cdot 3^{\frac{17}{3}}$

$= 2\cdot 3^{\frac{17}{3}} = 2\cdot\sqrt[3]{3^{17}} = 2\cdot\sqrt[3]{3^{3+3+3+3+3+2}} = 2\cdot\sqrt[3]{\left(3^3\cdot 3^3\cdot 3^3\cdot 3^3\cdot 3^3\right)\cdot 3^2} = 2\cdot(3\cdot3\cdot3\cdot3\cdot3)\sqrt[3]{3^2} = \left(2\cdot 3^5\right)\sqrt[3]{9} = \mathbf{486\sqrt[3]{9}}$

7. $x^2\cdot\left(x^{\frac{3}{4}}\right)^4\cdot\left(x^{\frac{1}{4}}\cdot y^{\frac{3}{5}}\right)^2 = x^2\cdot\left(x^{\frac{3}{4}\times4}\right)\cdot\left(x^{\frac{1}{4}\times2}\cdot y^{\frac{3}{5}\times2}\right) = x^2\cdot x^{\frac{12}{4}=\frac{3}{1}}\cdot x^{\frac{2}{4}=\frac{1}{2}}\cdot y^{\frac{6}{5}} = x^2\cdot x^{\frac{3}{1}}\cdot x^{\frac{1}{2}}\cdot y^{\frac{6}{5}} = \left(x^2\cdot x^3\right)\cdot x^{\frac{1}{2}}\cdot y^{\frac{6}{5}}$

$= \left(x^{2+3}\right)\cdot x^{\frac{1}{2}}\cdot y^{\frac{6}{5}} = x^5\cdot x^{\frac{1}{2}}\cdot y^{\frac{6}{5}} = \left(x^{\frac{5}{1}}\cdot x^{\frac{1}{2}}\right)\cdot y^{\frac{6}{5}} = \left(x^{\frac{5}{1}+\frac{1}{2}}\right)\cdot y^{\frac{6}{5}} = x^{\frac{(5\cdot2)+(1\cdot1)}{1\cdot2}}\cdot y^{\frac{6}{5}} = x^{\frac{10+1}{2}}\cdot y^{\frac{6}{5}} = x^{\frac{11}{2}}\cdot y^{\frac{6}{5}}$

$= \sqrt[2]{x^{11}}\cdot\sqrt[5]{y^6} = \sqrt[2]{x^{2+2+2+2+2+1}}\cdot\sqrt[5]{y^{5+1}} = \sqrt[2]{\left(x^2\cdot x^2\cdot x^2\cdot x^2\cdot x^2\right)x^1}\cdot\sqrt[5]{y^5\cdot y^1} = (x\cdot x\cdot x\cdot x\cdot x)\sqrt{x}\cdot\left(y\sqrt[5]{y}\right)$

$= \mathbf{x^5y\left(\sqrt{x}\sqrt[5]{y}\right)}$

8. $\left(x^0\cdot y^2\cdot z\right)^{\frac{3}{2}}\cdot\left(x^{\frac{1}{3}}\cdot y^{\frac{2}{5}}\cdot z\right) = \left(1\cdot y^2\cdot z\right)^{\frac{3}{2}}\cdot\left(x^{\frac{1}{3}}\cdot y^{\frac{2}{5}}\cdot z\right) = \left(y^{2\times\frac{3}{2}}\cdot z^{1\times\frac{3}{2}}\right)\cdot\left(x^{\frac{1}{3}}\cdot y^{\frac{2}{5}}\cdot z\right) = y^{\frac{6}{2}=\frac{3}{1}}\cdot z^{\frac{3}{2}}\cdot x^{\frac{1}{3}}\cdot y^{\frac{2}{5}}\cdot z$

$= x^{\frac{1}{3}}\left(y^{\frac{3}{1}}\cdot y^{\frac{2}{5}}\right)\cdot\left(z^{\frac{3}{2}}\cdot z^{\frac{1}{1}}\right) = x^{\frac{1}{3}}\left(y^{\frac{3}{1}+\frac{2}{5}}\right)\cdot\left(z^{\frac{3}{2}+\frac{1}{1}}\right) = x^{\frac{1}{3}}\left(y^{\frac{(3\cdot5)+(1\cdot2)}{1\cdot5}}\right)\cdot\left(z^{\frac{(3\cdot1)+(1\cdot2)}{2\cdot1}}\right) = x^{\frac{1}{3}}\left(y^{\frac{15+2}{5}}\right)\cdot\left(z^{\frac{3+2}{2}}\right)$

$= x^{\frac{1}{3}}\cdot y^{\frac{17}{5}}\cdot z^{\frac{5}{2}} = \sqrt[3]{x^1}\cdot\sqrt[5]{y^{17}}\cdot\sqrt[2]{z^5} = \sqrt[3]{x}\cdot\sqrt[5]{y^{5+5+5+2}}\cdot\sqrt[2]{z^{2+2+1}} = \sqrt[3]{x}\cdot\sqrt[5]{\left(y^5\cdot y^5\cdot y^5\right)\cdot y^2}\cdot\sqrt[2]{\left(z^2\cdot z^2\right)\cdot z^1}$

$= \sqrt[3]{x}\cdot\left[(y\cdot y\cdot y)\sqrt[5]{y^2}\right]\cdot\left[(z\cdot z)\sqrt[2]{z}\right] = \sqrt[3]{x}\cdot\left(y^3\sqrt[5]{y^2}\right)\cdot\left(z^2\sqrt{z}\right) = \mathbf{y^3z^2\left(\sqrt[3]{x}\sqrt[5]{y^2}\sqrt{z}\right)}$

9. $5^0\cdot\left(x^2\cdot y\right)^{\frac{2}{3}}\cdot\left(3^4\cdot x\cdot y^3\right)^{\frac{1}{3}} = 1\cdot\left(x^{2\times\frac{2}{3}}\cdot y^{1\times\frac{2}{3}}\right)\cdot\left(3^{4\times\frac{1}{3}}\cdot x^{1\times\frac{1}{3}}\cdot y^{3\times\frac{1}{3}}\right) = \left(x^{\frac{4}{3}}\cdot y^{\frac{2}{3}}\right)\cdot\left(3^{\frac{4}{3}}\cdot x^{\frac{1}{3}}\cdot y^{\frac{3}{3}}\right)$

$= 3^{\frac{4}{3}}\cdot\left(x^{\frac{4}{3}}\cdot x^{\frac{1}{3}}\right)\cdot\left(y^{\frac{2}{3}}\cdot y^{\frac{1}{1}}\right) = 3^{\frac{4}{3}}\cdot x^{\frac{4}{3}+\frac{1}{3}}\cdot y^{\frac{2}{3}+\frac{1}{1}} = 3^{\frac{4}{3}}\cdot x^{\frac{4+1}{3}}\cdot y^{\frac{(2\cdot1)+(1\cdot3)}{3\cdot1}} = 3^{\frac{4}{3}}\cdot x^{\frac{5}{3}}\cdot y^{\frac{2+3}{3}} = 3^{\frac{4}{3}}\cdot x^{\frac{5}{3}}\cdot y^{\frac{5}{3}} = 3^{\frac{4}{3}}\cdot(xy)^{\frac{5}{3}}$

$= \sqrt[3]{3^4}\sqrt[3]{(xy)^5} = \sqrt[3]{3^4(xy)^5} = \sqrt[3]{3^{3+1}(xy)^{3+2}} = \sqrt[3]{3^3\cdot3^1(xy)^3\cdot(xy)^2} = \mathbf{3(xy)\sqrt[3]{3(xy)^2}}$

10. $\left(x^{\frac{2}{3}}\right)^3\cdot\left[\left(2\cdot x\cdot y^2\right)^3\cdot x\right]^{\frac{1}{5}} = \left(x^{\frac{2}{3}\times3}\right)\cdot\left[\left(2^3\cdot x^3\cdot y^{2\times3}\right)\cdot x\right]^{\frac{1}{5}} = x^{\frac{6}{3}}\cdot\left[\left(2^3\cdot x^3\cdot y^6\right)\cdot x\right]^{\frac{1}{5}} = x^{\frac{2}{1}}\cdot\left[2^{3\times\frac{1}{5}}\cdot x^{3\times\frac{1}{5}}\cdot y^{6\times\frac{1}{5}}\cdot x^{1\times\frac{1}{5}}\right]$

$= x^{\frac{2}{1}}\cdot2^{\frac{3}{5}}\cdot x^{\frac{3}{5}}\cdot y^{\frac{6}{5}}\cdot x^{\frac{1}{5}} = 2^{\frac{3}{5}}\cdot\left(x^{\frac{3}{5}}\cdot x^{\frac{1}{5}}\cdot x^{\frac{2}{1}}\right)\cdot y^{\frac{6}{5}} = 2^{\frac{3}{5}}\cdot\left(x^{\frac{3}{5}+\frac{1}{5}}\cdot x^{\frac{2}{1}}\right)\cdot y^{\frac{6}{5}} = 2^{\frac{3}{5}}\cdot\left(x^{\frac{4}{5}}\cdot x^{\frac{2}{1}}\right)\cdot y^{\frac{6}{5}} = 2^{\frac{3}{5}}\cdot\left(x^{\frac{4}{5}+\frac{2}{1}}\right)\cdot y^{\frac{6}{5}}$

$= 2^{\frac{3}{5}}\cdot x^{\frac{(4\cdot1)+(2\cdot5)}{5\cdot1}}\cdot y^{\frac{6}{5}} = 2^{\frac{3}{5}}\cdot x^{\frac{4+10}{5}}\cdot y^{\frac{6}{5}} = 2^{\frac{3}{5}}\cdot x^{\frac{14}{5}}\cdot y^{\frac{6}{5}} = \sqrt[5]{2^3}\cdot\sqrt[5]{x^{14}}\cdot\sqrt[5]{y^6} = \sqrt[5]{8}\cdot\sqrt[5]{x^{5+5+4}}\cdot\sqrt[5]{y^{5+1}}$

$= \sqrt[5]{8}\cdot\sqrt[5]{\left(x^5\cdot x^5\right)\cdot x^4}\cdot\sqrt[5]{y^5\cdot y^1} = \sqrt[5]{8}\cdot(x\cdot x)\sqrt[5]{x^4}\cdot y\sqrt[5]{y} = x^2y\left(\sqrt[5]{8}\sqrt[5]{x^4}\sqrt[5]{y}\right) = \mathbf{x^2y\left(\sqrt[5]{8x^4y}\right)}$

Section 5.3 Case II Practice Problems

1. $\dfrac{x^{\frac{2}{3}}}{x^2} = \dfrac{x^{\frac{2}{3}}}{x^{\frac{2}{1}}} = \dfrac{1}{x^{\frac{2}{1}}x^{-\frac{2}{3}}} = \dfrac{1}{x^{\frac{2}{1}-\frac{2}{3}}} = \dfrac{1}{x^{\frac{(2\cdot3)-(1\cdot2)}{1\cdot3}}} = \dfrac{1}{x^{\frac{6-2}{3}}} = \dfrac{1}{x^{\frac{4}{3}}} = \dfrac{1}{\sqrt[3]{x^4}} = \dfrac{1}{\sqrt[3]{x^{3+1}}} = \dfrac{1}{\sqrt[3]{x^3\cdot x^1}} = \mathbf{\dfrac{1}{x\sqrt[3]{x}}}$

2. $\dfrac{a^2\cdot b^{\frac{1}{3}}}{a^{\frac{1}{4}}\cdot b} = \dfrac{a^{\frac{2}{1}}\cdot b^{\frac{1}{3}}}{a^{\frac{1}{4}}\cdot b^{\frac{1}{1}}} = \dfrac{a^{\frac{2}{1}}\cdot a^{-\frac{1}{4}}}{b^{\frac{1}{1}}\cdot b^{-\frac{1}{3}}} = \dfrac{a^{\frac{2}{1}-\frac{1}{4}}}{b^{\frac{1}{1}-\frac{1}{3}}} = \dfrac{a^{\frac{(2\cdot4)-(1\cdot1)}{1\cdot4}}}{b^{\frac{(1\cdot3)-(1\cdot1)}{1\cdot3}}} = \dfrac{a^{\frac{8-1}{4}}}{b^{\frac{3-1}{3}}} = \dfrac{a^{\frac{7}{4}}}{b^{\frac{2}{3}}} = \dfrac{\sqrt[4]{a^7}}{\sqrt[3]{b^2}} = \dfrac{\sqrt[4]{a^{4+3}}}{\sqrt[3]{b^2}} = \dfrac{\sqrt[4]{a^4\cdot a^3}}{\sqrt[3]{b^2}} = \mathbf{\dfrac{a\sqrt[4]{a^3}}{\sqrt[3]{b^2}}}$

3. $\left(\dfrac{a^2\cdot b^5\cdot c^2}{a^{\frac{2}{3}}}\right)^{\frac{0}{4}} = \left(\dfrac{a^2\cdot b^5\cdot c^2}{a^{\frac{2}{3}}}\right)^0 = \mathbf{1}$

4. $$\frac{y^{\frac{3}{4}}\cdot(z\cdot w)^{\frac{5}{2}}}{y^2\cdot w^{\frac{1}{2}}} = \frac{y^{\frac34}\cdot z^{1\times\frac52}\cdot w^{1\times\frac52}}{y^2\cdot w^{\frac12}} = \frac{y^{\frac34}\cdot z^{\frac52}\cdot w^{\frac52}}{y^{\frac21}\cdot w^{\frac12}} = \frac{z^{\frac52}\cdot w^{\frac52}\cdot w^{-\frac12}}{y^{\frac21}\cdot y^{-\frac34}} = \frac{z^{\frac52}\cdot w^{\frac52-\frac12}}{y^{\frac21-\frac34}} = \frac{z^{\frac52}\cdot w^{\frac{5-1}{2}}}{y^{\frac{(2\cdot4)-(1\cdot3)}{1\cdot4}}} = \frac{z^{\frac52}\cdot w^{\frac42}}{y^{\frac{8-3}{4}}} = \frac{z^{\frac52}\cdot w^{\frac42}}{y^{\frac54}}$$

$$= \frac{z^{\frac52}\cdot w^{\frac21}}{y^{\frac54}} = \frac{\sqrt[2]{z^5}\cdot w^2}{\sqrt[4]{y^5}} = \frac{\sqrt[2]{z^{2+2+1}}\cdot w^2}{\sqrt[4]{y^{4+1}}} = \frac{\sqrt[2]{z^2\cdot z^2\cdot z^1}\cdot w^2}{\sqrt[4]{y^4\cdot y^1}} = \frac{(z\cdot z)\sqrt{z}\cdot w^2}{y\sqrt[4]{y}} = \frac{w^2z^2\sqrt{z}}{y\sqrt[4]{y}}$$

5. $$\frac{(a\cdot b)^{\frac12}\cdot(x\cdot y)^{\frac23}}{a^3\cdot y^2\cdot x^{\frac12}} = \frac{a^{1\times\frac12}\cdot b^{1\times\frac12}\cdot x^{1\times\frac23}\cdot y^{1\times\frac23}}{a^3\cdot y^2\cdot x^{\frac12}} = \frac{a^{\frac12}\cdot b^{\frac12}\cdot x^{\frac23}\cdot y^{\frac23}}{a^{\frac31}\cdot y^{\frac21}\cdot x^{\frac12}} = \frac{b^{\frac12}\cdot x^{\frac23}\cdot x^{-\frac12}}{a^{\frac31}\cdot a^{-\frac12}\cdot y^{\frac21}\cdot y^{-\frac23}} = \frac{b^{\frac12}\cdot x^{\frac23-\frac12}}{a^{\frac31-\frac12}\cdot y^{\frac21-\frac23}}$$

$$= \frac{b^{\frac12}\cdot x^{\frac{(2\cdot2)-(1\cdot3)}{3\cdot2}}}{a^{\frac{(3\cdot2)-(1\cdot1)}{1\cdot2}}\cdot y^{\frac{(2\cdot3)-(1\cdot2)}{1\cdot3}}} = \frac{b^{\frac12}\cdot x^{\frac{4-3}{6}}}{a^{\frac{6-1}{2}}\cdot y^{\frac{6-2}{3}}} = \frac{b^{\frac12}\cdot x^{\frac16}}{a^{\frac52}\cdot y^{\frac43}} = \frac{\sqrt[2]{b^1}\cdot\sqrt[6]{x^1}}{\sqrt[2]{a^5}\cdot\sqrt[3]{y^4}} = \frac{\sqrt[2]{b}\cdot\sqrt[6]{x}}{\sqrt[2]{a^{2+2+1}}\cdot\sqrt[3]{y^{3+1}}} = \frac{\sqrt[2]{b}\cdot\sqrt[6]{x}}{\sqrt[2]{a^2\cdot a^2\cdot a^1}\cdot\sqrt[3]{y^3\cdot y^1}}$$

$$= \frac{\sqrt{b}\cdot\sqrt[6]{x}}{(a\cdot a)\sqrt{a}\cdot y\sqrt[3]{y}} = \frac{\sqrt{b}\,\sqrt[6]{x}}{a^2y\sqrt{a}\,\sqrt[3]{y}}$$

6. $$\frac{(x\cdot y)^{\frac23}\cdot x^2}{(x\cdot y)^{\frac32}} = \frac{x^{1\times\frac23}\cdot y^{1\times\frac23}\cdot x^2}{x^{1\times\frac32}\cdot y^{1\times\frac32}} = \frac{x^{\frac23}\cdot y^{\frac23}\cdot x^2}{x^{\frac32}\cdot y^{\frac32}} = \frac{x^{-\frac32}\cdot x^{\frac23}\cdot x^2}{y^{\frac32}\cdot y^{-\frac23}} = \frac{x^{-\frac32+\frac23}\cdot x^2}{y^{\frac32-\frac23}} = \frac{x^{\frac{-(3\cdot3)+(2\cdot2)}{2\cdot3}}\cdot x^2}{y^{\frac{(3\cdot3)-(2\cdot2)}{2\cdot3}}} = \frac{x^{\frac{-9+4}{6}}\cdot x^2}{y^{\frac{9-4}{6}}}$$

$$= \frac{x^{\frac{-5}{6}}\cdot x^{\frac21}}{y^{\frac56}} = \frac{x^{\frac{-5}{6}+\frac21}}{y^{\frac56}} = \frac{x^{\frac{-(5\cdot1)+(2\cdot6)}{1\cdot6}}}{y^{\frac56}} = \frac{x^{\frac{-5+12}{6}}}{y^{\frac56}} = \frac{x^{\frac76}}{y^{\frac56}} = \frac{\sqrt[6]{x^7}}{\sqrt[6]{y^5}} = \frac{\sqrt[6]{x^{6+1}}}{\sqrt[6]{y^5}} = \frac{\sqrt[6]{x^6\cdot x^1}}{\sqrt[6]{y^5}} = \frac{x\sqrt[6]{x}}{\sqrt[6]{y^5}} = x\left(\sqrt[6]{\frac{x}{y^5}}\right)$$

7. $$\frac{2\cdot a^{\frac32}\cdot b^2\cdot c^{\frac15}\cdot a}{b\cdot c^{\frac25}} = \frac{2\cdot a^{\frac32}\cdot b^2\cdot c^{\frac15}\cdot a^1}{b^1\cdot c^{\frac25}} = \frac{2\cdot a^{\frac32}\cdot a^{\frac11}\cdot b^2\cdot b^{-1}}{c^{\frac25}\cdot c^{-\frac15}} = \frac{2\cdot a^{\frac32+\frac11}\cdot b^{2-1}}{c^{\frac25-\frac15}} = \frac{2\cdot a^{\frac{(1\cdot3)+(1\cdot2)}{1\cdot2}}\cdot b^1}{c^{\frac{2-1}{5}}} = \frac{2\cdot a^{\frac{3+2}{2}}\cdot b}{c^{\frac15}}$$

$$= \frac{2\cdot a^{\frac52}\cdot b}{c^{\frac15}} = \frac{2b\cdot\sqrt[2]{a^5}}{\sqrt[5]{c^1}} = \frac{2b\cdot\sqrt[2]{a^{2+2+1}}}{\sqrt[5]{c}} = \frac{2b\cdot\sqrt{a^2\cdot a^2\cdot a^1}}{\sqrt[5]{c}} = \frac{2b\cdot(a\cdot a)\sqrt{a}}{\sqrt[5]{c}} = \frac{\mathbf{2a^2b\sqrt{a}}}{\mathbf{\sqrt[5]{c}}}$$

8. $$\left(\frac{z^{\frac12}\cdot w^3\cdot(a\cdot b)^{\frac32}}{a^2\cdot w^0\cdot z}\right)^{\frac23} = \left(\frac{z^{\frac12}\cdot w^3\cdot a^{1\times\frac32}\cdot b^{1\times\frac32}}{a^2\cdot1\cdot z}\right)^{\frac23} = \left(\frac{z^{\frac12}\cdot w^3\cdot a^{\frac32}\cdot b^{\frac32}}{a^2\cdot z}\right)^{\frac23} = \left(\frac{w^3\cdot b^{\frac32}}{a^2\cdot a^{-\frac32}\cdot z\cdot z^{-\frac12}}\right)^{\frac23} = \left(\frac{w^3\cdot b^{\frac32}}{a^{\frac21}\cdot a^{-\frac32}\cdot z^{\frac11}\cdot z^{-\frac12}}\right)^{\frac23}$$

$$= \left(\frac{w^3\cdot b^{\frac32}}{a^{\frac21-\frac32}\cdot z^{\frac11-\frac12}}\right)^{\frac23} = \left(\frac{w^3\cdot b^{\frac32}}{a^{\frac{(2\cdot2)-(1\cdot3)}{1\cdot2}}\cdot z^{\frac{(1\cdot2)-(1\cdot1)}{1\cdot2}}}\right)^{\frac23} = \left(\frac{w^3\cdot b^{\frac32}}{a^{\frac{4-3}{2}}\cdot z^{\frac{2-1}{2}}}\right)^{\frac23} = \left(\frac{w^3\cdot b^{\frac32}}{a^{\frac12}\cdot z^{\frac12}}\right)^{\frac23} = \frac{w^{3\times\frac23}\cdot b^{\frac32\times\frac23}}{a^{\frac12\times\frac23}\cdot z^{\frac12\times\frac23}} = \frac{w^{\frac63=\frac21}\cdot b^{\frac66=\frac11}}{a^{\frac26=\frac13}\cdot z^{\frac26=\frac13}}$$

$$= \frac{w^{\frac21}\cdot b^{\frac11}}{a^{\frac13}\cdot z^{\frac13}} = \frac{w^2\cdot b}{a^{\frac13}\cdot z^{\frac13}} = \frac{w^2\cdot b}{\sqrt[3]{a^1}\cdot\sqrt[3]{z^1}} = \frac{w^2\cdot b}{\sqrt[3]{a}\cdot\sqrt[3]{z}} = \frac{\mathbf{bw^2}}{\mathbf{\sqrt[3]{az}}}$$

9. $\frac{2^3 \cdot y^4 \cdot b^{\frac{3}{5}}}{2^2 \cdot b^2 \cdot y^{\frac{1}{2}}} = \frac{2^3 \cdot 2^{-2} \cdot y^4 \cdot y^{-\frac{1}{2}}}{b^2 \cdot b^{-\frac{3}{5}}} = \frac{2^{3-2} \cdot y^{\frac{4}{1}} \cdot y^{-\frac{1}{2}}}{b^{\frac{2}{1}} \cdot b^{-\frac{3}{5}}} = \frac{2^1 \cdot y^{\frac{4}{1}-\frac{1}{2}}}{b^{\frac{2}{1}-\frac{3}{5}}} = \frac{2 \cdot y^{\frac{(2\cdot4)-(1\cdot1)}{1\cdot2}}}{b^{\frac{(2\cdot5)-(1\cdot3)}{1\cdot5}}} = \frac{2 \cdot y^{\frac{8-1}{2}}}{b^{\frac{10-3}{5}}} = \frac{2 \cdot y^{\frac{7}{2}}}{b^{\frac{7}{5}}} = \frac{2 \cdot \sqrt[2]{y^7}}{\sqrt[5]{b^7}}$

$= \frac{2 \cdot \sqrt[2]{y^{2+2+2+1}}}{\sqrt[5]{b^{5+2}}} = \frac{2 \cdot \sqrt[2]{y^2 \cdot y^2 \cdot y^2 \cdot y^1}}{\sqrt[5]{b^5 \cdot b^2}} = \frac{2 \cdot (y \cdot y \cdot y)\sqrt[2]{y^1}}{b\sqrt[5]{b^2}} = \mathbf{\frac{2y^3\sqrt{y}}{b\sqrt[5]{b^2}}}$

10. $\left(\frac{a^3 \cdot b^{\frac{3}{2}} \cdot (c \cdot d)^6}{a \cdot b^4 \cdot c^3 \cdot d}\right)^{\frac{2}{3}} = \left(\frac{a^3 \cdot b^{\frac{3}{2}} \cdot c^{1\times6} \cdot d^{1\times6}}{a \cdot b^4 \cdot c^3 \cdot d}\right)^{\frac{2}{3}} = \left(\frac{a^3 \cdot b^{\frac{3}{2}} \cdot c^6 \cdot d^6}{a^1 \cdot b^4 \cdot c^3 \cdot d^1}\right)^{\frac{2}{3}} = \left(\frac{a^3 \cdot a^{-1} \cdot c^6 \cdot c^{-3} \cdot d^6 \cdot d^{-1}}{b^4 \cdot b^{-\frac{3}{2}}}\right)^{\frac{2}{3}}$

$= \left(\frac{a^{3-1} \cdot c^{6-3} \cdot d^{6-1}}{b^{\frac{4}{1}} \cdot b^{-\frac{3}{2}}}\right)^{\frac{2}{3}} = \left(\frac{a^2 \cdot c^3 \cdot d^5}{b^{\frac{4}{1}-\frac{3}{2}}}\right)^{\frac{2}{3}} = \left(\frac{a^2 \cdot c^3 \cdot d^5}{b^{\frac{(2\cdot4)-(1\cdot3)}{1\cdot2}}}\right)^{\frac{2}{3}} = \left(\frac{a^2 \cdot c^3 \cdot d^5}{b^{\frac{8-3}{2}}}\right)^{\frac{2}{3}} = \left(\frac{a^2 \cdot c^3 \cdot d^5}{b^{\frac{5}{2}}}\right)^{\frac{2}{3}}$

$= \frac{a^{2\times\frac{2}{3}} \cdot c^{3\times\frac{2}{3}} \cdot d^{5\times\frac{2}{3}}}{b^{\frac{5}{2}\times\frac{2}{3}}} = \frac{a^{\frac{4}{3}} \cdot c^{\frac{6}{3}=\frac{2}{1}} \cdot d^{\frac{10}{3}}}{b^{\frac{10}{6}=\frac{5}{3}}} = \frac{a^{\frac{4}{3}} \cdot c^{\frac{2}{1}} \cdot d^{\frac{10}{3}}}{b^{\frac{5}{3}}} = \frac{\sqrt[3]{a^4} \cdot c^2 \cdot \sqrt[3]{d^{10}}}{\sqrt[3]{b^5}} = \frac{\sqrt[3]{a^{3+1}} \cdot c^2 \cdot \sqrt[3]{d^{3+3+3+1}}}{\sqrt[3]{b^{3+2}}}$

$= \frac{\sqrt[3]{a^3 \cdot a^1} \cdot c^2 \cdot \sqrt[3]{d^3 \cdot d^3 \cdot d^3 \cdot d^1}}{\sqrt[3]{b^3 \cdot b^2}} = \frac{a\sqrt[3]{a} \cdot c^2 \cdot (d \cdot d \cdot d)\sqrt[3]{d}}{b\sqrt[3]{b^2}} = \frac{a\sqrt[3]{a} \cdot c^2 \cdot d^3\sqrt[3]{d}}{b\sqrt[3]{b^2}} = \frac{a \cdot c^2 \cdot d^3 \cdot \sqrt[3]{a}\sqrt[3]{d}}{b\sqrt[3]{b^2}} = \frac{ac^2d^3\sqrt[3]{a \cdot d}}{b\sqrt[3]{b^2}}$

$= \frac{ac^2d^3}{b}\left(\frac{\sqrt[3]{ad}}{\sqrt[3]{b^2}}\right) = \mathbf{\frac{ac^2d^3}{b}\left(\sqrt[3]{\frac{ad}{b^2}}\right)}$

Section 5.3 Case III Practice Problems

1. $\frac{4^{\frac{1}{2}} - 4^{\frac{3}{5}}}{5} + 2^{\frac{2}{3}} = \frac{4^{\frac{1}{2}} - 4^{\frac{3}{5}}}{5} + \frac{2^{\frac{2}{3}}}{1} = \frac{1 \cdot \left(4^{\frac{1}{2}} - 4^{\frac{3}{5}}\right) + \left(5 \cdot 2^{\frac{2}{3}}\right)}{1 \cdot 5} = \frac{4^{\frac{1}{2}} - 4^{\frac{3}{5}} + 5 \cdot 2^{\frac{2}{3}}}{5} = \frac{\sqrt[2]{4} - \sqrt[5]{4^3} + 5 \cdot \sqrt[3]{2^2}}{5}$

$= \frac{\sqrt[2]{2^2} - \sqrt[5]{64} + 5\sqrt[3]{4}}{5} = \mathbf{\frac{2 - \sqrt[5]{64} + 5\sqrt[3]{4}}{5}}$

2. $\frac{1}{x^{\frac{3}{5}} + x^{\frac{2}{3}}} + x^{\frac{2}{3}} = \frac{1}{x^{\frac{3}{5}} + x^{\frac{2}{3}}} + \frac{x^{\frac{2}{3}}}{1} = \frac{(1 \cdot 1) + \left[x^{\frac{2}{3}} \cdot \left(x^{\frac{3}{5}} + x^{\frac{2}{3}}\right)\right]}{1 \cdot \left(x^{\frac{3}{5}} + x^{\frac{2}{3}}\right)} = \frac{1 + \left[x^{\frac{2}{3}} \cdot x^{\frac{3}{5}} + x^{\frac{2}{3}} \cdot x^{\frac{2}{3}}\right]}{x^{\frac{3}{5}} + x^{\frac{2}{3}}} = \frac{1 + \left[x^{\frac{2}{3}+\frac{3}{5}} + x^{\frac{2}{3}+\frac{2}{3}}\right]}{x^{\frac{3}{5}} + x^{\frac{2}{3}}}$

$= \frac{1 + x^{\frac{(2\cdot5)+(3\cdot3)}{3\cdot5}} + x^{\frac{2+2}{3}}}{x^{\frac{3}{5}} + x^{\frac{2}{3}}} = \frac{1 + x^{\frac{10+9}{15}} + x^{\frac{4}{3}}}{x^{\frac{3}{5}} + x^{\frac{2}{3}}} = \frac{1 + x^{\frac{19}{15}} + x^{\frac{4}{3}}}{x^{\frac{3}{5}} + x^{\frac{2}{3}}} = \frac{1 + \sqrt[15]{x^{19}} + \sqrt[3]{x^4}}{\sqrt[5]{x^3} + \sqrt[3]{x^2}} = \frac{1 + \sqrt[15]{x^{15+4}} + \sqrt[3]{x^{3+1}}}{\sqrt[5]{x^3} + \sqrt[3]{x^2}}$

$= \frac{1 + \sqrt[15]{x^{15} \cdot x^4} + \sqrt[3]{x^3 \cdot x^1}}{\sqrt[5]{x^3} + \sqrt[3]{x^2}} = \mathbf{\frac{1 + x\sqrt[15]{x^4} + x\sqrt[3]{x}}{\sqrt[5]{x^3} + \sqrt[3]{x^2}}}$

3. $$\frac{2}{x^{\frac{2}{3}}}+x^{\frac{1}{3}} = \frac{2}{x^{\frac{2}{3}}}+\frac{x^{\frac{1}{3}}}{1} = \frac{(1\cdot 2)+\left(x^{\frac{2}{3}}\cdot x^{\frac{1}{3}}\right)}{1\cdot x^{\frac{2}{3}}} = \frac{2+x^{\frac{2}{3}+\frac{1}{3}}}{x^{\frac{2}{3}}} = \frac{2+x^{\frac{2+1}{3}}}{x^{\frac{2}{3}}} = \frac{2+x^{\frac{3}{3}=\frac{1}{1}}}{x^{\frac{2}{3}}} = \frac{2+x^{\frac{1}{1}}}{x^{\frac{2}{3}}} = \mathbf{\frac{2+x}{\sqrt[3]{x^2}}}$$

4. $$\frac{n^{\frac{2}{3}}}{n^{\frac{1}{3}}+n^{\frac{2}{3}}}+n^2 = \frac{n^{\frac{2}{3}}}{n^{\frac{1}{3}}+n^{\frac{2}{3}}}+\frac{n^2}{1} = \frac{\left(1\cdot n^{\frac{2}{3}}\right)+\left[n^2\cdot\left(n^{\frac{1}{3}}+n^{\frac{2}{3}}\right)\right]}{n^{\frac{1}{3}}+n^{\frac{2}{3}}} = \frac{n^{\frac{2}{3}}+\left[n^2\cdot n^{\frac{1}{3}}+n^2\cdot n^{\frac{2}{3}}\right]}{n^{\frac{1}{3}}+n^{\frac{2}{3}}} = \frac{n^{\frac{2}{3}}+\left[n^{\frac{2}{1}}\cdot n^{\frac{1}{3}}+n^{\frac{2}{1}}\cdot n^{\frac{2}{3}}\right]}{n^{\frac{1}{3}}+n^{\frac{2}{3}}}$$

$$= \frac{n^{\frac{2}{3}}+n^{\frac{2}{1}+\frac{1}{3}}+n^{\frac{2}{1}+\frac{2}{3}}}{n^{\frac{1}{3}}+n^{\frac{2}{3}}} = \frac{n^{\frac{2}{3}}+n^{\frac{(2\cdot 3)+(1\cdot 1)}{1\cdot 3}}+n^{\frac{(2\cdot 3)+(1\cdot 2)}{1\cdot 3}}}{n^{\frac{1}{3}}+n^{\frac{2}{3}}} = \frac{n^{\frac{2}{3}}+n^{\frac{6+1}{3}}+n^{\frac{6+2}{3}}}{n^{\frac{1}{3}}+n^{\frac{2}{3}}} = \frac{n^{\frac{2}{3}}+n^{\frac{7}{3}}+n^{\frac{8}{3}}}{n^{\frac{1}{3}}+n^{\frac{2}{3}}} = \frac{\sqrt[3]{n^2}+\sqrt[3]{n^7}+\sqrt[3]{n^8}}{\sqrt[3]{n^1}+\sqrt[3]{n^2}}$$

$$= \frac{\sqrt[3]{n^2}+\sqrt[3]{n^{3+3+1}}+\sqrt[3]{n^{3+3+2}}}{\sqrt[3]{n}+\sqrt[3]{n^2}} = \frac{\sqrt[3]{n^2}+\sqrt[3]{n^3\cdot n^3\cdot n^1}+\sqrt[3]{n^3\cdot n^3\cdot n^2}}{\sqrt[3]{n}+\sqrt[3]{n^2}} = \frac{\sqrt[3]{n^2}+(n\cdot n)\sqrt[3]{n}+(n\cdot n)\sqrt[3]{n^2}}{\sqrt[3]{n}+\sqrt[3]{n^2}} = \mathbf{\frac{\sqrt[3]{n^2}+n^2\sqrt[3]{n}+n^2\sqrt[3]{n^2}}{\sqrt[3]{n}+\sqrt[3]{n^2}}}$$

5. $$\frac{y^{\frac{2}{3}}}{y-y^{\frac{3}{4}}}+\frac{y^{\frac{1}{2}}}{2} = \frac{\left(2\cdot y^{\frac{2}{3}}\right)+\left[y^{\frac{1}{2}}\cdot\left(y-y^{\frac{3}{4}}\right)\right]}{2\cdot\left(y-y^{\frac{3}{4}}\right)} = \frac{2y^{\frac{2}{3}}+\left[y^{\frac{1}{2}}\cdot y-y^{\frac{1}{2}}\cdot y^{\frac{3}{4}}\right]}{2\cdot\left(y-y^{\frac{3}{4}}\right)} = \frac{2y^{\frac{2}{3}}+\left[y^{\frac{1}{2}}\cdot y^{\frac{1}{1}}-y^{\frac{1}{2}}\cdot y^{\frac{3}{4}}\right]}{2y-2y^{\frac{3}{4}}}$$

$$= \frac{2y^{\frac{2}{3}}+\left[y^{\frac{1}{2}+\frac{1}{1}}-y^{\frac{1}{2}+\frac{3}{4}}\right]}{2y-2y^{\frac{3}{4}}} = \frac{2y^{\frac{2}{3}}+\left[y^{\frac{(1\cdot 1)+(1\cdot 2)}{1\cdot 2}}-y^{\frac{(1\cdot 4)+(2\cdot 3)}{2\cdot 4}}\right]}{2y-2y^{\frac{3}{4}}} = \frac{2y^{\frac{2}{3}}+\left[y^{\frac{1+2}{2}}-y^{\frac{4+6}{8}}\right]}{2y-2y^{\frac{3}{4}}} = \frac{2y^{\frac{2}{3}}+y^{\frac{3}{2}}-y^{\frac{10}{8}=\frac{5}{4}}}{2y-2y^{\frac{3}{4}}}$$

$$= \frac{2\sqrt[3]{y^2}+\sqrt[2]{y^3}-\sqrt[4]{y^5}}{2y-2\sqrt[4]{y^3}} = \frac{2\sqrt[3]{y^2}+\sqrt[2]{y^{2+1}}-\sqrt[4]{y^{4+1}}}{2y-2\sqrt[4]{y^3}} = \frac{2\sqrt[3]{y^2}+\sqrt{y^2\cdot y^1}-\sqrt[4]{y^4\cdot y^1}}{2y-2\sqrt[4]{y^3}} = \mathbf{\frac{2\sqrt[3]{y^2}+y\sqrt{y}-y\sqrt[4]{y}}{2y-2\sqrt[4]{y^3}}}$$

6. $$\frac{w^{\frac{2}{5}}}{w^{\frac{1}{2}}+w^{\frac{2}{3}}}+\frac{1}{w} = \frac{w\cdot w^{\frac{2}{5}}+1\cdot\left(w^{\frac{1}{2}}+w^{\frac{2}{3}}\right)}{w\cdot\left(w^{\frac{1}{2}}+w^{\frac{2}{3}}\right)} = \frac{w^{\frac{1}{1}}\cdot w^{\frac{2}{5}}+\left(w^{\frac{1}{2}}+w^{\frac{2}{3}}\right)}{w^{\frac{1}{1}}\cdot\left(w^{\frac{1}{2}}+w^{\frac{2}{3}}\right)} = \frac{w^{\frac{1}{1}+\frac{2}{5}}+w^{\frac{1}{2}}+w^{\frac{2}{3}}}{w^{\frac{1}{1}}\cdot w^{\frac{1}{2}}+w^{\frac{1}{1}}\cdot w^{\frac{2}{3}}} = \frac{w^{\frac{(1\cdot 5)+(2\cdot 1)}{1\cdot 5}}+w^{\frac{1}{2}}+w^{\frac{2}{3}}}{w^{\frac{1}{1}+\frac{1}{2}}+w^{\frac{1}{1}+\frac{2}{3}}}$$

$$= \frac{w^{\frac{5+2}{5}}+w^{\frac{1}{2}}+w^{\frac{2}{3}}}{w^{\frac{(1\cdot 2)+(1\cdot 1)}{1\cdot 2}}+w^{\frac{(1\cdot 3)+(2\cdot 1)}{1\cdot 3}}} = \frac{w^{\frac{7}{5}}+w^{\frac{1}{2}}+w^{\frac{2}{3}}}{w^{\frac{2+1}{2}}+w^{\frac{3+2}{3}}} = \frac{w^{\frac{7}{5}}+w^{\frac{1}{2}}+w^{\frac{2}{3}}}{w^{\frac{3}{2}}+w^{\frac{5}{3}}} = \frac{\sqrt[5]{w^7}+\sqrt[2]{w^1}+\sqrt[3]{w^2}}{\sqrt[2]{w^3}+\sqrt[3]{w^5}} = \frac{\sqrt[5]{w^{5+2}}+\sqrt[2]{w}+\sqrt[3]{w^2}}{\sqrt[2]{w^{2+1}}+\sqrt[3]{w^{3+2}}}$$

$$= \frac{\sqrt[5]{w^5\cdot w^2}+\sqrt{w}+\sqrt[3]{w^2}}{\sqrt{w^2\cdot w^1}+\sqrt[3]{w^3\cdot w^2}} = \mathbf{\frac{w\sqrt[5]{w^2}+\sqrt{w}+\sqrt[3]{w^2}}{w\sqrt{w}+w\sqrt[3]{w^2}}}$$

7. $\dfrac{x+y^{\frac{2}{3}}}{x-y^{\frac{2}{5}}}+\dfrac{1}{x^{\frac{2}{3}}}=\dfrac{x^{\frac{2}{3}}\cdot\left(x+y^{\frac{2}{3}}\right)+1\cdot\left(x-y^{\frac{2}{5}}\right)}{x^{\frac{2}{3}}\cdot\left(x-y^{\frac{2}{5}}\right)}=\dfrac{x^{\frac{2}{3}}\cdot x+x^{\frac{2}{3}}\cdot y^{\frac{2}{3}}+x-y^{\frac{2}{5}}}{x^{\frac{2}{3}}\cdot x-x^{\frac{2}{3}}\cdot y^{\frac{2}{5}}}=\dfrac{x^{\frac{2}{3}}\cdot x^{\frac{1}{1}}+x^{\frac{2}{3}}\cdot y^{\frac{2}{3}}+x-y^{\frac{2}{5}}}{x^{\frac{2}{3}}\cdot x^{\frac{1}{1}}-x^{\frac{2}{3}}\cdot y^{\frac{2}{5}}}$

$=\dfrac{x^{\frac{2}{3}+\frac{1}{1}}+x^{\frac{2}{3}}\cdot y^{\frac{2}{3}}+x-y^{\frac{2}{5}}}{x^{\frac{2}{3}+\frac{1}{1}}-x^{\frac{2}{3}}\cdot y^{\frac{2}{5}}}=\dfrac{x^{\frac{(1\cdot2)+(1\cdot3)}{1\cdot3}}+x^{\frac{2}{3}}\cdot y^{\frac{2}{3}}+x-y^{\frac{2}{5}}}{x^{\frac{(1\cdot2)+(1\cdot3)}{1\cdot3}}-x^{\frac{2}{3}}\cdot y^{\frac{2}{5}}}=\dfrac{x^{\frac{2+3}{3}}+x^{\frac{2}{3}}\cdot y^{\frac{2}{3}}+x-y^{\frac{2}{5}}}{x^{\frac{2+3}{3}}-x^{\frac{2}{3}}\cdot y^{\frac{2}{5}}}=\dfrac{x^{\frac{5}{3}}+x^{\frac{2}{3}}y^{\frac{2}{3}}+x-y^{\frac{2}{5}}}{x^{\frac{5}{3}}-x^{\frac{2}{3}}y^{\frac{2}{5}}}$

$=\dfrac{\sqrt[3]{x^5}+\sqrt[3]{x^2y^2}+x-\sqrt[5]{y^2}}{\sqrt[3]{x^5}-\sqrt[3]{x^2}\sqrt[5]{y^2}}=\dfrac{\sqrt[3]{x^{3+2}}+\sqrt[3]{x^2y^2}+x-\sqrt[5]{y^2}}{\sqrt[3]{x^{3+2}}-\sqrt[3]{x^2}\sqrt[5]{y^2}}=\dfrac{\sqrt[3]{x^3\cdot x^2}+\sqrt[3]{x^2y^2}+x-\sqrt[5]{y^2}}{\sqrt[3]{x^3\cdot x^2}-\sqrt[3]{x^2}\sqrt[5]{y^2}}=\dfrac{\mathbf{x\sqrt[3]{x^2}+\sqrt[3]{x^2y^2}+x-\sqrt[5]{y^2}}}{\mathbf{x\sqrt[3]{x^2}-\sqrt[3]{x^2}\sqrt[5]{y^2}}}$

8. $\dfrac{1}{a^{\frac{1}{2}}-b^{\frac{1}{2}}}+\dfrac{b^{\frac{3}{4}}}{3}=\dfrac{(1\cdot3)+b^{\frac{3}{4}}\cdot\left(a^{\frac{1}{2}}-b^{\frac{1}{2}}\right)}{3\cdot\left(a^{\frac{1}{2}}-b^{\frac{1}{2}}\right)}=\dfrac{3+a^{\frac{1}{2}}\cdot b^{\frac{3}{4}}-b^{\frac{3}{4}}\cdot b^{\frac{1}{2}}}{3a^{\frac{1}{2}}-3b^{\frac{1}{2}}}=\dfrac{3+a^{\frac{1}{2}}\cdot b^{\frac{3}{4}}-b^{\frac{3}{4}+\frac{1}{2}}}{3a^{\frac{1}{2}}-3b^{\frac{1}{2}}}=\dfrac{3+a^{\frac{1}{2}}\cdot b^{\frac{3}{4}}-b^{\frac{(2\cdot3)+(1\cdot4)}{2\cdot4}}}{3a^{\frac{1}{2}}-3b^{\frac{1}{2}}}$

$=\dfrac{3+a^{\frac{1}{2}}\cdot b^{\frac{3}{4}}-b^{\frac{6+4}{8}}}{3a^{\frac{1}{2}}-3b^{\frac{1}{2}}}=\dfrac{3+a^{\frac{1}{2}}\cdot b^{\frac{3}{4}}-b^{\frac{10}{8}=\frac{5}{4}}}{3a^{\frac{1}{2}}-3b^{\frac{1}{2}}}=\dfrac{3+a^{\frac{1}{2}}\cdot b^{\frac{3}{4}}-b^{\frac{5}{4}}}{3a^{\frac{1}{2}}-3b^{\frac{1}{2}}}=\dfrac{3+\sqrt[2]{a^1}\sqrt[4]{b^3}-\sqrt[4]{b^5}}{3\sqrt[2]{a^1}-3\sqrt[2]{b^1}}=\dfrac{3+\sqrt{a}\sqrt[4]{b^3}-\sqrt[4]{b^{4+1}}}{3\sqrt{a}-3\sqrt{b}}$

$=\dfrac{3+\sqrt{a}\sqrt[4]{b^3}-\sqrt[4]{b^4\cdot b^1}}{3\sqrt{a}-3\sqrt{b}}=\dfrac{\mathbf{3+\sqrt{a}\sqrt[4]{b^3}-b\sqrt[4]{b}}}{\mathbf{3\sqrt{a}-3\sqrt{b}}}$

9. $\dfrac{x^{\frac{2}{5}}}{x^{\frac{2}{3}}-3}+\dfrac{1}{y^{\frac{1}{3}}}=\dfrac{\left(x^{\frac{2}{5}}\cdot y^{\frac{1}{3}}\right)+1\cdot\left(x^{\frac{2}{3}}-3\right)}{y^{\frac{1}{3}}\cdot\left(x^{\frac{2}{3}}-3\right)}=\dfrac{x^{\frac{2}{5}}\cdot y^{\frac{1}{3}}+x^{\frac{2}{3}}-3}{x^{\frac{2}{3}}\cdot y^{\frac{1}{3}}-3\cdot y^{\frac{1}{3}}}=\dfrac{\sqrt[5]{x^2}\cdot\sqrt[3]{y^1}+\sqrt[3]{x^2}-3}{\sqrt[3]{x^2}\cdot\sqrt[3]{y^1}-3\cdot\sqrt[3]{y^1}}=\dfrac{\sqrt[5]{x^2}\sqrt[3]{y}+\sqrt[3]{x^2}-3}{\sqrt[3]{x^2}\sqrt[3]{y}-3\sqrt[3]{y}}$

$=\dfrac{\mathbf{\sqrt[5]{x^2}\sqrt[3]{y}+\sqrt[3]{x^2}-3}}{\mathbf{\sqrt[3]{x^2y}-3\sqrt[3]{y}}}$

10. $\dfrac{a^{\frac{2}{3}}+b^{\frac{1}{2}}}{a-b}-a^{\frac{2}{5}}=\dfrac{a^{\frac{2}{3}}+b^{\frac{1}{2}}}{a-b}-\dfrac{a^{\frac{2}{5}}}{1}=\dfrac{1\cdot\left(a^{\frac{2}{3}}+b^{\frac{1}{2}}\right)-a^{\frac{2}{5}}\cdot(a-b)}{1\cdot(a-b)}=\dfrac{a^{\frac{2}{3}}+b^{\frac{1}{2}}-a\cdot a^{\frac{2}{5}}+a^{\frac{2}{5}}\cdot b}{a-b}=\dfrac{a^{\frac{2}{3}}+b^{\frac{1}{2}}-a^{\frac{1}{1}}\cdot a^{\frac{2}{5}}+a^{\frac{2}{5}}b}{a-b}$

$=\dfrac{a^{\frac{2}{3}}+b^{\frac{1}{2}}-a^{\frac{1}{1}+\frac{2}{5}}+a^{\frac{2}{5}}b}{a-b}=\dfrac{a^{\frac{2}{3}}+b^{\frac{1}{2}}-a^{\frac{(1\cdot5)+(1\cdot2)}{1\cdot5}}+a^{\frac{2}{5}}b}{a-b}=\dfrac{a^{\frac{2}{3}}+b^{\frac{1}{2}}-a^{\frac{5+2}{5}}+a^{\frac{2}{5}}b}{a-b}=\dfrac{a^{\frac{2}{3}}+b^{\frac{1}{2}}-a^{\frac{7}{5}}+a^{\frac{2}{5}}b}{a-b}$

$=\dfrac{\sqrt[3]{a^2}+\sqrt[2]{b^1}-\sqrt[5]{a^7}+b\sqrt[5]{a^2}}{a-b}=\dfrac{\sqrt[3]{a^2}+\sqrt{b}-\sqrt[5]{a^{5+2}}+b\sqrt[5]{a^2}}{a-b}=\dfrac{\sqrt[3]{a^2}+\sqrt{b}-\sqrt[5]{a^5\cdot a^2}+b\sqrt[5]{a^2}}{a-b}$

$=\dfrac{\mathbf{\sqrt[3]{a^2}+\sqrt{b}-a\sqrt[5]{a^2}+b\sqrt[5]{a^2}}}{\mathbf{a-b}}$

Section 5.4 Case I Practice Problems

1. $\left(a^2 \cdot a^3\right)^{-\frac{1}{3}} = a^{2\times-\frac{1}{3}} \cdot a^{3\times-\frac{1}{3}} = a^{-\frac{2}{3}} \cdot a^{-\frac{3}{3}} = a^{-\frac{2}{3}-\frac{3}{3}} = a^{\frac{-2-3}{3}} = a^{-\frac{5}{3}} = \frac{1}{a^{\frac{5}{3}}} = \frac{1}{\sqrt[3]{a^5}} = \frac{1}{\sqrt[3]{a^{3+2}}} = \frac{1}{\sqrt[3]{a^3 \cdot a^2}} = \mathbf{\frac{1}{a\sqrt[3]{a^2}}}$

2. $\left(a^2 \cdot y^{-1}\right)^{-\frac{1}{4}} \cdot y^{-3} = \left(a^{2\times-\frac{1}{4}} \cdot y^{-1\times-\frac{1}{4}}\right) \cdot y^{-3} = \left(a^{-\frac{2}{4}} \cdot y^{+\frac{1}{4}}\right) \cdot y^{-3} = a^{-\frac{2}{4}=-\frac{1}{2}} \cdot \left(y^{\frac{1}{4}} \cdot y^{-\frac{3}{1}}\right) = a^{-\frac{1}{2}} \cdot \left(y^{\frac{1}{4}-\frac{3}{1}}\right)$

$= a^{-\frac{1}{2}} y^{\frac{(1\cdot1)-(3\cdot4)}{4\cdot1}} = a^{-\frac{1}{2}} y^{\frac{1-12}{4}} = a^{-\frac{1}{2}} y^{-\frac{11}{4}} = \frac{1}{a^{\frac{1}{2}} y^{\frac{11}{4}}} = \frac{1}{\sqrt[2]{a^1}\,\sqrt[4]{y^{11}}} = \frac{1}{\sqrt[2]{a^1}\,\sqrt[4]{y^{4+4+3}}} = \frac{1}{\sqrt{a}\,\sqrt[4]{\left(y^4 \cdot y^4\right) \cdot y^3}}$

$= \frac{1}{\sqrt{a}\,(y \cdot y)\sqrt[4]{y^3}} = \mathbf{\frac{1}{\sqrt{a}\, y^2 \sqrt[4]{y^3}}}$

3. $\left(a^{-2} \cdot b^{-3}\right)^{-\frac{1}{2}} \cdot \left(a \cdot b^2\right) = \left(a^{-2\times-\frac{1}{2}} \cdot b^{-3\times-\frac{1}{2}}\right) \cdot \left(a \cdot b^2\right) = a^{\frac{2}{2}=\frac{1}{1}} \cdot b^{\frac{3}{2}} \cdot a \cdot b^2 = a^{1} \cdot b^{\frac{3}{2}} \cdot a \cdot b^2 = (a \cdot a) \cdot \left(b^{\frac{3}{2}} \cdot b^2\right)$

$= a^2 \cdot \left(b^{\frac{3}{2}} \cdot b^{\frac{2}{1}}\right) = a^2 \cdot b^{\frac{3}{2}+\frac{2}{1}} = a^2 \cdot b^{\frac{(3\cdot1)+(2\cdot2)}{2\cdot1}} = a^2 \cdot b^{\frac{3+4}{2}} = a^2\, b^{\frac{7}{2}} = a^2 \sqrt[2]{b^7} = a^2 \sqrt[2]{b^{2+2+2+1}} = a^2 \sqrt[2]{\left(b^2 \cdot b^2 \cdot b^2\right) \cdot b^1}$

$= a^2 (b \cdot b \cdot b)\sqrt{b} = \mathbf{a^2 b^3 \sqrt{b}}$

4. $512^{-\frac{1}{3}} \cdot \left(x^{2a} \cdot x^{3a}\right)^{-\frac{1}{5}} = 512^{-\frac{1}{3}} \cdot \left(x^{2a\times-\frac{1}{5}} \cdot x^{3a\times-\frac{1}{5}}\right) = 512^{-\frac{1}{3}} \cdot \left(x^{-\frac{2a}{5}} \cdot x^{-\frac{3a}{5}}\right) = 512^{-\frac{1}{3}} \cdot x^{-\frac{2a}{5}-\frac{3a}{5}} = 512^{-\frac{1}{3}} \cdot x^{\frac{-2a-3a}{5}}$

$= 512^{-\frac{1}{3}} \cdot x^{-\frac{5a}{5}=-\frac{a}{1}} = 512^{-\frac{1}{3}} \cdot x^{-\frac{a}{1}} = 512^{-\frac{1}{3}} \cdot x^{-a} = \frac{1}{512^{\frac{1}{3}} \cdot x^a} = \frac{1}{\sqrt[3]{512}\, x^a} = \frac{1}{\sqrt[3]{8^3}\, x^a} = \mathbf{\frac{1}{8x^a}}$

5. $\left(x \cdot z^2\right)^{-\frac{3}{2}} \cdot \left(x^2\right)^{-\frac{2}{3}} = \left(x^{1\times-\frac{3}{2}} \cdot z^{2\times-\frac{3}{2}}\right) \cdot \left(x^{2\times-\frac{2}{3}}\right) = x^{-\frac{3}{2}} \cdot z^{-\frac{6}{2}} \cdot x^{-\frac{4}{3}} = \left(x^{-\frac{3}{2}} \cdot x^{-\frac{4}{3}}\right) \cdot z^{-\frac{6}{2}} = x^{-\frac{3}{2}-\frac{4}{3}} \cdot z^{-\frac{6}{2}}$

$= x^{\frac{-(3\cdot3)-(4\cdot2)}{2\cdot3}} \cdot z^{-\frac{6}{2}=-\frac{3}{2}} = x^{\frac{-9-8}{6}} \cdot z^{-\frac{3}{1}} = x^{-\frac{17}{6}} \cdot z^{-3} = \frac{1}{x^{\frac{17}{6}} \cdot z^3} = \frac{1}{\sqrt[6]{x^{17}} \cdot z^3} = \frac{1}{\sqrt[6]{x^{6+6+5}} \cdot z^3} = \frac{1}{\sqrt[6]{\left(x^6 \cdot x^6\right) \cdot x^5} \cdot z^3}$

$= \frac{1}{(x \cdot x)\sqrt[6]{x^5}\, z^3} = \mathbf{\frac{1}{x^2 z^3 \sqrt[6]{x^5}}}$

6. $27^{-\frac{2}{3}} \cdot 4^2 \cdot 4^{-\frac{1}{2}} \cdot 4^0 = 27^{-\frac{2}{3}} \cdot 4^{\frac{2}{1}} \cdot 4^{-\frac{1}{2}} \cdot 1 = 27^{-\frac{2}{3}} \cdot 4^{\frac{2}{1}} \cdot 4^{-\frac{1}{2}} = 27^{-\frac{2}{3}} \cdot 4^{\frac{2}{1}-\frac{1}{2}} = 27^{-\frac{2}{3}} \cdot 4^{\frac{(2\cdot2)-(1\cdot1)}{1\cdot2}} = 27^{-\frac{2}{3}} \cdot 4^{\frac{4-1}{2}}$

$= 27^{-\frac{2}{3}} \cdot 4^{\frac{3}{2}} = \frac{1}{27^{\frac{2}{3}}} \cdot 4^{\frac{3}{2}} = \frac{4^{\frac{3}{2}}}{27^{\frac{2}{3}}} = \frac{\sqrt[2]{4^3}}{\sqrt[3]{27^2}} = \frac{\sqrt{4^{2+1}}}{\sqrt[3]{729}} = \frac{\sqrt{4^2 \cdot 4^1}}{\sqrt[3]{9^3}} = \frac{4\sqrt{4}}{9} = \frac{4\sqrt{2^2}}{9} = \frac{4 \cdot 2}{9} = \mathbf{\frac{8}{9}}$

7. $y^2\cdot x^{-\frac{2}{3}}\cdot\left(x\cdot y^{-1}\right)^{-\frac{2}{3}} = y^2\cdot x^{-\frac{2}{3}}\cdot\left(x^{1\times-\frac{2}{3}}\cdot y^{-1\times-\frac{2}{3}}\right) = y^{\frac{2}{1}}\cdot x^{-\frac{2}{3}}\cdot x^{-\frac{2}{3}}\cdot y^{\frac{2}{3}} = \left(x^{-\frac{2}{3}}\cdot x^{-\frac{2}{3}}\right)\cdot\left(y^{\frac{2}{1}}\cdot y^{\frac{2}{3}}\right)$

$= x^{-\frac{2}{3}-\frac{2}{3}}\cdot y^{\frac{2}{1}+\frac{2}{3}} = x^{\frac{-2-2}{3}}\cdot y^{\frac{(2\cdot3)+(1\cdot2)}{1\cdot3}} = x^{-\frac{4}{3}}\cdot y^{\frac{6+2}{3}} = x^{-\frac{4}{3}}\cdot y^{\frac{8}{3}} = \frac{y^{\frac{8}{3}}}{x^{\frac{4}{3}}} = \frac{\sqrt[3]{y^8}}{\sqrt[3]{x^4}} = \frac{\sqrt[3]{y^{3+3+2}}}{\sqrt[3]{x^{3+1}}} = \frac{\sqrt[3]{\left(y^3\cdot y^3\right)\cdot y^2}}{\sqrt[3]{x^3\cdot x^1}}$

$= \frac{(y\cdot y)\sqrt[3]{y^2}}{x\sqrt[3]{x}} = \frac{y^2\sqrt[3]{y^2}}{x\sqrt[3]{x}} = \mathbf{\frac{y^2}{x}\left(\sqrt[3]{\frac{y^2}{x}}\right)}$

8. $\left(a^2\cdot b^2\right)^{-\frac{3}{4}}\cdot\left(a^5\cdot b^2\right)^{-\frac{1}{4}} = \left(a^{2\times-\frac{3}{4}}\cdot b^{2\times-\frac{3}{4}}\right)\cdot\left(a^{5\times-\frac{1}{4}}\cdot b^{2\times-\frac{1}{4}}\right) = \left(a^{-\frac{6}{4}=-\frac{3}{2}}\cdot b^{-\frac{6}{4}=-\frac{3}{2}}\right)\cdot\left(a^{-\frac{5}{4}}\cdot b^{-\frac{2}{4}=-\frac{1}{2}}\right)$

$= a^{-\frac{3}{2}}\cdot b^{-\frac{3}{2}}\cdot a^{-\frac{5}{4}}\cdot b^{-\frac{1}{2}} = \left(a^{-\frac{3}{2}}\cdot a^{-\frac{5}{4}}\right)\cdot\left(b^{-\frac{3}{2}}\cdot b^{-\frac{1}{2}}\right) = a^{-\frac{3}{2}-\frac{5}{4}}\cdot b^{-\frac{3}{2}-\frac{1}{2}} = a^{\frac{-(3\cdot4)-(2\cdot5)}{2\cdot4}}\cdot b^{\frac{-3-1}{2}} = a^{\frac{-12-10}{8}}\cdot b^{\frac{-4}{2}}$

$= a^{-\frac{22}{8}=-\frac{11}{4}}\cdot b^{-\frac{4}{2}=-\frac{2}{1}} = a^{-\frac{11}{4}}\cdot b^{-\frac{2}{1}} = a^{-\frac{11}{4}}\cdot b^{-2} = \frac{1}{a^{\frac{11}{4}}\cdot b^2} = \frac{1}{\sqrt[4]{a^{11}}\cdot b^2} = \frac{1}{\sqrt[4]{a^{4+4+3}}\cdot b^2} = \frac{1}{\sqrt[4]{\left(a^4\cdot a^4\right)\cdot a^3}\cdot b^2}$

$= \frac{1}{(a\cdot a)\sqrt[4]{a^3}\cdot b^2} = \mathbf{\frac{1}{a^2b^2\sqrt[4]{a^3}}}$

9. $x^{-\frac{2}{3}}\cdot\left(64^{-1}\cdot x\cdot y^3\right)^{-\frac{2}{3}} = x^{-\frac{2}{3}}\cdot\left(64^{-1\times-\frac{2}{3}}\cdot x^{1\times-\frac{2}{3}}\cdot y^{3\times-\frac{2}{3}}\right) = x^{-\frac{2}{3}}\cdot 64^{+\frac{2}{3}}\cdot x^{-\frac{2}{3}}\cdot y^{-\frac{6}{3}} = 64^{\frac{2}{3}}\cdot\left(x^{-\frac{2}{3}}\cdot x^{-\frac{2}{3}}\right)\cdot y^{-\frac{6}{3}}$

$= 64^{\frac{2}{3}}\cdot x^{-\frac{2}{3}-\frac{2}{3}}\cdot y^{-\frac{6}{3}=-\frac{2}{1}} = 64^{\frac{2}{3}}\cdot x^{\frac{-2-2}{3}}\cdot y^{-\frac{2}{1}} = 64^{\frac{2}{3}}\cdot x^{-\frac{4}{3}}\cdot y^{-2} = \frac{64^{\frac{2}{3}}}{x^{\frac{4}{3}}\cdot y^2} = \frac{\sqrt[3]{64^2}}{\sqrt[3]{x^4}\cdot y^2} = \frac{\sqrt[3]{\left(4^3\right)^2}}{\sqrt[3]{x^{3+1}}\cdot y^2}$

$= \frac{\sqrt[3]{4^{3\times2}}}{\sqrt[3]{x^3\cdot x^1}\cdot y^2} = \frac{\sqrt[3]{4^6}}{x\sqrt[3]{x}\,y^2} = \frac{\sqrt[3]{4^{3+3}}}{xy^2\sqrt[3]{x}} = \frac{\sqrt[3]{4^3\cdot4^3}}{xy^2\sqrt[3]{x}} = \frac{4\cdot4}{xy^2\sqrt[3]{x}} = \mathbf{\frac{16}{xy^2\sqrt[3]{x}}}$

10. $\left(4\cdot m^2\right)^{-\frac{1}{2}}\cdot\left[\left(m\cdot p^2\right)^{-1}\cdot m\right]^{\frac{1}{2}} = \left(4^{-\frac{1}{2}}\cdot m^{2\times-\frac{1}{2}}\right)\cdot\left[\left(m^{1\times-1}\cdot p^{2\times-1}\right)\cdot m^1\right]^{\frac{1}{2}} = 4^{-\frac{1}{2}}\cdot m^{-\frac{2}{2}}\cdot\left[m^{-1}\cdot p^{-2}\cdot m^1\right]^{\frac{1}{2}}$

$= 4^{-\frac{1}{2}}\cdot m^{-\frac{2}{2}}\cdot\left[m^{-1}\cdot m^1\cdot p^{-2}\right]^{\frac{1}{2}} = 4^{-\frac{1}{2}}\cdot m^{-1}\cdot\left[m^{-1+1}\cdot p^{-2}\right]^{\frac{1}{2}} = 4^{-\frac{1}{2}}\cdot m^{-1}\cdot\left[m^0\cdot p^{-2}\right]^{\frac{1}{2}} = 4^{-\frac{1}{2}}\cdot m^{-1}\cdot\left[1\cdot p^{-2}\right]^{\frac{1}{2}}$

$= 4^{-\frac{1}{2}}\cdot m^{-1}\cdot\left(p^{-2}\right)^{\frac{1}{2}} = 4^{-\frac{1}{2}}\cdot m^{-1}\cdot p^{-2\times\frac{1}{2}} = 4^{-\frac{1}{2}}\cdot m^{-1}\cdot p^{-\frac{2}{2}=-\frac{1}{1}} = 4^{-\frac{1}{2}}\cdot m^{-1}\cdot p^{-\frac{1}{1}} = 4^{-\frac{1}{2}}\cdot m^{-1}\cdot p^{-1} = \frac{1}{4^{\frac{1}{2}}\cdot m^1\cdot p^1}$

$= \frac{1}{\sqrt{4}\cdot m\cdot p} = \frac{1}{\sqrt{2^2}\,mp} = \mathbf{\frac{1}{2mp}}$

Section 5.4 Case II Practice Problems

1. $\frac{a^{-\frac{1}{2}}}{a^2} = \frac{1}{a^2 \cdot a^{\frac{1}{2}}} = \frac{1}{a^{\frac{2}{1}} \cdot a^{\frac{1}{2}}} = \frac{1}{a^{\frac{2}{1}+\frac{1}{2}}} = \frac{1}{a^{\frac{(2\cdot2)+(1\cdot1)}{1\cdot2}}} = \frac{1}{a^{\frac{4+1}{2}}} = \frac{1}{a^{\frac{5}{2}}} = \frac{1}{\sqrt[2]{a^5}} = \frac{1}{\sqrt[2]{a^{2+2+1}}} = \frac{1}{\sqrt[2]{a^2 \cdot a^2 \cdot a^1}}$

$= \frac{1}{(a \cdot a)\sqrt{a}} = \mathbf{\frac{1}{a^2\sqrt{a}}}$

2. $\frac{a^{-\frac{2}{3}} \cdot b^3}{a^{-\frac{1}{3}}} = \frac{b^3}{a^{-\frac{1}{3}} \cdot a^{\frac{2}{3}}} = \frac{b^3}{a^{-\frac{1}{3}+\frac{2}{3}}} = \frac{b^3}{a^{\frac{-1+2}{3}}} = \frac{b^3}{a^{\frac{1}{3}}} = \frac{b^3}{\sqrt[3]{a^1}} = \mathbf{\frac{b^3}{\sqrt[3]{a}}}$

3. $\left(\frac{a^3 \cdot b^2}{a^2 \cdot b^5}\right)^{-\frac{2}{3}} = \left(\frac{a^3 \cdot a^{-2}}{b^5 \cdot b^{-2}}\right)^{-\frac{2}{3}} = \left(\frac{a^{3-2}}{b^{5-2}}\right)^{-\frac{2}{3}} = \left(\frac{a^1}{b^3}\right)^{-\frac{2}{3}} = \frac{a^{1\times-\frac{2}{3}}}{b^{3\times-\frac{2}{3}}} = \frac{a^{-\frac{2}{3}}}{b^{-\frac{6}{3}=-\frac{2}{1}}} = \frac{a^{-\frac{2}{3}}}{b^{-\frac{2}{1}}} = \frac{b^{\frac{2}{1}}}{a^{\frac{2}{3}}} = \frac{b^2}{a^{\frac{2}{3}}} = \mathbf{\frac{b^2}{\sqrt[3]{a^2}}}$

4. $\frac{d^3 \cdot (a \cdot b \cdot c)^{-\frac{1}{4}}}{d^{\frac{2}{3}} \cdot b^{-3}} = \frac{d^3 \cdot a^{1\times-\frac{1}{4}} \cdot b^{1\times-\frac{1}{4}} \cdot c^{1\times-\frac{1}{4}}}{d^{\frac{2}{3}} \cdot b^{-3}} = \frac{d^3 \cdot a^{-\frac{1}{4}} \cdot b^{-\frac{1}{4}} \cdot c^{-\frac{1}{4}}}{d^{\frac{2}{3}} \cdot b^{-3}} = \frac{d^3 \cdot d^{-\frac{2}{3}} \cdot b^3 \cdot b^{-\frac{1}{4}}}{a^{\frac{1}{4}} \cdot c^{\frac{1}{4}}} = \frac{d^{\frac{3}{1}} \cdot d^{-\frac{2}{3}} \cdot b^{\frac{3}{1}} \cdot b^{-\frac{1}{4}}}{a^{\frac{1}{4}} \cdot c^{\frac{1}{4}}}$

$= \frac{d^{\frac{3}{1}-\frac{2}{3}} \cdot b^{\frac{3}{1}-\frac{1}{4}}}{a^{\frac{1}{4}} \cdot c^{\frac{1}{4}}} = \frac{d^{\frac{(3\cdot3)-(2\cdot1)}{1\cdot3}} \cdot b^{\frac{(3\cdot4)-(1\cdot1)}{1\cdot4}}}{a^{\frac{1}{4}} \cdot c^{\frac{1}{4}}} = \frac{d^{\frac{9-2}{3}} \cdot b^{\frac{12-1}{4}}}{a^{\frac{1}{4}} \cdot c^{\frac{1}{4}}} = \frac{d^{\frac{7}{3}} \cdot b^{\frac{11}{4}}}{(a\,c)^{\frac{1}{4}}} = \frac{\sqrt[3]{d^7} \cdot \sqrt[4]{b^{11}}}{\sqrt[4]{ac}} = \frac{\sqrt[3]{d^{3+3+1}} \cdot \sqrt[4]{b^{4+4+3}}}{\sqrt[4]{ac}}$

$= \frac{\sqrt[3]{(d^3 \cdot d^3) \cdot d^1} \cdot \sqrt[4]{(b^4 \cdot b^4) \cdot b^3}}{\sqrt[4]{ac}} = \frac{(d \cdot d)\sqrt[3]{d^1} \cdot (b \cdot b)\sqrt[4]{b^3}}{\sqrt[4]{ac}} = \frac{d^2\sqrt[3]{d^1} \cdot b^2\sqrt[4]{b^3}}{\sqrt[4]{ac}} = \mathbf{\frac{b^2 d^2\left(\sqrt[3]{d}\sqrt[4]{b^3}\right)}{\sqrt[4]{ac}}}$

5. $\frac{(a \cdot b)^{-\frac{2}{3}} \cdot (x \cdot y)^{-\frac{1}{4}}}{a^2 \cdot y^{-\frac{3}{4}}} = \frac{\left(a^{-\frac{2}{3}} \cdot b^{-\frac{2}{3}}\right) \cdot \left(x^{-\frac{1}{4}} \cdot y^{-\frac{1}{4}}\right)}{a^2 \cdot y^{-\frac{3}{4}}} = \frac{a^{-\frac{2}{3}} \cdot b^{-\frac{2}{3}} \cdot x^{-\frac{1}{4}} \cdot y^{-\frac{1}{4}}}{a^2 \cdot y^{-\frac{3}{4}}} = \frac{y^{\frac{3}{4}} \cdot y^{-\frac{1}{4}}}{a^2 \cdot a^{\frac{2}{3}} \cdot b^{\frac{2}{3}} \cdot x^{\frac{1}{4}}} = \frac{y^{\frac{3}{4}-\frac{1}{4}}}{a^{\frac{2}{1}+\frac{2}{3}} \cdot b^{\frac{2}{3}} \cdot x^{\frac{1}{4}}}$

$= \frac{y^{\frac{3-1}{4}}}{a^{\frac{(2\cdot3)+(1\cdot2)}{1\cdot3}} \cdot b^{\frac{2}{3}} \cdot x^{\frac{1}{4}}} = \frac{y^{\frac{2}{4}}}{a^{\frac{6+2}{3}} \cdot b^{\frac{2}{3}} \cdot x^{\frac{1}{4}}} = \frac{y^{\frac{2}{4}=\frac{1}{2}}}{a^{\frac{8}{3}} \cdot b^{\frac{2}{3}} \cdot x^{\frac{1}{4}}} = \frac{y^{\frac{1}{2}}}{a^{\frac{8}{3}} \cdot b^{\frac{2}{3}} \cdot x^{\frac{1}{4}}} = \frac{\sqrt{y}}{\sqrt[3]{a^8} \cdot \sqrt[3]{b^2} \cdot \sqrt[4]{x}} = \frac{\sqrt{y}}{\sqrt[3]{a^{3+3+2}} \cdot \sqrt[3]{b^2} \cdot \sqrt[4]{x}}$

$= \frac{\sqrt{y}}{\sqrt[3]{(a^3 \cdot a^3) \cdot a^2} \cdot \sqrt[3]{b^2} \cdot \sqrt[4]{x}} = \frac{\sqrt{y}}{(a \cdot a)\sqrt[3]{a^2} \cdot \sqrt[3]{b^2} \cdot \sqrt[4]{x}} = \mathbf{\frac{\sqrt{y}}{a^2\sqrt[3]{a^2b^2}\sqrt[4]{x}}}$

6. $\frac{(x \cdot y)^0 \cdot x^2}{(x \cdot y^{-1})^{-\frac{3}{4}}} = \frac{1 \cdot x^2}{(x \cdot y^{-1})^{-\frac{3}{4}}} = \frac{x^2}{(x \cdot y^{-1})^{-\frac{3}{4}}} = \frac{x^2}{x^{1\times-\frac{3}{4}} \cdot y^{-1\times-\frac{3}{4}}} = \frac{x^2}{x^{-\frac{3}{4}} \cdot y^{\frac{3}{4}}} = \frac{x^2 \cdot x^{\frac{3}{4}}}{y^{\frac{3}{4}}} = \frac{x^{\frac{2}{1}} \cdot x^{\frac{3}{4}}}{y^{\frac{3}{4}}} = \frac{x^{\frac{2}{1}+\frac{3}{4}}}{y^{\frac{3}{4}}}$

$= \frac{x^{\frac{(2\cdot4)+(1\cdot3)}{1\cdot4}}}{y^{\frac{3}{4}}} = \frac{x^{\frac{8+3}{4}}}{y^{\frac{3}{4}}} = \frac{x^{\frac{11}{4}}}{y^{\frac{3}{4}}} = \frac{\sqrt[4]{x^{11}}}{\sqrt[4]{y^3}} = \frac{\sqrt[4]{x^{4+4+3}}}{\sqrt[4]{y^3}} = \frac{\sqrt[4]{(x^4 \cdot x^4) \cdot x^3}}{\sqrt[4]{y^3}} = \frac{(x \cdot x)\sqrt[4]{x^3}}{\sqrt[4]{y^3}} = \frac{x^2\sqrt[4]{x^3}}{\sqrt[4]{y^3}} = \mathbf{x^2\left(\sqrt[4]{\frac{x^3}{y^3}}\right)}$

7. $$\frac{1000^{-\frac{1}{3}}\cdot b^{-2}\cdot c^{-\frac{1}{3}}}{(b\cdot c)^{-\frac{2}{3}}} = \frac{1000^{-\frac{1}{3}}\cdot b^{-2}\cdot c^{-\frac{1}{3}}}{b^{-\frac{2}{3}}\cdot c^{-\frac{2}{3}}} = \frac{c^{\frac{2}{3}}\cdot c^{-\frac{1}{3}}}{1000^{\frac{1}{3}}\cdot b^{2}\cdot b^{-\frac{2}{3}}} = \frac{c^{\frac{2}{3}-\frac{1}{3}}}{1000^{\frac{1}{3}}\cdot b^{\frac{2}{1}-\frac{2}{3}}} = \frac{c^{\frac{2-1}{3}}}{1000^{\frac{1}{3}}\cdot b^{\frac{(2\cdot 3)-(1\cdot 2)}{1\cdot 3}}} = \frac{c^{\frac{1}{3}}}{1000^{\frac{1}{3}}\cdot b^{\frac{6-2}{3}}}$$

$$= \frac{c^{\frac{1}{3}}}{1000^{\frac{1}{3}}\cdot b^{\frac{4}{3}}} = \frac{\sqrt[3]{c^{1}}}{\sqrt[3]{1000^{1}}\cdot\sqrt[3]{b^{4}}} = \frac{\sqrt[3]{c}}{\sqrt[3]{1000}\cdot\sqrt[3]{b^{3+1}}} = \frac{\sqrt[3]{c}}{\sqrt[3]{10^{3}}\cdot\sqrt[3]{b^{3}\cdot b^{1}}} = \frac{\sqrt[3]{c}}{10\cdot b\sqrt[3]{b^{1}}} = \frac{\sqrt[3]{c}}{10b\sqrt[3]{b}} = \mathbf{\frac{1}{10b}\left(\sqrt[3]{\frac{c}{b}}\right)}$$

8. $$\left(\frac{z^{2}\cdot w^{-3}\cdot(a\cdot b)^{-2}}{a^{2}\cdot w^{-3}\cdot z^{0}}\right)^{-\frac{1}{2}} = \left(\frac{z^{2}\cdot w^{-3}\cdot a^{-2}\cdot b^{-2}}{a^{2}\cdot w^{-3}\cdot 1}\right)^{-\frac{1}{2}} = \left(\frac{z^{2}\cdot b^{-2}}{a^{2}\cdot a^{2}\cdot w^{-3}\cdot w^{3}}\right)^{-\frac{1}{2}} = \left(\frac{z^{2}\cdot b^{-2}}{a^{2+2}\cdot w^{-3+3}}\right)^{-\frac{1}{2}} = \left(\frac{z^{2}\cdot b^{-2}}{a^{4}\cdot w^{0}}\right)^{-\frac{1}{2}}$$

$$= \left(\frac{z^{2}\cdot b^{-2}}{a^{4}\cdot 1}\right)^{-\frac{1}{2}} = \left(\frac{z^{2}\cdot b^{-2}}{a^{4}}\right)^{-\frac{1}{2}} = \frac{z^{2\times-\frac{1}{2}}\cdot b^{-2\times-\frac{1}{2}}}{a^{4\times-\frac{1}{2}}} = \frac{z^{-\frac{2}{2}}\cdot b^{\frac{2}{2}}}{a^{-\frac{4}{2}}} = \frac{z^{-\frac{2}{2}=-\frac{1}{1}}\cdot b^{\frac{2}{2}=\frac{1}{1}}}{a^{-\frac{4}{2}=-\frac{2}{1}}} = \frac{z^{-\frac{1}{1}}\cdot b^{\frac{1}{1}}}{a^{-\frac{2}{1}}} = \frac{z^{-1}\cdot b}{a^{-2}} = \mathbf{\frac{a^{2}b}{z}}$$

9. $$\frac{2^{-\frac{3}{4}}\cdot 2^{\frac{1}{4}}\cdot(y\cdot z)^{-3}}{2^{-\frac{1}{4}}\cdot b^{-1}\cdot y^{-\frac{2}{3}}} = \frac{2^{-\frac{3}{4}+\frac{1}{4}}\cdot\left(y^{-3}\cdot z^{-3}\right)}{2^{-\frac{1}{4}}\cdot b^{-1}\cdot y^{-\frac{2}{3}}} = \frac{2^{\frac{-3+1}{4}}\cdot y^{-3}\cdot z^{-3}}{2^{-\frac{1}{4}}\cdot b^{-1}\cdot y^{-\frac{2}{3}}} = \frac{2^{-\frac{2}{4}=-\frac{1}{2}}\cdot y^{-3}\cdot z^{-3}}{2^{-\frac{1}{4}}\cdot b^{-1}\cdot y^{-\frac{2}{3}}} = \frac{b}{2^{\frac{1}{2}}\cdot 2^{-\frac{1}{4}}\cdot y^{3}\cdot y^{-\frac{2}{3}}\cdot z^{3}}$$

$$= \frac{b}{2^{\frac{1}{2}-\frac{1}{4}}\cdot y^{\frac{3}{1}-\frac{2}{3}}\cdot z^{3}} = \frac{b}{2^{\frac{(1\cdot 4)-(1\cdot 2)}{2\cdot 4}}\cdot y^{\frac{(3\cdot 3)-(1\cdot 2)}{1\cdot 3}}\cdot z^{3}} = \frac{b}{2^{\frac{4-2}{8}}\cdot y^{\frac{9-2}{3}}\cdot z^{3}} = \frac{b}{2^{\frac{2}{8}=\frac{1}{4}}\cdot y^{\frac{7}{3}}\cdot z^{3}} = \frac{b}{2^{\frac{1}{4}}\cdot y^{\frac{7}{3}}\cdot z^{3}}$$

$$= \frac{b}{\sqrt[4]{2}\cdot\sqrt[3]{y^{7}}\cdot z^{3}} = \frac{b}{\sqrt[4]{2}\cdot\sqrt[3]{y^{3+3+1}}\cdot z^{3}} = \frac{b}{\sqrt[4]{2}\cdot\sqrt[3]{y^{3}\cdot y^{3}\cdot y^{1}}\cdot z^{3}} = \frac{b}{\sqrt[4]{2}\cdot(y\cdot y)\sqrt[3]{y}\cdot z^{3}} = \mathbf{\frac{b}{\sqrt[4]{2}\,y^{2}\sqrt[3]{y}\,z^{3}}}$$

10. $$\frac{a^{\frac{1}{2}}\cdot b^{\frac{4}{3}}\cdot\left(c^{5}\cdot d\right)^{-\frac{4}{5}}}{c^{3}\cdot d^{\frac{1}{5}}} = \frac{a^{\frac{1}{2}}\cdot b^{\frac{4}{3}}\cdot c^{5\times-\frac{4}{5}}\cdot d^{1\times-\frac{4}{5}}}{c^{3}\cdot d^{\frac{1}{5}}} = \frac{a^{\frac{1}{2}}\cdot b^{\frac{4}{3}}\cdot c^{-\frac{20}{5}=-\frac{4}{1}}\cdot d^{-\frac{4}{5}}}{c^{3}\cdot d^{\frac{1}{5}}} = \frac{a^{\frac{1}{2}}\cdot b^{\frac{4}{3}}\cdot c^{-\frac{4}{1}}\cdot d^{-\frac{4}{5}}}{c^{3}\cdot d^{\frac{1}{5}}} = \frac{a^{\frac{1}{2}}\cdot b^{\frac{4}{3}}}{c^{3}\cdot c^{4}\cdot d^{\frac{1}{5}}\cdot d^{\frac{4}{5}}}$$

$$= \frac{a^{\frac{1}{2}}\cdot b^{\frac{4}{3}}}{c^{3+4}\cdot d^{\frac{1}{5}+\frac{4}{5}}} = \frac{a^{\frac{1}{2}}\cdot b^{\frac{4}{3}}}{c^{7}\cdot d^{\frac{1+4}{5}}} = \frac{a^{\frac{1}{2}}\cdot b^{\frac{4}{3}}}{c^{7}\cdot d^{\frac{5}{5}=\frac{1}{1}}} = \frac{a^{\frac{1}{2}}\cdot b^{\frac{4}{3}}}{c^{7}\cdot d^{\frac{1}{1}}} = \frac{\sqrt[2]{a^{1}}\cdot\sqrt[3]{b^{4}}}{c^{7}\cdot d} = \frac{\sqrt{a}\cdot\sqrt[3]{b^{3+1}}}{c^{7}\cdot d} = \frac{\sqrt{a}\cdot\sqrt[3]{b^{3}\cdot b^{1}}}{c^{7}\cdot d} = \mathbf{\frac{b\sqrt{a}\sqrt[3]{b}}{c^{7}d}}$$

Section 5.4 Case III Practice Problems

1. $2^{-\frac{1}{2}} - 2^3 = \frac{1}{2^{\frac{1}{2}}} - 8 = \frac{1}{\sqrt[2]{2^1}} - 8 = \frac{1}{\sqrt{2}} - \frac{8}{1} = \frac{(1\cdot1)-(8\cdot\sqrt{2})}{1\cdot\sqrt{2}} = \frac{1-8\sqrt{2}}{\sqrt{2}} = \frac{1-8\sqrt{2}}{\sqrt{2}}\times\frac{\sqrt{2}}{\sqrt{2}} = \frac{\sqrt{2}\cdot(1-8\sqrt{2})}{\sqrt{2}\cdot\sqrt{2}}$

$= \frac{\sqrt{2}-8\sqrt{2}\cdot\sqrt{2}}{\sqrt{2\cdot2}} = \frac{\sqrt{2}-8\sqrt{2\cdot2}}{\sqrt{2\cdot2}} = \frac{\sqrt{2}-8\sqrt{2^2}}{\sqrt{2^2}} = \frac{\sqrt{2}-8\cdot2}{2} = \mathbf{\frac{\sqrt{2}-16}{2}}$

2. $\left(x^3+x^2\right)^{-\frac{1}{3}} - 3 = \frac{1}{\left(x^3+x^2\right)^{\frac{1}{3}}} - 3 = \frac{1}{\sqrt[3]{x^3+x^2}} - \frac{3}{1} = \frac{(1\cdot1)-\left(3\cdot\sqrt[3]{x^3+x^2}\right)}{1\cdot\sqrt[3]{x^3+x^2}} = \mathbf{\frac{1-3\sqrt[3]{x^3+x^2}}{\sqrt[3]{x^3+x^2}}}$

3. $\frac{2}{a^{-\frac{1}{2}}} + a^{-\frac{1}{2}} = \frac{2}{\frac{1}{\frac{1}{a^{\frac{1}{2}}}}} + \frac{1}{a^{\frac{1}{2}}}$...

$\frac{2}{a^{-\frac{1}{2}}} + a^{-\frac{1}{2}} = \frac{2}{\frac{1}{a^{\frac{1}{2}}}} + \frac{1}{a^{\frac{1}{2}}} = \frac{\frac{2}{1}}{\frac{1}{a^{\frac{1}{2}}}} + \frac{1}{a^{\frac{1}{2}}} = \frac{\frac{2}{1}}{\frac{1}{a^{\frac{1}{2}}}} + \frac{1}{a^{\frac{1}{2}}} = \frac{2\cdot a^{\frac{1}{2}}}{1\cdot1} + \frac{1}{a^{\frac{1}{2}}} = \frac{2\sqrt{a}}{1} + \frac{1}{\sqrt{a}} = \frac{\left(2\sqrt{a}\cdot\sqrt{a}\right)+(1\cdot1)}{1\cdot\sqrt{a}}$

$= \frac{2\sqrt{a\cdot a}+1}{\sqrt{a}} = \frac{2\sqrt{a^2}+1}{\sqrt{a}} = \frac{2a+1}{\sqrt{a}} = \frac{2a+1}{\sqrt{a}}\times\frac{\sqrt{a}}{\sqrt{a}} = \frac{\sqrt{a}\cdot(2a+1)}{\sqrt{a}\cdot\sqrt{a}} = \frac{\sqrt{a}(2a+1)}{\sqrt{a\cdot a}} = \frac{\sqrt{a}(2a+1)}{\sqrt{a^2}} = \mathbf{\frac{\sqrt{a}(2a+1)}{a}}$

4. $\frac{\frac{3}{5}}{x^2+x^{-\frac{2}{5}}} = \frac{\frac{3}{5}}{x^2+\frac{1}{x^{\frac{2}{5}}}} = \frac{\frac{3}{5}}{x^2+\frac{1}{\sqrt[5]{x^2}}} = \frac{\frac{3}{5}}{\frac{x^2}{1}+\frac{1}{\sqrt[5]{x^2}}} = \frac{\frac{3}{5}}{\frac{\left(x^2\cdot\sqrt[5]{x^2}\right)+(1\cdot1)}{1\cdot\sqrt[5]{x^2}}} = \frac{\frac{3}{5}}{\frac{x^2\sqrt[5]{x^2}+1}{\sqrt[5]{x^2}}} = \frac{3\cdot\sqrt[5]{x^2}}{5\cdot\left(x^2\sqrt[5]{x^2}+1\right)}$

$= \mathbf{\frac{3}{5}\left(\frac{\sqrt[5]{x^2}}{x^2\sqrt[5]{x^2}+1}\right)}$

5. $\frac{y^{-\frac{1}{2}}}{y-y^{-\frac{1}{2}}} = \frac{\frac{1}{y^{\frac{1}{2}}}}{y-\frac{1}{y^{\frac{1}{2}}}} = \frac{\frac{1}{\sqrt{y}}}{\frac{y}{1}-\frac{1}{\sqrt{y}}} = \frac{\frac{1}{\sqrt{y}}}{\frac{(y\cdot\sqrt{y})-(1\cdot1)}{1\cdot\sqrt{y}}} = \frac{\frac{1}{\sqrt{y}}}{\frac{y\sqrt{y}-1}{\sqrt{y}}} = \frac{1\cdot\sqrt{y}}{\sqrt{y}\cdot\left(y\sqrt{y}-1\right)} = \frac{\sqrt{y}}{\sqrt{y}\cdot\left(y\sqrt{y}-1\right)} = \frac{1}{1\cdot\left(y\sqrt{y}-1\right)}$

$= \frac{1}{y\sqrt{y}-1} = \frac{1}{y\sqrt{y}-1}\times\frac{y\sqrt{y}+1}{y\sqrt{y}+1} = \frac{1\times y\sqrt{y}+1}{\left(y\sqrt{y}-1\right)\times\left(y\sqrt{y}+1\right)} = \frac{y\sqrt{y}+1}{y\sqrt{y}\cdot y\sqrt{y}+y\sqrt{y}-y\sqrt{y}-1} = \frac{y\sqrt{y}+1}{y\cdot y\sqrt{y\cdot y}-1}$

$= \frac{y\sqrt{y}+1}{y^2\sqrt{y^2}-1} = \frac{y\sqrt{y}+1}{y^2\cdot y-1} = \mathbf{\frac{y\sqrt{y}+1}{y^3-1}}$

6. $\frac{b^{-\frac{2}{3}}}{b^{-1}+b^{-2}} = \frac{\frac{1}{b^{\frac{2}{3}}}}{\frac{1}{b^1}+\frac{1}{b^2}} = \frac{\frac{1}{\sqrt[3]{b^2}}}{\frac{\left(1\cdot b^2\right)+(1\cdot b)}{b\cdot b^2}} = \frac{\frac{1}{\sqrt[3]{b^2}}}{\frac{b^2+b}{b^3}} = \frac{\frac{1}{\sqrt[3]{b^2}}}{\frac{b(b+1)}{b^3}} = \frac{\frac{1}{\sqrt[3]{b^2}}}{\frac{(b+1)}{b^3\cdot b^{-1}}} = \frac{\frac{1}{\sqrt[3]{b^2}}}{\frac{(b+1)}{b^{3-1}}} = \frac{\frac{1}{\sqrt[3]{b^2}}}{\frac{(b+1)}{b^2}} = \frac{1\cdot b^2}{\sqrt[3]{b^2}(b+1)}$

$= \frac{b^2}{(b+1)\sqrt[3]{b^2}} = \frac{b^2}{(b+1)}\left(\frac{1}{\sqrt[3]{b^2}}\right) = \frac{b^2}{(b+1)}\left(\frac{1}{\sqrt[3]{b^2}}\times\frac{\sqrt[3]{b^1}}{\sqrt[3]{b^1}}\right) = \frac{b^2}{(b+1)}\left(\frac{\sqrt[3]{b}}{\sqrt[3]{b^2\cdot b^1}}\right) = \frac{b^2}{(b+1)}\left(\frac{\sqrt[3]{b}}{\sqrt[3]{b^{2+1}}}\right) = \frac{b^2}{(b+1)}\left(\frac{\sqrt[3]{b}}{\sqrt[3]{b^3}}\right)$

$$= \frac{b^2}{(b+1)}\left(\frac{\sqrt[3]{b}}{b}\right) = \frac{b^2\sqrt[3]{b}}{b(b+1)} = \mathbf{\frac{b\sqrt[3]{b}}{b+1}}$$

7. $$\frac{x^{-1}+y^{\frac{2}{3}}}{x^{-\frac{2}{3}}} = \frac{\frac{1}{x^1}+y^{\frac{2}{3}}}{\frac{1}{x^{\frac{2}{3}}}} = \frac{\frac{1}{x}+\sqrt[3]{y^2}}{\frac{1}{\sqrt[3]{x^2}}} = \frac{\frac{1}{x}+\frac{\sqrt[3]{y^2}}{1}}{\frac{1}{\sqrt[3]{x^2}}} = \frac{\frac{(1\cdot 1)+\left(x\cdot\sqrt[3]{y^2}\right)}{1\cdot x}}{\frac{1}{\sqrt[3]{x^2}}} = \frac{\frac{1+x\sqrt[3]{y^2}}{x}}{\frac{1}{\sqrt[3]{x^2}}} = \frac{\sqrt[3]{x^2}\cdot\left(1+x\sqrt[3]{y^2}\right)}{1\cdot x}$$

$$= \frac{\left(1\cdot\sqrt[3]{x^2}\right)+\left(x\sqrt[3]{y^2}\cdot\sqrt[3]{x^2}\right)}{x} = \mathbf{\frac{\sqrt[3]{x^2}+x\sqrt[3]{x^2y^2}}{x}}$$

8. $$\frac{\frac{1}{2}}{a^{-\frac{1}{2}}-b^{-\frac{1}{2}}} = \frac{\frac{1}{2}}{\frac{1}{a^{\frac{1}{2}}}-\frac{1}{b^{\frac{1}{2}}}} = \frac{\frac{1}{2}}{\frac{1}{\sqrt[2]{a^1}}-\frac{1}{\sqrt[2]{b^1}}} = \frac{\frac{1}{2}}{\frac{1}{\sqrt{a}}-\frac{1}{\sqrt{b}}} = \frac{\frac{1}{2}}{\frac{\left(1\cdot\sqrt{b}\right)-\left(1\cdot\sqrt{a}\right)}{\sqrt{a}\cdot\sqrt{b}}} = \frac{\frac{1}{2}}{\frac{\sqrt{b}-\sqrt{a}}{\sqrt{a\cdot b}}} = \frac{1\cdot\sqrt{ab}}{2\cdot\left(\sqrt{b}-\sqrt{a}\right)}$$

$$= \frac{\sqrt{ab}}{2\left(\sqrt{b}-\sqrt{a}\right)} = \frac{1}{2}\frac{\sqrt{ab}}{\sqrt{b}-\sqrt{a}}\times\frac{\sqrt{b}+\sqrt{a}}{\sqrt{b}+\sqrt{a}} = \frac{1}{2}\frac{\sqrt{ab}\cdot\left(\sqrt{b}+\sqrt{a}\right)}{\left(\sqrt{b}-\sqrt{a}\right)\cdot\left(\sqrt{b}+\sqrt{a}\right)} = \frac{1}{2}\frac{\sqrt{ab\cdot b}+\sqrt{ab\cdot a}}{\sqrt{b\cdot b}-\sqrt{a\cdot b}+\sqrt{a\cdot b}-\sqrt{a\cdot a}}$$

$$= \frac{1}{2}\frac{\sqrt{ab^2}+\sqrt{a^2b}}{\sqrt{b^2}-\sqrt{ab}+\sqrt{ab}-\sqrt{a^2}} = \frac{1}{2}\frac{b\sqrt{a}+a\sqrt{b}}{\sqrt{b^2}-\sqrt{a^2}} = \mathbf{\frac{1}{2}\frac{b\sqrt{a}+a\sqrt{b}}{b-a}}$$

9. $$\frac{x^{-1}}{(x-3)^{-\frac{1}{4}}}+y^{-\frac{1}{3}} = \frac{\frac{1}{x^1}}{\frac{1}{(x-3)^{\frac{1}{4}}}}+\frac{1}{y^{\frac{1}{3}}} = \frac{\frac{1}{x}}{\frac{1}{\sqrt[4]{(x-3)^1}}}+\frac{1}{\sqrt[3]{y^1}} = \frac{1\cdot\sqrt[4]{x-3}}{1\cdot x}+\frac{1}{\sqrt[3]{y}} = \frac{\sqrt[4]{x-3}}{x}+\frac{1}{\sqrt[3]{y}} = \frac{\sqrt[3]{y}\cdot\left(\sqrt[4]{x-3}\right)+(1\cdot x)}{x\cdot\sqrt[3]{y}}$$

$$= \mathbf{\frac{\sqrt[3]{y}\left(\sqrt[4]{x-3}\right)+x}{x\sqrt[3]{y}}}$$

10. $$\frac{\left(a^2\right)^{-\frac{2}{3}}+\left(b^{-2}\right)^{-\frac{1}{3}}}{(ab)^{\frac{1}{5}}} = \frac{a^{2\times-\frac{2}{3}}+b^{-2\times-\frac{1}{3}}}{(ab)^{\frac{1}{5}}} = \frac{a^{-\frac{4}{3}}+b^{\frac{2}{3}}}{(ab)^{\frac{1}{5}}} = \frac{\frac{1}{a^{\frac{4}{3}}}+b^{\frac{2}{3}}}{(ab)^{\frac{1}{5}}} = \frac{\frac{1}{a^{\frac{4}{3}}}+\frac{b^{\frac{2}{3}}}{1}}{(ab)^{\frac{1}{5}}} = \frac{\frac{1}{\sqrt[3]{a^4}}+\frac{\sqrt[3]{b^2}}{1}}{\sqrt[5]{(ab)^1}}$$

$$= \frac{\frac{(1\cdot 1)+\left(\sqrt[3]{a^4}\cdot\sqrt[3]{b^2}\right)}{1\cdot\sqrt[3]{a^4}}}{\sqrt[5]{ab}} = \frac{\frac{1+\sqrt[3]{a^4\cdot b^2}}{\sqrt[3]{a^4}}}{\frac{\sqrt[5]{ab}}{1}} = \frac{1\cdot\left(1+\sqrt[3]{a^4\cdot b^2}\right)}{\sqrt[3]{a^4}\cdot\sqrt[5]{ab}} = \frac{1+\sqrt[3]{a^{3+1}\cdot b^2}}{\sqrt[3]{a^{3+1}}\cdot\sqrt[5]{ab}} = \frac{1+\sqrt[3]{a^3\cdot a^1\cdot b^2}}{\sqrt[3]{a^3\cdot a^1}\cdot\sqrt[5]{ab}} = \mathbf{\frac{1+a\sqrt[3]{ab^2}}{a\sqrt[3]{a}\sqrt[5]{ab}}}$$

Chapter 5 Appendix Case I Practice Problems

1. $0.00047 = \mathbf{4.7\times 10^{-4}}$
2. $1245.78 = \mathbf{1.24578\times 10^{3}}$
3. $0.000000456 = \mathbf{4.56\times 10^{-7}}$
4. $45789.456 = \mathbf{4.5789456\times 10^{4}}$
5. $23 = \mathbf{2.3\times 10^{1}}$
6. $45.8 = \mathbf{4.58\times 10^{1}}$

7. $3445.38 = \mathbf{3.44538 \times 10^3}$

8. $51244 = \mathbf{5.1244 \times 10^4}$

9. $0.0058 = \mathbf{5.8 \times 10^{-3}}$

10. $456794324.0 = \mathbf{4.56794324 \times 10^8}$

Chapter 5 Appendix Case II Practice Problems

1. $3.8 \times 10^{+4} = \mathbf{38000.0}$

2. $12.783 \times 10^{-3} = \mathbf{0.012783}$

3. $2.36796 \times 10^{+7} = \mathbf{23679600.0}$

4. $1.0035 \times 10^{-5} = \mathbf{0.000010035}$

5. $2.5 \times 10^{+2} = \mathbf{250.0}$

6. $3.7865 \times 10^{+5} = \mathbf{378650.0}$

7. $1.00004 \times 10^{-3} = \mathbf{0.00100004}$

8. $2.04506 \times 10^{-1} = \mathbf{0.204506}$

9. $9.34587654 \times 10^{+9} = \mathbf{9345876540.}$

10. $3.0500 \times 10^{+2} = \mathbf{305.0}$

Chapter 5 Appendix Case III Practice Problems

1. $\left(5.4 \times 10^{-3}\right) \cdot \left(1.2 \times 10^{+3}\right) = (5.4 \times 1.2) \cdot \left(10^{-3} \times 10^{+3}\right) = \mathbf{6.48 \times 10^0}$

2. $\left(12.564 \times 10^{+3}\right) \cdot \left(9 \times 10^{+2}\right) = (12.564 \times 9) \cdot \left(10^{+3} \times 10^{+2}\right) = (113.076) \cdot \left(10^{+3+2}\right) = \left(1.13076 \times 10^{+2}\right) \cdot 10^{+5}$

$= 1.13076 \cdot \left(10^{+2} \cdot 10^{+5}\right) = 1.13076 \times 10^{+2+5} = \mathbf{1.13076 \times 10^7}$

3. $\left(2.002 \times 10^{-2}\right) \cdot \left(3 \times 10^{-2}\right) = (2.002 \times 3) \cdot \left(10^{-2} \times 10^{-2}\right) = (6.006) \cdot \left(10^{-2-2}\right) = \mathbf{6.006 \times 10^{-4}}$

4. $\left(5 \times 10^{-2}\right) \cdot \left(8 \times 10^{+6}\right) = (5 \times 8) \cdot \left(10^{-2} \times 10^{+6}\right) = (40) \cdot \left(10^{-2+6}\right) = \left(4.0 \times 10^{+1}\right) \cdot 10^{+4} = 4.0 \cdot \left(10^{+1} \cdot 10^{+4}\right) = 4.0 \times 10^{+1+4}$

$= \mathbf{4.0 \times 10^5}$

5. $\left(22.34 \times 10^{-4}\right) \cdot \left(39.4 \times 10^{-3}\right) = (22.34 \times 39.4) \cdot \left(10^{-4} \times 10^{-3}\right) = (880.196) \cdot \left(10^{-4-3}\right) = \left(8.80196 \times 10^{+2}\right) \cdot 10^{-7}$

$= 8.80196 \cdot \left(10^{+2} \cdot 10^{-7}\right) = 8.80196 \times 10^{+2-7} = \mathbf{8.80196 \times 10^{-5}}$

6. $\left(4.334 \times 10^{-2}\right) \cdot \left(2.94 \times 10^{+4}\right) = (4.334 \times 2.94) \cdot \left(10^{-2} \times 10^{+4}\right) = (12.742) \cdot \left(10^{-2+4}\right) = \left(1.2742 \times 10^{+1}\right) \cdot 10^{+2}$

$= 1.2742 \cdot \left(10^{+1} \cdot 10^{+2}\right) = 1.2742 \times 10^{+1+2} = \mathbf{1.2742 \times 10^3}$

7. $\left(2 \times 10^{-4}\right) \cdot \left(9 \times 10^{+5}\right) = (2 \times 9) \cdot \left(10^{-4} \times 10^{+5}\right) = (18) \cdot \left(10^{-4+5}\right) = \left(1.8 \times 10^{+1}\right) \cdot 10^{+1} = 1.8 \cdot \left(10^{+1} \cdot 10^{+1}\right) = 1.8 \cdot 10^{+1+1}$

$= \mathbf{1.8 \times 10^2}$

8. $\left(8.01 \times 10^{-5}\right) \cdot \left(3.4 \times 10^{-1}\right) = (8.01 \times 3.4) \cdot \left(10^{-5} \times 10^{-1}\right) = (27.234) \cdot \left(10^{-5-1}\right) = \left(2.7234 \times 10^{+1}\right) \cdot 10^{-6}$

$= 2.7234 \cdot \left(10^{+1} \times 10^{-6}\right) = 2.7234 \times 10^{+1-6} = \mathbf{2.7234 \times 10^{-5}}$

9. $\left(4.4 \times 10^{+1}\right) \cdot \left(5.4 \times 10^{+1}\right) = (4.4 \times 5.4) \cdot \left(10^{+1} \times 10^{+1}\right) = (23.76) \cdot \left(10^{+1+1}\right) = \left(2.376 \times 10^{+1}\right) \cdot 10^{+2} = 2.376 \cdot \left(10^{+1} \times 10^{+2}\right)$

$= 2.376 \times 10^{+1+2} = \mathbf{2.376 \times 10^3}$

10. $\left(2.889\times10^{-1}\right)\cdot\left(9\times10^{-2}\right)=\left(2.889\times9\right)\cdot\left(10^{-1}\times10^{-2}\right)=\left(26.001\right)\cdot\left(10^{-1-2}\right)=\left(2.6001\times10^{+1}\right)\cdot10^{-3}$

$=2.6001\cdot\left(10^{+1}\cdot10^{-3}\right)=2.6001\times10^{+1-3}=\mathbf{2.6001\times10^{-2}}$

Chapter 5 Appendix Case IV Practice Problems

1. $\dfrac{48.4\times10^{+4}}{12\times10^{+3}}=\left(\dfrac{48.4}{12}\right)\cdot\left(\dfrac{10^{+4}}{10^{+3}}\right)=4.033\cdot\left(10^{+4}\times10^{-3}\right)=4.033\cdot\left(10^{+4-3}\right)=\mathbf{4.033\times10^{1}}$

2. $\dfrac{8.4\times10^{-3}}{2.2\times10^{+5}}=\left(\dfrac{8.4}{2.2}\right)\cdot\left(\dfrac{10^{-3}}{10^{+5}}\right)=3.818\cdot\left(10^{-3}\times10^{-5}\right)=3.818\cdot\left(10^{-3-5}\right)=\mathbf{3.818\times10^{-8}}$

3. $\dfrac{235.5\times10^{+3}}{5\times10^{+3}}=\left(\dfrac{235.5}{5}\right)\cdot\left(\dfrac{10^{+3}}{10^{+3}}\right)=47.1\cdot\left(10^{+3}\times10^{-3}\right)=47.1\cdot\left(10^{+3-3}\right)=\left(4.71\times10^{+1}\right)\cdot10^{0}=\left(4.71\times10^{+1}\right)\cdot1$

$=\mathbf{4.71\times10^{1}}$

4. $\dfrac{1.84\times10^{+2}}{0.2\times10^{-1}}=\left(\dfrac{1.84}{0.2}\right)\cdot\left(\dfrac{10^{+2}}{10^{-1}}\right)=9.2\cdot\left(10^{+2}\times10^{+1}\right)=9.2\cdot\left(10^{+2+1}\right)=9.2\times10^{+3}=\mathbf{9.2\times10^{3}}$

5. $\dfrac{14.484\times10^{-3}}{2\times10^{-2}}=\left(\dfrac{14.484}{2}\right)\cdot\left(\dfrac{10^{-3}}{10^{-2}}\right)=7.242\cdot\left(10^{-3}\times10^{+2}\right)=7.242\cdot\left(10^{-3+2}\right)=\mathbf{7.242\times10^{-1}}$

6. $\dfrac{2444.4\times10^{+5}}{4.4\times10^{+4}}=\left(\dfrac{2444.4}{4.4}\right)\cdot\left(\dfrac{10^{+5}}{10^{+4}}\right)=555.54\cdot\left(10^{+5}\times10^{-4}\right)=555.54\cdot\left(10^{+5-4}\right)=\left(5.5554\times10^{+2}\right)\cdot10^{+1}$

$=5.5554\cdot\left(10^{+2}\times10^{+1}\right)=5.5554\cdot\left(10^{+2+1}\right)=\mathbf{5.5554\times10^{3}}$

7. $\dfrac{35.745\times10^{+5}}{0.35\times10^{+2}}=\left(\dfrac{35.745}{0.35}\right)\cdot\left(\dfrac{10^{+5}}{10^{+2}}\right)=102.128\cdot\left(10^{+5}\times10^{-2}\right)=102.128\cdot\left(10^{+5-2}\right)=\left(1.02128\times10^{+2}\right)\cdot10^{+3}$

$=1.02128\cdot\left(10^{+2}\times10^{+3}\right)=1.02128\cdot\left(10^{+2+3}\right)=\mathbf{1.02128\times10^{5}}$

8. $\dfrac{8.45\times10^{-4}}{5.5\times10^{-3}}=\left(\dfrac{8.45}{5.5}\right)\cdot\left(\dfrac{10^{-4}}{10^{-3}}\right)=1.536\cdot\left(10^{-4}\times10^{+3}\right)=1.536\cdot\left(10^{-4+3}\right)=\mathbf{1.536\times10^{-1}}$

9. $\dfrac{2.4\times10^{0}}{4\times10^{+1}}=\left(\dfrac{2.4}{4}\right)\cdot\left(\dfrac{10^{0}}{10^{+1}}\right)=0.6\cdot\left(10^{0}\times10^{-1}\right)=0.6\cdot\left(10^{0-1}\right)=\left(6.0\times10^{-1}\right)\cdot10^{-1}=6.0\cdot\left(10^{-1}\times10^{-1}\right)=6.0\cdot\left(10^{-1-1}\right)$

$=\mathbf{6.0\times10^{-2}}$

10. $\dfrac{22.45\times10^{+6}}{2.2\times10^{0}}=\left(\dfrac{22.45}{2.2}\right)\cdot\left(\dfrac{10^{+6}}{10^{0}}\right)=10.205\cdot\left(\dfrac{10^{+6}}{1}\right)=10.205\cdot10^{+6}=\left(1.0205\times10^{+1}\right)\cdot10^{+6}=1.0205\cdot\left(10^{+1}\times10^{+6}\right)$

$=1.0205\cdot\left(10^{+1+6}\right)=\mathbf{1.0205\times10^{7}}$

Chapter 6 Solutions:

Section 6.1 Case I Practice Problems

Polynomials	Standard Form	Type	Degree	No. of Terms
1. $3x+2x^3-6$	$2x^3+3x-6$	trinomial	3	3
2. $-6y^8+2$	$-6y^8+2$	binomial	8	2
3. $2w+6w^2+8w^5$	$8w^5+6w^2+2w$	trinomial	5	3
4. $6y$	$6y$	monomial	1	1
5. $\sqrt{72}$	$\sqrt{72}$	monomial	0	1
6. $-16+2x^4$	$2x^4-16$	binomial	4	2
7. $x^5+8x^4+2x-x^3-5$	$x^5+8x^4-x^3+2x-5$	polynomial	5	5
8. $x^4+2x^2-x^{-1}+1$		not a polynomial		
9. $\frac{2}{3}y+\frac{5}{8}y^2+\frac{1}{4}y^3-\frac{1}{2}$	$\frac{1}{4}y^3+\frac{5}{8}y^2+\frac{2}{3}y-\frac{1}{2}$	polynomial	3	4
10. $x^4-2x^3+\frac{5}{x}-4$		not a polynomial		

Section 6.1 Case II Practice Problems

1. $-x^3+4x-8x^2+3x-5x^3-5x = \left(-x^3-5x^3\right)+(4x+3x-5x)-8x^2 = (-1-5)x^3+(4+3-5)x-8x^2$
   $= -6x^3+2x-8x^2 = \mathbf{-6x^3-8x^2+2x}$

2. $2y+2y^3-5+4y-5y^3+1+y = (2y+4y+y)+\left(2y^3-5y^3\right)+(-5+1) = (2+4+1)y+(2-5)y^3-4 = 7y-3y^3-4$
   $= \mathbf{-3y^3+7y-4}$

3. $2a^5+2a^2-3+4a^5+a^2 = \left(2a^5+4a^5\right)+\left(2a^2+a^2\right)-3 = (2+4)a^5+(2+1)a^2-3 = \mathbf{6a^5+3a^2-3}$

4. $3x+2x^4+2x^3-7x-5x^4 = (3x-7x)+\left(2x^4-5x^4\right)+2x^3 = (3-7)x+(2-5)x^4+2x^3 = -4x-3x^4+2x^3$
   $= \mathbf{-3x^4+2x^3-4x}$

5. $2rs+4r^3s^3-20+2rs-5r^3s^3-3 = (2rs+2rs)+\left(4r^3s^3-5r^3s^3\right)+(-20-3) = (2+2)rs+(4-5)r^3s^3-23$
   $= 4rs-r^3s^3-23 = \mathbf{-r^3s^3+4rs-23}$

6. $2xyz+2x^3y^3z^3+10-4xyz-4 = (2xyz-4xyz)+2x^3y^3z^3+(10-4) = (2-4)xyz+2x^3y^3z^3+6 = -2xyz+2x^3y^3z^3+6$
   $= \mathbf{2x^3y^3z^3-2xyz+6}$

7. $-8+2u^2v^2+6uv-5+2uv-8u^2v^2 = (-8-5)+\left(2u^2v^2-8u^2v^2\right)+(6uv+2uv) = -13+(2-8)u^2v^2+(6+2)uv$
   $= -13-6u^2v^2+8uv = \mathbf{-6u^2v^2+8uv-13}$

8. $2x+7x-8+2x^2-3-5x+2 = (2x+7x-5x)+(-8-3+2)+2x^2 = (2+7-5)x-9+2x^2 = 4x-9+2x^2$

$= \mathbf{2x^2+4x-9}$

9. $-2y^8+5y^5-8+12y^8-5y^8+y^5-3 = \left(-2y^8+12y^8-5y^8\right)+\left(5y^5+y^5\right)+(-3-8) = (-2+12-5)y^8+(5+1)y^5-11$

$= \mathbf{5y^8+6y^5-11}$

10. $2m^3+4m^3-4+2m-5m+3 = \left(2m^3+4m^3\right)+(-4+3)+(2m-5m) = (2+4)m^3-1+(2-5)m = 6m^3-1-3m$

$= \mathbf{6m^3-3m-1}$

Section 6.2 Case I a Practice Problems

1. $(2ax)\cdot\left(3a^2x^2\right) = (2\cdot3)\cdot\left(a\cdot a^2\right)\cdot\left(x\cdot x^2\right) = 6\cdot a^{1+2}\cdot x^{1+2} = \mathbf{6a^3x^3}$

2. $\left(5x^2y^2\right)\cdot(2x)\cdot(4y) = (5\cdot2\cdot4)\cdot\left(x^2\cdot x\right)\cdot\left(y^2\cdot y\right) = 40\cdot x^{2+1}\cdot y^{2+1} = \mathbf{40x^3y^3}$

3. $\left(6x^2\right)^0\cdot\left(3x^2\right)\cdot(-2x) = 1\cdot\left(3x^2\right)\cdot(-2x) = -(3\cdot2)\cdot\left(x^2\cdot x\right) = -6\cdot x^{2+1} = \mathbf{-6x^3}$

4. $\left(x^2y\right)\cdot(3xy)\cdot\left(4x^3y^2\right) = (3\cdot4)\left(x^2\cdot x\cdot x^3\right)\cdot\left(y\cdot y\cdot y^2\right) = 12\left(x^{2+1+3}\right)\cdot\left(y^{1+1+2}\right) = \mathbf{12x^6y^4}$

5. $\left(3x^2y^2\right)\cdot\left(2xy^0\right)\cdot\left(5x^0y\right) = \left(3x^2y^2\right)\cdot(2x)\cdot(5y) = (3\cdot2\cdot5)\left(x^2\cdot x\right)\cdot\left(y^2\cdot y\right) = 30x^{2+1}\cdot y^{2+1} = \mathbf{30x^3y^3}$

6. $\left(8a^2b^2\right)\cdot(2a)\cdot\left(3a^2b^3\right) = (8\cdot2\cdot3)\left(a^2\cdot a\cdot a^2\right)\cdot\left(b^2\cdot b^3\right) = 48a^{2+1+2}\cdot b^{2+3} = \mathbf{48a^5b^5}$

7. $\left(4m^2n^2\right)\cdot\left(3n^2\right) = (4\cdot3)\cdot m^2\cdot\left(n^2\cdot n^2\right) = 12\cdot m^2\cdot n^{2+2} = \mathbf{12m^2n^4}$

8. $\left(3m^2n^3\right)\cdot\left(2mn^2\right)\cdot(4n) = (3\cdot2\cdot4)\cdot\left(m^2\cdot m\right)\cdot\left(n^3\cdot n^2\cdot n\right) = 24\cdot m^{2+1}\cdot n^{3+2+1} = \mathbf{24m^3n^6}$

9. $\left(6x^3y^6z^2\right)\cdot\left(3x^5y^0z\right)\cdot\left(z^3\right) = \left(6x^3y^6z^2\right)\cdot\left(3x^5z\right)\cdot\left(z^3\right) = (6\cdot3)\cdot\left(x^3\cdot x^5\right)\cdot y^6\cdot\left(z^2\cdot z\cdot z^3\right) = 18\cdot x^{3+5}\cdot y^6\cdot z^{2+1+3}$

$= \mathbf{18x^8y^6z^6}$

10. $\left(2w^2z^3\right)\cdot\left(2w^3z\right)^0\cdot\left(5wz^2\right) = \left(2w^2z^3\right)\cdot1\cdot\left(5wz^2\right) = \left(2w^2z^3\right)\cdot\left(5wz^2\right) = (2\cdot5)\left(w^2\cdot w\right)\cdot\left(z^3\cdot z^2\right) = 10\cdot w^{2+1}\cdot z^{3+2}$

$= \mathbf{10w^3z^5}$

Section 6.2 Case I b Practice Problems

1. $2\cdot\left(5x^2+6x-2x^2-x+5\right) = 2\cdot\left[\left(5x^2-2x^2\right)+(6x-x)+5\right] = 2\cdot\left[(5-2)x^2+(6-1)x+5\right] = 2\cdot\left[3x^2+5x+5\right]$

$= (2\cdot3)x^2+(2\cdot5)x+(2\cdot5) = \mathbf{6x^2+10x+10}$

2. $\left(2x^2y-5y^2+3x^2y-2y^2+3\right)\cdot\left(3x^2y^2\right) = \left[\left(2x^2y+3x^2y\right)+\left(-5y^2-2y^2\right)+3\right]\cdot\left(3x^2y^2\right)$

$= \left[(2+3)x^2y+(-5-2)y^2+3\right]\cdot\left(3x^2y^2\right) = \left[5x^2y-7y^2+3\right]\cdot\left(3x^2y^2\right)$

$= (5\cdot3)\cdot\left(x^2\cdot x^2\right)\cdot\left(y\cdot y^2\right)-(7\cdot3)\cdot x^2\cdot\left(y^2\cdot y^2\right)+(3\cdot3)x^2y^2 = 15\cdot x^{2+2}\cdot y^{1+2}-21\cdot x^2\cdot y^{2+2}+9x^2y^2$

$= \mathbf{15x^4y^3-21x^2y^4+9x^2y^2}$

3. $\left(5x^3+2x^2-5+3x-2x^3\right)\cdot(-2x)^2 = \left[\left(5x^3-2x^3\right)+2x^2-5+3x\right]\cdot 4x^2 = \left[(5-2)x^3+2x^2-5+3x\right]\cdot 4x^2$

$= \left[3x^3+2x^2+3x-5\right]\cdot 4x^2 = (3\cdot 4)\cdot\left(x^3\cdot x^2\right)+(2\cdot 4)\cdot\left(x^2\cdot x^2\right)+(3\cdot 4)\cdot\left(x\cdot x^2\right)-(5\cdot 4)x^2$

$= 12\cdot x^{3+2}+8\cdot x^{2+2}+12\cdot x^{1+2}-20x^2 = \mathbf{12x^5+8x^4+12x^3-20x^2}$

4. $6w\cdot\left(4w+2w^2+2-3w+w^2\right) = 6w\cdot\left[(4w-3w)+\left(2w^2+w^2\right)+2\right] = 6w\cdot\left[(4-3)w+(2+1)w^2+2\right] = 6w\cdot\left[w+3w^2+2\right]$

$= 6w\cdot\left[3w^2+w+2\right] = (6\cdot 3)\cdot\left(w^2\cdot w\right)+6(w\cdot w)+(2\cdot 6)w = 18w^{2+1}+6w^2+12w = \mathbf{18w^3+6w^2+12w}$

5. $2x\cdot\left(2x^2\right)^2\cdot\left(5x^2+3x-2x^2+x-2\right) = 2x\cdot 4x^4\cdot\left[\left(5x^2-2x^2\right)+(3x+x)-2\right] = (2\cdot 4)\left(x\cdot x^4\right)\cdot\left[(5-2)x^2+(3+1)x-2\right]$

$= 8x^5\cdot\left[3x^2+4x-2\right] = (8\cdot 3)\cdot\left(x^5\cdot x^2\right)+(8\cdot 4)\cdot\left(x^5\cdot x\right)-(8\cdot 2)x^5 = 24\cdot x^{5+2}+32\cdot x^{5+1}-16x^5 = \mathbf{24x^7+32x^6-16x^5}$

6. $\left(\sqrt{162}+\sqrt{9}x-2x^2+\sqrt{16}x^3\right)\cdot\left(2x^3\right) = \left(\sqrt{81\cdot 2}+\sqrt{3^2}x-2x^2+\sqrt{4^2}x^3\right)\cdot\left(2x^3\right) = \left(\sqrt{9^2\cdot 2}+3x-2x^2+4x^3\right)\cdot\left(2x^3\right)$

$= \left(9\sqrt{2}+3x-2x^2+4x^3\right)\cdot\left(2x^3\right) = (9\cdot 2)\sqrt{2}x^3+(3\cdot 2)\cdot\left(x\cdot x^3\right)-(2\cdot 2)\cdot\left(x^2\cdot x^3\right)+(4\cdot 2)\cdot\left(x^3\cdot x^3\right)$

$= 18\sqrt{2}x^3+6x^{1+3}-4x^{2+3}+8x^{3+3} = 18\sqrt{2}x^3+6x^4-4x^5+8x^6 = \mathbf{8x^6-4x^5+6x^4+18\sqrt{2}x^3}$

7. $\left(5y-3y^2+2y-4\right)\cdot\left(3y^2\right) = \left[(5y+2y)-3y^2-4\right]\cdot\left(3y^2\right) = \left[(5+2)y-3y^2-4\right]\cdot\left(3y^2\right) = \left[7y-3y^2-4\right]\cdot\left(3y^2\right)$

$= \left[-3y^2+7y-4\right]\cdot\left(3y^2\right) = -(3\cdot 3)\cdot\left(y^2\cdot y^2\right)+(7\cdot 3)\cdot\left(y\cdot y^2\right)-(4\cdot 3)y^2 = -9\cdot y^{2+2}+21\cdot y^{1+2}-12y^2$

$= \mathbf{-9y^4+21y^3-12y^2}$

8. $9x\cdot\left(2x^2+5x-5x^2+6\right)\cdot\left(3x^3\right) = 9x\cdot\left[\left(2x^2-5x^2\right)+5x+6\right]\cdot\left(3x^3\right) = 9x\cdot\left[(2-5)x^2+5x+6\right]\cdot\left(3x^3\right)$

$= 9x\cdot\left[-3x^2+5x+6\right]\cdot\left(3x^3\right) = \left[-(3\cdot 9)\cdot\left(x^2\cdot x\right)+(5\cdot 9)\cdot(x\cdot x)+(6\cdot 9)x\right]\cdot\left(3x^3\right) = \left[-27x^{2+1}+45x^{1+1}+54x\right]\cdot\left(3x^3\right)$

$= \left[-27x^3+45x^2+54x\right]\cdot\left(3x^3\right) = -(27\cdot 3)\cdot\left(x^3\cdot x^3\right)+(45\cdot 3)\cdot\left(x^2\cdot x^3\right)+(54\cdot 3)\cdot\left(x\cdot x^3\right) = -81x^{3+3}+135x^{2+3}+162x^{1+3}$

$= \mathbf{-81x^6+135x^5+162x^4}$

9. $\left(5x^2\right)\cdot\left(2x^3-4x+2+x^3-x\right)\cdot 2x = 2x\cdot\left(5x^2\right)\cdot\left[\left(2x^3+x^3\right)+(-4x-x)+2\right] = (2\cdot 5)\cdot\left(x\cdot x^2\right)\cdot\left[(2+1)x^3-(4+1)x+2\right]$

$= 10x^3\cdot\left[3x^3-5x+2\right] = (10\cdot 3)\cdot\left(x^3\cdot x^3\right)-(10\cdot 5)\cdot\left(x^3\cdot x\right)+(10\cdot 2)x^3 = 30x^{3+3}-50x^{3+1}+20x^3 = \mathbf{30x^6-50x^4+20x^3}$

10. $\left(\sqrt[3]{8}x^2-4x-2x^2+8x-\sqrt[3]{125}+28\right)\cdot\left(2x^3\right) = \left(\sqrt[3]{2^3}x^2-4x-2x^2+8x-\sqrt[3]{5^3}+28\right)\cdot\left(2x^3\right)$

$= \left(2x^2-4x-2x^2+8x-5+28\right)\cdot\left(2x^3\right) = \left[\left(2x^2-2x^2\right)+(-4x+8x)+(-5+28)\right]\cdot\left(2x^3\right)$

$= \left[(2-2)x^2+(-4+8)x+23\right]\cdot\left(2x^3\right) = (4x+23)\cdot 2x^3 = (4\cdot 2)\cdot\left(x\cdot x^3\right)+(23\cdot 2)x^3 = 8x^{1+3}+46x^3 = \mathbf{8x^4+46x^3}$

Section 6.2 Case II Practice Problems

1. $(x+3)(x-2) = (x\cdot x)-(2\cdot x)+(3\cdot x)-(2\cdot 3) = x^2-2x+3x-6 = x^2+(-2x+3x)-6 = x^2+(-2+3)x-6$
$= \mathbf{x^2+x-6}$

2. $(-y+8)(y-6) = -(y\cdot y)+(6\cdot y)+(8\cdot y)-(8\cdot 6) = -y^2+6y+8y-48 = -y^2+(6y+8y)-48 = -y^2+(6+8)y-48$
$= \mathbf{-y^2+14y-48}$

3. $(x^2-2xy)(-y^2+2xy) = -(x^2\cdot y^2)+2(x^2\cdot x)y+2x(y\cdot y^2)-(2\cdot 2)(x\cdot x)(y\cdot y) = -x^2y^2+2x^3y+2xy^3-4x^2y^2$
$= (-x^2y^2-4x^2y^2)+2x^3y+2xy^3 = (-1-4)x^2y^2+2x^3y+2xy^3 = -5x^2y^2+2x^3y+2xy^3 = \mathbf{2x^3y-5x^2y^2+2xy^3}$

4. $(a^3-a^2)(a-6) = (a^3\cdot a)-(6\cdot a^3)-(a^2\cdot a)+(6\cdot a^2) = a^4-6a^3-a^3+6a^2 = a^4+(-6a^3-a^3)+6a^2$
$= a^4+(-6-1)a^3+6a^2 = \mathbf{a^4-7a^3+6a^2}$

5. $(\sqrt{x^3}-2x\sqrt{x^5})(\sqrt{x}-4) = (\sqrt{x^{2+1}}-2x\sqrt{x^{2+2+1}})(\sqrt{x}-4) = (\sqrt{x^2\cdot x^1}-2x\sqrt{x^2\cdot x^2\cdot x^1})(\sqrt{x}-4)$
$= [x\sqrt{x}-2x(x\cdot x)\sqrt{x}](\sqrt{x}-4) = [x\sqrt{x}-2(x\cdot x^2)\sqrt{x}](\sqrt{x}-4) = [x\sqrt{x}-2x^3\sqrt{x}](\sqrt{x}-4)$
$= x(\sqrt{x}\cdot\sqrt{x})-(4\cdot x\sqrt{x})-(2x^3\cdot\sqrt{x}\cdot\sqrt{x})+(2\cdot 4)x^3\cdot\sqrt{x} = x(\sqrt{x\cdot x})-4x\sqrt{x}-2x^3\sqrt{x\cdot x}+8x^3\sqrt{x}$
$= x\sqrt{x^2}-4x\sqrt{x}-2x^3\sqrt{x^2}+8x^3\sqrt{x} = x\cdot x-4x\sqrt{x}-2x^3\cdot x+8x^3\sqrt{x} = x^2-4x\sqrt{x}-2x^4+8x^3\sqrt{x}$
$= \mathbf{-2x^4+8x^3\sqrt{x}+x^2-4x\sqrt{x}}$

6. $(\sqrt[3]{y^5}-\sqrt[3]{y^2})(\sqrt[3]{y^7}-\sqrt[3]{y}) = (\sqrt[3]{y^{3+2}}-\sqrt[3]{y^2})(\sqrt[3]{y^{3+3+1}}-\sqrt[3]{y}) = (\sqrt[3]{y^3\cdot y^2}-\sqrt[3]{y^2})(\sqrt[3]{y^3\cdot y^3\cdot y^1}-\sqrt[3]{y})$
$= (y\sqrt[3]{y^2}-\sqrt[3]{y^2})[(y\cdot y)\sqrt[3]{y}-\sqrt[3]{y}] = (y\sqrt[3]{y^2}-\sqrt[3]{y^2})[y^2\sqrt[3]{y}-\sqrt[3]{y}]$
$= (y\cdot y^2)(\sqrt[3]{y^2}\cdot\sqrt[3]{y})-y(\sqrt[3]{y^2}\cdot\sqrt[3]{y})-y^2(\sqrt[3]{y^2}\cdot\sqrt[3]{y})+(\sqrt[3]{y^2}\cdot\sqrt[3]{y}) = y^3\sqrt[3]{y^2\cdot y}-y\sqrt[3]{y^2\cdot y}-y^2\sqrt[3]{y^2\cdot y}+\sqrt[3]{y^2\cdot y}$
$= y^3\sqrt[3]{y^3}-y\sqrt[3]{y^3}-y^2\sqrt[3]{y^3}+\sqrt[3]{y^3} = (y^3\cdot y)-(y\cdot y)-(y^2\cdot y)+y = y^4-y^2-y^3+y = \mathbf{y^4-y^3-y^2+y}$

7. $(\sqrt{81}-\sqrt{72})(\sqrt{36}+\sqrt{18}) = (\sqrt{9^2}-\sqrt{36\cdot 2})(\sqrt{6^2}+\sqrt{9\cdot 2}) = (9-\sqrt{6^2\cdot 2})(6+\sqrt{3^2\cdot 2}) = (9-6\sqrt{2})(6+3\sqrt{2})$
$= (9\cdot 6)+(9\cdot 3)\sqrt{2}-(6\cdot 6)\sqrt{2}-(6\cdot 3)(\sqrt{2}\cdot\sqrt{2}) = 54+27\sqrt{2}-36\sqrt{2}-18(\sqrt{2\cdot 2}) = 54+27\sqrt{2}-36\sqrt{2}-18\sqrt{2^2}$
$= 54+(27\sqrt{2}-36\sqrt{2})-(18\cdot 2) = 54+(27-36)\sqrt{2}-36 = (54-36)-9\sqrt{2} = \mathbf{18-9\sqrt{2}}$

8. $(5-3\sqrt{x^4})(\sqrt{x^2}-2) = (5-3\sqrt{x^{2+2}})(x-2) = (5-3\sqrt{x^2\cdot x^2})(x-2) = [5-3(x\cdot x)](x-2) = (5-3x^2)(x-2)$
$= (5\cdot x)-(5\cdot 2)-3(x^2\cdot x)+(3\cdot 2)x^2 = 5x-10-3x^3+6x^2 = \mathbf{-3x^3+6x^2+5x-10}$

9. $(u^2-\sqrt[3]{27})(u^3-\sqrt[3]{8}) = (u^2-\sqrt[3]{3^3})(u^3-\sqrt[3]{2^3}) = (u^2-3)(u^3-2) = (u^2\cdot u^3)-(2\cdot u^2)-(3\cdot u^3)+(2\cdot 3)$
$= u^5-2u^2-3u^3+6 = \mathbf{u^5-3u^3-2u^2+6}$

10. $(\sqrt{a^2c}+2)(-\sqrt{a^4c}+8) = (a\sqrt{c}+2)(-\sqrt{a^{2+2}c}+8) = (a\sqrt{c}+2)(-\sqrt{(a^2\cdot a^2)c}+8) = (a\sqrt{c}+2)[-(a\cdot a)\sqrt{c}+8]$
$= (a\sqrt{c}+2)[-a^2\sqrt{c}+8] = -(a\cdot a^2)\cdot(\sqrt{c}\cdot\sqrt{c})+8a\sqrt{c}-2a^2\sqrt{c}+(2\cdot 8) = -a^3(\sqrt{c\cdot c})+8a\sqrt{c}-2a^2\sqrt{c}+16$
$= -a^3\sqrt{c^2}+8a\sqrt{c}-2a^2\sqrt{c}+16 = \mathbf{-ca^3-2\sqrt{c}a^2+8\sqrt{c}a+16}$

Section 6.2 Case III Practice Problems

1. Multiply $\left(6x^2-5x-2\right)$ by $(3x+5)$.

$$\begin{array}{l} 6x^2-5x-2 \\ \underline{3x+5} \\ 18x^3-15x^2-6x \\ \quad +30x^2-25x-10 \\ \hline \mathbf{18x^3+15x^2-31x-10} \end{array}$$

2. Multiply $\left(b^3-b^2+6\right)$ by $\left(b^2-b+2\right)$.

$$\begin{array}{l} b^3-b^2+6 \\ \underline{b^2-b+2} \\ b^5-b^4+6b^2 \\ \quad -b^4+b^3-6b \\ \quad\quad +2b^3-2b^2+12 \\ \hline \mathbf{b^5-2b^4+3b^3+4b^2-6b+12} \end{array}$$

3. Multiply $\left(u^6-3u^4-2u^2+2\right)$ by $\left(u^3-2u^2+1\right)$.

$$\begin{array}{l} u^6-3u^4-2u^2+2 \\ \underline{u^3-2u^2+1} \\ u^9-3u^7-2u^5+2u^3 \\ \quad -2u^8+6u^6+4u^4-4u^2 \\ \quad\quad +u^6-3u^4-2u^2+2 \\ \hline \mathbf{u^9-2u^8-3u^7+7u^6-2u^5+u^4+2u^3-6u^2+2} \end{array}$$

4. Multiply $\left(x^5-3x^4+x^3-2x\right)$ by $(x+3)$.

$$\begin{array}{l} x^5-3x^4+x^3-2x \\ \underline{x+3} \\ x^6-3x^5+x^4-2x^2 \\ \quad +3x^5-9x^4+3x^3-6x \\ \hline \mathbf{x^6+0x^5-8x^4+3x^3-2x^2-6x} \end{array}$$

5. Multiply $\left(x^2-3x-8\right)$ by $\left(x^3-2x\right)$.

$$\begin{array}{l} x^2-3x-8 \\ \underline{x^3-2x} \\ x^5-3x^4-8x^3 \\ \quad -2x^3+6x^2+16x \\ \hline \mathbf{x^5-3x^4-10x^3+6x^2+16x} \end{array}$$

6. Multiply $\left(-3x^3+x^2-2x+1\right)$ by $\left(x^2-x+1\right)$.

$$\begin{array}{l} -3x^3+x^2-2x+1 \\ \underline{x^2-x+1} \\ -3x^5+x^4-2x^3+x^2 \\ \quad +3x^4-x^3+2x^2-x \\ \quad\quad -3x^3+x^2-2x+1 \\ \hline \mathbf{-3x^5+4x^4-6x^3+4x^2-3x+1} \end{array}$$

7. Multiply $\left(v^5-3v^3+2v-1\right)$ by $\left(v^2+v-2\right)$.

$$\begin{array}{l} v^5-3v^3+2v-1 \\ \underline{v^2+v-2} \\ v^7-3v^5+2v^3-v^2 \\ \quad +v^6-3v^4+2v^2-v \\ \quad\quad -2v^5+6v^3-4v+2 \\ \hline \mathbf{v^7+v^6-5v^5-3v^4+8v^3+v^2-5v+2} \end{array}$$

8. Multiply $\left(x^3+5x^2-5x+6\right)$ by $(x-5)$.

$$\begin{array}{l} x^3+5x^2-5x+6 \\ \underline{x-5} \\ x^4+5x^3-5x^2+6x \\ \quad -5x^3-25x^2+25x-30 \\ \hline \mathbf{x^4+0x^3-30x^2+31x-30} \end{array}$$

9. Multiply $(3y^3+2y^2-2)$ by $(-2y^2+1)$.

$$\begin{array}{r} 3y^3+2y^2-2 \\ \underline{-2y^2+1} \\ -6y^5-4y^4 \qquad +4y^2 \\ \underline{+3y^3+2y^2-2} \\ \mathbf{-6y^5-4y^4+3y^3+6y^2-2} \end{array}$$

10. Multiply $(-3a^3-2a^2+1)$ by $(-a^2+a-2)$.

$$\begin{array}{r} -3a^3-2a^2+1 \\ \underline{-a^2+a-2} \\ +3a^5+2a^4 \qquad -a^2 \\ -3a^4-2a^3 \qquad +a \\ \underline{+6a^3+4a^2 \quad -2} \\ \mathbf{3a^5-a^4+4a^3+3a^2+a-2} \end{array}$$

Section 6.3 Case I a Practice Problems

1. $\dfrac{-4xyz}{-8xyz} = +\dfrac{4xyz}{8xyz} = \dfrac{\overset{1}{\cancel{4}}\cancel{xyz}}{\underset{2}{\cancel{8}}\cancel{xyz}} = \mathbf{\dfrac{1}{2}}$

2. $\dfrac{u^2v^3w}{-uw^4} = -\dfrac{u^2v^3w^1}{u^1w^4} = -\dfrac{(u^2u^{-1})v^3}{w^4w^{-1}} = -\dfrac{(u^{2-1})v^3}{w^{4-1}} = -\dfrac{u^1v^3}{w^3} = -\mathbf{\dfrac{uv^3}{w^3}}$

3. $\dfrac{\sqrt{72x^2y^4}}{-12xy^2} = -\dfrac{x\sqrt{(36\cdot2)y^4}}{12xy^2} = -\dfrac{\cancel{x}\sqrt{(6^2\cdot2)y^{2+2}}}{12\cancel{x}y^2} = -\dfrac{6\sqrt{2(y^2\cdot y^2)}}{12y^2} = -\dfrac{6(y\cdot y)\sqrt2}{12y^2} = -\dfrac{6y^2\sqrt2}{12y^2} = -\dfrac{\overset{1}{\cancel{6}}\sqrt2}{\underset{2}{\cancel{12}}(y^2y^{-2})}$

$= -\dfrac{\sqrt2}{2y^{2-2}} = -\dfrac{\sqrt2}{2y^0} = -\mathbf{\dfrac{\sqrt2}{2}}$

4. $\dfrac{-36x^3y^3z^4}{-\sqrt{25}xyz^3} = +\dfrac{36x^3y^3z^4}{\sqrt{5^2}xyz^3} = \dfrac{36x^3y^3z^4}{5x^1y^1z^3} = \dfrac{36(x^3x^{-1})\cdot(y^3y^{-1})\cdot(z^4z^{-3})}{5} = \dfrac{36}{5}\dfrac{(x^{3-1})\cdot(y^{3-1})\cdot(z^{4-3})}{1} = \mathbf{7\frac{1}{5}(x^2y^2z)}$

5. $\dfrac{-9a^2b^2c^3}{\sqrt[3]{27a^6b^3c^3}} = -\dfrac{9a^2b^2c^3}{bc\sqrt[3]{3^3a^6}} = -\dfrac{\overset{3}{\cancel{9}}a^2b^2c^3}{\underset{1}{\cancel{3}}bc\sqrt[3]{a^{3+3}}} = -\dfrac{3a^2b^2c^3}{b^1c^1\sqrt[3]{a^3\cdot a^3}} = -\dfrac{a^2(b^2b^{-1})\cdot(c^3c^{-1})}{a\cdot a} = -\dfrac{3a^2b^{2-1}c^{3-1}}{a^2} = -\dfrac{3b^1c^2}{a^2a^{-2}}$

$= -\dfrac{3bc^2}{a^{2-2}} = -\dfrac{3bc^2}{a^0} = -\dfrac{3bc^2}{1} = \mathbf{-3bc^2}$

6. $\dfrac{-24lm^3n^2}{12l^2mn} = -\dfrac{\overset{2}{\cancel{24}}lm^3n^2}{\underset{1}{\cancel{12}}l^2mn} = -\dfrac{2l^1m^3n^2}{l^2m^1n^1} = -\dfrac{2(m^3m^{-1})\cdot(n^2n^{-1})}{l^2l^{-1}} = -\dfrac{2(m^{3-1})\cdot(n^{2-1})}{l^{2-1}} = -\dfrac{2m^2n^1}{l^1} = -\mathbf{\dfrac{2m^2n}{l}}$

7. $\dfrac{xyz^2}{-xy^2z^3} = -\dfrac{x^1y^1z^2}{x^1y^2z^3} = -\dfrac{x^1x^{-1}}{(y^2y^{-1})\cdot(z^3z^{-2})} = -\dfrac{x^{1-1}}{(y^{2-1})\cdot(z^{3-2})} = -\dfrac{x^0}{y^1z^1} = -\mathbf{\dfrac{1}{yz}}$

8. $\dfrac{\sqrt[4]{3^4x^5y^4z^6}}{\sqrt{18x^6y^2}} = \dfrac{3y\sqrt[4]{x^5z^6}}{y\sqrt{9\cdot2x^6}} = \dfrac{3y\sqrt[4]{x^{4+1}z^{4+2}}}{y\sqrt{(3^2\cdot2)x^{2+2+2}}} = \dfrac{3\sqrt[4]{(x^4\cdot x^1)\cdot(z^4\cdot z^2)}}{3\sqrt{2(x^2\cdot x^2\cdot x^2)}} = \dfrac{\overset{1}{\cancel{3}}xz\sqrt[4]{xz^2}}{\underset{1}{\cancel{3}}(x\cdot x\cdot x)\sqrt2} = \dfrac{x^1z\sqrt[4]{xz^2}}{x^3\sqrt2} = \dfrac{z\sqrt[4]{xz^2}}{\sqrt2(x^3x^{-1})}$

$= \dfrac{z\sqrt[4]{xz^2}}{\sqrt2(x^{3-1})} = \dfrac{z\sqrt[4]{xz^2}}{\sqrt2x^2} = \dfrac{z\sqrt[4]{xz^2}}{\sqrt2x^2}\times\dfrac{\sqrt2}{\sqrt2} = \dfrac{\sqrt2\times z\sqrt[4]{xz^2}}{\sqrt2\times\sqrt2x^2} = \dfrac{\sqrt2z\sqrt[4]{xz^2}}{\sqrt{2\cdot2}x^2} = \dfrac{\sqrt2z\sqrt[4]{xz^2}}{\sqrt{2^2}x^2} = \mathbf{\dfrac{\sqrt2z\sqrt[4]{xz^2}}{2x^2}}$

9. $\frac{\sqrt[3]{27x^3y^6}}{-9x^2y} = -\frac{x\sqrt[3]{3^3y^{3+3}}}{9x^2y} = -\frac{\overset{1}{\cancel{3}}x\sqrt[3]{y^3\cdot y^3}}{\underset{3}{\cancel{9}}x^2y} = -\frac{x(y\cdot y)}{3x^2y} = -\frac{xy^2}{3x^2y} = -\frac{x^1y^2}{3x^2y^1} = -\frac{y^2y^{-1}}{3x^2x^{-1}} = -\frac{y^{2-1}}{3x^{2-1}} = -\frac{\mathbf{y}}{\mathbf{3x}}$

10. $\frac{\sqrt{x^2y^3}}{-\sqrt[4]{2^4x^4y^8}} = -\frac{x\sqrt{y^{2+1}}}{2x\sqrt[4]{y^{4+4}}} = -\frac{\cancel{x}\sqrt{y^2\cdot y^1}}{2\cancel{x}\sqrt[4]{y^4\cdot y^4}} = -\frac{y\sqrt{y}}{2(y\cdot y)} = -\frac{y^1\sqrt{y}}{2y^2} = -\frac{\sqrt{y}}{2y^2y^{-1}} = -\frac{\sqrt{y}}{2y^{2-1}} = -\frac{\sqrt{y}}{2y^1} = -\frac{\sqrt{\mathbf{y}}}{\mathbf{2y}}$

Section 6.3 Case I b Practice Problems

1. $\frac{98-46}{-12} = -\frac{\overset{13}{\cancel{52}}}{\underset{3}{\cancel{12}}} = -\frac{13}{3} = \mathbf{-4\frac{1}{3}}$

2. $\frac{x^3y^3z+4x^2y^2}{-2xy^2z} = \frac{x^3y^3z}{-2xy^2z}+\frac{4x^2y^2}{-2xy^2z} = -\frac{x^3y^3z^1}{2x^1y^2z^1} - \frac{\overset{2}{\cancel{4}}x^2y^2}{\underset{1}{\cancel{2}}x^1y^2z} = -\frac{(x^3x^{-1})\cdot(y^3y^{-2})}{2(z^1z^{-1})} - \frac{2(x^2x^{-1})}{(y^2y^{-2})z}$

$= -\frac{x^{3-1}\cdot y^{3-2}}{2z^{1-1}} - \frac{2x^{2-1}}{y^{2-2}z} = -\frac{x^2\cdot y^1}{2z^0} - \frac{2x^1}{y^0z} = -\frac{\mathbf{x^2y}}{\mathbf{2}} - \frac{\mathbf{2x}}{\mathbf{z}}$

3. $\frac{-a^3b^3c+a^2bc^2}{-a^2b^2c^2} = \frac{-a^3b^3c}{-a^2b^2c^2}+\frac{a^2bc^2}{-a^2b^2c^2} = +\frac{a^3b^3c^1}{a^2b^2c^2} - \frac{a^2b^1c^2}{a^2b^2c^2} = \frac{(a^3a^{-2})\cdot(b^3b^{-2})}{c^2c^{-1}} - \frac{(a^2a^{-2})\cdot(c^2c^{-2})}{b^2b^{-1}}$

$= \frac{a^{3-2}\cdot b^{3-2}}{c^{2-1}} - \frac{a^{2-2}\cdot c^{2-2}}{b^{2-1}} = \frac{a^1\cdot b^1}{c^1} - \frac{a^0\cdot c^0}{b^1} = \frac{ab}{c} - \frac{1\cdot 1}{b} = \frac{\mathbf{ab}}{\mathbf{c}} - \frac{\mathbf{1}}{\mathbf{b}}$

4. $\frac{\sqrt[4]{a^5b^4c^3}-\sqrt[3]{a^3b^6c}}{\sqrt{a^2b^4c^6}} = \frac{b\sqrt[4]{a^{4+1}c^3}-a\sqrt[3]{b^{3+3}c}}{a\sqrt{b^{2+2}c^{2+2+2}}} = \frac{b\sqrt[4]{(a^4\cdot a^1)\cdot c^3}-a\sqrt[3]{(b^3\cdot b^3)\cdot c}}{a\sqrt{(b^2\cdot b^2)\cdot(c^2\cdot c^2\cdot c^2)}} = \frac{ab\sqrt[4]{ac^3}-a(b\cdot b)\sqrt[3]{c}}{a(b\cdot b)(c\cdot c\cdot c)}$

$= \frac{ab\sqrt[4]{ac^3}-ab^2\sqrt[3]{c}}{ab^2c^3} = \frac{ab\sqrt[4]{ac^3}}{ab^2c^3}+\frac{-ab^2\sqrt[3]{c}}{ab^2c^3} = \frac{\cancel{a}b\sqrt[4]{ac^3}}{\cancel{a}b^2c^3} - \frac{\cancel{a}b^2\sqrt[3]{c}}{\cancel{a}b^2c^3} = \frac{\sqrt[4]{ac^3}}{(b^2b^{-1})\cdot c^3} - \frac{\sqrt[3]{c}}{(b^2b^{-2})\cdot c^3} = \frac{\sqrt[4]{ac^3}}{b^{2-1}\cdot c^3} - \frac{\sqrt[3]{c}}{b^{2-2}\cdot c^3}$

$= \frac{\sqrt[4]{ac^3}}{b^1c^3} - \frac{\sqrt[3]{c}}{b^0c^3} = \frac{\sqrt[4]{\mathbf{ac^3}}}{\mathbf{bc^3}} - \frac{\sqrt[3]{\mathbf{c}}}{\mathbf{c^3}}$

5. $\frac{m^3n^2l+ml^2}{mnl} = \frac{m^3n^2l}{mnl}+\frac{ml^2}{mnl} = \frac{m^3n^2l}{mnl}+\frac{\cancel{m}l^2}{\cancel{m}nl} = \frac{m^3n^2}{m^1n^1}+\frac{l^2}{nl^1} = \frac{(m^3m^{-1})\cdot(n^2n^{-1})}{1}+\frac{l^2l^{-1}}{n} = \frac{m^{3-1}n^{2-1}}{1}+\frac{l^{2-1}}{n}$

$= \frac{m^2n^1}{1}+\frac{l^1}{n} = \mathbf{m^2n}+\frac{\mathbf{l}}{\mathbf{n}}$

6. $\frac{36y^2-18y^3}{-9y} = \frac{36y^2}{-9y}+\frac{-18y^3}{-9y} = -\frac{\overset{4}{\cancel{36}}y^2}{\underset{1}{\cancel{9}}y}+\frac{\overset{2}{\cancel{18}}y^3}{\underset{1}{\cancel{9}}y} = -\frac{4y^2}{y^1}+\frac{2y^3}{y^1} = -\frac{4y^2y^{-1}}{1}+\frac{2y^3y^{-1}}{1} = -\frac{4y^{2-1}}{1}+\frac{2y^{3-1}}{1}$

$= -\frac{4y^1}{1}+\frac{2y^2}{1} = \mathbf{2y^2-4y}$

7. $\frac{-5\sqrt{150}+10\sqrt{125}}{\sqrt{25}} = \frac{-5\sqrt{25\cdot 6}+10\sqrt{25\cdot 5}}{\sqrt{5^2}} = \frac{-5\sqrt{5^2\cdot 6}+10\sqrt{5^2\cdot 5}}{5} = \frac{-(5\cdot 5)\sqrt{6}+(10\cdot 5)\sqrt{5}}{5} = \frac{-25\sqrt{6}+50\sqrt{5}}{5} = \frac{-25\sqrt{6}}{5}+\frac{50\sqrt{5}}{5} = \frac{\overset{5}{\cancel{25}}\sqrt{6}}{\underset{1}{\cancel{5}}}+\frac{\overset{10}{\cancel{50}}\sqrt{5}}{\underset{1}{\cancel{5}}}$

$= \frac{5\sqrt{6}}{1}+\frac{10\sqrt{5}}{1} = \frac{-5\sqrt{6}+10\sqrt{5}}{1} = \mathbf{-5\sqrt{6}+10\sqrt{5}}$

8. $$\frac{99x^2-18x^3}{\sqrt{81x^4}}=\frac{99x^2-18x^3}{\sqrt{9^2x^{2+2}}}=\frac{99x^2-18x^3}{9\sqrt{x^2\cdot x^2}}=\frac{99x^2-18x^3}{9(x\cdot x)}=\frac{99x^2-18x^3}{9x^2}=\frac{99x^2}{9x^2}+\frac{-18x^3}{9x^2}=\frac{\overset{11}{\cancel{99}}x^2}{\underset{1}{\cancel{9}}x^2}-\frac{\overset{2}{\cancel{18}}x^3}{\underset{1}{\cancel{9}}x^2}$$

$$=\frac{11x^2}{x^2}-\frac{2x^3}{x^2}=\frac{11x^2x^{-2}}{1}-\frac{2x^3x^{-2}}{1}=\frac{11x^{2-2}}{1}-\frac{2x^{3-2}}{1}=\frac{11x^0}{1}-\frac{2x^1}{1}=\frac{11}{1}-\frac{2x}{1}=\frac{11-2x}{1}=\mathbf{-2x+11}$$

9. $$\frac{\sqrt{8x^2y^3}-\sqrt{9xy^2}}{\sqrt{4x^4y^6}}=\frac{x\sqrt{(2^2\cdot2)y^{2+1}}-y\sqrt{3^2x}}{\sqrt{2^2x^{2+2}y^{2+2+2}}}=\frac{2x\sqrt{2(y^2\cdot y^1)}-3y\sqrt{x}}{2\sqrt{(x^2\cdot x^2)\cdot(y^2\cdot y^2\cdot y^2)}}=\frac{2xy\sqrt{2y}-3y\sqrt{x}}{2(x\cdot x)(y\cdot y\cdot y)}=\frac{2xy\sqrt{2y}-3y\sqrt{x}}{2x^2y^3}$$

$$=\frac{2xy\sqrt{2y}}{2x^2y^3}+\frac{-3y\sqrt{x}}{2x^2y^3}=\frac{\overset{1}{\cancel{2}}xy\sqrt{2y}}{\underset{1}{\cancel{2}}x^2y^3}-\frac{3y\sqrt{x}}{2x^2y^3}=\frac{x^1y^1\sqrt{2y}}{x^2y^3}-\frac{3y^1\sqrt{x}}{2x^2y^3}=\frac{\sqrt{2y}}{(x^2x^{-1})\cdot(y^3y^{-1})}-\frac{3\sqrt{x}}{2x^2(y^3y^{-1})}$$

$$=\frac{\sqrt{2y}}{x^{2-1}y^{3-1}}-\frac{3\sqrt{x}}{2x^2y^{3-1}}=\frac{\sqrt{2y}}{x^1y^2}-\frac{3\sqrt{x}}{2x^2y^2}=\frac{\sqrt{2y}}{xy^2}-\frac{3}{2}\left(\frac{\sqrt{x}}{x^2y^2}\right)=\mathbf{\frac{\sqrt{2y}}{xy^2}-1\frac{1}{2}\left(\frac{\sqrt{x}}{x^2y^2}\right)}$$

10. $$\frac{\sqrt{100}w^5+\sqrt{75}w^3}{-\sqrt{25}w^2}=\frac{\sqrt{10\cdot10}w^5+\sqrt{25\cdot3}w^3}{-\sqrt{5^2}w^2}=\frac{\sqrt{10^2}w^5+\sqrt{5^2\cdot3}w^3}{-5w^2}=\frac{10w^5+5\sqrt{3}w^3}{-5w^2}=\frac{10w^5}{-5w^2}+\frac{5\sqrt{3}w^3}{-5w^2}$$

$$=-\frac{\overset{2}{\cancel{10}}w^5}{\underset{1}{\cancel{5}}w^2}-\frac{\overset{1}{\cancel{5}}\sqrt{3}w^3}{\underset{1}{\cancel{5}}w^2}=-\frac{2w^5}{w^2}-\frac{\sqrt{3}w^3}{w^2}=-\frac{2w^5w^{-2}}{1}-\frac{\sqrt{3}w^3w^{-2}}{1}=-\frac{2w^{5-2}}{1}-\frac{\sqrt{3}w^{3-2}}{1}=\frac{-2w^3}{1}-\frac{\sqrt{3}w^1}{1}$$

$$=\frac{-2w^3-\sqrt{3}w}{1}=\mathbf{-2w^3-\sqrt{3}w}$$

Section 6.3 Case I c Practice Problems

1. $$\frac{x^6-x^4+2x^2-6}{4}=\frac{x^6}{4}+\frac{-x^4}{4}+\frac{+2x^2}{4}+\frac{-6}{4}=\frac{x^6}{4}-\frac{x^4}{4}+\frac{\overset{1}{\cancel{2}}x^2}{\underset{2}{\cancel{4}}}-\frac{\overset{3}{\cancel{6}}}{\underset{2}{\cancel{4}}}=\mathbf{\frac{x^6}{4}-\frac{x^4}{4}+\frac{x^2}{2}-\frac{3}{2}}$$

2. $$\frac{y^3-2y^2+4y-12}{-2}=\frac{y^3}{-2}+\frac{-2y^2}{-2}+\frac{+4y}{-2}+\frac{-12}{-2}=-\frac{y^3}{2}+\frac{\overset{1}{\cancel{2}}y^2}{\underset{1}{\cancel{2}}}-\frac{\overset{2}{\cancel{4}}y}{\underset{1}{\cancel{2}}}+\frac{\overset{6}{\cancel{12}}}{\underset{1}{\cancel{2}}}=-\frac{y^3}{2}+\frac{y^2}{1}-\frac{2y}{1}+\frac{6}{1}=\mathbf{-\frac{1}{2}y^3+y^2-2y+6}$$

3. $$\frac{x^2y-2xy+4xy^3-8}{6xy}=\frac{x^2y}{6xy}+\frac{-2xy}{6xy}+\frac{+4xy^3}{6xy}+\frac{-8}{6xy}=\frac{x^2y}{6xy}-\frac{\overset{1}{\cancel{2}}xy}{\underset{3}{\cancel{6}}xy}+\frac{\overset{2}{\cancel{4}}xy^3}{\underset{3}{\cancel{6}}xy}-\frac{\overset{4}{\cancel{8}}}{\underset{3}{\cancel{6}}xy}=\frac{x^2y}{6xy}-\frac{\cancel{xy}}{3\cancel{xy}}+\frac{2\cancel{x}y^3}{3\cancel{x}y}-\frac{4}{3xy}$$

$$=\frac{x^2}{6x}-\frac{1}{3}+\frac{2y^3}{3y}-\frac{4}{3xy}=\frac{x^2x^{-1}}{6}-\frac{1}{3}+\frac{2y^3y^{-1}}{3}-\frac{4}{3xy}=\frac{x^{2-1}}{6}-\frac{1}{3}+\frac{2y^{3-1}}{3}-\frac{4}{3xy}=\mathbf{\frac{x}{6}-\frac{1}{3}+\frac{2y^2}{3}-\frac{4}{3xy}}$$

4. $$\frac{\sqrt[3]{375}y^3-\sqrt[3]{108}y^2-3y+5}{15y}=\frac{\sqrt[3]{125\cdot3}y^3-\sqrt[3]{27\cdot4}y^2-3y+5}{15y}=\frac{\sqrt[3]{5^3\cdot3}y^3-\sqrt[3]{3^3\cdot4}y^2-3y+5}{15y}$$

$$=\frac{5\sqrt[3]{3}y^3-3\sqrt[3]{4}y^2-3y+5}{15y}=\frac{5\sqrt[3]{3}y^3}{15y}+\frac{-3\sqrt[3]{4}y^2}{15y}+\frac{-3y}{15y}+\frac{+5}{15y}=\frac{\overset{1}{\cancel{5}}\sqrt[3]{3}y^3}{\underset{3}{\cancel{15}}y}-\frac{\overset{1}{\cancel{3}}\sqrt[3]{4}y^2}{\underset{5}{\cancel{15}}y}-\frac{\overset{1}{\cancel{3}}y}{\underset{5}{\cancel{15}}y}+\frac{\overset{1}{\cancel{5}}}{\underset{3}{\cancel{15}}y}$$

$$=\frac{\sqrt[3]{3}y^3}{3y}-\frac{\sqrt[3]{4}y^2}{5y}-\frac{y}{5y}+\frac{1}{3y}=\frac{\sqrt[3]{3}y^3y^{-1}}{3}-\frac{\sqrt[3]{4}y^2y^{-1}}{5}-\frac{1}{5}+\frac{1}{3y}=\frac{\sqrt[3]{3}y^{3-1}}{3}-\frac{\sqrt[3]{4}y^{2-1}}{5}-\frac{1}{5}+\frac{1}{3y}=\mathbf{\frac{\sqrt[3]{3}}{3}y^2-\frac{\sqrt[3]{4}}{5}y-\frac{1}{5}+\frac{1}{3y}}$$

5. $$\frac{-w^6+2w^3+4w^2-8}{-2w} = \frac{-w^6}{-2w}+\frac{+2w^3}{-2w}+\frac{+4w^2}{-2w}+\frac{-8}{-2w} = +\frac{w^6}{2w}-\frac{\overset{1}{\cancel{2}}w^3}{\underset{1}{\cancel{2}}w}-\frac{\overset{2}{\cancel{4}}w^2}{\underset{1}{\cancel{2}}w}+\frac{\overset{4}{\cancel{8}}}{\underset{1}{\cancel{2}}w} = \frac{w^6}{2w}-\frac{w^3}{w}-\frac{2w^2}{w}+\frac{4}{w}$$

$$= \frac{w^6w^{-1}}{2}-\frac{w^3w^{-1}}{1}-\frac{2w^2w^{-1}}{1}+\frac{4}{w} = \frac{w^{6-1}}{2}-\frac{w^{3-1}}{1}-\frac{2w^{2-1}}{1}+\frac{4}{w} = \frac{w^5}{2}-\frac{w^2}{1}-\frac{2w}{1}+\frac{4}{w} = \mathbf{\frac{1}{2}w^5-w^2-2w+\frac{4}{w}}$$

6. $$\frac{3\sqrt{a^{10}}-6a^4-\sqrt{64}a^2-12}{4a^2} = \frac{3\sqrt{(a^5)^2}-6a^4-\sqrt{8^2}a^2-12}{4a^2} = \frac{3a^5-6a^4-8a^2-12}{4a^2} = \frac{3a^5}{4a^2}+\frac{-6a^4}{4a^2}+\frac{-8a^2}{4a^2}+\frac{-12}{4a^2}$$

$$= \frac{3a^5}{4a^2}-\frac{\overset{3}{\cancel{6}}a^4}{\underset{2}{\cancel{4}}a^2}-\frac{\overset{2}{\cancel{8}}a^2}{\underset{1}{\cancel{4}}a^2}-\frac{\overset{3}{\cancel{12}}}{\underset{1}{\cancel{4}}a^2} = \frac{3a^5}{4a^2}-\frac{3a^4}{2a^2}-\frac{2a^2}{a^2}-\frac{3}{a^2} = \frac{3a^5a^{-2}}{4}-\frac{3a^4a^{-2}}{2}-\frac{2a^2a^{-2}}{1}-\frac{3}{a^2}$$

$$= \frac{3a^{5-2}}{4}-\frac{3a^{4-2}}{2}-\frac{2a^{2-2}}{1}-\frac{3}{a^2} = \frac{3a^3}{4}-\frac{3a^2}{2}-\frac{2a^0}{1}-\frac{3}{a^2} = \mathbf{\frac{3a^3}{4}-\frac{3a^2}{2}-2-\frac{3}{a^2}}$$

7. $$\frac{-m^6-3m^3+2m-8}{-m^2} = \frac{-m^6}{-m^2}+\frac{-3m^3}{-m^2}+\frac{+2m}{-m^2}+\frac{-8}{-m^2} = +\frac{m^6}{m^2}+\frac{3m^3}{m^2}-\frac{2m}{m^2}+\frac{8}{m^2} = \frac{m^6m^{-2}}{1}+\frac{3m^3m^{-2}}{1}-\frac{2}{m^2m^{-1}}+\frac{8}{m^2}$$

$$= \frac{m^{6-2}}{1}+\frac{3m^{3-2}}{1}-\frac{2}{m^{2-1}}+\frac{8}{m^2} = \frac{m^4}{1}+\frac{3m^1}{1}-\frac{2}{m^1}+\frac{8}{m^2} = \mathbf{m^4+3m-\frac{2}{m}+\frac{8}{m^2}}$$

8. $$\frac{m^3n^4-m^3n^2+mn-6n}{2m^2n^2} = \frac{m^3n^4}{2m^2n^2}+\frac{-m^3n^2}{2m^2n^2}+\frac{+mn}{2m^2n^2}+\frac{-6n}{2m^2n^2} = \frac{m^3n^4}{2m^2n^2}-\frac{m^3n^2}{2m^2n^2}+\frac{mn}{2m^2n^2}-\frac{6n}{2m^2n^2}$$

$$= \frac{(m^3m^{-2})(n^4n^{-2})}{2}-\frac{m^3m^{-2}}{2n^2n^{-2}}+\frac{1}{2(m^2m^{-1})(n^2n^{-1})}-\frac{6}{2m^2(n^2n^{-1})} = \frac{m^{3-2}n^{4-2}}{2}-\frac{m^{3-2}}{2n^{2-2}}+\frac{1}{2m^{2-1}n^{2-1}}-\frac{6}{2m^2n^{2-1}}$$

$$= \frac{m^1n^2}{2}-\frac{m^1}{2n^0}+\frac{1}{2m^1n^1}-\frac{\overset{3}{\cancel{6}}}{\underset{1}{\cancel{2}}m^2n^1} = \mathbf{\frac{mn^2}{2}-\frac{m}{2}+\frac{1}{2mn}-\frac{3}{m^2n}}$$

9. $$\frac{y^7-3y^5-5y^3+6y-12}{15y^2} = \frac{y^7}{15y^2}+\frac{-3y^5}{15y^2}+\frac{-5y^3}{15y^2}+\frac{+6y}{15y^2}+\frac{-12}{15y^2} = \frac{y^7}{15y^2}-\frac{\overset{1}{\cancel{3}}y^5}{\underset{5}{\cancel{15}}y^2}-\frac{\overset{1}{\cancel{5}}y^3}{\underset{3}{\cancel{15}}y^2}+\frac{\overset{2}{\cancel{6}}y}{\underset{5}{\cancel{15}}y^2}-\frac{\overset{4}{\cancel{12}}}{\underset{5}{\cancel{15}}y^2}$$

$$= \frac{y^7}{15y^2}-\frac{y^5}{5y^2}-\frac{y^3}{3y^2}+\frac{2y}{5y^2}-\frac{4}{5y^2} = \frac{y^7y^{-2}}{15}-\frac{y^5y^{-2}}{5}-\frac{y^3y^{-2}}{3}+\frac{2}{5y^2y^{-1}}-\frac{4}{5y^2}$$

$$= \frac{y^{7-2}}{15}-\frac{y^{5-2}}{5}-\frac{y^{3-2}}{3}+\frac{2}{5y^{2-1}}-\frac{4}{5y^2} = \frac{y^5}{15}-\frac{y^3}{5}-\frac{y^1}{3}+\frac{2}{5y^1}-\frac{4}{5y^2} = \mathbf{\frac{y^5}{15}-\frac{y^3}{5}-\frac{y}{3}+\frac{2}{5y}-\frac{4}{5y^2}}$$

10. $$\frac{\sqrt[3]{32}-\sqrt{72}+\sqrt[3]{108}+\sqrt{162}}{-\sqrt{81}} = \frac{\sqrt[3]{8\cdot4}-\sqrt{36\cdot2}+\sqrt[3]{27\cdot4}+\sqrt{81\cdot2}}{-\sqrt{9\cdot9}} = \frac{\sqrt[3]{2^3\cdot4}-\sqrt{6^2\cdot2}+\sqrt[3]{3^3\cdot4}+\sqrt{9^2\cdot2}}{-\sqrt{9^2}}$$

$$= \frac{2\sqrt[3]{4}-6\sqrt{2}+3\sqrt[3]{4}+9\sqrt{2}}{-9} = -\frac{(2\sqrt[3]{4}+3\sqrt[3]{4})+(-6\sqrt{2}+9\sqrt{2})}{9} = -\frac{(2+3)\sqrt[3]{4}+(-6+9)\sqrt{2}}{9} = -\frac{5\sqrt[3]{4}+3\sqrt{2}}{9}$$

$$= \frac{-5\sqrt[3]{4}-3\sqrt{2}}{9} = -\frac{5\sqrt[3]{4}}{9}-\frac{\overset{1}{\cancel{3}}\sqrt{2}}{\underset{3}{\cancel{9}}} = \mathbf{-\frac{5\sqrt[3]{4}}{9}-\frac{\sqrt{2}}{3}}$$

Section 6.3 Case II Practice Problems

1. Divide $3x^2 + 10x + 7$ by $x + 3$.

$$\begin{array}{l} \qquad\quad 3x + 1 \\ x + 3\ \overline{)\,+3x^2 + 10x + 7} \\ \qquad\ \overset{-}{+}3x^2 \overset{-}{+} 9x \\ \qquad\qquad\quad + x + 7 \\ \qquad\qquad\quad \overset{-}{+} x \overset{-}{+} 3 \\ \qquad\qquad\qquad + 4 \end{array}$$

The answer is $\mathbf{3x + 1}$ with remainder of $\mathbf{+4}$, or

$\mathbf{3x + 1 + \frac{4}{x + 3}}$.

2. Divide $x^4 + 7x^3 + 13x^2 + 17x + 10$ by $x + 5$.

$$\begin{array}{l} \qquad\quad x^3 + 2x^2 + 3x + 2 \\ x + 5\ \overline{)\,+x^4 + 7x^3 + 13x^2 + 17x + 10} \\ \qquad\ \overset{-}{+}x^4 \overset{-}{+}5x^3 \\ \qquad\qquad +2x^3 + 13x^2 \\ \qquad\qquad \overset{-}{+}2x^3 \overset{-}{+}10x^2 \\ \qquad\qquad\qquad + 3x^2 + 17x \\ \qquad\qquad\qquad \overset{-}{+} 3x^2 \overset{-}{+} 15x \\ \qquad\qquad\qquad\qquad + 2x + 10 \\ \qquad\qquad\qquad\qquad \overset{-}{+} 2x \overset{-}{+}10 \\ \qquad\qquad\qquad\qquad\qquad 0 \end{array}$$

The answer is $\mathbf{x^3 + 2x^2 + 3x + 2}$ with remainder of zero.

3. Divide $x^6 - x^5 - 2x^4 - x^3 + 2x^2 + 5x - 10$ by $x - 2$.

$$\begin{array}{l} \qquad\quad x^5 + x^4 - x^2 + 5 \\ x - 2\ \overline{)\,+x^6 - x^5 - 2x^4 - x^3 + 2x^2 + 5x - 10} \\ \qquad\ \overset{-}{+}x^6 \pm 2x^5 \\ \qquad\qquad + x^5 - 2x^4 \\ \qquad\qquad \overset{-}{+} x^5 \pm 2x^4 \\ \qquad\qquad\qquad -x^3 + 2x^2 \\ \qquad\qquad\qquad \pm x^3 \overset{-}{+}2x^2 \\ \qquad\qquad\qquad\qquad +5x - 10 \\ \qquad\qquad\qquad\qquad \overset{-}{+}5x \pm 10 \\ \qquad\qquad\qquad\qquad\qquad 0 \end{array}$$

The answer is $\mathbf{x^5 + x^4 - x^2 + 5}$ with remainder of zero.

4. Divide $-2x^4 + 5x^3 - 4x^2 + 16x - 15$ by $-2x + 5$.

$$\begin{array}{l} \qquad\qquad x^3 + 2x - 3 \\ -2x + 5\ \overline{)\,-2x^4 + 5x^3 - 4x^2 + 16x - 15} \\ \qquad\quad \pm 2x^4 \overset{-}{+}5x^3 \\ \qquad\qquad\qquad -4x^2 + 16x \\ \qquad\qquad\qquad \pm 4x^2 \overset{-}{+}10x \\ \qquad\qquad\qquad\qquad + 6x - 15 \\ \qquad\qquad\qquad\qquad \overset{-}{+} 6x \pm 15 \\ \qquad\qquad\qquad\qquad\qquad 0 \end{array}$$

The answer is $\mathbf{x^3 + 2x - 3}$ with remainder of zero.

5. Divide $2x^4 - 13x^3 + 13x^2 + 15x - 35$ by $x - 5$.

$$\begin{array}{l} \qquad\quad 2x^3 - 3x^2 - 2x + 5 \\ x - 5\ \overline{)\,+2x^4 - 13x^3 + 13x^2 + 15x - 35} \\ \qquad\ \overset{-}{+}2x^4 \pm 10x^3 \\ \qquad\qquad - 3x^3 + 13x^2 \\ \qquad\qquad \pm 3x^3 \overset{-}{+}15x^2 \\ \qquad\qquad\qquad -2x^2 + 15x \\ \qquad\qquad\qquad \pm 2x^2 \overset{-}{+}10x \\ \qquad\qquad\qquad\qquad + 5x - 35 \\ \qquad\qquad\qquad\qquad \overset{-}{+} 5x \pm 25 \\ \qquad\qquad\qquad\qquad\qquad - 10 \end{array}$$

The answer is $\mathbf{2x^3 - 3x^2 - 2x + 5}$ with remainder of $\mathbf{-10}$,

or $\mathbf{2x^3 - 3x^2 - 2x + 5 - \frac{10}{x - 5}}$.

6. Divide $-2x^4 + 7x^3 - 6x^2 - 2x + 3$ by $-2x + 3$.

$$\begin{array}{l} \qquad\qquad x^3 - 2x^2 + 1 \\ -2x + 3\ \overline{)\,-2x^4 + 7x^3 - 6x^2 - 2x + 3} \\ \qquad\quad \pm 2x^4 \overset{-}{+}3x^3 \\ \qquad\qquad\quad +4x^3 - 6x^2 \\ \qquad\qquad\quad \overset{-}{+}4x^3 \pm 6x^2 \\ \qquad\qquad\qquad\qquad -2x + 3 \\ \qquad\qquad\qquad\qquad \pm 2x \overset{-}{+} 3 \\ \qquad\qquad\qquad\qquad\qquad 0 \end{array}$$

The answer is $\mathbf{x^3 - 2x^2 + 1}$ with remainder of zero.

7. Divide $3x^7 + 3x^6 - 2x^3 - 2x^2 + 5x + 5$ by $x + 1$.

$$\begin{array}{r|l} & 3x^6 - 2x^2 + 5 \\ x+1 & +3x^7 + 3x^6 - 2x^3 - 2x^2 + 5x + 5 \\ & \mp 3x^7 \mp 3x^6 \\ \hline & -2x^3 - 2x^2 \\ & \pm 2x^3 \pm 2x^2 \\ \hline & +5x + 5 \\ & \mp 5x \mp 5 \\ \hline & 0 \end{array}$$

The answer is $\mathbf{3x^6 - 2x^2 + 5}$ with remainder of zero.

8. Divide $2y^3 + 5y^2 - 4y - 12$ by $y + 2$.

$$\begin{array}{r|l} & 2y^2 + y - 6 \\ y+2 & +2y^3 + 5y^2 - 4y - 12 \\ & \mp 2y^3 \mp 4y^2 \\ \hline & + y^2 - 4y \\ & \mp y^2 \mp 2y \\ \hline & -6y - 12 \\ & \pm 6y \pm 12 \\ \hline & 0 \end{array}$$

The answer is $\mathbf{2y^2 + y - 6}$ with remainder of zero.

9. Divide $x^3 + 2x^2 - 18x + 14$ by $x - 3$.

$$\begin{array}{r|l} & x^2 + 5x - 3 \\ x-3 & +x^3 + 2x^2 - 18x + 14 \\ & \mp x^3 \pm 3x^2 \\ \hline & +5x^2 - 18x \\ & \mp 5x^2 \pm 15x \\ \hline & -3x + 14 \\ & \pm 3x \mp 9 \\ \hline & +5 \end{array}$$

The answer is $\mathbf{x^2 + 5x - 3}$ with remainder of $+\ \mathbf{5}$, or

$$\mathbf{x^2 + 5x - 3 + \frac{5}{x-3}}.$$

10. Divide $x^4 + 2x^3 + 2x^2 + 5x + 2$ by $x + 2$.

$$\begin{array}{r|l} & x^3 + 2x + 1 \\ x+2 & +x^4 + 2x^3 + 2x^2 + 5x + 2 \\ & \mp x^4 \mp 2x^3 \\ \hline & +2x^2 + 5x \\ & \mp 2x^2 \mp 4x \\ \hline & + x + 2 \\ & \mp x \mp 2 \\ \hline & 0 \end{array}$$

The answer is $\mathbf{x^3 + 2x + 1}$ with remainder of zero.

Section 6.4 Case I Practice Problems

1. $(x^3 + 2x^5 - 3x + 2) + (3x^3 + x - x^5) = (2x^5 + x^3 - 3x + 2) + (-x^5 + 3x^3 + x) = (2x^5 - x^5) + (x^3 + 3x^3) + (-3x + x) + 2$

$= (2-1)x^5 + (1+3)x^3 + (-3+1)x + 2 = \mathbf{x^5 + 4x^3 - 2x + 2}$

2. $(y - y^2 + 2y^4 + 3y^2 - 3) + (2y^4 + y^3 + 5 - y^2) = (2y^4 + 3y^2 - y^2 + y - 3) + (2y^4 + y^3 - y^2 + 5)$

$= (2y^4 + 2y^4) + y^3 + (3y^2 - y^2 - y^2) + y + (-3+5) = (2+2)y^4 + y^3 + (3-1-1)y^2 + y + 2 = \mathbf{4y^4 + y^3 + y^2 + y + 2}$

3. $(3x - 3x^2 + 5x - 3) - (-2x + 5 - x^2 + 2) = (3x - 3x^2 + 5x - 3) + (2x - 5 + x^2 - 2) = (-3x^2 + 3x + 5x - 3) + (x^2 + 2x - 5 - 2)$

$= (-3x^2 + x^2) + (3x + 5x + 2x) + (-3 - 5 - 2) = (-3+1)x^2 + (3+5+2)x - 10 = \mathbf{-2x^2 + 10x - 10}$

4. $(xyz + 2x^2yz + 4xyz) + (4x^2yz - x^2yz + 2xyz) = (2x^2yz + xyz + 4xyz) + (4x^2yz - x^2yz + 2xyz)$

$= (2x^2yz + 4x^2yz - x^2yz) + (xyz + 4xyz + 2xyz) = (2+4-1)x^2yz + (1+4+2)xyz = \mathbf{5x^2yz + 7xyz}$

5. $(-2ab - 3 + 2a^2b^2) + (-3ab + a^2b^2 + 2(ab)^0) = (-2ab - 3 + 2a^2b^2) + (-3ab + a^2b^2 + 2)$

$= (2a^2b^2 - 2ab - 3) + (a^2b^2 - 3ab + 2) = (2a^2b^2 + a^2b^2) + (-2ab - 3ab) + (-3 + 2) = (2 + 1)a^2b^2 + (-2 - 3)ab - 1$

$= \mathbf{3a^2b^2 - 5ab - 1}$

6. $(5x^6 - x^5 - 4x^4 + 3x + x^2) - (x - 3x^2 + x^4 - 3x^6) = (5x^6 - x^5 - 4x^4 + 3x + x^2) + (-x + 3x^2 - x^4 + 3x^6)$

$= (5x^6 + 3x^6) - x^5 + (-4x^4 - x^4) + (x^2 + 3x^2) + (3x - x) = (5 + 3)x^6 - x^5 + (-4 - 1)x^4 + (1 + 3)x^2 + (3 - 1)x$

$= \mathbf{8x^6 - x^5 - 5x^4 + 4x^2 + 2x}$

7. $-(w^2 + w + 2w^4 + 8) + (w - 3w^4 - 2w^2 - 8) = (-w^2 - w - 2w^4 - 8) + (w - 3w^4 - 2w^2 - 8)$

$= (-2w^4 - w^2 - w - 8) + (-3w^4 - 2w^2 + w - 8) = (-2w^4 - 3w^4) + (-w^2 - 2w^2) + (-w + w) + (-8 - 8)$

$= (-2 - 3)w^4 + (-1 - 2)w^2 + (-1 + 1)w - 16 = -5w^4 - 3w^2 + 0w - 16 = \mathbf{-5w^4 - 3w^2 - 16}$

8. $(u^2v + 2uv + 5u^2v - 6 + uv) + (uv + 2 - 3u^2v) = (u^2v + 5u^2v + 2uv + uv - 6) + (-3u^2v + uv + 2)$

$= (u^2v + 5u^2v - 3u^2v) + (2uv + uv + uv) + (-6 + 2) = (1 + 5 - 3)u^2v + (2 + 1 + 1)uv - 4 = \mathbf{3u^2v + 4uv - 4}$

9. $(x^3 + x) - (3x^2 + x^3 + 5x) + (-x^2 - 6x - 4x^3 + x) = (x^3 + x) + (-3x^2 - x^3 - 5x) + (-x^2 - 6x - 4x^3 + x)$

$= (x^3 + x) + (-x^3 - 3x^2 - 5x) + (-4x^3 - x^2 - 6x + x) = (x^3 - x^3 - 4x^3) + (-3x^2 - x^2) + (x - 5x - 6x + x)$

$= (1 - 1 - 4)x^3 + (-3 - 1)x^2 + (1 - 5 - 6 + 1)x = \mathbf{-4x^3 - 4x^2 - 9x}$

10. $(x^5 + x^4 + 2) - (x^4 + 3 - 2x^5 + x^3) + (-4x^4 - 8) = (x^5 + x^4 + 2) + (-x^4 - 3 + 2x^5 - x^3) + (-4x^4 - 8)$

$= (x^5 + x^4 + 2) + (+2x^5 - x^4 - x^3 - 3) + (-4x^4 - 8) = (x^5 + 2x^5) + (x^4 - x^4 - 4x^4) - x^3 + (2 - 3 - 8)$

$= (1 + 2)x^5 + (1 - 1 - 4)x^4 - x^3 - 9 = \mathbf{3x^5 - 4x^4 - x^3 - 9}$

Section 6.4 Case II Practice Problems

1. $(x^2 + 2x + x^3) + (3x - 2x^3) = (x^3 + x^2 + 2x) + (-2x^3 + 3x) =$

$$\begin{array}{r} +\ x^3 + x^2 + 2x \\ -\ 2x^3 \qquad\ + 3x \\ \hline \mathbf{-\ x^3 + x^2 + 5x} \end{array}$$

2. $(y + y^2 + 3y^3 + 4) + (-2 + y^2 + 3y^2 + 2y) = (3y^3 + y^2 + y + 4) + [(y^2 + 3y^2) + 2y - 2]$

$= (3y^3 + y^2 + y + 4) + [(1 + 3)y^2 + 2y - 2] = (3y^3 + y^2 + y + 4) + (4y^2 + 2y - 2) =$

$$\begin{array}{r} 3y^3 + y^2 + y + 4 \\ + 4y^2 + 2y - 2 \\ \hline \mathbf{3y^3 + 5y^2 + 3y + 2} \end{array}$$

3. $(x^3 + x^2 - 3 + 3x^2) - (-2x^3 - 5x + 5) = [x^3 + (x^2 + 3x^2) - 3] + (2x^3 + 5x - 5) = [x^3 + (1 + 3)x^2 - 3] + (2x^3 + 5x - 5)$

$= (x^3 + 4x^2 - 3) + (2x^3 + 5x - 5) =$

$$\begin{array}{r} x^3 + 4x^2 \qquad - 3 \\ 2x^3 \qquad + 5x - 5 \\ \hline \mathbf{3x^3 + 4x^2 + 5x - 8} \end{array}$$

4. $\left(z^5+3z^2+z-2z^2-4z+2\right)+\left(z^2+4z^5+z^0\right) = \left[z^5+\left(3z^2-2z^2\right)+(z-4z)+2\right]+\left(4z^5+z^2+1\right)$

$= \left[z^5+(3-2)z^2+(1-4)z+2\right]+\left(4z^5+z^2+1\right) = \left(z^5+z^2-3z+2\right)+\left(4z^5+z^2+1\right) =$

$$\begin{array}{r} z^5+z^2-3z+2 \\ 4z^5+z^2+1 \\ \hline \mathbf{5z^5+2z^2-3z+3} \end{array}$$

5. $-\left(a^3-2a+a+2-3a^3\right)+\left(-2a^3-4a-3\right) = \left(-a^3+2a-a-2+3a^3\right)+\left(-2a^3-4a-3\right)$

$= \left[\left(3a^3-a^3\right)+(2a-a)-2\right]+\left(-2a^3-4a-3\right) = \left[(3-1)a^3+(2-1)a-2\right]+\left(-2a^3-4a-3\right)$

$= \left(2a^3+a-2\right)+\left(-2a^3-4a-3\right) =$

$$\begin{array}{r} +2a^3+a-2 \\ -2a^3-4a-3 \\ \hline \mathbf{0a^3-3a-5} \end{array}$$

6. $\left(u^2+2u+u+5\right)+\left(-2u^2-3-5u-8\right) = \left[u^2+(2u+u)+5\right]+\left[-2u^2-5u+(-3-8)\right] = \left[u^2+(2+1)u+5\right]+\left[-2u^2-5u-11\right]$

$= \left(u^2+3u+5\right)+\left(-2u^2-5u-11\right) =$

$$\begin{array}{r} +u^2+3u+5 \\ -2u^2-5u-11 \\ \hline \mathbf{-u^2-2u-6} \end{array}$$

7. $\left(x-x^2+3x^4-5\right)-\left(-2x^4+x-3+4x^2\right) = \left(x-x^2+3x^4-5\right)+\left(2x^4-x+3-4x^2\right)$

$= \left(3x^4-x^2+x-5\right)+\left(2x^4-4x^2-x+3\right) =$

$$\begin{array}{r} 3x^4-x^2+x-5 \\ 2x^4-4x^2-x+3 \\ \hline \mathbf{5x^4-5x^2+0x-2} \end{array}$$

8. $\left(a+3a^2-60+5a\right)+\left(-4a-5a+5^0+2a^2\right) = \left(3a^2+a+5a-60\right)+\left(2a^2-4a-5a+1\right)$

$= \left[3a^2+(a+5a)-60\right]+\left[2a^2+(-4a-5a)+1\right] = \left[3a^2+(1+5)a-60\right]+\left[2a^2+(-4-5)a+1\right]$

$= \left(3a^2+6a-60\right)+\left(2a^2-9a+1\right) =$

$$\begin{array}{r} 3a^2+6a-60 \\ 2a^2-9a+1 \\ \hline \mathbf{5a^2-3a-59} \end{array}$$

9. $\left(x^2y^2-2xy-8+2x^2y^2\right)+\left(3x^2y^2+4xy+3+2xy\right) = \left[\left(x^2y^2+2x^2y^2\right)-2xy-8\right]+\left[3x^2y^2+(4xy+2xy)+3\right]$

$= \left[(1+2)x^2y^2-2xy-8\right]+\left[3x^2y^2+(4+2)xy+3\right] = \left(3x^2y^2-2xy-8\right)+\left(3x^2y^2+6xy+3\right) =$

$$\begin{array}{r} 3x^2y^2-2xy-8 \\ 3x^2y^2+6xy+3 \\ \hline \mathbf{6x^2y^2+4xy-5} \end{array}$$

10. $\left(3a^3+2a+a^6+1\right)+\left(8+3a^3+a\right)-\left(2a^6-4+4a+2a^3\right) = \left(3a^3+2a+a^6+1\right)+\left(8+3a^3+a\right)+\left(-2a^6+4-4a-2a^3\right)$

$= \left(a^6+3a^3+2a+1\right)+\left(3a^3+a+8\right)+\left(-2a^6-2a^3-4a+4\right) =$

$$\begin{array}{r} +a^6+3a^3+2a+1 \\ +3a^3+a+8 \\ -2a^6-2a^3-4a+4 \\ \hline \mathbf{-a^6+4a^3-a+13} \end{array}$$

Glossary

The following glossary terms are used throughout this book:

Absolute value - The numerical value or magnitude of a quantity, as of a negative number, without regard to its sign. The symbol for absolute value is two parallel lines "$|\ |$". For instance, $|-2| = |2| = 2$, $|-35| = |35| = 35$, $|-0.23| = |0.23| = 0.23$, and $|-5.13| = |5.13| = 5.13$ are some examples of how absolute value is used.

Addend - Any of a set of numbers to be added.

Addition - The process of adding two or more numbers to get a number called the sum.

Algebraic approach - An approach in which only numbers, letters, and arithmetic operations are used.

Algebraic expression - Designating an expression, equation, or function in which only numbers, letters, and arithmetic operations are contained or used.

Apply - To put on. To put to or adapt for particular use. To use.

Associative - Pertaining to an operation in which the result is the same regardless of the way the elements are grouped, as, in addition, $2+(4+5)$ = $(2+4)+5$ = 11 and, in multiplication, $2\times(4\times 5)$ = $(2\times 4)\times 5$ = 40.

Base - *a.* The number on which a system of numeration is based. For example, the base of the decimal system is 10. Computers use the binary system, which has the base 2. *b.* A number that is to be multiplied by itself the number of times indicated by an exponent or logarithm. For example, in 2^5, 2 is the base and 5 is the exponent.

Binomial - An expression consisting of two terms connected by a plus or minus sign. For example, $a+b$, $\sqrt{x^3}-\sqrt{y}$, x^3+3x, and a^2b^3-3ab are referred to as binomials.

Brackets [ ] - A pair of symbols used to enclose a mathematical expression.

Case - Supporting facts offered in justification of a statement.

Change - To replace by another; alter; transform.

Class - A group of persons or things that have something in common, a set, collection, group.

Classification - The act, process, or result of classifying.

Classify - To put or divide into classes or groups.

Coefficient - A number placed in front of an algebraic expression and multiplying it; factor. For example, in the expression $3x^2 + 5x = 2$, 3 is the coefficient of x^2, and 5 is the coefficient of x.

Combine - To bring together; unite; join; merge.

Common denominator - A common multiple of the denominators of two or more fractions. For example, 10 is a common denominator of $\frac{1}{2}$ and $\frac{3}{5}$.

Common divisor - A number or quantity that can evenly divide two or more other numbers or quantities. For example, 4 is a common divisor of 12 and 20.

Common factor - Another name for common divisor.

Common fraction - A fraction whose numerator and denominator are both integers (whole numbers). In this book a common fraction is the same as an integer fraction.

Commutative - Pertaining to an operation in which the order of the elements does not affect the result, as, in addition, $5+3 = 3+5$ and, in multiplication, $5 \times 3 = 3 \times 5$.

Conjugate - Inversely or oppositely related to one of a group of otherwise identical properties.

Conversion - A change in the form of a quantity or an expression without a change in the value.

Convert - To change from one form or use to another; transform.

Cube - The third power of a number or quantity.

Cube root $\left(\sqrt[3]{\ }\right)$ - A number which, cubed, equals the number given. For example, the cube root of 216 is 6.

Decimal number - Any number written using base 10; a number containing a decimal point.

Decimal point - A period placed to the left of a decimal.

Decrease - Reduce; make less; lessen usually refers to decrease in numbers.

Degree - The greatest sum of the exponents of the variables in a term of a polynomial or polynomial equation. For example, the polynomial $w^3 + 3w + 5$ is a third degree polynomial.

Denominator - The term below the line in a fraction; the divisor of the numerator. For example, in the fraction $\frac{3}{5}$, 5 is the denominator.

Descend - To move from a higher to a lower place. To go down.

Descending order - Decreasing order.

Difference - The amount by which one quantity differs from another; remainder left after subtraction.

Digit - Any of the numerals from 0 through 9 - in the base-ten system.

Distributive - Of the principle in multiplication that allows the multiplier to be used separately with each term of the multiplicand.

Dividend - A quantity to be divided. For example, in the problem $14 \div 2$, 14 is called the dividend.

Division - The process of finding how many times a number (the divisor) is contained in another number (the dividend). The number of times equals the quotient.

Divisor - The quantity by which another quantity, the dividend, is to be divided. For example, in the problem $14 \div 2$, 2 is called the divisor.

Enhance - To add to; to increase or make greater.

Equal - Exactly the same. Of the same quantity, size, number, value, degree, intensity, or quality.

Equation - A mathematical sentence involving the use of an equal sign. For example, $x^3 + 3x^2 + 5x = 3$ is referred to as an equation.

Equivalent fractions - Fractions that are numerically the same.

Even number - A number which is exactly divisible by two; not odd. For example, $(0, 2, 4, 6, 8, 10, \ldots)$ are even numbers.

Exact order - Not deviating in form or content; precise.

Example - One that is representative of a group as a whole; a sample.

Expanded form - To write, a quantity, as a sum of terms, as a continued product, or as another extended form.

Exponent - A number placed as a superscript to show how many times another number is to be placed as a factor. For example, in the problem $5^3 = 5 \times 5 \times 5 = 125$, 3 is an exponent.

Exponential - Containing, involving, or expressed as an exponent.

Exponential notation - A way of expressing a number as the product of the factor and 10 raised to some power. The factor is either a whole number or a decimal number. For example, the exponential notation form of 0.0353, 0.048, 489, 3987 are 35.3×10^{-3}, 48×10^{-3}, 48.9×10^1, and 398.7×10^1, respectively.

Expression - A designation of any symbolic mathematical form, such as an equation. The means by which something is expressed.

Factor - One of two or more quantities having a designated product. For example, 3 and 5 are factors of 15.

Form - A specific type; kind.

Fraction - A number which indicates the ratio between two quantities in the form of $\frac{a}{b}$ such that a is any real number and b is any real number not equal to zero.

Fractional - Having to do with or making up a fraction.

Fractional exponent - An exponential expression of the form $x^{\frac{a}{b}}$.

General - Not precise or detailed. Not limited to one class of things. Relating to all.

Greater than $(\rangle)$ - A symbol used to compare two numbers with the greater number given first. For example, $5 \rangle 2$, $23 \rangle 20$, $50 \rangle 10$.

Greatest common factor - A greatest number that divides two or more numbers without a remainder. For example, 6 is the greatest common factor among 6, 12, and 36.

Group - An assemblage of objects or numbers.

Horizontal - Flat. Parallel to the horizon. Something that is horizontal, as a line, plane, or bar.

Identify - To recognize. To establish the identity of.

Imaginary number - The positive square root of a negative number. For example, $\sqrt{-5}$, $\sqrt{-3}$, and $\sqrt{-1}$ are imaginary numbers. Not real number.

Improper fraction - A fraction in which the numerator is larger than or equal to the denominator. For example, $\frac{6}{5}$, $\frac{10}{9}$, and $\frac{23}{7}$ are improper fractions.

Index - A number or symbol, often written as a subscript or superscript to a mathematical expression, that indicates an operation to be performed on. For example, in the problem $\sqrt[3]{x^2}$, 3 is referred to as an index.

Inequality $(\neq)$ - A relation indicating that the two numbers are not the same.

Instance - A case or example.

Integer fraction - A fraction having positive or negative whole numbers in the numerator and the denominator.

Integer number - Any member of the set of positive whole numbers $(1, 2, 3, 4, \ldots)$, negative whole numbers $(-1, -2, -3, -4, \ldots)$, and zero is an integer number.

Introduction - To inform of something for the first time. The act of introducing.

Irrational number - A number not capable of being expressed by an integer (a whole number) or an integer fraction (quotient of an integer). For example, $\sqrt{3}$, π, and $\sqrt[4]{7}$ are irrational numbers.

Law - A general principle or rule that is obeyed in all cases to which it is applicable.

Less than $(\langle)$ - A symbol used to compare two numbers with the lesser number given first. For example, $5 \langle 8$, $23 \langle 30$, $12 \langle 25$.

Like terms - Similar terms.

Lowest term - Smallest value.

Match - A person or thing that is exactly like another, counterpart.

Mathematical operation - The process of performing addition, subtraction, multiplication, and division in a specified sequence.

Method - A way of doing or accomplishing something.

Minimize - To reduce to the least possible amount; reduce to a minimum.

Mixed fraction - A fraction made up of a positive or negative whole number and an integer fraction.

Mixed operation - Combining addition, subtraction, multiplication, and division in a math process is defined as a mixed operation.

Monomial - An expression consisting of only one term. Being a simple algebraic term. For example, 5, $\sqrt{xy}$, x^3, and $2ab$ are referred to as monomials.

Multiplicand - The number that is or is to be multiplied by another.

Multiplication - The process of finding the number obtained by repeated additions of a number a specified number of times: Multiplication is symbolized in various ways, i.e., $3 \times 4 = 12$ or $3 \cdot 4 = 12$, which means $3+3+3+3 = 12$, to add the number three together four times.

Multiplier - The number by which the multiplicand is multiplied. For example, if 3 is multiplied by 4, 3 is the multiplicand, 4 is the multiplier, and 12 is the product.

Negative number - A quantity less than zero.

Not Applicable - In this book *Not Applicable* implies to a *step* that can not be put to a specific use. A *Step* that is not relevant.

Not real number - Imaginary number.

Numerator - The term above the line in a fraction. For example, in the fraction $\frac{3}{5}$, 3 is the numerator.

Numerical coefficient - Coefficients represented by a number or numbers rather than by letter or symbol.

Objective - A goal or end.

Odd number - A number having a remainder of one when divided by two; not even. For example, $(1, 3, 5, 7, 9, 11, \ldots)$ are even numbers.

Operation - A process or action, such as addition, subtraction, multiplication, or division, performed in a specified sequence and in accordance with specific rules of procedure.

Parentheses ( ) - A pair of symbols used to enclose a sum, product, or other mathematical expressions.

Polynomial - An algebraic function of two or more summed terms, each term consisting of a constant multiplier and one or more variables raised to a power. For example, the general form of a polynomial of degree n in a single real variable x is $P(x) = a_n x^n + a_{n-1} x^{n-1} + a_{n-2} x^{n-2} + \cdots + a_0$.

Positive number - A quantity greater than zero.

Power - An exponent. The result of a number multiplied by itself a given number of times. For example, the third power of 3 is 27.

Practice - To exercise or perform repeatedly in order to acquire or polish a skill.

Primary - Something that is first in degree, quality, or importance. Occurring first in time or sequence. Original.

Prime factorization - A factorization that shows only prime factors. For example, $21 = 1 \times 3 \times 7$.

Prime number - A number that has itself and unity as its only factors. For example, 2, 3, 5, 7, and 11 are prime numbers since they have no common divisor except unity.

Principal - First, highest, or foremost in importance.

Problem - Something to be done or solved.

Proceed - To go on. To continue. To begin and carry on an activity.

Process - A series of operations or a method for producing something. A series of actions, changes, or functions that bring about an end or result.

Product - The quantity obtained by multiplying two or more quantities together.

Proper fraction - A fraction in which the numerator is smaller than the denominator.

Quality - That which makes something the way it is; distinctive feature or characteristic.

Quantity - An amount or number.

Quotient - The quantity resulting from division of one quantity by another.

Radical - The root of a quantity as indicated by the radical sign. Indicating or having to do with a square root or cube root.

Radical expression - A mathematical expression or form in which radical signs appear.

Radical sign $\left(\sqrt{\ }\right)$ - A sign that indicates a specified root of the number written under it. For example, $\sqrt[3]{27}$ = the cube root of 27, which is, 3.

Radicand - The quantity under a radical sign. For example, 27 is the radicand of $\sqrt[3]{27}$.

Rationalization - The act, process, or practice of rationalizing.

Rationalize - To remove radicals without changing the value of an expression or roots of an equation.

Rational number - A number that can be represented as an integer (a whole number) or an integer fraction (quotient of integers). For example, $\frac{1}{5}$, $-\frac{2}{15}$, $12=\frac{12}{1}$, $-230=-\frac{230}{1}=\frac{230}{-1}=\ldots$, $-10=-\frac{10}{1}=-\frac{100}{10}=-\frac{50}{5}=\frac{350}{-35}=\ldots$, and $0.13=\frac{13}{100}=\frac{130}{1000}=\frac{26}{200}=\ldots$ are rational numbers.

Real number - A number that is either a rational number or an irrational number. For example, $\frac{3}{5}$, $-\frac{4}{13}$, -23, 0.13, $\sqrt{5}$, and π are real numbers.

Reference - The directing of attention to a person or thing.

Re-group - A repeated assemblage of objects or numbers.

Remainder - *a.* What is left when a smaller number is subtracted from a larger number. *b.* What is left undivided when one number is divided by another that is not one of its factors.

Result - To end in a particular way. The consequence of a particular action. An outcome.

Resultant - That which results. Consequence.

Revise - To change or modify. To read carefully so as to correct errors or make improvements and changes.

Revision - The result of revising. Something that has been revised.

Root - A quantity that, multiplied by itself a specified number of times, produces a given quantity. For example, 5 is the square root (5×5) of 25 and the cube root $(5\times 5\times 5)$ of 125.

Rule - A method or procedure prescribed for computing or solving a problem.

Scientific notation - A way of expressing a number as the product of the factor and 10 raised to some power. The factor is always of the form where the decimal point is to the right of the first non-zero digit. For example, the scientific notation form of 0.0353, 0.048, 489, 3987 are 3.53×10^{-2}, 4.8×10^{-2}, 4.89×10^{2}, 3.987×10^{3}, respectively.

Section - One of several component parts of something; piece; portion.

Sequence - The order in which one thing comes after another. A number of things following each other; series.

Show - Demonstrate; to point out; indicate.

Sign - A mark or symbol having an accepted and specific meaning. For example, the sign $+$ implies addition.

Signed number - A number which can have a positive or negative value as designated by $+$ or $-$ symbol. A signed number with no accompanying symbol is understood to be positive.

Similar - Alike but not completely the same.

Similar radicals - Radical expressions with the same index and the same radicand. For example, $\sqrt[3]{x^2}$, $5\sqrt[3]{x^2}$, and $3\sqrt[3]{x^2}$ are referred to as similar radicals.

Simplify - Make easier; less complex.

Solution - The act, method, or process of solving a problem. The answer to a problem.

Solve - To find a solution to; answer.

Special - Exceptional. Surpassing what is common or usual.

Specific example - An example that is precise and explicit.

Square root $\left(\sqrt{\ }\right)$ - The factor of a number which, multiplied by itself, gives the original number. For example, the square root of 36 is 6.

Standard - Any type, model, or example for comparison. Serving as a gauge or model.

Step - One of a series of actions or measures taken toward some end.

Sub-group - A distinct group within a group.

Subject - A topic discussed in writing.

Subscript - A number, letter, or a symbol, written below and to the right or left of a character. For example, 2 is the subscript in x_2.

Subtraction - The mathematical process of finding the difference between two numbers.

Sum - The amount obtained as a result of adding two or more numbers together.

Summary - Reduced into few words; concise.

Superscript - A number, letter, or a symbol, written above a character. For example, 5 is the superscript in y^5.

Symbol - A sign used to represent a mathematical operation.

Technique - A special method of doing something. The systematic procedure by which a complex or scientific task is accomplished.

Term - The parts of a mathematical expression that are added or subtracted. For example, in the equation $ax^3 + bx^2 + cx - d$, ax^3, bx^2, cx, and d are referred to as terms.

Trinomial - An expression consisting of three terms connected by a plus or minus sign. For example, $a^2 + a + 3$, $\sqrt[3]{x^2} + \sqrt[3]{x} - 5$, and $x^3 + 3x^2 + 2$ are referred to as trinomials.

Type - An example or model; kind.

Variable - A quantity capable of assuming any of a set of values. Having no fixed quantitative value.

Vertical - Upright. At right angles to the horizon. Straight up and down.

Whole number - A whole number is defined as an integer number.

With - Having as a possession, attribute, or characteristic.

Without - In the absence of; with no or none of.

Zero - The symbol or numeral 0. The point, marked 0, from which positive or negative quantities are reckoned on a graduated scale.

The following references were used in developing this glossary:

1) *The Webster's New World Dictionary of American English, Victoria E. Neufeldt, editor in chief, third college edition, 1995.*

2) *The American Heritage Dictionary of the English Language, William Morris, editor, third edition, 1994.*

3) *HBJ School Dictionary, Harcourt Brace Jovanovich publishing, fourth edition, 1985.*

Index

D

E

F

G

H

I

L

M

N

O

P

T

V

W

Z

About the Author

Said Hamilton received his B.S. degree in Electrical Engineering from Oklahoma State University and Master's degree in Electrical Engineering from the University of Texas at Austin. He has taught a number of math and engineering courses as a visiting lecturer at the University of Oklahoma, Department of Mathematics, and as a faculty member at Rose State College, Department of Engineering Technology, at Midwest City, Oklahoma. He is currently working in the field of aerospace technology and has published numerous technical papers.

About the Editor

Pat Eblen received his Bachelor of Science degree in Electrical Engineering from the University of Kentucky where he was a member of Eta Kappa Nu Electrical Engineering Honor Society. He has worked in the aerospace industry for nearly twenty years where he has received numerous awards for contributions to spacecraft technology programs. Mr. Eblen enjoys studying mathematical theories in probability and quantum mechanics and has developed several original concepts in these fields.

Watch for these other Hamilton Education Guides:

- Mastering Fractions
- Mastering Algebra: Intermediate Level
- Mastering Algebra: Advanced Level

Order Form

call 1-800-209-8186 to order

Or: Use the order form

Last Name ______________________ First Name ______________ M.I. ____

Address ____

City ________________ State __________ Zip Code _________________

No. of Books	Book Price ($49.95)	Total Price
		$
	Subtotal	$
	Shipping (Add $3.50 for the first book $2.00 for each additional book)	$
	Sales Tax Va. residents add 4.5% ($49.95 × 0.045 = $2.25)	$
	Total Payment	$

Enclosed is:

A check ☐ Master Card ☐

VISA ☐ American Express ☐

Account Number ______________________ Expiration Date ______________

Signature __________________________

Make checks payable to Hamilton Education Guides.

Please send completed form to:

Hamilton Education Guides

P.O. Box 681

Vienna, Va. 22183

~ *Notes* ~

~ Notes ~